CONIC SECTIONS

Parabola, vertical axis: $y = ax^2 + bx + c,\ a \neq 0$
Parabola, horizontal axis: $x = ay^2 + by + c,\ a \neq 0$
Ellipse, center at origin: $ax^2 + by^2 = c$, a, b, c all positive
Hyperbola, center at origin: $ax^2 - by^2 = c$, a, b, c all positive, asymptotes $ax \pm by = 0$
$ay^2 - bx^2 = c$, a, b, c all positive, asymptotes $ay \pm bx = 0$
$xy = c$, asymptotes $x = 0$, $y = 0$

CRAMER'S RULE

For the system
$$\begin{aligned} a_{11}x + a_{12}y + a_{13}z &= c_1 \\ a_{21}x + a_{22}y + a_{23}z &= c_2 \\ a_{31}x + a_{32}y + a_{33}z &= c_3 \end{aligned}$$

Let $D = \begin{vmatrix} a_{11} & a_{12} & a_{13} \\ a_{21} & a_{22} & a_{23} \\ a_{31} & a_{32} & a_{33} \end{vmatrix}$

and let D_x, D_y, D_z be the determinants formed from D by replacing the coefficients of x, y, z, respectively, by the corresponding c's. Then if $D \neq 0$, the solution is $x = D_x/D$, $y = D_y/D$, $z = D_z/D$.

COMPLEX NUMBERS

Polar form: $a + bi = r(\cos\theta + i\sin\theta) = r \operatorname{cis}\theta$, where $r = \sqrt{a^2 + b^2}$, $\cos\theta = a/r$, $\sin\theta = b/r$
De Moivre's Theorem: $(r \operatorname{cis}\theta)^n = r^n \operatorname{cis} n\theta$

Roots: $\sqrt[n]{r \operatorname{cis}\theta} = \sqrt[n]{r} \operatorname{cis} \dfrac{\theta + k \cdot 360^\circ}{n}$, $k = 0, 1, 2, \ldots, (n-1)$

FIBONACCI SEQUENCE

$a_1 = 1$, $a_2 = 1$, and for $n > 2$, $a_n = a_{n-1} + a_{n-2}$

ARITHMETIC PROGRESSION

$a_n = a_1 + (n-1)d$, $S_n = n\left(\dfrac{a_1 + a_n}{2}\right)$

GEOMETRIC PROGRESSION

$a_n = a_1 r^{n-1}$, $S_n = \dfrac{a_1(1 - r^n)}{1 - r}$, $r \neq 1$

INFINITE GEOMETRIC SERIES

With $|r| < 1$: $S = \dfrac{a_1}{1 - r}$

PERMUTATIONS

Of n things, r at a time: $P(n, r) = n(n-1)\cdots(n-r+1) = \dfrac{n!}{(n-r)!}$, $r \leq n$

COMBINATIONS

Of n things, r at a time: $C(n, r) = \dfrac{P(n, r)}{r!} = \dfrac{n!}{r!(n-r)!}$

BINOMIAL THEOREM

$(a + b)^n = \displaystyle\sum_{k=0}^{n} \binom{n}{k} a^{n-k} b^k$, where $\dbinom{n}{k} = \dfrac{n!}{k!(n-k)!}$

CONTEMPORARY COLLEGE ALGEBRA AND TRIGONOMETRY

CONTEMPORARY COLLEGE ALGEBRA AND TRIGONOMETRY

JACK R. BRITTON
University of South Florida

IGNACIO BELLO
Hillsborough Community College

HARPER & ROW, PUBLISHERS, New York
Cambridge, Philadelphia, San Francisco,
London, Mexico City, São Paulo, Sydney

Sponsoring editor: Fred Henry
Project editor: Cynthia L. Indriso
Designer: Gayle Jaeger
Manager, new book production: Kewal K. Sharma/Jacqui Brownstein
Compositor: Syntax International Pte. Ltd.
Printer and binder: R. R. Donnelly & Sons Company
Art studio: J & R Art Services, Inc.
Cover: Painting by Frances Torbert Tilley

Library of Congress Cataloging in Publication Data
Britton, Jack Rolf, Date-
Contemporary college algebra and trigonometry.
Includes index.
1. Algebra. 2. Trigonometry, Plane. I. Bello, Ignacio. II. Title
QA152.2.B76 512′.13 81-7113
ISBN 0-06-040989-4 AACR2

CONTENTS

PREFACE

Some years ago, we started with the idea of creating a new and different mathematics book. But more than that, we wanted to do so in a manner that was comprehensible and relevant to students. In other words, we wanted a book that was innovative in many ways: motivation, presentation, format, and testing.

PREREQUISITES

We assume that the student has the algebraic skills that can be obtained in a high school or a post high school algebra course such as intermediate algebra.

MOTIVATION

It is well known to any sensitive mathematics instructor that mathematics proper does not appeal to most students. They constantly ask, "Why do we have to learn this? What good is it?" To answer these students, we have given, whenever possible, the natural motivation by means of real problems taken from newspapers and advertisements, as well as from other sources. We have always presented this motivation along with the topics to be taught. It will not do to assure students that they will appreciate *later* the value of the mathematics that they are asked to learn *now.*

PRESENTATION

Our presentation is made through carefully constructed discussion, examples, and exercises that gradually lead to the desired results. We have tried to avoid the traditional definition-theorem-proof approach. Instead, we have given the students an intuitive idea of *why* the theorem or the definition is needed, and then combined this with examples and with the available pertinent applications. Definitions and theorems are set in the margin, and important concepts are set apart from the text by color rules. Moreover, we have kept the terminology to a minimum. Common words have been used to introduce new concepts. Last, but not least, we have tried to keep symbolism to a minimum.

USING YOUR KNOWLEDGE

These special problems are presented after the exercises at the end of each section. They try to answer the question: "Why do we need this material and how can we use it?"

TESTING

A Self-Test is provided at the end of each chapter. In addition, the *Instructor's Manual* contains two test sequences for each chapter, one independent one and one patterned after the Self-Test.

TOPICS

The topics presented are the usual ones for courses of this type. We have included as many topics as possible to enable instructors to tailor the course to their needs.

A NOTE ABOUT CALCULATORS

Although the basic mathematics in this text does not require a calculator, a great deal of time that would be spent on numerical computations can be saved if students have calculators. Problems that are intended to be worked with a calculator have been labeled with the symbol C.

Scientific calculators use mainly two types of logic: algebraic logic and RPN logic. We suggest using a calculator with one of these types of logic. RPN calculators are characterized by an ENTER or equivalent key. The difference between algebraic logic and RPN logic is illustrated by the example

$3 \bullet 4 + 2 \bullet 7$

Algebraic logic [3] [×] [4] [+] [2] [×] [7] [=]

RPN logic [3] [ENTER] [4] [×] [2] [ENTER] [7] [×] [+]

The calculations in this text have been done on a Texas Instruments TI-30 to illustrate algebraic logic and on a Hewlett-Packard HP-32E to illustrate RPN logic. However, students should consult the owner's manual for their own calculators.

ACKNOWLEDGMENTS

This text has been class-tested several times, not only by the authors but also by the following persons to whom we are extremely grateful: Professors James R. Gard, George Kosan, Arthur Price, Donald C. Rose, and David A. Rose.

We also wish to express our thanks for the helpful comments and suggestions made by the following reviewers: Arthur Dull, Richard Langlie, Peter Lindstrom, Kathleen Salter, Ara Sullenberger, and Edward Wright.

Jack R. Britton
Ignacio Bello

CHAPTER 1

THE REAL NUMBERS

1.1 NUMBERS

In this chapter, we shall review the set of real numbers and the properties of these numbers that are used in their arithmetic. Because we shall use a small amount of set language, you should first recall the definition of a set. (See Definition 1.1a.) Capital letters such as A, B, C, X, Y, and Z are usually used to denote sets, and lower-case letters such as a, b, c, x, y, and z are used to denote elements of sets. When feasible, it is customary to "list" the elements of a set in braces, $\{ \quad \}$, and to separate the elements by commas. Thus, we may write

▼ DEFINITION 1.1a

A set is a collection of objects called the *elements* or *members* of the set.

$$A = \{a, b, c\}$$

for the set that has a, b, and c for its elements;

$$B = \{2, 5, 7, 11, 13\}$$

for the set that has 2, 5, 7, 11, and 13 for its elements; and so on. If a set has no elements, it is called an *empty set* or a *null set*. Because one empty set cannot be distinguished from another empty set, we agree that there is just one empty set, which we denote by the symbol $\varnothing$ (read "the empty set").

An important set of numbers, which is undoubtedly familiar to you, is the set of *natural* numbers (*counting* numbers), denoted by the letter N and defined in Definition 1.1b. The three dots inside the braces indicate that the enumeration continues without end. (If these dots were omitted, we would understand the set to consist of only the three numbers 1, 2, and 3, and *not* the entire set of natural

▼ DEFINITION 1.1b

The set of *natural* numbers is the set

$$N = \{1, 2, 3, \ldots\}$$

numbers.) The set N can also be described by using *set-builder* notation and a *defining property* for the set. With this notation, we write:

$N = \{x \mid x$ is a natural number$\}$

which is read "N is the set of all x's such that x is a natural number." (The vertical stroke | is read "such that.")

EXAMPLE 1

Write the set of natural numbers less than 7 by:

a. listing the elements in braces;
b. using set-builder notation.

SOLUTION:

a. $\{1, 2, 3, 4, 5, 6\}$
b. $\{x \mid x$ is a natural number less than $7\}$

You might have noticed that the number zero, which plays an extremely important role in mathematics, is *not* an element of the set N. To include the number zero, we construct a new set by adjoining zero to the set of natural numbers, thus forming the set of *whole numbers*, denoted by W. (See Definition 1.1c.)

▼ DEFINITION 1.1c

The set of *whole numbers* is the set

$W = \{0, 1, 2, 3, \ldots\}$

Note that every element of N is also an element of W. We express this fact by saying that *N is a subset of W*. In general, we define a subset as in Definition 1.1d. For example, if $A = \{1, 2, 5, 7\}$ and $B = \{1, 2, 3, 5, 6, 7\}$, then A is a subset of B. Similarly, if C is the set of students in your class and S is the set of students in your school, then C is a subset of S, because every student in your class is also a student in your school.

▼ DEFINITION 1.1d

A set A is a *subset* of a set B if every element of A is also an element of B.

We are sometimes interested in knowing whether two given sets are equal. Equality of sets is defined in Definition 1.1e. With this definition of equality, we see that

$$\{1, 2, 3\} = \{3, 1, 2\} \quad \text{and} \quad \{2, 2, 4\} = \{2, 4\}$$

▼ DEFINITION 1.1e

Two sets A and B are equal, denoted by $A = B$, if and only if every element of A is an element of B and every element of B is an element of A.

These illustrations indicate that the order in which the elements of a set are listed is immaterial as is the repetition of elements. The listing without repetition is preferred.

It is possible to state the definition of equality in terms of subsets. Thus, *$A = B$ if and only if A is a subset of B and B is a subset of A*. Note that the definition of subset implies that *every set is a subset of itself*.

Besides being a subset of the set of whole numbers W, the set N of natural numbers is also a subset of the set I of integers. As you can see from Definition 1.1f, the set of integers includes three important subsets:

▼ DEFINITION 1.1f

The set of *integers* is the set

$I = \{\ldots, -2, -1, 0, 1, 2, \ldots\}$

1. the set of positive integers, $I^+ = \{1, 2, 3, \ldots\}$,
2. the set of negative integers, $I^- = \{-1, -2, -3, \ldots\}$,
3. the singleton set $\{0\}$, with the single element zero.

Notice that, by the definition of equality of sets, $I^+ = N$. You should also notice that no two of the three sets I^+, I^-, and $\{0\}$ have any elements in common. Two sets with no common elements are said to be *disjoint*.

None of the sets of numbers considered so far has contained any fractions. However, the set of integers is a subset of the very important set of *rational numbers*, which contains these fractions. Recall that a rational number is a number that can be expressed as the ratio of two integers. Definition 1.1g says that the set Q consists of all possible ratios of integers with nonzero denominators. Here are some rational numbers:

▼ DEFINITION 1.1g

The set of *rational numbers* Q is the set given by

$$Q = \left\{ r \mid r = \frac{a}{b}, \text{ where } a \text{ and } b \text{ are integers, } b \neq 0 \right\}.$$

$$\frac{3}{4}, \quad \frac{-1}{5}, \quad \frac{8}{3}, \quad \frac{9}{-4}, \quad 4 = \frac{4}{1}, \quad 0 = \frac{0}{1}, \quad 2.73 = \frac{273}{100}, \quad \text{and so on}$$

Of course, there are some numbers that are *not* rational; that is, they cannot be written as ratios of two integers. Such numbers are called *irrational*, meaning *not* ratios of integers. One of the simplest examples of an irrational number is the number s such that $s^2 = 2$. (Recall that s^2 means $s \cdot s$.) We write this irrational number as $\sqrt{2}$ (read "the square root of 2"). The nonnegative integers that are perfect squares, such as $0 = 0 \cdot 0$, $1 = 1 \cdot 1$, $4 = 2 \cdot 2$, $9 = 3 \cdot 3$, and so on, all have *rational* square roots, but all other nonnegative integers have *irrational* square roots. We shall denote the set of all irrational numbers by the letter H. (See Definition 1.1h.) A famous irrational number that you are probably acquainted with is the number π (pi), $\pi = 3.1415926\ldots$.

▼ DEFINITION 1.1h

The set of *irrational numbers* is the set H of all numbers that *cannot* be expressed as ratios of integers.

The set consisting of all the rational and all the irrational numbers is called the set of *real* numbers and is denoted by the letter R. This set is the basis for most of our work in algebra. (See Definition 1.1i.)

▼ DEFINITION 1.1i

The set R of *real numbers* consists of all the rational and all the irrational numbers.

It is important to know that every real number can be written in *decimal* form. If the number is rational, say a/b with a and b both integers, we simply divide a by b. When this division is performed, the result is either a terminating decimal (as in the case of $\frac{1}{4} = 0.25$) or a nonterminating, repeating decimal (as in the case of $\frac{1}{3} = 0.333\ldots$). We often put a bar over the set of repeating digits in a nonterminating repeating decimal so that

$$\tfrac{1}{3} = 0.333\ldots = 0.\overline{3}$$

$$\tfrac{2}{11} = 0.181818\ldots = 0.\overline{18}$$

It can be shown not only that every rational number is equal to either a terminating or a nonterminating, repeating decimal, but also that every such decimal is equal to a rational number. Thus, the set Q is the set of all numbers that can be expressed as decimals that are either terminating or nonterminating and repeating. As a consequence, we see that the set H of irrational numbers expressed in decimal form must consist of all nonterminating, nonrepeating decimals. For instance, 0.101001000 . . . , where the *n*th 1 is followed by *n* zeros, is an irrational number because it neither terminates nor repeats.

EXAMPLE 2

Write as a decimal:

a. $\frac{4}{5}$

b. $\frac{3}{11}$

SOLUTION:

a. Dividing 4 by 5, we obtain

$$\begin{array}{r} 0.8 \\ 5\,\overline{)4.0} \\ 4\,0 \\ \hline 0 \end{array}$$

$\frac{4}{5} = 0.8$

b. Dividing 3 by 11, we have

$$\begin{array}{r} 0.27\ldots \\ 11\,\overline{)3.0} \\ 2\,2 \\ \hline 80 \\ 77 \\ \hline 3 \\ \vdots \end{array}$$

After the second step, the remainder is 3, the original dividend. Thus, the digits 27 will repeat without end to give

$\frac{3}{11} = 0.272727\ldots = 0.\overline{27}$

Example 2 indicates how we know that a rational number must yield either a terminating or a nonterminating, repeating decimal. In the division of a natural number a by a natural number b, the only possible remainders are the numbers 0, 1, 2, . . . , $b - 1$. If the remainder 0 occurs, the division terminates, and the result is a terminating decimal. Otherwise, after not more than $b - 1$ steps, a remainder must repeat, and the result is a nonterminating, repeating decimal.

Figure 1.1a shows the relationships among the sets of numbers that we have discussed in this section.

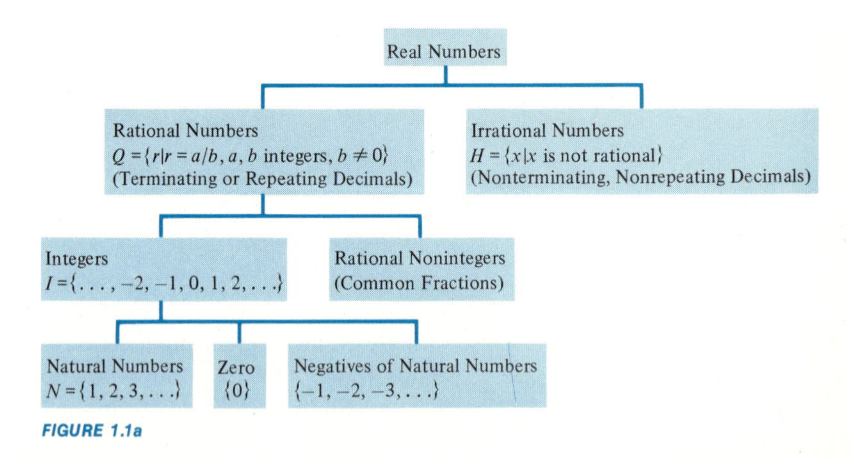

FIGURE 1.1a

EXERCISE 1.1

In Problems 1–10, write each set by: (a) listing its elements in braces and (b) using set-builder notation.

1 The set of natural numbers less than 2.

2 The set of natural numbers between 4 and 10, inclusive.

3 The set of natural numbers between 8 and 11.

4 The set of natural numbers greater than 6.

5 The set of natural numbers less than 9 and greater than 4.

6 The set of whole numbers less than 3.

7 The set of integers less than 0.

8 The set of integers between -3 and 2.

9 The set of nonnegative integers.

10 The set of positive integers.

11 The set of letters in the word *root*.

12 The set of letters in the word *sleeper*.

In Problems 13–20, let $A = \{1, 3, 5, 7, 9\}$, $B = \{2, 4, 6, 8, 10\}$, and $C = \{1, 4, 5, 6, 9\}$.

13 A set D consists of all the elements common to A and B. Describe D by listing its elements in braces.

14 A set E consists of all the elements common to A and C. Describe E by listing its elements in braces.

15 A set F consists of all the elements that are in A or B. Describe F by listing its elements in braces.

16 A set G consists of all the elements that are in A or C. Describe G by listing its elements in braces.

17 A set X consists of all the elements that are in both the set C and the set F of Problem 15. Describe X by listing its elements in braces.

18 A set Y consists of all the elements that are in both the set B and the set E of Problem 14. Describe Y by listing its elements in braces.

19 A set Z consists of all the elements that are in the set C or in the set F of Problem 15. Describe Z by listing its elements in braces.

20 Which of the sets A, B, and C are disjoint?

In Problems 21–26, fill in the blank with the appropriate letter denoting one of the following sets:
N, the set of natural numbers
W, the set of whole numbers
I, the set of integers
Q, the set of rational numbers
H, the set of irrational numbers
R, the set of real numbers

21 The set of rational numbers is a subset of the set ____.

22 The set of whole numbers is constructed by adjoining the number zero to the set ____.

23 The set H is a subset of the set ____.

24 The set consisting of all the elements of W and all the negative integers is the set ____.

25 The set of all the numbers that are in either Q or H is the set ____.

26 The sets Q and ____ are disjoint.

In Problems 27–36, classify the given number as rational or irrational. If the number is rational, express it as a decimal.

27 $\frac{5}{11}$

28 $\frac{9}{5}$

29 $\frac{8}{27}$

30 $\frac{7}{20}$

31 $\frac{7}{6}$

32 $\frac{17}{6}$

33 $0.1223334444\ldots$

34 $0.1211311141111\ldots$

35 $\sqrt{0.64}$

36 $\sqrt{13}$

USING YOUR KNOWLEDGE 1.1

In this section, we learned how to change rational numbers in the form a/b into decimals. This idea is used in many areas outside of mathematics. Use your knowledge to do the following problems.

1 An optometrist must grind an optical lens to $\frac{3}{64}$ of an inch. Write $\frac{3}{64}$ as a decimal. (The machine that grinds lenses is graduated in decimals.)

2 Parts suppliers usually list the dimensions of parts in decimals. A mechanic needs to replace a piston ring that is $3\frac{7}{8}$ inches in diameter. How will this size be listed by the supplier?

3 The total permissible variation in a dimension of a machine part is called its *tolerance*. For example, if the nominal size of a certain screw is 0.437 inch in diameter and it can be larger or smaller by 0.002 inch, then the tolerance is ± 0.002 inch. This means that an acceptable diameter is between 0.435 and 0.439 inch, inclusive. If the nominal size of the diameter of a certain pin is $\frac{1}{4}$ inch with a tolerance of ± 0.002 inch, what is the largest acceptable diameter for this pin? (Express your answer as a single decimal.)

4 If the nominal diameter of the pin in Problem 3 is $\frac{3}{4}$ inch, what is the smallest acceptable diameter for this pin? (Express your answer as a single decimal.)

1.2 SOME BASIC PROPERTIES OF THE REAL NUMBERS

In the preceding section, we used set language to discuss some sets of real numbers. We are now ready to proceed with the discussion using the *language of algebra.*

As you know, a language often uses *pronouns*, such as "he," "she," and "it," in place of nouns. In algebra, we often use *symbols*, such as $x, y, z, \square$, and ——, in place of numbers. A language uses *verbs* to indicate actions. In algebra, actions are expressed by using *operations*. Finally, a language has certain rules that have evolved through its development. In algebra, these fundamental rules or assumptions are called *axioms*. What are these axioms?

Before answering this question, let us consider a very simple puzzle, in which a magician asks a person to think of a number and to make several calculations involving this number. Then, without knowing the original number, the magician reveals the number that the person finally obtains. How is this done? Here is one of these puzzles, along with the operations performed at each step.

Think of a number.	n
Add 1 to it.	$n + 1$
Multiply the result by 5.	$5(n + 1)$
Subtract 5 times the number you started with.	$5(n + 1) - 5n$
Your result is 5.	

Now, you undoubtedly see how the puzzle works, but how do you justify the results? To do this, you must make certain basic assumptions (axioms) about the operations of addition, multiplication, and subtraction. You must also understand the meaning of the phrase "is equal to."

Let us start with the idea of *equality*. In mathematics, an equality is simply a statement that two symbols are names for the same thing. Thus, if a and b are real numbers, the statement

$$a = b \qquad \text{(Read "}a\text{ equals }b\text{.")}$$

(left member = right member)

means that a and b represent the same number. For example, the statement $5 + 2 = 3 + 4$ means that $5 + 2$ and $3 + 4$ represent the same number.

The axioms in Table 1.2a are basic to the equality (=) relationship. As an example of the use of the Substitution Law (Axiom E-4), we know that $3 + 4 = 7$ and $5 + 2 = 7$, so that the axiom justifies our writing $5 + 2 = 3 + 4$. (We substituted the $3 + 4$ for the 7 in the second equation.)

TABLE 1.2a

Axioms for equality		Name
If a, b, and c are real numbers, then		
E-1	$a = a$	Reflexive Law
E-2	If $a = b$, then $b = a$.	Symmetric Law
E-3	If $a = b$ and $b = c$, then $a = c$.	Transitive Law
E-4	If $a = b$, then a may be replaced by b (or b by a) in any statement without changing the truth or falsity of the statement.	Substitution Law

AXIOMS FOR ADDITION AND MULTIPLICATION

The next assumption we must make is that if we start with a real number and perform certain operations involving this number, the result will be a real number. We say that the set R is *closed* under the operations being performed, and we call this the *closure* property for these operations. With reference to our puzzle, this means that each operation that was performed starting with the number n simply leads to another number.

Because we used the fundamental operations of addition and multiplication in our little puzzle, we now list as axioms in Table 1.2b the properties of the *sum*, $\boldsymbol{a + b}$, and the *product*, $\boldsymbol{ab}$, of two numbers a and b. A set of numbers satisfying these axioms under the two given operations is called a *field*. Thus, we speak of the *field R of real numbers*.

TABLE 1.2b *Field Axioms for R*

Addition axioms	Multiplication axioms	Name
A-1 $a + b$ is a unique real number.	M-1 ab is a unique real number.	Closure
A-2 $a + b = b + a$	M-2 $ab = ba$	Commutative Laws
A-3 $(a + b) + c = a + (b + c)$	M-3 $(ab)c = a(bc)$	Associative Laws
A-4 There exists a unique real number 0, such that $a + 0 = a$ and $0 + a = a$.	M-4 There exists a unique real number 1, such that $a \cdot 1 = a$ and $1 \cdot a = a$.	Identity Laws
A-5 For each real number a, there exists a unique real number $-a$ (the *additive inverse* or the *opposite* of a) such that $a + (-a) = 0$ and $(-a) + a = 0$.	M-5 For each *nonzero* real number a, there exists a unique real number $1/a$ (the *multiplicative inverse* or the *reciprocal* of a) such that $a \cdot (1/a) = 1$ and $(1/a) \cdot a = 1$.	Inverses
D-1 $a(b + c) = ab + ac$ D-2 $(b + c)a = ba + ca$		Distributive Laws

EXAMPLE 1

Indicate which of the preceding axioms justifies the given statement:

a. $a + (b + 3) = a + (3 + b)$

b. $4(x + 2) = 4x + (4)(2)$

c. $(x + 5) \cdot \dfrac{1}{x + 5} = 1$ if $x + 5 \neq 0$

d. $-1 \cdot 1 = -1$

e. $4(xy) = (4x)y$

SOLUTION:

a. $a + (b + 3) = a + (3 + b)$ because the addition of numbers is commutative (Axiom A-2).

b. $4(x + 2) = 4x + (4)(2)$ because of the Distributive Law (Axiom D-1).

c. If $x + 5 \neq 0$, $(x + 5) \cdot 1/(x + 5) = 1$ because $x + 5$ and $1/(x + 5)$ are reciprocals (Axiom M-5).

d. $-1 \cdot 1 = -1$ because 1 is the multiplicative identity element (Axiom M-4).

e. $4(xy) = (4x)y$ because multiplication of numbers is associative (Axiom M-3).

The preceding discussion is concerned with only two operations, addition and multiplication. You know, of course, that the inverse operations, subtraction and division, are also very important. We define these operations in Definitions 1.2a and 1.2b.

▼ DEFINITION 1.2a

If a and b are real numbers, the *difference* of a and b is defined by

$$a - b = a + (-b)$$

The operation of finding a difference is called *subtraction*.

▼ DEFINITION 1.2b

If a and b are real numbers and $b \neq 0$, the *quotient* of a and b is defined by

$$\frac{a}{b} = a \cdot \frac{1}{b}$$

The operation of finding a quotient is called *division*.

The quotient $\dfrac{a}{b}$ is often written a/b or $a \div b$, and the symbol $\dfrac{a}{b}$ is sometimes called "the fraction, a over b." The numbers a and b are called the *numerator* and the *denominator*, respectively, of the fraction. Moreover, because $1/b$ is defined for $b \neq 0$ *only*, *division by zero is not defined;* that is, $a/0$ *is meaningless.*

EXAMPLE 2

Let us return to the puzzle introduced earlier in this section, and justify the final result.

SOLUTION:

We noted before that the closure properties of the operations being performed guarantee that all the steps will lead to real numbers. In the first two calculations, adding 1 and multiplying the result by 5, the Closure Axioms A-1 and M-1 apply. Because subtraction is defined in terms of addition, the Closure Axiom A-1 tells us that the result of subtracting 5 times the original number is again a real number. Now, we have

$5(n+1) - 5n = (5n + 5 \cdot 1) - 5n$	Distributive Law, D-1
$= (5n + 5) - 5n$	Identity Element for Multiplication, M-4
$= (5 + 5n) - 5n$	Commutative Law, A-2
$= (5 + 5n) + [-(5n)]$	Definition of Subtraction
$= 5 + \{5n + [-(5n)]\}$	Associative Law, A-3
$= 5 + 0$	Additive Inverse Axiom, A-5, and the Substitution Law of Equality
$= 5$	Identity Element for Addition, A-4

This completes the required justification.

Although the construction of proofs such as that in Example 2 is not a main objective of this book, you should be aware of the fact that all the procedures we shall study can be justified by using the basic definitions and axioms of algebra. This should reassure you that correctly performed procedures will lead to reasonable results.

THE NUMBER LINE

It is quite helpful to associate the real numbers with the points on a line. This provides a geometrical picture of the real numbers that enables us to visualize many of the important properties of these numbers and the operations we perform on them. To obtain this association, we first draw a horizontal straight line as in Figure 1.2a, select a point on it to represent the number 0, and label this point, the *origin*. We then choose a unit length and measure successive unit intervals to the right of the origin, labeling the points of division with the successive positive integers. Similarly, we measure successive unit intervals to the left of the origin and label the points of division in order $-1, -2, -3, \ldots$. We call this line the *number line*, because it allows us to set up a one-to-one correspondence between the real numbers and the points on the line. (This means that there will correspond exactly one point for each number and exactly one number for each point.) As an illustration, let the line

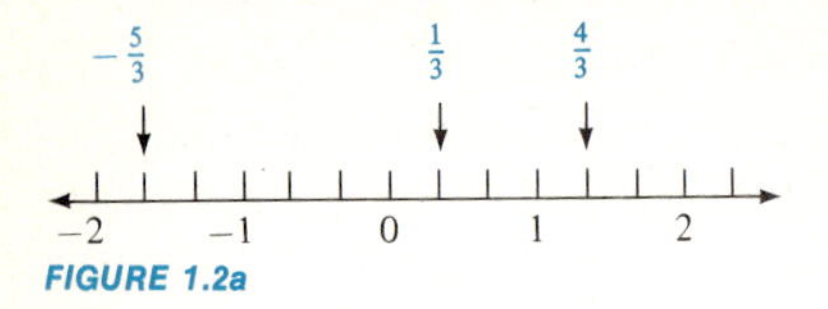

FIGURE 1.2a

segment from 0 to 1 be divided into three equal parts. Then the point of division closest to the origin represents the rational number $\frac{1}{3}$ (see Figure 1.2a). By counting off four such segments to the right of the origin, we locate the point that corresponds to the number $\frac{4}{3}$, and by counting off five such segments to the left of the origin, we find the point that corresponds to the number $-\frac{5}{3}$. In general, if m and n are positive integers, we can locate the points corresponding to m/n and $-m/n$ by a procedure similar to that used for the numbers $\frac{4}{3}$ and $-\frac{5}{3}$. We divide the segment from 0 to 1 into n equal parts and count off m of these segments to the right of the origin for the point corresponding to m/n and to the left for $-m/n$. Thus, we can locate the point that corresponds to any given rational number.

Except for certain special cases, it is much more difficult to locate exactly the point that corresponds to a given irrational number. However, such points can be located as accurately as desired by using the appropriate decimal approximations. It seems plausible now that each real number corresponds to a single point on the line, and we shall simply state the axiom that is needed as a basis for working with the number line. This axiom is called the *Completeness Axiom.* (See Axiom C-1.)

▼ AXIOM C-1

There is a one-to-one correspondence between the set of real numbers and the set of points on the number line.

The number that is associated with a point on the number line is called the *coordinate* of the point, and the point itself is called the *graph* of the number. Thus, the coordinate of point A in Figure 1.2b is $\frac{5}{2}$. The heavy dot at B indicates that this point is the graph of the number -3

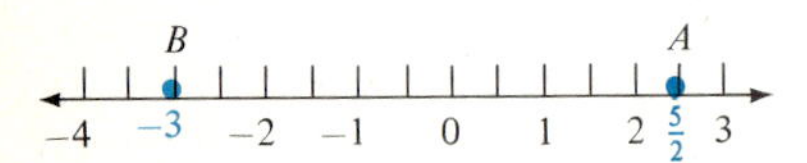

FIGURE 1.2b

ORDERING THE REAL NUMBERS

As the preceding discussion shows, the points to the *right* of the origin on the number line correspond to the *positive* real numbers, whereas those to the *left* correspond to the *negative* real numbers. The *origin* corresponds to the number *zero*, which is neither positive nor negative. The following *order axiom* states more precisely what has just been said. (See Axiom O-1.)

▼ AXIOM O-1

If a is a real number, then exactly one of the following statements is true:

I. a is positive.
II. a is zero.
III. a is negative.

Another order axiom simply states something we already know, namely, that the set of positive real numbers is closed under the operations of addition and multiplication. Thus, we make the formal statement in Axiom O-2.

▼ AXIOM O-2

If a and b are positive real numbers, then $a + b$ and ab are also positive.

We can now order the real numbers as follows: Suppose that a and b are two real numbers. We say that a is *greater than* b, denoted by $a > b$, or b is *less than* a, denoted by $b < a$, if $a - b$ is positive. This principle is restated in Definition 1.2c.

▼ DEFINITION 1.2c

If a and b are real numbers, then $a > b$ (or $b < a$) if and only if $a - b$ is positive.

The symbols $>$ and $<$ are called *inequality* signs, and statements such as $a > b$ or $b < a$ are called *inequalities.* From the definition, we can see that $1 < 2$ because $2 - 1 = 1$, which is positive. Similarly, $-5 < -2$ because $-2 - (-5) = 3$, which is positive.

It is helpful to note that this definition orders the real numbers to correspond to the order of their graphs on the number line. Thus, $a > b$ means that the graph of a is to the *right* of the graph of b. Similarly, $c < d$ means that the graph of c is to the *left* of the graph of d. This idea will be useful to us in working with inequalities.

The symbols $\leq$ and $\geq$ (read "less than or equal to" and "greater than or equal to") are also used in algebra. Thus, $x \leq 3$ means that $x < 3$ or $x = 3$. Similarly, $x \geq 2$ means that $x > 2$ or $x = 2$. Sometimes, we can combine inequalities into a chain as follows:

If $a < b$ and $b < c$, we write $a < b < c$.

Thus the inequalities $0 \leq x$ *and* $x \leq 3$ are written as $0 \leq x \leq 3$.

GRAPHS OF INEQUALITIES

The correspondence between the points on the number line and the real numbers makes it easy for us to graph simple inequalities as in the following examples.

EXAMPLE 3

Make a graph of the inequality $0 \leq x \leq 3$.

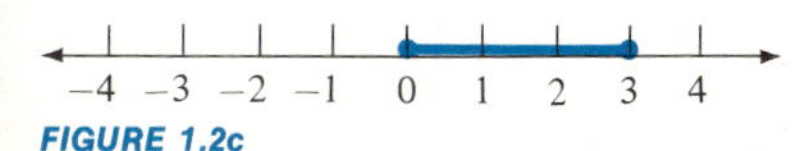

FIGURE 1.2c

SOLUTION:

The graph of this inequality is the set of points on the number line whose coordinates are real numbers between 0 and 3, inclusive. This graph is shown in Figure 1.2c. As a consequence of the *completeness axiom*, the graph is a solid line segment. The solid dots at 0 and 3 indicate that these points are part of the graph.

As just noted in the preceding example, we use a solid dot at the end of a line segment to show that this end point is a part of the graph. If we wish to indicate that a point is *not* part of a graph, we draw a small open circle about this point. Thus, the graph of the inequality $-2 < x < 3$ is shown in Figure 1.2d. In this figure, the points -2 and 3 have open circles about them to indicate that these points are *not* part of the graph.

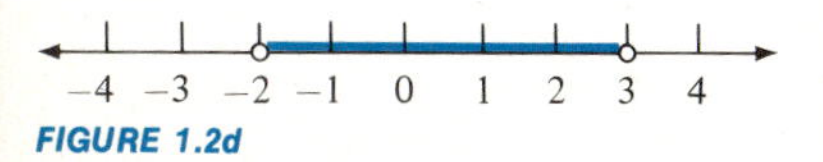

FIGURE 1.2d

EXAMPLE 4

Graph the inequalities:

a. $-1 \leq x < 2$

b. $x > 2$

SOLUTION:

a. The desired graph is shown in Figure 1.2e. You should note the heavy dot at -1 and the open circle at 2 to indicate that the first of these points is part of the graph and the second is not.

b. The graph of the inequality $x > 2$ is the set of all points to the right of point 2, as indicated in Figure 1.2f.

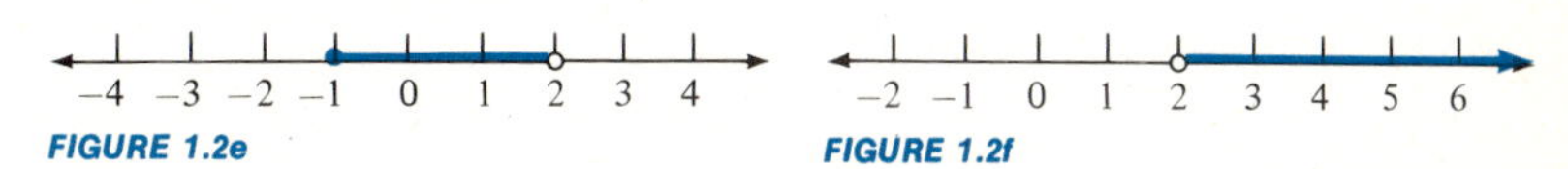

FIGURE 1.2e **FIGURE 1.2f**

In the preceding examples, we graphed inequalities involving x, and we included all real numbers that satisfied the given inequalities. Sometimes, there might be an additional restriction on x, as in the first two parts of Example 5.

EXAMPLE 5

Graph:

a. $-3 < x < 4$, x an integer;

b. $-3 < x < 4$, x a whole number;

c. $-3 < x < 4$

SOLUTION:

a. Here, we must graph the integers between -3 and 4. The graph is shown in Figure 1.2g. Notice that there are no solid segments of the number line included.

b. Because the whole numbers include no negative integers, the whole numbers between -3 and 4 are 0, 1, 2, and 3. The graph is shown in Figure 1.2h.

c. Because no additional restriction on x is given, we assume that x can be any real number satisfying the inequality. Thus, the graph is the line segment between -3 and 4, not including the end points, as shown in Figure 1.2i.

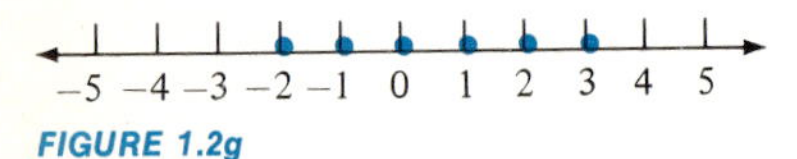

FIGURE 1.2g

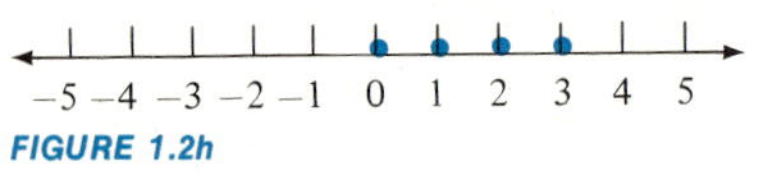

FIGURE 1.2h

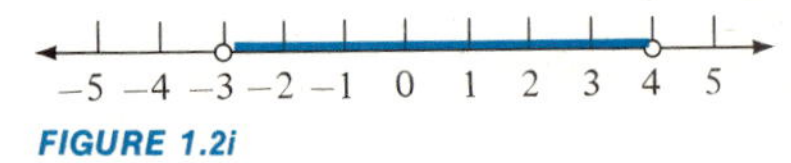

FIGURE 1.2i

EXERCISE 1.2

In Problems 1–6, state which of the axioms for equality justifies the given statement. Assume all letters represent real numbers.

1 $a + b = a + b$

2 If $18 = 2x + 6$, then $2x + 6 = 18$.

3 If $P = 4S$ and $S = 5$, then $P = 4 \cdot 5$.

4 If $P = I(I \cdot R)$ and $I \cdot R = V$, then $P = I \cdot V$.

5 If $d = rt$ and $d = (r + 3)(t - 2)$, then $rt = (r + 3)(t - 2)$.

6 If $I = Prt$ and $A = P + I$, then $A = P + Prt$.

In Problems 7–20, state which of the field axioms for the set R of real numbers justifies the given statement. The letters x, y, a, b, c, and d represent real numbers.

7 $-3 + 9$ is a real number.

8 $5x$ is a real number.

9 $3x + y = y + 3x$

10 $ad = da$

11 $4 + (x + 5) = (4 + x) + 5$

12 $3(4y) = (3 \cdot 4)y$

13 $(a + 8) \cdot 5 = 5 \cdot (a + 8)$

14 $4(x + y) = 4x + 4y$

15 $(ab)(cd) = (cd)(ab)$

16 $(5x) + (-5x) = 0$

17 $(x + y)(a + b) = (x + y)a + (x + y)b$

18 $1 \cdot (x + y) = x + y$

19 $(2 + x) \cdot \dfrac{1}{2 + x} = 1$ if $2 + x \neq 0$

20 $(x + 3) + 0 = x + 3$

In Problems 21–36, graph the given inequality.

21 $x < 4$, x a natural number

22 $x \leq 6$, x a natural number

23 $0 \leq x < 3$, x an integer

24 $-2 < x < 1$, x an integer

25 $-4 \leq x \leq 1$, x an integer

26 $-4 < x \leq 1$, x an integer

27 $x > -2$

28 $x < 2$

29 $1 < x < 3$

30 $-2 \leq x < 4$

31 $5 < x \leq 7$

32 $4 \leq x \leq 6$

33 $x < -\frac{1}{2}$

34 $x > \frac{3}{2}$

35 $\frac{1}{2} < x \leq \frac{7}{2}$

36 $-1 \leq x < \frac{9}{2}$

In Problems 37–46, write the inequality that corresponds to the given statement.

37 x is between 0 and 3. **38** x is between -1 and 4. **39** x is positive.

40 x is negative. **41** x is nonnegative. **42** x is between 0 and 1, inclusive.

43 x is between -3 and 4, inclusive. **44** $x - y$ is greater than 1 and less than 2.

45 $2x$ is between 3 and 5. **46** $x + y$ is less than or equal to 7.

USING YOUR KNOWLEDGE 1.2

In calculus and other more advanced mathematics, inequalities are used to define *intervals*. There are several types of intervals. See Definitions 1.2d and 1.2e. These two types of intervals are illustrated in Figure 1.2j. Notice that the end points *are not* included in an *open* interval, and they *are* included in a *closed* interval. Sometimes an interval includes one of its end points and not the other. Such an interval is called *half-open*. Figure 1.2k illustrates the two types of half-open intervals that are possible, $(a, b]$ and $[a, b)$.

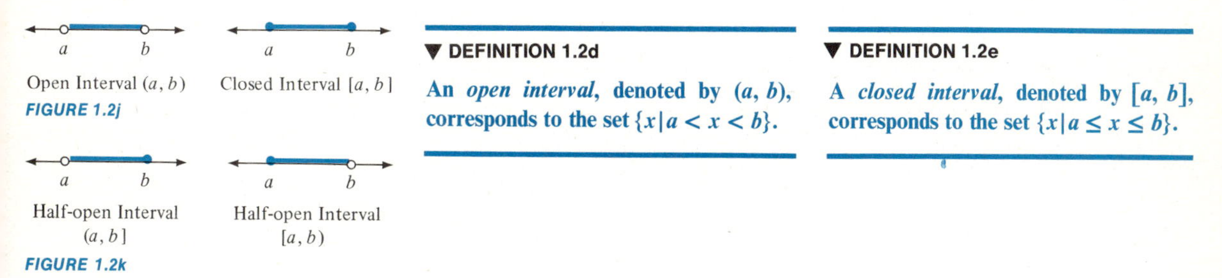

Open Interval (a, b) Closed Interval $[a, b]$

FIGURE 1.2j

Half-open Interval $(a, b]$ Half-open Interval $[a, b)$

FIGURE 1.2k

▼ **DEFINITION 1.2d**

An *open interval*, denoted by (a, b), corresponds to the set $\{x \mid a < x < b\}$.

▼ **DEFINITION 1.2e**

A *closed interval*, denoted by $[a, b]$, corresponds to the set $\{x \mid a \leq x \leq b\}$.

All the above intervals are said to be *bounded*. An interval that is not bounded is said to be *unbounded*. For example, the interval corresponding to the set of positive real numbers, $\{x \mid x > 0\}$, is *unbounded*. To describe this interval, we use the symbol ∞ (read "infinity") to denote that there is no right-hand end point. The interval corresponding to the positive real numbers is symbolized by $(0, \infty)$ In general,

An unbounded interval with a left-hand end point a is denoted by $[a, \infty)$ or (a, ∞) depending on whether end point a is included or not.

To describe an interval that is unbounded to the left, we use the symbol $-\infty$ (read "minus infinity"). For instance, we may write $(-\infty, 3]$ or $(-\infty, 3)$ for the interval that has 3 as its right-hand end point and is unbounded to the left. A bracket or a parenthesis is used on the right depending on whether the 3 is included or not.

Use this knowledge to do the following problems.

1 Use interval notation for the interval described by: **a** $3 < x < 8$ **b** $-2 \leq x < 0$ **c** $-1 \leq x \leq 1$ **d** $2 < x \leq 5$

2 Use interval notation for the interval described by: **a** the set of negative real numbers, $\{x \mid x < 0\}$; **b** the set of real numbers greater than -2; **c** the set of real numbers less than 2.

3 Use interval notation for the interval corresponding to: **a** the real numbers between 2 and 3; **b** the real numbers between -1 and 5, inclusive.

4 Graph the interval: **a** $(2, 4]$ **b** $[-1, 3)$ **c** $[-4, 0]$ **d** $(-\infty, 1)$ **e** $[2, \infty)$ **f** $(-\infty, -2]$

1.3 THE ARITHMETIC OF THE REAL NUMBERS

Even though the fundamental operations involving the rational numbers have been used for many centuries, they are still often misunderstood. For this reason, we will review some basic ideas in arithmetic before going on in algebra.

In the preceding section, we stated axioms for the real numbers but said nothing about the *rules* governing the operations of addition, subtraction, multiplication, and division. We shall now state several properties of the real numbers that can be derived from the axioms and that can help us with the fundamental operations. We shall number some of these derived properties, which are called *laws* or *theorems*, for easy reference.

▼ DEFINITION 1.3a

The *absolute value* of a number x, denoted by $|x|$, is the distance of the number from the origin on the number line.

As you may recall, one of the order axioms (Axiom O-2) states that the sum of two positive numbers is positive. What happens if at least one of the numbers is negative? The answer is stated in terms of *absolute values.* According to Definition 1.3a, $|5| = 5$, $|-3| = 3$, $|0.8| = 0.8$, $|-\frac{1}{5}| = \frac{1}{5}$, and $|0| = 0$. (Note that you can obtain the absolute value of a specific real number simply by disregarding the sign of the number.)

The following rule is used to add two numbers with the same sign.

To add two real numbers with *like* signs, add their absolute values and prefix the result with their common sign.

Thus, $(+8) + (+2) = +(8 + 2) = +10$, which we write simply

$8 + 2 = 10$

Similarly, $(-3) + (-5) = -(3 + 5) = -8$, which we may write

$-3 + (-5) = -8$

The rule for adding numbers with unlike signs is:

To add two numbers with *unlike* signs, subtract the smaller absolute value from the larger absolute value and prefix the result with the sign of the number having the larger absolute value.

Thus, $(-3) + (+7) = +(7 - 3) = +4$, which we usually write

$-3 + 7 = 4$

Similarly,

$2 + (-5) = -(5 - 2) = -3$

EXAMPLE 1

Perform the following additions:

a. $-7+(-9)$
b. $8+(-2)$
c. $-9+2$

SOLUTION:

a. The numbers have *like* signs. Hence,

$$-7+(-9) = -(7+9) = -16$$

b. The numbers have *unlike* signs. Thus,

$$8+(-2) = +(8-2) = 6$$

c. Here the numbers also have *unlike* signs, so that

$$-9+2 = -(9-2) = -7$$

▼ $a - b = a + (-b)$

For your convenience, we repeat the definition of subtraction (Definition 1.2a). This means that to *subtract* b from a, we simply *add* the *opposite* (the *additive inverse*) of b to a. For instance,

$$3-9 = 3+(-9) = -6$$
$$-5-2 = -5+(-2) = -7$$
$$-4-(-6) = -4+6 = 2$$

Note, in the last illustration, that we added 6, the opposite of -6, to the -4.

EXAMPLE 2

Perform the following subtractions:

a. $-9-(-2)$
b. $-1-5$
c. $4-9$

SOLUTION:

a. $-9-(-2) = -9+2 = -7$
b. $-1-5 = -1+(-5) = -6$
c. $4-9 = 4+(-9) = -5$

▼ **THEOREM 1.3a**

If a and b are real numbers, then

I. $-(-a) = a$
II. $(-a)(b) = (a)(-b) = -(ab)$
III. $(-a)(-b) = ab$
IV. $(-1)(a) = -a$

The multiplication of signed numbers is based on Theorem 1.3a, the different parts of which can be derived from the axioms for the real numbers in Section 1.2. What does this theorem really say? The first statement says that the opposite of the opposite of a number is the number itself; that is, $-(-a)$ is just a. (We used this idea earlier when we wrote $-4-(-6) = -4+6$.) The second statement says that the product of b and the opposite of a, or of a and the opposite of b, is the opposite of ab. The third part says that the product of the opposites of a and b is the same as the product of a and b. You may think of the fourth part as the special case of the second part, in which a is replaced by 1 and b by a.

Although most students have no difficulty accepting Parts I, II, and IV, a great many students do not understand the reasons for III. Here is a proof of Theorem 1.3a, Part III in case you need to be convinced.

Statement	*Reason*
1. $-b + b = 0$	b and $-b$ are additive inverses (Axiom A-5).
2. $-a(-b + b) = -a \cdot 0 = 0$	by the Substitution Axiom (E-4) and the fact that $-a \cdot 0 = 0$
3. $(-a)(-b) + (-a)(b) = 0$	by the Distributive Law (D-1)
4. $(-a)(-b) + [-(ab)] = 0$	Theorem 1.3a, Part II
5. $(-a)(-b) = -[-(ab)]$	By step 4, $(-a)(-b)$ and $-(ab)$ are additive inverses (Axiom A-5).
6. $(-a)(-b) = ab$	Theorem 1.3a, Part I

Fortunately, you do not need to memorize the preceding proof. The information that is needed to multiply (or divide) signed numbers can be derived from Theorem 1.3a. This information is stated in the Rule of Signs.

THE RULE OF SIGNS

1. **The product (or quotient) of two numbers of *like* sign is *positive*.**
2. **The product (or quotient) of two numbers of *unlike* sign is *negative*.**

EXAMPLE 3

Perform the indicated operations:

a. $(+8)(-3)$

b. $\dfrac{-16}{4}$ c. $\dfrac{-25}{-5}$

SOLUTION:

a. $+8$ and -3 have unlike signs. Thus, $(+8)(-3) = -24$.

b. -16 and 4 have unlike signs. Hence, $-16/4 = -4$.

c. Because -25 and -5 are of the same sign, $-25/-5 = 5$.

Now that we know how to handle the signs in dealing with signed numbers, we will turn to the rules for operating with fractions. All the required rules are stated in Theorem 1.3b, *where all denominators are assumed to be nonzero.*

THEOREM 1.3b

If a, b, c, and d are real numbers, then

I. $\dfrac{a}{b} = \dfrac{c}{d}$ **if and only if** $ad = bc$.

II. $\dfrac{a}{b} = \dfrac{ad}{bd}$ III. $\dfrac{a}{-b} = \dfrac{-a}{b} = -\dfrac{a}{b}$

IV. $\dfrac{a}{b} + \dfrac{c}{b} = \dfrac{a+c}{b}$

V. $\dfrac{a}{b} + \dfrac{c}{d} = \dfrac{ad+bc}{bd}$ VI. $\dfrac{a}{b} \cdot \dfrac{c}{d} = \dfrac{ac}{bd}$

VII. $\dfrac{a}{b} \div \dfrac{c}{d} = \dfrac{a}{b} \cdot \dfrac{d}{c} = \dfrac{ad}{bc}$

Part I of this theorem tells us that $\frac{2}{5}$ and $\frac{8}{20}$ represent the same rational number because $2 \cdot 20 = 8 \cdot 5$. However, we prefer the $\frac{2}{5}$ as being the *simpler* of the two.

A common fraction a/b, where a and b are integers, is in *simplest form* if a and b have no common divisor except 1 or -1.

Because $\frac{8}{20} = (2 \cdot 4)/(5 \cdot 4)$, we can *reduce* the fraction to $\frac{2}{5}$ by Part II of Theorem 1.3b.

EXAMPLE 4

Reduce to simplest form:

a. $\dfrac{48}{54}$ b. $\dfrac{-42}{63}$

SOLUTION:

a. $\dfrac{48}{54} = \dfrac{8 \cdot 6}{9 \cdot 6} = \dfrac{8}{9}$

b. $\dfrac{-42}{63} = \dfrac{-2 \cdot 21}{3 \cdot 21} = \dfrac{-2}{3}$ or $-\dfrac{2}{3}$

Part IV of Theorem 1.3b tells us how to add fractions with the same denominator. Thus,

$$\frac{3}{5}+\frac{8}{5}=\frac{3+8}{5}=\frac{11}{5} \quad \text{and} \quad \frac{4}{9}+\frac{3}{9}=\frac{7}{9}$$

But what about something like $\frac{1}{6}+\frac{1}{9}$? We could, of course, use Part V of the theorem, but it is usually more convenient to change the fractions to equivalent fractions with the least common denominator (LCD), that is, the smallest integer that is divisible by all the denominators, 6 and 9 in this case. The smallest positive integer that is divisible by both 6 and 9 is 18; so we proceed as follows:

$$\frac{1}{6}+\frac{1}{9}=\frac{1\cdot 3}{6\cdot 3}+\frac{1\cdot 2}{9\cdot 2}=\frac{3}{18}+\frac{2}{18}=\frac{5}{18}$$

Subtraction is done in a similar way.

EXAMPLE 5

Perform the indicated operations:

a. $\frac{3}{4}+\frac{7}{10}$

b. $\frac{5}{6}-\frac{9}{10}$

SOLUTION:

a. We first write the fractions as equivalent ones with a denominator of 20, the LCD of the two fractions. Thus,

$$\frac{3}{4}+\frac{7}{10}=\frac{3\cdot 5}{4\cdot 5}+\frac{7\cdot 2}{10\cdot 2}=\frac{15}{20}+\frac{14}{20}=\frac{29}{20}$$

b. The LCD for $\frac{5}{6}$ and $\frac{9}{10}$ is 30. Hence, we write

$$\frac{5}{6}-\frac{9}{10}=\frac{5\cdot 5}{6\cdot 5}-\frac{9\cdot 3}{10\cdot 3}=\frac{25}{30}-\frac{27}{30}=-\frac{2}{30}=-\frac{1}{15}$$

Notice that the answer was reduced to simplest form.

The operations remaining to be discussed are multiplication and division of fractions. Parts VI and VII of Theorem 1.3b show exactly what must be done. To multiply two fractions, we multiply numerator by numerator and denominator by denominator. To divide one fraction by another, we invert the divisor and multiply. There is just one warning: To keep the numbers as small as possible, make sure that all fractions are reduced to simplest form. Example 6 illustrates the procedure.

EXAMPLE 6

Perform the indicated operations:

a. $\frac{-5}{9}\cdot\frac{18}{27}$

b. $\frac{3}{4}\div\frac{-9}{16}$

SOLUTION:

a. Because $\frac{18}{27}=\frac{2}{3}$, we have

$$\frac{-5}{9}\cdot\frac{18}{27}=-\frac{5}{9}\cdot\frac{2}{3}=-\frac{10}{27}$$

b. Here, we must invert the divisor and multiply. We note that the quotient is negative, so that

$$\frac{3}{4} \div \frac{-9}{16} = -\frac{\overset{1}{\cancel{3}}}{\underset{1}{\cancel{4}}} \cdot \frac{\overset{4}{\cancel{16}}}{\underset{3}{\cancel{9}}} = -\frac{4}{3}$$

Be certain to notice the simplification in the last step.

EXERCISE 1.3

In Problems 1–50, perform the indicated operations.

1 $-13 + (-2)$ **2** $-8 + (-3)$ **3** $-14 + 5$ **4** $-18 + 7$ **5** $-\frac{1}{5} + (-\frac{2}{5})$ **6** $-\frac{2}{7} + (-\frac{3}{7})$

7 $-3 - (-5)$ **8** $-8 - (-2)$ **9** $-7 - 9$ **10** $-9 - 6$ **11** $-\frac{1}{5} - (-\frac{3}{5})$ **12** $-\frac{3}{8} - (-\frac{1}{8})$

13 $-0.8 - (-0.1)$ **14** $-0.6 - (-0.3)$ **15** $-0.4 - 0.7$ **16** $-0.2 - 0.9$ **17** $(9)(-4)$ **18** $8(-6)$

19 $(-3)(-6)$ **20** $(-7)(-9)$ **21** $(-3)(4)(-2)$ **22** $(-5)(2)(-3)$ **23** $(-2)(-3)(-4)$ **24** $(-4)(-2)(-3)$

25 $\frac{-18}{2}$ **26** $\frac{30}{-5}$ **27** $\frac{-8}{-4}$ **28** $\frac{-12}{-4}$ **29** $3 - (8 - 4)$ **30** $(2 - 7 + 1) - (8 - 9 - 1)$

31 $(6)(-2) - (2)(-5)$ **32** $(3)(-5) - (5)(-2)$ **33** $-\frac{1}{2} + \frac{2}{5}$ **34** $\frac{1}{4} + (-\frac{1}{6})$

35 $-\frac{7}{8} + (-\frac{1}{12})$ **36** $-\frac{1}{3} + (-\frac{5}{6})$ **37** $\frac{5}{8} - \frac{7}{12}$ **38** $\frac{8}{15} - \frac{2}{25}$ **39** $\frac{3}{4} + \frac{1}{12} - \frac{1}{6}$ **40** $\frac{5}{6} - \frac{1}{9} - \frac{1}{3}$

41 $\frac{-3}{4} \cdot \frac{8}{27}$ **42** $\frac{5}{6} \cdot \frac{6}{-15}$ **43** $\frac{9}{-7} \cdot \frac{-14}{27}$ **44** $\frac{8}{-5} \cdot \frac{-15}{16}$ **45** $\frac{4}{5} \div \frac{16}{-15}$ **46** $\frac{7}{8} \div \frac{21}{-16}$

47 $\frac{-9}{4} \div \frac{-45}{16}$ **48** $\frac{-30}{19} \div \frac{15}{-38}$ **49** $\frac{3}{4} \cdot \frac{5}{6} \div \frac{25}{18}$ **50** $(\frac{5}{6} \div \frac{25}{12})(\frac{18}{25})$

51 Prove that if a is a real number, then $-(-a) = a$.

Hint: Start with $(-a) + a = 0$ and use the Additive Inverse Axiom.

52 Prove that if a and b are real numbers, then $(-a)(b) = -(ab)$.

Hint: Start with $ab + (-a)b = [a + (-a)]b$ and use the fact that $0 \cdot b = 0$.

53 Prove that if a is a real number, then $(-1)a = -a$.

54 According to Theorem 1.3b,

$$\frac{a}{b} \div \frac{c}{d} = \frac{ad}{bc}$$

Show that this checks with the requirement that quotient times divisor equals dividend.

USING YOUR KNOWLEDGE 1.3

The oxidation number (valence) of a chemical compound is the sum of the oxidation numbers of the individual atoms in the compound. For example, the oxidation number of hydrogen is $+1$ and that of oxygen is -2. Hence, the oxidation number of water, H_2O, is found as follows:

$$\overset{H_2O}{\overbrace{2(\text{oxid. no. H}) + (\text{oxid. no. O})}} = 2(+1) + (-2) = 2 + (-2) = 0$$

Use this idea to find the oxidation numbers of the compounds in Problems 1–4.

1 Sodium selenide, Na_2Se, given that (oxid. no. Na) = +1, (oxid. no. Se) = −2.
2 Phosphate, PO_4, given that (oxid. no. P) = +5, (oxid. no. O) = −2.
3 Sulphuric acid, H_2SO_4, given that (oxid. no. H) = +1, (oxid. no. S) = +6, (oxid. no. O) = −2.
4 Cryolite, Na_3AlF_6, given that (oxid. no. Na) = +1, (oxid. no. Al) = +3, (oxid. no. F) = −1.
5 The compound MnO_4^- is known to have an oxidation number of −1. If (oxid. no. O) = −2, what must be the oxidation number of manganese, Mn?

SELF-TEST

1 Write the set of whole numbers less than 3 by:
a listing the elements in braces
b using set-builder notation.

2 Let $A = \{1, 3, 5, 6, 8\}$, $B = \{3, 7, 8\}$, and $C = \{1, 7, 9\}$. Find:
a the set D consisting of all elements in A or B;
b the set E consisting of all elements common to A and C;
c the set of all elements common to C and the set D of part (a).

3 Write as a decimal:
a $\frac{2}{5}$ b $\frac{10}{11}$

4 Classify each number as rational or irrational:
a $\frac{8}{9}$ b $2.\overline{7}$ c $\sqrt{15}$ d 9.21

5 State which axiom for equality justifies the given statement:
a If $E = IR$ and $R = r_1 + r_2$, then $E = I(r_1 + r_2)$.
b If $5 = 3x - 1$, then $3x - 1 = 5$.

6 State which field axiom for the real numbers justifies the given statement:
a $a + (b + c) = (b + c) + a$
b $-5 + 0 = -5$

7 State which field axiom for the real numbers justifies the given statement:
a $(x+2) \cdot \dfrac{1}{x+2} = 1$, if $x + 2 \neq 0$
b $a(bc) = a(cb)$

8 Graph the inequalities:
a $-2 \leq x < 1$ b $x > 1$

9 Graph the inequality $-\frac{3}{2} \leq x \leq \frac{5}{2}$.

10 Graph the set of numbers described by:
a $-2 < x < 3$, x an integer
b the integers between $\frac{1}{2}$ and 4.

11 Write the inequality that corresponds to the statement:
a x is greater than 2 and less than 5
b x is negative.

12 Perform the indicated additions:
a $(-8) + (-4)$ b $9 + (-1)$
c $-8 + 6$ d $(-3) + 5$
e $(-2) + 7$ f $(-5) + 9 + (-6)$

13 Subtract as indicated:
a $8 - 1$ b $9 - (-3)$
c $-8 - (-2)$ d $-8 - (-10)$
e $8 - 9$

14 Multiply as indicated:
a $(+9)(-2)$ b $(-8)(-3)$
c $(-5)(7)$ d $(4)(-9)$ e $(-3)(-3)$

15 Divide as indicated:
a $\dfrac{16}{-8}$ b $\dfrac{-9}{3}$ c $\dfrac{0}{-7}$ d $\dfrac{-24}{-8}$
e $-\dfrac{-14}{-7}$

16 Reduce to simplest form:
a $\dfrac{64}{72}$ b $\dfrac{-56}{48}$

17 Fill in the correct number for the numerator on the right:
a $\dfrac{5}{9} = \dfrac{\quad}{54}$ b $\dfrac{-4}{7} = \dfrac{\quad}{49}$

18 Perform the indicated operations:
a $\frac{3}{5} + \frac{7}{10}$ b $\frac{5}{6} + \frac{3}{10}$ c $\frac{7}{8} - \frac{1}{5}$
d $\frac{5}{9} - \frac{7}{11}$ e $\frac{3}{4} - \frac{1}{5} + \frac{1}{8}$
f $\frac{5}{6} - \frac{2}{5} - \frac{3}{10}$

19 Multiply as indicated:
a $\dfrac{-3}{4} \cdot \dfrac{16}{-9}$ b $\dfrac{8}{9} \cdot \dfrac{-27}{16}$

20 Divide as indicated:
a $\dfrac{9}{5} \div \dfrac{-9}{5}$ b $\dfrac{4}{5} \div \dfrac{32}{-15}$

CHAPTER 2

POLYNOMIALS AND RATIONAL FRACTIONS

2.1 POLYNOMIALS: ADDITION AND SUBTRACTION

ALGEBRAIC EXPRESSIONS

Suppose that a man jumps from a height of 80 feet into a safety mattress. Do you know how long it takes him to hit the mattress? It is shown in physics that after t seconds ($0 \le t \le \sqrt{5}$), the man's height above the mattress is given by the expression

$$-16t^2 + 80$$

In this expression, the letter t, which may be replaced by any number between 0 and $\sqrt{5}$, inclusive, is called a *variable*. In general, we can define a variable as in Definition 2.1a. Notice that the expression $-16t^2 + 80$ has a meaningful value for each permissible value of the variable t. In particular, the expression has the value zero for $t = \sqrt{5}$; thus, it takes $\sqrt{5}$ seconds for the man to hit the mattress. (This is the reason for the restriction $t \le \sqrt{5}$.)

▼ DEFINITION 2.1a

A *variable* is a symbol that may be replaced by any one of a specified set of numbers.

The expression $-16t^2 + 80$ is an example of an *algebraic expression*. (See Definition 2.1b.) For instance,

▼ DEFINITION 2.1b

An *algebraic expression* is formed by starting with a collection of variables (usually denoted by letters of the alphabet) and real numbers and applying a finite number of the operations of addition, subtraction, multiplication, division, and the extraction of roots.

$$s^2 + t^2 + 80, \quad \frac{4x + \sqrt{5x + 8}}{2x^2 - 9}, \quad xy + 5\sqrt{z} - w, \quad \text{and } 5$$

are all examples of algebraic expressions.

If we substitute specific numbers for the variables in an algebraic expression, the resulting number is called the *value* of the expression for the given numerical values of the variables. For example, refer to the jumper who was mentioned at the beginning of this section and suppose we wish to know how high above the mattress he is after one second. To find out, we substitute $t = \mathbf{1}$ in the expression $-16t^2 + 80$, obtaining

$$-16(\mathbf{1})^2 + 80 = -16 + 80 = 64$$

Thus, after one second, the man is 64 feet above the mattress.

Sometimes we represent algebraic expressions in one variable by symbols such as

$$P(x), \quad D(y), \quad \text{and } Q(z)$$

(read "P of x", "D of y", and "Q of z"). The letter inside the parentheses is the variable being used. Thus, we might write

$$P(x) = x^2 - 3x$$
$$D(y) = \sqrt{y} + 1$$
$$Q(z) = -z + 2$$

With this notation, we can indicate the value of an expression for a specific value of the variable. Thus, $P(\mathbf{-1})$ represents the value of the expression $P(x)$ when x is replaced by $\mathbf{-1}$. For example, if

$$P(x) = x^2 - 3x$$

then

$$P(\mathbf{-1}) = (\mathbf{-1})^2 - 3(\mathbf{-1}) = 1 + 3 = 4$$

Similarly, because $D(y) = \sqrt{y} + 1$,

$$D(4) = \sqrt{4} + 1 = 2 + 1 = 3$$

Note that the radical sign $\sqrt{}$ *always* means the positive square root.

EXAMPLE 1

a. If $H(t) = -16t^2 + 80$, find $H(0)$, $H(\frac{3}{2})$, $H(2)$, $H(\sqrt{5})$, $H(a)$.

b. If $Q(x) = (2 + x)/x^2$, find $Q(-2)$, $Q(1)$, $Q(2)$, $Q(x + h)$.

SOLUTION:

a.
$$H(0) = -16(0)^2 + 80 = -16 \cdot 0 + 80 = 80$$
$$H(\tfrac{3}{2}) = -16(\tfrac{3}{2})^2 + 80 = -16 \cdot \tfrac{9}{4} + 80 = 44$$
$$H(2) = -16 \cdot 2^2 + 80 = 16$$
$$H(\sqrt{5}) = -16(\sqrt{5})^2 + 80 = -16 \cdot 5 + 80 = 0$$
$$H(a) = -16a^2 + 80$$

b.
$$Q(-2) = \frac{2 + (-2)}{(-2)^2} = 0$$

$$Q(1) = \frac{2 + (1)}{1^2} = 3$$

$$Q(2) = \frac{2 + (2)}{2^2} = 1$$

$$Q(x + h) = \frac{2 + (x + h)}{(x + h)^2}$$

POSITIVE INTEGER EXPONENTS

In the expression $-16t^2$, the t^2 is called a *power* of t; the t is called the *base*, and the 2 is the *exponent* of the power. In general, we have Definition 2.1c. In view of this definition, t^2 is the second power of t (often called the *square of* t); x^3 is the third power of x (often called the *cube of* x); y^4 is the fourth power of y, and so on.

▼ DEFINITION 2.1c

If n is a positive integer and x is a real number, then the *nth power of* x is denoted by x^n, where

$$x^n = \underbrace{x \cdot x \cdot x \cdots x}_{n \text{ factors}}$$

x is called the base and n the *exponent* of the power. We agree to write x for x^1.

POLYNOMIALS

Algebraic expressions that are built up from the variables and the real numbers by using only the three operations of addition, subtraction, and multiplication are called *polynomials*. Thus,

$$4, \quad 3x^2 + 7x - 8, \quad \tfrac{5}{3}y^3 - \tfrac{1}{3}x^2z, \quad \text{and} \quad \pi xz^2 + \sqrt{2}$$

are all polynomials, but $1/x$, $2 + \sqrt{x}$, and $(1 - 2x)/x^2$ are not polynomials ($1/x$ and $(1 - 2x)/x^2$ involve division by a variable, and $2 + \sqrt{x}$ involves the square root of a variable).

The parts of a polynomial that are separated by plus signs are called *terms* of the expression. Thus, the polynomial $-16t^2 + 80$ has two terms, $-16t^2$ and 80; the polynomial $x - 2y + 7$ has the three terms x, $-2y$, and 7. Notice that $x - 2y + 7$ is regarded as $x + (-2y) + 7$; the minus sign is taken to be part of the multiplier of y.

Polynomials can be classified according to the number of terms they have. Thus, polynomials of *one* term are called *monomials*; polynomials with *two* terms are called *binomials*; and polynomials with *three* terms are called *trinomials*. Special names are generally not used if there are more than three terms. For example, $-16t^2 + 80$ is a binomial; $2x^n$ is a monomial, and $x + 3y^4 - 2z$ is a trinomial.

In a monomial, the numerical multiplier is called the *numerical coefficient* (or just the *coefficient*) of the other factor of the monomial. Thus, in $3xy^2$, the 3 is the coefficient of xy^2; in x^3, the coefficient is 1 (because $x^3 = 1 \cdot x^3$); in $-2x^4$, the coefficient is -2 Sometimes we speak of the coefficient of some particular variable factor of the monomial, so that $3x$ is the coefficient of y^2 in $3xy^2$, and $3y^2$ is the coefficient of x in the same monomial. Coefficients that involve variables as in the last illustration are often called *literal* coefficients.

Polynomials in one variable can also be classified according to the highest power of the variable. The exponent of this power is called the *degree* of the polynomial. For example, the monomial $-2x^4$ is of the *fourth* degree; the trinomial $x^2 - 10x^5 - 2$ is of the *fifth* degree (because the highest exponent of the variable is 5); and the binomial $1 - 2x$ is of the *first* degree. By convention, a nonzero number, such as 3 or -17 is called a polynomial of degree zero. The number 0 is called the *zero polynomial* and is not assigned a degree.

The degree of a monomial with more than one variable is the sum of the exponents of the variable factors. Thus, the degree of $5xy^2$ is $1 + 2 = 3$. The degree of a polynomial with more than one variable is the degree of the term of highest degree. For example, the degree of $3x^2 - y + 6$ is 2, the degree of the first term. The degree of $3x^2 - y + xyz$ is $1 + 1 + 1 = 3$, the degree of the third term. The degree of $-x^2y^3z + x^4y^2$ is 6, because the degree of each term is 6. Sometimes it is desirable to consider the degree with respect to a particular variable. Thus, the preceding polynomial is of degree 4 in x, of degree 3 in y, and of degree 1 in z, although it is of degree 6 in the three variables combined.

EXAMPLE 2

Classify each polynomial as a monomial, binomial, or trinomial. Give the degree and name the coefficients:

a. $5x^4 - 3x^2 - 8$ **b.** x
c. $xy^2z^3 + 9z^7$ **d.** 0

SOLUTION:

a. A trinomial of degree 4. The coefficient of x^4 is 5; of x^2 is -3; of the constant term (the term with no variable) is -8.
b. A monomial of degree 1. The coefficient is 1.
c. A binomial of degree 7. The coefficient of xy^2z^3 is 1; of z^7 is 9.
d. The zero polynomial. No degree is assigned. The coefficient is 0.

ADDITION AND SUBTRACTION OF POLYNOMIALS

We are now able to perform the basic operations of addition and subtraction of polynomials. We shall consider multiplication and division in the next section. In doing these calculations, we keep in mind that the variables in the polynomials represent real numbers, so that the polynomials become real numbers when such substitutions are made for the variables. This means that all the rules given in Chapter 1 may be applied. For example, to add $2x^2 + 5x - 4$ and $3x^2 - 2x + 7$, we use the Associative, Commutative, and Distributive Laws to write

$$\begin{aligned}(2x^2 + 5x - 4) + (3x^2 - 2x + 7) &= (2x^2 + 3x^2) + (5x - 2x) + (-4 + 7)\\ &= (2 + 3)x^2 + (5 - 2)x + 3\\ &= 5x^2 + 3x + 3\end{aligned}$$

This procedure is often shortened by writing the terms of the polynomial in order of descending degree and then placing the terms with like powers in the same column as shown next.

$$\begin{array}{rr} & 2x^2 + 5x - 4\\ (+) & 3x^2 - 2x + 7\\ \hline & 5x^2 + 3x + 3\end{array}$$

To subtract polynomials, we first recall that $a - b = a + (-b)$. Consequently,

$$\begin{aligned} a - (b + c) &= a + [-(b + c)] \\ &= a + [-1(b + c)] \\ &= a + [-b - c] \\ &= a - b - c \end{aligned}$$

This shows that to subtract $(b + c)$ from a, we simply change the sign of each term inside the parentheses and add. Hence,

$$\begin{aligned} &(7x^3 + 2x^2 - 3x + 2) - (x^3 + 6x^2 - 3x - 4) \\ &\quad = (7x^3 + 2x^2 - 3x + 2) + (-x^3 - 6x^2 + 3x + 4) \\ &\quad = (7 - 1)x^3 + (2 - 6)x^2 + (-3 + 3)x + (2 + 4) \\ &\quad = 6x^3 - 4x^2 + 6 \end{aligned}$$

We can shorten the procedure by writing

$$\begin{array}{rr} & 7x^3 + 2x^2 - 3x + 2 \\ (-) & x^3 + 6x^2 - 3x - 4 \\ \hline \end{array} \quad \text{as} \quad \begin{array}{rr} & 7x^3 + 2x^2 - 3x + 2 \\ (+) & -x^3 - 6x^2 + 3x + 4 \\ \hline & 6x^3 - 4x^2 \qquad + 6 \end{array}$$

EXAMPLE 3

Subtract $4 + 3x - 7x^3$ from $9x^3 + 3x^2 - 2x + 1$.

SOLUTION:

We follow the foregoing shortened procedure, writing each polynomial in order of descending degree with like powers in the same column and changing the signs of all the terms in the polynomial to be subtracted.

$$\begin{array}{rr} & 9x^3 + 3x^2 - 2x + 1 \\ (+) & 7x^3 \qquad\quad - 3x - 4 \\ \hline & 16x^3 + 3x^2 - 5x - 3 \end{array}$$

Note that all signs have been changed.

EXAMPLE 4

If $P(x) = 3x^3 + 4x - 7$ and $Q(x) = x^3 - x^2 - x + 2$, find $P(x) - 3Q(x)$.

SOLUTION:

We first multiply the second polynomial by 3 in order to form $3Q(x)$. Thus, $3Q(x) = 3(x^3 - x^2 - x + 2) = 3x^3 - 3x^2 - 3x + 6$. We then follow the procedure of Example 3 to obtain

$$\begin{array}{rr} & 3x^3 \qquad\quad + 4x - 7 \\ (+) & -3x^3 + 3x^2 + 3x - 6 \\ \hline & 3x^2 + 7x - 13 \end{array}$$

Hence, $P(x) - 3Q(x) = 3x^2 + 7x - 13$. Notice that we leave a blank space for the missing power in the first polynomial.

EXERCISE 2.1

In Problems 1–10, classify each polynomial as a monomial, binomial, or trinomial. Then give the degree of the polynomial.

1 $x^2 - 5z + 7y$ **2** $9xyz - z^2 + 4y^2$ **3** $x^4 - 28yz^3$ **4** $3 - 17x^2$ **5** -9 **6** x^8 **7** 0 **8** $xy - xz + yz$

9 $7x^8 + 10$ **10** $3xyz$

11 What is the coefficient of the z term in Problem 1?

12 What is the coefficient of the highest degree term in Problem 2?

13 What is the coefficient of the yz^3 term in Problem 3?

14 What is the coefficient of the xz term in Problem 8?

In Problems 15–20, find (a) $P(x) + Q(x)$ and (b) $P(x) - Q(x)$.

15 $P(x) = 2x^2 + 5x - 3$, $Q(x) = 4x^2 - 3x + 2$

16 $P(x) = 5x^2 + 4x + 1$, $Q(x) = -3x^2 - 5x - 8$

17 $P(x) = 8x^3 + 7x^2 + 5x - 5$, $Q(x) = 8x^3 + 7x - 6$

18 $P(x) = x^3 - 4x^2 + 2x - 5$, $Q(x) = 3x^3 - x^2 + x + 2$

19 $P(x) = 4x^2 - 7x - 5$, $Q(x) = x^2 - x - 2$

20 $P(x) = 3x^2 - 2x - 3$, $Q(x) = 3x^2 + 2x + 3$

In Problems 21–28, perform the indicated operations.

21 $(2x^2 + x - 3) - 2(x^2 - 2x + 2)$

22 $(x^3 + y^3 - 8xy + 3) - (2x^3 - y^3 + 10xy - 6)$

23 $(2x^3 + x^2 - 3) - 2(x^2 - 2x + 2)$

24 $(x^3 + 3x^2 + 2x - 3) - 3(x^2 + 2x + 1)$

25 $(a^2 + a) + (9a - 4a^2) + (a^2 - 5a)$

26 $(2x + 5x^2) - (x + x^2) + (x^2 - 7x)$

27 $(3x^2 + y^2 - 3x^2y^2) + (x^3 + 3y^2 + 3x^2y^2)$

28 $(x + x^3 - 3) - (3x^2 + x - 3)$

29 If $P(x) = 2x^2 + 5x - 3$, find (a) $P(0)$, (b) $P(-1)$, (c) $P(2)$.

30 If $Q(x) = 4x^2 - 3x + 2$, find (a) $Q(0)$, (b) $Q(-1)$, (c) $Q(\frac{1}{2})$.

31 If $H(t) = t - 2/t$, find (a) $H(1)$, (b) $H(2)$, (c) $H(-2)$.

32 If $G(y) = y^3 - 2y + 3$, find (a) $G(2)$, (b) $G(-2)$.

33 If $P(x) = 2x^2 + 5x - 3$ and $Q(x) = 4x^2 - 3x + 2$, find $P(2) + Q(2)$.

34 If $P(x) = 5x^2 + 4x + 1$ and $Q(x) = -3x^2 - 5x - 8$, find $P(-1) - Q(-1)$.

USING YOUR KNOWLEDGE 2.1

As you will see in the work of some later chapters, it is frequently necessary to evaluate a polynomial in x for some given values of x. If these values of x are not small, simple numbers, then the arithmetic can become quite tedious. However, if you have a small, hand calculator that can do addition, subtraction, and multiplication, all of the tiresome arithmetic can easily be done on your calculator. Let's illustrate this with an example.

Suppose that $P(x) = 2x^3 - 7x^2 + 5x - 10$ and we wish to calculate $P(3.1)$. First, we use the Distributive and Associative Laws to rewrite $P(x)$ as follows:

$$P(x) = (2x^3 - 7x^2 + 5x) - 10 = x(2x^2 - 7x + 5) - 10 = x[x(2x - 7) + 5] - 10$$

Although it looks as if we have complicated matters, we have actually broken the computation down into the individual small steps that we need in order to use the calculator. Follow these steps as we do the calculation starting from the innermost parentheses and working out.

1. Multiply $\underline{2}$ by the value of x, 3.1, to get 6.2.
2. Subtract $\underline{7}$ to get -0.8.
3. Multiply -0.8 by the value of x, 3.1, to get -2.48.
4. Add $\underline{5}$ to -2.48 to get 2.52.
5. Multiply 2.52 by the value of x, 3.1, to get 7.812.
6. Subtract $\underline{10}$ from 7.812 to get -2.188. This is the value of $P(3.1)$.

The underlined numbers are obtained from the coefficients of $P(x)$. In the even-numbered steps, these numbers are added or subtracted according to the sign of the coefficient.

If you have storage and recall keys on your calculator, you can store the value of x and recall it as needed. Here are the procedures for finding $P(3.1)$ following the steps just outlined.

Algebraic logic

[3.1] [STO] [2] [×] [RCL] [−] [7] [=] [×] [RCL] [+] [5] [=] [×] [RCL] [−] [10] [=]

RPN logic

[3.1] [STO] [2] [ENTER] [RCL] [×] [7] [−] [RCL] [×] [5] [+] [RCL] [×] [10] [−]

Try it. You should get the answer -2.188. A little practice will enable you to do calculations such as the preceding one in just a few seconds. Use the procedure to find $P(3.21)$ for the same polynomial. You should come out with 0.073622.

Notice that the important numbers in the calculation are the numerical coefficients of the polynomial. For the given $P(x)$, we can list the coefficients as

$$2, \quad -7, \quad +5, \quad -10$$

The first of these is the one we multiply by the value of x in the first step. The others are added or subtracted according to their signs in every even-numbered step before multiplying by the value of x in the succeeding step. If a power of x is missing, be sure to supply the coefficient zero for this term.

[C][1] Use your calculator to find the indicated values of the polynomials in the following problems.

1 $P(x) = x^3 - 4x^2 + 2x - 5$; $P(2.56)$ **2** $Q(x) = 3x^3 - x^2 + x + 2$; $Q(-0.24)$
3 $H(t) = 8t^3 + 7t^2 + 5t + 30$; $H(-1.68)$ **4** $G(y) = 8y^3 - 7y - 80$; $G(2.35)$
5 $R(x) = x^4 + 5x^3 + 3x - 20$; $R(1.38)$ **6** $S(t) = 5t^4 + 8t^3 - t^2 + t - 52$; $S(-2.44)$

2.2 MULTIPLICATION AND DIVISION OF POLYNOMIALS

In algebra, when we multiply together two variables, such as x and y, we simply write xy. However, with specific numbers, such as four times four, we write $4 \cdot 4$ with a multiplication dot, and not 44.

EXPONENTS IN MULTIPLICATION

In this section, we shall study multiplication and division of polynomials, but we must first consider the product of the monomials x^m and x^n, where m and n are positive integers. By the definition of positive integer coefficients (Definition 2.1c),

$$x^m \cdot x^n = \underbrace{(x \cdot x \cdot x \cdots x)}_{m \text{ factors}}\underbrace{(x \cdot x \cdot x \cdots x)}_{n \text{ factors}} = \underbrace{x \cdot x \cdot x \cdots x}_{m+n \text{ factors}} = x^{m+n}$$

For example,

$$x^2 \cdot x^3 = x^{2+3} = x^5 \quad \text{and} \quad x^6 \cdot x^9 = x^{6+9} = x^{15}.$$

[1] This symbol designates exercises and problems that are to be done using a calculator.

▼ **THEOREM 2.2a**

Rules of exponents in multiplication

If x and y are real numbers and m and n are positive integers, then

I. $x^m \cdot x^n = x^{m+n}$
II. $(x^m)^n = x^{mn}$ **III.** $(xy)^n = x^n y^n$

The preceding discussion proves Part I of Theorem 2.2a, which states two additional results that we need in the following work. The proofs of these two results follow directly from the definition of positive integer exponents and are left for the exercises.

MULTIPLICATION OF POLYNOMIALS

To multiply $x^3 - 2x$ by $x^2 + 3x$, we use the Distributive, Commutative, and Associative Laws together with Theorem 2.2a to write

$$\begin{aligned}
&(x^3-2x)(x^2+3x)\\
&\quad=(x^3-2x)(x^2)+(x^3-2x)(3x) && \text{Distributive Law}\\
&\quad=x^3\cdot x^2-2x\cdot x^2+x^3\cdot 3x-2x\cdot 3x && \text{Distributive Law}\\
&\quad=x^5-2x^3+3x^4-6x^2 && \text{Theorem 2.2a, Part I}\\
&\quad=x^5+3x^4-2x^3-6x^2 && \text{Commutative and Associative Laws}
\end{aligned}$$

Again, to multiply $3x^2 - x + 5$ and $2x^3 + 3x^2 - 1$, we can follow a similar procedure to write

$$\begin{aligned}
&(3x^2 - x + 5)(2x^3 + 3x^2 - 1)\\
&\quad= (3x^2 - x + 5)(2x^3) + (3x^2 - x + 5)(3x^2) + (3x^2 - x + 5)(-1)\\
&\quad= 6x^5 - 2x^4 + 10x^3 + 9x^4 - 3x^3 + 15x^2 - 3x^2 + x - 5\\
&\quad= 6x^5 + (-2 + 9)x^4 + (10 - 3)x^3 + (15 - 3)x^2 + x - 5\\
&\quad= 6x^5 + 7x^4 + 7x^3 + 12x^2 + x - 5
\end{aligned}$$

We can use a form similar to that used in arithmetic to obtain the more convenient procedure that follows.

$$\begin{array}{rrrrrrl}
 & & & 3x^2 & - x & + 5 & \\
(\times) & 2x^3 & & + 3x^2 & & - 1 & \\
\hline
6x^5 & - 2x^4 & + 10x^3 & & & & \text{Multiply first polynomial by } 2x^3.\\
 & 9x^4 & - 3x^3 & + 15x^2 & & & \text{Multiply first polynomial by } 3x^2.\\
 & & & - 3x^2 & + x & - 5 & \text{Multiply first polynomial by } -1.\\
\hline
6x^5 & + 7x^4 & + 7x^3 & + 12x^2 & + x & - 5 & \text{Add.}
\end{array}$$

EXAMPLE 1

Find the product of $x^2 + xy + y^2$ and $x - y$.

SOLUTION:

$$\begin{array}{lrrrr}
 & x^2 & + xy & + y^2 & \\
(\times) & x & - y & & \\
\hline
x^3 & + x^2y & + xy^2 & & \\
 & - x^2y & - xy^2 & - y^3 & \\
\hline
x^3 & & & - y^3 &
\end{array}$$

As you can see, both the middle terms drop out, and the answer is $x^3 - y^3$.

SPECIAL PRODUCTS

Certain "special products" occur so frequently in algebra that we list the results obtained when the multiplication is carried out. You should verify these results.

SPECIAL POLYNOMIAL PRODUCTS

P-1 $(x + a)(x + b) = x^2 + (a + b)x + ab$
P-2 $(x + a)^2 = x^2 + 2ax + a^2$
P-3 $(x - a)^2 = x^2 - 2ax + a^2$
P-4 $(x + a)(x - a) = x^2 - a^2$
P-5 $(ax + b)(cx + d) = acx^2 + (ad + bc)x + bd$

EXAMPLE 2

Multiply:

a. $(x + 3)(x - 4)$
b. $(x - 2y)^2$
c. $(3x + 4y)(3x - 4y)$

SOLUTION:

a. Letting $a = 3$ and $b = -4$ in P-1, we obtain

$$\begin{aligned}(x + 3)(x - 4) &= x^2 + (3 - 4)x + (3)(-4)\\ &= x^2 - x - 12\end{aligned}$$

b. Letting $a = 2y$ in P-3, we obtain

$$(x - 2y)^2 = x^2 - 2(2yx) + (2y)^2 = x^2 - 4xy + 4y^2$$

c. Replacing x by $3x$ and a by $4y$ in P-4, we have

$$(3x + 4y)(3x - 4y) = (3x)^2 - (4y)^2 = 9x^2 - 16y^2$$

EXPONENTS IN DIVISION

Before discussing division of polynomials in general, we will look at the division of the monomial x^m by the monomial x^n, where m and n are natural numbers. For example, to divide x^5 by x^3, we may write

$$\begin{aligned}x^5 \div x^3 = \frac{x^5}{x^3} &= \frac{x \cdot x \cdot x \cdot x \cdot x}{x \cdot x \cdot x} = \frac{x \cdot x \cdot x}{x \cdot x \cdot x} \cdot x \cdot x\\ &= 1 \cdot x \cdot x = x^2 = x^{5-3}\end{aligned}$$

The procedure in this illustration is quite general and can be used to prove Theorem 2.2b. For example,

$$\frac{x^{10}}{x^4} = x^{10-4} = x^6 \qquad \frac{x^{12}}{x^{12}} = 1 \qquad \frac{x^4}{x^{10}} = \frac{1}{x^{10-4}} = \frac{1}{x^6}$$

▼ THEOREM 2.2b

Rules of exponents in division

If x is a real nonzero number and m and n are positive integers, then

I. $\dfrac{x^m}{x^n} = x^{m-n}$ **for $m > n$**

II. $\dfrac{x^m}{x^n} = 1$ **for $m = n$**

III. $\dfrac{x^m}{x^n} = \dfrac{1}{x^{n-m}}$ **for $m < n$**

With the aid of Theorem 2.2b, we can divide a polynomial by a monomial using a term-by-term procedure as in Example 3.

EXAMPLE 3

Divide $x^3y^2 + 2x^2y - 3x^3$ by x^2y.

SOLUTION:

$(x^3y^2 + 2x^2y - 3x^3) \div x^2y$

$$= (x^3y^2 + 2x^2y - 3x^3) \cdot \frac{1}{x^2y} \qquad \text{Definition of Division}$$

$$= \frac{x^3y^2}{x^2y} + \frac{2x^2y}{x^2y} - \frac{3x^3}{x^2y} \qquad \text{Distributive Law}$$

$$= x^{3-2}y^{2-1} + 2 \cdot 1 \cdot 1 - \frac{3x^{3-2}}{y} \qquad \text{Laws of Exponents}$$

$$= xy + 2 - \frac{3x}{y}$$

DIVISION OF POLYNOMIALS

Can we do more complicated divisions? Of course we can. We use a procedure similar to that used in arithmetic long division, as Example 4 illustrates.

EXAMPLE 4

Divide $3x^2 - 7x - 7$ by $x - 3$.

SOLUTION:

We write the division in the usual way, with each polynomial arranged in order of descending powers of x.

$$\begin{array}{r l}
 & 3x \;\; + 2 \\
x - 3 \,\big) & \overline{3x^2 - 7x - 7} \\
 & \underline{3x^2 - 9x} \\
 & \qquad 2x - 7 \\
 & \qquad \underline{2x - 6} \\
 & \qquad\quad - 1
\end{array}$$

Divide the first term in the dividend, $3x^2$, by x, obtaining $3x$. Multiply the divisor, $x - 3$, by $3x$ and write the result under the dividend. Then subtract. Repeat the procedure, using $2x - 7$ as the new dividend. The result shows that the quotient is $3x + 2$, the remainder is -1.

Example 5 illustrates the division procedure for a divisor of degree greater than one.

EXAMPLE 5

Divide $2x^4 + 7x^3 - 4x + 2$ by $2x^2 + 3x - 2$.

SOLUTION:

$$\begin{array}{r l}
 & x^2 + 2x \;\; - 2 \\
2x^2 + 3x - 2 \,\big) & \overline{2x^4 + 7x^3 \qquad\quad - 4x + 2} \\
 & \underline{2x^4 + 3x^3 - 2x^2} \\
 & \qquad 4x^3 + 2x^2 - 4x + 2 \\
 & \qquad \underline{4x^3 + 6x^2 - 4x} \\
 & \qquad\quad - 4x^2 \qquad + 2 \\
 & \qquad\quad \underline{- 4x^2 - 6x + 4} \\
 & \qquad\qquad\qquad 6x - 2
\end{array}$$

Divide the first term of the dividend, $2x^4$, by the first term of the divisor, $2x^2$, obtaining x^2. Multiply the divisor by x^2 and subtract the result from the dividend. Repeat the procedure, using the result of the subtraction as the new dividend. Continue until the remainder is either zero or a polynomial of lower degree than that of the divisor. Notice that, to keep the exponents in proper descending order, we leave a space for the missing x^2 term in the dividend.

We have found the quotient to be $x^2 + 2x - 2$ and the remainder to be $6x - 2$. You can check the answer by multiplying the quotient by the divisor and then adding the remainder. The result should be the dividend. Thus,

$$\begin{array}{rrrrrrl}
 & 2x^2 & + 3x & - 2 & & & \text{Divisor} \\
(\times) & x^2 & + 2x & - 2 & & & \text{Quotient} \\
\hline
 & 2x^4 + 3x^3 & - 2x^2 & & & & \\
 & + 4x^3 & + 6x^2 & - 4x & & & \\
 & & - 4x^2 & - 6x & + 4 & & \\
\hline
 & 2x^4 + 7x^3 & & - 10x & + 4 & & \\
(+) & & & 6x & - 2 & & \text{Remainder} \\
\hline
 & 2x^4 + 7x^3 & & - 4x & + 2 & & \text{Dividend}
\end{array}$$

EXERCISE 2.2

In Problems 1–12, perform the indicated operations.

1 $3x(x + 7)$
2 $5x^2(x^2 + x + 1)$
3 $(x^2)^3(x + 1)$
4 $(x^2y)^3(x - 2)$
5 $(3xy + x^2y + 2)(x - y)$
6 $(x^2y + 3x - 7)(x - y + 1)$
7 $4(x^2 + x - 2) - 3x(x + 1)$
8 $5x(x^2+2x-2)-x^2(x+1)+3(x-7)$
9 $(a^2 + a - 2)(2a^2 - 3a + 1)$
10 $(3b^2 + 2b - 5)(2b^2 - b + 2)$
11 $(2t^2 - 5t - 12)(t^2 + t - 2)$
12 $(4y^2 - y - 3)(3y^2 + y - 4)$

In Problems 13–32, use the special products P-1 through P-5 to perform the indicated operations.

13 $(x + 9)(x - 2)$
14 $(x - 4)(x + 5)$
15 $(x + 3)(x + 4)$
16 $(x - 5)(x - 4)$
17 $(2x + 4)^2$
18 $(3x + 2y)^2$
19 $(x - 5)^2$
20 $(x - 3y)^2$
21 $(2x - 5y)^2$
22 $(3x - 4y)^2$
23 $\left(\frac{x}{2} - \frac{y}{3}\right)^2$
24 $\left(\frac{x}{3} + \frac{y}{2}\right)^2$
25 $(3x - y)(3x + y)$
26 $(4y - 5x)(4y + 5x)$
27 $(3x + 5y)(x - 2y)$
28 $(3x - 2y)(x + 3y)$
29 $(2x - 3y)(3x - 5y)$
30 $(4x + 5y)(2x + 7y)$
31 $(4x + 3y)(2x - 3y)$
32 $(7x - 3y)(4x + 7y)$

In Problems 33–42, use the special products P-1 through P-5 to multiply.

33 $-3x(x - 2)(x - 4)$
34 $-5x(x - 3)(x - 7)$
35 $(x - 1)^2 2x$
36 $-3x(x + 1)^2$
37 $(x + 2)^2(x + 1)$
38 $(x - 3)^2(x - 1)$
39 $-3x(x + 1)(x + 2)$
40 $-5x(x - 1)(x - 3)$
41 $(x + 1)(3x + 1)(x + 4)$
42 $(x - 2)(3x + 2)(3x - 2)$

In Problems 43–58, perform the indicated operations.

43 $(6x^3 + 3x^2 - 8x + 12) \div 12x$

44 $(30y^5 - 10y^4 + 18y^3 - 9) \div 15y^4$

45 $(8u^2 + 12v^2) \div 4uv$

46 $(7a^4b - 35ab^4) \div a^2b^2$

47 $(a^8 - 6a^6b^4 + 6a^4b^6 - b^8) \div a^4b^4$

48 $(2x^6 + x^4y^2 - x^2y^4 + 4y^8) \div 4x^2y^2$

49 $(x^2 - 5x + 6) \div (x - 3)$

50 $(y^2 + 5y + 5) \div (y + 1)$

51 $(y^2 - 2y - 14) \div (y + 3)$

52 $(x^3 - x^2 - 10x - 8) \div (x - 4)$

53 $(x^4 - y^4) \div (x + y)$

54 $(t^8 - 16u^{16}) \div (t^2 - 2u^4)$

55 $(x^3 + 8x^2 + 9x + 2) \div (x^2 + 7x + 2)$

56 $(s^4 - 5s^2 + 9) \div (s^2 - s + 3)$

57 $(x^3 - 8y^3 - 8z^3 - 12xyz) \div (x - 2y - 2z)$

58 $(a^3 + 27b^3 - c^3 + 9abc) \div (a + 3b - c)$

59 Prove Theorem 2.2a, Part II.

60 Prove Theorem 2.2a, Part III.

61 Prove Theorem 2.2b, Part I.

62 Prove Theorem 2.2b, Parts II and III.

USING YOUR KNOWLEDGE 2.2

If you let $x = 10$ in the polynomial $4x^3 + 3x^2 + 9x + 5$, you get the decimal integer 4395. In fact, you can see that the set of digits in any decimal integer consists of exactly the set of coefficients in the corresponding polynomial with $x = 10$. Thus, the integer **25,308** corresponds to the polynomial

$$\mathbf{2}x^4 + \mathbf{5}x^3 + \mathbf{3}x^2 + \mathbf{0}x + \mathbf{8}$$

with $x = 10$. This correspondence is the basis for all our arithmetic operations in the decimal system. The fact that we do not write the powers of ten means that we are operating with "detached coefficients." For instance, we can multiply $x^3 - 5x + 7$ by $x^2 + 3x - 1$ using detached coefficients as follows:

1	0	−5	+7			Coefficients of $x^3 - 5x + 7$
	1	+3	−1			Coefficients of $x^2 + 3x - 1$
1	0	−5	+7			Multiply first row by 1.
	+3	0	−15	+21		Multiply first row by +3, moving right one column.
		−1	0	+5	−7	Multiply first row by −1, moving right one column.
1	+3	−6	−8	+26	−7	Add separate columns to get coefficients of answer.

The result is $x^5 + 3x^4 - 6x^3 - 8x^2 + 26x - 7$. If you carry out the multiplication in the usual way and then erase all the powers of x, the result will be exactly as the foregoing.

Use detached coefficients to do the multiplications in Exercise 2.2 as indicated below.

1 Problem 9 **2** Problem 10 **3** Problem 11 **4** Problem 12

[C] **5** Use your calculator to multiply $47x^3 + 92x^2 - 31x + 17$ by $12x^3 - 19x^2 + 23x - 14$. Try to devise a scheme for calculating each final coefficient without writing down the numbers in the intermediate rows. You should obtain the answer

$$564x^6 + 211x^5 - 1039x^4 + 2251x^3 - 2324x^2 + 825x - 238$$

2.3 FACTORING POLYNOMIALS

As you might recall, the numbers that are multiplied together to form a product are called *factors* of that product. Before discussing the factors of a polynomial, we need to review briefly the factoring of natural numbers.

A natural number is said to be *prime* if it has *exactly* two *distinct* divisors, itself and 1.

Thus, 2, 3, 5, 7, 11, 13, and so on are primes.

A natural number that has *more* than two *distinct* divisors is said to be *composite.*

For example, 4 is composite because it has the divisors 1, 2, and 4. Every natural number greater than 1 that is *not* prime is composite. The numbers 4, 6, 8, 9, 10, 12, and so on, are composite.

To *factor* a composite number means to write the number as a product of primes.

For instance, the factorization of 90 is

$$90 = 2 \cdot 3 \cdot 3 \cdot 5 = 2 \cdot 3^2 \cdot 5$$

As in the case of numbers, a polynomial can be written as a product of other polynomials called *factors* of the original polynomial. The process of writing a polynomial as the product of its factors is called *factoring* the polynomial. For example, we saw in the preceding section that

$$(x + a)(x + b) = x^2 + (a + b)x + ab$$

Hence, $(x + a)$ and $(x + b)$ are *factors* of $x^2 + (a + b)x + ab$.

In the remainder of this section, we will study the factorization of polynomials with integer coefficients, and we will require each factor to have integer coefficients. Thus, even though

$$x + 2 = x\left(1 + \frac{2}{x}\right)$$

and

$$x + 2 = 2\left(\frac{x}{2} + 1\right)$$

we say that $x + 2$ *cannot* be factored *over the integers*, because $x + 2$ is *not* a product of two *polynomials* with *integer* coefficients. A polynomial with integer coefficients that cannot be factored into a product of polynomials with integer coefficients (without using either 1 or -1 as a factor) is called *irreducible*. Thus, $x + 2$ is irreducible.

FACTORING OUT A COMMON FACTOR

The easiest type of factorization makes use of the generalized Distributive Law:

$$ax + ay + az = a(x + y + z)$$

For instance, to factor the polynomial $9x^2 + 6x + 3$, we observe that 3 is a factor of each term and write

$$9x^2 + 6x + 3 = 3(\qquad)$$

We then insert in the parentheses the result of dividing the original polynomial by 3, that is, $3x^2 + 2x + 1$. The final result is written as

$$9x^2 + 6x + 3 = 3(3x^2 + 2x + 1)$$

Later in this section (Example 3b) you will see that the trinomial $3x^2 + 2x + 1$ is irreducible.

Similarly, to factor $4x^3 + 6x^2 + 10x$, we first note that $2x$ is a factor common to $4x^3$, $6x^2$, and $10x$. Thus,

$$4x^3 + 6x^2 + 10x = 2x(2x^2 + 3x + 5)$$

FACTORING BY GROUPING

In some cases where there is no factor common to all the terms of a polynomial, we may be able to group the terms in such a way that the groups have a common factor. For example, $ax + ay + bx + by$ does not have a factor common to all the terms. However, we can group the terms to obtain

$$(ax + ay) + (bx + by) = a(x + y) + b(x + y)$$

and we can now factor out the common factor $x + y$. Thus,

$$ax + ay + bx + by = (x + y)(a + b)$$

EXAMPLE 1

Factor:

a. $21x^2y^3 + 18xy^2$

b. $6xy - 6yz + 5x - 5z$

SOLUTION:

a. We see that $3xy^2$ is a common factor. Hence,

$$21x^2y^3 + 18xy^2 = 3xy^2(7xy + 6)$$

b. There is no factor common to all the terms. Therefore, we try factoring by grouping as follows:

$$\begin{aligned} 6xy - 6yz + 5x - 5z &= (6xy - 6yz) + (5x - 5z) \\ &= 6y(x - z) + 5(x - z) \\ &= (x - z)(6y + 5) \end{aligned}$$

THE TRINOMIAL $x^2 + px + q$

The special products given in the preceding section, rewritten with the right and left members of the equations interchanged, provide the following factoring formulas:

FACTORS OF SPECIAL POLYNOMIALS

F-1 $\mathbf{x^2 + (a + b)x + ab = (x + a)(x + b)}$
F-2 $\mathbf{x^2 + 2ax + a^2 \quad = (x + a)^2}$
F-3 $\mathbf{x^2 - 2ax + a^2 \quad = (x - a)^2}$

Equation F-1 tells us that we can factor any polynomial of the form $x^2 + px + q$ if the last term, q, is the product of two numbers a and b and the coefficient, p, of the middle term is the sum of these two numbers. For example, to factor $x^2 + 5x + 6$, we need to factor 6 as a product of two factors whose sum is 5. These factors are **2** and **3**. Thus,

$$x^2 + 5x + 6 = (x + 2)(x + 3)$$

Note that F-2 and F-3 are really special cases of F-1 with $a = b$. For example, to factor $x^2 - 8x + 16$, we can find two numbers, a and b, whose product is 16 and whose sum is -8. These numbers are $a = -4$ and $b = -4$, so that we get

$$\begin{aligned} x^2 - 8x + 16 &= (x - 4)(x - 4) \\ &= (x - 4)^2 \end{aligned}$$

Similarly,

$$x^2 + 6x + 9 = (x + 3)(x + 3) = (x + 3)^2$$

EXAMPLE 2

Factor:

a. $x^2 - 6x + 8$
b. $b^4 + 8b^2 + 16$
c. $9b^2 - 12bc + 4c^2$

SOLUTION:

a. We need two numbers whose product is 8 and whose sum is -6. These numbers are $\mathbf{-4}$ and $\mathbf{-2}$. Thus,

$$x^2 - 6x + 8 = (x - 2)(x - 4)$$

b. Letting $x = b^2$ and $a = 4$ in F-2, we have

$$b^4 + 8b^2 + 16 = (b^2)^2 + 2(4)(b^2) + 4^2 = (b^2 + 4)^2$$

c. Letting $x = 3b$ and $a = 2c$ in F-3, we have

$$9b^2 - 12bc + 4c^2 = (3b)^2 - 2(2c)(3b) + (2c)^2 = (3b - 2c)^2$$

THE TRINOMIAL $ax^2 + bx + c$

Except for a possible common factor, a trinomial of the form $ax^2 + bx + c$ is factorable if and only if it can be written in the form

$(dx + e)(fx + g)$

However, $(dx + e)(fx + g) = dfx^2 + (dg + ef)x + eg$, so that if the two trinomials are identical, then $a = df$, $c = eg$, and $b = dg + ef$. Thus, $ac = defg$, which shows that ac must be the product of dg and ef, whose sum is b. We state this result in Theorem 2.3a. For instance, $4x^2 - 11x + 6$ is factorable, because (4)(6) or 24 is the product of -8 and -3, and $(-8) + (-3) = -11$. We can now factor the trinomial as follows: Use the factors -8 and -3 to write

▼ THEOREM 2.3a

The trinomial $ax^2 + bx + c$ is factorable as the product of two binomials if and only if ac is the product of two factors whose sum is b.

$$4x^2 - 11x + 6 = 4x^2 - 8x - 3x + 6$$

Then the factoring can be completed by grouping the first two and the last two terms thus:

$$\begin{aligned} 4x^2 - 11x + 6 &= (4x^2 - 8x) + (-3x + 6) \\ &= 4x(x - 2) - 3(x - 2) \\ &= (x - 2)(4x - 3) \end{aligned}$$

EXAMPLE 3

Factor into lower-degree factors if possible:

a. $3x^2 - 5x - 2$
b. $3x^2 + 2x + 1$

SOLUTION:

a. To factor $3x^2 - 5x - 2$, we need to find factors of $(3)(-2) = -6$ whose sum is -5. These factors are -6 and 1. Thus,

$$\begin{aligned} 3x^2 - 5x - 2 &= 3x^2 - 6x + x - 2 \\ &= (3x^2 - 6x) + (x - 2) \\ &= 3x(x - 2) + 1(x - 2) \\ &= (x - 2)(3x + 1) \end{aligned}$$

b. To factor $3x^2 + 2x + 1$, we need to find factors of (3)(1) whose sum is 2. Because such factors do not exist, the polynomial $3x^2 + 2x + 1$ is irreducible.

THE DIFFERENCE OF TWO SQUARES

We consider next the factorization of some important *binomials*. The first of these binomials, called the *difference of two squares*, has already been mentioned in Formula P-4. From that formula we have:

F-4 $\quad x^2 - a^2 = (x + a)(x - a)$

Thus, using Formula F-4 with $a = 4$, we obtain

$$x^2 - 16 = x^2 - 4^2 = (x + 4)(x - 4)$$

Similarly, letting $x = 3b$ and $a = 5c$ in this formula, we obtain

$$9b^2 - 25c^2 = (3b)^2 - (5c)^2 = (3b + 5c)(3b - 5c)$$

EXAMPLE 4

Factor:

a. $9b^2 - 1$

b. $(y + z)^2 - (y - w)^2$

c. $b^4 - c^4$

SOLUTION:

a. $9b^2 - 1 = (3b)^2 - 1^2 = (3b + 1)(3b - 1)$

b. Using Formula F-4 with $x = y + z$, $a = y - w$, we obtain

$$(y + z)^2 - (y - w)^2 = [(y + z) + (y - w)][(y + z) - (y - w)]$$
$$= (2y + z - w)(z + w)$$

c. Letting $x = b^2$ and $a = c^2$ in Formula F-4, we have

$$b^4 - c^4 = (b^2)^2 - (c^2)^2 = (b^2 + c^2)(b^2 - c^2)$$
$$= (b^2 + c^2)(b + c)(b - c)$$

THE SUM AND DIFFERENCE OF TWO CUBES

The following formulas serve to factor the sum and the difference of two cubes. The formulas are easily verified by multiplying out the right-hand members.

F-5 $x^3 + a^3 = (x + a)(x^2 - ax + a^2)$

F-6 $x^3 - a^3 = (x - a)(x^2 + ax + a^2)$

EXAMPLE 5

Factor:

a. $8b^3 + 27c^3$

b. $64b^3 - 1$

c. $b^6 - c^6$

SOLUTION:

a. Letting $x = 2b$ and $a = 3c$ in F-5, we obtain

$$8b^3 + 27c^3 = (2b)^3 + (3c)^3$$
$$= (2b + 3c)[(2b)^2 - (2b)(3c) + (3c)^2]$$
$$= (2b + 3c)(4b^2 - 6bc + 9c^2)$$

b. Letting $x = 4b$ and $a = 1$ in F-6, we see that

$$64b^3 - 1 = (4b)^3 - 1^3$$
$$= (4b - 1)[(4b)^2 + (4b)(1) + 1^2]$$
$$= (4b - 1)(16b^2 + 4b + 1)$$

c. Letting $x = b^3$ and $a = c^3$ in F-4, we obtain

$$b^6 - c^6 = (b^3)^2 - (c^3)^2$$
$$= (b^3 + c^3)(b^3 - c^3)$$
$$= (b + c)(b^2 - bc + c^2)(b - c)(b^2 + bc + c^2)$$

Note that if you let $x = b^2$ and $a = c^2$ in F-6, your factorization of $b^6 - c^6$ will be incomplete. Can you see why? (See Using Your Knowledge 2.3.)

EXERCISE 2.3

In Problems 1–40, factor completely.

1 $6x + 10$

2 $ax + 5x$

3 $21x^2 + 7x^4$

4 $2ax^2 + 6bx^3 - cx^2$

5 $5a^3b^2 - 10a^2b^3 + 15a^4b^4$

6 $18a^6y^4 - 24a^3b^3 + 12a^4b$

7 $(a + b)x + (a + b)y$

8 $(a - c)x - (a - c)y$

9 $ax + 3x + 2ay + 6y$

10 $4x - bx - 24y + 6by$
11 $2bx + 15n - 5b - 6nx$
12 $12b^2 + 3y - 4b^2z^3 - yz^3$
13 $x^2 + 6x + 8$
14 $x^2 + 10x + 21$
15 $c^2 - 8cd + 15d^2$
16 $x^2 - 5mx + 6m^2$
17 $x^2 - 14xy + 49y^2$
18 $x^2 - 16xy + 63y^2$
19 $y^2 + 2y + 1$
20 $x^2 + 14x + 49$
21 $9x^2 + 30xy + 25y^2$
22 $36y^2 + 48yz + 16z^2$
23 $x^2 - 14x + 49$
24 $9y^2 - 60y + 100$
25 $a^2 - 2ab + b^2$
26 $a^2 - 10a + 25$
27 $49x^2 - 28xy + 4y^2$
28 $16x^2 - 24xy + 9y^2$
29 $m^4 - 12m^2 + 36$
30 $a^4 - 2a^2 + 1$
31 $x^2/4 - xy + y^2$
32 $x^2/9 + 2x/3 + 1$
33 $2x^2 + 3x - 2$
34 $3x^2 - 5x - 2$
35 $5y^2 + 13y - 6$
36 $6y^2 - 5y - 6$
37 $6x^4 + 5x^2 - 6$
38 $5x^6 + 4x^3 - 12$
39 $15x^2 - ax - 2a^2$
40 $20x^2 + ax - a^2$

In Problems 41–60, factor as the difference of two squares or as the sum or difference of two cubes.

41 $x^2 - 49$
42 $9x^2 - 36y^2$
43 $9x^2y^2 - 49z^2$
44 $a^2x^4 - 49$
45 $a^2 - (b + c)^2$
46 $a^2 - (b - c)^2$
47 $b^8 - y^{16}$
48 $m^4 - 16v^4$
49 $16x^4 - 81y^4$
50 $9x^2y^2 - 49z^6$
51 $x^3 + 125$
52 $a^3 + 27$
53 $x^6 + 1$
54 $x^9 - y^{18}$
55 $(a + b)^3 + (a - b)^3$
56 $(a + 2b)^3 + (c + 3d)^3$
57 $x^3 - (x + y)^3$
58 $x^{12} - y^6$
59 $27a^6 - (b + 3c)^3$
60 $(m + v)^3 - (c - d)^3$

USING YOUR KNOWLEDGE 2.3

Sometimes we add and subtract a perfect square so that a given polynomial can be written as a difference of two squares. For example, $a^4 + 64$ is *not* written as the difference of two squares. However, because

$$a^4 + 64 = (a^2)^2 + (8)^2$$

we add and subtract $2(8)(a^2) = 16a^2$, obtaining

$$\begin{aligned} a^4 + 64 &= a^4 + \underline{16a^2} + 64 - \underline{16a^2} \\ &= (a^2 + 8)^2 - (4a)^2 \qquad \text{Which is the difference of two squares} \\ &= (a^2 + 8 + 4a)(a^2 + 8 - 4a) \\ &= (a^2 + 4a + 8)(a^2 - 4a + 8) \end{aligned}$$

We now return to Example 5c and let $x = b^2$, $a = c^2$ in F-6 to obtain

$$\begin{aligned} b^6 - c^6 &= (b^2 - c^2)[(b^2)^2 + (b^2)(c^2) + (c^2)^2] \\ &= (b^2 - c^2)(b^4 + b^2c^2 + c^4) \end{aligned}$$

By inspection, we can factor $b^2 - c^2 = (b + c)(b - c)$. To factor $b^4 + b^2c^2 + c^4$, we use the fact that $b^4 + 2b^2c^2 + c^4$ is a perfect square. Hence, we add and subtract b^2c^2 to obtain the difference of two squares as follows:

$$b^4 + b^2c^2 + c^4 = b^4 + 2b^2c^2 + c^4 - b^2c^2$$
$$= (b^2 + c^2)^2 - (bc)^2$$
$$= (b^2 + c^2 + bc)(b^2 + c^2 - bc)$$

This completes the factorization, giving

$$b^6 - c^6 = (b + c)(b - c)(b^2 + bc + c^2)(b^2 - bc + c^2)$$

which agrees with the result obtained in Example 5c.

Use the preceding ideas to factor:

1 $x^4 + 4$ **2** $64x^4 + 1$ **3** $x^4 + x^2 + 1$ **4** $64y^4 + a^8$ **5** $x^4 + 3x^2 + 4$ **6** $x^4 - 7x^2 + 9$
7 $b^8 + b^4 + 1$ **8** $a^4 + 2a^2 + 9$ **9** $c^4 - 45c^2 + 100$ **10** $a^4 - 3a^2b^2 + b^4$

2.4 ADDING AND SUBTRACTING RATIONAL EXPRESSIONS

When a tennis ball is hit by a racket, the ball is distorted by the impact. The average force of this impact on the ball is

$$\frac{mv - mv_0}{t}$$

where m is the mass of the ball, v_0 is the initial velocity of the racket, v is the terminal velocity of the racket, t is the time the ball and racket are in contact. The expression

$$\frac{mv - mv_0}{t}$$

is an example of a *rational expression.* A *rational expression* is simply a quotient of two polynomials. For example,

$$\frac{x}{x-3}, \frac{8xy - z}{x^2 + y^2}, \frac{y^2 - 3y + 9}{y}, \text{ and } \frac{x}{y}$$

are rational expressions. Because the fraction itself is a real number for each replacement of the variable(s) for which the numerator is a real number and the denominator is a nonzero real number, the rules of operation for rational expressions are the same as those for the fractions of arithmetic. Theorem 2.4a is a direct consequence of Theorem 1.3b, which gives the rules for operating with arithmetic fractions. **It must be emphasized that in this theorem and *in all subsequent work*, there will be allowed no values of the variables for which any denominator is zero.**

▼ THEOREM 2.4a

If A, B, C, and D are polynomials, then

I. $\dfrac{A}{B} = \dfrac{C}{D}$ **if and only if** $AD = BC$

II. $\dfrac{-A}{B} = -\dfrac{A}{B} = \dfrac{A}{-B} = -\dfrac{-A}{-B}$ **and**

$\dfrac{A}{B} = \dfrac{-A}{-B} = -\dfrac{-A}{B} = -\dfrac{A}{-B}$

III. $\dfrac{AC}{BC} = \dfrac{A}{B}$ **(Fundamental Principle of Fractions)**

Part I of the preceding theorem allows us to verify the equality of two fractions. For instance,

$$\frac{x-a}{x^2 - a^2} = \frac{1}{x+a}$$

because $(x - a)(x + a) = (x^2 - a^2) \cdot 1$. Part II shows the various permissible arrangements of signs in a fraction. An illustration of this is as follows:

$$\frac{1}{2 - x} = \frac{-1}{-(2 - x)} = \frac{-1}{x - 2}$$

Part III is used to reduce a rational expression to lowest terms. A rational expression is in *lowest terms* when the numerator and the denominator have no common factors other than 1 or -1. Thus, to reduce

$$\frac{8x^3y^2}{12xy^5}$$

to lowest terms, we may write

$$\frac{8x^3y^2}{12xy^5} = \frac{2x^2 \cdot \mathbf{4xy^2}}{3y^3 \cdot \mathbf{4xy^2}} = \frac{2x^2}{3y^3}$$

If the given rational expression contains polynomials in the numerator and denominator, it is often useful to factor these polynomials. For example, to reduce the rational expression

$$\frac{b^2 - a^2}{b^3 - a^3}$$

we first write the numerator and denominator in factored form and then use the Fundamental Principle of Fractions to obtain

$$\frac{b^2 - a^2}{b^3 - a^3} = \frac{(b + a)\,\mathbf{(b - a)}}{\mathbf{(b - a)}(b^2 + ab + a^2)} = \frac{b + a}{b^2 + ab + a^2}$$

EXAMPLE 1

Reduce to lowest terms:

a. $\dfrac{32x^5y^7}{-4x^2y^9}$

b. $\dfrac{2 - x}{3x^2 - 5x - 2}$

SOLUTION:

a. $\dfrac{32x^5y^7}{-4x^2y^9} = -\dfrac{8x^3 \cdot \mathbf{4x^2y^7}}{y^2 \cdot \mathbf{4x^2y^7}} = -\dfrac{8x^3}{y^2}$

b. $\dfrac{2 - x}{3x^2 - 5x - 2} = \dfrac{2 - x}{(3x + 1)(x - 2)} = \dfrac{(-1)\,\mathbf{(x - 2)}}{(3x + 1)\mathbf{(x - 2)}} = \dfrac{-1}{3x + 1}$

▼ THEOREM 2.4b

If A, B, and C are polynomials, then for all values of the variables such that $C \neq 0$,

$$\frac{A}{C} + \frac{B}{C} = \frac{A + B}{C}$$

Addition of rational expressions is defined in exactly the same way as addition of common fractions in arithmetic. (See Theorem 2.4b.) Thus,

$$\frac{3x^2 + 4}{y^3} + \frac{x^2 - 1}{y^3} = \frac{(3x^2 + 4) + (x^2 - 1)}{y^3}$$

$$= \frac{4x^2 + 3}{y^3}$$

Similarly,

$$\frac{x^2}{x+2}+\frac{x-1}{x+2}+\frac{x+9}{x+2}=\frac{x^2+(x-1)+(x+9)}{x+2}$$

$$=\frac{x^2+2x+8}{x+2}$$

If the expressions involved do not have the same denominator, then we may replace the expressions by equivalent ones with a common denominator and complete the addition as in the preceding illustrations. In arithmetic, this procedure is usually accomplished by using the least common denominator (LCD) of the fractions. In many cases, the LCD is found by inspection. However, the LCD of a set of fractions can be found by the following procedure:

1. Write each denominator as a product of primes.
2. Select each different prime factor raised to the highest power to which it occurs in the products of step 2.
3. The product of the factors selected in step 2 is the LCD.

For instance, to find the LCD of $\frac{1}{12}$, $\frac{5}{18}$, and $\frac{7}{24}$, we follow these three steps:

1. $12 = 2^2 \cdot 3$
   $18 = 2 \cdot 3^2$
   $24 = 2^3 \cdot 3$
2. From the first column (the 2's) we select 2^3, the highest power of the prime 2 in this factorization. From the second column we pick 3^2.
3. The LCD is the product of these factors, that is, $2^3 \cdot 3^2 = 72$.

To add rational expressions, a similar procedure is employed. Thus, to add

$$\frac{a}{a^2-b^2}+\frac{b}{(a+b)^2}$$

we first find the LCD as follows:

1. Factor each denominator, writing common factors in the same column.

   $a^2 - b^2 = (a+b) \quad | \quad (a-b)$

   $(a+b)^2 = (a+b)^2 \quad |$

2. Select the different factors with the greatest exponent in each column, that is, $(a+b)^2$ and $(a-b)$
3. The LCD is $(a+b)^2(a-b)$

Writing the expressions $a/(a^2-b^2)$ and $b/(a+b)^2$ as equivalent ones with the LCD $(a+b)^2(a-b)$ as denominator, we obtain

$$\frac{a}{a^2-b^2}=\frac{a}{(a+b)(a-b)}=\frac{a}{(a+b)(a-b)}\cdot\frac{a+b}{a+b}=\frac{a(a+b)}{(a+b)^2(a-b)}$$

and

$$\frac{b}{(a+b)^2} = \frac{b(a-b)}{(a+b)^2(a-b)}$$

Thus,

$$\frac{a}{a^2-b^2} + \frac{b}{(a+b)^2} = \frac{a(a+b)}{(a+b)^2(a-b)} + \frac{b(a-b)}{(a+b)^2(a-b)}$$

$$= \frac{a^2+ab+ba-b^2}{(a+b)^2(a-b)} = \frac{a^2+2ab-b^2}{(a+b)^2(a-b)}$$

You should note that, in replacing a fraction by an equivalent fraction with the LCD, we use the Fundamental Principle (Theorem 2.4a, Part III) to go in the direction opposite that used in reducing fractions. Of course, the principle works both ways.

Because the subtraction of a number is the same as the addition of the opposite of that number, we see that subtraction of rational expressions must proceed just as addition.

EXAMPLE 2

Perform the indicated operation.

$$\frac{1}{x^2+3xy+2y^2} - \frac{1}{x^2-y^2}$$

SOLUTION:

We first find the LCD.

1. $x^2 + 3xy + 2y^2 = (x+2y)\ (x+y)$
   $x^2 - y^2 = \qquad\qquad (x+y)\ (x-y)$
2. The different factors with the greatest exponents are $(x+2y)$, $(x+y)$, and $(x-y)$.
3. The LCD is the product $(x+2y)(x+y)(x-y)$.

We next write the given fractions using the LCD.

$$\frac{1}{x^2+3xy+2y^2} = \frac{1}{(x+2y)(x+y)} = \frac{1\cdot(x-y)}{(x+2y)(x+y)(x-y)}$$

$$\frac{1}{x^2-y^2} = \frac{1}{(x+y)(x-y)} = \frac{1\cdot(x+2y)}{(x+y)(x-y)(x+2y)}$$

Thus,

$$\frac{1}{x^2+3xy+2y^2} - \frac{1}{x^2-y^2}$$

$$= \frac{x-y}{(x+2y)(x+y)(x-y)} - \frac{x+2y}{(x+2y)(x+y)(x-y)}$$

$$= \frac{x-y-x-2y}{(x+2y)(x+y)(x-y)}$$

$$= \frac{-3y}{(x+2y)(x+y)(x-y)}$$

EXAMPLE 3

Find:

$$\frac{a}{a^2+2a-3} - \frac{1}{a-2} + \frac{1}{1-a}$$

SOLUTION:

We first find the LCD.

$a^2+2a-3 =$	$(a+3)$	$(a-1)$	
$a-2 =$			$a-2$
$1-a =$		$-(a-1)$	

Note that because a^2+2a-3 has $a-1$ as a factor and

$$\frac{1}{1-a} = \frac{-1}{-(1-a)} = -\frac{1}{a-1}$$

we write the last fraction as

$$-\frac{1}{a-1}$$

Thus, the LCD is $(a+3)(a-1)(a-2)$. Now

$$\frac{a}{a^2+2a-3} - \frac{1}{a-2} + \frac{1}{1-a}$$

$$= \frac{a}{(a+3)(a-1)} - \frac{1}{a-2} - \frac{1}{a-1}$$

$$= \frac{a(a-2)}{(a+3)(a-1)(a-2)} - \frac{(a+3)(a-1)}{(a+3)(a-1)(a-2)}$$

$$- \frac{(a+3)(a-2)}{(a+3)(a-1)(a-2)}$$

$$= \frac{a^2-2a-a^2-2a+3-a^2-a+6}{(a+3)(a-1)(a-2)}$$

$$= \frac{-a^2-5a+9}{(a+3)(a-1)(a-2)} = \frac{-(a^2+5a-9)}{(a+3)(a-1)(a-2)}$$

Because a^2+5a-9 is irreducible, the answer is in lowest terms.

EXERCISE 2.4

In Problems 1–10, reduce the given fraction to lowest terms.

1 $\dfrac{8x^2y^2}{12x^4y}$

2 $\dfrac{28a^3b^2}{14a^2b^4}$

3 $\dfrac{16a^6b^7}{12a^3b^5+20a^5b^4}$

4 $\dfrac{18m^5v^3}{27m^4v^8-36m^6v^6}$

5 $\dfrac{5c-5d}{c^2-d^2}$

6 $\dfrac{x^2-y^2}{3y-3x}$

7 $\dfrac{r^4-r}{r^2+r+1}$

8 $\dfrac{a^3+8}{a^2-4}$

9 $\dfrac{a^2b^4-a^4b^2}{a^2-b^2}$

10 $\dfrac{2x^2-7x-15}{25-x^2}$

In Problems 11–16, replace the question mark with an appropriate expression.

11 $\dfrac{1}{2-x}=\dfrac{?}{x-2}$

12 $-\dfrac{-3}{x-1}=\dfrac{?}{1-x}$

13 $\dfrac{x-y}{y-x}=(?)1$

14 $\dfrac{x}{3b^2}=\dfrac{?}{6xb^2+3b^3}$

15 $\dfrac{3x+y}{x-y}=\dfrac{?}{3x^2-2xy-y^2}$

16 $\dfrac{2x}{3x+2}=\dfrac{?}{3x^2-x-2}$

In Problems 17–36, perform the indicated operation and simplify.

17 $\dfrac{3}{5x+10}+\dfrac{2x}{5(x+2)}$

18 $\dfrac{2x+1}{2x+2}-\dfrac{x-1}{2(x+1)}$

19 $\dfrac{3y}{y^2-25}+\dfrac{2}{y-5}$

20 $\dfrac{ab}{9a^2-b^2}-\dfrac{a}{3a+b}$

21 $\dfrac{8x}{x^2-4y^2}-\dfrac{2x}{x^2-5xy+6y^2}$

22 $\dfrac{x+3}{x^2-x-2}+\dfrac{x-1}{x^2+2x+1}$

23 $\dfrac{a-b}{a+b}+\dfrac{a+b}{a-b}+\dfrac{4ab}{a^2-b^2}$

24 $\dfrac{a}{a-x}+\dfrac{a}{a+x}+\dfrac{2a^2}{a^2-x^2}$

25 $\dfrac{a-b}{a+b}-\dfrac{a+b}{a-b}-\dfrac{4a^2}{a^2-b^2}$

26 $\dfrac{1}{x-1}-\dfrac{2x}{x^2-1}-\dfrac{3x^2}{x^3-1}$

27 $\dfrac{1}{x-1}-\dfrac{2}{x}+\dfrac{1}{x+1}$

28 $\dfrac{1}{a}-\dfrac{2}{a+1}+\dfrac{1}{a+2}$

29 $\dfrac{1}{x-3}+\dfrac{x+1}{(3-x)(x-2)}+\dfrac{2}{(2-x)(1-x)}$

30 $\dfrac{1}{x-1}+\dfrac{2x^3+2x^2}{1-x^3}+\dfrac{1}{x^2+x+1}$

31 $\dfrac{x}{4x^2-1}-\dfrac{3}{2x-1}-x+1$

32 $\dfrac{1}{x^2+2x+1}-\dfrac{3}{x+1}-2x+3$

33 $\dfrac{1}{(a-b)(a-c)}+\dfrac{1}{(b-a)(b-c)}+\dfrac{1}{(c-a)(c-b)}$

34 $\dfrac{x}{(x-y)(2-x)}+\dfrac{y}{(y-x)(x-2)}-\dfrac{y}{(x-y)(2-x)}$

35 $\dfrac{x+2}{3x-1}+\dfrac{x+1}{3-2x}+\dfrac{4x^2+6x+3}{6x^2-11x+3}$

36 $\dfrac{2x}{x-1}-\dfrac{2x^3-2x^2}{1-x^3}+\dfrac{1}{x^2+x+1}$

37 Use Theorem 2.4a, Part II to write the fraction $a/(a-b)$ in three other equivalent forms.

38 Is the value of $(1-x)/(x^2-1)$ equal to that of $-1/(x+1)$ for all values of x? If not, what value of x would make the fractions *not* equal?

39 The equation

$$\frac{x^2+2x+4}{x^3-8}=\frac{1}{x-2}$$

is true for all real values of x except one. What is this value? (As you will see later, there are no real values of x for which $x^2+2x+4=0$.)

40 Repeat Problem 39 for the equation

$$\frac{x^2-1}{(x+1)^2}=\frac{x-1}{x+1}$$

USING YOUR KNOWLEDGE 2.4

Certain basic problems in calculus require you to simplify expressions of the type

$$\frac{1}{(x+h)^2}-\frac{1}{x^2}$$

and to display h as a factor of the final numerator. We can do this as follows:

$$\frac{1}{(x+h)^2} - \frac{1}{x^2} = \frac{x^2 - (x+h)^2}{x^2(x+h)^2}$$

$$= \frac{(x - x - h)(x + x + h)}{x^2(x+h)^2}$$

$$= \frac{-h(2x+h)}{x^2(x+h)^2}$$

Carry out the same procedure for the following expressions.

1 $\dfrac{1}{x+h} - \dfrac{1}{x}$ **2** $\dfrac{1}{(x+h)^3} - \dfrac{1}{x^3}$ **3** $\dfrac{1}{1-(x+h)^2} - \dfrac{1}{1-x^2}$ **4** $\dfrac{x+h}{(x+h)^2+4} - \dfrac{x}{x^2+4}$

2.5 MULTIPLICATION AND DIVISION OF RATIONAL EXPRESSIONS

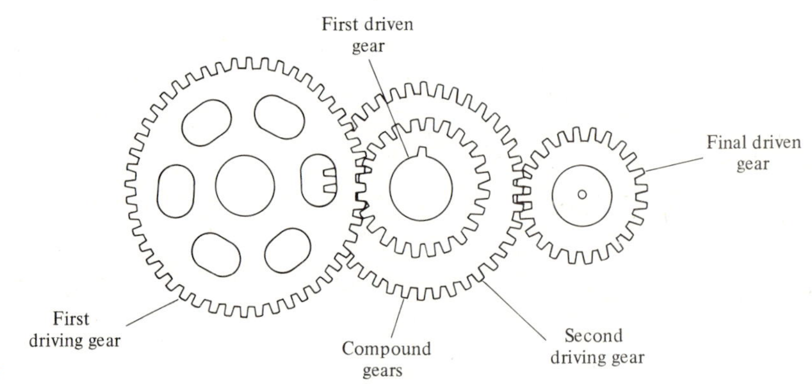

FIGURE 2.5a

Figure 2.5a shows a compound gear train. To find the speed in rpm (revolutions per minute) of the final driven gear of a compound gear train, we use the formula

$$\frac{T_1}{t_1} \cdot \frac{T_2}{t_2} \cdot R$$

where T_1 and T_2 represent the number of teeth in the first and second driving gears, t_1 and t_2 represent the number of teeth in the first and second driven gears, and R is the speed of the first driving gear.

In designing a gear train, an engineer assumes that

$$T_1 = x^2, \qquad T_2 = x^2 - 64, \qquad t_1 = x + 8, \qquad t_2 = x, \qquad R = 12$$

Find a formula for the speed of the final driven gear in terms of x. To solve this problem or to simplify the given formula, we need to know how to find the *product* of two rational expressions. Because a rational expression becomes an arithmetic fraction when the variables are replaced by permissible numbers, we see that such expressions are multiplied just as fractions are multiplied in arithmetic. Hence we have Theorem 2.5a. Thus,

THEOREM 2.5a

If $B, D \neq 0$, then $\frac{A}{B} \cdot \frac{C}{D} = \frac{AC}{BD}$.

$$\frac{T_1}{t_1} \cdot \frac{T_2}{t_2} \cdot R = \frac{T_1}{t_1} \cdot \frac{T_2}{t_2} \cdot \frac{R}{1} \qquad \text{Because } R = \frac{R}{1}$$

$$= \frac{T_1 \cdot T_2 \cdot R}{t_1 \cdot t_2}$$

We now substitute the given expressions for the T's, the t's, and the R and then simplify to obtain the desired formula. Thus,

$$\frac{T_1 T_2 R}{t_1 t_2} = \frac{x^2(x^2 - 64)(12)}{(x + 8)(x)}$$

$$= \frac{x^2(x - 8)(x + 8)(12)}{x(x + 8)}$$

$$= \frac{x(x + 8)}{x(x + 8)} \cdot 12x(x - 8)$$

$$= 12x(x - 8)$$

This results in a formula for the number of rpm of the final driven gear. For instance, if the first driving gear has 144 teeth, so that $x = 12$, then the final driven gear turns at the rate of (12)(12)(4) or 576 rpm.

Example 1 illustrates further the procedures used in the multiplication of rational expressions.

EXAMPLE 1

Find:

$$\frac{a^3 - b^3}{a + b} \cdot \frac{a^2 + 2ab + b^2}{a^2 - b^2}$$

SOLUTION:

We first factor each expression, obtaining

$$\frac{a^3 - b^3}{a + b} \cdot \frac{a^2 + 2ab + b^2}{a^2 - b^2} = \frac{(a - b)(a^2 + ab + b^2)}{a + b} \cdot \frac{(a + b)^2}{(a - b)(a + b)}$$

$$= \frac{(a^2 + ab + b^2)\,(a + b)^2(a - b)}{(a + b)^2(a - b)}$$

$$= a^2 + ab + b^2$$

This procedure is usually shortened by writing the expressions in factored form and dividing common factors from the numerator and denominator according to the Fundamental Principle of Fractions as follows:

$$\frac{a^3 - b^3}{a + b} \cdot \frac{a^2 + 2ab + b^2}{a^2 - b^2} = \frac{\overset{1}{\cancel{(a - b)}}(a^2 + ab + b^2)\overset{1}{\cancel{(a + b)^2}}}{\underset{1}{\cancel{(a + b)}}\underset{1}{\cancel{(a - b)}}\underset{1}{\cancel{(a + b)}}}$$

$$= a^2 + ab + b^2$$

▼ THEOREM 2.5b

If x and y are real numbers, $y \neq 0$, and n is a positive integer, then

$$\left(\frac{x}{y}\right)^n = \frac{x^n}{y^n}$$

Theorem 2.5b is a simple consequence of Theorem 2.5a. For example,

$$\left(\frac{2}{5}\right)^3 = \frac{2^3}{5^3} = \frac{8}{125} \quad \text{and} \quad \left(\frac{x}{y}\right)^{10} = \frac{x^{10}}{y^{10}}$$

▼ THEOREM 2.5c

If $B, C, D \neq 0$, then

$$\frac{A}{B} \div \frac{C}{D} = \frac{A}{B} \cdot \frac{D}{C} = \frac{AD}{BC}$$

The division of rational expressions is exactly like the division of fractions in arithmetic. Theorem 2.5c is the corresponding theorem for division. This theorem says that to divide a rational expression A/B by C/D, we *multiply* A/B by the *reciprocal* of C/D, that is, by D/C, to obtain AD/BC. For instance, to perform the following division, we invert the divisor and multiply. Thus,

$$\frac{2-x}{x+1} \div \frac{x^2-4}{x^2+4x+3} = \frac{2-x}{x+1} \cdot \frac{x^2+4x+3}{x^2-4} \qquad \text{Inverting the divisor and multiplying}$$

$$= \frac{2-x}{x+1} \cdot \frac{(x+3)(x+1)}{(x+2)(x-2)} \qquad \text{Factoring}$$

$$= \frac{\overset{-1}{\cancel{(2-x)}}(x+3)\overset{1}{\cancel{(x+1)}}}{\underset{1}{\cancel{(x+1)}}(x+2)\underset{1}{\cancel{(x-2)}}} \qquad \text{Simplifying and using the fact that } 2-x=-(x-2)$$

$$= \frac{-(x+3)}{x+2}$$

EXAMPLE 2

Simplify:

$$\frac{a^2+4a}{2a+6} \div \frac{a^2-16}{3a+9}$$

SOLUTION:

Following the procedure in Example 1, we have

$$\frac{a^2+4a}{2a+6} \div \frac{a^2-16}{3a+9} = \frac{a^2+4a}{2a+6} \cdot \frac{3a+9}{a^2-16} \qquad \text{Inverting the divisor and multiplying}$$

$$= \frac{a\overset{1}{\cancel{(a+4)}}}{2\underset{1}{\cancel{(a+3)}}} \cdot \frac{3\overset{1}{\cancel{(a+3)}}}{\underset{1}{\cancel{(a+4)}}(a-4)} \qquad \text{Factoring and simplifying}$$

$$= \frac{3a}{2(a-4)} \quad \text{or} \quad \frac{3a}{2a-8}$$

A fraction whose numerator or denominator, or both, contains other fractions is called a *complex fraction.* A fraction that is *not* complex is called a *simple fraction.* For example,

$$\frac{1/2a + \frac{1}{2}}{\frac{1}{4} + 1/2a}$$

is a complex fraction, but

$$\frac{a+2}{a-2}$$

is a simple fraction.

To simplify a complex fraction, we can proceed in either of two ways:

1. We can multiply the numerator and the denominator of the complex fraction by the LCD of the simple fractions appearing.
2. We can perform the operations indicated in the numerator and the denominator of the complex fraction, and then divide the numerator by the denominator.

Both these methods are illustrated in Example 3.

EXAMPLE 3

Simplify:

$$\frac{1/2a + \frac{1}{2}}{\frac{1}{4} + 1/2a}$$

SOLUTION:

Method 1 We first multiply the numerator and the denominator of the complex fraction by the LCD, $4a$, of the simple fractions, obtaining

$$\frac{1/2a + \frac{1}{2}}{\frac{1}{4} + 1/2a} = \frac{\mathbf{4a}(1/2a + \frac{1}{2})}{\mathbf{4a}(\frac{1}{4} + 1/2a)} = \frac{2 + 2a}{a + 2} \quad \text{or} \quad \frac{2(1 + a)}{a + 2}$$

Method 2 We first perform the additions in the numerator and the denominator. Then we invert the divisor and multiply.

$$\frac{1/2a + \frac{1}{2}}{\frac{1}{4} + 1/2a} = \frac{1/2a + a/2a}{a/4a + 2/4a}$$

$$= \frac{(1 + a)/2a}{(a + 2)/4a}$$

$$= \frac{1 + a}{2a} \cdot \frac{4a}{a + 2}$$

$$= \frac{2(1 + a)}{a + 2} \quad \text{or} \quad \frac{2 + 2a}{a + 2}$$

If the complex fraction is very complicated, we may simplify small portions of the fraction at a time. For example, to simplify

$$1 + \frac{a}{1 + 1/a}$$

we start with the lowermost terms and simplify $1 + 1/a$ first to obtain

$$1 + \frac{1}{a} = \frac{a}{a} + \frac{1}{a} = \frac{a + 1}{a}$$

Hence,

$$1 + \frac{a}{1 + 1/a} = 1 + \frac{a}{(a + 1)/a}$$

Next,

$$\frac{a}{(a+1)/a} = a \cdot \frac{a}{a+1}$$

Thus,

$$1 + \frac{a}{1 + 1/a} = 1 + \frac{a^2}{a+1}$$

$$= \frac{a+1}{a+1} + \frac{a^2}{a+1}$$

$$= \frac{a^2 + a + 1}{a+1}$$

EXAMPLE 4

Simplify:

$$\frac{x^3 + 1}{x - 1/[1 + 1/(x-1)]}$$

SOLUTION:

We work from the bottom up, first simplifying $1 + 1/(x-1)$ to obtain

$$1 + \frac{1}{x-1} = \frac{x-1}{x-1} + \frac{1}{x-1} = \frac{x}{x-1}$$

Next,

$$\frac{1}{1 + 1/(x-1)} = \frac{1}{x/(x-1)} = \frac{x-1}{x}$$

Then,

$$x - \frac{1}{1 + 1/(x-1)} = x - \frac{x-1}{x} = \frac{x^2 - x + 1}{x}$$

Finally,

$$\frac{x^3 + 1}{x - 1/[1 + 1/(x-1)]} = \frac{x^3 + 1}{(x^2 - x + 1)/x}$$

$$= (x^3 + 1) \div \frac{x^2 - x + 1}{x}$$

$$= (x+1)\cancel{(x^2 - x + 1)} \cdot \frac{x}{\cancel{x^2 - x + 1}}$$

$$= x(x+1)$$

Note: The denominator of the complex fraction means

$$x - \left[1 \div \left(1 + \frac{1}{x-1}\right)\right]$$

so that our procedure is equivalent to starting with the innermost parentheses and working out.

EXERCISE 2.5

In Problems 1–34, perform the indicated operations and reduce to lowest terms. Assume that the variables have no values resulting in a zero denominator.

1 $\frac{4a^2b^2}{9a^3b^2} \cdot \frac{3a^3b^5}{2ab^3}$

2 $\frac{27x^3y}{72xy^3} \cdot \frac{16x}{15y^2}$

3 $\frac{a+b}{a+2b} \cdot \frac{a^2+3ab+2b^2}{a^2-b^2}$

4 $\frac{x^2-1}{x^2+3x-10} \cdot \frac{x^2-25}{x^2-3x-4}$

5 $\frac{35xy^5}{24x^3y^9} \div \frac{15x^4y^7}{84x^3y^8}$

6 $\frac{15c^4d^4}{54cd^2} \div \frac{105c^5d^6}{14c^3d^2}$

7 $\frac{6x^3y^4}{x^2-16} \div \frac{3x^2y}{x^2-5x+4}$

8 $\frac{3a^4b^5}{a^2-4} \div \frac{9a^3b^3}{a^2-a-2}$

9 $\frac{y^2+2y-3}{y-5} \cdot \frac{y^2-3y-10}{y^2+5y-6}$

10 $\frac{x^2+2x-8}{x^2+7x+12} \cdot \frac{x^2+2x-3}{x^2-3x+2}$

11 $\frac{x^2-3x+2}{x^2-5x+6} \div \frac{x^2-5x+4}{x^2-7x+12}$

12 $\frac{x^2-6x+5}{x^2-15x+50} \div \frac{x^2+6x-7}{x^2+2x-35}$

13 $\frac{x^3-a^3}{x^2-4a^2} \cdot \frac{x+2a}{x-a}$

14 $\frac{x^3+8y^3}{x^2-64y^2} \cdot \frac{x^3+512y^3}{x^2+xy-2y^2}$

15 $\frac{a^3-x^3}{a^3+x^3} \div \frac{(a-x)^2}{a^2-x^2}$

16 $\frac{a^3+y^3}{a^3-y^3} \div \frac{(a+y)^2}{a^2-y^2}$

17 $\frac{(a-b)^2-c^2}{(a-c)^2-b^2} \cdot \frac{b^2-(c-a)^2}{c^2-(a-b)^2}$

18 $\frac{c^2-(a-b)^2}{b^2-(a-c)^2} \cdot \frac{(c-a)^2-b^2}{(a-b)^2-c^2}$

19 $\frac{x^2-(y+z)^2}{x^2-(y-z)^2} \div \frac{y^2-(x+z)^2}{y^2-(x-z)^2}$

20 $\frac{(y+z)^2-x^2}{(y-z)^2-x^2} \div \frac{(x+z)^2-y^2}{(x-z)^2-y^2}$

21 $\frac{64a^2-81b^2}{x^2-81} \cdot \frac{(x-9)^2}{8a-9b} \div \frac{8a^2+9ab}{(x+9)^2}$

22 $\frac{x^2-x-12}{x^2-49} \cdot \frac{x^2-x-56}{x^2+x-20} \div \frac{x^2-5x-24}{x+5}$

23 $\frac{x+x/y}{y-1/y}$

24 $\frac{x/y-y/x}{1+y/x}$

25 $\frac{1/a+1/b}{1/a-1/b}$

26 $\frac{x/y-y/x}{1-y^2/x^2}$

27 $\frac{1/x-1/a}{x-a}$

28 $\frac{1/x^2-1/a^2}{x-a}$

29 $\frac{1/(x-1)-1/(a-1)}{x-a}$

30 $\frac{1/(x^2-a^2)-1/(h^2-a^2)}{x-h}$

31 $\frac{1+(x+1)/(x-1)}{1/(x-1)-1/(x+1)}$

32 $\frac{4/(x+1)-1/x}{(x-2)/x+(2x+6)/(x+1)}$

33 $\frac{2}{1+2/(1+2/x)}$

34 $\frac{1}{x-x/[x-x^2/(x+1)]}$

35 The formula for the Doppler Effect in light is

$$f = f_s\sqrt{\frac{1+v/c}{1-v/c}}$$

where f_s is the frequency of the source, v is the relative velocity between source and observer, and c is the velocity of light. Write the fraction under the radical as a simple fraction in lowest terms.

36 Balmer's Formula for the wavelength λ (lambda) of the hydrogen spectrum lines gives

$$\lambda = \frac{1}{R(1/m^2-1/n^2)}$$

Write this fraction in simplest form.

37 If x dollars are invested at 5 percent simple interest and y dollars are invested at 7 percent simple interest, what rational fraction represents the average rate of interest on the total of $x+y$ dollars?

38 A motorist drives 45 miles per hour for x hours, 50 miles per hour for y hours, and 55 miles per hour for z hours. Write the rational fraction that represents his average speed for the total time $x+y+z$ hours.

USING YOUR KNOWLEDGE 2.5

1 In physics, the formula for the speed of a particle with energy eV is

$$v = c\sqrt{1 - \frac{1}{(1 + eV/m_0c^2)^2}}$$ Write the quantity under the radical in simplest form.

2 An observer in a closed vehicle that is traveling at a speed V throws a ball, which this observer sees as moving at the speed v'. According to the theory of relativity, an outside observer, at rest with respect to the vehicle, will see the ball moving at the speed

$$v = \frac{v' + V}{1 + v'V/c^2}$$ Write the complex fraction in simplest form.

SELF-TEST

1 Classify the given polynomials as monomials, binomials, or trinomials, and give the degree of each:
a $x^2 - 3x + 5$ **b** $3x^2 + x^3 - 25x$
c 0 **d** $5x^2y^3 - x^7$ **e** $-9x$

2 Add or subtract the polynomials as indicated:
a $(4x^2 + 6x - 3) + (3x^2 - 8x - 7)$
b $(5x^3 + 2x - 4) - 2(3x^3 - 2x^2 + 4x - 5)$

3 If $P(x) = x^2 + x - 3$ and $Q(x) = 3x^2 - 5x + 5$, find:
a $P(x) + Q(x)$ **b** $Q(x) - P(x)$
c $P(0)$ **d** $Q(-1)$ **e** $P(0) + Q(-1)$

4 Multiply $x^2 + 3x - 2$ by $x^3 + x - 3$.

5 Perform the indicated multiplications:
a $(x - 5)(x + 2)$ **b** $(2x + 3)(3x - 4)$
c $(5x - 2y)^2$ **d** $(5x + 2y)(5x - 2y)$
e $(x + y)(x^2 - xy + y^2)$

6 Perform the indicated divisions:
a $(3x^3 - 10x - 4) \div (x - 2)$
b $(2a^3 + 2a - 2) \div (a^2 + a - 1)$

7 Factor:
a $12a^3b^2 - 8a^2b$
b $8ac + 3b - 8bc - 3a$
c $b^2 - 25$ **d** $64 - z^3$ **e** $27b^3 + 8c^3$

8 Factor:
a $x^2 - 6x + 9$ **b** $x^2 - 8x + 16$
c $36u^2 - 60uv + 25v^2$

9 Factor:
a $2x^2 + x - 15$ **b** $6x^2 - x - 12$

10 Factor:
a $x^6 + y^{12}$ **b** $a^4 - 16b^4$

11 Reduce to simplest terms:
a $\frac{24x^4y^7}{6x^3y^9}$ **b** $\frac{15 + 7x - 2x^2}{x^2 - 25}$

12 Add the fractions as indicated:
a $\frac{2}{3x + 6} + \frac{4}{3(x + 2)}$
b $\frac{x + 2}{x^2 - 4} + \frac{x - 3}{x^2 - 9}$

13 Perform the indicated subtractions:
a $\frac{2x + 3}{2x + 2} - \frac{x - 3}{2(x + 1)}$
b $\frac{2}{(2 - x)(x + 2)} - \frac{x}{x^2 - 4}$

14 Combine and simplify as indicated:
$\frac{x}{4x^2 - 1} + \frac{3}{2x - 1} - x + 1$

15 Combine and simplify as indicated:
$\frac{x}{2x^2 - 3x + 1} + \frac{2}{2x^2 + 3x - 2} - \frac{2x + 1}{x^2 + x - 2}$

16 Multiply as indicated and simplify:
a $\frac{6a^2b^2}{9a^3b^2} \cdot \frac{3a^4b^5}{2ab^3}$
b $\frac{a^3 + 8b^3}{a - 2b} \cdot \frac{a^2 - 4ab + 4b^2}{a^2 - 4b^2}$

17 Divide as indicated and simplify:
$\frac{1 - x}{x + 2} \div \frac{x^2 - 1}{x^2 + 3x + 2}$

18 Divide as indicated and simplify:
$\frac{c^2 - (a - b)^2}{b^2 - (a - c)^2} \div \frac{(c + a)^2 - b^2}{(a + b)^2 - c^2}$

19 Simplify: $\frac{1/a + \frac{1}{3}}{\frac{1}{9} + 1/3a}$

20 Simplify: $\frac{a + 2/(1 - a)}{a + (4a + 2)/(a - 1)}$

CHAPTER 3

EXPONENTS AND RADICALS

3.1 INTEGRAL EXPONENTS

▼

Laws of exponents

I. $x^m \cdot x^n = x^{m+n}$

II. $(x^m)^n = x^{mn}$

III. $(xy)^n = x^n y^n$

IV. $\dfrac{x^m}{x^n} = \begin{cases} x^{m-n} & \text{if } m > n \\ 1 & \text{if } m = n \\ \dfrac{1}{x^{n-m}} & \text{if } m < n \end{cases} \quad x \neq 0$

V. $\left(\dfrac{x}{y}\right)^n = \dfrac{x^n}{y^n}, \; y \neq 0$

There are many places in science and technology where numbers are expressed in terms of *negative exponents*. For example, the diameter of a DNA molecule is 10^{-8} meter, and the time it takes for an electron to go from source to screen in a TV tube is 10^{-6} second. What do 10^{-8} and 10^{-6} mean? In Chapter 2, we defined positive integer exponents and obtained the important laws that they obey. In this section, we shall extend the definition of an exponent so as to give meaning to such symbols as 10^{-8}, 10^{-6}, x^{-4}, and x^0. In doing this, we observe a very basic principle; we make sure that the Laws of Exponents, which were previously obtained, continue to hold. We list these laws for ready reference.

THE ZERO EXPONENT

One of the simplest ways to arrive at x^0 is to require that the first line of Law IV apply for $m = n$. Thus, for $x \neq 0$, we should have

$$\frac{x^m}{x^m} = x^{m-m} = x^0$$

However, by the second line of Law IV, the result must be 1. This leads us to make Definition 3.1a. By this definition, $8^0 = 1$, $(-5)^0 = 1$, and $(-\frac{4}{9})^0 = 1$, but 0^0 is *not* defined; when using Law IV, remember that $x \neq 0$.

▼ **DEFINITION 3.1a**

If x is a real nonzero number, then

$$x^0 = 1$$

NEGATIVE EXPONENTS

We are now ready to find a meaning for such symbols as 10^{-8}. To do this, we reason that if $n > 0$ and Law I is to apply to negative exponents, then

$$x^{-n} \cdot x^n = x^{-n+n} = x^0 = 1$$

Thus, x^{-n} should be defined to be the reciprocal (the multiplicative inverse) of x^n; that is,

$$x^{-n} = \frac{1}{x^n}$$

▼ DEFINITION 3.1b

If x is a nonzero real number and n is any positive integer, then

$$x^{-n} = \frac{1}{x^n}$$

This idea motivates Definition 3.1b. For example,

$$3^{-2} = 1/3^2 = \tfrac{1}{9}$$
$$10^{-8} = 1/10^8 = 1/100{,}000{,}000$$

and

$$(ab)^{-4} = \frac{1}{(ab)^4} = \frac{1}{a^4b^4}$$

Because $a^{-4}b^{-4} = 1/a^4 \cdot 1/b^4 = 1/a^4b^4$, we conclude that $(ab)^{-4} = a^{-4}b^{-4}$. This indicates that Law III applies to negative as well as positive exponents. It can be shown, although we shall not do so, that all five Laws of Exponents listed, where m and n now represent *any* integers, are valid. Moreover, the three parts of Law IV may be stated simply as:

$$\frac{x^m}{x^n} = x^{m-n}, \qquad x \neq 0$$

For instance,

$$\frac{x^4}{x^7} = x^{4-7} = x^{-3} = \frac{1}{x^3}$$

which is in complete agreement with what we would get by using the third line of Law IV.

EXAMPLE 1

Write the given expression in simplified form using positive exponents. In (d) and (e) assume that n is a positive integer.

a. $(x^{-5} \cdot x^8)^{-2}$
b. $(x^2y^{-1})^{-3}$

SOLUTION:

a. $(x^{-5} \cdot x^8)^{-2} = (x^{-5+8})^{-2}$ By Law I

$= (x^3)^{-2}$

$= x^{-6}$ By Law II

$= \dfrac{1}{x^6}$ By Definition 3.1b

b. $(x^2y^{-1})^{-3} = (x^2)^{-3}(y^{-1})^{-3}$ By Law III

$= x^{-6}y^3$ By Law II

c. $\left(\dfrac{x^2y^2}{x^{-1}y^3}\right)^{-2}$

d. $x^{2n} \cdot x^{-3n}$

e. $\left(\dfrac{x^n}{x^{n-2}}\right)^3$

$$= \frac{1}{x^6} \cdot y^3 \qquad \text{By Definition 3.1b}$$

$$= \frac{y^3}{x^6}$$

c. $\left(\dfrac{x^2y^2}{x^{-1}y^3}\right)^{-2} = (x^{2-(-1)}y^{2-3})^{-2}$ By Law IV

$$= (x^3y^{-1})^{-2}$$

$$= (x^3)^{-2}(y^{-1})^{-2} \qquad \text{By Law III}$$

$$= x^{-6}y^2 \qquad \text{By Law II}$$

$$= \frac{1}{x^6} \cdot y^2 \qquad \text{By Definition 3.1b}$$

$$= \frac{y^2}{x^6}$$

d. $x^{2n} \cdot x^{-3n} = x^{2n+(-3n)}$ By Law I

$$= x^{-n}$$

$$= \frac{1}{x^n} \qquad \text{By Definition 3.1b}$$

e. $\left(\dfrac{x^n}{x^{n-2}}\right)^3 = (x^{n-(n-2)})^3$ By Law IV

$$= (x^{n-n+2})^3$$

$$= (x^2)^3$$

$$= x^6 \qquad \text{By Law II}$$

EXAMPLE 2

Simplify:

$$\frac{y^{-1}}{x + y^{-1}}$$

SOLUTION:

First method

$$\frac{y^{-1}}{x + y^{-1}} = \frac{1/y}{x + 1/y} \qquad \text{By Definition 3.1b}$$

$$= \frac{1/y}{(xy + 1)/y}$$

$$= \frac{1}{y} \cdot \frac{y}{xy + 1}$$

$$= \frac{1}{xy + 1}$$

Second method Multiply numerator and denominator by y to obtain

$$\frac{y^{-1}}{x+y^{-1}} = \frac{y \cdot y^{-1}}{y(x+y^{-1})} = \frac{y^0}{xy+y^0} \qquad \text{By Law I}$$

$$= \frac{1}{xy+1} \qquad \text{By Definition 3.1a}$$

SCIENTIFIC NOTATION

In science and in other areas, there frequently occur very large or very small numbers. For example, a red cell of human blood contains 270,000,000 hemoglobin molecules, and the mass of a single carbon atom is 0.000 000 000 000 000 000 000 019 9 gram. Numbers in this form are difficult to write and to work with and, because of this, are written in *scientific notation.* (See Definition 3.1c.) To find n, we simply count the number of places that the decimal point must be moved to give the coefficient m. If the decimal point must be moved to the *left*, n is positive; if it must be moved to the *right*, n is *negative.* Thus,

▼ DEFINITION 3.1c

A number is said to be in *scientific notation* if it is written in the form

$$m \times 10^n$$

where m is a number such that $1 \le m < 10$ and n is an integer.

$$3 = 3 \times 10^0$$

$$87 = 8.7 \times 10^1 = 8.7 \times 10$$

$$68{,}000 = 6.8 \times 10^4$$

and

$$0.4 = 4 \times 10^{-1}$$

$$0.07 = 7 \times 10^{-2}$$

$$0.00038 = 3.8 \times 10^{-4}$$

EXAMPLE 3

Write in scientific notation:

a. 270,000,000

b. 0.000 000 000 000 000 000 000 019 9

SOLUTION:

a. $270{,}000{,}000 = 2.7 \times 10^8$

b. $0.000\,000\,000\,000\,000\,000\,000\,019\,9 = 1.99 \times 10^{-23}$

Scientific notation is also used to simplify calculations involving very large or very small quantities. For example, our solar system is 3×10^4 light years away from the center of the Milky Way. Because a light year is 6×10^{12} miles, our distance from the center of the galaxy is

$$(3 \times 10^4) \times (6 \times 10^{12}) = (3 \times 6) \times (10^4 \times 10^{12})$$

$$= 18 \times 10^{16}$$

$$= 1.8 \times 10 \times 10^{16}$$

$$= 1.8 \times 10^{17} \text{ miles}$$

EXAMPLE 4

Compute:

$$\frac{(6 \times 10^4) \times (8 \times 10^{-4})}{12 \times 10^{-3}}$$

SOLUTION:

$$\frac{(6 \times 10^4) \times (8 \times 10^{-4})}{12 \times 10^{-3}} = \frac{48 \times 10^0}{12 \times 10^{-3}}$$

$$= \frac{48}{12} \cdot \frac{10^0}{10^{-3}}$$

$$= 4 \cdot 10^{0-(-3)}$$

$$= 4 \times 10^3$$

If you have a scientific calculator, you might have noticed that if a calculation results in a very large or a very small number, the display shows the answer in scientific notation. For example, if you multiply 9,832,000 by 4,567,000, the display may show

4.4903 13

which means

4.4903×10^{13}

(Of course, the 4.4903 is only approximate; it is correct to four decimal places.) Similarly, if you divide 0.04217 by 3,128,500, the display may show

1.3479 −08

which means

1.3479×10^{-8}

(Again, the number multiplying the power of ten is correct to four decimal places.)

EXAMPLE 5

C Use your calculator to evaluate the following fraction and express the answer in scientific notation.

$$\frac{(6.32 \times 10^8)(8.17 \times 10^{-4})}{1.26 \times 10^{-3}}$$

SOLUTION:

We can collect the powers of ten and rewrite the fraction as

$$\frac{(6.32)(8.17)}{1.26} \times 10^{8-4+3} = \frac{(6.32)(8.17)}{1.26} \times 10^7$$

Evaluating the first factor on the calculator, we obtain approximately 40.98. Thus, the answer is 4.098×10^8.

EXERCISE 3.1

In Problems 1–8, evaluate the given expression.

1 5^{-3} **2** $(-3)^{-2}$ **3** $\left(\frac{-2}{3}\right)^{-3}$ **4** $\frac{3^{-1} + 2^0}{2^{-1} + 2^2}$ **5** $\frac{3^0 + 2^0}{2^{-3}}$ **6** $\frac{4^{-2} \cdot 2^3}{2}$ **7** $\frac{3^{-2} \cdot 9^2}{3^4}$

8 $\frac{4^2 \cdot 2^{-3}}{4^{-1}}$

In Problems 9–34, write the given expression in simplified form with positive exponents. In Problems 23–30, assume that m and n are positive integers.

9 $(x^{-1}y^2)^{-2}$

10 $(3x^{-2}y^3)^{-2}$

11 $\left(\dfrac{a^{-4}}{b^2}\right)^{-2}$

12 $\left(\dfrac{a^{-2}}{b^{-3}}\right)^{-3}$

13 $2m^{-6}v^7$

14 $3h^{-1}m^{-4}$

15 $\dfrac{1}{a^{-2}b^3}$

16 $\dfrac{4}{c^{-2}d^8}$

17 $(m^{-5}v^7)(m^{-2}v^{-7})$

18 $(x^2y^3)(x^{-1}y^{-4})$

19 $\left(\dfrac{a^{-1}b^{-2}}{ab^{-1}}\right)\left(\dfrac{a^{-4}b^2}{a^{-6}b^3}\right)^{-2}$

20 $\left(\dfrac{-8a^2}{b^{-5}}\right)^2\left(\dfrac{16b^3}{a^{-3}}\right)^{-2}$

21 $\left(\dfrac{3x}{2y}\right)^2\left(\dfrac{y}{x}\right)^{-1}\left(\dfrac{x^{-2}}{3^{-1}}\right)$

22 $\left(\dfrac{x}{y}\right)^3\left(\dfrac{-2x}{3}\right)^{-2}\left(\dfrac{-y}{x}\right)^{-3}$

23 $x^{3n}\cdot x^{-5n}$

24 $(x^m)^n\cdot x^{2mn}$

25 $\dfrac{x^{n+1}\cdot x^{2n}}{x^{n-1}}$

26 $\dfrac{x^{n-1}\cdot x^{-n}}{x^{n+1}}$

27 $\left(\dfrac{x^{2n}}{x^{2n-1}}\right)^{-3}$

28 $\left(\dfrac{x^n}{x^{-n+1}}\right)^{-2}$

29 $\dfrac{x^{2n}y^{n+2}}{x^ny^{n-1}}$

30 $\dfrac{x^{n+1}y^{2n-1}}{x^{n-1}y^n}, n>1$

31 $\dfrac{x^{-1}+y^{-1}}{x^{-1}-y^{-1}}$

32 $\dfrac{x^{-1}+y^{-1}}{x^{-1}}$

33 $a^{-1}b-ab^{-1}$

34 $\dfrac{(a+b)^{-1}}{a^{-1}+b^{-1}}$

In Problems 35–40, write the given number in scientific notation.

35 87

36 0.3

37 0.0012

38 37,400,000

39 0.000 000 29

40 3,451,000

In Problems 41–50, simplify the given expression and write the answer in scientific notation.

41 $\dfrac{(0.130)(2.2)(50)}{(2.60)(0.011)(0.05)}$

42 $\dfrac{(0.48)(3.8)(25)}{(0.12)(0.19)(10)}$

43 $\dfrac{(2\times 10^2)^3(6\times 10^{-5})}{4\times 10^3}$

44 $\dfrac{(8\times 10)^2(3\times 10^{-2})}{24\times 10^{-3}}$

45 $\dfrac{(210)(0.008)(800)}{(0.0014)(600)(20)}$

46 $\dfrac{(0.0024)(0.004)(350)}{(175)(0.001)(0.008)}$

47 The width of the asteroid belt is 2.8×10^8 kilometers. The speed of Pioneer 10 in passing through this belt was 1.4×10^5 kilometers per hour. Thus, Pioneer 10 took

$$\frac{2.8\times 10^8}{1.4\times 10^5}$$

hours to go through the belt. How many hours is that?

48 The mass of the earth is 6×10^{24} kilograms. If 1 kilogram is 1.1×10^{-3} tons, what is the mass of the earth in tons?

49 Sir Arthur Eddington claimed that the total number of electrons in the universe is 136×2^{256}. If you have a calculator with a y^x key, you can verify that 2^{256} is approximately 1.16×10^{77}. Use this result to write 136×2^{256} in scientific notation.

C **50** The most plentiful form of sea life known is the nematode sea worm. It is estimated that there are

40,000,000,000,000,000,000,000,000

of these sea worms in the world's oceans. There are about 3.16×10^9 cubic miles of ocean and about 1.10×10^{12} gallons of water per cubic mile. If these sea worms were uniformly dispersed through the oceans, about how many of them would there be per gallon?

In Problems 51–54, the number 6.023×10^{23}, known as Avogadro's Number, is the number of atoms of any element in one standard unit (a *gram atomic weight*) of the element. This number is also the number of molecules of any substance in the unit known as the *gram molecular weight*.

C **51** One gram atomic weight of hydrogen is known to weigh 1.008 grams. Thus, the mass of one atom of hydrogen is

$$\frac{1.008}{6.023 \times 10^{23}} \text{ gram}$$

Express this result in scientific notation, giving four digits in the answer.

C **52** One gram atomic weight of oxygen is known to weigh 16.00 grams. Hence, the mass of one atom of oxygen is

$$\frac{16.00}{6.023 \times 10^{23}} \text{ gram}$$

Express this result in scientific notation, giving four digits in the answer.

C **53** One gram molecular weight of water is known to weigh 18.02 grams. Thus, the mass of one molecule of water is

$$\frac{18.02}{6.023 \times 10^{23}} \text{ gram}$$

Express this result in scientific notation, giving four digits in the answer.

C **54** The smallest sample of magnesium that can be weighed on the most delicate analytical balance weighs 1.0×10^{-6} gram. One gram atomic weight of magnesium is known to weigh 24.31 grams, so that the number of atoms in one gram of magnesium is

$$\frac{6.023 \times 10^{23}}{24.31}$$

How many atoms are there in the given sample? Express your answer in scientific notation, giving two digits.

USING YOUR KNOWLEDGE 3.1

Have you had to wait in line for gasoline? Have you had sufficient heat in the winter? It is all a matter of energy, its production and consumption. The amounts involved are so enormous that we write them in scientific notation. Use your knowledge to answer the following questions.

1 In 1974, U.S. oil production was 3.2×10^9 barrels. In the same year, U.S. consumption amounted to 5.4×10^9 barrels. How many barrels had to be imported?

2 U.S. oil reserves are 3.5×10^{10} barrels. Production amounts to 3.2×10^9 barrels per year. At this rate, how long would U.S. oil reserves last? (Give your answer to the nearest year.)

3 Perhaps we can use other people's oil. Let's see what's happening in the world. The world's reserves of oil are 6.28×10^{11} barrels. Production is 2.0×10^{10} barrels per year. At this rate, how long would the world's oil reserves last? (Give your answer to the nearest year.)

C **4** Maybe we can use natural gas for energy. U.S. reserves of natural gas amount to 2.37×10^{14} cubic feet (ft^3). Production is 2.2648×10^{13} ft^3 per year. At this rate, how long would U.S. reserves of gas last? (Answer to the nearest year.)

C **5** The coal industry claims it can help. U.S. reserves of coal are 4.34×10^{11} long tons.

a If 1 long ton is 1.12 short tons, how many short tons of coal are in U.S. reserves?

b The United States consumes 6.12×10^8 short tons of coal annually. At this rate, how long would U.S. coal reserves last? (Answer to the nearest year.)

3.2 RADICALS

If a ball is dropped from some point above the earth's surface, the velocity of the ball (in meters per second) is given by the formula

$$v = \sqrt{2gh}$$

where $g = 9.8$ and h is the distance in meters through which the ball has fallen. The velocity here is said to be the "square root" of $2gh$. Now suppose we want to find the velocity of the ball after it has fallen 10 meters. Then, we need to find

$$\sqrt{(2)(9.8)(10)} = \sqrt{196}$$

In arithmetic, the roots of certain numbers are easy to find. For example,

1. A square root of 196 is 14 because $14^2 = 196$. (This gives the answer to our problem; the desired velocity is 14 meters per second.)
2. Another square root of 196 is -14 because $(-14)^2 = 196$.
3. A cube (third) root of 8 is 2 because $2^3 = 8$.

▼ **DEFINITION 3.2a**

If a and x are real numbers and n is a positive integer greater than 1, then x is an *nth root of a* if and only if $x^n = a$.

In general, we define an nth root of a number as in Definition 3.2a.

EVALUATION OF ROOTS

As you can see from items 1 and 2 in the foregoing list, there are *two* square roots of 196: 14 and -14. Similarly, $\frac{16}{81}$ has two real fourth roots, $\frac{2}{3}$ and $-\frac{2}{3}$, because $(-\frac{2}{3})^4 = (\frac{2}{3})^4 = \frac{16}{81}$. To be able to specify one of these roots and to avoid ambiguity, we introduce the notion of the "principal nth root of a." This root is denoted by $\sqrt[n]{a}$, where the symbol $\sqrt{\ }$ is a *radical sign*, n is the *index*, and a is the *radicand*. The entire expression $\sqrt[n]{a}$ is called a *radical*. By convention, the index 2 is omitted, and we write $\sqrt{2}$ rather than $\sqrt[2]{2}$. The principal root is defined in Definition 3.2b. In Table 3.2a we illustrate the cases in which $a \neq 0$.

TABLE 3.2a

n	$a = 64$	$a = -64$
$n = 2$	$\sqrt{64} = 8$ (because $8^2 = 64$)	$\sqrt{-64}$ is not a real number.
$n = 3$	$\sqrt[3]{64} = 4$ (because $4^3 = 64$)	$\sqrt[3]{-64} = -4$ (because $(-4)^3 = -64$)

You should make careful note of the fact that, although 8 and -8 are both square roots of 64, the radical $\sqrt{64}$ is 8, *not* ± 8. The radical sign denotes the *principal* root. If the negative root is desired, it must be denoted by $-\sqrt{64}$.

▼ DEFINITION 3.2b

If a is a real number and n is a positive integer greater than 1, then $\sqrt[n]{a}$ is defined in the following table.

n	$a > 0$	$a < 0$	$a = 0$
Even	$\sqrt[n]{a}$ is the *positive* nth root of a.	$\sqrt[n]{a}$ is not a real number.	$\sqrt[n]{a} = 0$
Odd	$\sqrt[n]{a}$ is the *positive* nth root of a.	$\sqrt[n]{a}$ is the *negative* nth root of a.	$\sqrt[n]{a} = 0$

EXAMPLE 1

Evaluate:
a. $\sqrt[3]{\frac{1}{8}}$ **b.** $\sqrt[5]{-32}$ **c.** $\sqrt{-16}$

SOLUTION:

a. $\sqrt[3]{\frac{1}{8}} = \frac{1}{2}$ because $(\frac{1}{2})^3 = \frac{1}{8}$.

b. $\sqrt[5]{-32} = -2$ because $(-2)^5 = -32$.

c. $\sqrt{-16}$ is not a real number.

SIMPLIFYING RADICALS

▼ THEOREM 3.2a

If n is a positive integer greater than 1 and if x and y are real numbers such that $\sqrt[n]{x}$ and $\sqrt[n]{y}$ are also real numbers, then

I. $(\sqrt[n]{x})^n = x$

II. $\sqrt[n]{x} \cdot \sqrt[n]{y} = \sqrt[n]{xy}$

III. $\dfrac{\sqrt[n]{x}}{\sqrt[n]{y}} = \sqrt[n]{\dfrac{x}{y}}$ **for** $y \neq 0$

IVa. **For n odd,** $\sqrt[n]{x^n} = x$

IVb. **For n even,** $\sqrt[n]{x^n} = |x|$

V. $\sqrt[kn]{x^{km}} = \sqrt[n]{x^m}$ **for k, a positive integer, and $x > 0$**

Theorem 3.2a states the important Laws of Radicals. Recall that $|x|$ denotes the *absolute value* of x, which was defined on page 14. Part IVb of Theorem 3.2a tells us, for instance, that

$$\sqrt[4]{(-2)^4} = |-2| = 2$$

We may regard the following as a proof of Part III of the theorem. (Proofs of the remaining parts are left for the problems.) If both sides of the equation in Part III are raised to the nth power, the result is x/y. Thus, both sides are nth roots of x/y. It remains to be verified that $\sqrt[n]{x}/\sqrt[n]{y}$ is the principal root. This is so if n is even because both $\sqrt[n]{x}$ and $\sqrt[n]{y}$ are nonnegative. If n is odd, then the sign of $\sqrt[n]{x}/\sqrt[n]{y}$ agrees with the sign of x/y, which agrees with the sign of $\sqrt[n]{x/y}$. This completes the proof of Part III.

Theorem 3.2a can often be used to simplify expressions involving radicals. A radical is said to be in simplified form if the following three conditions are satisfied:

1. **No factor of the radicand appears to a power greater than or equal to the index of the radical.**
2. **The radicand contains no fractions.**
3. **The index of the radical is as small as possible.**

EXAMPLE 2

Assume that x, y, and z represent positive numbers and simplify:

a. $\sqrt[3]{81x^4y^6z^7}$

SOLUTION:

a.

$\sqrt[3]{81x^4y^6z^7} = \sqrt[3]{(3^3x^3y^6z^6)(3xz)}$	Laws of Exponents, Part I
$= \sqrt[3]{(3xy^2z^2)^3(3xz)}$	Laws of Exponents, Part III
$= \sqrt[3]{(3xy^2z^2)^3} \cdot \sqrt[3]{3xz}$	Theorem 3.2a, Part II
$= 3xy^2z^2 \cdot \sqrt[3]{3xz}$	Theorem 3.2a, Part IVa

b. $\dfrac{\sqrt{8x^5}}{\sqrt{3y^7}}$

b. $\dfrac{\sqrt{8x^5}}{\sqrt{3y^7}} = \sqrt{\dfrac{8x^5}{3y^7}}$ Theorem 3.2a, Part III

$= \sqrt{\dfrac{2^3x^5}{3y^7} \cdot \dfrac{3y}{3y}}$ We multiplied by $3y/3y$ in order to make the denominator a perfect square.

$= \sqrt{\dfrac{2^2x^4}{3^2y^8}(6xy)}$ Laws of Exponents, Part I

$= \dfrac{2x^2}{3y^4}\sqrt{6xy}$ Theorem 3.2a, Part IVb

The procedure used in simplifying Example 2b is usually called "rationalizing the denominator," because the result is a fraction with no radicals in the denominator.

EXAMPLE 3

Simplify $\sqrt[4]{9x^2y^{10}}$, where x and y represent positive numbers.

SOLUTION:

$\sqrt[4]{9x^2y^{10}} = \sqrt[4]{3^2x^2y^{10}}$

Notice that the index 4 and the exponents 2, 2, and 10 under the radical have the common factor 2. By Theorem 3.2a, Part V, this common factor may be divided out. Thus,

$$\sqrt[4]{9x^2y^{10}} = \sqrt{3xy^5} = \sqrt{3xy(y^2)^2} = y^2\sqrt{3xy}$$

EXAMPLE 4

Simplify $\sqrt{a^2 - 6ab + 9b^2}$.

SOLUTION:

$\sqrt{a^2 - 6ab + 9b^2} = \sqrt{(a-3b)^2}$

$= |a - 3b|$ By Theorem 3.2a, Part IVb

Note that the absolute value bars are needed because we do not know whether $a - 3b$ is nonnegative and the principal root must be nonnegative.

EXERCISE 3.2

In Problems 1–10, evaluate if the radical is a real number.

1 $\sqrt[3]{-125}$ **2** $\sqrt[4]{625}$ **3** $\sqrt[5]{\dfrac{-1}{243}}$

4 $\sqrt[4]{-81}$ **5** $\sqrt{x^8y^2}$, $y > 0$ **6** $\sqrt[3]{x^6y^3}$

7 $\sqrt{\frac{16}{9}x^6y^8}$, $x > 0$ **8** $\sqrt[5]{\dfrac{-32x^5y^5}{243}}$ **9** $\sqrt[3]{-64a^9b^3}$

10 $\sqrt[7]{128r^{14}s^{21}}$

In Problems 11–24, simplify. Assume that all the letters represent positive numbers.

11 $\sqrt{125xy^6}$ **12** $\sqrt{360x^4y^5}$ **13** $\sqrt[3]{1024x^7y^9}$

14 $2\sqrt[3]{16x^2y^7}$ **15** $\dfrac{1}{3a}\sqrt{27a^3m^7}$ **16** $\dfrac{3}{5x}\cdot\sqrt[3]{375x^8y}$

17 $\sqrt{(x+y)(x^2-y^2)},\ x>y$

18 $\sqrt{9x^3-36x^2+36x},\ x>2$

19 $-\sqrt{\dfrac{3}{8ab^3}}$

20 $\sqrt[3]{\dfrac{3}{16x^2}}$

21 $\sqrt{\dfrac{7x^5}{5y^3}}$

22 $\sqrt{\dfrac{16z^8}{x^3y^5}}$

23 $\sqrt[3]{\dfrac{-6a^5}{49b^2}}$

24 $\sqrt[3]{\dfrac{3b^{12}}{4a^4}}$

In Problems 25–34, use Theorem 3.2a to simplify. Assume that all the letters represent positive numbers.

25 $\sqrt[4]{x^8y^4}$

26 $\sqrt{a^2+2ab+b^2}$

27 $\sqrt[3]{7xy}\cdot\sqrt[3]{49x^2y^4}$

28 $\sqrt[4]{8a^6b^7c^3}\cdot\sqrt[4]{2a^2bc^7}$

29 $\dfrac{\sqrt{a^3b}}{\sqrt{ab^4}}$

30 $\dfrac{\sqrt{x^4y^5}}{\sqrt{2xy^8}}$

31 $\sqrt[8]{x^{12}}$

32 $\sqrt[9]{x^6}$

33 $(\sqrt[3]{x^2y^7})^3$

34 $(\sqrt[5]{-3xy^5})^5$

In Problems 35–42, rationalize the denominator and simplify. Assume that all the letters represent positive numbers.

35 $\dfrac{\sqrt{2}}{\sqrt{7}}$

36 $\dfrac{\sqrt{y}}{\sqrt{x}}$

37 $\dfrac{\sqrt[3]{3y}}{\sqrt[3]{16x^2}}$

38 $\dfrac{\sqrt[3]{xz^4}}{\sqrt[3]{9xy^2}}$

39 $\dfrac{\sqrt[3]{4x^4}}{\sqrt[3]{t^4}}$

40 $\dfrac{\sqrt[5]{a^2b^2}}{\sqrt[5]{a^4b}}$

41 $\dfrac{\sqrt[5]{xy^3}}{\sqrt[5]{x^2y^2}}$

42 $\dfrac{\sqrt[3]{9/x^2}}{\sqrt[3]{1/x}}$

43 The radius r (meters) of the orbit of a satellite that makes one revolution around the earth every 24 hours is given by the expression

$$r=\sqrt[3]{75.3\times 10^{21}}$$

Simplify this radical.

44 The root-mean-square velocity, $\bar{v}$, of a gas particle is given by the formula

$$\bar{v}=\frac{\sqrt{3kT}}{\sqrt{m}}$$

where k is a constant, T is the temperature (in degrees Kelvin), and m is the mass of the particle. Rationalize the denominator of the expression on the right side.

45 The mass m of an object depends on its speed, v, and the speed of light, c. The relationship is given by the formula

$$m=\frac{m_0}{\sqrt{1-v^2/c^2}}$$

where m_0 is the "rest mass," the mass when $v=0$. Simplify the expression on the right side and rationalize the denominator.

46 Have you heard of supersonic airplanes with a speed of Mach 2? Mach 2 means that the speed of the plane is *twice* the speed of sound. The Mach number can be found from the formula

$$M=\sqrt{\frac{2}{\gamma}}\sqrt{\frac{P_2-P_1}{P_1}}$$

where γ is a constant and P_1 and P_2 are air pressures that are measured by an instrument called a Pitot tube. Write the expression on the right in simplified form with a rationalized denominator.

In Problems 47–50, use absolute values when needed to simplify the given expression.

Hint: See Theorem 3.2a, Part IVb.

47 $\sqrt{2x^2 - 8x + 8}$

48 $\sqrt{9y - 6y^3 + y^5}$

49 $\sqrt{3z^2 - 24z + 48}$

50 $\sqrt{75y^3 - 30y^4 + 3y^5}$

51 Prove Theorem 3.2a, Part I.

52 Prove Theorem 3.2a, Part II.

53 Prove Theorem 3.2a, Parts IVa and b.

Hint: Make careful note of the definition of *principal root.*

54 Prove Theorem 3.2a, Part V.

USING YOUR KNOWLEDGE 3.2

From Theorem 3.2a, Part IVb, it follows that if x is any real number, then

$$\sqrt{x^2} = |x|$$

1 Use this result to prove that $|x|\,|y| = |xy|$.

Hint: Use Theorem 3.2a and the Laws of Exponents.

2 Prove that $|x|/|y| = |x/y|$.

Hint: See Problem 1.

3.3 RATIONAL EXPONENTS

It is shown in physics that the time T (seconds) that it takes a pendulum to swing from left to right and then back to its original position is given by the formula

$$T = 2(L/g)^{1/2}$$

where L is the length (meters) of the string holding the bob and $g = 9.8$ m/sec^2 (meters per second per second). The expression $(L/g)^{1/2}$ displays the *rational exponent* $\frac{1}{2}$, and we shall now use radicals to arrive at a reasonable meaning for such exponents.

DEFINING $x^{1/n}$

We assume that $x \geq 0$, and we use the Laws of Exponents, Part II to multiply $x^{1/2}$ by itself:

$$x^{1/2} \cdot x^{1/2} = (x^{1/2})^2 = x^{(1/2)\cdot 2} = x^1 = x$$

This means that if $x^{1/2}$ is to have a meaning consistent with the Laws of Exponents, it should be taken to be the *principal* square root of x; that is, $x^{1/2} = \sqrt{x}$, provided x is nonnegative.

In general, for $x \geq 0$ and n a positive integer, we see that

$$(x^{1/n})^n = x^{(1/n)\cdot n} = x^1 = x$$

if we assume that Law II applies. This result means that $x^{1/n}$ should be defined to be an nth root of x and leads to Definition 3.3a. For example,

$$16^{1/2} = \sqrt{16} = 4$$

and

$$(1/81)^{1/4} = \sqrt[4]{1/81} = \tfrac{1}{3}$$

▼ DEFINITION 3.3a

If x is a nonnegative number and n is a positive integer, then

$$x^{1/n} = \sqrt[n]{x}$$

DEFINING $x^{m/n}$

If $x > 0$, we can arrive at a definition of $x^{m/n}$, where m and n are integers and $n > 0$. Again, we assume that Law II applies, so that

$$(\sqrt[n]{x})^m = (x^{1/n})^m = (x^m)^{1/n}$$

However,

$$(x^{1/n})^m = \underbrace{x^{1/n} \cdot x^{1/n} \cdot x^{1/n} \cdots x^{1/n}}_{m \text{ factors}}$$

$= \sqrt[n]{x} \cdot \sqrt[n]{x} \cdot \sqrt[n]{x} \cdots \sqrt[n]{x}$	By Definition 3.3a
$= \sqrt[n]{x \cdot x \cdot x \cdots x}$	By Theorem 3.2a, Part II
$= \sqrt[n]{x^m}$	Definition of x^m
$= (x^m)^{1/n}$	By Definition 3.3a

This result shows that, under the stated conditions,

$$x^{m/n} = (x^m)^{1/n} = (x^{1/n})^m$$
$$= \sqrt[n]{x^m} = (\sqrt[n]{x})^m$$

If m and n are both positive integers and $x > 0$, a similar discussion shows that we should define

$$x^{-m/n} = \frac{1}{x^{m/n}}$$

As a consequence of these results, we make Definition 3.3b.

▼ DEFINITION 3.3b

If m and n are integers with $n > 0$ and $x > 0$, then

$$x^{m/n} = (x^m)^{1/n} = (x^{1/n})^m$$
$$= \sqrt[n]{x^m} = (\sqrt[n]{x})^m$$

EXAMPLE 1

Evaluate:

a. $8^{2/3}$
b. $32^{-3/5}$
c. $-64^{-1/3}$

SOLUTION:

a. We can use Definition 3.3b to evaluate $8^{2/3}$ by either of the following two methods.

Method 1 Taking the cube root first,

$$8^{2/3} = (\sqrt[3]{8})^2 = 2^2 = 4$$

Method 2 Squaring first,

$$8^{2/3} = \sqrt[3]{8^2} = \sqrt[3]{64} = 4$$

Note that it is usually simpler to extract the root first, because this keeps the numbers smaller. Try using the second method on Part b of this example if you have any doubts.

b. $32^{-3/5} = \dfrac{1}{32^{3/5}} = \dfrac{1}{(\sqrt[5]{32})^3} = \dfrac{1}{2^3} = \dfrac{1}{8}$

c. $-64^{-1/3} = -\dfrac{1}{64^{1/3}} = -\dfrac{1}{\sqrt[3]{64}} = -\dfrac{1}{4}$

It is easy to prove, although we shall not do it here, that all the Laws of Exponents are valid for rational exponents as defined in Definition 3.3b. Notice carefully that *the base has been required to be positive.* If this restriction is removed, then we have to give up some of the nice properties of exponents. For instance, if the base $x = 0$, then we cannot allow zero or negative exponents. However, if m and n are restricted to be positive, then Definition 3.3b may be extended to allow a zero base.

We shall avoid the use of negative bases as much as possible because of certain inherent difficulties that arise. If it is necessary to work with a negative base, then in place of Definition 3.3b we must use Definition 3.3c. For example,

▼ DEFINITION 3.3c

If $n > 0$, m/n is a rational number *in lowest terms*, and $\sqrt[n]{x}$ is a *real nonzero number*, then

$$x^{m/n} = (x^m)^{1/n} = (x^{1/n})^m = \sqrt[n]{x^m} = (\sqrt[n]{x})^m$$

$$(-8)^{2/3} = \sqrt[3]{(-8)^2} = \sqrt[3]{64} = 4$$

or

$$(-8)^{2/3} = (\sqrt[3]{-8})^2 = (-2)^2 = 4$$

However, $(-8)^{4/6}$ is *not* defined because the exponent is not in lowest terms. Notice that $[(-8)^4]^{1/6} = (8^4)^{1/6} = 8^{2/3} = 4$, but $[(-8)^{1/6}]^4$ is not defined because $\sqrt[6]{-8}$ is not a real number. Thus, we have the undesirable result

$$[(-8)^4]^{1/6} \neq [(-8)^{1/6}]^4$$

This does not contradict Definition 3.3c because $\sqrt[6]{-8}$ is *not* a real number, but it does give a reason for requiring the exponent to be in lowest terms in this definition.

SIMPLIFICATION INVOLVING RATIONAL EXPONENTS

The following examples show you how the Laws of Exponents are used in handling rational exponents.

EXAMPLE 2

Simplify each of the following. Assume that the letters represent positive numbers.

a. $\dfrac{x^{2/3}}{x^{1/4}}$

b. $\left(\dfrac{3x^{1/2}}{y^{2/3}}\right)^2$

SOLUTION:

a. $\dfrac{x^{2/3}}{x^{1/4}} = x^{2/3-1/4} = x^{5/12}$

b. $\left(\dfrac{3x^{1/2}}{y^{2/3}}\right)^2 = \dfrac{3^2(x^{1/2})^2}{(y^{2/3})^2} = \dfrac{9x}{y^{4/3}}$

EXAMPLE 3

Simplify each of the following. Assume that the letters represent positive numbers.

a. $(y^6)^{1/2}$

b. $x^{3/4}(x^{1/2} - 2x^{-1/2})$

SOLUTION:

a. $(y^6)^{1/2} = y^{6 \cdot (1/2)} = y^3$

b.
$$\begin{aligned} x^{3/4}(x^{1/2} - 2x^{-1/2}) &= x^{3/4} \cdot x^{1/2} - 2x^{3/4} \cdot x^{-1/2} \\ &= x^{3/4+1/2} - 2x^{3/4-1/2} \\ &= x^{5/4} - 2x^{1/4} \end{aligned}$$

Note that if y is not restricted to be a positive number in Example 3a, then we must interpret $(y^6)^{1/2}$ as $\sqrt{y^6}$, which means the principal square root of y^6, that is, $|y^3|$. (See Definition 3.3a.)

EXAMPLE 4

Simplify the expression $[(x^2 - 4x + 4)(x + 1)]^{1/2}$. Assume that x may be replaced by only those numbers for which the expression is defined.

SOLUTION:

$$\begin{aligned} [(x^2 - 4x + 4)(x + 1)]^{1/2} &= [(x - 2)^2(x + 1)]^{1/2} \\ &= [(x - 2)^2]^{1/2}(x + 1)^{1/2} \\ &= |x - 2|(x + 1)^{1/2} \end{aligned}$$

Notice that the absolute value bars are essential here because we do not know whether $x - 2$ is positive, negative, or zero. You can see that the original expression is defined for all values such that $x \geq -1$, because $(x - 2)^2(x + 1)$ is nonnegative for these values of x. Because $x - 2$ is negative for x between -1 and 2 and is positive for x greater than 2, the absolute value bars insure that we have written down the *principal* root.

If you have a scientific calculator with a y^x (or x^y) key, you can calculate rational powers of numbers. For example, to calculate $25^{15/32}$, proceed as follows:

Algebraic logic

[15] [÷] [32] [=] [STO] [25] [y^x] [RCL] [=] ≈ 4.521519

RPN logic

[25] [ENTER] [15] [ENTER] [32] [÷] [y^x] ≈ 4.521519

EXERCISE 3.3

In Problems 1–12, evaluate the given expression.

1 $4^{1/2}$ **2** $8^{2/3}$ **3** $4^{-1/2}$

4 $64^{-1/3}$ **5** $(-27)^{2/3}$ **6** $27^{-2/3}$

7 $(\frac{1}{8})^{2/3}$ **8** $(-\frac{1}{8})^{-2/3}$ **9** $(\frac{9}{16})^{3/2}$

10 $(\frac{9}{16})^{-3/2}$ **11** $[(-11)^6]^{1/6}$ **12** $[(-16)^6]^{1/3}$

In Problems 13–32, simplify the given expression and write the answer using positive exponents only. Assume that all the letters represent positive numbers.

13 $(3x^{3/2})(2x^{1/2})$ **14** $(-4c^{2/3})(2c^{1/6})$ **15** $(-4a^{-2})(a^{-1/2})$

16 $(5x^{-2/3}y^{1/3})(x^{-1/3}y^{2/3})$

17 $\dfrac{x^{3/4}y^{-3/4}}{x^{-1/2}y^{3/4}}$

18 $\dfrac{-8x^{-2}y^{2/5}}{4xy^{-1/5}}$

19 $\dfrac{(2x^2y^{1/2})^2}{(3xy^{2/3})^3}$

20 $\dfrac{(x^4y^6)^{-1/2}}{(x^6y^3)^{-1/3}}$

21 $\left(\dfrac{x^{1/2}y^{3/4}z^{-2}}{x^{-3/2}y^{1/4}}\right)^{-2}$

22 $\left(\dfrac{x^{3/4}y^{1/2}z^{3/2}}{x^{-5/4}y^{7/2}z^{9/2}}\right)^{-1/2}$

In Problems 23–26, n is a positive integer.

23 $\dfrac{(x^n)^{5/4}}{x^{n/4}}$

24 $(a^{2n}b^{n/3})^6$

25 $x^{3n/2} \cdot x^{n/2}$

26 $\left(\dfrac{x^n y^{3n}}{y^{6n}}\right)^{1/n}$

27 $\dfrac{x^{1/6} \cdot x^{-5/6}}{x^{1/3}}$

28 $\dfrac{x^{-8/3}y^{-1}}{x^{-1/6}}$

29 $x^{1/3}(x^{2/3} + x)$

30 $y^{3/4}(y - 3y^{1/2})$

31 $x^{-4/5}(x + x^{-1/5})$

32 $y^{-2/3}(y^{-1/2} - y)$

In Problems 33–36, simplify the given expression. Assume that the letters represent real numbers for which the expression is defined.

33 $[x^2(x-5)]^{1/2}$

34 $[16x^2(x-1)]^{1/2}$

35 $\dfrac{1}{[(x^2+6x+9)(x-1)]^{1/2}}$

36 $\left[\dfrac{4}{x^4(x+1)}\right]^{1/2}$

37 What is wrong with the following argument?

$[(-2)^6]^{1/2} = 64^{1/2} = 8$ and $[(-2)^6]^{1/2} = (-2)^{6 \cdot (1/2)} = (-2)^3 = -8$

Therefore, $8 = -8$.

C **38** Use your calculator to evaluate: **a** $(12.7)^{2/3}$ **b** $36^{5/6}$

C **39** Use your calculator to evaluate: **a** $108^{-2/5}$ **b** $32^{-5/9}$

C **40** Use your calculator to evaluate $10^{0.1}$, $100^{0.01}$, $1000^{0.001}$, and $10{,}000^{0.0001}$. What seems to be happening to the results as you take a larger and larger base with a correspondingly smaller and smaller exponent?

USING YOUR KNOWLEDGE 3.3

C **1** If a satellite is revolving in a circular orbit of radius r about the earth, the speed of the satellite is given by the formula $v = (rg)^{1/2}$. For a satellite that is 150 kilometers above the surface of the earth, the value of g is 9.33 m/sec² (meters per second per second). The radius of the earth is about 6.38×10^6 meters, so that r, the distance of the satellite from the center of the earth, is about 6.53×10^6 meters. **a** Find the speed of the satellite. **b** Find how long it takes the satellite to make one orbit about the earth.

Hint: The distance the satellite must go is $2\pi r$.

C **2** For a very low-intensity sound (10 decibels at 1000 cycles per second), the amplitude A of the vibration of the air is

$$A = \frac{1}{\pi \times 10^3}\left[\frac{10^{-11}}{2 \times 3.31 \times 10^2 \times 1.29}\right]^{1/2} \text{ meter}$$

Evaluate A.

C **3** It is a consequence of one of Kepler's laws of planetary motion that the radius R of a planet's orbit about the sun is given in terms of its period P (the time it takes to complete one orbit) by the formula $R = P^{2/3}$. In this formula, the unit of time is the earth year, and the unit of distance is the astronomical unit, A.U., the average distance from earth to sun—about 9.3×10^7 miles. The planet Jupiter takes 11.86 earth years to orbit the sun. What is the radius of Jupiter's orbit: **a** in A.U.? **b** in miles?

C **4** The period of the planet Venus is about 6.15×10^{-1} earth years. Answer the same questions as in Problem 3 for Venus.

3.4 OPERATIONS WITH RADICALS

You probably recall from arithmetic that you can add only the same kinds of things (apples and apples, for example). In algebra, we have learned to add like terms, an idea we can extend to radicals. If radicals agree in index and radicand, then they may be added or subtracted by using the Distributive Law to factor out the common radical; this amounts to adding like terms. For instance,

$$4\sqrt{7} + 2\sqrt{7} = (4+2)\sqrt{7} = 6\sqrt{7}$$
$$9\sqrt{5} - 7\sqrt{5} = (9-7)\sqrt{5} = 2\sqrt{5}$$

Of course, it may be necessary to simplify some of the terms in an expression before they can be combined, as Example 1 shows.

EXAMPLE 1

Simplify:

a. $\sqrt{45} + \sqrt{125}$

b. $3\sqrt[3]{8x^4y^5} - xy\sqrt[3]{64xy^2}$

SOLUTION:

a. $\sqrt{45} + \sqrt{125} = \sqrt{9 \cdot 5} + \sqrt{25 \cdot 5} = 3\sqrt{5} + 5\sqrt{5} = 8\sqrt{5}$

b.
$$\begin{aligned} 3\sqrt[3]{8x^4y^5} - xy\sqrt[3]{64xy^2} &= 3\sqrt[3]{(2xy)^3xy^2} - xy\sqrt[3]{4^3xy^2} \\ &= 6xy\sqrt[3]{xy^2} - 4xy\sqrt[3]{xy^2} \\ &= 2xy\sqrt[3]{xy^2} \end{aligned}$$

As we saw in Section 3.2, if two radicals agree in index, they can be multiplied or divided by using Theorem 3.2a, Parts II and III. In some cases, however, we must also use the Distributive Law, as Example 2 illustrates.

EXAMPLE 2

Simplify:

a. $\dfrac{1}{\sqrt{3}}(2 - 3\sqrt{3})$

b. $(\sqrt{x} + 3)(\sqrt{x} - 5)$

SOLUTION:

a.
$$\begin{aligned} \frac{1}{\sqrt{3}}(2 - 3\sqrt{3}) &= \frac{2}{\sqrt{3}} - \frac{3\sqrt{3}}{\sqrt{3}} \\ &= \frac{2\sqrt{3}}{(\sqrt{3})^2} - 3 \\ &= \tfrac{2}{3}\sqrt{3} - 3 \end{aligned}$$

b.
$$\begin{aligned} (\sqrt{x} + 3)(\sqrt{x} - 5) &= (\sqrt{x})^2 + (3-5)\sqrt{x} - 15 \\ &= x - 2\sqrt{x} - 15 \end{aligned}$$

SIMPLIFICATION OF EXPRESSIONS INVOLVING RADICALS

Fractional exponents are often convenient to use in simplifying expressions involving radicals. The radicals can be changed to equivalent forms using rational exponents. Then the simplification can be carried out by using the Laws of Exponents. The procedure is illustrated in Example 3.

EXAMPLE 3

Simplify:

a. $\dfrac{\sqrt[3]{9}}{\sqrt{3}}$

b. $(\sqrt{2x})(\sqrt[4]{2x^3}),\ x > 0$

SOLUTION:

a. $\dfrac{\sqrt[3]{9}}{\sqrt{3}} = \dfrac{9^{1/3}}{3^{1/2}} = \dfrac{(3^2)^{1/3}}{3^{1/2}} = \dfrac{3^{2/3}}{3^{1/2}} = 3^{2/3-1/2} = 3^{1/6} = \sqrt[6]{3}$

b. $(\sqrt{2x})(\sqrt[4]{2x^3}) = (2x)^{1/2}(2x^3)^{1/4} = (2^{1/2}x^{1/2})(2^{1/4}x^{3/4})$

$= 2^{3/4}x^{5/4} = 2^{3/4} \cdot x \cdot x^{1/4}$

$= x(2^3x)^{1/4} = x\sqrt[4]{8x}$

Suppose you wish to approximate $1/(\sqrt{2} + 1)$. To avoid this division, you can rationalize the denominator of this fraction by multiplying the numerator and denominator of the fraction by $\sqrt{2} - 1$, the *conjugate* of $\sqrt{2} + 1$. We then have

$$\frac{1}{\sqrt{2}+1} = \frac{1}{\sqrt{2}+1} \cdot \frac{\sqrt{2}-1}{\sqrt{2}-1}$$

$$= \frac{\sqrt{2}-1}{2-1}$$

$$= \sqrt{2} - 1$$

$$\approx 1.414 - 1$$

$$= 0.414$$

In general, the expressions $\sqrt{a} + \sqrt{b}$ and $\sqrt{a} - \sqrt{b}$ are called *conjugates* of each other; their product, $a - b$, is free of radicals. A denominator (or numerator) consisting of one of these expressions can be rationalized by multiplying by the conjugate expression, as in Example 4.

EXAMPLE 4

Assume that all the letters represent positive numbers.

a. Rationalize the denominator: $\dfrac{x}{\sqrt{x}+\sqrt{y}}$

b. Rationalize the numerator and simplify:

$\dfrac{\sqrt{x}-\sqrt{a}}{x-a}$

SOLUTION:

a. $\dfrac{x}{\sqrt{x}+\sqrt{y}} = \dfrac{x}{\sqrt{x}+\sqrt{y}} \cdot \dfrac{\sqrt{x}-\sqrt{y}}{\sqrt{x}-\sqrt{y}} = \dfrac{x(\sqrt{x}-\sqrt{y})}{x-y}, \quad (x \neq y)$

b. $\dfrac{\sqrt{x}-\sqrt{a}}{x-a} = \dfrac{\sqrt{x}-\sqrt{a}}{x-a} \cdot \dfrac{\sqrt{x}+\sqrt{a}}{\sqrt{x}+\sqrt{a}} = \dfrac{x-a}{(x-a)(\sqrt{x}+\sqrt{a})}$

$= \dfrac{1}{\sqrt{x}+\sqrt{a}}$

The procedure of Example 4 can also be used if the expression to be rationalized is of the form $a + \sqrt{b}$ or $a - \sqrt{b}$, because these two are conjugates. For example,

$$\frac{1}{4+\sqrt{7}} \cdot \frac{4-\sqrt{7}}{4-\sqrt{7}} = \frac{4-\sqrt{7}}{16-7} = \frac{4-\sqrt{7}}{9}$$

EXERCISE 3.4

In Problems 1–16, simplify the given expression. Assume that all the letters represent positive numbers.

1 $\sqrt{48} - \sqrt{27} - 5\sqrt{3}$

2 $\sqrt{8} + \sqrt{32} - \sqrt{50}$

3 $\sqrt[3]{54} + \sqrt[3]{16} - \sqrt[3]{128}$

4 $\sqrt[3]{24} - \sqrt[3]{81} + \sqrt[3]{192}$

5 $4c\sqrt{k^3} - 2\sqrt{c^2k^3} + 5k\sqrt{c^2k}$

6 $3b\sqrt{m^5} - m\sqrt{4b^2m^3} - m^2\sqrt{16b^2m}$

7 $\sqrt[3]{-3a^4} - \sqrt[3]{-24a} + \sqrt[3]{375a}$

8 $ab^2\sqrt[3]{4a^2} + ab\sqrt[3]{108a^2b^3} + \sqrt[3]{-32a^5b^6}$

9 $\sqrt[3]{\frac{27}{36}} - \sqrt[3]{\frac{3}{32}} - \sqrt[3]{\frac{-2}{9}}$

10 $\sqrt[5]{\frac{5}{16}} - \sqrt[5]{\frac{-2}{625}} + \sqrt[5]{\frac{-1}{10{,}000}}$

11 $\sqrt{\frac{3e}{f}} - \frac{5}{3}\sqrt{\frac{3f}{e}} - \sqrt{\frac{f}{3e}}$

12 $\sqrt{\frac{h^3}{2d}} + hd\sqrt{\frac{h}{2d^3}} + \frac{d^2}{2}\sqrt{\frac{2h^3}{d^5}}$

13 $\sqrt{x}(\sqrt{x} - \sqrt{y})$

14 $\sqrt{y}(\sqrt{y} + \sqrt{xy})$

15 $(\sqrt{x} + \sqrt{5y})(\sqrt{x} - \sqrt{5y})$

16 $(4\sqrt{5} - 3\sqrt{x})(4\sqrt{5} + 3\sqrt{x})$

17 Find the value of $x^2 + 2x - 2$ for $x = 1 + \sqrt{3}$.

18 Find the value of $y^2 - 3y + 1$ for $y = (3 - \sqrt{5})/2$.

In Problems 19–28, simplify the given expression. Assume all the letters represent positive numbers and write the answer in radical form.

19 $\frac{\sqrt{16}}{\sqrt[3]{16}}$

20 $\frac{\sqrt[3]{16}}{\sqrt{2}}$

21 $\frac{\sqrt{9x}}{\sqrt[3]{3x^2}}$

22 $\frac{\sqrt[3]{3x^4}}{\sqrt[9]{27x^3}}$

23 $\frac{\frac{4}{5}\sqrt[3]{4xy}}{\frac{1}{10}\sqrt{2x^2}}$

24 $(\sqrt{2x})(\sqrt[3]{4x^2y})$

25 $(\sqrt[3]{4x^2})(\sqrt{x})$

26 $(\frac{2}{3}\sqrt[3]{4x^2})(\frac{3}{4}\sqrt[5]{16x^4y})$

27 $(\sqrt[3]{x^2y^2})(-3\sqrt[4]{3x^3y})$

28 $\left(-\frac{2}{3}\sqrt{\frac{2x}{y}}\right)\left(\frac{3}{4}\sqrt[3]{\frac{x^2}{4y^2}}\right)$

In Problems 29–32, rationalize the denominator and simplify.

29 $\frac{\sqrt{5} - \sqrt{3}}{\sqrt{5} + \sqrt{3}}$

30 $\frac{4 + \sqrt{5}}{4 - \sqrt{5}}$

31 $\frac{x - \sqrt{3y}}{x + \sqrt{3y}}$

32 $\frac{\sqrt{x} + \sqrt{2y}}{\sqrt{x} - \sqrt{2y}}$

USING YOUR KNOWLEDGE 3.4

There are certain calculus problems that require the rationalization of the numerator of a fraction, such as

$$\frac{\sqrt{x} - \sqrt{a}}{x - a} \qquad x \neq a$$

We can do this rationalization by multiplying numerator and denominator by the conjugate of the numerator, that is, $\sqrt{x} + \sqrt{a}$. This gives

$$\frac{\sqrt{x} - \sqrt{a}}{x - a} = \frac{(\sqrt{x} - \sqrt{a})(\sqrt{x} + \sqrt{a})}{(x - a)\ (\sqrt{x} + \sqrt{a})}$$

$$= \frac{x - a}{(x - a)(\sqrt{x} + \sqrt{a})}$$

$$= \frac{1}{\sqrt{x} + \sqrt{a}}$$

which is the required form. Follow this procedure in the problems below. Assume $x \neq a$.

1 $\dfrac{\sqrt{2x + 1} - \sqrt{2a + 1}}{x - a}$ **2** $\dfrac{x^{3/2} - a^{3/2}}{x - a}$ **3** $\dfrac{\sqrt{x^2 + b^2} - \sqrt{a^2 + b^2}}{x - a}$ **4** $\dfrac{x^{-1/2} - a^{-1/2}}{x - a}$

SELF-TEST

1 Write in simplified form with positive exponents:
a $(x^{-3} \cdot x^6)^{-3}$ **b** $(x^{-3}y^2)^{-4}$

2 Write in simplified form with positive exponents:

$$\left(\frac{xy^{-2}}{x^{-3}y}\right)^{-3}$$

3 Simplify and write your answer with positive exponents:

$$\frac{x - y^{-1}}{x^{-1} + y}$$

4 Simplify:

$$\frac{(4 \times 10^4)(6 \times 10^{-3})}{8 \times 10^{-6}}.$$

5 At atmospheric pressure at 0°C, there are about 6.02×10^{23} molecules in 22.4 liters of any gas. How many molecules are there in 22,400 liters of gas? Give answer in scientific notation.

6 Evaluate:

a $\sqrt{\dfrac{16}{25}}$ **b** $\sqrt[5]{\dfrac{-1}{32}}$

7 Simplify:
a $\sqrt{9x^4y^7}$ **b** $\sqrt[3]{16x^5y^6z^3}$

8 Simplify:

a $\sqrt{\dfrac{3x^5}{8y^3}}$ **b** $\sqrt[3]{\dfrac{54x^4}{y^2z}}$

9 Rationalize the denominator and give answer in simplest form:

$$\frac{1}{\sqrt[3]{(x + y)^2}}$$

10 Simplify:
$\sqrt{x^3 - 4x^2y + 4xy^2}$

11 Evaluate:
a $27^{2/3}$ **b** $32^{-3/5}$

12 Assume that x and y represent positive numbers and simplify:

$$\left(\frac{8x^{3/2}}{y^{-1/2}}\right)^{2/3}$$

13 Assume that x represents a positive number and simplify:

$$x^{1/2}(x^{3/4} - 3x^{-1/2})$$

14 Assume that x represents a positive number and simplify:

$$\frac{x^{3/4}}{x^{1/6}}$$

15 Combine like terms and simplify.
a $\sqrt{12} + 2\sqrt{27}$
b $6\sqrt[3]{27x^5y^4} - xy\sqrt[3]{8x^2y}$

16 Assume x and y represent positive numbers and simplify:

a $\dfrac{1}{\sqrt{2}}(4 + 3\sqrt{2})$

b $(\sqrt{x} + 7\sqrt{y})(\sqrt{x} - 3\sqrt{y})$

17 Simplify:

$$\frac{\sqrt[3]{16}}{\sqrt{2}}$$

18 Simplify:

$$(\sqrt{x^2y})(\sqrt[3]{xy^2})$$

19 Rationalize the denominator and simplify:

$$\frac{y}{\sqrt{x} - \sqrt{y}}$$

20 Rationalize the numerator and simplify:

$$\frac{\sqrt{x} + \sqrt{y}}{x - y}$$

CHAPTER 4

EQUATIONS AND INEQUALITIES IN ONE VARIABLE

4.1 FIRST-DEGREE EQUATIONS IN ONE VARIABLE

Recently, National Car Rental Systems introduced the New York businessman's rate for renting a car, \$8.50 per day plus 13 cents per mile. Suppose Mr. Smith rented one of these cars for one day and drove 76 miles. How much would he have had to pay National? Of course, we would multiply 0.13 by 76 to get the mileage charge and add 8.50 to get the total amount:

$$\underbrace{(0.13)(76)}_{\text{Mileage charge}} + \underbrace{8.50}_{\text{Per day charge}} = \underbrace{18.38}_{\text{Total charge}}$$

Now suppose that Mr. Smith rented the car for a second day and at the end of that day paid \$15.00. How many miles did he drive the second day? To answer this question, we first let m be the number of miles he drove. To get his cost for the day, we proceed as before to multiply 0.13 by m to get the mileage charge and add 8.50 to get the total charge, \$15.00. Thus, we can write

$$\underbrace{(0.13)m}_{\text{Mileage charge}} + \underbrace{8.50}_{\text{Per day charge}} = \underbrace{15}_{\text{Total charge}}$$

The expression $0.13m + 8.50 = 15$ is an example of an *equation*. To find how many miles Mr. Smith drove, we must find the value of m that makes the equation true. We shall solve this problem shortly.

CONDITIONAL EQUATIONS

▼ **DEFINITION 4.1a**

An *equation* is a statement of equality between two quantities.

In general, an equation is defined as in Definition 4.1a. Other examples of equations are

$$x + 8 = 9, \quad x^2 - 7x + 3 = 5, \quad \text{and} \quad x + y^2 = 10$$

Some equations can immediately be classified as true or false. For instance, $4 + 2 = 6$ is true, but $3 \cdot 0 = 3$ is false. On the other hand, the equation

$$0.13m + 8.50 = 15 \tag{1}$$

cannot be classified as either true or false until we replace m by some specific number. Of course, we want this number to be such that the equation is true, and this imposes a condition on m (m must be such that 0.13 times it added to 8.50 gives 15). If the requirement that an equation be true imposes a *condition* on the variable(s) involved, the equation is called a *conditional equation.* In effect, we regard such an equation as saying, "Here is a condition that only certain numbers can satisfy. Find these numbers." Thus, Equation 1 is an example of a *conditional equation.*

LINEAR EQUATIONS

In this section, we shall consider conditional equations involving only one variable and with that variable to the first power. Such equations are called *linear* or *first-degree* equations and can always be written in the form

$$\mathbf{ax + b = 0 \qquad a \text{ and } b \text{ real numbers, } a \neq 0} \tag{2}$$

In Equation 2, the *replacement set* of the variable x, that is, the set of numbers that may replace x in the equation, is assumed to be the set of all real numbers. However, the replacement set for the equation

$$\frac{1}{x-2} = 5$$

is the set of all real numbers *except* $x = 2$, because for $x = 2$,

$$\frac{1}{x-2} = \frac{1}{0}$$

which is *not defined.* Whenever a problem involves no other restrictions, *the replacement set is taken to be the set of all real numbers for which the terms in the equation are defined.* The replacement set is simply the set of numbers that we have in mind when seeking those numbers (if there are any) that make the equation true.

A number that is in the replacement set and that, upon substitution for the variable, makes the equation true is called a *solution* or a *root* of the equation. For example, if x is replaced by **2** in the equation $x + 5 = 7$, the result is the true statement $2 + 5 = 7$. Hence, 2 is a solution of the equation $x + 5 = 7$. A solution of an equation is said to *satisfy* the equation. Obviously, 2 is the only solution of the equation $x + 5 = 7$. On the other hand, the equation

$$(x - 2)(x + 3) = 0$$

is satisfied by $x = 2$ (because $0 \cdot 5 = 0$ is true) and also by $x = -3$ (because $-5 \cdot 0 = 0$ is true). Because at least one of the factors $(x - 2)$ and $(x + 3)$ must be zero to satisfy this equation, there can be no other solutions. The set of all solutions of an equation is called its *solution set*. The solution set of the equation $x + 5 = 7$ consists of the single number 2; the solution set of $(x - 2)(x + 3) = 0$ is the set $\{-3, 2\}$.

SOLVING LINEAR EQUATIONS

We now consider the problem of solving linear equations. If the equation is simple enough, we can solve it by inspection. Thus, the equation $x - 5 = 0$ has the obvious solution $x = 5$, and the equation $x + 1 = 3$ has the obvious solution $x = 2$. Equations that are not as simple are solved by performing a sequence of operations on the equation until an equation with an obvious solution is produced. Of course, we require that the solutions of the final equation be identical to those of the given equation. Two equations that have identical solutions are called *equivalent equations*. For example, the equations

$$x + 7 = 10 \quad \text{and} \quad x = 3$$

are equivalent equations, because both have exactly the one solution, 3.

The usual simple operations that are used to obtain equivalent equations from a given equation are adding or subtracting the same quantity on both sides and multiplying or dividing by the same quantity on both sides. These operations are described more precisely by Theorem 4.1a, which is a direct consequence of the addition and multiplication laws for equality.

▼ THEOREM 4.1a

If $P(x)$, $Q(x)$, and $R(x)$ are algebraic expressions, then for all values of x for which $P(x)$, $Q(x)$, and $R(x)$ are real numbers, the equation

$$P(x) = Q(x)$$

is equivalent to each of the following:

I. $P(x) + R(x) = Q(x) + R(x)$

II. $P(x) - R(x) = Q(x) - R(x)$

III. $P(x) \cdot R(x) = Q(x) \cdot R(x)$ **for all values of x such that $R(x) \neq 0$**

IV. $\dfrac{P(x)}{R(x)} = \dfrac{Q(x)}{R(x)}$ **for all values of x such that $R(x) \neq 0$**

We can apply this theorem to solve the mileage problem, Equation 1, as follows:

$0.13m + 8.50 = 15$	Given
$0.13m + 8.50 - 8.50 = 15 - 8.50$	Subtract 8.50 from both sides.
$0.13m = 6.50$	Simplify.
$\dfrac{0.13m}{0.13} = \dfrac{6.50}{0.13}$	Divide both sides by 0.13
$m = 50$	Simplify.

Because 50 is the only solution of the last equation, the theorem assures us that 50 is also the only solution of the original equation.

That **50** is a solution of that equation can be checked by direct substitution of 50 for m:

$$(0.13)(\mathbf{50}) + 8.50 = 6.50 + 8.50 = 15,$$

which is a true statement.

It is very important for you to realize that Theorem 4.1a does not insure against making mistakes in the course of solving an equation. For this reason, proposed solutions should always be checked in the original equation.

EXAMPLE 1

Solve:
$3x - 5 = x + 7$

SOLUTION:

$3x - 5 = x + 7$	
$3x - 5 + \mathbf{5} = x + 7 + \mathbf{5}$	Add **5** to both sides.
$3x = x + 12$	Simplify
$3x - \boldsymbol{x} = x + 12 - \boldsymbol{x}$	Subtract $\boldsymbol{x}$ from both sides.
$2x = 12$	Simplify.
$x = 6$	Divide both sides by **2**.

CHECK:
For $x = \mathbf{6}$, the original equation becomes $(3)(\mathbf{6}) - 5 = 6 + 7$, or $13 = 13$, which is true.

The procedure used in Example 1 is quite general. Hence, we summarize it here for your convenience:

PROCEDURE FOR SOLVING LINEAR EQUATIONS

1. **Add or subtract the same numbers on both sides to bring all the specific numbers on one side and simplify.**
2. **Add or subtract terms involving the variable to bring all such terms on the other side and simplify.**
3. **If the coefficient of the variable is not 1, divide both sides by this coefficient to produce an equation such as $x = k$, where k is a specific number.**
4. **Check the solution in the original equation.**

SOLVING LINEAR EQUATIONS INVOLVING FRACTIONS

When fractions are present in an equation, we can first use Theorem 4.1a, Part III to multiply both sides by the LCD of the fractions involved. This procedure "clears" the equation of the fractions and usually makes it a little more pleasant to work with. For instance, to solve the equation

$$\frac{x}{2} + \frac{1}{4} = \frac{x}{6}$$

we first multiply each term by **12**, the LCD of the three fractions, obtaining

$$\mathbf{12}\left(\frac{x}{2} + \frac{1}{4}\right) = \mathbf{12}\left(\frac{x}{6}\right)$$

$$(12)\left(\frac{x}{2}\right) + (12)\left(\frac{1}{4}\right) = (12)\left(\frac{x}{6}\right)$$

or

$$\begin{aligned} 6x + 3 &= 2x \\ 6x &= 2x - 3 \\ 4x &= -3 \\ x &= -\tfrac{3}{4} \end{aligned}$$

The check is left for you.

Sometimes, simplification of a nonlinear equation leads to a linear equation, as in Examples 2 and 3.

EXAMPLE 2

Solve:

$$\frac{3}{x^2 - 4} = \frac{2}{x + 2}$$

SOLUTION:

Because $x^2 - 4 = (x - 2)(x + 2)$ is the LCD of the fractions, we assume $x^2 \neq 4$ and multiply by $x^2 - 4$ to obtain

$$(x^2 - 4) \cdot \frac{3}{x^2 - 4} = (x - 2)(x + 2) \cdot \frac{2}{x + 2}$$

or

$$\begin{aligned} 3 &= 2(x - 2) \\ 3 &= 2x - 4 \\ -2x &= -7 \\ x &= \tfrac{7}{2} \end{aligned}$$

CHECK:

For $x = \frac{7}{2}$,

$$\frac{3}{x^2 - 4} = \frac{3}{\frac{49}{4} - 4} = \frac{12}{49 - 16} = \frac{12}{33} = \frac{4}{11}$$

and

$$\frac{2}{x + 2} = \frac{2}{\frac{7}{2} + 2} = \frac{4}{7 + 4} = \frac{4}{11}$$

Because both sides reduce to the same number, the proposed solution checks.

EXAMPLE 3

Solve:

$$\frac{x}{x-2} = 1 + \frac{8}{x^2-4}$$

SOLUTION:

As in the preceding example, $x^2 - 4$ is the LCD, and we multiply through by it, assuming that $x^2 \neq 4$. This gives

$$(x+2)(x-2)\frac{x}{x-2} = (x^2-4)\cdot 1 + (x^2-4)\frac{8}{x^2-4}$$

or

$$x^2 + 2x = x^2 - 4 + 8$$

or

$$2x = 4$$
$$x = 2$$

Now we have a contradiction because we had to assume that $x^2 \neq 4$ in order to multiply through by a nonzero quantity, but, if $x = 2$, then $x^2 = 4$. Furthermore, if we put $x = 2$ into the given equation, we obtain

$$\frac{2}{0} = 1 + \frac{8}{0}$$

which is completely *meaningless*. Hence, the given equation has no solution, because that equation is equivalent to the final equation *except* for $x = \pm 2$, and 2 is the *only* solution of this final equation. Of course, -2 does not satisfy the original equation because it gives a zero denominator on the right side.

LITERAL EQUATIONS

The ideas that we have discussed can also be used to solve literal equations, that is, equations involving one or more letters besides the one whose value is required. Thus, to solve $ax - 2b = c$ for x, we follow the usual procedure for linear equations, treating a, b, and c as if they were specific known numbers.

$ax - 2b = c$	Given
$ax = 2b + c$	Adding $2b$ to both sides
$x = \dfrac{2b+c}{a}$	Dividing both sides by a (This assumes $a \neq 0$.)

The check is left for you.

EXAMPLE 4

The formula for converting degrees Fahrenheit to degrees Celsius (centigrade) is

$$C = \tfrac{5}{9}(F - 32)$$

Solve this formula for F.

SOLUTION:

$C = \frac{5}{9}(F - 32)$	Given
$\frac{9}{5}C = F - 32$	Multiplying both sides by $\frac{9}{5}$, the reciprocal of $\frac{5}{9}$
$\frac{9}{5}C + 32 = F$	Adding 32 to both sides
$F = \frac{9}{5}C + 32$	By the Symmetric Law of Equality

The check is left for you.

EXAMPLE 5

Solve for x:

$c(x - c) = 2x + c(1 - c), c \neq 2$

SOLUTION:

$c(x - c) = 2x + c(1 - c), c \neq 2$

$cx - c^2 = 2x + c - c^2$ Simplify.

$cx = 2x + c$ Adding c^2 to both sides

$cx - 2x = c$ Subtracting $2x$ from both sides

$(c - 2)x = c$ Factoring the left side

$x = \dfrac{c}{c - 2}$ Dividing both sides by $c - 2$

CHECK:

For $x = c/(c - 2)$, the left side becomes

$$c\left(\frac{c}{c-2} - c\right) = c\left(\frac{c - c^2 + 2c}{c - 2}\right) = \frac{3c^2 - c^3}{c - 2}$$

and the right side becomes

$$\frac{2c}{c-2} + c(1 - c) = \frac{2c + c(1 - c)(c - 2)}{c - 2}$$

$$= \frac{2c + c(-2 + 3c - c^2)}{c - 2}$$

$$= \frac{2c - 2c + 3c^2 - c^3}{c - 2}$$

$$= \frac{3c^2 - c^3}{c - 2}$$

EXERCISE 4.1

In Problems 1–24, solve the given equation.

1 $5x - 4 = 6$

2 $-3x + 1 = -9$

3 $2x - 1 = 3x - 7$

4 $-5x + 1 = -13$

5 $-4x + \frac{1}{2} = 6(x - \frac{1}{8})$

6 $-6x + \frac{2}{3} = 4(x - \frac{1}{5})$

7 $\frac{1}{4}x - \frac{7}{12} = \frac{11}{12} - \frac{5}{4}x$

8 $\frac{2}{3}v - \frac{3}{5} = \frac{4}{3}v + \frac{2}{15}$

9 $\dfrac{2x + 7}{3} = 5$

10 $\dfrac{3y - 1}{7} = 2$

11 $\dfrac{2m - 17}{3} = \dfrac{8m + 11}{6}$

12 $\dfrac{2t - 3}{7} - \dfrac{6t - 11}{5} = -\dfrac{2}{35}$

13 $2y(y + 1) - 4 = y(2y + 5) + 6$

14 $4x(3x - 1) + 19 = 2x(6x + 5) - 9$

15 $(m - 8)(m + 3) = (m - 7)(m + 5)$

16 $(9y - 8)(4y + 5) = (12y + 7)(3y - 2)$

17 $\dfrac{2}{4x - 1} = \dfrac{3}{4x + 1}$

18 $\dfrac{2}{x + 1} = \dfrac{1}{x - 1}$

19 $\dfrac{5}{x^2 - 1} = \dfrac{1}{x + 1}$

20 $\dfrac{3}{x + 1} = \dfrac{1}{x^2 - 1}$

21 $\dfrac{-x}{3 - x} = 2 - \dfrac{3}{3 - x}$

22 $\dfrac{-1}{3 - x} = 2 - \dfrac{3}{3 - x}$

23 $\dfrac{1}{x + 1} = 1 + \dfrac{2}{x + 1}$

24 $\dfrac{x}{x + 1} = 1 + \dfrac{2}{x + 1}$

In Problems 25–35, solve for the indicated variable.

25 $A = ab/2$ for b

26 $V = \pi r^2 h$ for h

27 $V = gt$ for g

28 $S = \frac{1}{2}gt^2 + S_0$ for g

29 $P = 2L + 2W$ for L

30 $I = \dfrac{E}{R + nr}$ for R

31 $f = \dfrac{ab}{a+b}$, $f \neq b$, for a

32 $s = \dfrac{a}{1-r}$ for r

33 $a(x + a) - x = a(a + 1) + 1$, $a \neq 1$, for x

34 $x(3 - 2b) - 1 = x(2 - 3b) - b^2$, $b \neq -1$, for x

35 $\dfrac{x+m}{x-n} = \dfrac{n+x}{m+x}$, $m \neq 0$, for x

36 For what value of c does the equation $3x + c = 2x - 2c$ have -6 for its solution?

37 For what value of c does the equation $2x + 1 = (3 + x)/c$ have 2 as its solution?

38 Consider the equation $ax + b = 0$.
a Find values of a and b so that the equation has the solution $-\frac{5}{4}$.
b Are the values found in part a unique, or can you find another pair of values satisfying the requirement in a?

39 Which pair of the following equations are not equivalent?

$x = 2 \qquad x^2 = 4 \qquad x = \sqrt{4}$

40 Which pair of the following equations are not equivalent if a represents any positive number?

$x = a \qquad x^2 = a^2 \qquad x = \sqrt{a^2}$

USING YOUR KNOWLEDGE 4.1

In this section, you learned how to tell if a specific number is a solution of a given equation. This idea can sometimes be used to do some detective work. Suppose that a femur bone measuring 20 inches in length is found. The relationship between the length L of the femur bone of a man and his height H is given by

$H = 1.88L + 32$

where H and L are both measured in inches. The corresponding equation for a woman is

$H = 1.95L + 29$

1 Can the femur belong to a man whose height was 6 feet?
2 Can the bone belong to a man whose height was 69.6 inches?
3 Can a femur measuring 20 inches in length belong to a woman whose height was 5 feet 8 inches?
C 4 A police pathologist wants to solve the two equations relating L, the length of the femur, and H, the height of the person for L. Do this and then use your calculator to find L for: **a** a man 5 feet 6 inches in height; **b** a man 5 feet 8 inches in height; **c** a woman 5 feet 6 inches in height; **d** a woman 5 feet 10 inches in height.
C 5 Find the height to the nearest tenth of an inch of a woman whose femur is the same length as the femur of a 5-foot, 10-inch man.

4.2 FIRST-DEGREE INEQUALITIES IN ONE VARIABLE

Suppose that the unemployment rate for a certain year was between 7 and 9 percent. If r is the percentage of people unemployed, we can write

$$7 < r < 9$$

which is read "7 is less than r and r is less than 9" or, more briefly, "r is between 7 and 9." If the main verb in a statement that compares two quantities is "is greater than," "is less than," or a combination of one of these with the verb "equals," the statement is called an *inequality*. The statement $7 < r < 9$ is an example of an inequality; other examples are

$$x - 4 < 10 \quad \text{and} \quad \frac{3 - y}{5} \geq 7$$

The symbols $\geq$ and $\leq$ are read "is greater than or equal to" and "is less than or equal to," respectively.

CONDITIONAL AND UNCONDITIONAL INEQUALITIES

In this section, we shall discuss the solution of inequalities. As in the case of equations, any element of the replacement set that, when substituted for the variable, makes the inequality true is a *solution* of the inequality. Inequalities that are true for all elements of the replacement set are called *absolute* or *unconditional* inequalities. For example, if x is a real number, then $x^2 \geq 0$ is an unconditional inequality. Inequalities that are not true for all numbers in the replacement set are called *conditional* inequalities. Thus,

$$x - 4 < 10 \quad \text{and} \quad \frac{3 - y}{5} \geq 7$$

are conditional inequalities.

SOLVING INEQUALITIES

Inequalities are solved in a manner similar to that used for equations. Again, the main idea is to perform certain simple operations on the given inequality until an equivalent one with an obvious solution is obtained. These operations are the same as those used for equations but require a little more care when it comes to multiplication and division, as Theorem 4.2a indicates. This theorem holds if $<$ is replaced by $\leq$, and $>$ by $\geq$. You must observe the important difference between Theorems 4.1a and 4.2a when it comes to multiplication and division (Parts IIIb and IVb). In the case of an inequality, the sense (direction) of the inequality sign is reversed if both sides are multiplied or divided by a negative quantity. Thus, if

$$-3x < 6$$

then

$$\frac{-3x}{-3} > \frac{6}{-3}$$

or

$$x > -2$$

Note the reversal of the inequality sign upon division by -3

Just as for an equation, the *solution set* of an inequality is the set of all its solutions. For instance, the inequality $-3x < 6$ has for its solution set all numbers x such that $x > -2$. We think of an inequality as solved when we arrive at a simple inequality that provides an obvious description of the solution set.

▼ THEOREM 4.2a

If $P(x)$, $Q(x)$, and $R(x)$ are algebraic expressions, then for all values of x for which $P(x)$, $Q(x)$, and $R(x)$ are real numbers, the inequality $P(x) < Q(x)$ is equivalent to each of the following:

I. $P(x) + R(x) < Q(x) + R(x)$

II. $P(x) - R(x) < Q(x) - R(x)$

IIIa. $P(x)R(x) < Q(x)R(x)$ **for all x such that $R(x) > 0$**

IIIb. $P(x)R(x) > Q(x)R(x)$ **for all x such that $R(x) < 0$**

IVa. $\frac{P(x)}{R(x)} < \frac{Q(x)}{R(x)}$ **for all x such that $R(x) > 0$**

IVb. $\frac{P(x)}{R(x)} > \frac{Q(x)}{R(x)}$ **for all x such that $R(x) < 0$**

EXAMPLE 1

Solve:

$$\frac{1-x}{3} \leq \frac{3}{2}$$

SOLUTION:

$\frac{1-x}{3} \leq \frac{3}{2}$	Given
$(6)\left(\frac{1-x}{3}\right) \leq (6)\left(\frac{3}{2}\right)$	Multiply both sides by 6, the LCD of the two fractions.
$2 - 2x \leq 9$	Simplify.
$-2x \leq 7$	Subtract 2 from both sides.
$x \geq -3.5$	Divide both sides by -2 and reverse the sense of the inequality.

The solution set consists of all numbers x such that $x \geq -3.5$.

GRAPHING INEQUALITIES

The solution set that was obtained in Example 1 can be graphed on the number line as shown in Figure 4.2a. Note that the point -3.5 is part of the graph, as indicated by the heavy dot.

Sometimes inequalities appear in the form $-6 < 2x \leq 8$. As you may recall from Chapter 1, this notation means that

$$-6 < 2x \quad \textit{and} \quad 2x \leq 8$$

The solution of such inequalities is obtained in a manner similar to that used in Example 1. In the case of $-6 < 2x \leq 8$, we simply divide each term by 2, obtaining $-3 < x \leq 4$, which describes the solution set. The graph of this set is shown in Figure 4.2b. Note that 4 is part of the graph and -3 is *not*, as is indicated by the heavy dot at 4 and the open circle at -3.

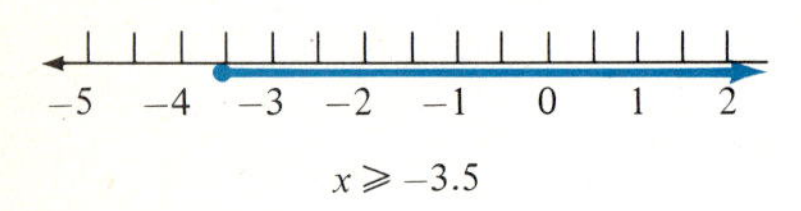

FIGURE 4.2a

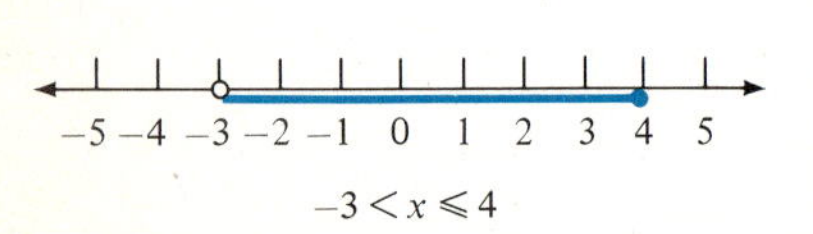

FIGURE 4.2b

EXAMPLE 2

Graph the solution set of $-4 \le -3x - 1 < 5$.

SOLUTION:

We first solve the inequality as follows:

$-4 \le -3x - 1 < 5$	Given
$-4 + 1 \le -3x - 1 + 1 < 5 + 1$	Add 1 to each member.
$-3 \le -3x < 6$	Simplify
$1 \ge x > -2$	Divide each member by -3 and reverse the sense of the inequality.
$-2 < x \le 1$	Rewrite the inequality.

The required graph is shown in Figure 4.2c.

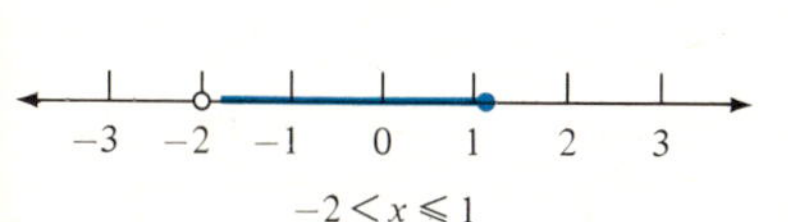

FIGURE 4.2c

EXAMPLE 3

Solve the inequality $-2x + 8 < x - 7$.

SOLUTION:

$-2x + 8 < x - 7$	Given
$-2x + 8 - 8 < x - 7 - 8$	Subtract 8 from both members.
$-2x < x - 15$	Simplify.
$-2x - x < x - 15 - x$	Subtract x from both members.
$-3x < -15$	Simplify.
$x > 5$	Divide by -3 and reverse the sense of the inequality.

Thus, the required solution is $x > 5$.

EXAMPLE 4

The revenue (in dollars) of a certain business based on the sale of x units of a product is given by

$4x + 200$

How many units must be sold so that the revenue exceeds \$13,000?

SOLUTION:

For the revenue to exceed \$13,000, we must have

$4x + 200 > 13{,}000$

or

$4x > 12{,}800$

$x > 3200$

Thus, more than 3200 units must be sold.

EXERCISE 4.2

In Problems 1–20, solve and graph the solution of the given inequality.

1 $4x - 1 < 19$

2 $3x - 14 > 1$

3 $2x + 5 \le 3$

4 $2x - 2 \ge 3$

5 $\dfrac{9x - 2}{4} \le 4$

6 $\dfrac{10x + 5}{3} \le 5$

7 $\frac{1}{4}(1 - x) < 2x + \frac{5}{2}$

8 $3x - 4 + \dfrac{x}{4} > \dfrac{5x}{2} + 2$

9 $\frac{1}{4}(4 - x) < \frac{1}{6}x$

10 $\frac{1}{3}(x - 2) > \frac{1}{2}x$

11 $-4 \le 6x - 1 < 11$

12 $-21 < 11x + 1 < 34$

13 $-1 \le \frac{x}{2} < 2$

14 $-2 \le \frac{x+2}{3} < 1$

15 $4 < x + 3 \le 9$

16 $1 < 5 - 3x < 17$

17 $-1 \le \frac{12 - 2x}{2} \le 1$

18 $-5 \le \frac{-12 - 6x}{3} \le 2$

19 $-1 \le \frac{x-1}{3} < 2$

20 $-2 < \frac{-2x + 1}{3} < 1$

21 A car saleswoman earns \$150 per week plus \$50 for every car she sells. Her total salary for the week is $50x + 150$, where x is the number of cars sold. How many cars does she have to sell in order to earn more than \$550 per week?

22 A car rental agency charges \$15 per day plus 10¢ per mile. Suppose a person travels x miles. Write an inequality indicating the number of miles that can be traveled so that the cost is under \$70.

23 The total revenue (in dollars) from selling x units of a product is $3x + 200$. Find how many units have to be sold so that the revenue exceeds \$700.

24 A commissioned saleswoman earns \$200 per week plus 10% of gross sales. If gross sales amount to x dollars, her weekly income is $0.10x + 200$. Write an inequality indicating the amount of gross sales necessary for her to earn more than \$500 per week?

25 Show that

$$\frac{b}{3a} + \frac{3a}{b} > 2$$

if $a > 0$, $b > 0$, and $3a \ne b$

26 Show that

$$\frac{3d}{4c} > 1 - \frac{c}{3d}$$

if $c > 0$, $d > 0$, and $2c \ne 3d$

27 Show that

$$\frac{\sqrt{n}}{\sqrt{v}} + \frac{\sqrt{v}}{\sqrt{n}} > 2$$

if $n > 0$, $v > 0$, and $n \ne v$

28 $\sqrt{(x-1)/3}$ is a real number if $(x - 1)/3 \ge 0$. Find the values of x for which $\sqrt{(x-1)/3}$ is a real number.

29 Find the values of x for which $\sqrt{(x+3)/x^2}$ is a real number. (See Problem 28.)

30 Find the values of x for which $\sqrt{x^3/(x+1)}$ is a real number. (See Problem 28.)

USING YOUR KNOWLEDGE 4.2

Sometimes you may want to find the common solution (if there is one) of a pair of inequalities such as

$x + 2 > 3$ *and* $x - 1 < 2$

You can often find the solution by solving each of the given inequalities separately and then combining the solutions in the proper way. For instance, you can solve the given two inequalities as follows:

$$\begin{aligned} x + 2 &> 3 \qquad & x - 1 &< 2 \\ x &> 1 & x &< 3 \end{aligned}$$

Thus, you must have $x > 1$ (or its equivalent $1 < x$) *and* $x < 3$. Hence, the common solution is $1 < x < 3$.

See if you can find the common solutions of the following pairs of inequalities.

1 $x + 2 < 1$ *and* $x + 3 > 0$

2 $x + 1 < 3$ *and* $x + 1 > 0$

3 $\frac{x+1}{3} \le 1$ *and* $\frac{x-1}{2} > -1$

4 $\frac{x-2}{2} > 1$ *and* $\frac{x-3}{2} \le 1$

5 $\frac{x-1}{2} \le 1$ *and* $\frac{x-3}{2} \ge 0$

6 $\frac{x-1}{2} < -1$ *and* $\frac{x-4}{3} \ge -1$

4.3 ABSOLUTE VALUE EQUATIONS AND INEQUALITIES

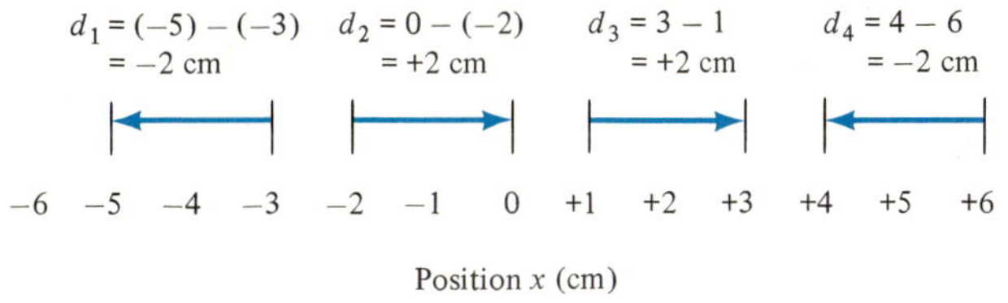

$d_1 = (-5) - (-3) = -2$ cm $d_2 = 0 - (-2) = +2$ cm $d_3 = 3 - 1 = +2$ cm $d_4 = 4 - 6 = -2$ cm

−6 −5 −4 −3 −2 −1 0 +1 +2 +3 +4 +5 +6

Position x (cm)

FIGURE 4.3a

Figure 4.3a pictures the *directed* distance from one point to another on the number line in four different cases. For instance, if we take the first point to be $x_1 = -3$ and the second point to be $x_2 = -5$, then the directed distance *from* x_1 *to* x_2 is clearly -2 units, because the distance is 2 units and the direction is the negative direction on the number line. This directed distance can be obtained by writing $x_2 - x_1 = (-5) - (-3) = -2$. Similarly, if $x_1 = -2$ and $x_2 = 0$, the directed distance from x_1 to x_2 is correctly given by $x_2 - x_1 = 0 - (-2) = 2$. You can check in the remaining two cases that the directed distance is given by the formula

$$d = x_2 - x_1$$

where x_1 is the first point and x_2 is the second point. By taking various points on the number line, you can convince yourself that the formula always gives the directed distance *from* the point x_1 *to* the point x_2.

ABSOLUTE VALUE

Frequently, we are interested in the distance itself and wish to disregard the direction. In all four cases in the figure, the *undirected* distance between the pairs of points is 2 units. In mathematics, we are especially interested in the undirected distance between the origin, the zero point on the number line, and any other point x. This *undirected distance* is called the *absolute value* of x and is denoted by $|x|$. (Compare Section 1.3.) Thus,

$$|-3| = 3, \quad |3| = 3, \quad |\tfrac{1}{7}| = \tfrac{1}{7}, \quad \text{and } |-\tfrac{1}{7}| = \tfrac{1}{7}$$

In terms of x itself the absolute value can be defined as in Definition 4.3a. It is important for you to note that an absolute value is *never* negative.

▼ DEFINITION 4.3a

$$|x| = \begin{cases} x & \text{if } x \geq 0 \\ -x & \text{if } x < 0 \end{cases}$$

EXAMPLE 1

Find:

a. $\left|\dfrac{8}{-2}\right|$ **b.** $|(-4)(5)|$

SOLUTION:

a. $\left|\dfrac{8}{-2}\right| = |-4| = 4$

b. $|(-4)(5)| = |-20| = 20$

Using Definition 4.3a, we can now see that the *undirected* distance d between two points x_1 and x_2 on the number line is given by

$$d = |x_2 - x_1|$$

For instance, if we return to Figure 4.3a, we can verify that the undirected distance between $x_1 = -3$ and $x_2 = -5$ is

$$|(-5) - (-3)| = |-2| = 2$$

EQUATIONS INVOLVING ABSOLUTE VALUES

We now discuss the solution of equations and inequalities that involve absolute values. For example, to solve the equation

$$|x| = 2$$

we reason that $|x| = 2$ means that the undirected distance from the origin to the point x is 2 units. There are two points that satisfy this requirement, 2 and -2. Hence, the solutions of

$$|x| = 2$$

are

$$x = \pm 2$$

where ± 2 means $+2$ or -2.

In general, the solutions of

$$|x| = a \qquad \text{for } a > 0$$

are

$$x = \pm a$$

The solution of $|x| = 0$ is, of course, simply $x = 0$.

If $E(x)$ is an algebraic expression in x, then $|E(x)|$ is determined by Definition 4.3a, the definition of absolute value. This means that

$$|E(x)| = \begin{cases} E(x) \text{ for all values of } x \text{ such that } E(x) \geq 0 \\ -E(x) \text{ for all values of } x \text{ such that } E(x) < 0 \end{cases}$$

Thus, to solve an equation such as $|E(x)| = a$, where $a > 0$, we must solve two equations, $E(x) = a$ and $-E(x) = a$. This is equivalent to solving $E(x) = a$ and $E(x) = -a$. This idea is illustrated in Example 2.

EXAMPLE 2

Solve the equation $|x + 1| = 2$.

SOLUTION:

Because the equation $|x + 1| = 2$ is equivalent to the two equations $x + 1 = 2$ and $x + 1 = -2$, we must solve these two equations. Thus,

$x + 1 = 2$ gives $x = 1$

and

$x + 1 = -2$ gives $x = -3$

Hence, there are two solutions, -3 and 1.

Alternatively, we can interpret $|x + 1| = |x - (-1)|$ as the undirected distance between x and -1. If this distance is to be 2 units, then x must be -3 or 1.

INEQUALITIES INVOLVING ABSOLUTE VALUES

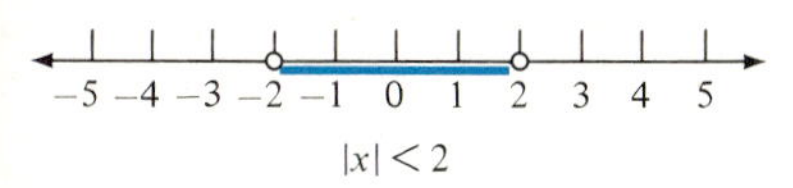

FIGURE 4.3b

We have seen that the solution set of $|x| = 2$ consists of the two points that are exactly 2 units away from 0. It follows that the solution set of $|x| < 2$ must consist of all points that are *less than* 2 units away from 0. Thus, the solution set of $|x| < 2$ consists of all x such that $-2 < x < 2$. The graph of this solution set is shown in Figure 4.3b.

Exactly the same reasoning as that used for the inequality $|x| < 2$ leads to Theorem 4.3a, which we shall accept without formal proof. A similar result holds for the inequality $|x| \le a$, which is equivalent to $-a \le x \le a$. Furthermore, the x in these inequalities may be replaced by an expression in x, say $E(x)$, so that $|E(x)| < a$ is equivalent to $-a < E(x) < a$, and $|E(x)| \le a$ is equivalent to $-a \le E(x) \le a$.

▼ THEOREM 4.3a

For $a > 0$, $|x| < a$ if and only if $-a < x < a$.

EXAMPLE 3

Find the graph the solution set of $|3x + 2| < 4$.

SOLUTION:

We see from the preceding discussion that $|3x + 2| < 4$ is equivalent to

$$-4 < 3x + 2 < 4$$

$$-6 < 3x < 2$$

$$-2 < x < \tfrac{2}{3}$$

The graph of the solution set is shown in Figure 4.3c.

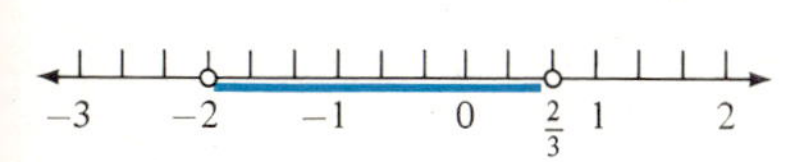

FIGURE 4.3c

To find the solution set of $|x| > 2$, we argue that because the solution set of $|x| = 2$ consists of all points *exactly* 2 units away from 0, the solution of $|x| > 2$ must consist of all points that are *more than* 2 units away from 0. The graph of this solution set appears in Figure 4.3d. As you can see, the solution set of $|x| > 2$ consists of all x such that $x < -2$ *or* $x > 2$. Notice that there are two *separate* unbounded intervals in this solution set that must be described by two *separate* inequalities.

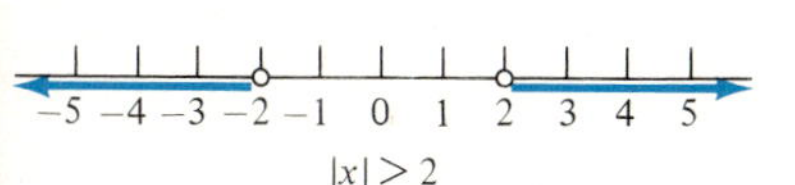

FIGURE 4.3d

▼ THEOREM 4.3b

For $a > 0$, $|x| > a$ if and only if $x < -a$ *or* $x > a$.

The preceding result is generalized in Theorem 4.3b, and we again omit any formal proof. Note carefully that, unlike $|x| < a$, the inequality $|x| > a$ is equivalent to two inequalities that must be separated by the word "or" and must *not* be run together.

As before, similar results may be stated for $|x| \ge a$ and for $|E(x)| \ge a$. Example 4 illustrates the idea.

EXAMPLE 4

Find and graph the solution set of $|2x + 3| \geq 5$.

SOLUTION:

The given inequality is equivalent to

$$2x + 3 \leq -5 \quad or \quad 2x + 3 \geq 5$$

that is

$$2x \leq -8 \quad or \quad 2x \geq 2$$

$$x \leq -4 \quad or \quad x \geq 1$$

The solution set consists of all x such that $x \leq -4$ *or* $x \geq 1$. The graph of this set is shown in Figure 4.3e.

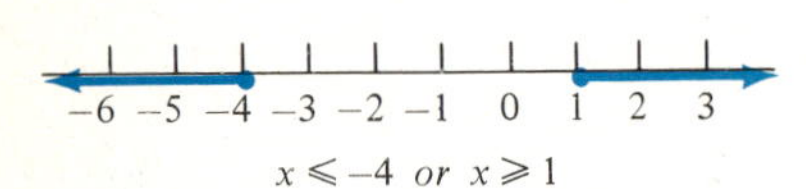

FIGURE 4.3e

Examples 3 and 4 illustrate the following general result. The solution set of an inequality of the form $|ax + b| < c$, where $c > 0$, consists of a *single* interval, but the solution set of an inequality of the form $|ax + b| > c$ consists of *two separate* intervals. You should graph your solution to check that it meets this requirement.

EXAMPLE 5

Solve the equation $|2x + 3| = 3x + 7$.

SOLUTION:

If this equation has any solution, it must be a solution of one or the other of the equations $2x + 3 = 3x + 7$ and $2x + 3 = -(3x + 7)$. Hence, we solve these two equations and check in the given equation.

$$2x + 3 = 3x + 7 \qquad 2x + 3 = -(3x + 7) = -3x - 7$$

$$-x = 4 \qquad 5x = -10$$

$$x = -4 \qquad x = -2$$

CHECK:

For $x = -4$,

$$|2x + 3| = |-8 + 3| = |-5| = 5$$

and

$$3x + 7 = -12 + 7 = -5$$

Because $5 \neq -5$, $x = -4$ does not check; it is not a solution of the given equation.

For $x = -2$,

$$|2x + 3| = |-4 + 3| = |-1| = 1$$

and

$$3x + 7 = -6 + 7 = 1$$

Therefore, $x = -2$ checks and is the only solution of the given equation.

EXERCISE 4.3

In Problems 1–6, find and graph the solution set of the given equation or inequality.

1 $|x| = 5$ **2** $|x| = 4$ **3** $|x| \le 1$ **4** $|x| < \frac{5}{2}$ **5** $|x| > 3$ **6** $|x| \ge \frac{7}{2}$

In Problems 7–12, replace the question marks with the appropriate quantities.

7 $-4 \le x \le 4$ if and only if $|?| \le ?$ **8** $-2 < x + 1 < 2$ if and only if $|?| < ?$

9 $x > 3$ or $x < -3$ if and only if $|?| > ?$ **10** $x + 1 \ge 2$ or $x + 1 \le -2$ if and only if $|?| \ge ?$

11 $x \ge 4$ or $x \le -2$ if and only if $|?| \ge ?$

Hint: The answer should be of the form $|x - x_1| \ge a$, where x_1 is the midpoint of the interval from -2 to 4.

12 $2 < x < 8$ if and only if $|?| < ?$ *Hint:* See Problem 11.

In Problems 13–36, find and graph the solution set of the given equation or inequality.

13 $|y - 2| = 3$ **14** $|y - 3| = 2$ **15** $|x + 2| < 3$

16 $|x + 5| \le 2$ **17** $|2x + 3| \le 7$ **18** $|2x - 1| \le 3$

19 $|3 - 4x| < 7$ **20** $|5 - 3x| \le 11$ **21** $|x - 1| \ge 2$

22 $|x - 3| > 4$ **23** $|2x - 1| > 3$ **24** $|3x - 1| \ge 2$

25 $|3 - 4x| > 7$ **26** $|5 - 3x| > 11$ **27** $|3t - 2| = 7$

28 $|4t - 1| = 3$ **29** $|\frac{1}{2}x - 1| \le 2$ **30** $|\frac{1}{3}x - 1| < 1$

31 $|\frac{1}{4}x - 1| > \frac{1}{8}$ **32** $|\frac{1}{3}x - 2| \le \frac{1}{3}$ **33** $|\frac{1}{3}x + 1| \ge \frac{1}{6}$

34 $|\frac{1}{2}x + 1| < \frac{1}{2}$ **35** $|1 - \frac{1}{4}x| < \frac{1}{8}$ **36** $|2 - \frac{1}{3}x| > \frac{1}{6}$

In Problems 37–46, solve the given equation.

37 $|2x - 3| = -(2x - 3)$ **38** $|5x - 1| = -(5x - 1)$ **39** $|3x - 2| = 3x - 2$

40 $|-2x - 1| = -2x - 1$ **41** $|4x - 2| = 2x$ **42** $|3x + 1| = -3x$

43 $|3 - 2x| = -x$ **44** $|5 - 2x| = 3x$ **45** $|4 - 3x| = -3x + 6$ **46** $|5 - 2x| = x + 5$

47 Prove that if x and y are real numbers, then $|x + y| \le |x| + |y|$.

Hint: One way to do this is to square both sides, use the fact that the square of the absolute value of a number is equal to the square of the number itself, and simplify. You should end with $xy \le |x| \cdot |y|$, which is obviously true by the definition of the absolute value. Now start with this last inequality, and proceed backwards through your analysis to construct the proof.

48 Use the result of Problem 47 to prove that $||x| - |y|| \le |x - y|$.

Hint: Consider two cases: (1) Assume $|x| \ge |y|$ and write $x = (x - y) + y$. (2) Assume $|y| > |x|$ and write $y = (y - x) + x$.

USING YOUR KNOWLEDGE 4.3

It is proved in calculus that $x^2 - 9$ is arbitrarily close to 16 for all values of x sufficiently close to 5 but not equal to 5. The proof depends on the fact that

$$\begin{aligned} |(x^2 - 9) - 16| &= |x^2 - 25| \\ &= |(x + 5)(x - 5)| \\ &= |x + 5|\,|x - 5| \end{aligned}$$

Now, if x is restricted to an interval of unit radius centered at 5, then $|x - 5| < 1$; that is, $-1 < x - 5 < 1$ or $4 < x < 6$, so that $9 < x + 5 < 11$. Therefore, $|x + 5| < 11$ and $|(x^2 - 9) - 16| < 11|x - 5|$. Thus, you can keep $x^2 - 9$ within $\frac{1}{10}$ of a unit from 16 by requiring $11|x - 5| < \frac{1}{10}$; that is, $|x - 5| < \frac{1}{110}$. This means that for all values of x such that $5 - \frac{1}{110} < x < 5 + \frac{1}{110}$, $x^2 - 9$ will surely be between 15.9 and 16.1.

1 Show that if x is restricted to an interval of unit radius centered at $x = 2$, then $|(x^2 + 4) - 8| < 5|x - 2|$ for $x \neq 2$.

2 Show that if x is restricted to an interval of unit radius centered at $x = 4$, then $|(x^2 - 2x - 3) - 5| < 7|x - 4|$ for $x \neq 4$.

3 Show that if x is restricted to an interval of unit radius centered at $x = 5$, then

$$\left|\frac{10}{x + 5} - 1\right| < \frac{1}{9}|x - 5| \qquad \text{for } x \neq 5$$

4 Show that if x is restricted to an interval of unit radius centered at $x = 3$, then

$$\left|\frac{x}{x + 1} - \frac{3}{4}\right| < \frac{1}{12}|x - 3| \qquad \text{for } x \neq 3$$

4.4 QUADRATIC EQUATIONS

An airplane is at an altitude s_0 when a parachutist jumps from the plane with an initial downward velocity v_0. His height, h feet above the ground at the end of t seconds after he jumps and before his parachute opens, is given by

$$h = -16t^2 - v_0 t + s_0$$

Suppose that $s_0 = 4800$ and $v_0 = 20$. Then

$$h = -16t^2 - 20t + 4800$$

If the man intends to open his parachute at a height of 3000 feet, how long should he wait before pulling the parachute release? To answer this question, you would have to solve the equation

$$3000 = -16t^2 - 20t + 4800$$

or

$$16t^2 + 20t - 1800 = 0 \tag{1}$$

Equation 1 is an example of a *quadratic* equation in t.

The equation

$$ax^2 + bx + c = 0 \qquad a \neq 0$$

is the *standard form* of a *second-degree* or *quadratic* equation in x.

SOLUTION BY FACTORING

▼ THEOREM 4.4a

Zero product principle

If the product of two numbers is zero, then at least one of the numbers is zero; that is

If $ab = 0$, then $a = 0$ or $b = 0$.

There are various ways of solving quadratic equations, one of these being the method of *factoring*. This method depends on the basic principle expressed in Theorem 4.4a, which is a consequence of the properties of the number system. Thus, according to this principle, to find the solution of Equation 1,

$$16t^2 + 20t - 1800 = 0$$

we can first simplify by dividing through by 4 to obtain

$$4t^2 + 5t - 450 = 0$$

and then factor the left side to obtain

$$(t - 10)(4t + 45) = 0$$

Now, the Zero Product Principle tells us that $t - 10 = 0$ or $4t + 45 = 0$. If $t - 10 = 0$, then $t = 10$, and if $4t + 45 = 0$, then $t = -\frac{45}{4}$. You can check to see that both these values of t satisfy Equation 1, but $t = 10$ is the only solution that has a practical significance in our problem. Thus, the man should wait 10 seconds before pulling the parachute release.

EXAMPLE 1

Solve:

a. $x^2 - x - 12 = 0$
b. $x^2 - 2x = 0$

SOLUTION:

a. $x^2 - x - 12 = 0$

$(x - 4)(x + 3) = 0$ — Factor the left side.

$x - 4 = 0$ or $x + 3 = 0$ — Zero Product Principle

$x = 4$ or $x = -3$ — Solve the two equations.

Thus, the equation has the two solutions, 4 and -3.

b. $x^2 - 2x = 0$

$x(x - 2) = 0$

$x = 0$ or $x - 2 = 0$

$x = 0$ or $x = 2$

Hence, the equation has the two solutions, 0 and 2. The checks are left for you to do.

SOLUTION BY EXTRACTING ROOTS

Obviously, the method of factoring is useful only with equations that are easily factored. If we try to solve the equation

$$x^2 - 2 = 0$$

we immediately find that $x^2 - 2$ is not factorable into factors with integer coefficients. However, because $x^2 - 2 = 0$ is equivalent to

$$x^2 = 2$$

we know that x must be either the positive or the negative square root of 2. Thus,

$$x = \pm\sqrt{2}$$

These solutions are easily checked because $(\pm\sqrt{2})^2 - 2 = 2 - 2 = 0$.

The method used in the preceding illustration is called *extraction of roots*. This method depends on putting the equation in the form

$$x^2 = a \qquad a > 0$$

which has the solution $x = \pm\sqrt{a}$.

EXAMPLE 2

Solve:

a. $x^2 - 3 = 0$
b. $9x^2 - 4 = 0$
c. $(x - 3)^2 = 4$

SOLUTION:

a. $x^2 - 3 = 0$

$x^2 = 3$

$x = \pm\sqrt{3}$ Extracting roots

b. $9x^2 - 4 = 0$

$9x^2 = 4$

$x^2 = \frac{4}{9}$

$x = \pm\sqrt{\frac{4}{9}} = \pm\frac{2}{3}$ Extracting roots

c. $(x - 3)^2 = 4$

$x - 3 = \pm 2$ Extracting roots

$x = 3 \pm 2$

$x = 1$ or $x = 5$

The checks are left for you to do.

SOLUTION BY COMPLETING THE SQUARE

In Example 2c, we solved an equation of the form

$$(x - a)^2 = b \qquad b > 0$$

by the method of *extraction of roots*. This method can be extended to quadratic equations that are *not* given in the form $(x - a)^2 = b$. The technique involves adding the proper term to both sides of the equation in order to make the left side a perfect trinomial square. This method is called *completing the square*.

The first step in this method is to determine what value to add to both sides to complete the square. First, consider the special quadratic equation in which the coefficient of x^2 is 1.

$$x^2 + bx + c = 0$$

We wish to write this equation in the form

$$(x + p)^2 = q$$

which we can solve by extracting roots. Because

$$(x + p)^2 = x^2 + 2px + p^2$$

we note that *the third term, p^2, is the square of one half of $2p$, the coefficient of x in the middle term.* This is the key to the following procedure.

$$x^2 + bx + c = 0$$

$$x^2 + bx = -c \qquad \text{Subtract } c \text{ from both sides.}$$

$$x^2 + bx + \left(\frac{b}{2}\right)^2 = -c + \left(\frac{b}{2}\right)^2 \qquad \text{Add } (b/2)^2\text{, the square of one half the coefficient of } x_1\text{, to both sides.}$$

$$\left(x + \frac{b}{2}\right)^2 = -c + \left(\frac{b}{2}\right)^2 \qquad \text{Factor the left side.}$$

The method of completing the square can now be summarized as follows:

TO SOLVE $x^2 + bx + c = 0$ BY COMPLETING THE SQUARE

1. **Subtract c from both sides to obtain the form**

   $$x^2 + bx = -c$$

2. **Add $(b/2)^2$ to both sides.**
3. **Factor the left side and simplify the right side.**
4. **Solve by extracting roots.**

It is extremely important to notice that this method assumes that the coefficient of x^2 is 1; if this is not the case, then you must first divide by that coefficient as in Example 3.

EXAMPLE 3

Solve $4x^2 - 4x - 7 = 0$ by completing the square.

SOLUTION:

We first divide by 4, so that the coefficient of x^2 becomes 1. This gives

$$x^2 - x - \tfrac{7}{4} = 0$$

1. $$x^2 - x = \tfrac{7}{4}$$ Subtract $-\frac{7}{4}$ (that is, add $\frac{7}{4}$) on both sides.

2. $$x^2 - x + (-\tfrac{1}{2})^2 = \tfrac{7}{4} + (-\tfrac{1}{2})^2$$ Add the square of one half the coefficient of x to both sides.

3. $$(x - \tfrac{1}{2})^2 = 2$$ Factor the left and simplify the right side.

4. $$x - \tfrac{1}{2} = \pm\sqrt{2}$$ Extract square roots.

Thus,

$$x = \tfrac{1}{2} \pm \sqrt{2}$$

CHECK:

For $x = \frac{1}{2} + \sqrt{2}$,

$$\begin{aligned} 4x^2 - 4x - 7 &= 4(\tfrac{1}{2} + \sqrt{2})^2 - 4(\tfrac{1}{2} + \sqrt{2}) - 7 \\ &= 1 + 4\sqrt{2} + 8 - 2 - 4\sqrt{2} - 7 \\ &= 0 \end{aligned}$$

which checks. The check for the other solution is left for you.

THE QUADRATIC FORMULA

The method of completing the square can be used to obtain a formula for solving any quadratic equation. The details are as follows:

$$ax^2 + bx + c = 0 \qquad a \neq 0$$

$$x^2 + \frac{b}{a}x + \frac{c}{a} = 0$$ Divide by a to make the coefficient of x^2 become 1.

$$x^2 + \frac{b}{a}x = -\frac{c}{a}$$ Subtract c/a from both sides.

$$x^2 + \frac{b}{a}x + \left(\frac{b}{2a}\right)^2 = -\frac{c}{a} + \left(\frac{b}{2a}\right)^2$$ Add the square of one half the coefficient of x to both sides.

$$\left(x + \frac{b}{2a}\right)^2 = \frac{b^2 - 4ac}{4a^2}$$ Factor the left and simplify the right side.

$$x + \frac{b}{2a} = \pm\sqrt{\frac{b^2 - 4ac}{4a^2}}$$ Extract square roots.

$$x = -\frac{b}{2a} \pm \frac{\sqrt{b^2 - 4ac}}{2a}$$ Subtract $b/2a$ from both sides and simplify the denominator on the right.

$$x = \frac{-b \pm \sqrt{b^2 - 4ac}}{2a}$$ Combine fractions on the right.

▼ THEOREM 4.4b

The solutions of the equation

$$ax^2 + bx + c = 0 \qquad a \neq 0$$

are

$$x = \frac{-b \pm \sqrt{b^2 - 4ac}}{2a}$$

This result is known as the *quadratic formula*, and we have the important conclusion stated in Theorem 4.4b. Of course, you *must be certain* that the equation is in the *standard form* before you apply the quadratic formula.

EXAMPLE 4

Solve:

$$\frac{x^2}{7} + \frac{x}{2} = \frac{3}{14}$$

SOLUTION:

We first write the equation in the *standard form*

$$\tfrac{1}{7}x^2 + \tfrac{1}{2}x - \tfrac{3}{14} = 0$$

$$2x^2 + 7x - 3 = 0$$ Multiply by 14 to clear the fractions.

For this equation, $a = 2$, $b = 7$, and $c = -3$. Therefore,

$$x = \frac{-7 \pm \sqrt{(7)^2 - 4(2)(-3)}}{2(2)}$$

$$= \frac{-7 \pm \sqrt{49 + 24}}{4}$$

$$= \frac{-7 \pm \sqrt{73}}{4}$$

Thus, the equation has the two solutions

$$\frac{-7 - \sqrt{73}}{4} \quad \text{and} \quad \frac{-7 + \sqrt{73}}{4}$$

The check is left for you.

NATURE OF THE ROOTS (THE DISCRIMINANT)

The nature of the roots of the equation $ax^2 + bx + c = 0$ depends on the quantity under the radical sign in the quadratic formula. This quantity, $b^2 - 4ac$, is called the *discriminant* of the quadratic equation. As you can see, if $b^2 - 4ac = 0$, the equation has just one root, $-b/2a$. If $b^2 - 4ac$ is a positive number, then its square root is a real number, and the equation has two real solutions. If $b^2 - 4ac$ is a negative number, then it has no real square root, and the equation has no real solutions. We shall consider this case in the chapter entitled "Complex Numbers and Theory of Equations."

If the coefficients a, b, and c are integers, then we can get additional information in the case where the discriminant is positive. If $b^2 - 4ac$ is a perfect square, then the roots are rational numbers; otherwise, the roots are irrational. Of course, if the coefficients are rational numbers, not integers, we can multiply through by the LCD of the denominators to make the coefficients integers. Thus, we can still determine whether or not the roots are rational.

The preceding information is summarized in the following manner.

Discriminant	***Nature of the roots***
1. $b^2 - 4ac = 0$	**There is exactly one root, the real number** $-b/2a$.
2. $b^2 - 4ac > 0$	**There are two distinct real roots.**
3. $b^2 - 4ac < 0$	**There are no real roots.**

In 4 and 5, it is assumed that a, b, and c are integers and $b^2 - 4ac > 0$.

4. $b^2 - 4ac$ ***is*** **a perfect square.**	**The roots are rational.**
5. $b^2 - 4ac$ ***is not*** **a perfect square.**	**The roots are irrational.**

EXAMPLE 5

a. Under what condition on k will $4x^2 + 4x + k = 0$ have two real roots?

b. If $k = \frac{3}{4}$, show that the roots are rational.

SOLUTION:

a. For this equation, $a = 4$, $b = 4$, and $c = k$. Thus, the discriminant

$$b^2 - 4ac = 16 - 16k$$

Because the discriminant must be positive for there to be two real roots, we require

$$16 - 16k > 0$$

or

$$-16k > -16$$

so that $k < 1$. Accordingly, the equation will have two real roots if k is any real number less than 1.

b. We see from a, that the roots are real if $k = \frac{3}{4}$. Also for $k = \frac{3}{4}$, the equation becomes

$$4x^2 + 4x + \frac{3}{4} = 0$$

or

$$16x^2 + 16x + 3 = 0$$

For this equation, $a = 16$, $b = 16$, and $c = 3$, so that

$$\begin{aligned} b^2 - 4ac &= (16)^2 - (4)(3)(16) \\ &= 16(16 - 12) \\ &= 64 \end{aligned}$$

which is a perfect square. Therefore, by item 4, the roots are rational numbers. (You can check this by solving the equation.)

If the coefficients of a quadratic equation are not convenient small numbers, then your calculator can be very helpful in finding the roots. Of course, the roots will usually be approximated by decimals (which might be just what you want). It is most convenient to start with the radical and to store its value so that you can evaluate both roots without having to copy down any intermediate numbers or to backtrack. Let's look at the equation

$$47x^2 + 83x + 13 = 0$$

Here is the procedure using the quadratic formula:

Algebraic logic

[83] [x^2] [−] [4] [×] [47] [×] [13] [=] [$\sqrt{x}$] [STO] [83] [+/−] [+] [RCL] [=] [÷] [2] [÷] [47] [=] (Display shows −0.17371454.)

[83] [+/−] [−] [RCL] [=] [÷] [2] [÷] [47] [=] (Display shows −1.5922429.)

RPN logic

[83] [ENTER] [×] [4] [ENTER] [47] [×] [13] [×] [−] [$\sqrt{x}$] [STO] [83] [CHS] [ENTER] [RCL] [+] [2] [÷] [47] [÷] (Display shows −0.1737145404.)
[83] [CHS] [ENTER] [RCL] [−] [2] [÷] [47] [÷] (Display shows −1.592242906.)

To three decimal places, the roots are −0.174 and −1.592.

EXERCISE 4.4

In Problems 1–12, solve by factoring. Check your solutions.

1 $x^2 + 4x + 3 = 0$ **2** $x^2 + 6x + 5 = 0$ **3** $x^2 + 4x + 4 = 0$

4 $x^2 + 8x + 16 = 0$ **5** $x^2 - 10x + 21 = 0$ **6** $x^2 - 2x - 143 = 0$

7 $x^2 + 64 = 16x$ **8** $x^2 + 121 = 22x$ **9** $2x^2 + 7x + 3 = 0$

10 $9x^2 + 9x + 2 = 0$ **11** $25y^2 = 25y - 6$ **12** $15 - 16y = -4y^2$

In Problems 13–20, solve by extracting square roots.

13 $(x - 1)^2 = 25$ **14** $(x + 7)^2 = 49$ **15** $x^2 - 2x + 1 = 2$

16 $x^2 - 8x + 16 = 25$ **17** $(x + \frac{3}{4})^2 = \frac{1}{16}$ **18** $(3x + 1)^2 = \frac{1}{4}$

19 $(x - a)^2 = 25$ **20** $(x + b)^2 = 36$

In Problems 21–24, solve by completing the square.

21 $y^2 + 10y + 18 = 0$ **22** $y^2 - 2y - 2 = 0$ **23** $3x^2 = 6 - 7x$

24 $2x^2 = 56 - 9x$

In Problems 25–32, solve by the quadratic formula.

25 $y^2 - 6y + 7 = 0$ **26** $y^2 + 6y + 7 = 0$ **27** $\frac{1}{4}x^2 + x = -\frac{1}{4}$

28 $\frac{1}{4}x^2 = -\frac{1}{4}x + \frac{1}{2}$ **29** $2t^2 - \sqrt{3}t = 3$ **30** $w^2 + 2\sqrt{3}w = 1$

31 $\frac{1}{2}x^2 + \sqrt{2}x = -1$ **32** $s^2 + \frac{5}{\sqrt{2}}s = -2$

In Problems 33–48, solve by any method.

33 $y^2 + 5y = 0$ **34** $7v^2 + 7v = 0$ **35** $x(6x + 7) = 3$

36 $z(3z + 11) = 20$ **37** $3x^2 - x + 12 = 10 + x(x + 3)$ **38** $(2x - 3)^2 - (x + 3)^2 = -6x^2$

39 $3(1 + y)^2 + y^2 = (y - 1)^2 - 3$ **40** $2(2 + y)^2 - 4(y + 1)^2 = (1 - y)^2 + 2$

41 $4w^2 + 4w = 1$ **42** $9z^2 + 6z - 8 = 0$

43 $x^2 + 2\sqrt{3}x = 13$ **44** $3x^2 - 2\sqrt{5}x = 5$

45 $\frac{x^2}{5} - \frac{x}{2} = \frac{3}{10}$ **46** $\frac{x^2}{4} - \frac{x}{5} = \frac{1}{20}$

47 $\sqrt{5}t^2 + 4t - \sqrt{5} = 0$ **48** $3s^2 + 2\sqrt{10}s - 2 = 0$

In Problems 49–52, use the discriminant to determine whether the roots are rational or irrational.

49 $2x^2 + 13x + 6 = 0$ **50** $2x^2 + 13x + 12 = 0$ **51** $3x^2 - 15x - 4 = 0$

52 $4x^2 - 15x - 4 = 0$

53 Find k so that $kx^2 + 2x + 1 = 0$ has just one root.

Hint: If there is just one root, the discriminant must be zero.

54 Repeat Problem 53 for the equation $x^2 + kx + k + 3 = 0$.

55 a Under what conditions on k will the equation $x^2 + 3x + k + 1 = 0$ have two real roots?
b Show that the equation has rational roots if $k = -89$.

56 Under what condition on k will the equation $kx^2 + (2k - 3)x + k = 0$ have no real roots?

In Problems 57–60, use your calculator to find the roots correct to three decimal places.

C **57** $29x^2 + 17x - 31 = 0$

C **58** $13x^2 - 45x + 23 = 0$

C **59** $2.73x^2 + 5.61x - 9.52 = 0$

C **60** $3.67x^2 - 7.32x + 2.16 = 0$

USING YOUR KNOWLEDGE 4.4

We have shown in this section that the equation $ax^2 + bx + c = 0$, $a \neq 0$, has the two roots r_1 and r_2 given by

$$r_1 = \frac{-b - \sqrt{b^2 - 4ac}}{2a} \quad \text{and} \quad r_2 = \frac{-b + \sqrt{b^2 - 4ac}}{2a}$$

1 Show that $r_1 + r_2 = -b/a$.

2 Show that $r_1 r_2 = c/a$.

3 Use the information in Problems 1 and 2 to find the sum and the product of the roots in the following equations: **a** $x^2 - 4x + 2 = 0$ **b** $3x^2 - 2x - 8 = 0$ **c** $2x^2 - 12x + 5 = 0$

4 What would you take for the value of b in the equation $5x^2 + bx + 7 = 0$ if you required the sum of the roots to be 4?

5 What would you take for the value of c in the equation $3x^2 + 5x + c = 0$ if you required the product of the roots to be -10?

4.5 EQUATIONS LEADING TO QUADRATICS

A ball is dropped from a height h (meters). The time t (seconds) that it takes the ball to hit the ground is

$$t = \sqrt{\frac{h}{5}} \tag{1}$$

If the ball takes 2 seconds to hit the ground, we can find the height from which the ball was dropped by solving Equation 1. Letting $t = 2$ in this equation, we obtain

$$\sqrt{\frac{h}{5}} = 2 \tag{2}$$

and, by squaring both sides,

$$\frac{h}{5} = 4 \tag{3}$$

$$h = 20$$

Thus, the ball was dropped from a height of 20 meters.

Equation 2 was solved by squaring both members to obtain the equivalent, Equation 3. In many cases, however, squaring both sides of an equation does *not* result in an equivalent equation. For example, if we square both sides of

$$x = 2 \tag{4}$$

we obtain

$$x^2 = 4 \tag{5}$$

which has the solutions $x = 2$ and $x = -2$. The number -2 is a solution of Equation 5 but *not* of 4. Such prospective solutions are called *extraneous roots;* such roots can be obtained in the solution process as roots of a derived equation but are not roots of the original equation, and their existence points out the *necessity of checking in the original equation.* From this example, you can see that all the solutions of Equation 4 are also solutions of Equation 5, which was obtained by squaring both sides of 4. On the other hand, not all solutions of 5 are solutions of 4. This is a particular illustration of Theorem 4.5a.

▼ THEOREM 4.5a

If $P(x)$ and $Q(x)$ are algebraic expressions, and n is a positive integer, then all solutions of

$$P(x) = Q(x)$$

are solutions of the nth power equation

$$[P(x)]^n = [Q(x)]^n$$

However, it may happen that the nth power equation has solutions that are not solutions of the original equation.

EQUATIONS INVOLVING RADICALS

We can take advantage of the preceding theorem to solve equations involving radicals. For example, we can solve the equation

$$\sqrt{x+3} = x + 1$$

by squaring both sides and transforming the equation into a quadratic equation. Theorem 4.5a assures us that the solutions of the original equation will be found among the solutions of the quadratic equation, although there might, of course, be solutions of the quadratic that do not satisfy the original equation.

EXAMPLE 1

Solve the equation $\sqrt{x+3} = x + 1$.

SOLUTION:

$\sqrt{x+3} = x + 1$	
$x + 3 = x^2 + 2x + 1$	Square both sides.
$x^2 + x - 2 = 0$	Write the equation in standard form.
$(x + 2)(x - 1) = 0$	Factor.
$x = -2 \quad \text{or} \quad x = 1$	Solve.

CHECK:

For $x = -2$, $\sqrt{x+3} = \sqrt{1} = 1$ and $x + 1 = -1$. Thus, -2 is *not* a root.

For $x = 1$, $\sqrt{x+3} = \sqrt{4} = 2$ and $x + 1 = 2$, which checks. Hence, $x = 1$ is the only solution of the equation $\sqrt{x+3} = x + 1$. By the definition of the principal square root, we see that the left side, $\sqrt{x+3}$, cannot be negative. Therefore, the right side cannot be negative, so that $x + 1 \geq 0$ and $x \geq -1$. This observation means that we could have discarded the extraneous solution $x = -2$ at once because it does not meet the condition $x \geq -1$.

In Example 1, we squared both sides to remove the radical and then solved the resulting equation. The fact that we found one root and one extraneous root shows the importance of the check in the original equation.

In some problems it is necessary to square both sides of an equation more than once, as Example 2 illustrates.

EXAMPLE 2

Solve the equation
$\sqrt{x-4} = \sqrt{x} - 2$

SOLUTION:

$\sqrt{x-4} = \sqrt{x} - 2$

$x - 4 = x - 4\sqrt{x} + 4$	Square both sides.
$4\sqrt{x} = 8$	Simplify.
$\sqrt{x} = 2$	Divide by 4.
$x = 4$	Square both sides.

CHECK:

For $x = 4$, $\sqrt{x-4} = 0$ and $\sqrt{x} - 2 = 2 - 2 = 0$, which checks. Thus, the solution is $x = 4$.

THE SUBSTITUTION METHOD

Although the equation $y + 2\sqrt{y} - 3 = 0$ can be solved by the preceding method, we recognize that this equation can be rewritten as

$$(\sqrt{y})^2 + 2\sqrt{y} - 3 = 0$$

which is actually a quadratic equation in $\sqrt{y}$. Hence, if we use the method of substitution and let $\sqrt{y} = x$, the equation becomes

$$x^2 + 2x - 3 = 0$$

which can be solved to obtain the solution of the original equation, as in Example 3.

EXAMPLE 3

Use the substitution method to solve the equation
$y + 2\sqrt{y} - 3 = 0$.

SOLUTION:

Let $\sqrt{y} = x$, so that $y = x^2$ and the equation becomes

$$x^2 + 2x - 3 = 0$$

or

$$(x + 3)(x - 1) = 0$$

which gives

$$x = -3 \quad \text{or} \quad x = 1$$

Because $x = \sqrt{y}$, $\sqrt{y} = -3$ or $\sqrt{y} = 1$, but $\sqrt{y}$, for real y, is *never* negative. Therefore, we must discard $\sqrt{y} = -3$, leaving $\sqrt{y} = 1$ as the only possibility. If

$$\sqrt{y} = 1$$

then

$$y = 1 \qquad \text{By squaring both sides}$$

CHECK:

For $y = 1$, $y + 2\sqrt{y} - 3 = 1 + 2\sqrt{1} - 3 = 1 + 2 - 3 = 0$, which checks. Thus, the solution of the original equation is $y = 1$.

The substitution method always applies to equations of the form

$$a[\mathbf{E(x)}]^2 + b\mathbf{E(x)} + c = 0$$

which is an equation in *quadratic form*, with the expression $\mathbf{E(x)}$ taking the place of x in the standard form. If we let $\mathbf{E(x)} = p$, the equation becomes

$$ap^2 + bp + c = 0$$

You might recognize that the equation in Example 3 is an equation of quadratic form. Example 4 shows you how to solve a more complicated problem of this type.

EXAMPLE 4

Solve:

$$\left(3x - \frac{2}{x}\right)^2 + 6\left(3x - \frac{2}{x}\right) + 5 = 0$$

SOLUTION:

Let $3x - 2/x = p$. Then the given equation becomes

$$p^2 + 6p + 5 = 0$$
$$(p + 5)(p + 1) = 0$$
$$p = -5 \quad \text{or } p = -1$$

Thus,

$$3x - \frac{2}{x} = -5 \quad \text{or} \quad 3x - \frac{2}{x} = -1$$
$$3x^2 - 2 = -5x \quad \text{or} \quad 3x^2 - 2 = -x$$
$$3x^2 + 5x - 2 = 0 \quad \text{or} \quad 3x^2 + x - 2 = 0$$
$$(3x - 1)(x + 2) = 0 \quad \text{or} \quad (3x - 2)(x + 1) = 0$$
$$x = \tfrac{1}{3}, \quad x = -2 \quad \text{or} \quad x = \tfrac{2}{3}, \quad x = -1$$

CHECK:

The check is left for you to do. All the preceding values check, so that the given equation has the four roots, $-2, -1, \frac{1}{3}, \frac{2}{3}$.

Example 5 illustrates the substitution method for an equation that is not in quadratic form.

EXAMPLE 5

Find the solution set of the equation $3\sqrt{y^2 - 1} = 2y^2 - 7$.

SOLUTION:

Because the radical in this equation is the bothersome term, we let

$$\sqrt{y^2 - 1} = x$$

so that

$$y^2 - 1 = x^2 \quad \text{and} \quad y^2 = x^2 + 1$$

By substituting into the given equation, we obtain

$$3x = 2(x^2 + 1) - 7$$

or

$$\begin{aligned} 2x^2 - 3x - 5 &= 0 \\ (2x - 5)(x + 1) &= 0 \\ x &= \tfrac{5}{2} \quad \text{or} \quad -1 \end{aligned}$$

Because $\sqrt{y^2 - 1}$ cannot be negative, we discard the value $x = -1$. This leaves us with

$$\begin{aligned} \sqrt{y^2 - 1} &= \tfrac{5}{2} \\ y^2 - 1 &= \tfrac{25}{4} \\ y^2 &= \tfrac{25}{4} + 1 = \tfrac{29}{4} \\ y &= \pm\frac{\sqrt{29}}{2} \end{aligned}$$

CHECK:

For $y = \pm\sqrt{29}/2$,

$$\begin{aligned} 3\sqrt{y^2 - 1} &= 3\sqrt{\tfrac{29}{4} - 1} \\ &= 3\sqrt{\tfrac{25}{4}} \\ &= \tfrac{15}{2} \end{aligned}$$

and

$$\begin{aligned} 2y^2 - 7 &= (2)(\tfrac{29}{4}) - 7 \\ &= \tfrac{29}{2} - \tfrac{14}{2} \\ &= \tfrac{15}{2} \end{aligned}$$

Thus, both values check, and the solutions are $\pm\sqrt{29}/2$.

EXERCISE 4.5

In Problems 1–14, solve the given equation.

1 $\sqrt{x+8}=4$

2 $\sqrt{x+4}=5$

3 $\sqrt{x+3}=x-3$

4 $\sqrt{x+9}=x-3$

5 $\sqrt{y+5}=y-1$

6 $\sqrt{y-5}=y-7$

7 $\sqrt{y+1}-y=1$

8 $\sqrt{y-1}-y=-3$

9 $\sqrt{x+9}=1+\sqrt{x}$

10 $\sqrt{x+5}=2+\sqrt{x}$

11 $\sqrt{y+3}=\sqrt{y}+\sqrt{3}$

12 $\sqrt{x+5}=\sqrt{x}+\sqrt{5}$

13 $\sqrt{2y-1}+\sqrt{y+3}=3$

14 $\sqrt{x-3}+\sqrt{2x+1}=2\sqrt{x}$

In Problems 15–38, use the substitution method to solve the given equation.

15 $y-4\sqrt{y}+3=0$

16 $y-5\sqrt{y}+4=0$

17 $x+2\sqrt{x}=15$

18 $x=-\sqrt{x}+20$

19 $\left(x-\frac{4}{x}\right)^2-\left(x-\frac{4}{x}\right)-12=0$

20 $\left(x-\frac{6}{x}\right)^2-4\left(x-\frac{6}{x}\right)=5$

21 $\left(x+\frac{5}{x}\right)^2-3\left(x+\frac{5}{x}\right)-54=0$

22 $\left(x+\frac{1}{x}\right)^2+6\left(x+\frac{1}{x}\right)=-8$

23 $x^2+8+\frac{16}{x^2}+x+\frac{4}{x}-20=0$

24 $y^2-12+\frac{36}{y^2}+4y-\frac{24}{y}=5$

25 $\frac{4x^2}{(x+2)^2}-\frac{4x}{x+2}-24=0$

26 $\frac{4}{(x+3)^2}+\frac{2}{x+3}=12$

27 $(x^2+2x)^2-14(x^2+2x)-15=0$

28 $(2y^2-y)^2-16(2y^2-y)+60=0$

29 $2w^2-3w-2\sqrt{2w^2-3w}-3=0$

30 $x^2+3x-\sqrt{x^2+3x}-6=0$

31 $14\sqrt{y^2-25}=y^2-1$

32 $16\sqrt{y^2+64}=47+y^2$

33 $x^4-29x^2+100=0$

34 $y^4-13y^2+36=0$

35 $y^{-8}-17y^{-4}+16=0$

36 $x^{-3/2}-26x^{-3/4}-27=0$

37 $y^{1/3}-1-2y^{-1/3}=0$

38 $u^{1/4}+2-8u^{-1/4}=0$

USING YOUR KNOWLEDGE 4.5

If an object falls through a distance h (feet), its velocity v (feet per second) is known from physics to be

$$v=\sqrt{2gh}$$

where g is the gravitational acceleration, approximately 32 feet per second per second. Using this value, we get

$$v=\sqrt{64h}=8\sqrt{h}$$

If water is discharged from a hole in the bottom of a large tank, its velocity is that of an object that falls through the distance h, where h is the depth of the water in the tank.

1 If the velocity of discharge, v, of water through a hole in the bottom of a tank is 32 feet per second, how deep is the water in the tank?

2 Answer the same question if the velocity is 5.5 feet per second.

3 If the pressure at the nozzle of a hose is p (pounds per square inch), then the velocity of discharge of the water is known to be

$$v=12.1\sqrt{p}$$

If a fire hose is discharging water at the rate of 96.8 feet per second, what is the pressure at the nozzle?

4.6 QUADRATIC INEQUALITIES

TABLE 4.6a *The Effect of Daily Compounding*

Nominal interest rate	Effective annual yield
5%	5.13%
6%	6.18%
7%	7.25%
8%	8.33%
9%	9.42%
10%	10.52%

Table 4.6a shows what happens for several nominal interest rates when interest is compounded daily. We will consider a simpler problem involving annual compounding. If P dollars are invested at the rate r, compounded annually, the interest for the first year is Pr, and the interest for the second year is $(P + Pr)r$. The total interest is $Pr + (P + Pr)r = 2Pr + Pr^2$. This interest is usually written as

$$I = Pr(2 + r)$$

If we wish to know the rate at which to invest $1000 in order to receive more than $210 in interest for the 2 years, then we must solve the inequality

$$1000r(2 + r) > 210 \tag{2}$$

or equivalently,

$$100r^2 + 200r - 21 > 0 \tag{3}$$

You should realize that solving this inequality is equivalent to finding the values of r for which the polynomial on the left,

$$100r^2 + 200r - 21$$

is positive. Let us write $P(r) = 100r^2 + 200r - 21$.

We shall make use of a fact that is proved in calculus, namely:

A polynomial $P(x)$ remains of constant sign between consecutive values of x for which $P(x) = 0$.

Thus, when we solve an inequality such as $P(x) > 0$, the roots of $P(x) = 0$ play a critical role. For this reason, we call these roots the *critical values* of x. To solve the inequality, all we need do is determine the sign of $P(x)$ for each of the intervals into which the critical values divide the number line. Keep in mind that $P(x)$ positive means $P(x) > 0$, and $P(x)$ negative means $P(x) < 0$.

We now return to the interest rate problem, where we had

$$P(r) = 100r^2 + 200r - 21$$

and we wanted to solve the inequality $P(r) > 0$. Following the preceding discussion, we first find the roots of $P(r) = 0$. Thus, we solve

$$100r^2 + 200r - 21 = 0$$

or

$$(10r - 1)(10r + 21) = 0$$

FIGURE 4.6a

$$r = \frac{1}{10} \quad \text{or} \quad r = -\frac{21}{10}$$

These are the critical values of r, which we graph on the number line as in Figure 4.6a. As the figure shows, these values divide the number line into three intervals:

$$r < -\frac{21}{10}, \quad -\frac{21}{10} < r < \frac{1}{10}, \text{ and } r > \frac{1}{10}$$

Now we use a convenient value of r in each of these intervals to determine the sign of $P(r)$; $r = -3$, $r = 0$, and $r = 1$ will do. Using the factored form,

$$P(r) = (10r - 1)(10r + 21)$$

we find

$$P(-3) = (-31)(-9) = +279$$
$$P(0) = (-1)(21) \quad = -21$$
$$P(1) = (9)(31) \quad = +279$$

These results show that $P(r)$ is positive for $r < -\frac{21}{10}$, $P(r)$ is negative for $-\frac{21}{10} < r < \frac{1}{10}$, and $P(r)$ is positive for $r > \frac{1}{10}$. The signs in Figure 4.6a correspond to these facts. We can conclude that

$$P(r) > 0 \qquad \text{for } r < -\frac{21}{10} \quad \text{or} \quad r > \frac{1}{10}$$

and

$$P(r) < 0 \qquad \text{for } -\frac{21}{10} < r < \frac{1}{10}$$

The heavy lines in the figure correspond to the values of r for which $P(r) > 0$.

In our investment problem, the result $r < -\frac{21}{10}$ is meaningless, because the interest rate must be positive. Hence, we have to obtain a rate greater than $\frac{1}{10} = 10$ percent to get more than \$210 total interest in 2 years.

EXAMPLE 1

Solve the inequality $(x - 1)(x + 3) < 0$.

SOLUTION:

Here we can read off the critical values by inspection, because

$$(x - 1)(x + 3) = 0$$

for

$$x = 1 \quad \text{or} \quad x = -3$$

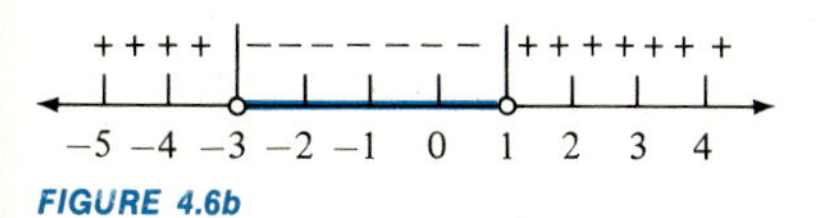

FIGURE 4.6b

We graph these on the number line shown in Figure 4.6b. The critical values divide the number line into the three intervals

$$x < -3, \qquad -3 < x < 1, \qquad x > 1$$

Writing $P(x) = (x - 1)(x + 3)$ and selecting the test values $x = -4$, $x = 0$, and $x = 2$, we find

$$P(-4) = (-5)(-1) = +5$$

$$P(0) = (-1)(3) \quad = -3$$
$$P(2) = (1)(5) \quad = +5$$

Accordingly, in Figure 4.6b, we mark plus signs on the intervals $x < -3$ and $x > 1$, and minus signs on the interval $-3 < x < 1$. The solution of the inequality $(x - 1)(x + 3) < 0$ is $-3 < x < 1$. This portion of the number line is heavy in the figure. The open circles at -3 and 1 mean that these points are not included.

EXAMPLE 2

Solve the inequality $2x^2 - 7x + 4 > 0$.

SOLUTION:

Here we obtain the critical values by using the quadratic formula to solve the equation

$$2x^2 - 7x + 4 = 0$$

The roots are

$$\frac{7 - \sqrt{17}}{4} \approx 0.72 \quad \text{and} \quad \frac{7 + \sqrt{17}}{4} \approx 2.78$$

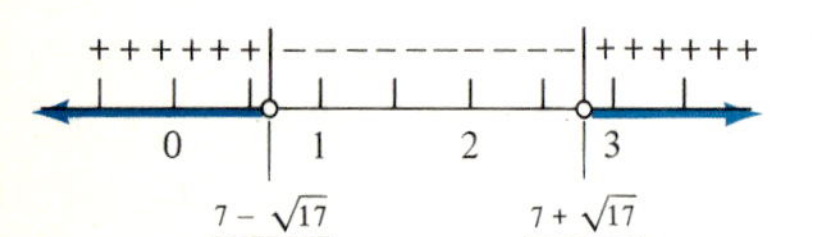

FIGURE 4.6c

These points are graphed on the number line in Figure 4.6c, which shows that the number line is divided into three intervals:

$$x < \frac{7 - \sqrt{17}}{4}, \quad \frac{7 - \sqrt{17}}{4} < x < \frac{7 + \sqrt{17}}{4}, \quad \text{and } x > \frac{7 + \sqrt{17}}{4}$$

Suitable test values are $x = 0$, $x = 2$, and $x = 3$. Writing $P(x)$ for the polynomial, we find $P(0) = +4$, $P(2) = -2$, and $P(3) = +1$. Accordingly, the left-hand and the right-hand intervals in the figure are marked with plus signs; the middle interval with minus signs. Thus, $P(x) > 0$ for all values of x in either the right-hand or the left-hand interval, and the solution of the inequality is

$$x < \frac{7 - \sqrt{17}}{4} \quad \text{or} \quad x > \frac{7 + \sqrt{17}}{4}$$

EXAMPLE 3

Solve:

$$\frac{2x - 3}{x + 2} \leq 0$$

SOLUTION:

First, we note that the given fraction is zero when its numerator, $2x - 3$, is zero, that is, when $x = \frac{3}{2}$; the fraction is undefined when its denominator, $x + 2$, is zero, that is, when $x = -2$. Next, we observe that the Rule of Signs tells us that a fraction has the same sign as the product of its numerator and its denominator. Consequently, the "less than" part of the problem can be solved by solving the inequality $(2x - 3)(x + 2) < 0$. Thus, the critical values are $x = -2$ and $x = \frac{3}{2}$, which divide the number line into three intervals as shown in Figure 4.6d. We write $P(x) = (2x - 3)(x + 2)$ and use -3, 0, and 3 as test values to get $P(-3) = (-9)(-1) = +9$, $P(0) = (-3)(2) = -6$, and $P(3) = (3)(5) = +15$. Hence, the interval $x < -2$ is marked with plus signs, $-2 < x < \frac{3}{2}$ with minus signs, and $x > \frac{3}{2}$

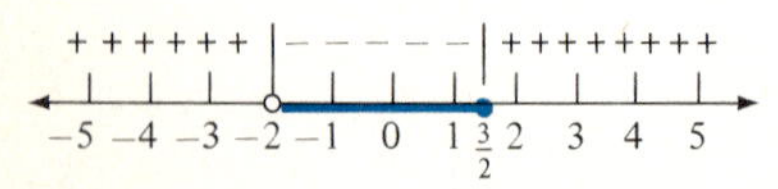

FIGURE 4.6d

with plus signs. The solution of the given inequality is $-2 < x \leq \frac{3}{2}$. Notice carefully that $x = -2$, the point where the fraction is undefined, is not included in the solution.

EXAMPLE 4

Solve:

$$\frac{x}{x-2} \leq 2$$

SOLUTION:

We first subtract 2 from both sides to obtain

$$\frac{x}{x-2} - 2 \leq 0$$

or, by combining the terms on the left,

$$\frac{4-x}{x-2} \leq 0$$

Thus, we have a problem similar to that in Example 3. We note that the fraction on the left is zero for $x = 4$ and is undefined for $x = 2$. To solve the "less than" part of the problem, we write $P(x) = (4 - x)(x - 2)$ and read off the critical values, 4 and 2. The numbers 0, 3, and 5 will do for testing. They give $P(0) = -8$, $P(3) = +1$, and $P(5) = -3$. The plus and minus signs in Figure 4.6e are in agreement with these results. The solution of the original inequality is $x < 2$ or $x \geq 4$.

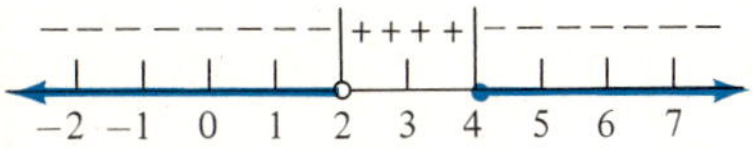

- - - - - - - - |+ + + +| - - - - - - -

−2 −1 0 1 2 3 4 5 6 7

FIGURE 4.6e

In the preceding examples, the quadratic equations we solved to find the critical values all had real solutions. If the equation has no real solutions, then the corresponding quadratic polynomial is either always positive or else always negative; it cannot change sign. Consequently, we can easily find whether the polynomial is positive or negative just by evaluating it for $x = 0$.

The procedure for solving a quadratic inequality is as follows:

PROCEDURE FOR SOLVING A QUADRATIC INEQUALITY

$P(x) = ax^2 + bx + c < 0$

1. **Solve $P(x) = 0$ to find the critical values.**
2. **If there are no real critical values, then evaluate $P(0)$. If this result is positive, the inequality has no solution; if it is negative, the solution is all values of x.**
3. **If the critical values are real, graph them on the number line and choose convenient test values, one in each interval determined by the critical values.**
4. **Evaluate $P(x)$ at the test points and mark plus and minus signs on the intervals to agree with the sign of $P(x)$ at the corresponding test point.**
5. **The solution of the inequality corresponds to the interval(s) where you have marked the minus signs.**

Of course, if the inequality to be solved reads $P(x) > 0$, you would use the intervals where you marked the plus signs.

EXERCISE 4.6

In Problems 1–24, solve the given inequality.

1 $(x + 1)(x - 3) > 0$

2 $(x - 1)(x + 2) < 0$

3 $x(x + 4) \leq 0$

4 $(x - 1)x \geq 0$

5 $x^2 - x - 2 \leq 0$

6 $x^2 - x - 6 \leq 0$

7 $x^2 - 3x \geq 0$

8 $x^2 + 2x \leq 0$

9 $x^2 - 3x + 3 < 0$

10 $x^2 - 2x - 2 > 0$

11 $x^2 + 2x - 4 < 0$

12 $x^2 + x + 1 < 0$

13 $\dfrac{1}{x} \leq 2$

14 $\dfrac{2}{x} \geq 3$

15 $\dfrac{2}{x - 2} \geq 0$

16 $\dfrac{3}{x - 1} \leq 0$

17 $\dfrac{x + 5}{x - 1} > 2$

18 $\dfrac{2x - 3}{x + 3} < 1$

19 $\dfrac{3x - 4}{2x - 1} < 1$

20 $\dfrac{x - 1}{x + 5} > 1$

21 $\dfrac{1}{x - 1} < \dfrac{1}{x + 2}$

22 $\dfrac{1}{x + 1} > \dfrac{1}{x - 2}$

23 $\dfrac{4}{x} + 6 > \dfrac{2}{x} + 2$

24 $\dfrac{3}{x} + 1 < \dfrac{1}{x} - 2$

In Problems 25–28, find all values of x for which the given expression is a real number.

Hint: $\sqrt{a}$ is a real number if $a \geq 0$.

25 $\sqrt{x^2 - 9}$

26 $\sqrt{x^2 - 4x + 4}$

27 $\sqrt{x^2 - 6x + 5}$

28 $\sqrt{3x - 8}$

29 Prove that if $a^2 > b^2$, and a and b are nonnegative numbers, then $a > b$.

30 Prove that $\sqrt{ab} \leq (a + b)/2$ for a and b nonnegative numbers. *Hint:* Start with $(\sqrt{a} - \sqrt{b})^2 \geq 0$.

31 Prove:

$$\frac{x^2 + y^2}{2} \geq \left(\frac{x + y}{2}\right)^2$$

Hint: $(x - y)^2 \geq 0$.

32 Prove that $(a + b)^2 > a^2 + b^2$ for a and b positive numbers.

Hint: $2ab > 0$.

33 A company figures that its profit P (in thousands of dollars) on a certain item is given by

$$P(x) = 2x + 8 - x^2 \qquad x > 0$$

where x is the number of thousands of this item that it makes. For what values of x will the company make a profit?

Hint: To make a profit, $P(x) > 0$.

34 Suppose you want to draw a triangle with one of its sides 12 inches long. Of the other sides, one is to be equal to the square of the other; that is, if one side is of length x inches, the other is of length x^2 inches. What inequality must x satisfy?

Hint: We know from geometry that the sum of the lengths of two sides of a triangle is greater than the length of the third side and the difference of the lengths of two sides is less than the length of the third side.

USING YOUR KNOWLEDGE 4.6

If v is the speed of a car in miles per hour, it is known by experiment that the stopping distance d, in feet, is given by

$$d = 0.044v^2 + 1.1v$$

This formula allows for the driver's reaction time, that is, the time it takes the driver to step on the brake.

C **1** Find the inequality that v would have to satisfy if $50 \le d < 60$.
C **2** Do the same for $80 \le d < 90$.
C **3** Do the same for $160 \le d < 170$.
C **4** Do the same for $190 \le d < 200$.

4.7 APPLICATIONS: WORD PROBLEMS

Do you know how the velocity of the wind is measured? One way is to use a wind pressure gauge. It is known that the pressure, in pounds per square foot, caused by a wind whose velocity is measured in miles per hour (mph) is found by multiplying the square of the wind's velocity by 0.003. If a wind pressure gauge registers 120 pounds per square foot during a gale, can we find the wind velocity?

The problem just stated is an example of a *word* or *story* problem. This kind of problem requires us to represent verbal sentences symbolically, that is, to translate word sentences into equations that correspond to the stated problem and then to solve these equations. The following procedure is often helpful in solving word problems.

PROCEDURE FOR SOLVING WORD PROBLEMS

1. **Read the problem carefully and decide what number is asked for.**
2. **Represent the number asked for by a letter and draw a picture (if possible) showing all the unknown and the known quantities.**
3. **Write an equation expressing the equality stated or implied by the problem.**
4. **Solve the equation you obtained in step 3.**
5. **Check your answer by making sure that it satisfies the requirements of the original problem.**

We will now use this procedure to solve the wind velocity problem.

1. After reading the problem carefully, we see that it is asking for the velocity of the wind.
2. Let the velocity of the wind be represented by v mph.
3. According to the problem:
4. The pressure caused by the wind | is found | by multiplying the square of the wind velocity by 0.003.

$$\underbrace{\text{The pressure caused by the wind}}_{120} \quad \underbrace{\text{is found}}_{=} \quad \underbrace{\text{by multiplying the square of the wind velocity by 0.003.}}_{0.003v^2}$$

We rewrite and solve this equation as follows:

$$0.003v^2 = 120$$

$$v^2 = \frac{120}{0.003} = 40{,}000$$

$$v = 200$$

Thus, the wind's velocity is **200** mph.

5. Because $(0.003)(\mathbf{200})^2 = 120$, which agrees with the requirements of the problem, our result is correct.

EXAMPLE 1

The world's strongest current is the Saltstraumen in Norway, which reaches a speed of 18 mph (Guinness). A speedboat can travel 36 miles downstream in the Saltstraumen in the same time that it takes the boat to go 12 miles upstream. What is the speed of the boat in still water? (Assume that the current stays steady at 18 mph.)

SOLUTION:

1. We are asked for the speed of the boat in still water.
2. Let this speed be R mph.
3. Going downstream, the boat's speed is increased by the speed of the current. Hence, the speed of the boat relative to the shore is $R + 18$ mph. Similarly, going upstream, the boat's speed is decreased by the speed of the current, so that, relative to the shore, the speed is $R - 18$. Because the time is the distance divided by the rate and the time is the same for both trips, we have the equation

$$\frac{36}{R + 18} = \frac{12}{R - 18}$$

4. To solve this equation, we first divide both sides by 12 and then multiply by the LCD $(R + 18)(R - 18)$ to obtain

$$3(R - 18) = R + 18$$

$$2R = 72$$

$$R = 36$$

Hence, the speed of the boat in still water is 36 mph.

5. Verification of this answer is left to you.

EXAMPLE 2

A woman had \$8000 to invest. She invested part at 10 percent and the rest at 12 percent. If her annual income from these investments totals \$840, how much does she have invested at each rate?

SOLUTION:

1. We are asked for the amount of money the woman has invested at each rate.
2. We let y be the number of dollars she has invested at 10 percent. The rest, $8000 - y$, is the amount she has invested at 12 percent.

We know that when P dollars are invested at a rate r, the annual interest I is given by $I = Pr$. It is helpful to display the information given in the problem in a chart such as the accompanying one.

r	P	$\times$ r	$=$ I
10%	y	0.10	$0.10y$
12%	$8000 - y$	0.12	$0.12(8000 - y)$

3. Because the total income from the two investments is \$840, we must have

$$0.10y + 0.12(8000 - y) = 840$$

4. Solving, we obtain

$$-0.02y = -120$$
$$y = 6000$$

Thus, the woman has invested \$6000 at 10 percent and \$2000 at 12 percent.

5. Because 10 percent of \$6000 is \$600, and 12 percent of \$2000 is \$240, the interest from the two investments does add up to \$840, as was required.

The following "mixture" problem can be handled in much the same way as the investment problem in Example 2. Again, a chart is helpful in organizing the solution of the problem.

EXAMPLE 3

How many liters of a 60-percent acid solution must be added to 2 liters of water to make a 40-percent solution?

SOLUTION:

1. We are asked how many liters of 60-percent solution to add to 2 liters of water.
2. Let x be the number of liters to be added. Then there will $x + 2$ liters of the new solution.

We chart the information in the problem as in the accompanying table.

Solution	Concentration ×	Liters of solution =	Liters of acid
60%	0.60	x	$0.60x$
40%	0.40	$x + 2$	$0.40(x + 2)$

3. Because the total amount of acid has not been changed, we have the equation

$$0.60x = 0.40(x + 2)$$

4. Solving, we obtain

$$0.2x = 0.8$$
$$x = 4$$

Thus, 4 liters of the 60-percent solution must be added to the 2 liters of water.

5. The amount of acid in the added 4 liters is (0.60)(4) or 2.4 liters. There are 6 liters of solution in the final mixture, and $2.4 \div 6 = 0.40$; that is, the final mixture is 40 percent acid as required. Therefore, the answer is correct.

EXAMPLE 4

A square box without a top is to be made by cutting a 2-inch square out of each corner of a square sheet of metal and folding up the sides. If the box is to hold 338 cubic inches, what must be the dimensions of the original square?

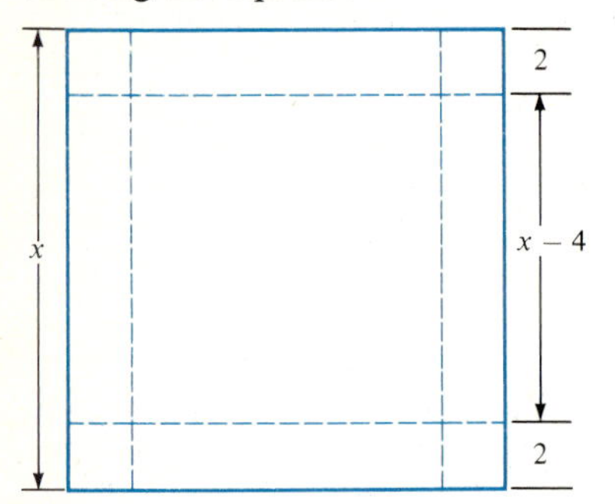

FIGURE 4.7a

SOLUTION:

1. We are asked for the dimensions of the original square.
2. Let x inches be the length of the side of the original square. Because a 2-inch square is to be cut from each corner, we make a diagram to show this as in Figure 4.7a. The volume of the box is obtained by multiplying the height by the length by the width, which in terms of x is

$$2(x-4)(x-4)$$

3. Because the volume of the box is to be 338 cubic inches, we have the equation

$$2(x-4)^2 = 338$$

or

$$(x-4)^2 = 169$$

4. Extracting roots and solving, we find $x = 17$ or $x = -9$. Because -9 is meaningless in this problem, we see that the original square must be 17 inches on a side.
5. You should verify that this result is correct.

EXAMPLE 5

If two parallel sides of a square are each increased by 2 meters and the other two sides are each increased by 4 meters, the area of the rectangle so formed exceeds twice the area of the original square by 12 square meters. Find the side of the original square.

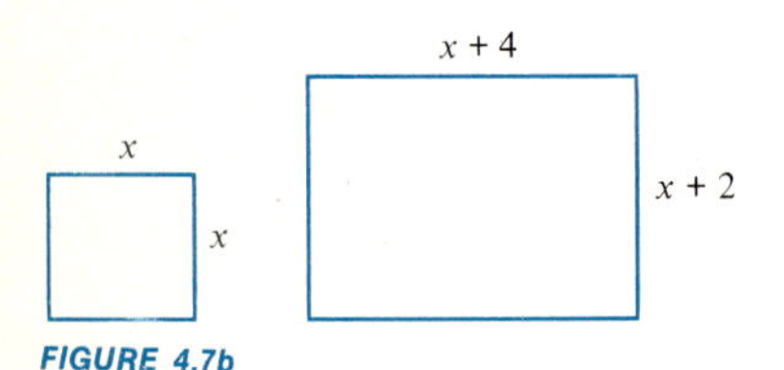

FIGURE 4.7b

SOLUTION:

1. We are asked for the side of the original square.
2. Let x meters be the length of the side of this square. Then the rectangle will have sides $x + 2$ meters and $x + 4$ meters.
3. The diagram in Figure 4.7b shows the square and the rectangle. The area of the square is x^2 square meters and the area of the rectangle is $(x + 2)(x + 4)$ square meters. Because the area of the rectangle exceeds twice the area of the square by 12 square meters, we can write the equation

$$(x+2)(x+4) = 2x^2 + 12$$

4. Simplifying this equation, we obtain

$$x^2 - 6x + 4 = 0$$

Now we use the quadratic formula to obtain

$$x = \frac{6 \pm \sqrt{20}}{2} = 3 \pm \sqrt{5}$$

Thus, the side of the square can be either $3 + \sqrt{5} \approx 5.24$ meters or $3 - \sqrt{5} \approx 0.764$ meter.

5. You can show that both solutions satisfy the conditions of the problem.

EXERCISE 4.7

1 The sum of three consecutive integers n, $n + 1$, and $n + 2$ is 17 less than four times the smallest of the three integers. Find the integers.

2 Three consecutive, positive, odd integers are such that the product of the larger two exceeds twice the square of the smallest by 1. Find the integers.

Hint: The difference between two consecutive odd integers is 2.

3 The denominator of a fraction exceeds twice the numerator by 2. If the numerator is increased by 10 and the denominator by 24, the value of the fraction is unaltered. Find the fraction.

4 Find two numbers whose sum is 190 such that the smaller is three sevenths of the larger.

5 The height of the Sears Tower and its antenna is 1800 feet. If the building is 1108 feet taller than the antenna, find the height of each.

6 A small plane, flying at top speed, goes 240 miles against the wind in the same time it takes to go 360 miles with the wind. If the wind velocity is constantly 30 mph, find the plane's top speed in still air.

7 Two trains, whose speeds differ by 15 mph, travel in oppositie directions. If the trains are 460 miles apart at the end of 4 hours, what is the speed of each?

8 A train leaves a station traveling at 65 mph. Two hours later, a second train leaves in the same direction but traveling at 75 mph. How long will it take the faster train to catch the slower one?

9 A man invested $15,000 in two types of stocks. Stock A earns 4.5 percent annually, whereas stock B earns 5.5 percent annually. If his annual income from the two stocks is $745, find how much he has invested in each stock.

10 Two sums of money totaling $25,000 were invested during a depression and earned 2.5 percent and 3.5 percent per year. Find the two amounts if together they earned $660 per year.

11 The amount of annual interest earned by $8000 is $200 less than that earned by $12,000 at $\frac{1}{2}$ percent less interest per year. What is the rate of interest on each amount?

12 A woman invested $2400 at an annual interest rate of 10 percent. What additional amount must she invest at 15 percent to obtain an overall annual return of 11 percent?

13 How many liters of a 10-percent salt solution must be added to 15 liters of a 20-percent solution to obtain a 16-percent solution?

14 How many ounces of regular vodka (40 percent alcohol) must be added to 30 ounces of Poland White Spirit vodka (80 percent alcohol) to obtain vodka that is 70 percent alcohol?

15 How many ounces of vermouth containing 10 percent alcohol should be added to 10 ounces of gin containing 40 percent alcohol so that the resulting pitcher of martinis will contain 30 percent alcohol?

16 How many parts of glacial acetic acid (99.5 percent acid) must a photographer add to 100 parts of a 10-percent solution of acetic acid to get a 28-percent solution?

17 How many pounds of coffee worth $3.50 per pound should be mixed with coffee worth $4.50 per pound to make 50 pounds of a blend worth $4.25 per pound?

18 The most expensive tea in the world is Oolong tea, recently retailing for $13 per pound. How many pounds of this Oolong tea should be added to 50 pounds of regular tea selling at $4 per pound to obtain a mixture selling for $7 per pound?

19 A rectangular box without a top is to be made by cutting a 2-inch square from each corner of a rectangular sheet of metal and folding up the sides. If the sheet of metal is to be twice as long as it is wide and the box is to have a volume of 320 cubic inches, what must be the dimensions of the sheet of metal?

20 Weighty Scales wishes to manufacture a bathroom scale whose top is made by cutting a 2-inch square from each corner of a rectangular sheet of metal and bending down the sides to form an inverted tray. If the face of the scale is to be 2 inches longer than it is wide and is to have an area of 255 square inches, what must be the dimensions of the sheet of metal?

21 Repeat Problem 20 if the face area is to be 195 square inches.

22 One leg of an isosceles right triangle (equal legs) is lengthened by 8 centimeters, and the other leg is shortened by 8 centimeters to make a new right triangle. If the hypotenuse of the new triangle is $2\sqrt{2}$ centimeters longer than the hypotenuse of the original triangle, how long was the leg of the original triangle?

23 A pool can be filled by an intake pipe in 4 hours and can be emptied by a drain pipe in 6 hours. How long would it take to fill the pool with both pipes open?

Hint: If T is the number of hours needed to fill the pool, then $T/4$ is the fraction of the pool filled in T hours by the intake pipe, and $T/6$ is the fraction emptied by the drain pipe in T hours. Because both pipes are open and one tank is to be filled, you must take the difference between $T/4$ and $T/6$ to be 1.

24 A faucet can fill a tank in 12 hours and a drain pipe can empty it in 18 hours. If the faucet and the drain pipe were both open, how long would it take to fill the tank?

Note: See Problem 23.

25 A can do a job in 6 days and B can do it in 8 days. How long would it take to do the job if both work on it together?

Note: See Problem 23.

26 A pipe can fill a tank in 9 hours, whereas a drain pipe can empty it in 7 hours. If the tank were full and both pipes were opened, how long would it take to empty the tank?

Note: See Problem 23.

27 A company finds that the cost of production of a certain item is $C = 4x$, where x is the number of thousands of this item produced. If the revenue R is given by $R = 4x^2 - 9x + 3$ for $x > 1$, for how many thousands of items produced will the company make a gross profit; that is, for how many thousands is $R > C$?

28 The cost of producing x thousands of an item is $C = 5x$. If the revenue is $R = 2x^2 - 6x + 5$ for $x \geq 2$, for how many thousands will the company make a gross profit on this item?

Note: See Problems 27.

29 A company sells its products for $6000 per unit. If the cost (in thousands of dollars) of producing x units is given by $3x^2 - 8x + 8$ for $x \geq 2$, for how many units produced will the company make a profit?

Note: See Problem 27.

30 The height h (feet) reached by a ball t seconds after it is thrown upward with an initial velocity of 96 feet per second is given by $h = -16t^2 + 96t$. During what interval of time will the ball be more than 80 feet above the ground?

31 A student in a mathematics class needs at least 360 points to obtain an A grade. If her scores on the first three tests were 85, 96, and 93, what range of scores on the last test would give her an A for the course?

32 A student must have an average of 70–79 percent on five tests in a course to receive a C grade. If his grades on the first four tests were 93, 62, 71, and 54 percent, what range of grades on the last test would give him a C for the course?

USING YOUR KNOWLEDGE 4.7

Let's see how good a detective you are. There is an island 8 miles due east of a point A on a strip of straight beach that runs due north and south. A murder was committed at a point P 5 miles due south of A. The coroner is able to determine that the crime occurred between 4:00 and 7:00 P.M. on September 1.

One of the prime suspects is Harry the Horse. Harry lives 11 miles due south of A, and he claims that he left his home at 4:30 P.M. on September 1 and jogged north on the beach to a point B, south of

P; just how far from P, he does not know. At B he was picked up by a friend in a motorboat and was taken directly to the island, which they reached at 6:30 P.M. Harry claims that he jogs at the rate of 5 mph and that the motor boat traveled at the rate of 10 mph.

Suppose that the times and the rates are all verified. Do you believe that Harry the Horse could have been at the scene of the crime? Make a diagram, locate the point B where Harry says he was picked up, and decide.

SELF TEST

1 Solve:

a $3x - 4 = 5x - 7$

b $\frac{3}{4} + \frac{x}{6} = x - \frac{7}{4}$

2 Solve:

a $\frac{5}{x+1} = 1 + \frac{3}{x+1}$

b $\frac{x}{x+1} = 1 + \frac{2}{x^2-1}$

3 a Solve for r: $A = P(1 + rt)$

b Solve for a: $S = \frac{n}{2}(a + b)$

4 Solve:
$(2y + 3)(6y - 1) = (3y + 2)(4y + 3)$

5 Solve:

a $-3x + 5 > x + 7$

b $\frac{1-2x}{2} \le \frac{2}{3}$

6 Find the graph the solution:
$-3 < -2x + 1 \le 7$

7 A car rental agency charges $16 per day plus 12 cents per mile. Suppose a person rents a car for one day and drives x miles. Write an inequality that indicates how many miles can be traveled to keep the total cost under $64.

8 Solve:

a $|4x - 1| = 3$

b $|2x + 1| = 2 + 6x$

9 Solve and graph the solution:
a $|3x + 1| < 5$ **b** $|3x + 1| \ge 2$

10 Solve by factoring:

a $3x^2 + 2x - 1 = 0$

b $6x^2 - 10 = 11x$

11 Solve:
a $5x^2 - 36 = 0$ **b** $3(x + 2)^2 = 6$

12 Solve:
a $3x^2 + 7x = 5$ **b** $\frac{x^2}{3} + \frac{7}{12} = \frac{x}{3}$

13 a Find the condition on k for the equation $kx^2 + 2x + 2 = 0$ to have no real roots.

b Without solving the equation, show that the roots are rational if $k = \frac{7}{32}$.

14 Solve:
$\sqrt{x + 8} = \sqrt{x} + 2$

15 Solve
$(x^2 - 1)^2 - 5(x^2 - 1) + 6 = 0$

16 Solve by substitution:
$3\sqrt{x^2 - 9} = x^2 - 13$

17 Solve:
a $x^2 + x - 6 \ge 0$ **b** $\frac{1}{x} \le \frac{1}{x-3}$

18 A small plane traveling at its top speed goes 220 miles with a tail wind in the same time that it takes to go 180 miles against a head wind. If the wind speed is 15 mph in both cases, what is the plane's top speed in still air?

19 A photographer wants to mix a 10-percent acetic acid solution with a 50-percent acetic acid solution to make 5 liters of a 28-percent solution. How many liters of each of the 10-percent and the 50-percent solutions should he use?

20 The cost (in thousands of dollars) of producing x thousands of an item is $6x$, for $x > 1$. The revenue (in thousands of dollars) is $R = 2x^2 - 5x + 5$. For how many units produced will the company make a gross profit?

CHAPTER 5

RELATIONS, FUNCTIONS, AND GRAPHS

5.1 ORDERED PAIRS AND THE DISTANCE FORMULA

The map in Figure 5.1a shows the position of hurricane Bob on July 11, 1979. As you can see, the storm's center was near 92° West longitude and 25.5° North latitude. (This means 92° west of Greenwich, England, and 25.5° north of the equator.) The numbers **92** and **25.5** describe the position of the hurricane if west longitude and north latitude are understood. For this reason, the numbers 92 and 25.5 are called the *coordinates* of the hurricane. If we agree to use an *ordered pair* of numbers, naming the longitude first and the latitude second, then we can write (**92**, **25.5**) for the coordinates of

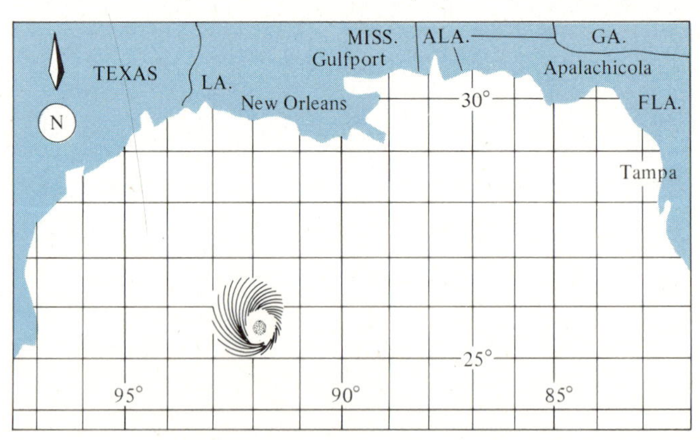

FIGURE 5.1a

the hurricane. Similarly, the city of New Orleans has the coordinates (90, 30).

In mathematics, we use a geometric picture similar to that of a map to represent *ordered pairs* of numbers. Such an ordered pair is denoted by (x, y), where the letter x stands for a real number called the *x-coordinate* or *abscissa*, and y stands for a real number called the *y-coordinate* or *ordinate*. Thus, in the ordered pair $(4, \sqrt{5})$, the 4 is the x-coordinate or the abscissa, and the $\sqrt{5}$ is the y-coordinate or the ordinate.

THE CARTESIAN COORDINATE SYSTEM

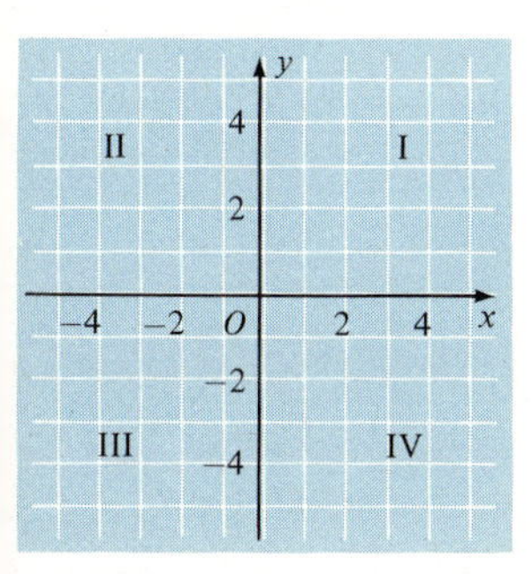

FIGURE 5.1b

In order to represent ordered pairs of numbers on a plane, we first draw two perpendicular lines, called the *x-axis* and the *y-axis*, intersecting at a point O, called the *origin*. (See Figure 5.1b.) As in this figure, the x-axis is usually taken as a horizontal line and the y-axis as a vertical line. The two axes divide the plane into four regions, called *quadrants*, which are numbered I, II, III, IV, in the counterclockwise direction, as in Figure 5.1b. We use convenient units of length and make each axis into a number line with its zero point at the origin. Notice in the figure that the positive direction on the x-axis is *to the right*, and on the y-axis it is *upward*. The configuration thus obtained is called a *Cartesian coordinate system*, a *rectangular coordinate system*, or simply a *coordinate plane*.

For the ordered pair (x, y), we agree that the x-coordinate denotes the *directed distance* of the point (x, y) from the y-axis, *to the right* if x is *positive* and *to the left* if x is *negative*. Similarly, the y-coordinate denotes the *directed distance* from the x-axis, *upward* if y is *positive* and *downward* if y is *negative*.

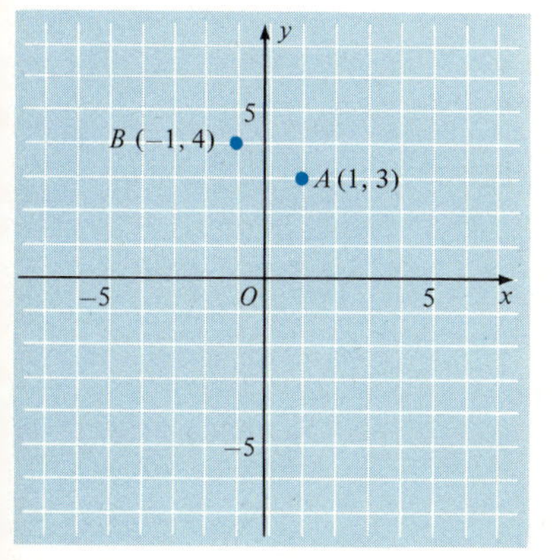

FIGURE 5.1c

The preceding conventions establish a one-to-one correspondence between the set of all ordered pairs of real numbers and the set of all points in the plane; that is, to each ordered pair (x, y) there corresponds a unique point P in the plane, and vice versa.

It is customary to refer to the point corresponding to the ordered pair (x, y) as "the point (x, y)" or "the graph of (x, y)." The notation $P(x, y)$ is often used for the point P with coordinates x and y. In Figure 5.1c, for instance, $A(1, 3)$ is the graph of the point A whose x-coordinate is 1 and whose y-coordinate is 3. Similarly, $B(-1, 4)$ is the graph of the point B with coordinates -1 and 4.

EXAMPLE 1

Graph the points $C(2, 4)$, $D(-2, 1)$, $E(-3, -2)$, $F(2, -3)$, and $G(0, -4)$.

SOLUTION:

The graphs of these points are shown in Figure 5.1d. Note that the points C, D, E, and F are in Quadrants I, II, III, and IV, respectively, but G is on the y-axis and not in any quadrant.

THE DISTANCE FORMULA

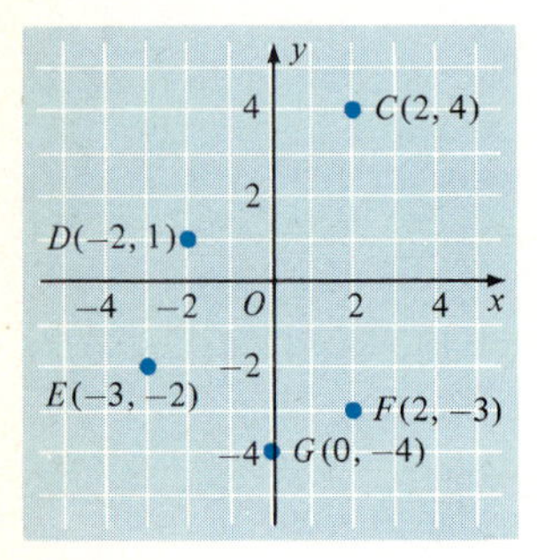

FIGURE 5.1d

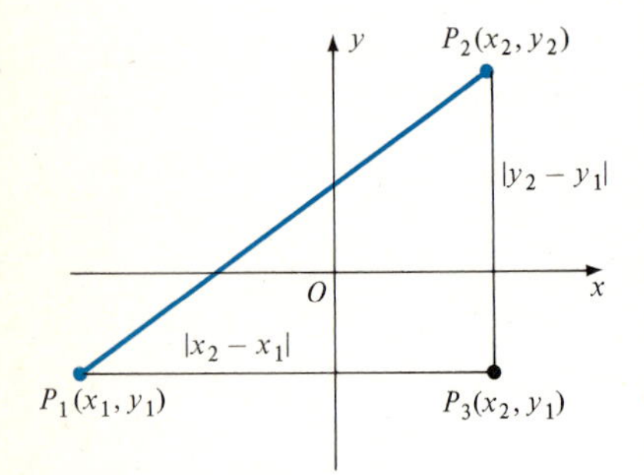

FIGURE 5.1e

Suppose that we wish to know the distance between the points $C(2, 4)$ and $E(-3, -2)$ in Figure 5.1d. To calculate this distance, it is convenient to have a formula for the distance, say $d(P_1, P_2)$, between any two points $P_1(x_1, y_1)$ and $P_2(x_2, y_2)$. First, suppose that P_1 and P_2 are any two points *not* on the same horizontal or vertical line, as shown in Figure 5.1e. The line P_1P_3 is drawn parallel to the x-axis, and P_2P_3 is drawn parallel to the y-axis. Thus, these two lines are the vertices of a right triangle $P_1P_2P_3$ with right angle at P_3. Because P_1 and P_3 are on the same horizontal line, they must have the same y-coordinate, y_1. Similarly, P_2 and P_3 are on the same vertical line and thus have the same x-coordinate, x_2. Hence, P_3 must be the point with coordinates (x_2, y_1).

The undirected distance between P_1 and P_3 is $|x_2 - x_1|$ (see Section 4.3), and the undirected distance between P_2 and P_3 is $|y_2 - y_1|$. We now use the Pythagorean theorem, which says that *the square of the length of the hypotenuse of a right triangle equals the sum of the squares of the lengths of the other two sides*. This gives

$$[d(P_1, P_2)]^2 = |x_2 - x_1|^2 + |y_2 - y_1|^2$$

Because $|x_2 - x_1|^2 = (x_2 - x_1)^2$ and $|y_2 - y_1|^2 = (y_2 - y_1)^2$, we may extract the square root and write

$$\mathbf{d(P_1, P_2) = \sqrt{(x_2 - x_1)^2 + (y_2 - y_1)^2}} \qquad \textbf{(1)}$$

Because the numbers $x_2 - x_1$ and $y_2 - y_1$ are squared in Formula 1, the order in which the given points are taken does not matter. Moreover, if the points P_1 and P_2 are on the same horizontal line, then $y_1 = y_2$ and

$$\begin{aligned} d(P_1, P_2) &= \sqrt{(x_2 - x_1)^2 + 0} \\ &= \sqrt{(x_2 - x_1)^2} \\ &= |x_2 - x_1| \end{aligned}$$

which agrees with our result in Section 4.3. A similar argument holds if P_1 and P_2 are on the same vertical line; the distance then is

$$d(P_1, P_2) = |y_2 - y_1|$$

EXAMPLE 2

Find the distance between the two given points:

a. (2, 3) and (−2, 5)
b. (−2, 1) and (−2, 3)

SOLUTION:

a. By Formula 1,

$$d = \sqrt{(-2-2)^2 + (5-3)^2}$$
$$= \sqrt{(-4)^2 + (2)^2}$$
$$= \sqrt{16+4} = \sqrt{20} = 2\sqrt{5}$$

b. By Formula 1,

$$d = \sqrt{[-2-(-2)]^2 + (3-1)^2}$$
$$= \sqrt{(-2+2)^2 + 2^2}$$
$$= \sqrt{0^2+4} = \sqrt{4} = 2$$

(Note that because these two points lie on a vertical line, their distance is $3 - 1 = 2$ units).

EXAMPLE 3

Use the distance formula to show that $A(-3, -6)$, $B(5, 0)$, and $C(1, 2)$ are the vertices of a right triangle.

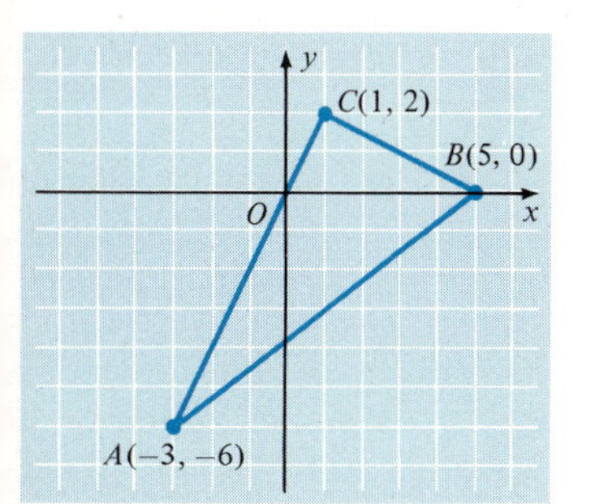

FIGURE 5.1f

SOLUTION:

A triangle is a right triangle if the square of one side equals the sum of the squares of the other two sides. Hence, we compute the squares of the lengths of the sides of the triangle ABC.

$$[d(A, B)]^2 = (-3-5)^2 + (-6-0)^2 = 64 + 36 = 100$$
$$[d(B, C)]^2 = (1-5)^2 + (2-0)^2 = 16 + 4 = 20$$
$$[d(A, C)]^2 = (-3-1)^2 + (-6-2)^2 = 16 + 64 = 80$$

Because $100 = 20 + 80$, $[d(A, B)]^2 = [d(B, C)]^2 + [d(A, C)]^2$, and the triangle ABC is a right triangle with AB as its hypotenuse. (See Figure 5.1f.)

THE CIRCLE

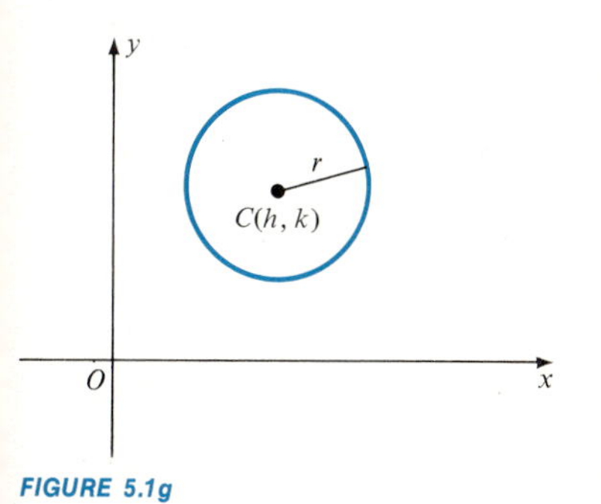

FIGURE 5.1g

We can use the distance formula to find the equation of the circle of radius r with center at $C(h, k)$ as in Figure 5.1g. Because every point on the circle is r units from the center, the distance from any point $P(x, y)$ on the circle to the center $C(h, k)$ is r. Hence, by the distance formula,

$$\sqrt{(x-h)^2 + (y-k)^2} = r$$

and, by squaring both sides, we obtain

$$(x-h)^2 + (y-k)^2 = r^2 \qquad (2)$$

Equation 2 is called the *standard form* of the equation of the circle. Notice that the equation displays the center $C(h, k)$ and the radius r.

EXAMPLE 4

Find the center and the radius of the circle represented by the equation $(x-1)^2+(y+2)^2=9$.

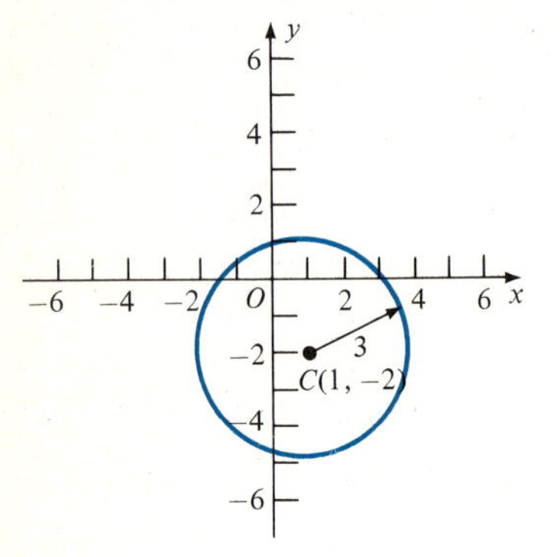

FIGURE 5.1h

SOLUTION:

By comparing the given equation with the standard equation

$$(x-h)^2+(y-k)^2=r^2$$

we see that $h=1$, $k=-2$, and $r=3$. Thus, the given equation represents a circle with radius 3 and center at $(1, -2)$. See Figure 5.1h.

THE MIDPOINT FORMULAS

It is frequently useful to have formulas for the coordinates of the midpoint of a line segment in terms of the coordinates of the end points. These formulas are given in Theorem 5.1a; the proof is left as an exercise. Notice that each coordinate is the simple average of the corresponding coordinates of the end points.

▼ THEOREM 5.1a

The midpoint formulas

The coordinates of the midpoint of the line segment from $P_1(x_1, y_1)$ to $P_2(x_2, y_2)$ are given by

$$x_0=\frac{x_1+x_2}{2} \qquad y_0=\frac{y_1+y_2}{2}$$

EXAMPLE 5

Find the midpoint of the line segment from $A(2, -4)$ to $B(-8, 9)$.

SOLUTION:

By the midpoint formulas, we obtain

$$x_0=\frac{2+(-8)}{2}=-3 \quad \text{and} \quad y_0=\frac{-4+9}{2}=\frac{5}{2}$$

Thus, the midpoint is the point $(-3, \frac{5}{2})$.

EXERCISE 5.1

1 Graph the given ordered pairs and state in which quadrant (if any) each point lies.
a $(-3, \frac{5}{2})$ **b** $(0, -3)$ **c** $(4, -\frac{2}{3})$ **d** $(-5, -\frac{1}{2})$

2 Graph the given ordered pairs and state in which quadrant (if any) each point lies.
a $(\frac{1}{3}, -2)$ **b** $(-\frac{1}{2}, -2)$ **c** $(-3, 0)$ **d** $(-\frac{7}{2}, -\frac{1}{2})$

In Problems 3–12, find (a) the distance between the points A and B; (b) the midpoint of the line segment AB.

3 $A(2, 2)$ and $B(6, 5)$

4 $A(-1, 2)$ and $B(-1, 4)$

5 $A(3, 5)$ and $B(0, 1)$

6 $A(-3, -4)$ and $B(0, 4)$

7 $A(-2, 3)$ and $B(4, 3)$

8 $A(-2, 5)$ and $B(-3, -4)$

9 $A(-2, -2)$ and $B(6, -4)$ **10** $A(7, 2)$ and $B(7, 8)$ **11** $A(-4, 5)$ and $B(-4, -3)$

12 $A(\frac{1}{2}, -\frac{1}{2})$ and $B(1, 1)$

In Problems 13–16, use the distance formula to find the equation of the circle with the given center and radius.

13 $C(2, 5)$, $r = 3$ **14** $C(-1, 3)$, $r = 2$ **15** $C(-3, -4)$, $r = 1$

16 $C(4, -2)$, $r = \frac{1}{2}$

In Problems 17–20, find the radius and the coordinates of the center of the circle represented by the given equation.

17 $(x-2)^2 + (y-4)^2 = 36$ **18** $(x+1)^2 + (y-2)^2 = 4$ **19** $(x+1)^2 + (y+3)^2 = 16$

20 $(x+\frac{1}{2})^2 + (y+\frac{1}{3})^2 = \frac{1}{25}$

In Problems 21–24, determine if the given points are the vertices of a right triangle. State whether the triangle is isosceles (two equal sides) or scalene (no equal sides).

21 $A(2, 2)$, $B(0, 5)$, $C(-20, 12)$ **22** $A(2, 2)$, $B(0, 5)$, $C(-19, -12)$ **23** $A(5, 5)$, $B(-1, -11)$, $C(-17, -5)$

24 $A(\frac{1}{2}, 0)$, $B(-\frac{1}{2}, 2)$, $C(3, \frac{5}{4})$

In Problems 25–28, determine if the given points are collinear.

Hint: Three points are collinear if they can be arranged in an order A, B, C so that $d(A, B) + d(B, C) = d(A, C)$.

25 $(0, 5)$, $(4, 1)$, $(12, -7)$ **26** $(-6, -1)$, $(-1, -3)$, $(-2, -5)$ **27** $(1, -1)$, $(3, 1)$, $(-2, -4)$

28 $(5, -3)$, $(\frac{5}{2}, 2)$, $(2, 3)$

In each of Problems 29–32, the given pair of points determines the hypotenuse (longest side) of a right triangle. Suppose each of the other two sides is parallel, one to either of the coordinate axes. Graph the given points and find the coordinates of the third vertex, say C (two answers). Find the lengths of the other two sides.

29 $A(2, 4)$, $B(7, 8)$ **30** $A(-2, -1)$, $B(2, 0)$ **31** $A(0, -4)$, $B(3, 1)$

32 $A(-5, -1)$, $B(-2, -5)$

33 The distance between the points $(x, -3)$ and $(-2, -6)$ is 5 units. Find x. (There are two answers.)

34 The distance between the points $(-2, 1)$ and $(-2, y)$ is 2 units. Find y. (There are two answers.)

35 Find an equation that must be satisfied by a point $P(x, y)$ that is equidistant from the points $A(1, 0)$ and $B(4, 3)$.

36 Consider the line segment joining the points $P_1(x_1, y_1)$ and $P_2(x_2, y_2)$ with midpoint $M(x, y)$ as shown in Figure 5.1i. As you can see, Q has coordinates (x, y_1), and R has coordinates (x_2, y). Also the directed distance P_1Q is equal to the directed distance MR, and the directed distance QM is equal to the directed distance RP_2. Use this information to prove Theorem 5.1a.

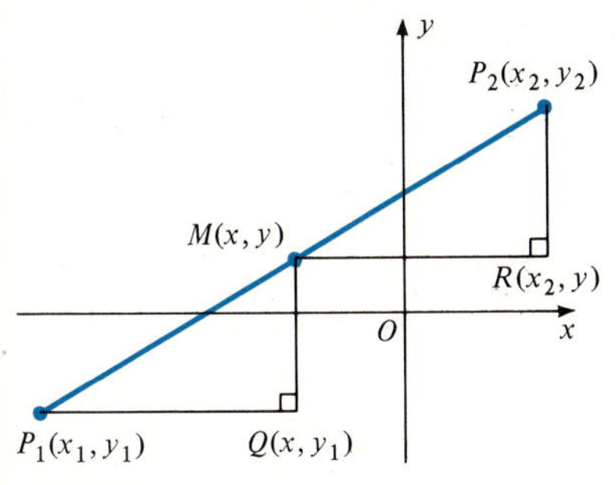

FIGURE 5.1i

Use your calculator in Problems 37–42.

C **37** Find the distance between the two points (2.6, 3.5) and (−5.6, 8.4). Answer to the nearest tenth of a unit.

C **38** Find the distance between the points (−5.63, 4.72) and (−25.69, 15.28). Answer to the nearest hundredth of a unit.

C **39** Determine (within the limits of accuracy of your calculator) whether or not the points (0, 2.13), (1.08, 7.53), and (−3.52, −15.47) lie on a straight line.

C **40** Find, to the nearest tenth of a unit, the radius of the circle with center at (1, −2) and passing through the point (13.5, 15.7).

C **41** If you refer to the map in Figure 5.1a at the beginning of this section, you will see that the coordinates of New Orleans are (90, 30) and the coordinates of Hurricane Bob are about (92, 25.5). Assume these are ordinary rectangular coordinates and find how far Bob was from New Orleans. Use the fact that 1° is about 69 miles to find the actual approximate distance in miles.

C **42** The coordinates of Tampa, Florida, are about (82.2, 28). Refer to Problem 41 and calculate about how far Bob was from Tampa.

USING YOUR KNOWLEDGE 5.1

The Soviet Union possesses the largest telescope in the world, a 6-meter reflector-type instrument that was opened for viewing in 1976 on Mt. Pastukhov in the Caucasus. The surface of the mirror in this kind of telescope is a special surface that reflects the parallel rays of light from a star into a single point. (See Using Your Knowledge 6.3 for a further discussion of this kind of surface.)

The first step in forming the required surface is to grind a spherical cavity out of a blank of the proper material. The surface of this cavity can then be further ground to give the required shape. An amateur can grind a small, concave, spherical mirror from a slab of glass. It takes about 50 hours to do this. Suppose that it is desired to grind a 6-inch spherical mirror that is 1 inch deep. Can we find the radius of the corresponding sphere? Yes. Here is one way to do it: Consider a cross section through the center of the mirror and the center of the sphere. Choose a rectangular coordinate system with the x-axis along the top of the circular arc (see Figure 5.1j) and the y-axis as the axis of symmetry. The center of the circle is then some point on the y-axis, say $(0, k)$, and the equation of the circle is

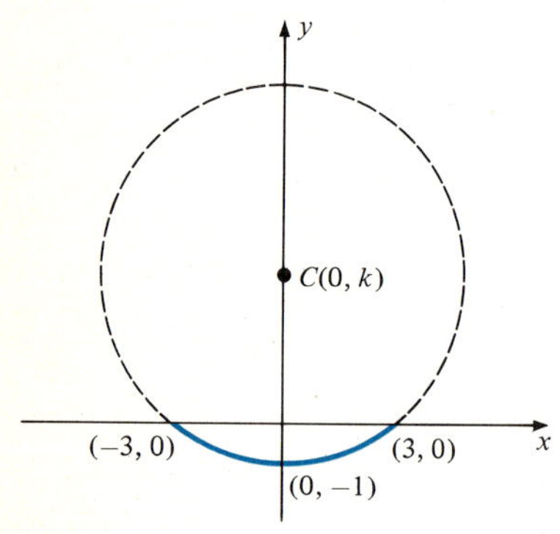

FIGURE 5.1j

$$x^2 + (y - k)^2 = r^2 \tag{1}$$

where r is the desired radius. Now we know that the circle passes through (3, 0) and through (0, −1), so that these two points must satisfy Equation 1. Substituting (3, 0) into the equation gives

$$9 + k^2 = r^2 \tag{2}$$

and substituting (0, −1) gives

$$(-1 - k)^2 = r^2 \tag{3}$$

Equating the two values of r^2 from 2 and 3 yields the equation

$$9 + k^2 = (-1 - k)^2$$

which we can solve to obtain $k = 4$. If we substitute $k = 4$ into Equation 3, we find $r = 5$. (The second solution, $r = -5$, is meaningless for our problem.) Thus, the desired radius is 5 inches.

1 Suppose that a reflector is to be made from a 10-inch spherical mirror with a depth of 1 inch. Find the radius of the sphere.

2 Suppose that the reflector is to be made from a 2-meter spherical mirror that is 10 centimeters deep. Find the radius of the sphere. (A meter is 100 centimeters.)

5.2 RELATIONS AND THEIR GRAPHS

Figure 5.2a suggests that there is a relationship between temperature and altitude. The exact relationship is indicated by the curve (graph) shown. If we agree to write the temperature as the first coordinate and the altitude (height) as the second coordinate, we can express the fact that the temperature at the top of Mt. Everest, which is about 11 kilometers high, is approximately $-56°C$ by using the ordered pair $(-56, 11)$. As you can see, each of the points on the curve can be identified by an ordered pair of numbers. Thus, we may regard the relationship between the temperature and the altitude to be expressed by a set of ordered pairs.

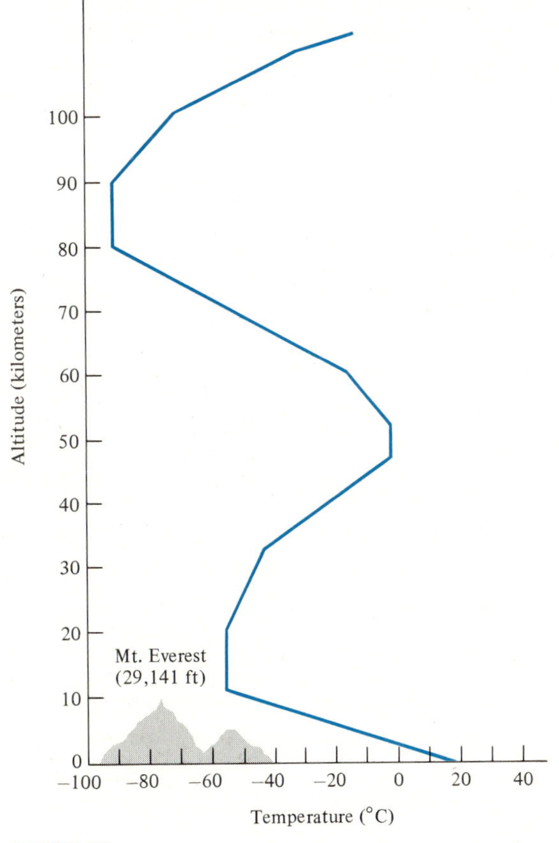

FIGURE 5.2a

RELATIONS

In mathematics, a given set of ordered pairs is called a relation as in Definition 5.2a.

▼ DEFINITION 5.2a

A *relation* is a set of ordered pairs. The set of all *first* components in a relation is called the *domain*, and the set of all *second* components is called the *range*.

EXAMPLE 1

Consider the relation $R = \{(A, 4), (B, 3), (C, 2), (D, 1), (F, 0)\}$, which pairs the grade obtained in a course with the number of grade points usually assigned to that grade.

a. Find the domain of R.

b. Find the range of R.

SOLUTION:

a. The domain of R is the set of all first components in R, that is, $\{A, B, C, D, F\}$.

b. The range of R is the set of all second components in R, that is, $\{0, 1, 2, 3, 4\}$.

GRAPHS OF RELATIONS

In this chapter, we shall be concerned with relations that are *ordered* pairs of numbers. Such a relation can be represented geometrically as a set of points in a plane. This set of points is called the *graph* of the relation. (See Definition 5.2b.)

▼ DEFINITION 5.2b

If S is a relation, then the *graph* of S is the set of all points in the plane corresponding to the ordered pairs in S.

EXAMPLE 2

Find the domain and range of the given relation and show its graph.

a. $S_1 = \{(0, 32), (37, 98.6), (100, 212)\}$

b. $S_2 = \{(x, y) | -3 \leq x \leq 2, -4 \leq y \leq 4\}$

SOLUTION:

a. The domain of S_1 is $\{0, 37, 100\}$, and the range is $\{32, 98.6, 212\}$. The graph of this relation is shown in Figure 5.2b.

b. The domain of S_2 is $\{x | -3 \leq x \leq 2\}$, and the range is $\{y | -4 \leq y \leq 4\}$. The graph of the relation is the set of all points inside and on the boundary of the rectangle shown in color in Figure 5.2c.

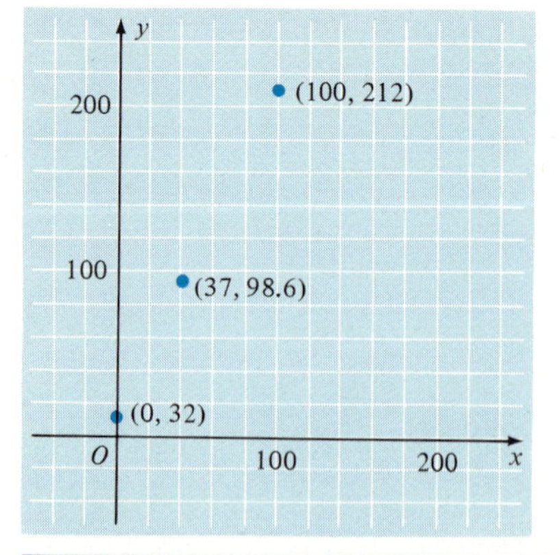

FIGURE 5.2b

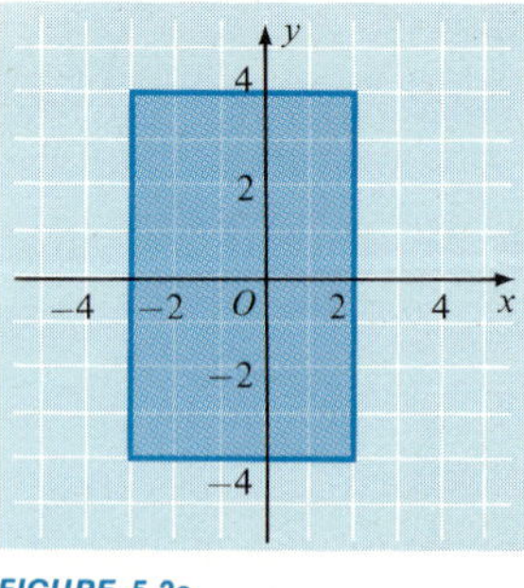

FIGURE 5.2c

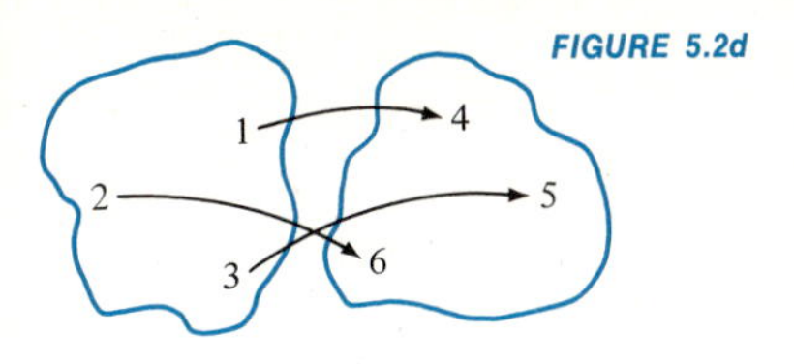

FIGURE 5.2d

There are many ways to describe a relation. In addition to using ordered pairs, we may use a diagram indicating how the elements of the domain and range are paired. In Figure 5.2d, for instance, 1 is paired with 4, 2 with 6, and 3 with 5.

A relation can also be described by stating a rule that defines the correspondence between x and y. If no restriction on the domain is specified, then the domain is assumed to be the set of all real numbers for which the pair (x, y) is real. For example, the relation

$$R = \{(x, y) \mid y = x + 1\}$$

consists of all ordered pairs in which x is a real number and $y = x + 1$; the domain and the range are both the set of all real numbers. Some of the pairs in this relation are (0, 1), (−1, 0), (8, 9), (6.3, 7.3), and so on. Of course, we cannot list all the ordered pairs in this relation, but we can make a picture of the relation by graphing some of its number pairs and connecting them with a smooth curve. To simplify matters, we list the x coordinates in a table and calculate the corresponding y-coordinates as shown in the accompanying table. The resulting graph is shown in Figure 5.2e.

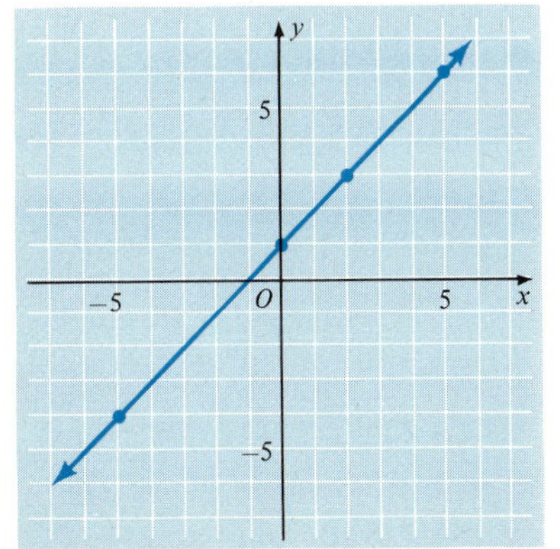

FIGURE 5.2e

x	$y = x + 1$	Pairs
−5	−5 + 1 = −4	(−5, −4)
0	0 + 1 = 1	(0, 1)
2	2 + 1 = 3	(2, 3)
5	5 + 1 = 6	(5, 6)

Notice that the graph in Figure 5.2e appears to be a straight line; the arrows indicate that the line continues without end in both directions. It can be shown[1] that every relation with a rule of the form $y = mx + b$, where m and b are constants, has a straight line for its graph. Such relations are called *linear* relations.

EXAMPLE 3

Graph the relation $\{(x, y) \mid y = 3x - 6\}$.

SOLUTION:

Because this is a linear relation, the graph is a straight line. Hence, we can draw the graph by finding two points on the line. A third point is useful as a check. The calculation is shown in the accompanying table. The graph appears in Figure 5.2f.

x	$y = 3x - 6$	Pairs
0	(3)(0) − 6 = −6	(0, −6)
2	(3)(2) − 6 = 0	(2, 0)
3	(3)(3) − 6 = 3	(3, 3)

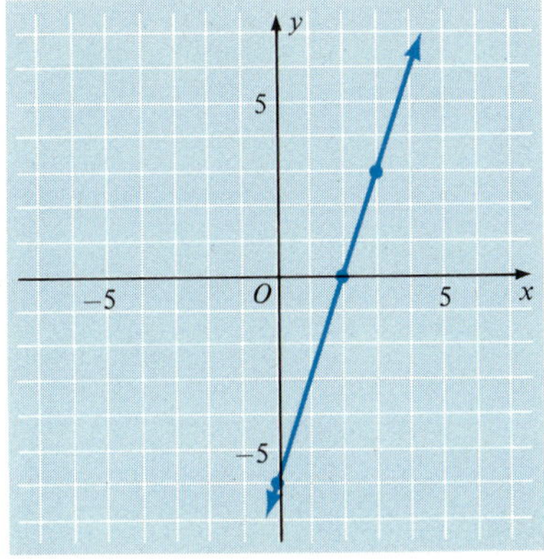

FIGURE 5.2f

[1] See Exercise 5.2, Problem 33.

EXAMPLE 4

Graph the relation $\{(x, y) | y < x + 1\}$.

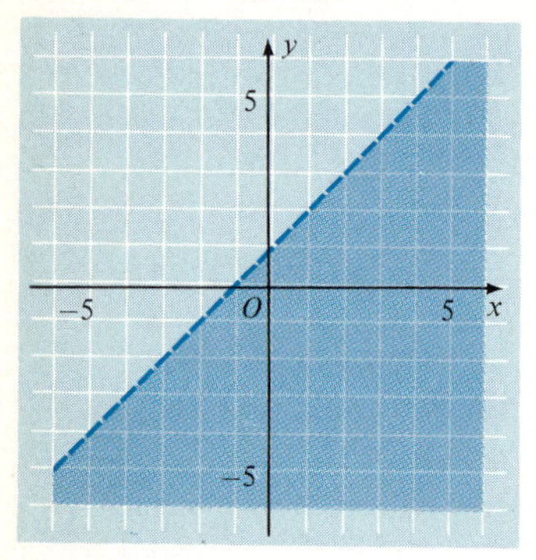

FIGURE 5.2g

SOLUTION:

We can show that all the points that satisfy the inequality $y < x + 1$ lie below the line $y = x + 1$. For example, the point (2, 3) is on the line, and all points $(2, y)$ such that $y < 3$ satisfy the inequality. Thus, for (2, 2), $(2, \frac{1}{2})$, (2, 0), (2, −5), and so on, $y < x + 1$. Similarly, if (x_1, y_1) is any point on the line, then, $y_1 = x_1 + 1$ and all points (x_1, y), where $y < y_1$, satisfy the inequality. You can also see by similar reasoning that no point above the line satisfies the inequality. The colored region in Figure 5.2g is the graph of the given relation. The line is shown dashed to indicate that it is not part of the graph.

It can be shown that the graph of any relation in which the rule is of the form $y < mx + b$ or $y > mx + b$ consists of the half plane on one side of the line $y = mx + b$. We shall not give the general proof but shall only point out that it proceeds as the analysis in Example 4. The result is that we have to test only one point on one side of the line. If this point satisfies the inequality, shade that side of the line; if the point does not satisfy the inequality, shade the other side of the line.

EXAMPLE 5

Graph the relation $\{(x, y) | y \geq 3x - 6\}$.

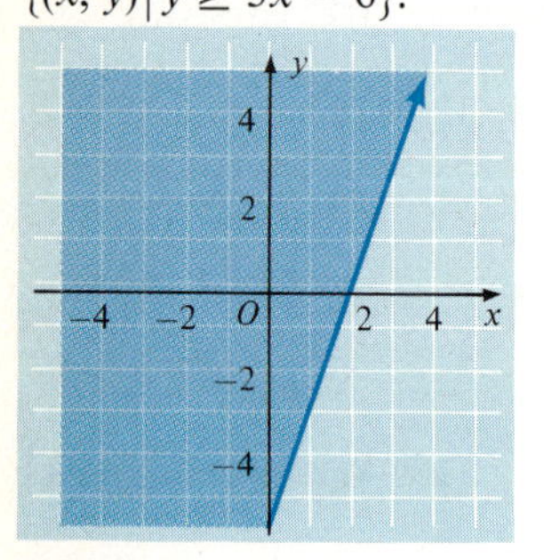

FIGURE 5.2h

SOLUTION:

We graphed the line $y = 3x - 6$ in Example 2. Let us use the test point (0, 0), which lies *above* the line. Substituting into the given inequality, we obtain $0 > (3)(0) - 6$ or $0 > -6$, which is true. Accordingly, we shade the half-plane containing the test point, that is, the half plane above the line, as in Figure 5.2h. Note that the line is shown solid to indicate that it is part of the graph.

FINDING RELATIONS FROM GRAPHS

So far, we have studied the problem of finding the *graph* of a given relation when the algebraic description for the relation is known. In some cases, it is necessary to find the algebraic description of a relation, given its *graph*. For instance, in Example 4 of Section 5.1, we showed that the circle of radius r centered at $C(h, k)$ has the equation

$$(x - h)^2 + (y - k)^2 = r^2$$

Thus, if the graph of a circle is given and we are asked to find the corresponding equation, we merely inspect the graph, determine the radius r and the center $C(h, k)$, and substitute the values in 1 to obtain the corresponding equation. This procedure is illustrated in Example 6.

EXAMPLE 6

Find the equation of the circle shown in Figure 5.2i.

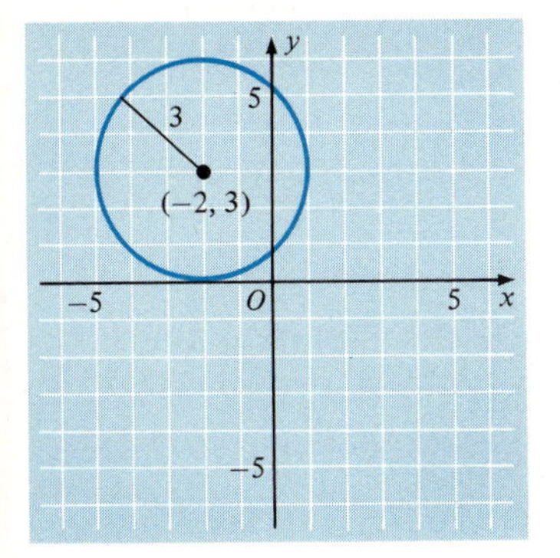

FIGURE 5.2i

SOLUTION:

The circle is centered at $(-2, 3)$ and has a radius $r = 3$. Thus, the equation of the circle is $(x + 2)^2 + (y - 3)^2 = 9$.

Of course, if we have to graph a relation of the type $\{(x, y) | (x - h)^2 + (y - k)^2 = r^2\}$, we know from Section 5.1 that the graph is a circle with radius r and center at (h, k). Thus, the circle in Figure 5.2i is the graph of the relation $\{(x, y) | (x + 2)^2 + (y - 3)^2 = 9\}$.

GRAPHING OTHER RELATIONS

In this section, we have thus far considered the graphs of two types of relations: *linear relations* (those whose graphs are straight lines) and those in which the defining relationship is the equation of a *circle*. There are many important relations that are *not* of these two types. For example, the relation $\{(x, y) | x = y^2\}$ has a graph that is neither a line nor a circle. To find the graph of this relation, we proceed as before; that is, we construct a table of corresponding values of x and y. In this case, however, since $x = y^2$, it is easier to give values to y and then to calculate the corresponding values of x. We enter the values of x and y in the accompanying table. Note that the domain of this relation is $\{x | x \geq 0\}$ because $y^2 \geq 0$ for all real values of y. The completed graph is shown in Figure 5.2j, which shows a type of curve called a *parabola*.

$x = y^2$	y	Pairs
0	0	(0, 0)
1	± 1	$(1, \pm 1)$
4	± 2	$(4, \pm 2)$
9	± 3	$(9, \pm 3)$

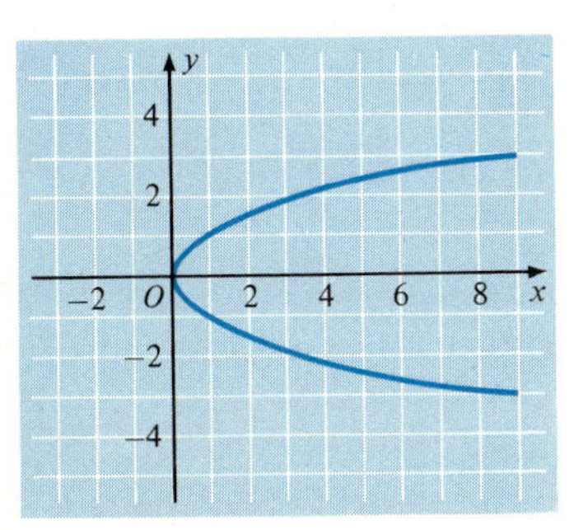

FIGURE 5.2j

Sometimes a problem asks for the graph of an equation. It is understood that this calls for the graph of the set of ordered pairs that satisfy the equation, as in Example 7.

EXAMPLE 7

Graph the equation $2x = y^2 - 2$.

SOLUTION:

In this problem, as in Example 6, it is easier to proceed by first solving the defining equation, $2x = y^2 - 2$, for x. Then we substitute in values of y and calculate x as in the accompanying table. The graph is shown in Figure 5.2k. Because $y^2 \geq 0$ for all real values of y, we see that $x \geq -1$.

$x = \frac{y^2 - 2}{2}$	y	**Pairs**
-1	0	$(-1, 0)$
$-\frac{1}{2}$	± 1	$(-\frac{1}{2}, \pm 1)$
1	± 2	$(1, \pm 2)$
$\frac{7}{2}$	± 3	$(\frac{7}{2}, \pm 3)$
7	± 4	$(7, \pm 4)$

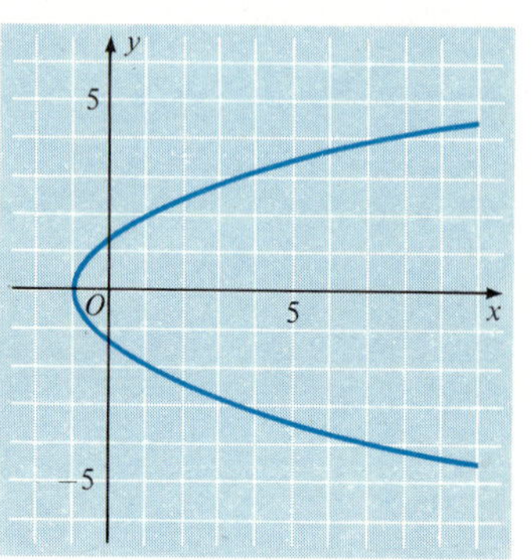

FIGURE 5.2k

A set of points is often described by a geometric condition. Sometimes the equation of the set of points can be found as described in Example 8.

EXAMPLE 8

Find the equation of the set of points such that each point $P(x, y)$ is equidistant from the line $y = -2$ and the point $F(0, 2)$

Hint: Use the distance formula and simplify the equation as much as possible.

SOLUTION:

Figure 5.2l shows one possible position of the point P. Because P is equidistant from the line $y = -2$ and the point $(0, 2)$, the distance DP is equal to the distance FP. Thus,

$$|y - (-2)| = \sqrt{(x - 0)^2 + (y - 2)^2}$$

We simplify this equation by squaring both sides to obtain

$$y^2 + 4y + 4 = x^2 + y^2 - 4y + 4$$

and, by collecting terms,

$$8y = x^2$$

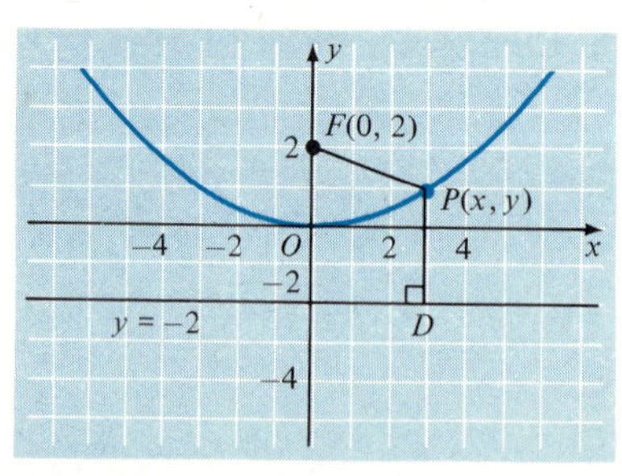

FIGURE 5.2l

EXERCISE 5.2

For the relations in Problems 1–20, (a) find the domain; (b) find the range; (c) sketch the graph.

1 $R = \{(1, 2), (1, 3), (1, 4)\}$

2 $R = \{(1, 1), (2, 1), (3, 1), (4, 1)\}$

3 $R = \{(x, y) | -1 < x < 2, y > 0\}$

4 $R = \{(x, y) | -1 \le x \le 0, 0 \le y \le 4\}$

5 $R = \{(x, y) | x = -2\}$

6 $R = \{(x, y) | y = 4\}$

7 $R = \{(x, y) | |x| < 4, |y| < 1\}$

8 $R = \{(x, y) | |x| \le 1, |y| \le 4\}$

9 $R = \{(x, y) | y = x - 2\}$

10 $R = \{(x, y) | y = x + 4\}$

11 $R = \{(x, y) | y = 5 - 3x\}$

12 $R = \{(x, y) | y = 2x - 6\}$

13 $R = \{(x, y) | y < x - 1\}$

14 $R = \{(x, y) | y > x + 3\}$

15 $R = \{(x, y) | y < 2x - 4\}$

16 $R = \{(x, y) | y > 3x - 9\}$

17 $R = \{(x, y) | y^2 = x + 1\}$

18 $R = \{(x, y) | y^2 = x - 1\}$

19 $R = \{(x, y) | y^2 = 4 - x\}$

20 $R = \{(x, y) | y^2 = 1 - x\}$

In Problems 21–24, find the standard equation of the circle with the given center and radius.

21 $C(2, 1), r = 4$

22 $C(-3, 4), r = 2$

23 $C(-4, 3), r = 1$

24 $C(-2, -3), r = 3$

In Problems 25–30, a set of points $\{P(x, y)\}$ is described by a geometric condition. Use the distance formula to find an equation for the graph. Simplify the equation in each case.

25 $P(x, y)$ is equidistant from $(-1, 3)$ and $(2, 5)$.

26 $P(x, y)$ is equidistant from $(0, 0)$ and $(2, 5)$.

27 $P(x, y)$ is equidistant from the line $y = 2$ and the point $(0, 4)$.

28 $P(x, y)$ is equidistant from the y-axis and the point $(2, 0)$

29 $P(x, y)$ is twice as far from $(2, 4)$ as from $(1, 2)$.

30 $P(x, y)$ is 6 units farther from $(-5, 0)$ than from $(5, 0)$.

In Problems 31 and 32, use your calculator to do the computations.

31 It is desired to construct an accurate graph of the equation $y^2 = x^2 + 4$ for the interval $0 \le x \le 5$. Make a table giving the y-values for values of x from 0 to 5 at half-unit intervals. Find each y-value correct to the nearest hundredth of a unit and draw the graph.

32 If $y^2 = x^3$, complete the accompanying table of values. Find each y-value correct to the nearest hundredth of a unit. Use a large scale and draw the graph of the equation for $0 \le x \le 1$.

x	0	0.1	0.2	0.3	0.4	0.5	0.6	0.7	0.8	0.9	1
y	0	± 0.03	± 0.09								± 1

33 It can be shown that the graph of the equation $y = mx + b$, where m and b are specified real numbers, is a straight line. Thus, let any three points $A(x_1, y_1)$, $B(x_2, y_2)$, $C(x_3, y_3)$, such that $x_1 < x_2 < x_3$, be chosen on the graph. Because $y_1 = mx_1 + b$, $y_2 = mx_2 + b$, and $y_3 = mx_3 + b$, we have

$$\begin{aligned} d(A, B) &= \sqrt{(x_2 - x_1)^2 + (y_2 - y_1)^2} \\ &= \sqrt{(x_2 - x_1)^2 + (mx_2 + b - mx_1 - b)^2} \\ &= \sqrt{(x_2 - x_1)^2 + m^2(x_2 - x_1)^2} \\ &= (x_2 - x_1)\sqrt{1 + m^2} \end{aligned}$$

Obtain similar expressions for $d(B, C)$ and $d(A, C)$. Then show that $d(A, C) = d(A, B) + d(B, C)$. This proves that the graph is a straight line. (Compare Exercise 5.1, Problems 25–28.)

USING YOUR KNOWLEDGE 5.2

Return to the beginning of this section and look at Figure 5.2a, which shows the relationship between temperature and altitude. Because we usually read a graph by looking first at the variable measured in the horizontal direction and then finding the corresponding value of the other variable, this graph makes it appear that the altitude depends on the temperature, which seems rather unnatural.

1 Redraw the graph with the altitude measured along the horizontal axis and the temperature measured along the vertical axis. The graph that you obtain is that of the relation in which the altitude is the first coordinate and the temperature is the second. This is exactly the reverse of the order used before. Reversing the order of the two members of the number pairs in a relation yields what is called the *inverse relation*. The topic of inverses is discussed in Section 5.4.
2 Read your graph and find the minimum temperature. At what altitude does this minimum occur?
3 What is the approximate temperature at the top of Mt. Rainier (4.4 kilometers). At what other altitudes below 100 kilometers is the temperature the same as that at the top of Rainier?

5.3 FUNCTIONS AND THEIR GRAPHS

A certain car rental firm advertises the cost of renting a Pinto at \$13.95 per day with no charge for mileage. For example, if you wish to rent a Pinto, the cost depends on how many days you keep the car. As you can see, if you rent the car for 1 day, the corresponding cost is \$13.95; for 2 days, the cost is \$27.90, and so on. In mathematics, we say that the *cost* of renting a Pinto is a *function* of the *number of days* you keep the car. The idea here is that the rental price corresponds to the number of days you keep the car in a very specific way. If we agree to let C be the cost of renting the car and d be the number of days you keep it, then we say that the equation

$$C = 13.95d \qquad (1)$$

defines a *function* that associates a unique value of C with each value of d. Equation 1 is often written in the form

$$C(d) = 13.95d \qquad \text{(Read "}C\text{ of }d\text{ equals }13.95d.\text{")} \qquad (2)$$

In many cases, a function is not defined by a formula as in Equations 1 or 2 but is defined by a table or perhaps a verbal description. For example, we could have displayed the relationship between the cost of renting the Pinto and the number of days the car is kept by means of the accompanying table.

Number of days	1	2	3	. . .
Cost	\$13.95	\$27.90	\$41.85	. . .

The important mathematical idea here is that we have two sets, $x = \{1, 2, 3, \ldots\}$ and $Y = \{\$13.95, \$27.90, \$41.85, \ldots\}$, and that there is an association between the elements of the two sets such that to each element x of X, there corresponds a *unique* element y of Y. Of course, y is the cost of keeping the car x days. Evidently, the association between the elements of X and Y defines a set of *ordered pairs*

{(1, $13.95), (2, $27.90), (3, $41.85), . . .}

in which each x is paired with *one and only one* value of y. Thus, we have just a special kind of relation, one is which *exactly one value of y corresponds to each value of x in the domain.* This kind of relation is especially important and is defined as in Definition 5.3a.

▼ DEFINITION 5.3a

A *function* f is a relation such that no two distinct ordered pairs have the same first component.

This definition means that if a is an element of the domain, and b and c are two *different* elements of the range, then (a, b) and (a, c) *cannot* both belong to the function. However, different elements of the domain may have the same element of the range corresponding to them. Thus, $\{(1, 3), (2, 5), (4, -1), (6, 3)\}$ is a function because each x in the domain $\{1, 2, 4, 6\}$ has exactly one element of the range $\{-1, 3, 5\}$ corresponding to it, but $\{(1, 3), (2, 5), (2, -1), (4, -1), (6, 3)\}$ is *not* a function because there are two different pairs $(\mathbf{2}, 5)$ and $(\mathbf{2}, -1)$ with the same first element of the domain. Of course, both sets of pairs are relations.

FUNCTIONAL NOTATION

As in the case of relations in general, functions can be represented diagrammatically. For example, Figure 5.3a shows several arrows indicating the correspondence between the elements of the domain X and the associated elements of the range Y. The x, which can be selected arbitrarily from the domain, is called the *independent variable*, and the y, which depends on x, is called the *dependent* variable. The value of y that corresponds to the value x is frequently denoted by $\boldsymbol{f(x)}$, which may be read "f of x" or "the value of f at x." This notation is used in Figure 5.3a, where $f(\boldsymbol{a})$ means the value of f for $x = \boldsymbol{a}$ and likewise for $f(b)$ and $f(c)$.

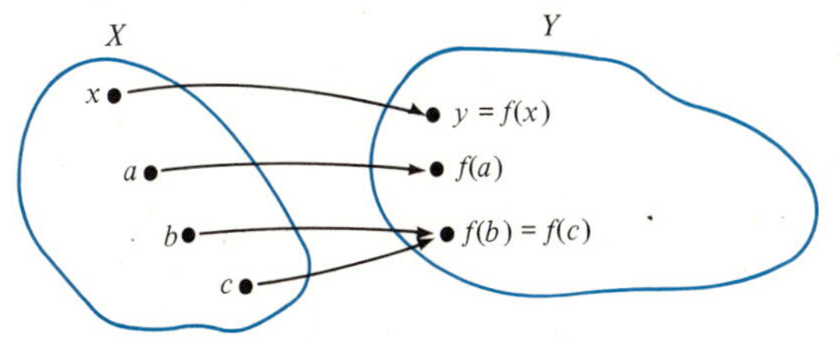

FIGURE 5.3a

Since the graph of a function f is the graph of the relation f, you can tell by a glance at the graph whether or not a relation is a function. For a relation to be a function, there must be not more than one y value for each x value. Therefore, every vertical line that cuts the graph intersects it at one point only. Thus, in Figure 5.3b, (I) and (III) are graphs of functions, but (II), which shows a circle,

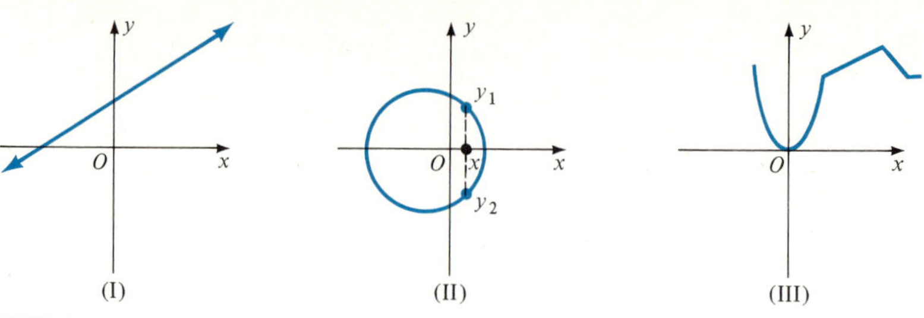

FIGURE 5.3b

is the graph of a relation that is *not* a function. Notice that a vertical line can cut the circle in two points, so that the x shown in the figure has two distinct values of y corresponding to it.

As in the case of relations in general, *if the domain of a function is not specified, we assume that the domain is the set of all real numbers for each of which the function has a real value.* For example, if

$$f = \{(x, y) \mid y = x^2 + 2\}$$

the domain is the set of all real numbers because $x^2 + 2$ is a real number for each real value of x. However, if

$$f = \left\{(x, y) \mid y = \frac{1}{x}\right\}$$

then the domain of f is the set of all real numbers *except* 0, because for $x = 0$, y is not defined, but y is a real number for each other value of x.

EXAMPLE 1

Let $f = \{(x, y) \mid y = 2x + 4\}$.

a. Find the domain of f.

b. Find the range of f.

c. Draw the graph of f.

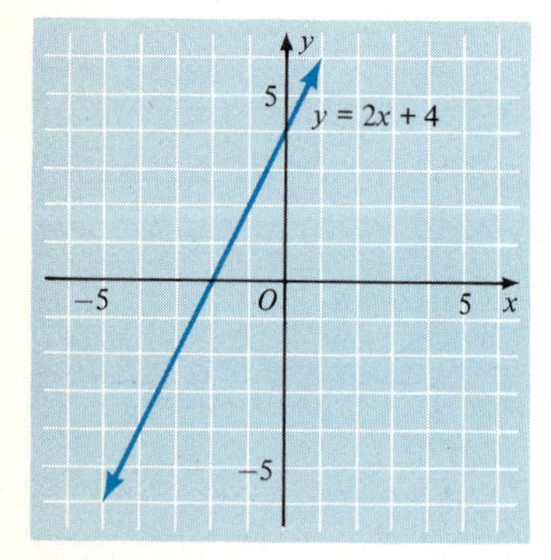

FIGURE 5.3c

SOLUTION:

a. Because the domain is not specified and $2x + 4$ is a real number for each real value of x, the domain is the set of all real numbers.

b. Because x can be any real number, $2x + 4$ can take on all real values. Thus, the range of f is the set of all real numbers.

c. We see that f is a linear relation, so that the graph can be obtained as before. This graph is the straight line shown in Figure 5.3c.

We often wish to know the value of a function for a given value of the independent variable. For example, Equation 2 gives us the cost of renting a Pinto for d days. If we wish to know the cost for **3** days, we substitute $d = \mathbf{3}$ into Equation 2 and obtain

$$C(\mathbf{3}) = (13.95)(\mathbf{3}) = 41.85$$

showing that the cost for 3 days is \$41.85. The cost for 4 days is found by letting $d = 4$ in Equation 2, which gives

$$C(4) = (13.95)(4) = 55.80$$

Thus, the cost for 4 days is $55.80.

In general, to find the value of a function f for a specified value of x, we substitute the value for x into the expression for $f(x)$. Thus, if

$$f(x) = x^2 + 1$$

then

$$f(1) = (1)^2 + 1 = 2$$
$$f(0) = (0)^2 + 1 = 1$$
$$f(-2) = (-2)^2 + 1 = 5$$

and

$$f(a + b) = (a + b)^2 + 1 = a^2 + 2ab + b^2 + 1$$

You should notice the great convenience of the *functional notation*, $f(x)$. It indicates the value of the function for the specified value of x. We could think of x as a blank and write

$$f(\quad) = (\quad)^2 + 1$$

and then fill the blank with whatever is specified for x. Thus, if x is to be replaced by -3, we have

$$f(-3) = (-3)^2 + 1 = 10$$

If x is to be replaced by $x - 1$, we have

$$f(x - 1) = (x - 1)^2 + 1 = x^2 - 2x + 2$$

Although f is the most commonly used letter to denote a function, other letters such as g, h, F, G, and ϕ are also used in exactly the same way. For example, if

$$G(x) = x^2 + \frac{1}{x}$$

then

$$G(x + 1) = (x + 1)^2 + \frac{1}{x + 1}$$

As in the case of a relation, a function can be defined by a formula indicating how the ordered pairs are to be obtained. Thus, the function

$$f\{(x, y) | y = x^2 + 1\}$$

is frequently regarded as defined by the equation

$$f(x) = x^2 + 1$$

However, you should not confuse the function f, which is the entire set of ordered pairs, with the functional value $f(x)$, which is simply the value corresponding to x.

EXAMPLE 2

Let a function g be defined by $g(x) = x^2$.

a. Find the domain of g.
b. Find the range of g.
c. Draw the graph of g.

SOLUTION:

a. Because the domain of g is not specified and x^2 is real for all real values of x, the domain is the set of all real numbers.

b. Because $g(x) = x^2$ and x^2 can take on all nonnegative real values, the range of g is the set of all nonnegative numbers; that is

$\{y \mid y \geq 0\}$

c. We first calculate a few values of $g(x)$ as in the accompanying table. Then we graph the corresponding number pairs and join them with a smooth curve, as shown in Figure 5.3d. This curve is a parabola that has the y-axis for a line of symmetry.

x	$g(x) = x^2$	Pairs
0	0	(0, 0)
± 1	1	$(\pm 1, 1)$
± 2	4	$(\pm 2, 4)$
± 3	9	$(\pm 3, 9)$

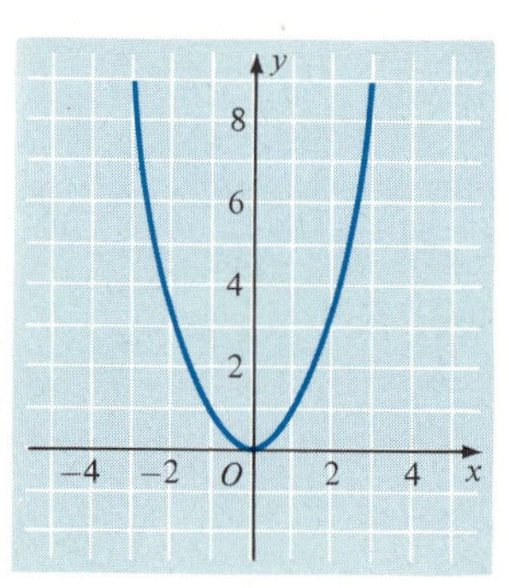

FIGURE 5.3d

EXAMPLE 3

Consider the absolute value function defined by

$$F(x) = \begin{cases} x & \text{if } x \geq 0 \\ -x & \text{if } x < 0 \end{cases}$$

This function is usually denoted by

$F(x) = |x|$ (Read "F of x is the absolute value of x.")

a. Find $F(2)$, $F(0)$, $F(-3)$.
b. What is the domain of F?
c. What is the range of F?

SOLUTION:

a. $F(2) = 2$, because $2 > 0$.
$F(0) = 0$, because $0 = 0$.
$F(-3) = -(-3) = 3$, because $-3 < 0$.

b. The domain of F is the set of all real numbers.

c. Because F can take on all nonnegative values, the range is the set of all nonnegative real numbers; that is, $\{y \mid y \geq 0\}$.

d. Draw the graph of F.

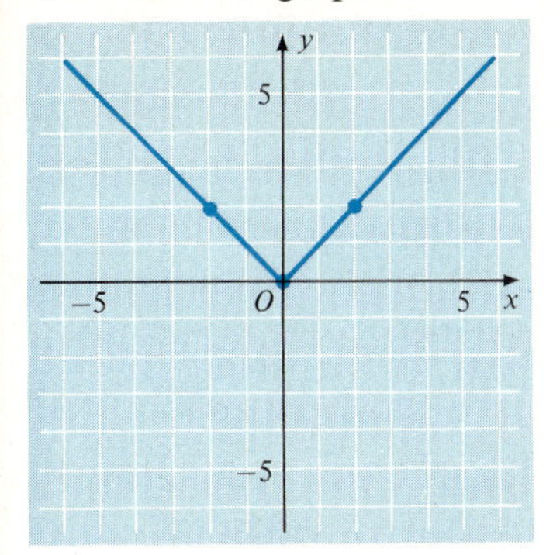

FIGURE 5.3e

d. Note that $F(x) = x$ for $x \geq 0$, and $F(x) = -x$ for $x < 0$. Thus, the graph consists of the straight line given by $y = x$ for $x \geq 0$ and the straight line given by $y = -x$ for $x < 0$. The graph is the V-shaped configuration shown in Figure 5.3e.

As we noted before, not all functions are defined by a simple formula. An interesting example is the function defined by the accompanying table, which gives the cost of mailing a letter.[2] This table

Weight (x ounces)	Cost (y cents)
$0 < x \leq 1$	15
$1 < x \leq 2$	28
$2 < x \leq 3$	41
$3 < x \leq 4$	54
...	...

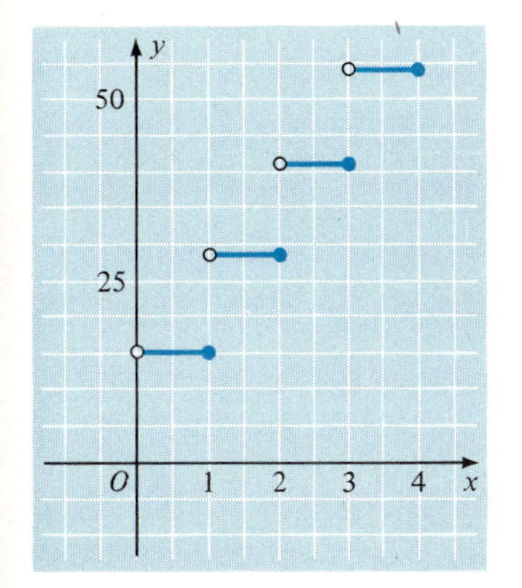

FIGURE 5.3f

defines the cost, y cents, in terms of the weight of the letter, x ounces. The graph of this function for $0 < x \leq 4$ is shown in Figure 5.3f. As usual in many problems, different scales have been used on the two axes. Notice that the open circle at the point (0, 15) in the graph indicates that this point *is not* part of the graph, whereas the solid dot at (1, 15) indicates that this point *is* part of the graph. For obvious reasons, the function graphed here is called a *step function.*

An important *step function* in mathematics is the *greatest integer* function, which is denoted by $[\![x]\!]$ and which assigns to each real number x the greatest integer less than or equal to x. For example,

$$[\![3]\!] = 3, \qquad [\![2.5]\!] = 2, \qquad [\![0.786]\!] = 0, \qquad [\![-1.25]\!] = -2$$

EXAMPLE 4

Graph the function defined by $f(x) = [\![x]\!]$ and give its domain and range.

SOLUTION:

As usual, we make a table showing some corresponding values of x and $f(x)$. The values in the table are graphed in Figure 5.3g. As you can see from the definition of the function, the domain is the set of all real numbers, and the range is the set of integers.

x	$f(x) = [\![x]\!]$
$-3 \leq x < -2$	-3
$-2 \leq x < -1$	-2
$-1 \leq x < 0$	-1
$0 \leq x < 1$	0
$1 \leq x < 2$	1
$2 \leq x < 3$	2

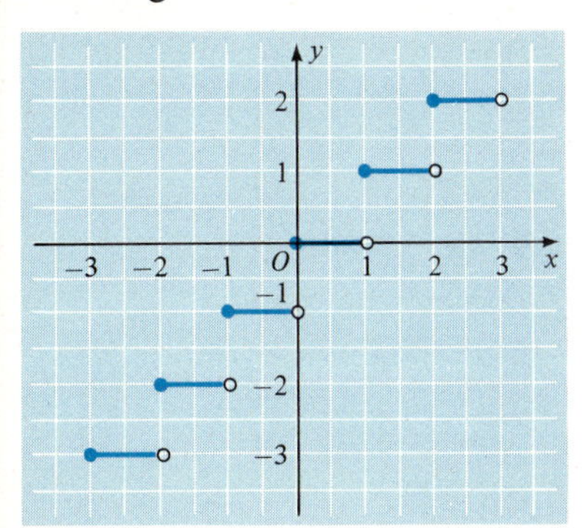

FIGURE 5.3g

[2] Check with your local post office for current prices.

In Problem 49, Exercise 5.3, you are asked to verify that the first-class postage function can be formulated in terms of the greatest integer function.

EXERCISE 5.3

In Problems 1–22, a function is defined. (a) Find the domain; (b) find the range; (c) draw the graph.

1 $f = \{(x, y) | y = x + 5\}$ **2** $f = \{(x, y) | y = -x + 3\}$ **3** $g = \{(x, y) | y = 2x + 5\}$

4 $g = \{(x, y) | y = -2x + 4\}$ **5** $F = \{(x, y) | y = x^2 + 3\}$ **6** $F = \{(x, y) | y = 1 - 2x^2\}$

7 $h = \{(x, y) | y = 4 - x^2\}$ **8** $H = \{(x, y) | y = -x^2 + 3\}$ **9** $G(x) = -3x + 6$

10 $f(x) = -4x + 8$ **11** $f(x) = 5$ **12** $f(x) = -2$

13 $F = \{(x, y) | y = |x| - 1\}$ **14** $G = \{(x, y) | y = |x| + 2\}$ **15** $H(x) = |x - 1|$

16 $g(x) = |x + 1|$ **17** $f(x) = \begin{cases} 1 & \text{if } x \geq 0 \\ -1 & \text{if } x < 0 \end{cases}$ **18** $g(x) = \begin{cases} -1 & \text{if } x < -1 \\ x & \text{if } -1 \leq x \leq 1 \\ 1 & \text{if } x > 1 \end{cases}$

19 $h(x) = \begin{cases} 2 & \text{if } x < -2 \\ -x & \text{if } -2 \leq x \leq 2 \\ -2 & \text{if } x > 2 \end{cases}$ **20** $f(x) = |2x|$ **21** $g(x) = [\![x]\!] + 1$ **22** $h(x) = [\![-x]\!] + 1$

In Problems 23–30, a relation is given. (a) Find the domain; (b) determine whether the relation is a function; (c) draw the graph.

23 $R = \{(1, 3), (2, 3), (5, 3)\}$ **24** $S = \{(3, 1), (3, 2), (3, 3)\}$ **25** $T = \{(2, 6), (3, -5), (-4, 6), (2, 0)\}$

26 $S = \{(-2, 0), (2, 0), (-3, 1), (3, 1)\}$ **27** $R = \{(x, y) | y = 2x^2 + 4\}$ **28** $R = \{(x, y) | y^2 = x + 4\}$

29 $S = \{(x, y) | x = |y|\}$ **30** $S = \{(x, y) | x^2 + y^2 = 9\}$

31 Let $f(x) = 1 - x$. Find:
a $f(a - 1)$ **b** $f(a) - 1$ **c** $f(x - y)$ **d** $f(x) - y$

32 Let $G(y) = y^2$. Find:
a $\dfrac{1}{G(b)}$ **b** $G(\sqrt{b})$ **c** $\sqrt{G(b)}$

33 Let $F(x) = x + \dfrac{1}{x}$. Find:
a $F(y)$ **b** $F\left(\dfrac{1}{y}\right)$ **c** $\dfrac{1}{F(y)}$

34 Let $H(x) = \sqrt{x^2 - 1}$. Find:
a $H(a + 1)$ **b** $H(a^2 + 1)$ **c** $H\left(\dfrac{1}{a}\right)$

35 Suppose that $g(x) = \sqrt{x - 2}$.
a What is the domain of g? **b** Find a value in the domain such that $g(x) = 4$.

36 Suppose that $f(x) = \sqrt{x^2 - 1}$.
a What is the domain of f? **b** Find a value in the domain such that $f(x) = 8$.

37 Let $f(x) = x + 2$. Find:
a $f(x + h)$ **b** $\dfrac{f(x + h) - f(x)}{h}, h \neq 0$

38 Let $g(x) = x^2$. Find:
a $g(x + h)$ **b** $\dfrac{g(x + h) - g(x)}{h}, h \neq 0$

39 Let $H(x) = \sqrt{x + 1}$. Find:
a $H(x + h)$ **b** $\dfrac{H(x + h) - H(x)}{h}$

Hint: To simplify this fraction, rationalize the numerator.

40 Let $G(x) = \sqrt{x^2 + 1}$. Find:

a $G(x + h)$ **b** $\dfrac{G(x + h) - G(x)}{h}$ (See Problem 35.)

41 Suppose that the cost of renting a car is \$14.95 per day plus 15¢ a mile. Let m be the number of miles traveled and $C(m)$ the daily cost in dollars.
a Find $C(50)$, the price paid by a customer who traveled 50 miles in one day. **b** Find $C(100)$.
c Find a formula for $C(m)$.

42 Repeat Problem 41 if the cost is \$16.95 per day plus 17¢ a mile.

43 A market analysis shows that the profit P (dollars) that a store will make on a certain item depends on the price x (dollars) as follows:

$$P(x) = 230x - x^2 + 5000, \qquad 100 \le x \le 130$$

Make a table of values of $P(x)$ for prices from \$100 to \$130 at intervals of \$5. Which of these prices gives the greatest profit?

44 A market analysis shows that the sales of a certain item depend on the price as follows: If x (dollars) is the price, then the number $N(x)$ sold is given by

$$N(x) = 240x - 10x^2 - \left[\!\left[\frac{1000}{x}\right]\!\right], \qquad 10 \le x \le 15$$

Make a table of values of $N(x)$ for $10 \le x \le 15$ at unit intervals. Which of these prices gives the greatest sales volume?

45 Torricelli's law states that the square of the velocity v of a stream of liquid flowing from a hole in a tank is equal to $2gh$, where g is the gravitational acceleration and h is the height of the surface of the liquid above the hole. Use $g = 32$ ft/sec^2 and express v as a function of h.

46 Express the radius of a circle as a function of its area. Find a formula that gives the increase I in the radius of any circle if the area is doubled.

47 A motorboat, $\frac{1}{4}$ mile out from shore, is traveling parallel to the shoreline at a speed of 15 miles per hour. At 10:00 A.M. the boat passes directly in front of an observer on the shoreline. Let t denote the time in *minutes* after 10:00 A.M., and express the distance d (in miles) from the observer to the boat as a function of t.

48 A ship leaves port at 1:00 P.M. and sails east at the rate of 15 mph. A second ship leaves the same port at 2:00 P.M. (the same day) and sails north at the rate of 20 mph. Let t be the number of hours after 2:00 P.M., and express the distance between the two ships as a function of t.

49 Show that domestic, first-class postage (in cents) is given by the formula

$$p(x) = 15 - 13[\![1 - x]\!]$$

where x is the weight in ounces.

50 Make a graph of the function defined by $f(x) = [\![x]\!] + [\![-x]\!]$.

USING YOUR KNOWLEDGE 5.3

Symmetry properties of functions are quite helpful in making the corresponding graphs. For example, we have seen that the function $g(x) = x^2$ has for its graph a parabola that is symmetric to the y-axis. This is a consequence of the fact that $(-x_1)^2 = (x_1)^2$, so that if (x_1, y_1) is on the graph, then $(-x_1, y_1)$ is also on the graph. A function that behaves in sign like x^2 is called an *even* function. Thus, an even function is characterized by the equation

$$f(-x) = f(x) \qquad (1)$$

An even function has a graph that is symmetric with respect to the y-axis. If you make the portion of the graph that is on one side of the y-axis, then you can graph the portion on the other side by using this symmetry without having to calculate any more points.

Similarly, a function that behaves in sign like x^3 is called an *odd* function. An odd function is characterized by the equation

$$f(-x) = -f(x) \tag{2}$$

For such a function, we see that if (x_1, y_1) is on the graph, then $(-x_1, -y_1)$ is also on the graph. This means that the graph is symmetric with respect to the origin; that is, a line through the origin that cuts the curve in a point on one side of the origin must cut it in another point on the opposite side of the origin, and the two points are equidistant from the origin. Figure 5.3h shows the two types of symmetry that we have discussed.

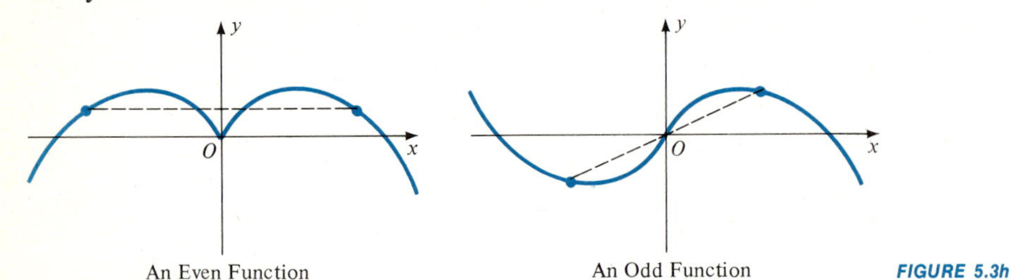

An Even Function An Odd Function FIGURE 5.3h

The next illustrations show how to apply Equations 1 and 2.

a. $f(x) = x^2 - 5x^4$ is an even function because $f(-x) = (-x)^2 - 5(-x)^4 = x^2 - 5x^4 = f(x)$.
b. $g(x) = x^3 + 2x$ is an odd function because $g(-x) = (-x)^3 + 2(-x) = -x^3 - 2x = -g(x)$.
c. $h(x) = (x + x^2)/(1 - x)$ is neither even nor odd because

$$h(-x) = \frac{(-x) + (-x)^2}{1 - (-x)} = \frac{-x + x^2}{1 + x}$$

which does not agree with either $h(x)$ or $-h(x)$.

Classify each of the following functions as even, odd, or neither.

1 $f(x) = x^2 + 4x^4 - x^6 + 3$ **2** $f(x) = -x + 4x^3 - 6x^5$ **3** $f(x) = |x|$ **4** $f(x) = x^5 + 3$

5 $f(x) = x^5 - 8x$ **6** $f(x) = |x^3 - x|$ **7** $f(x) = \dfrac{x^2 + 1}{x^2 - 1}$ **8** $f(x) = \dfrac{1}{x^3 + 2x}$

5.4 INVERSE FUNCTIONS AND OPERATIONS ON FUNCTIONS

THE INVERSE OF A FUNCTION

Table 5.4a shows the relationship between the number n of chirps that a cricket makes in a minute and the temperature in degrees Fahrenheit. The information in this table can also be written as the function

$$f = \{(40, 0), (50, 40), (60, 80), (70, 120), (80, 160)\}$$

or as

$$f = \{(t, n) | n = 4(t - 40), t = 40, 50, 60, 70, 80\}$$

TABLE 5.4a

Temperature °F	Number of chirps
40	0
50	40
60	80
70	120
80	160

In this function, t is the independent variable, and n is the dependent variable. We would naturally expect the number of chirps to depend on the temperature and not the other way around.

It is said that farmers use the information in Table 5.4a to tell the temperature by counting the number of chirps per minute that the cricket makes. Suppose that you wanted to do this. Then it would be convenient to have the temperature t expressed in terms of n rather than as we have it. We can accomplish this by solving the equation

$$n = 4(t - 40)$$

for t to obtain

$$t = \frac{n}{4} + 40$$

The effect of this procedure is to set up a new function, say g, where

$$g = \left\{(n, t) \middle| t = \frac{n}{4} + 40, n = 0, 40, 80, 120, 160\right\}$$

An equivalent procedure would be to interchange the members of each ordered pair in the function f to obtain

$$g = \{(0, 40), (40, 50), (80, 60), (120, 70), (160, 80)\}$$

The new function g, which has been so constructed, is called the *inverse* of f and is often denoted by $\boldsymbol{f^{-1}}$ (read "f inverse"). Note that the superscript -1 is *not* an exponent; it is simply the notation used to indicate the inverse of f.

***To find the inverse of a function*, do either of the following:**

1. **Solve the defining relationship for the independent variable and then make the original dependent variable the new independent one.**
2. **Interchange the members of each ordered pair in the function.**

Thus, if

$$f = \{(\mathbf{1}, 3), (\mathbf{2}, 7), (\mathbf{3}, -10)\}$$

then the inverse of f is the relation

$$S = \{(3, \mathbf{1}), (7, \mathbf{2}), (-10, \mathbf{3})\}$$

obtained by interchanging the members of each ordered pair in f. Of course, the domain of f has become the range of S, and the range of f has become the domain of S. Notice that S is a function, so that we may write f^{-1} in place of S.

As a second illustration, if

$$f\{(x, y) | y = 2x + 5\}$$

then the inverse is obtained by solving the equation $y = 2x + 5$ for x to obtain $x = (y - 5)/2$. Because we prefer to keep the letter x in its usual role as the independent variable, we interchange the x and y in the last equation, and write the inverse of f as

$$f^{-1} = \left\{(x, y) \mid y = \frac{x - 5}{2}\right\}$$

Notice that this relation is a function, so that the notation f^{-1} is justified.

In the preceding illustrations, the inverses turned out to be functions; however, this is not always the case. For instance, if

$$f = \{(1, 2), (3, 2)\}$$

then the inverse is the relation

$$S = \{(2, 1), (2, 3)\}$$

which is *not* a function. Obviously, for the relation obtained by interchanging the members of the ordered pairs of a function f to be a function, f must consist of ordered pairs with *all different second* components. A function in which no two ordered pairs have the same second component is called a *one-to-one* function; such a function associates exactly one y for each x and one x for each y. This idea is formalized in Definition 5.4a.

▼ DEFINITION 5.4a

A function is *one-to-one* if and only if no two ordered pairs in the function have the same second component.

Just as the graph of a function must pass the "vertical-line test," the graph of a one-to-one function must pass the "horizontal-line test." No horizontal line can cut the graph of a one-to-one function in more than one point. (Otherwise, there would be two pairs with the same second member in the function.) If the graph of a one-to-one function is a continuous (unbroken) curve, then the graph cannot go up and down; it must be a rising (or a falling) curve as we proceed from left to right over the entire domain of the function. The idea is illustrated in Figure 5.4a.

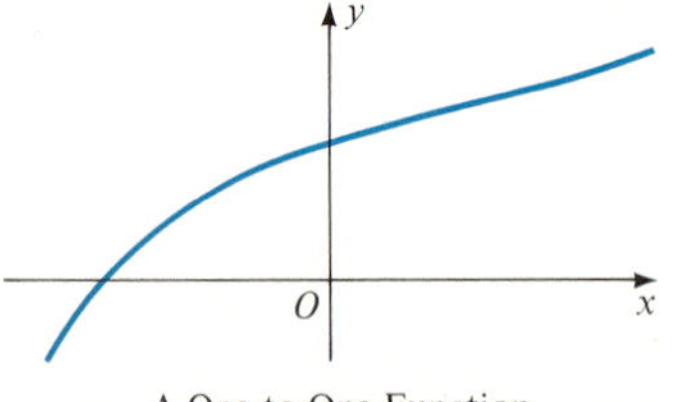

A One-to-One Function

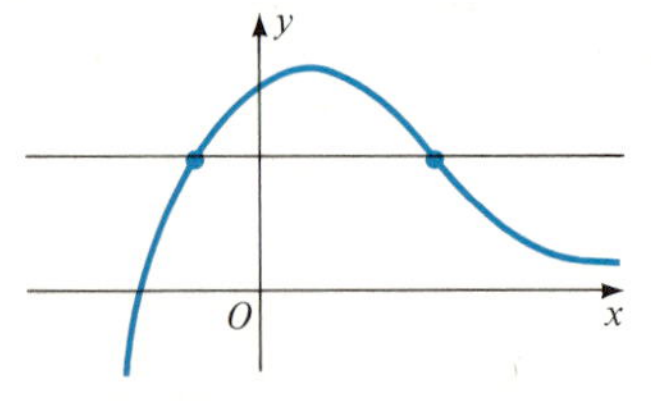

Not a One-to-One Function

FIGURE 5.4a

If a function f with domain X and range Y is one-to-one, then the inverse is also a function; the domain of the inverse is Y, and the range is X. Thus, if the function f is defined by

$$y = f(x)$$

and we solve for x to obtain

$$x = g(y)$$

then

$$f(x) = f(g(y)) = y$$

for every y in Y. Similarly,

$$g(y) = g(f(x)) = x$$

for every x in X. In view of these results, we can state Definition 5.4b. According to the definition, if g satisfies 1 and 2, then $g = f^{-1}$, and we may rewrite these conditions as follows:

$$f(f^{-1}(x)) = x \qquad \text{for every } x \text{ in } Y \tag{1'}$$

and

$$f^{-1}(f(x)) = x \qquad \text{for every } x \text{ in } X \tag{2'}$$

▼ DEFINITION 5.4b

If f is a one-to-one function with domain X and range Y, then a function g with domain Y and range X is the inverse function f^{-1} of f if and only if

$$f(g(x)) = x \qquad \text{for every } x \text{ in } Y \tag{1}$$

and

$$g(f(x)) = x \qquad \text{for every } x \text{ in } X \tag{2}$$

You should be warned that the preceding definition acts as a test to determine whether or not g is the inverse of f, but it is not a good tool for finding the inverse. For the function f defined by $f(x) = 2x + 5$, we found the inverse function f to be defined by $f^{-1}(x) = (x - 5)/2$. Both of these functions have the set of all real numbers for domain and range. Applying Definition 5.4b, we find

1. $f(f^{-1}(x)) = 2\left(\dfrac{x-5}{2}\right) + 5 = x$ for all real values of x, and

2. $f^{-1}(f(x)) = \dfrac{(2x+5)-5}{2} = x$ for all real values of x.

This verifies that the two functions are inverses of each other.

EXAMPLE 1

Find the inverse of the given function and then draw the graphs of the function and its inverse on the same coordinate axes.

a. $f = \{(-2, 0), (-1, 2), (0, 5), (1, 6)\}$

b. $g = \{(x, y) \mid y = 2x + 4\}$

SOLUTION:

a. The inverse of f is $f^{-1} = \{(0, -2), (2, -1), (5, 0), (6, 1)\}$. The graphs of f (in black) and f^{-1} (**in color**) are shown in Figure 5.4b. Notice that the graphs of f and f^{-1} are symmetric with

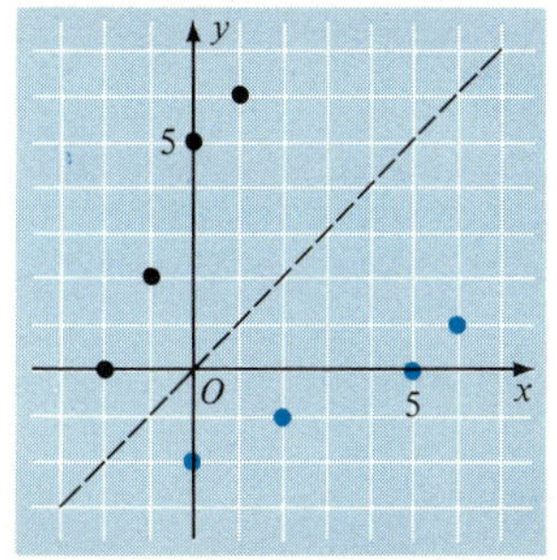

FIGURE 5.4b

respect to the line $y = x$ (shown dashed). This means that if the paper is folded along the line $y = x$, the two graphs will coincide.

b. Because $y = 2x + 4$, we solve for x, obtaining

$$x = \frac{y - 4}{2}$$

Then, we interchange the x and the y, and write the inverse function as

$$g^{-1} = \left\{(x, y) \middle| y = \frac{x - 4}{2}\right\}$$

The graphs of g (in black) and g^{-1} (**in color**) are shown in Figure 5.4c. Note again that the graphs of g and g^{-1} are symmetric with respect to the line $y = x$.

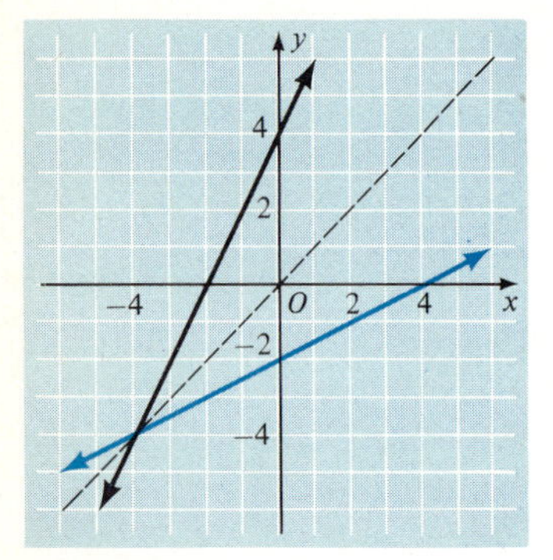

FIGURE 5.4c

The symmetry obtained in the graphs of the function and its inverse in Example 1 is not an accident. If the pair (a, b) belongs to f, then the pair (b, a) belongs to f^{-1}. If you graph the points (a, b) and (b, a) for $a \neq b$, you will see that these points must be symmetric with respect to the line $y = x$. (See Figure 5.4d.) Of course, for $a = b$, the two points coincide and lie on the line $y = x$.

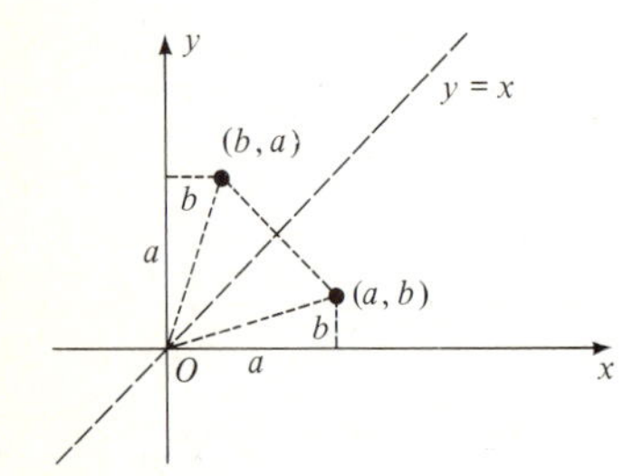

FIGURE 5.4d

EXAMPLE 2

Let $f(x) = x^2 + 2$.

a. What restrictions must be imposed on the domain of f so that the inverse of f is a function?

b. Find the inverse of the restricted f and verify that it satisfies Equations 1′ and 2′.

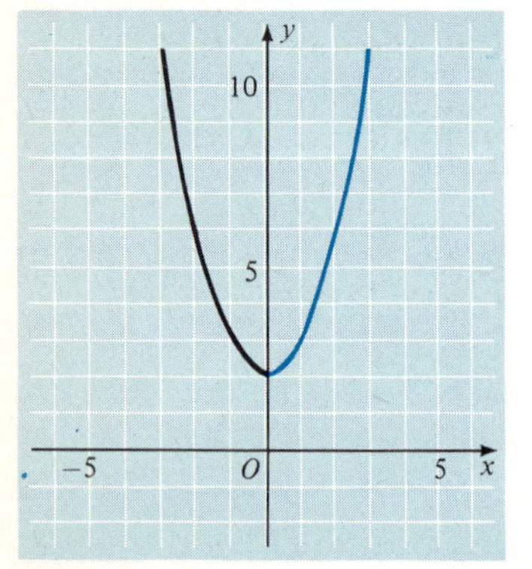

FIGURE 5.4e

SOLUTION:

a. The graph of f is shown in Figure 5.4e. Because f must be a one-to-one function for the inverse to be a function, f cannot have two ordered pairs with the same second member. This difficulty can be avoided by restricting the domain of f. For instance, we could restrict x to take on nonpositive values only, or we could restrict x to take on nonnegative values only. These restrictions give us the simplest one-to-one functions. Notice that the graphs in both cases pass the horizontal-line test. Let us restrict x to nonnegative values and call the restricted function f_r, so that

$$f_r(x) = x^2 + 2, \qquad x \geq 0$$

The graph of f_r is the portion of the graph in Figure 5.4e that lies to the right of the y-axis and includes the origin (shown **in color**).

b. If we write $y = f_r(x)$, then

$$y = x^2 + 2, \qquad x \geq 0$$

and we can solve for x to obtain

$$x = \sqrt{y - 2}$$

Note that it is not $\pm$ because $x \geq 0$. Now, we interchange the x and the y to write

$$f_r^{-1}(x) = \{(x, y) | y = \sqrt{x - 2}\}$$

The domain of the inverse function is $x \geq 2$, so that y is a real number for each x in the domain. We can verify that Equations 1′ and 2′ are satisfied by f_r and f_r^{-1} as follows:

$$\begin{aligned} f_r(f_r^{-1}(x)) &= (\sqrt{x - 2})^2 + 2 \\ &= x - 2 + 2 \\ &= x \qquad \text{for all } x \geq 2 \qquad \text{This is the range of } f_r. \qquad (1') \\ f_r^{-1}(f_r(x)) &= \sqrt{(x^2 + 2) - 2} \\ &= \sqrt{x^2} \\ &= x \qquad \text{for all } x \geq 0 \qquad \text{This is the domain of } f_r. \qquad (2') \end{aligned}$$

We have thus verified that f_r and f_r^{-1} satisfy Equations 1′ and 2′.

ELEMENTARY OPERATIONS ON FUNCTIONS

If we are given two functions, f and g, we can form from them other functions by the operations of addition, subtraction, multiplication, and division. We define these operations as in Definition 5.4c. Note that the domain of the sum, difference, and product is simply the set of numbers common to the domains of f and g, but the domain of the quotient must exclude all values of x for which the denominator $g(x)$ is zero.

▼ DEFINITION 5.4c

Let f and g be functions with domains D_f and D_g, respectively.

1. **The sum of f and g, denoted by $f + g$, is given by $(f + g)(x) = f(x) + g(x)$.**
2. **The difference of f and g, denoted by $f - g$, is given by $(f - g)(x) = f(x) - g(x)$.**
3. **The product of f and g, denoted by fg, is given by $(fg)(x) = f(x)g(x)$.**
4. **The quotient of f and g, denoted by f/g, is given by $(f/g)(x) = f(x)/g(x)$.**

EXAMPLE 3

Let $f(x) = 4x^2 - 4$ and $g(x) = 3x + 5$. Find:

a. $(f + g)(x)$
b. $(f - g)(x)$
c. $(g - f)(x)$
d. $(fg)(x)$

SOLUTION:

a. $(f + g)(x) = f(x) + g(x) = (4x^2 - 4) + (3x + 5) = 4x^2 + 3x + 1$
b. $(f - g)(x) = f(x) - g(x) = (4x^2 - 4) - (3x + 5) = 4x^2 - 3x - 9$
c. $(g - f)(x) = g(x) - f(x) = (3x + 5) - (4x^2 - 4) = -4x^2 + 3x + 9$
d. $(fg)(x) = f(x)g(x) = (4x^2 - 4)(3x + 5) = 12x^3 + 20x^2 - 12x - 20$

e. $\left(\dfrac{f}{g}\right)(x)$

f. $\left(\dfrac{g}{f}\right)(x)$

e. $\left(\dfrac{f}{g}\right)(x) = \dfrac{f(x)}{g(x)} = \dfrac{4x^2 - 4}{3x + 5}, x \neq -\dfrac{5}{3}$

f. $\left(\dfrac{g}{f}\right)(x) = \dfrac{g(x)}{f(x)} = \dfrac{3x + 5}{4x^2 - 4}, x \neq \pm 1$

COMPOSITION OF FUNCTIONS

There is another operation on two functions that is frequently used in calculus to break a complicated function down into a composite of simpler functions. For example,

$$g(x) = \sqrt{x^2 + 1}$$

may be thought of as $g(f(x))$, where

$$g(x) = \sqrt{x} \quad \text{and} \quad f(x) = x^2 + 1$$

A function formed in this manner, that is, by putting $f(x)$ in place of x in $g(x)$, is called a *composite function.* We used the idea of composite functions in defining the inverse of a function (Definition 5.4b). Sometimes the composite function given by $g(f(x))$ is denoted by $g \circ f$ (read "g circle f") and $g(f(x))$ is written $(g \circ f)(x)$. Of course, we can also form $f(g(x))$ to define $f \circ g$. Definition 5.4d describes $g \circ f$ more precisely and is illustrated in Figure 5.4f.

▼ **DEFINITION 5.4d**

If f is a function with domain X and range Y, and g is a function with domain Y and range Z, then $g \circ f$ is the function defined by $g(f(x))$ with domain X and range Z.

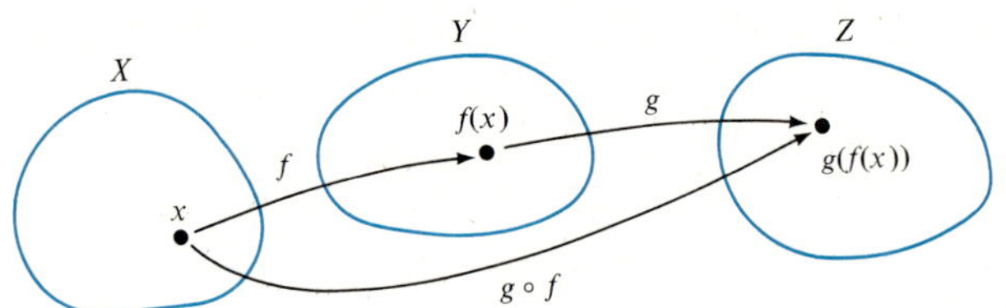

FIGURE 5.4f

In many applications, the domain of g might be only a subset of Y rather than all of it. Moreover, it might be necessary to restrict x to some subset of X so that $f(x)$ falls into the domain of g, as illustrated in Example 4. In general, the domain of $g \circ f$ is just that part of the domain of f that yields values of $f(x)$ that are in the domain of g. Thus, the domain of $g \circ f$ is always a subset of the domain of f.

EXAMPLE 4

Let $f(x) = x + 5$ and $g(x) = \sqrt{x + 1}$

a. Find $(f \circ g)(x)$. What is the domain of $f \circ g$?

b. Find $(g \circ f)(x)$. What is the domain of $g \circ f$?

SOLUTION:

a.
$$\begin{aligned}(f \circ g)(x) &= f(g(x)) \\ &= f(\sqrt{x + 1}) \\ &= \sqrt{x + 1} + 5\end{aligned}$$

Here, the domain of f is the set of all real numbers, and the domain of g is $\{x \mid x \geq -1\}$. The domain of $f \circ g$ is $\{x \mid x \geq -1\}$ also, so that the domain of $f \circ g$ is the same as that of g.

b. $(g \circ f)(x) = g(f(x))$

$$= g(x + 5)$$
$$= \sqrt{(x + 5) + 1}$$
$$= \sqrt{x + 6}$$

The domains of f and g are as before. However, the domain of $g \circ f$ is $\{x \mid x \geq -6\}$, which is a subset of the domain of f. Note that $f(g(x)) \neq g(f(x))$.

EXAMPLE 5

Let F be defined by $F(x) = \sqrt{x^3 + 3x - 1}$. Write F as a composite of two other functions, f and g.

SOLUTION:

We can let $f(x) = \sqrt{x}$ and $g(x) = x^3 + 3x - 1$. Then,

$$F = f \circ g$$

because

$$(f \circ g)(x) = f(g(x)) = \sqrt{x^3 + 3x - 1}$$

In case you are wondering how to go about making the choice of the f and the g in a problem such as that in Example 5, you should know that the answer is not unique. There are many ways of writing a given function as a composite of other functions. For instance, we could have used

$$f(x) = \sqrt[4]{x} \quad \text{and} \quad g(x) = (x^3 + 3x - 1)^2$$

in Example 5. The choice is usually indicated by the type of application that is being made.

EXAMPLE 6

The area of a circle of radius r is given by the formula $A = \pi r^2$, and the circumference by the formula $C = 2\pi r$. Let

$$f(r) = \pi r^2 \quad \text{and} \quad g(r) = 2\pi r$$

a. Write a formula for $f(g^{-1}(C))$.

b. What can the formula $A = f(g^{-1}(C))$ be used for?

SOLUTION:

a. To find $g^{-1}(C)$, we solve the equation $C = 2\pi r$ for r. The result is

$$r = \frac{C}{2\pi} = g^{-1}(C)$$

Substituting this result in the function f, we find

$$f(g^{-1}(C)) = \pi\left(\frac{C}{2\pi}\right)^2 = \frac{C^2}{4\pi}$$

b. The formula $A = f(g^{-1}(C))$ expresses the area of the circle as a function of the circumference C. Thus, the formula

$$A = \frac{C^2}{4\pi}$$

could be used to find the area if the circumference is known.

EXERCISE 5.4

In Problems 1–12, find the inverse of the given function.

1 $f = \{(4, 3), (3, 2), (2, 1)\}$

2 $g = \{(-1, 4), (-2, 3), (-3, 2)\}$

3 $f = \{(x, y) \mid y = 3x - 2\}$

4 $g = \{(x, y) \mid y = (3x + 2)/5\}$

5 $f = \{(x, y) \mid y = x^2 - 2, x \geq 0\}$

6 $F = \{(x, y) \mid y = 2x^2 - 1, x \geq 0\}$

7 $f(x) = \dfrac{1}{2 - x}, x \neq 2$

8 $G(x) = \dfrac{1}{1 - x}, x \neq 1$

9 $g(x) = x^3 - 1$

10 $f(x) = x^{1/3} + 2$

11 $h(x) = \sqrt{3x + 4}, x \geq -\frac{4}{3}$

12 $f(x) = \sqrt{1 + 2x^2}, x \leq 0$

In Problems 13–18, use Definition 5.4b to prove that the functions *f* and *g* are inverses of each other. Then graph the two functions on the same coordinate axes.

13 $f(x) = \dfrac{x + 2}{4}, g(x) = 4x - 2$

14 $f(x) = \sqrt{\dfrac{x}{2}}, g(x) = 2x^2, x \geq 0$

15 $f = \{(x, y) \mid y = 3x + 6\}, g = \left\{(x, y) \mid y = \dfrac{x - 6}{3}\right\}$

16 $f = \{(x, y) \mid y = x^3\}, g = \{(x, y) \mid y = \sqrt[3]{x}\}, x \geq 0$

17 $f = \{(x, y) \mid y = x\}, g = \{(x, y) \mid y = x\}$

18 $f = \{(x, y) \mid y = 1/x\}, g = \{(x, y) \mid y = 1/x\}$

In Problem 19–22, two functions are given. Find (a) $(f + g)(x)$; (b) $(f - g)(x)$; (c) $(fg)(x)$; (d) $(f/g)(x)$.

19 $f(x) = x^2 - 7, g(x) = -x + 4$

20 $f(x) = x + 3, g(x) = -x^2 - 1$

21 $f(x) = -x^2 + 2x + 1, g(x) = x + 3$

22 $f(x) = \dfrac{1}{x + 1}, g(x) = \dfrac{x}{x + 1}$

In Problems 23–28, find (a) $(f \circ g)(x)$ and (b) $(g \circ f)(x)$.

23 $f(x) = 3x - 2, g(x) = x + 1$

24 $f(x) = x^2, g(x) = x - 1$

25 $f(x) = \sqrt{x + 1}, g(x) = x^2 - 1$

26 $f(x) = \sqrt{x^2 + 1}, g(x) = 2x + 1$

27 $f(x) = 3, g(x) = -1$

28 $f(x) = ax, g(x) = bx$

In Problems 29–32, find two functions *f* and *g*, such that $F(x) = (f \circ g)(x)$.

29 $F(x) = (x^2 - x - 1)^3$

30 $F(x) = \sqrt{x^2 + 2x}$

31 $F(x) = \sqrt{x + 1} + 5$

32 $F(x) = \sqrt{x^2 + 1} - 3$

33 The formula for changing inches (i) to centimeters (c) may be written as

$c = f(i)$ where $f(i) = 2.54i$

The formula for changing centimeters (c) to meters (m) may be written as

$m = g(c)$ where $g(c) = 0.01c$

a Write a formula for changing inches to meters. **b** What composite function does the answer to part (a) define?

34 The formula for changing degrees Fahrenheit (F) to degrees Celsius (C) may be written as

$C = g(F)$ where $g(F) = \frac{5}{9}(F - 32)$

The formula or changing degrees Celsius (C) to degrees Kelvin (K) may be written as

$K = h(C)$ where $h(C) = C + 273$

a Write a formula for changing degrees Fahrenheit to degrees Kelvin.
b What composite function does the answer to part (a) define?

35 Find the inverse of the composite function found in Problem 33b. What can the defining equation of this inverse function be used for?

36 Find the inverse of the composite function found in Problem 34b. What can the defining equation of this inverse function be used for?

37 The volume of a sphere is given by the formula

$$V = \tfrac{4}{3}\pi r^3 = F(r)$$

where r is the radius. The surface area of the sphere is

$$S = 4\pi r^2 = G(r)$$

a Find $F(G^{-1}(S))$. **b** What can the formula $V = F(G^{-1}(S))$ be used for?

38 The accompanying table shows some experimental data on the thermal conductivity of copper.

Absolute temperature	10	15	20	30	40	60	80
Thermal conductivity	1.5	2.4	4.1	3.8	3.1	2.4	1.8

a Graph these points, taking temperature along the horizontal axis and conductivity along the vertical axis. Draw a smooth curve that displays the thermal conductivity as a function of the absolute temperature.
b If we write $C = f(T)$ for the function in part (a), is the inverse of f also a function? Explain. For what value of T (approximately) is $C = 3$?

39 Let $f_1(x) = \sqrt[4]{x} = x^{0.25}$, $f_2(x) = f_1(f_1(x))$, $f_3(x) = f_1(f_2(x))$, and so on, for $n > 1$, $f_n(x) = f_1(f_{n-1}(x))$. Take $x = 10$, and evaluate $f_1(10)$, $f_2(10)$, $f_3(10)$, and $f_4(10)$. (You can do this by using the y^x key on your calculator, or if you don't have such a key, recall that the fourth root is the square root of the square root.) What happens to the outcome as you take more and more compositions?

40 Repeat Problem 39 taking $x = 0.1$.

USING YOUR KNOWLEDGE 5.4

At the beginning of this section, you were told that the rate at which a house cricket chirps depends on the temperature. The relationship can be written as

$$n = 4(\mathrm{F} - 40)$$

where n is the number of chirps per minute and F is the temperature in degrees Fahrenheit.

It has also been claimed that the speed at which a certain kind of ant crawls depends on the temperature, and that relationship has been given as

$$d = \tfrac{1}{6}(C - 4)$$

where d is the speed in centimeters per second and C is the temperature in degrees Celsius. (We don't pretend to know why ants prefer the Celsius scale, whereas crickets prefer the Fahrenheit scale.)

It is also known that the relationship between the Fahrenheit and the Celsius scales is given by

$$\mathrm{F} = \tfrac{9}{5}\mathrm{C} + 32$$

If we use functional notation, all this information can be written as follows:

$$n = g(\mathrm{F}) = 4(\mathrm{F} - 40)$$
$$d = f(\mathrm{C}) = \tfrac{1}{6}(\mathrm{C} - 4)$$
$$\mathrm{F} = h(\mathrm{C}) = \tfrac{9}{5}\mathrm{C} + 32$$

Use this knowledge to do the following problems.

1 Find: **a** $g^{-1}(n)$ **b** $f^{-1}(d)$ **c** $h^{-1}(F)$

2 a Express d as a composite of f, h^{-1}, and g^{-1}. **b** Express d in terms of n. **c** If the cricket is chirping at the rate of 112 times per minute, how fast is the ant crawling?

SELF-TEST

1 a Graph $A(-1, -3)$ and $B(2, -5)$, and name the quadrant containing each point.
b Find the distance between the two points in part (a).

2 a Write the equation of the circle of radius 4 and with center at $(2, -3)$.
b Use the distance formula to determine whether the point $(5, -6)$ is inside or outside of the circle in (a).

3 Use the distance formula to determine whether or not the points $A(-3, 4)$, $B(2, -1)$, and $C(8, 6)$ are the vertices of a right triangle.

4 Find the distance from the point $P(1, -2)$ to the midpoint of the line segment from $A(4, -2)$ to $B(-6, 8)$.

5 a State the domain and the range of the relation $\{(x, y) | -1 < x < 2, -2 < y < 3\}$.
b Graph the relation in (a).

6 Graph the relation $\{(x, y) | y > x + 1\}$.

7 Find the equation of the circle in the accompanying figure.

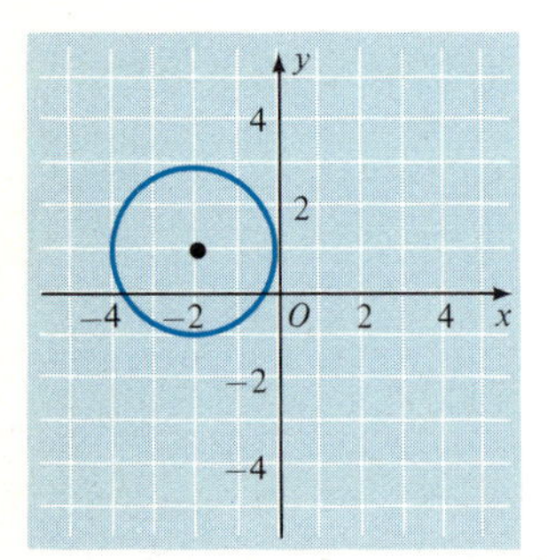

8 a Graph the relation $\{(x, y) | 3x = y^2\}$.
b Find the domain and the range of the relation in (a).

9 A set of points is such that each point $P(x, y)$ in the set is three times as far from $(0, 0)$ as from $(4, 0)$. Find and simplify the equation of the graph.

10 a Find the domain and the range of the function $f = \{(x, y) | y = 3x + 3\}$.
b Graph the function in (a).

11 a Let $f(x) = -x^2$. Find the domain and the range of f.
b Graph the function in (a).

12 Let $g(x) = \sqrt{x^2 - 9}$.
a Find the domain of g.
b If $g(x) = 4$, what is the value of x?

13 a Find the domain and the range of f if $f(x) = |x| + 1$.
b Graph the function in (a).

14 Suppose that the cost of renting a car is \$16 per day plus 15 cents per mile. Let x be the number of miles driven and $C(x)$ the daily cost in dollars.
a Find $C(50)$.
b Find a formula for $C(x)$ in terms of x.

15 Let $f(x) = \sqrt{x^2 + 16}$. Find:
a $f(3)$ **b** $f(x + h)$

16 Find the inverse of the function $f = \{(x, y) | y = x - 4\}$.

17 a If $f(x) = x^2 - 1$, what restriction should be put on x so that the restricted function has an inverse function?
b What is the inverse of the restricted f in (a)?

18 Let $f(x) = x^2 + 1$ and $g(x) = 2x - 1$. Find:
a $(f + g)(x)$ **b** $(f - g)(x)$
c $(fg)(x)$ **d** $(f/g)(x)$

19 Let $f(x) = x^2 + 2$ and $g(x) = \sqrt{x - 1}$.
a Find $(g \circ f)(x)$, that is, $g(f(x))$.
b What is the domain of $(g \circ f)(x)$?

20 Find two functions f and g, so that $F(x) = \sqrt[3]{x^2 - 1}$ can be written as the composite $f(g(x))$.

CHAPTER 6

SOME EQUATIONS IN TWO VARIABLES

6.1 LINEAR EQUATIONS

SLOPE

The solid line in Figure 6.1a shows the percentage of Fli-Hi flights arriving on time. As you can see, the on-time performance of Fli-Hi rose faster between 1979 and 1980 than between 1980 and 1981. This is evident because the line segment from 1979 to 1980 is "steeper" than the segment from 1980 to 1981. In mathematics, we measure the steepness of a line by comparing the *rise* of a segment with its *run*. The rise and the run as well as the rise and run ratio of the segments of the Fli-Hi graph for 1979–80 and 1980–81 are shown in Figure 6.1b. The ratio of rise to run is called the *slope* of the line segment and is denoted by the letter *m*. Thus,

$$m = \frac{\textbf{rise}}{\textbf{run}}$$

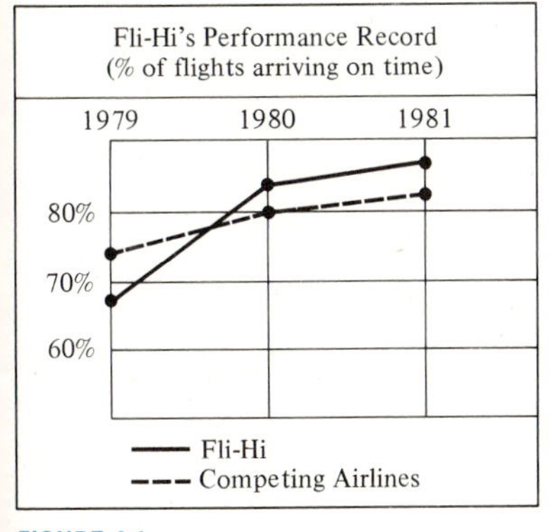

FIGURE 6.1a

Rise = 17.5%

1979 Run 1980

$\frac{\text{Rise}}{\text{Run}} = \frac{17.5}{1}$

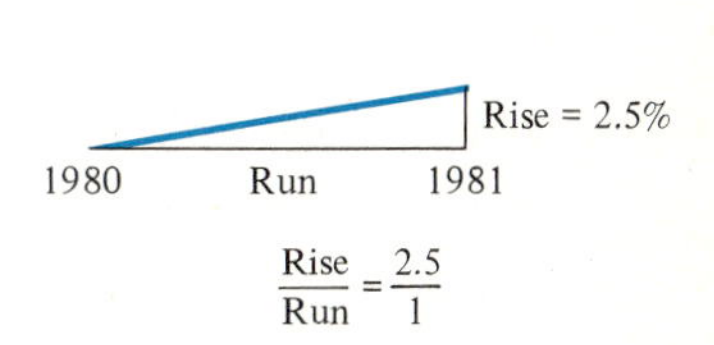

FIGURE 6.1b

▼ **DEFINITION 6.1a**

If $P_1(x_1, y_1)$ and $P_2(x_2, y_2)$ are any two distinct points on a line L that is not parallel to the y-axis, then the slope of L is

$$m = \frac{y_2 - y_1}{x_2 - x_1} = \frac{\text{change in } y}{\text{change in } x} \qquad (1)$$

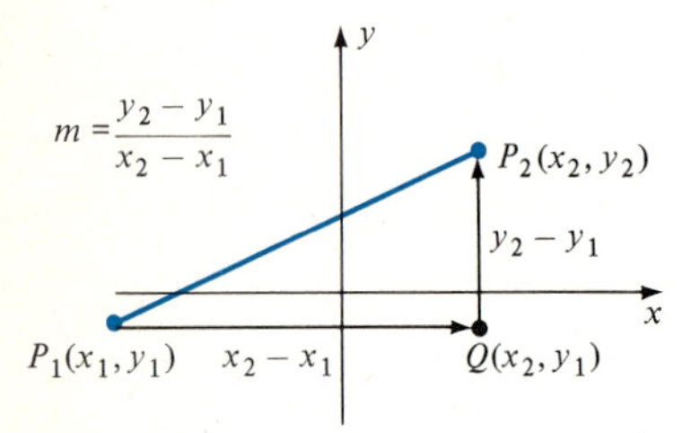

FIGURE 6.1c

This concept of slope is slightly generalized in Definition 6.1a, which allows for falling as well as rising lines. Figure 6.1c illustrates this definition. Note carefully that $x_2 - x_1$ and $y_2 - y_1$ are both *directed* distances. (In calculus it is customary to denote the change in x by Δx and the corresponding change in y by Δy. Thus, $\Delta x = x_2 - x_1$, $\Delta y = y_2 - y_1$, and $m = \Delta y/\Delta x$. Here are some remarks about Definition 6.1a:

1. Note that

$$\frac{y_2 - y_1}{x_2 - x_1} = \frac{y_1 - y_2}{x_1 - x_2}$$

so that it does not matter which point is named P_1 and which P_2. For instance, the slope of the line through $P_1(0, -6)$ and $P_2(3, 3)$ is

$$m = \frac{3 - (-6)}{3 - 0} = \frac{9}{3} = 3$$

If we choose P_1 to be (3, 3) and P_2 to be (0, −6), then

$$m = \frac{-6 - 3}{0 - 3} = \frac{-9}{-3} = 3$$

as before.

2. The slope given in Definition 6.1a is independent of the two points that are chosen on the line L. For example, if in Figure 6.1d we select the points P_3 and P_4 rather than P_1 and P_2, the resulting value of m is the same, because the triangles P_1AP_2 and P_3BP_4 are similar, so that the ratios of corresponding sides are equal; that is,

$$\frac{y_4 - y_3}{x_4 - x_3} = \frac{y_2 - y_1}{x_2 - x_1} = m$$

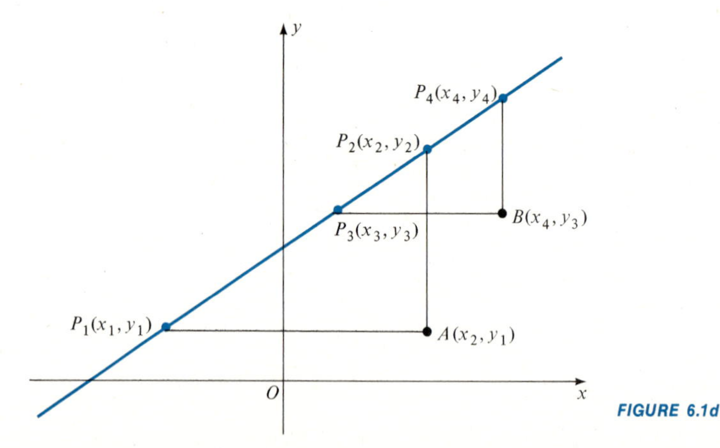

FIGURE 6.1d

3. The slope of a line is positive, negative, zero, or undefined as illustrated in Figures 6.1e, 6.1f, 6.1g, and 6.1h, respectively.

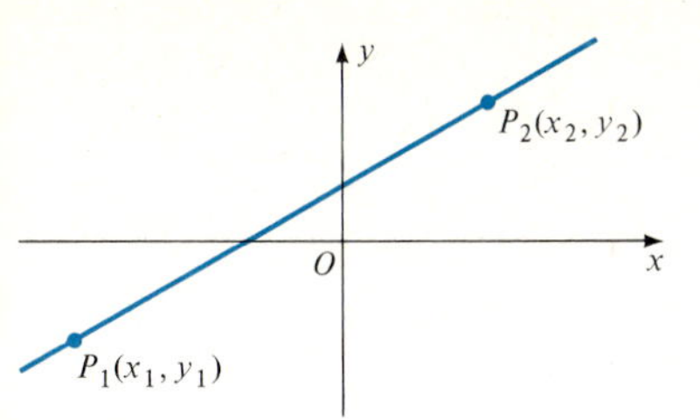

FIGURE 6.1e

A line with positive slope: $x_2 > x_1$ *and* $y_2 > y_1$, *so that* $x_2 - x_1$ *and* $y_2 - y_1$ *are both positive. Thus, m is positive.*

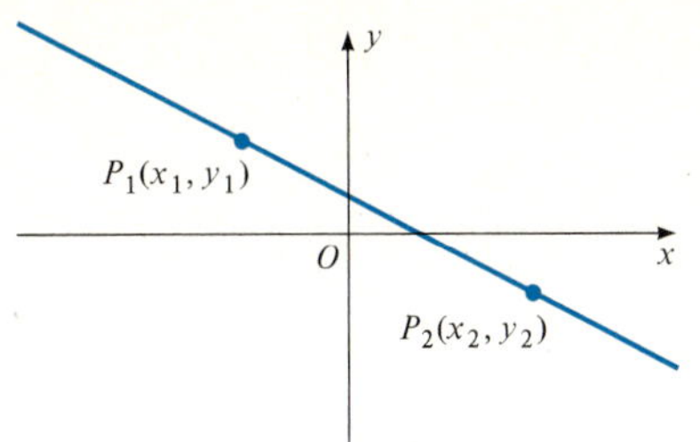

FIGURE 6.1f

A line with negative slope: $x_2 > x_1$ *and* $y_2 < y_1$, *so that* $x_2 - x_1$ *is positive and* $y_2 - y_1$ *is negative. Thus, m is negative.*

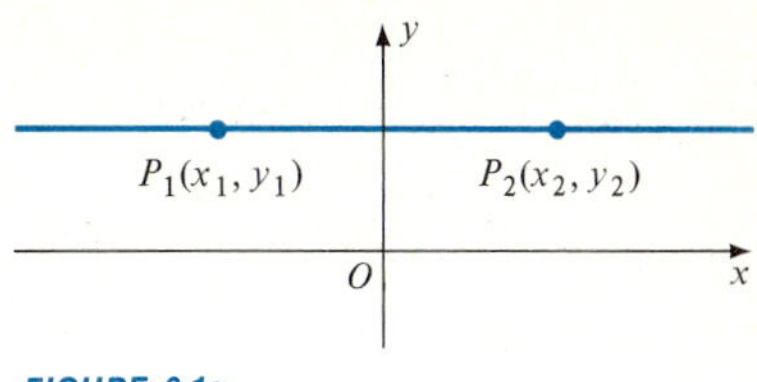

FIGURE 6.1g

A line with zero slope: $x_2 > x_1$ *and* $y_2 = y_1$. *Thus,* $m = 0$.

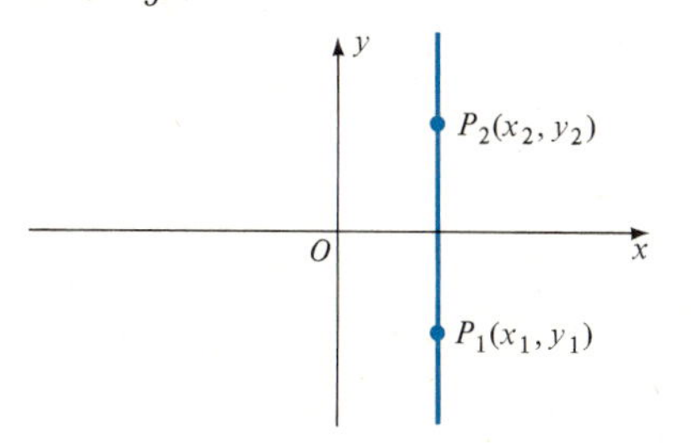

FIGURE 6.1h

A line whose slope is not defined: $x_2 = x_1$ *and* $y_2 > y_1$. *Thus,* $(y_2 - y_1)/(x_2 - x_1) = (y_2 - y_1)/0$, *which is meaningless. Definition 6.1a excludes this case.*

EXAMPLE 1

Find the slope of the line through the two points.

a. (2, 7) and (−5, −7)
b. (−3, 2) and (5, 2)
c. (3, −5) and (3, 2)

SOLUTION:

a. Using Formula 1, we find

$$m = \frac{-7-7}{-5-2} = \frac{-14}{-7} = 2$$

b. Note that the two points have the same y-coordinates, so that this is a horizontal line (zero rise) and the slope is zero. The formula agrees, giving

$$m = \frac{2-2}{5-(-3)} = \frac{0}{8} = 0$$

c. The slope of this line is *not defined* because it is a vertical line. Definition 6.1a does not apply, and if we try to use Formula 1, we obtain

$$\frac{2-(-5)}{3-3} = \frac{7}{0}$$

which is meaningless.

POINT-SLOPE EQUATION OF A LINE

If the slope and a point on the line are given, we can find an equation of the line. Thus, let $P_1(x_1, y_1)$ be a point on a line L with slope m. Then, another point $P(x, y)$ is on this line if and only if the slope of the line segment joining the two points is m; that is, if and only if

$$\frac{y - y_1}{x - x_1} = m$$

▼ DEFINITION 6.1b

The point-slope equation of a line

$$y - y_1 = m(x - x_1) \qquad (2)$$

Multiplying both sides of this equation by $x - x_1$, we obtain the equation in Definition 6.1b. Notice that (x_1, y_1) satisfies Equation 2.

EXAMPLE 2

Find an equation of the line with slope -2 and passing through $(-3, 4)$.

SOLUTION:

Using Equation 2 with $m = -2$, $x = -3$, and $y = 4$, we obtain

$$y - 4 = -2[x - (-3)]$$

or

$$y = -2x - 2$$

EXAMPLE 3

Find an equation of the line through the two points (3, 5) and (9, 2).

SOLUTION:

We first use Equation 1 to find the slope. Thus,

$$m = \frac{2 - 5}{9 - 3} = \frac{-3}{6} = -\frac{1}{2}$$

Now, we can substitute this value of m and one of the points, say (3, 5), into Equation 2 to obtain

$$y - 5 = -\tfrac{1}{2}(x - 3)$$

or

$$2y = -x + 13$$

as an equation of the line through the given points. Note that if we use the other point, (7, 3), in the point-slope equation, we obtain

$$y - 2 = -\tfrac{1}{2}(x - 9)$$

or

$$2y = -x + 13$$

again.

SLOPE-INTERCEPT EQUATION OF A LINE

An important form of an equation for a straight line can be obtained by using the slope and the point of intersection of the line and the y-axis. Let this point be labeled $(0, b)$, where b is the directed distance from the origin to the point of intersection; this distance is called the *y-intercept* of the line. If we substitute $(0, b)$ for (x_1, y_1) in the point-slope equation, we have

$$y - b = m(x - 0)$$

▼ DEFINITION 6.1c

The slope-intercept equation of a line

$$y = mx + b \qquad (3)$$

Adding b to both sides gives us the equation in Definition 6.1c. You should be sure to keep in mind that m is the slope and b is the y-intercept in Equation 3.

EXAMPLE 4

Find an equation of the line with slope $-\frac{5}{2}$ and y-intercept -6.

SOLUTION:

We substitute $m = -\frac{5}{2}$ and $b = -6$ into the slope-intercept equation and obtain

$$y = -\frac{5}{2}x - 6$$

EXAMPLE 5

Find the slope and the y-intercept of the line with equation $3x - 2y - 4 = 0$.

SOLUTION:

The best way to solve this problem is to write the equation in the slope-intercept form, which is simply done by solving for y. If we solve the given equation for y, we obtain

$$y = \frac{3}{2}x - 2$$

Thus, the slope is $\frac{3}{2}$, and the y-intercept is -2.

STANDARD FORM OF THE EQUATION OF A LINE

▼ DEFINITION 6.1d

The standard form of the equation of a line is

$$ax + by + c = 0 \tag{4}$$

where a, b, and c are real numbers, and a and b are not both zero.

As you might have noticed, both Equations 2 and 3 can be written in what is generally considered as standard form. Equation 4 in Definition 6.1d is the general *linear* or *first-degree* equation in x and y.

We can show that a linear equation in x and y always has a straight line for its graph. Indeed, if $b \neq 0$, we can solve Equation 4 for y to obtain

$$y = -\frac{a}{b}x - \frac{c}{b} \tag{5}$$

which is an equation of a line with slope $-a/b$ and y-intercept $-c/b$. In case $a = 0$ in Equation 5, we obtain the horizontal line

$$y = -\frac{c}{b} \tag{6}$$

If $b = 0$ and $a \neq 0$ in Equation 4, we can solve for x to obtain

$$x = -\frac{c}{a} \tag{7}$$

which is an equation for a vertical line.

In the preceding paragraph, we proved that the graph of Equation 4 is always a straight line, and it is interesting to know that every straight line is the graph of a linear equation. We combine this information in Theorem 6.1a.

▼ THEOREM 6.1a

The graph of (4), $ax + by + c = 0$, where a, b, and c are real numbers, and a and b are not both zero, is a straight line, and, conversely, every straight line is the graph of an equation of this form.

We can prove the second part of Theorem 6.1a as follows: We know that every vertical line has an equation of the form $x = k$, where k is a real number, and we can write this equation in the form $x + 0 \cdot y - k = 0$. If a line is not vertical, then it has a slope, say m, and a y-intercept, say b. Hence, the line has a slope-intercept equation $y = mx + b$. Again, we can write the equation in standard linear form, $mx - y + b = 0$. This completes the proof of the theorem.

If we have a linear function defined by $f(x) = mx + b$, we usually write $y = f(x)$, so that the graph of f coincides with the graph of $y = mx + b$. Thus, in proving Theorem 6.1a, we have incidentally proved that the graph of a linear function is a straight line.

Because the graph of any linear equation in x and y is a straight line and any two points determine a unique line, it is enough to locate two points in order to graph the equation. The easiest points to calculate are those involving the x- *and y-intercepts*, that is, the points where the line crosses the axes. Thus, to graph a linear equation, we simply let $x = 0$ and find the y-intercept, then let $y = 0$ and find the x-intercept. The line through the corresponding points is the desired line. It is a good idea to find a convenient third point as a check on the graph.

EXAMPLE 6

Graph the equation $2x + 3y - 9 = 0$.

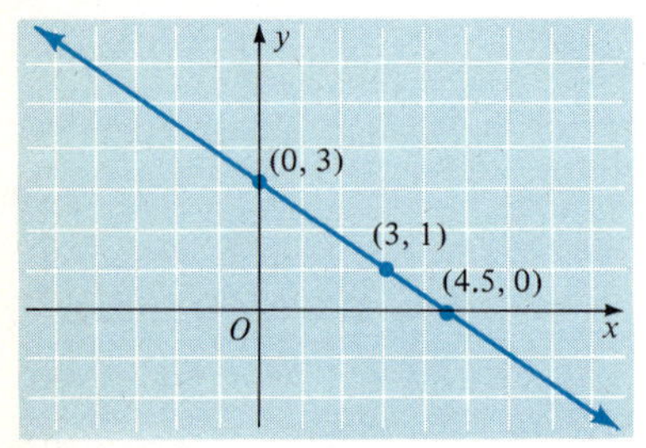

FIGURE 6.1i

SOLUTION:

If $x = 0$ in this equation, then $y = 3$, the y-intercept. If $y = 0$, then $x = 4.5$, the x-intercept. We graph the points (0, 3) and (4.5, 0), and draw the straight line through them. This line, which is shown in Figure 6.1i, is the desired graph. For a check, we substitute $x = 3$ into the equation and find $y = 1$. Hence, the point (3, 1) must be on the graph, and the figure shows that it is.

If the intercepts are inconvenient numbers to use or if they are too close together to permit an accurate drawing, then other points can be used, as in Example 7.

EXAMPLE 7

Graph the equation $6x + 5y - 4 = 0$.

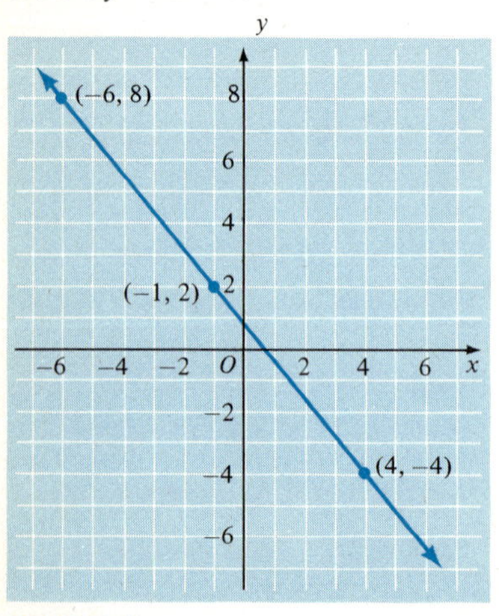

FIGURE 6.1j

SOLUTION:

For this line, the x- and y-intercepts are $\frac{2}{3}$ and $\frac{4}{5}$, respectively. The corresponding points are quite close together and difficult to plot accurately, so that it is desirable to graph the line by using other points. If we let $x = -1$, we find $y = 2$, and if we let $x = 4$, we find $y = -4$. Thus, the points $(-1, 2)$ and $(4, -4)$ are on the line. We graph these two points and draw the line as shown in Figure 6.1j. As a check, we note that for $x = -6$, $y = 8$, and the figure shows that $(-6, 8)$ is on the line.

If an equation is in the slope-intercept form, then it can be graphed by using the m and the b, as in Example 8.

EXAMPLE 8

Graph the equation $y = -\frac{3}{2}x + 2$.

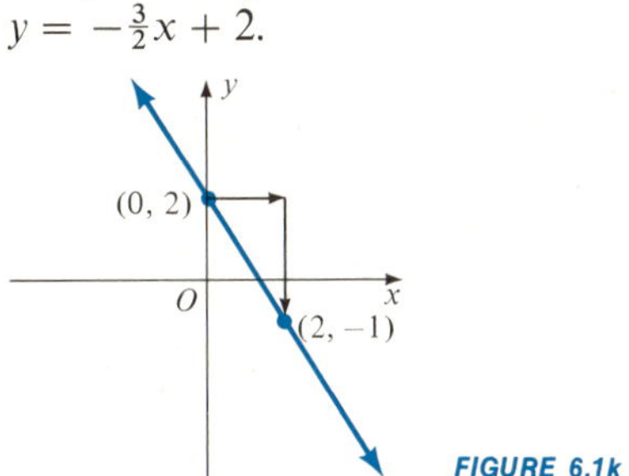

FIGURE 6.1k

SOLUTION:

Because the y-intercept, b, is 2, we graph the point (0, 2). Then, because the slope, m, is $-\frac{3}{2}$, we can locate a second point by going any convenient distance in the positive x-direction from (0, 2) and then $\frac{3}{2}$ this distance in the negative y-direction. In Figure 6.1k, we have gone 2 units to the right and 3 units down to locate the point $(2, -1)$ on the line. The graph is the line through the two points (0, 2) and $(2, -1)$.

PARALLEL AND PERPENDICULAR LINES

Sometimes, it is necessary to find equations for lines that are parallel or perpendicular to a given line. Theorem 6.1b is indispensable for solving such problems. The proof of this theorem is left for the exercises. Note that the condition given in Part II of the theorem is equivalent to

$$m_2 = -\frac{1}{m_1}$$

Hence, the slopes of two perpendicular lines are negative reciprocals of each other.

▼ THEOREM 6.1b

If L_1 and L_2 are two lines with slopes m_1 and m_2, respectively, then

I. L_1 and L_2 are *parallel* if and only if

$$m_1 = m_2$$

II. L_1 and L_2 are *perpendicular* if and only if

$$m_1 m_2 = -1$$

Using these ideas, you can verify that if $b \neq 0$, then

$$ax + by + c = 0$$

and

$$ax + by + k = 0 \qquad k \neq c$$

are equations of two *parallel* lines; both have the slope $-a/b$ (which you can see by writing the equations in slope-intercept form). Similarly, if $a \neq 0$ and $b \neq 0$, then

$$ax + by + c = 0$$

and

$$bx - ay + k = 0$$

are equations of two *perpendicular* lines; the first has slope $-a/b$ and the second has slope b/a, which are negative reciprocals of each other. Notice in this case that you can write the second equation by interchanging the coefficients of x and y in the first equation and changing the sign of one of them.

EXAMPLE 9

Write in standard form the equation of the line that passes through the point $(2, -3)$ and is

a. parallel to the line $6x - 3y + 2 = 0$;

b. perpendicular to the line $6x - 3y + 2 = 0$.

SOLUTION:

a. We use the preceding discussion and write $6x - 3y + k = 0$ for the equation of a line parallel to the given line. We evaluate k by substituting the point $(2, -3)$ into this equation. Thus,

$$(6)(2) - (3)(-3) + k = 0$$

so that $k = -21$, and the required equation is

$$6x - 3y - 21 = 0$$

or

$$2x - y - 7 = 0$$

b. We follow a similar procedure and write $3x + 6y + k = 0$ for the equation of a line perpendicular to the given line. Notice that we interchanged the 6 and the -3 and changed the sign of the -3. Again, we evaluate k by substituting the point $(2, -3)$ into the equation. Thus,

$$(3)(2) + (6)(-3) + k = 0$$

so that $k = 12$. Hence, the required equation is

$$3x + 6y + 12 = 0$$

or

$$x + 2y + 4 = 0$$

You can check the answers in this example by noting that $(2, -3)$ satisfies both equations. The slope of the line $2x - y - 7 = 0$ is 2, the slope of the given line. The slope of $x + 2y + 4 = 0$ is $-\frac{1}{2}$, the negative reciprocal of the slope of the given line. This completes the check.

EXERCISE 6.1

In Problems 1–8, find the slope of the line that passes through the given points.

1 $(-3, 5)$ and $(2, 0)$

2 $(7, -1)$ and $(-2, 5)$

3 $(-2, -4)$ and $(-2, 1)$

4 $(-2, -4)$ and $(-5, -4)$

5 $(\frac{2}{5}, \frac{1}{2})$ and $(1, 2)$

6 $(1.5, 0.2)$ and $(0.5, -0.2)$

7 $(\sqrt{8}, \sqrt{27})$ and $(\sqrt{2}, 5\sqrt{3})$

8 $(\sqrt{32}, \sqrt{75})$ and $(\sqrt{18}, \sqrt{3})$

In Problems 9–26, find an equation in standard form for the line that satisfies the given conditions.

9 through $(4, -2)$, slope $\frac{1}{2}$

10 through $(-3, 5)$, slope -3

11 slope 5, y-intercept 3

12 slope -2, y-intercept -4

13 through $(2, 0)$ and $(10, 3)$

14 through $(-3, 5)$ and $(-4, 4)$

15 through $(1, 2)$ and parallel to the line $3x - 2y + 5 = 0$

16 through $(-1, 2)$ and parallel to the line $x + 2y + 14 = 0$

17 through $(-2, 3)$ and perpendicular to the line $x + 3y - 6 = 0$

18 through $(-4, -2)$ and perpendicular to the line $x - 3y - 8 = 0$

19 with y-intercept -2 and parallel to the line $3y = 6x - \frac{1}{2}$

20 with y-intercept 3 and parallel to the x-axis

21 with y-intercept -1 and parallel to the x-axis

22 through (3, 5) and parallel to (a) the x-axis; (b) the y-axis

23 through $(-2, 5)$ and perpendicular to (a) the x-axis; (b) the y-axis

24 through $(4, -1)$ and with no slope (slope undefined)

25 through $(-2, 3)$ and with no slope (slope undefined)

26 through (5, 3) and with slope 0

In Problems 27–32, an equation of a line is given. (a) Find the x-intercept; (b) find the y-intercept; (c) find the slope; (d) draw the graph.

27 $3y = 6x + 9$

28 $2y = -4x + 8$

29 $x - 2y + 4 = 0$

30 $-x + 3y + 6 = 0$

31 $x = 3$

32 $y = -1$

In Problems 33–40, find the slope-intercept equation of a line that passes through (1, 0) and is (a) parallel to the given line; (b) perpendicular to the given line.

33 $2y = 4x + 6$

34 $-3y = 6x - 2$

35 $y = 5$

36 $x = 2$

37 $-x + y - 3 = 0$

38 $2x + 4y + 5 = 0$

39 $x = 4y + 3$

40 $-3x = 5y + 2$

41 A line has slope $m = 2$ and passes through the points (3, 5) and $(4, a)$. Find a.

42 A line has slope $m = -3$ and passes through the points $(-2, 4)$ and $(a, 10)$. Find a.

43 A line has slope $m = \frac{1}{2}$ and passes through the points $(8, b)$ and $(b, 2)$. Find b.

44 A line has slope $m = -3$ and passes through the points $(b, 2b)$ and $(-5, 5)$. Find b.

45 Refer to Figure 6.1a and draw a straight line from the Fli-Hi point for 1979 to the Fli-Hi point for 1981. What is the slope of this line? Interpret this slope in terms of percentage increase in on-time performance per year.

46 Refer to Figure 6.1a and notice that the dashed line from 1979 to 1981 is nearly straight. If this were a straight line, what would be its slope? Interpret this slope in terms of percentage increase in on-time performance per year.

Use your calculator in Problems 47–50.

47 The x-intercept of a line is 2.34, and the slope is 1.56. Find the y-intercept correct to the nearest hundredth.

48 Write the equation $6.92x - 3.76y - 2.39 = 0$ in the slope-intercept form with the coefficients correct to the nearest hundredth.

49 The equation of a line is $0.697x - 3.125y + 1.608 = 0$. Find x- and y-intercepts correct to the nearest thousandth.

50 A line passes through the two points (2.317, 4.526) and (0.523, -6.371). Find the slope-intercept equation of the line with coefficients correct to the nearest thousandth.

51 A line passes through the two points (x_1, y_1), (x_2, y_2), $x_1 \neq x_2$. Show that an equation of the line may be written in the *two-point form*

$$\frac{y - y_1}{x - x_1} = \frac{y_2 - y_1}{x_2 - x_1}$$

52 a Show that if a line has nonzero x- and y-intercepts, a and b, respectively, then an equation of the line may be written in the *intercept form*

$$\frac{x}{a} + \frac{y}{b} = 1$$

b If a line cuts the x-axis at (25, 0) and the y-axis at (0, 40), find an equation of the line by using the result in (a).

53 Consider the two lines

L_1: $Ax + By = C$ and L_2: $Ax + By = D$

Show that L_1 and L_2 are parallel if $C \neq D$.

54 If $AB \neq 0$, show that the lines

L_1: $Ax + By = C$ and L_2: $Bx - Ay = D$

are perpendicular.

55 Prove Theorem 6.1b, Part I.

Hint: In Figure 6.11, if the lines L_1 and L_2 are parallel, then the indicated angles are equal, so that the two right triangles are similar. Use this idea to show that the slopes are equal.

On the other hand, if the slopes are equal, then the triangles are similar. Use this idea to show that the lines are parallel.

Hint: Two triangles are similar if and only if their corresponding angles are equal. Furthermore, two triangles are similar if and only if their corresponding sides are proportional.

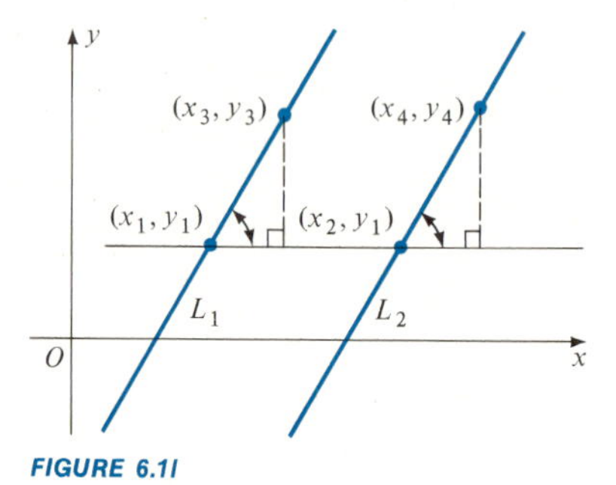

FIGURE 6.1l

56 Prove Theorem 6.1b, Part II.

Hint: Let $y = m_1x$ and $y = m_2x$ be lines through the origin and parallel to L_1 and L_2, respectively. Then L_1 is perpendicular to L_2 if and only if the triangle with vertices at $A(1, m_1)$, $B(1, m_2)$, $C(0, 0)$ is a right triangle with right angle at C; that is, if and only if

$(AC)^2 + (BC)^2 = (AB)^2$

USING YOUR KNOWLEDGE 6.1

In economics, we are interested in the total cost $C(x)$ of producing x units of a product whose producing x units of a product whose production cost is m dollars per unit and whose fixed cost is b dollars. This total cost is given by

$C(x) = mx + b$

(Notice that m and b are what we have called the slope and the y-intercept, respectively.)

1 Find $C(x)$ for a company that produces radios at \$3 per unit and with fixed costs of \$8000.
2 Find $C(x)$ for a company that produces sport coats at \$17 per unit and with fixed costs of \$7800.
3 The total cost $C(x)$, in dollars, of producing x units of a certain product is

$C(x) = 4x + 9000$

a What is the production cost per unit? **b** What are the fixed costs?
c What is the cost of producing 1000 units?

4 The total cost $C(x)$, in dollars, of producing x television sets is

$C(x) = 60x + 12{,}000$

a What is the production cost per set? **b** What are the fixed costs?
c If 1000 sets are produced, what is the actual average cost per set?
d If x sets are produced, write an equation giving the actual average cost per set.

6.2 THE EQUATION $y = ax^2 + bx + c$

Figure 6.2a shows the path followed by a ball that is projected horizontally. This path is in the shape of a curve that is called a *parabola*. (See Definition 6.2a.)

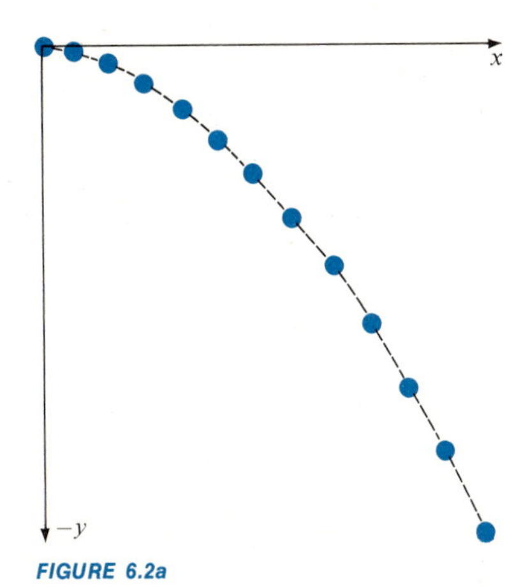

FIGURE 6.2a

▼ DEFINITION 6.2a

If a, b, c are real numbers, $a \neq 0$, the graph of

$$f(x) = ax^2 + bx + c \quad (1)$$

or

$$y = ax^2 + bx + c \quad (2)$$

is a *parabola*.

GRAPHING PARABOLAS

We graphed the parabola $y = x^2$ in Figure 5.3d, Section 5.3. The graph of Equation 1 or 2 has a similar shape, the exact shape depending on the number a. The numbers b and c, along with a, locate the curve in the xy-plane. In order to see how this happens, first change the form of the quadratic polynomial by completing the square (compare Section 4.4). We can write

$$\begin{aligned} y = f(x) &= ax^2 + bx + c \\ &= a\left(x^2 + \frac{b}{a}x\right) + c \\ &= a\left(x^2 + \frac{b}{a}x + \frac{b^2}{4a^2}\right) + c - \frac{b^2}{4a} \end{aligned}$$

Thus, we obtain

$$y = a\left(x + \frac{b}{2a}\right)^2 + c - \frac{b^2}{4a}$$

Because a may be a positive or a negative number, we discuss these two cases.

Case 1. $a > 0$

The minimum value of $y = f(x)$ occurs for $x = -b/(2a)$, because then $[x + b/(2a)]^2 = 0$, and for all other values of x, $a[x + b/(2a)]^2 > 0$. Therefore,

$$f\left(-\frac{b}{2a}\right) = c - \frac{b^2}{4a} < f(x) \qquad \text{for all } x \neq -\frac{b}{2a}$$

It follows that

$$\left(-\frac{b}{2a}, c - \frac{b^2}{4a}\right)$$

is the *lowest* point on the curve. This point is called the *vertex* of the parabola.

Case 2. $a < 0$
As before, if $x = -b/(2a)$, then $[x + b/(2a)]^2 = 0$, but for all other values of x, the quantity $a[x + b/(2a)]^2 < 0$, because a is negative. Therefore,

$$f\left(-\frac{b}{2a}\right) = c - \frac{b^2}{4a} > f(x) \qquad \text{for all } x \neq \frac{b}{2a}$$

It follows that

$$\left(-\frac{b}{2a}, c - \frac{b^2}{4a}\right)$$

the *vertex* of the parabola, is the *highest* point on the curve.

Knowing that the vertex is the lowest point when a is positive and is the highest point when a is negative, we can conclude that the parabola given by

$$y = ax^2 + bx + c$$

opens *upward* if a is *positive* and opens *downward* if a is *negative.*

Note that you do not need to remember both coordinates of the vertex. The x-coordinate is $-b/(2a)$, and you can find the other coordinate by substituting into the equation. For example, if the equation is

$$y = 2x^2 + 8x + 7$$

then the x-coordinate of the vertex is $-b/2a = -\frac{8}{4} = -2$ and for $x = -2$,

$$y = 2(-2)^2 + (8)(-2) + 7 = -1$$

Thus, the vertex is at $(-2, -1)$.

We can find the y-intercept of the parabola by letting $x = 0$ in the equation. This gives $y = c$. The x-intercepts (if there are any) are found by substituting $y = 0$ in the equation to obtain

$$ax^2 + bx + c = 0$$

If this equation has real solutions, these give us the intercepts. Of course, if the solutions are not real numbers, then the parabola does not intersect the x-axis.

Another important result follows from the completed-square form of equation

$$y = a\left(x + \frac{b}{2a}\right)^2 + c - \frac{b^2}{4a}$$

We note that two values $x = -b/(2a) \pm k$, where k is any real number, give the same value for y because

$$y = a\left(-\frac{b}{2a} \pm k + \frac{b}{2a}\right)^2 + c - \frac{b^2}{4a}$$

$$= ak^2 + c - \frac{b^2}{4a}$$

This means that the curve is symmetric to the line $x = -b/(2a)$ which is called the *axis* of the parabola.

The preceding discussion can be summarized by the following procedure for graphing a parabola given by Equation 1 or 2.

PROCEDURE FOR GRAPHING THE PARABOLA
$y = ax^2 + bx + c, a \neq 0$

1. **Find and graph the vertex and the axis of the parabola. Remember that the x-coordinate of the vertex is $-b/(2a)$ and the axis is the line $x = -b/(2a)$.**
2. **Find and graph the x-intercepts by lettering $y = 0$ and finding the corresponding values of x (if they are real numbers).**
3. **Find and graph the y-intercept by letting $x = 0$ and finding the corresponding value of y.**
4. **Find one or two other points on the curve by substituting convenient values of x and calculating the corresponding values of y. Note that each point leads to a symmetric point on the other side of the axis of the parabola.**
5. **Connect all the points found in the preceding steps with a smooth curve, keeping in mind that the parabola opens upward if a is positive and downward if a is negative.**

EXAMPLE 1

Graph the following two parabolas on the same set of axes.

a. $y = 2x^2 + 4x - 6$
b. $y = 2x^2 - 8x + 8$

SOLUTION:

a. **1.** The x-coordinate of the vertex of the parabola

$$y = 2x^2 + 4x - 6$$

is $-\frac{4}{4} = -1$. For $x = -1$, $y = -8$. Hence, the vertex is at $(-1, -8)$, and the axis is the line $x = -1$.

2. If $y = 0$, then $2x^2 + 4x - 6 = 0$. By factoring, we obtain

$$2(x - 1)(x + 3) = 0$$

giving $x = -3$ and $x = 1$. Notice that $(-3, 0)$ and $(1, 0)$ are symmetric with respect to the axis of the parabola.

3. If $x = 0$, then $y = -6$, and, by symmetry with respect to the line $x = -1$, we know that $(-2, -6)$ is also on the graph.

4. Letting $x = 2$, we obtain $y = 10$. Thus, $(2, 10)$ is on the graph, and, by symmetry, $(-4, 10)$ is also on the graph.

5. The graph is the lower of the two parabolas shown in Figure 6.2b.

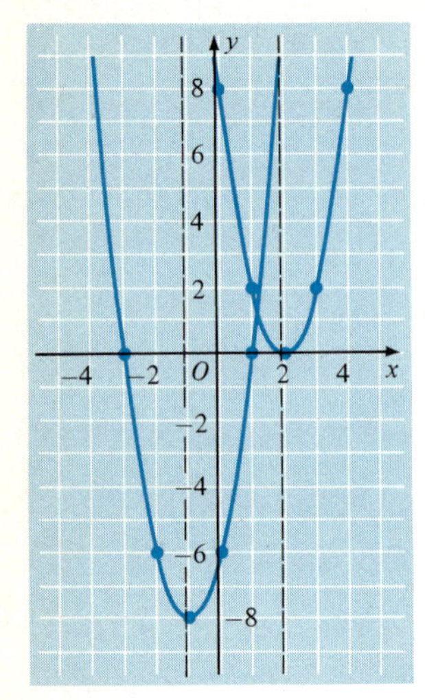

FIGURE 6.2b

b. 1. The x-coordinate of the vertex of the parabola

$$y = 2x^2 - 8x + 8$$

is $-(-8/4) = 2$. For $x = 2$, $y = 0$, so that the vertex is at $(2, 0)$ and the axis is the line $x = 2$.

2. Because the vertex is at $(2, 0)$, there can be no other x-intercept.

3. If $x = 0$, then $y = 8$, so that $(0, 8)$ is on the graph. By symmetry with respect to the line $x = 2$, we find that $(4, 8)$ is also on the graph.

4. If $x = 1$, then $y = 2$. Thus, the point $(1, 2)$ is on the curve, and, by symmetry, the point $(3, 2)$ is also on the curve.

5. The graph is the upper parabola in Figure 6.2b.

Notice that the two parabolas in Figure 6.2b have exactly the same shape and both equations have the same coefficient, $a = 2$. The two curves are simply in different positions in the plane.

EXAMPLE 2

Graph the following two equations on the same set of axes.

a. $y = -4x^2 + 1$

b. $y = -\frac{1}{4}x^2 + 1$

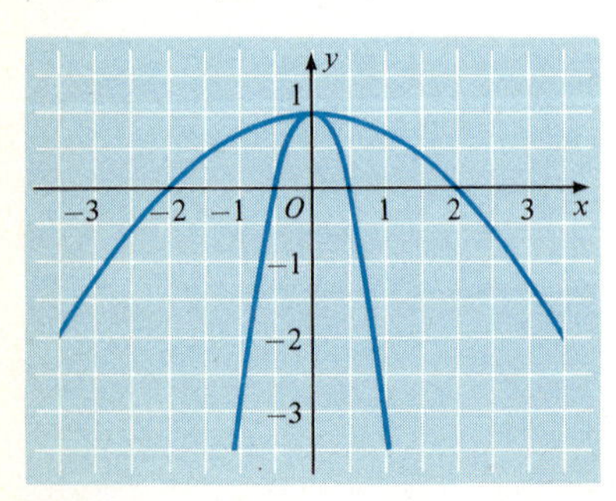

FIGURE 6.2c

SOLUTION:

a. 1. The x-coordinate of the vertex of the parabola

$$y = -4x^2 + 1$$

is 0 because there is no first-degree term in x. For $x = 0$, we have $y = 1$. Thus, the vertex is at $(0, 1)$, and the line $x = 0$ (the y-axis) is the axis of the parabola.

2. For $y = 0$, we have $-4x^2 + 1 = 0$, so that $x = \pm\frac{1}{2}$.

3. The parabola cuts the y-axis at $(0, 1)$.

4. For $x = 1$, we find $y = -3$. Thus, $(1, -3)$ is on the graph, and, by symmetry, $(-1, -3)$ is also on the graph.

5. The graph is the narrower of the parabolas shown in Figure 6.2c.

b. 1. Just as in part (a), we find the vertex is at $(0, 1)$, and the y-axis is the axis of the parabola.

2. For $y = 0$, we have $-\frac{1}{4}x^2 + 1 = 0$, so that $x = \pm 2$.

3. The curve cuts the y-axis at $(0, 1)$
4. For $x = 4$, we get $y = -3$, so that $(4, -3)$ is on the graph. By symmetry, $(-4, -3)$ is also on the curve.
5. The graph is the wider of the two parabolas in Figure 6.2c.

Notice that the coefficient a is negative in both equations in Example 2 and that both parabolas open downward. Also notice that the absolute value of a in the equation $y = -4x^2 + 1$ is greater than the absolute value of a in the equation $y = -\frac{1}{4}x^2 + 1$. The effect of this is to make the first parabola *narrower* than the second. In general, the *larger* the absolute value of a, the *narrower* is the parabola.

QUADRATIC INEQUALITIES

Our knowledge of parabolas is useful in solving quadratic inequalities of the type $f(x) > 0$ or $f(x) < 0$, where

$$f(x) = ax^2 + bx + c.$$

Let us write

$$y = ax^2 + bx + c.$$

Then, we keep in mind that $y > 0$ when the parabola is above the x-axis, and $y < 0$ when the parabola is below the x-axis. If we are interested only in solving the inequality, then a very rough sketch of the curve will settle the question. Example 3 illustrates the idea.

EXAMPLE 3

Solve the inequality $x^2 - 6x + 8 > 0$.

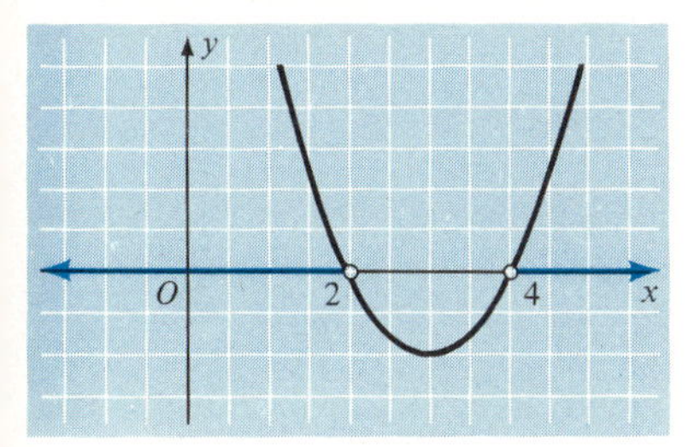

FIGURE 6.2d

SOLUTION:

We write $y = x^2 - 6x + 8$. Then, we see that the parabola opens upward and has its vertex at $(3, -1)$. For $y = 0$, we have

$$x^2 - 6x + 8 > 0.$$

or

$$(x - 2)(x - 4) = 0$$

which gives

$$x = 2 \quad \text{or} \quad x = 4$$

We now make the sketch shown in Figure 6.2d, from which we can read the solution of the inequality. This corresponds to the portions of the x-axis where the curve lies above the axis:

$$x < 2 \quad \text{or} \quad x > 4$$

EXAMPLE 4

Show that $x^2 + 2x + 2 > 0$ for all real values of x.

SOLUTION:

We write $y = x^2 + 2x + 2$. Then, we see that the parabola opens *upward*. Next, we find the vertex to be at the point $(-1, 1)$, which is *above* the x-axis. These two results show that the entire curve must lie above the x-axis, and therefore,

$$x^2 + 2x + 2 > 0$$

for all real values of x.

For the parabola $y = ax^2 + bx + c$, we have seen that the vertex is the maximum (highest) point if the curve opens downward and is the minimum (lowest) point if the curve opens upward. There are certain problems in which this information is quite useful, as Example 5 illustrates.

EXAMPLE 5

A long rectangular piece of tin, 10 centimeters wide, is to be bent into a rectangular channel, as shown in Figure 6.2e. How deep should the channel be if its cross-sectional area is to be as large as possible?

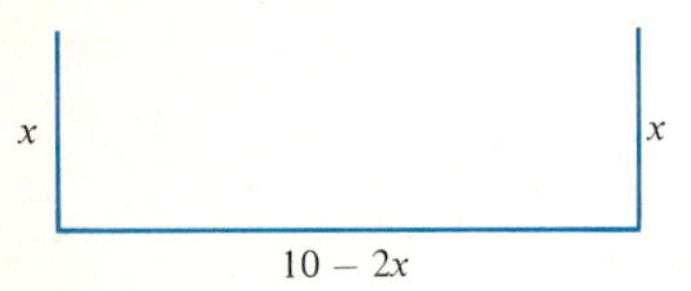

FIGURE 6.2e

SOLUTION:

As indicated in the figure, we let the depth of the channel be x centimeters. Then, the width is $(10 - 2x)$ centimeters. If the cross-sectional area is y square centimeters.

$$y = x(10 - 2x) = -2x^2 + 10x$$

This is the equation of a parabola that opens downward and hence has its vertex as the highest point. The x-coordinate of the vertex is $-(10/-4) = \frac{5}{2}$, and, for $x = \frac{5}{2}$, we obtain $y = \frac{25}{2}$. Accordingly, taking $x = 2.5$ gives the largest y-value possible, $y = 12.5$. Thus, a depth of 2.5 centimeters gives the maximum cross-sectional area, 12.5 square centimeters.

PARABOLAS AND ROOTS OF QUADRATIC EQUATIONS

As another important application, we note that the graph of the parabola $y = ax^2 + bx + c$ indicates the nature of the roots of the equation $ax^2 + bx + c = 0$. As you may recall, a quadratic equation in one variable may have *one* real solution, *two* real solutions, or *no* real solution. Corresponding to *one* real solution, the graph of the parabola touches the x-axis at just *one* point; corresponding to *two* real solutions, the graph cuts the x-axis at *two* distinct points; and, corresponding to *no* real solutions, the graph has *no* points in common with the x-axis. For example, the lower parabola in Figure 6.2b intersects the x-axis at two points. Hence, the corresponding quadratic equation $2x^2 + 4x - 6 = 0$ has two real solutions. The upper parabola in Figure 6.2b touches the x-axis at just one point, and the corresponding equation, $2x^2 - 8x + 8 = 0$ has exactly one solution. Finally, the graph of the equation $y = x^2 + 2x + 2$ (see Example 4) has its vertex at $(-1, 1)$ and opens upward. Thus, this graph has no points in common with the x-axis, and the corresponding equation $x^2 + 2x + 2 = 0$, has no real roots, as you can easily check.

EXAMPLE 6

Determine the nature of the roots of the quadratic equation $2x^2 + 4x + 1 = 0$

SOLUTION:

We let $y = 2x^2 + 4x + 1$, which we see to be the equation of a parabola that opens upward. The vertex of this parabola is at the point $(-1, -1)$, which is *below* the x-axis. Hence, the curve cuts the x-axis in two distinct points, so that the quadratic equation $2x^2 + 4x + 1 = 0$ has two distinct real roots. You can check this by solving the equation.

EXAMPLE 7

Return to Figure 6.2a and suppose that the ball is projected horizontally with an initial velocity of 10 feet per second. If air resistance is neglected, then the horizontal distance (x feet) of the ball from its starting point at the end of t seconds is given by

$$x = 10t$$

It is known that the ball falls so that the y distance is

$$y = -16t^2$$

Find the equation of the path of the ball.

SOLUTION:

The required equation of the path is the direct relationship between x and y, which we can obtain by eliminating the variable t from the pair of equations $x = 10t$ and $y = -16t^2$. By solving the first of these equations for t, we obtain

$$t = \frac{x}{10}$$

We then substitute this expression for t in the second equation to obtain

$$y = -16\left(\frac{x}{10}\right)^2 = -\frac{16}{100}x^2$$

or

$$y = -\frac{4}{25}x^2$$

the equation of a simple parabola with its vertex at the origin and opening downward. Note that if $y = f(t) = -16t^2$ and $t = g(x) = x/10$, then $y = f(g(x))$ will be the required answer.

EXERCISE 6.2

In Problems 1–12, use the procedure in the text to graph the given equation or function.

1 $y = x^2 + 3$
2 $y = -x^2 + 1$
3 $y = -x^2 - 2x$
4 $y = x^2 - 3x$
5 $f(x) = x^2 - 4x - 4$
6 $f(x) = x^2 + 3x + 2$
7 $g(x) = 2x^2 + 3x - 1$
8 $h(x) = -3x^2 - 5x - 2$
9 $y = \frac{1}{3}x^2 - \frac{1}{2}x + \frac{3}{16}$
10 $y = \frac{1}{3}x^2 + \frac{1}{3}x - \frac{1}{4}$
11 $y = -x^2 + 2x - 1$
12 $2y = -x^2 - 2x + 8$

In Problems 13–24, use the results of Problems 1–12 to solve the given inequality.

13 $x^2 + 3 > 0$
14 $-x^2 + 1 < 0$
15 $-x^2 - 2x \leq 0$
16 $x^2 - 3x \geq 0$
17 $x^2 - 4x > 4$
18 $x^2 + 3x \geq -2$
19 $2x^2 \leq -3x + 1$
20 $-3x^2 \geq 5x + 2$
21 $\frac{1}{3}x^2 < \frac{1}{2}x - \frac{3}{16}$
22 $\frac{1}{3}x^2 + \frac{1}{3}x \geq \frac{1}{4}$
23 $2x - x^2 \geq 1$
24 $-x^2 - 2x + 8 \geq 0$

In Problems 25–30, use the graph constructed in the indicated preceding problem to determine the nature of the roots of the corresponding quadratic equation.

25 Problem 1 **26** Problem 2 **27** Problem 3

28 Problem 5 **29** Problem 7 **30** Problem 9

In each of Problems 31–40, graph the two given equations on the same coordinate axes.

31 a $y = x^2 + 1$ **b** $y = x^2 - 1$ **32 a** $y = x^2 + 3$ **b** $y = x^2 + 5$ **33 a** $y = -x^2$ **b** $y = -x^2 + 1$

34 a $y = -x^2 - 1$ **b** $y = -x^2 - 3$ **35 a** $x = y^2$ **b** $3x = y^2$ **36 a** $x = y^2$ **b** $4x = y^2$

37 a $x + 1 = y^2$ **b** $x + 3 = y^2$ **38 a** $x = -y^2$ **b** $x + 1 = -y^2$ **39 a** $2x + 4 = -y^2$ **b** $2x + 6 = -y^2$

40 a $3x + 9 = y^2$ **b** $6x + 9 = y^2$

41 An open trough is to be made by bending a long, flat piece of tin, 20 inches wide, into a rectangular shape. What must be the depth of the trough if the cross-sectional area is to be a maximum?

42 A farmer wishes to fence in a rectangular piece of land that is adjacent to a straight river, so that only three sides need to be fenced. (The river side needs no fence.) What is the maximum area that he can fence with 400 feet of fencing?

43 A rectangular plot is to be fenced and then divided into four rectangular lots by running fences across the middle both ways. If 600 meters of fencing are available, what must be the dimensions of the plot to give maximum total fenced area?

44 A rectangular field is to be fenced and then divided into five lots by fences running parallel to one of the sides. What is the maximum total area that can be enclosed with 540 meters of fencing?

45 In Example 7 of this section, suppose that the ball is projected from a building 100 feet high. How far from the foot of the building will the ball hit the ground?

46 A rifle fires a bullet with an initial velocity of 3000 feet per second in a horizontal direction from the top of the building in Problem 45. If air resistance is neglected and there are no obstructions, how far from the foot of the building will the bullet strike the ground?

Use your calculator to do Problems 47–50.

C **47** Find the vertex and the x-intercepts of the parabola

$$y = 46x^2 - 94x - 125$$

(Give answers to the nearest hundredth.)

C **48** Find the equation of the parabola of the form $y = ax^2$ if the curve passes through the point (3.67, 15.75). Obtain a to the nearest hundredth.

C **49** A merchant finds that the profit on a certain article depends on the price x (dollars) according to the equation

$$P = 25 + 16x - 4x^2$$

Find the profit for the following prices: \$1.55, \$1.75, \$2.15, and \$2.25.

C **50** If a bullet is fired straight up with a muzzle velocity of 1500 feet per second and air resistance is neglected, the height y (feet) of the bullet at the end of t seconds is given by

$$y = -16t^2 + 1500t$$

How high will the bullet go? Express your answer in miles to the nearest tenth of a mile (1 mile = 5280 feet).

USING YOUR KNOWLEDGE 6.2

Because of their esthetic as well as their engineering properties, parabolic arches are often used in buildings and bridges. If the height and span of a parabolic arch are known, we can find the equation of the parabola. This is equivalent to the problem of finding the equation of a parabola with a vertical axis if the origin is taken at the vertex and another point on the curve is known. The required equation is of the form $y = ax^2$, and the value of a can be found by substituting the values of x and y at the known point into the equation and then solving for a. For example, suppose the curve passes through the point $(2, -1)$. Then

$$-1 = (a)(2^2)$$

Thus, $a = -\frac{1}{4}$, and the desired equation is $y = -\frac{1}{4}x^2$.

1 Suppose that a parabolic arch is 10 feet high and 20 feet wide at the base. Find the equation of the parabola if the origin is taken at the highest point.

2 A parabolic arch is 20 feet high and 10 feet wide at the base. **a** Find the equation of the parabola if the origin is taken at the highest point. **b** How high is the arch at a horizontal distance of 2 feet from the center line?

6.3 THE CONIC SECTIONS

Figure 6.3a shows the various curves that can be cut from a cone by a plane. Because they can be obtained in this way, these curves, the *ellipse*, the *parabola*, and the *hyperbola*, are called *conic sections*. It is shown in analytic geometry that the equation of a conic section can always be put in the form

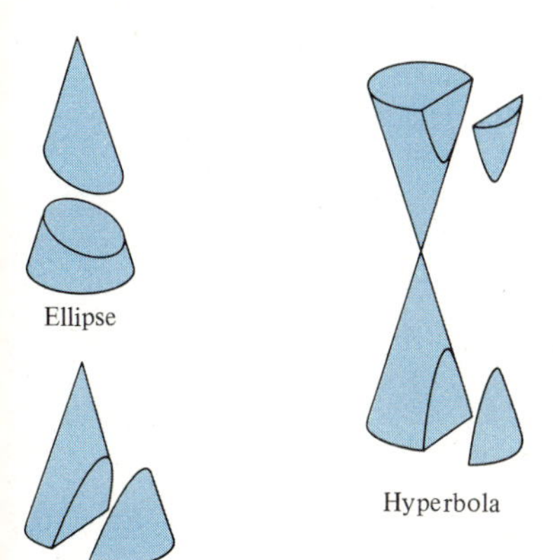

FIGURE 6.3a

$$Ax^2 + By^2 + Cx + Dy + E = 0 \tag{1}$$

where the coefficients A, B, C, D, and E are real numbers, and A and B are not both zero. Conversely, if A and B are not both zero and Equation 1 is satisfied by more than one pair of real numbers, then the graph of the equation is a conic section. (We include as conic sections, the lines that are obtained if the cutting plane passes through the vertex of the cone.)

We have already discussed two special cases of Equation 1, namely, the circle

$$(x - h)^2 + (y - k)^2 = r^2 \tag{2}$$

and the parabola with vertical axis

$$y = ax^2 + bx + c \tag{3}$$

To obtain the circle, Equation 2, we take $A = 1$, $B = 1$, $C = -2h$, $D = -2k$, and $E = h^2 + k^2 - r^2$ in Equation 1. Similarly, to obtain the parabola, Equation 3, we take $A = a$, $B = 0$, $C = b$, $D = -1$, and $E = c$.

In the remainder of this section, we shall discuss some other important special cases of Equation 1.

THE ELLIPSE

The equation

$$ax^2 + by^2 = c \tag{4}$$

where a, b, and c are *positive* numbers, has an *ellipse* for its graph. Notice that x and y enter this equation only as squares. Consequently, we see that if (x_1, y_1) is a point on the graph, then $(-x_1, y_1)$, $(x_1, -y_1)$, and $(-x_1, -y_1)$ are also on the graph. This means that the graph is *symmetric* with respect to both axes. By putting $y = 0$, we find the x-intercepts to be $\pm\sqrt{c/a}$; similarly, by putting $x = 0$, we find the y-intercepts to be $\pm\sqrt{c/b}$. A rough sketch of an ellipse can be made by using the intercepts and drawing a smooth oval curve through them.

EXAMPLE 1

Graph the ellipse whose equation is $9x^2 + 25y^2 = 225$.

SOLUTION:

By putting $y = 0$, we find the x-intercepts to be ± 5, and by putting $x = 0$, we find the y-intercepts to be ± 3. To find coordinates of other points on the graph, we solve the equation for y to obtain

$$y = \pm\tfrac{3}{5}\sqrt{25 - x^2}$$

This result shows that x cannot be greater than 5 in absolute value; otherwise, y will not be a real number. Thus, we see that $-5 \leq x \leq 5$. By solving the equation for x, we find in the same way that $-3 \leq y \leq 3$. By substituting values of x in the preceding solution for y, we can find additional points on the graph. For example, if $x = \pm 3$, we have

$$y = \pm\tfrac{3}{5}\sqrt{25 - 9} = \pm\tfrac{3}{5}\sqrt{16} = \pm\tfrac{12}{5} = \pm 2.4$$

Thus, $(3, 2.4)$, $(3, -2.4)$, $(-3, 2.4)$, and $(-3, -2.4)$ are all points on the graph. The ellipse is shown in Figure 6.3b.

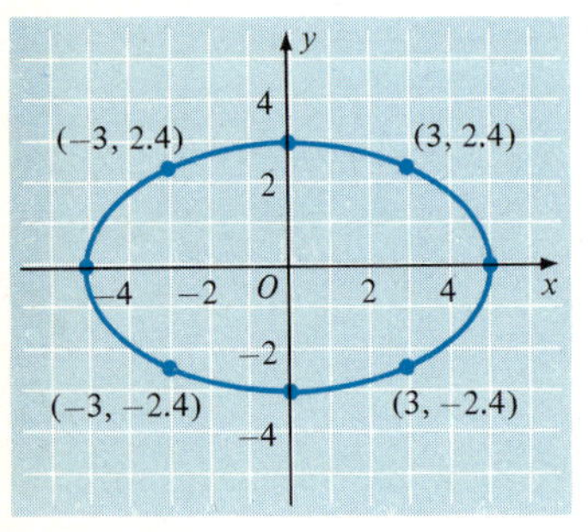

FIGURE 6.3b

Notice that if $a = b$ in Equation 4, we can divide by a to obtain $x^2 + y^2 = c/a$. Because a and c are both positive, c/a is also positive, and we may write $c/a = r^2$. Thus, we have

$$x^2 + y^2 = r^2 \tag{5}$$

the equation of a circle of radius r with center at the origin. This shows that we may regard the circle as a special case of the ellipse.

THE PARABOLA

We have seen that the equation

$$y = ax^2 + bx + c, a \neq 0 \tag{6}$$

has for its graph a parabola with a vertical axis whose equation is $x = -b/(2a)$. The x-coordinate of the vertex is $-b/(2a)$, and the y-coordinate can be found by substituting this value of x into the equation (see Section 6.2).

If we interchange x and y in Equation 6, we obtain

$$x = ay^2 + by + c \tag{7}$$

which also has a parabola for its graph, this time with a horizontal axis, the line $y = -b/(2a)$. The y-coordinate of the vertex is $-b/(2a)$, and the x-coordinate can be found by substituting this value of y into Equation 7.

EXAMPLE 2

Graph the equation $x = -\frac{1}{2}y^2 - \frac{3}{2}y + 2$.

SOLUTION:

We proceed as in Section 6.2, keeping in mind that the roles of x and y have been interchanged. The y-coordinate of the vertex is now

$$-\frac{b}{2a} = -\frac{-\frac{3}{2}}{2(-\frac{1}{2})} = -\frac{3}{2}$$

For $y = -\frac{3}{2}$, we obtain

$$x = (-\tfrac{1}{2})(-\tfrac{3}{2})^2 - (\tfrac{3}{2})(-\tfrac{3}{2}) + 2 = \tfrac{25}{8}$$

so that the vertex of the parabola is at $(\frac{25}{8}, -\frac{3}{2})$. The axis of the curve is the line $y = -\frac{3}{2}$. For $y = 0$, we obtain $x = 2$, and for $x = 0$, we find $y = 1$ or $y = -4$. Other points on the graph may be found by substituting values for y and calculating corresponding values for x from the given equation. The graph appears in Figure 6.3c. Notice that the coefficient of y^2 in the given equation is negative, so that the curve opens to the left.

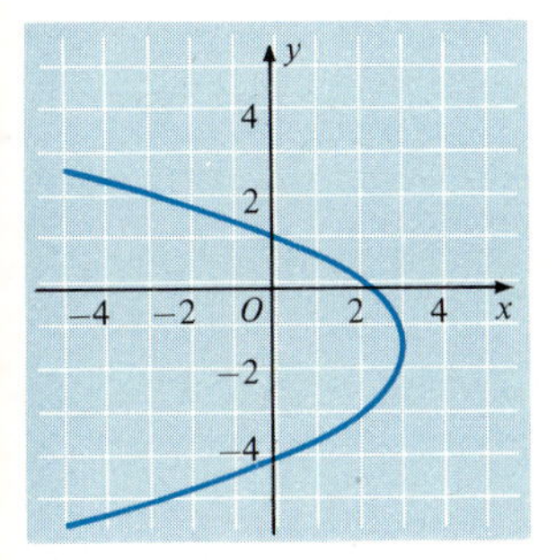

FIGURE 6.3c

THE HYPERBOLA

The equation

$$ax^2 - by^2 = c \tag{8}$$

where a, b, and c are positive real numbers, has for its graph the curve called a *hyperbola*.

EXAMPLE 3

Graph the hyperbola whose equation is $16x^2 - 9y^2 = 144$.

SOLUTION:

Solving the given equation for y, we obtain

$$y = \pm\tfrac{4}{3}\sqrt{x^2 - 9}$$

This result shows that $y = 0$ for $x = \pm 3$, that y is not a real number for $-3 < x < 3$, and that y has two real values for each x such that $|x| > 3$. Thus, the graph consists of two separate pieces, called *branches*, one to the left of the point $(-3, 0)$ and the other to the right of the point $(3, 0)$. As in the case of the ellipse, the hyperbola is symmetric to both coordinate axes. For $x = \pm 5$, we find

$$y = \pm\tfrac{4}{3}\sqrt{25 - 9} = \pm\tfrac{16}{3}$$

Hence, $(5, \frac{16}{3})$, $(5, -\frac{16}{3})$, $(-5, \frac{16}{3})$, and $(-5, -\frac{16}{3})$ are all on the graph.

If we return to the given equation and factor the left member, we obtain

$$(4x - 3y)(4x + 3y) = 144$$

so that

$$4x - 3y = \frac{144}{4x + 3y}$$

For very large positive values of x and y, the fraction $144/(4x + 3y)$ is close to zero. Thus, if $x = 1000$, then $y \approx 1333$ and $144/(4x + 3y) \approx 0.018$ which is close to 0. This tells us that the point (1000, 1333) on the curve is close to the line $4x - 3y = 0$. A similar argument applies if x and y are both negative but large in absolute value. As we go farther and farther from the origin along the curve in either the first or the third quadrant, we get closer and closer to the line $4x - 3y = 0$. We can show in the same way that as we recede from the origin in either the second or the fourth quadrant, we get closer and closer to the line $4x + 3y = 0$. The two lines

$$4x + 3y = 0 \quad \text{and} \quad 4x - 3y = 0$$

which the curve approaches, are called *asymptotes* to the curve. The asymptotes are quite helpful in drawing the graph. Figure 6.3d shows the hyperbola along with its asymptotes.

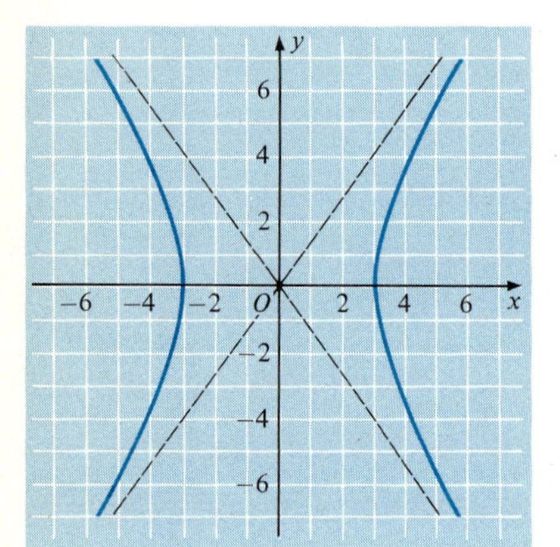

FIGURE 6.3d

The asymptotes of the hyperbola given by Equation 8 are easily found by replacing the c by zero, factoring the left side, and putting each factor equal to zero. This is better than trying to memorize formulas for the asymptotes.

If x and y are interchanged in Equation 8, the result is

$$ay^2 - bx^2 = c \tag{9}$$

where a, b, and c are positive real numbers. The graph of Equation 9 is the hyperbola given by Equation 8, but turned through an angle of 90°. This is illustrated by Figure 6.3e, which shows the graph of

$$16y^2 - 9x^2 = 144$$

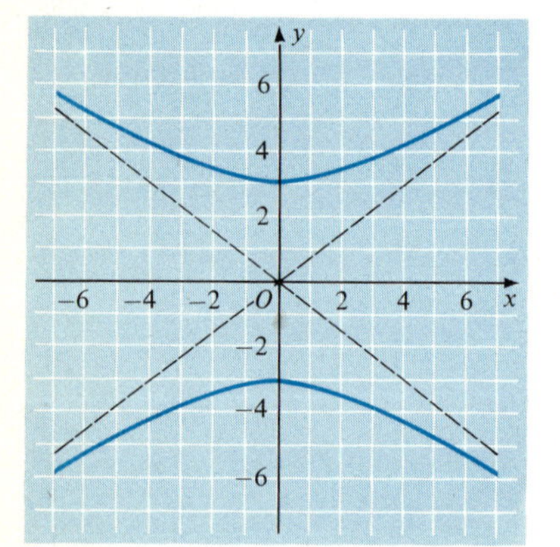

FIGURE 6.3e

the equation obtained by interchanging x and y in the equation of Example 3. Note that the left side $16y^2 - 9x^2 = (4y + 3x)(4y - 3x)$ and the asymptotes $4y + 3x = 0$ and $4y - 3x = 0$ can be obtained by putting each factor of the left side equal to zero.

Another form of the equation of the hyperbola that occurs in many applied problems is

$$xy = c \tag{10}$$

where c is a nonzero real number. For example, Boyle's Law, which is important in physics and chemistry, states that if a gas is held at a constant temperature, then the product of its volume V and its pressure P is a constant; that is,

$$PV = k$$

an equation in the exact form of Equation 10.

EXAMPLE 4

Graph the equation $xy = 12$.

SOLUTION:

We can find points on the graph by solving for y to obtain

$$y = \frac{12}{x}$$

and then substituting values for x. For example, (2, 6), (3, 4), (4, 3), (6, 2), (−2, −6), (−3, −4), (−4, −3), and (−6, −2) are all points on the graph. As you can see from the equation $xy = 12$, neither x nor y can be zero, so that the graph does not cross the coordinate axes. In fact, these axes are the asymptotes of this hyperbola. (This can be arrived at in a manner similar to that used to get the asymptotes of the hyperbola $16x^2 - 9y^2 = 144$ in Example 3.) The graph of $xy = 12$ appears in Figure 6.3f.

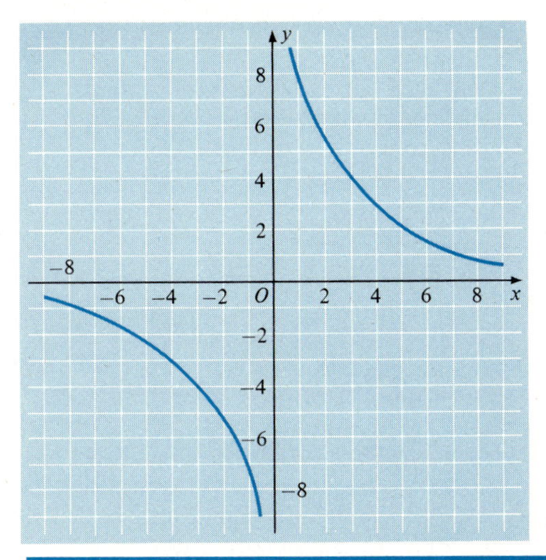

FIGURE 6.3f

The conic sections occur in a wide variety of applied problems. For example, the planets in our solar system travel in elliptical orbits about the sun. Comets that visit the solar system travel in elliptical, parabolic, or hyperbolic orbits. (Of course, those traveling in parabolic or hyperbolic orbits visit us only once.) Of those with elliptical orbits, Halley's comet is perhaps the best known. Machine gears are sometimes cut in the shape of an ellipse, and cams of an elliptical shape are necessary for certain kinds of machinery. The path of a projectile or a baseball is closely approximated by a parabola. Parabolic arches are used in buildings. Loaded cables such as those on a suspension bridge are parabolic in shape. The hyperbola is the basis of a method of *long range* navigation known as "Loran."

EXERCISE 6.3

In Problems 1–12, identify the curve and draw its graph.

1 $9x^2 + 16y^2 = 144$ **2** $4x^2 + 9y^2 = 36$ **3** $x^2 + y^2 = 25$

4 $4x^2 + 4y^2 = 25$ **5** $y^2 = 9x$ **6** $4y^2 = x$

7 $y = x^2 - 2x - 15$ **8** $x = 2y^2 + 8y + 9$ **9** $x^2 - y^2 = 16$

10 $y^2 - 4x^2 = 4$ **11** $xy = 6$ **12** $xy + 6 = 0$

In Problems 13–16, use your calculator to obtain answers in decimal form.

C **13** Find the x- and y-intercepts of the ellipse $42x^2 + 29y^2 = 145$. Give answers to the nearest tenth.

C **14** Find the asymptotes to the hyperbola $27y^2 - 85x^2 = 107$. Find the coefficients to the nearest tenth.

C **15** Find to the nearest tenth the coordinates of the vertex of the parabola $x = 5y^2 + 48y + 16$.

C **16** An arch is to be constructed in the shape of the upper half of an ellipse. The width at the base of the arch is to be 12 feet, and the height at the center is to be 10 feet.
a Find the equation of the ellipse.
b Make a table giving the height of the arch to the nearest hundredth of a foot at 1-foot intervals along the base.

The conic sections can be defined without reference to a cone, and the following problems are concerned with such definitions.

17 A parabola is a set of points in the plane such that each point is equidistant from a fixed point (called the *focus*) and a fixed line (called the *directrix*). Let the focus be the point $(0, p)$ and let the directrix be the line $y = -p$. (See Figure 6.3g.) Then, if $P(x, y)$ is a point on the curve, we must have $FP = DP$; that is,

$$\sqrt{x^2 + (y - p)^2} = |y + p|$$

Now square both sides and simplify to obtain the equation

$$x^2 = 4py$$

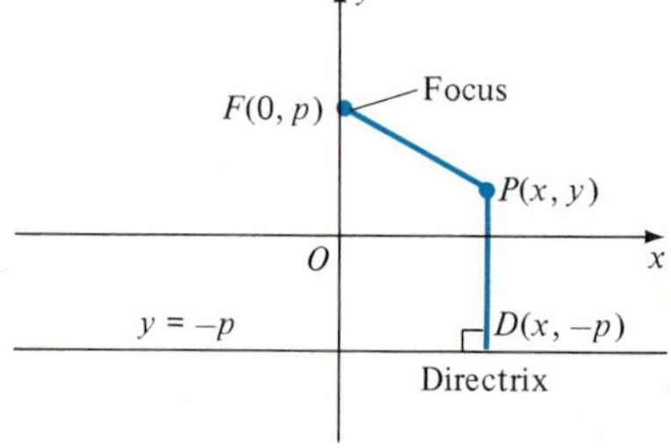

FIGURE 6.3g

18 Refer to Problem 17 and find the coordinates of the focus of the parabola:
a $x^2 = 32y$ **b** $x^2 + 18y = 0$

19 An ellipse is a set of points in the plane such that the sum of the distances of each point from two fixed points (called the *foci*) is a constant. (Note that "foci" is the plural of "focus.") Let the foci be located at the points $(c, 0)$ and $(-c, 0)$, and let the sum of the distances PF_1 and PF_2 (see Figure 6.3h) be denoted by $2a$, where $a > c$. If $P(x, y)$ is a point on the curve, then we must have

$$\sqrt{(x - c)^2 + y^2} + \sqrt{(x + c)^2 + y^2} = 2a$$

Rewrite this equation with one radical on each side. Then square both sides and simplify. Isolate the remaining radical and square both sides again. Now collect the terms and let $a^2 - c^2 = b^2$. You should end up with the equation of an ellipse in the form

$$b^2x^2 + a^2y^2 = a^2b^2$$

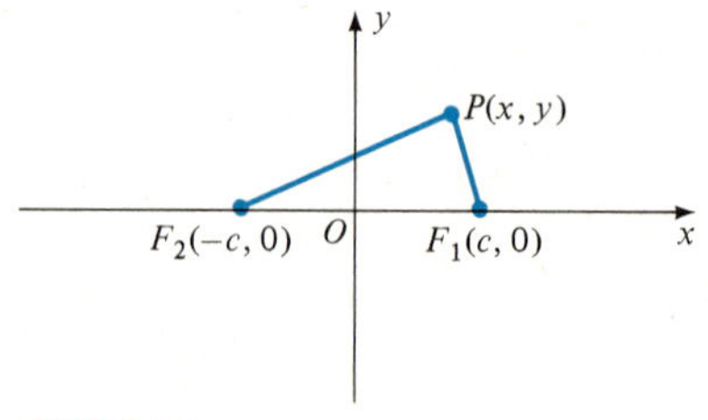

FIGURE 6.3h

20 A hyperbola is a set of points in the plane such that the difference of the distances of each point from two fixed points (called the *foci*) is a constant. Let the foci be located at $(c, 0)$ and $(-c, 0)$, and let the difference of the distances PF_1 and PF_2 (see Figure 6.3h) be denoted by $2a$, where $a < c$. Then, if $P(x, y)$ is a point on the hyperbola, we must have

$$\sqrt{(x-c)^2+y^2}-\sqrt{(x+c)^2+y^2}=\pm 2a$$

(The plus sign on the right applies if $PF_2 < PF_1$, and the minus sign applies if $PF_2 > PF_1$.) Use the same procedure as in Problem 19, but at the end, let $c^2 - a^2 = b^2$. You should come out with the equation of a hyperbola in the form

$$b^2x^2 - a^2y^2 = a^2b^2$$

USING YOUR KNOWLEDGE 6.3

The parabola has many applications, most of which depend on a very elegant geometric property of the curve. It can be shown (by using calculus methods) that a line tangent to the parabola at a point P makes equal angles with the focal radius FP and a line through P parallel to the axis of the parabola (see Figure 6.3i). If the parabola were a mirror, then a ray of light parallel to the axis of the curve would be reflected to the focus, and a ray originating at the focus would be reflected parallel to the axis.

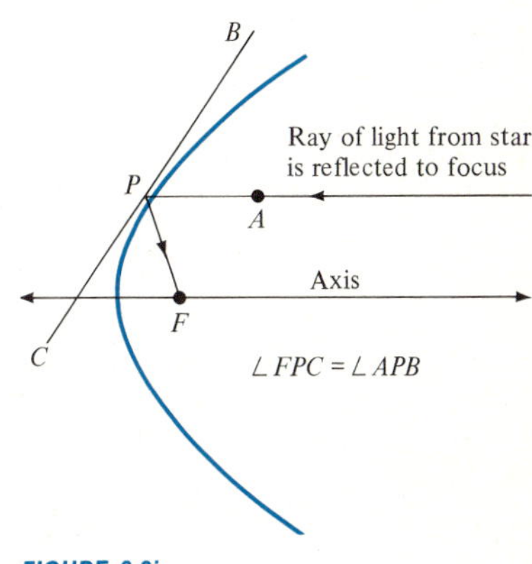

FIGURE 6.3i

A surface, called a *paraboloid of revolution*, is formed by revolving a parabola around its axis. This is the shape that is used for automobile headlights and for searchlights that throw a parallel beam of light when the light source is placed at the focus; it is also the shape of a radar dish or a reflecting telescope mirror that collects parallel rays of energy (light) and reflects them to the focus.

In Problem 17, Exercise 6.3, we found that the parabola $x^2 = 4py$ has its focus at the point $(0, p)$. Because this parabola has its vertex at the origin and the y-axis for its axis, it follows that the focus of the curve is on its axis, p units from the vertex. For example, if the equation is $x^2 = 6y$, then by comparison with the equation $x^2 = 4py$, we see that $4p = 6$, so that $p = \frac{6}{4} = \frac{3}{2}$. Thus, the focus is at $(0, \frac{3}{2})$, $\frac{3}{2}$ units from the vertex.

It is not difficult to find the equation of the parabola needed to generate a paraboloid of given dimensions. Thus, if a parabolic mirror is to have a diameter of 6 feet and a depth of 6 inches, we can use a parabola with vertex at the origin and equation of the form $x^2 = 4py$, and make the curve pass through the point $(3, \frac{1}{2})$. This means that we substitute $(3, \frac{1}{2})$ for (x, y) and determine p as follows:

$$x^2 = 4py$$
$$3^2 = (4p)(\tfrac{1}{2})$$
$$= 2p$$

Hence, $p = 4.5$ and the equation of the parabola is $x^2 = 18y$; the focus is 4.5 feet from the vertex.

1 A radar dish has a diameter of 10 feet and a depth of 2 feet. The dish is in the shape of a paraboloid of revolution. Find an equation for a parabola that would generate this dish and locate the focus of the dish.

2 The cables of a suspension bridge hang very nearly in the shape of parabolas. A cable on such a bridge spans a distance of 1000 feet and sags 50 feet in the middle. Find an equation for this parabola.

6.4 GRAPHS OF RATIONAL FUNCTIONS

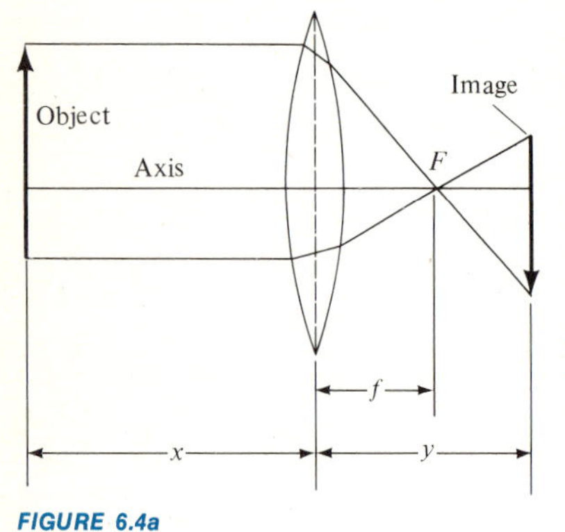

FIGURE 6.4a

Figure 6.4a shows a simple converging optical lens. If an object is placed in front of such a lens, the lens will form an image as shown in the diagram. Rays of light that are parallel to the axis are bent so that they pass through a point (F in the diagram) called the focal point of the lens. It is shown in the study of optics that the object distance (x in the diagram) and the image distance (y in the diagram) are related by the lens equation

$$\frac{1}{x}+\frac{1}{y}=\frac{1}{f} \tag{1}$$

where f is the distance from the center of the lens to the focal point.

If we wanted to study the behavior of the image distance as a function of the object distance, then we would solve Equation 1 for y and find

$$y=\frac{fx}{x-f} \tag{2}$$

The right side of Equation 2 is a simple example of a *rational fraction*. In general, we use Definition 6.4a. For example,

▼ **DEFINITION 6.4a**

A *rational fraction* is the ratio of two polynomials. A function that is defined by such a ratio is called a *rational function*.

$$F(x)=\frac{2x^2}{x-2}$$

defines a rational function. Other examples are

$$g(x)=\frac{x(x-1)}{(x-2)(x-3)} \qquad G(x)=\frac{x^2+2x-3}{x^3+x^2+1}$$

and, in general,

$$f(x)=\frac{a_nx^n+a_{n-1}x^{n-1}+\cdots+a_1x+a_0}{b_mx^m+b_{m-1}x^{m-1}+\cdots+b_1x+b_0} \qquad a_n\neq 0,\ b_m\neq 0$$

where the a's and b's are real numbers. A polynomial function is a special case of a rational function in which the denominator is simply 1.

There are many applications of algebra in which rational functions occur and in which it is helpful to have a rough sketch of the graph of the function. We shall not be concerned with the exact shape of the graph because it takes calculus methods to determine this. Frequently, all we need to know is the set of values of x for which the function is positive (or negative), and we shall consider the problem of sketching rapidly the graphs of some simple rational functions.

FACTORED POLYNOMIALS

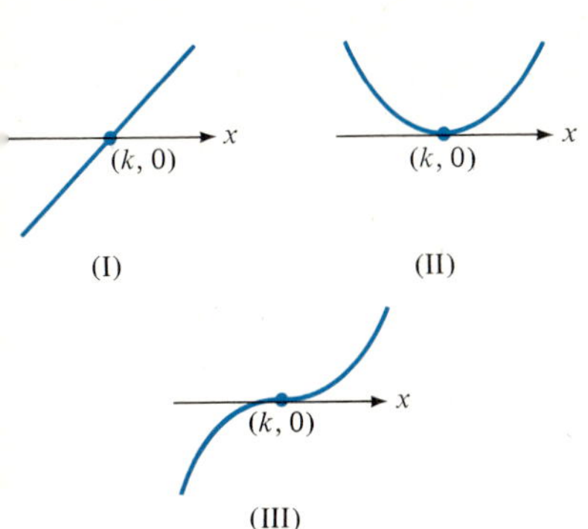

FIGURE 6.4b

First, we consider the graph of a factored polynomial. Figure 6.4b shows the graphs of the functions

I. $f(x) = x - k$
II. $f(x) = (x - k)^2$
III. $f(x) = (x - k)^3$

all for values of x close to the value $x = k$. In each case, the value $x = k$ gives a zero value for the function, so that the graph has the point $(k, 0)$ in common with the x-axis. In Equations I and III, $f(x)$ changes sign as x changes from values less than k to values greater than k. Thus, the curve crosses the x-axis at $(k, 0)$. On the other hand, the function in II is a perfect square; it cannot have any negative values. Hence, the curve in Figure 6.4b(II) remains above the x-axis except at the point $(k, 0)$, where it touches (is tangent to) the axis.

In case the functions are the negatives of those graphed in Figure 6.4b, you can see their graphs by holding your book upside down and looking at Figure 6.4b through the back of the page. Thus, the graphs of $f(x) = -(x - k)$ and $f(x) = -(x - k)^3$ lie above the x-axis to the left of $(k, 0)$ and below the x-axis to the right of $(k, 0)$. The graph of $f(x) = -(x - k)^2$ lies below the x-axis on both sides of $(k, 0)$.

For exactly the same reasons as in the cases discussed for Figure 6.4b, it is evident that the graph of

$$f(x) = (x - k)^n \qquad \text{for } n = 5, 7, 9, \ldots$$

will be similar to that in Figure 6.4b(III), and the graph of

$$f(x) = (x - k)^n \qquad \text{for } n = 4, 6, 8, \ldots$$

will resemble that in Figure 6.4b(II). The actual change in the appearance caused by increasing the value of n is a "flattening" of the curve near the point $(k, 0)$.

If several factors of the preceding type are combined into a single product, the graph will display all the types of behavior corresponding to the exponents that occur. For example, Figure 6.4c shows the graph of

$$f(x) = x^2(x - 1)(x - 2)^3$$

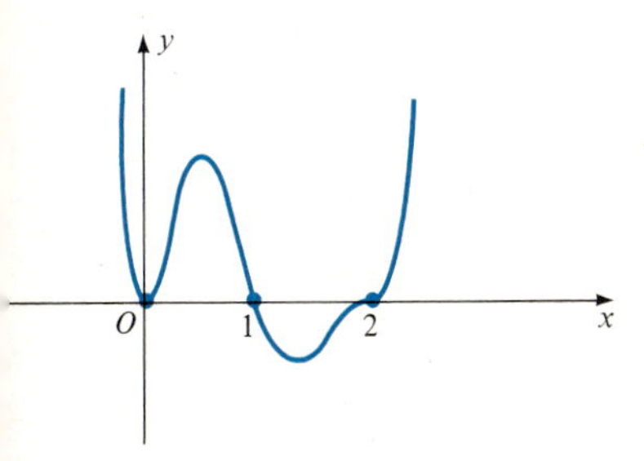

FIGURE 6.4c

At $(0, 0)$, we see the typical intersection for a squared factor; at $(1, 0)$ the typical intersection for a linear factor; and at $(2, 0)$ for a cubic factor. You can check the general behavior of the curve at points of intersection with the x-axis by noting that if x is greater than 1, then x^2 and $x - 1$ are both positive and thus cannot affect the general way in which the curve crosses the x-axis at $x = 2$. You can make similar statements about the points $(0, 0)$ and $(1, 0)$ by considering values of x less than 1 and values of x between 0 and 2, respectively. Because $f(x)$ is positive for x greater than 2, the graph lies above the x-axis to the right of $x = 2$. With this in mind, we see that the graph must lie *below* the x-axis for x between 1 and 2, because the factor $x - 2$ occurs to an odd power; the curve must lie above the x-axis for x between 0 and 1, because the factor $x - 1$ is to an odd power; and the curve must lie above the x-axis for x less than 0, because the factor x occurs to an even power. Recall that an *odd power* means a *change of sign* and an *even power* means *no change of sign* as we pass through the corresponding point of contact with the x-axis.

The preceding discussion may be generalized to fit any function of the form

$$f(x) = A(x - x_1)^{p_1}(x - x_2)^{p_2} \cdots (x - x_n)^{p_n}$$

where the exponents $p_1, p_2, \ldots, p_n$ are all positive integers. The graph of $f(x)$ will have the points $(x_1, 0), (x_2, 0), \ldots, (x_n, 0)$ in common with the x-axis. At each such point, the curve will cross in the general manner of a straight line if the corresponding exponent is 1 [Figure 6.4b(I)], and it will resemble the cubic if the exponent is odd and greater than 1 [Figure 6.4b(III)]. If the exponent is even, the curve touches the axis in the general manner of a second power [Figure 6.4b(II)]. The exact shape of the curve between consecutive points where it touches or crosses the x-axis will not concern us here and may be indicated schematically. In order to find whether the curve lies above or below the axis for values of x greater than the largest x_i, notice that all the binomial factors will be positive for such values of x; hence, the sign of $f(x)$ is that of the constant coefficient A.

EXAMPLE 1

Make a schematic drawing showing how the graph of $f(x) = -6(x - 1)^5(x - 2)^2(x - 3)^4$ meets the x-axis.

SOLUTION:

Because $f(x) = 0$ for $x = 1$, 2, and 3, the graph meets the x-axis at the corresponding points. The factor $x - 1$ has the exponent 5, so that at (1, 0) the curve crosses the axis in the general manner of a cubic. The factor $x - 2$ has the exponent 2; thus, the graph is tangent to the axis at (2, 0). Finally, the factor $x - 3$ has the exponent 4, so that the graph is also tangent to the axis at (3, 0). For values of x greater than 3, the binomials $x - 1$, $x - 2$, and $x - 3$ are all positive, so that $f(x)$ is negative because of the coefficient -6. The required schematic graph is shown in Figure 6.4d.

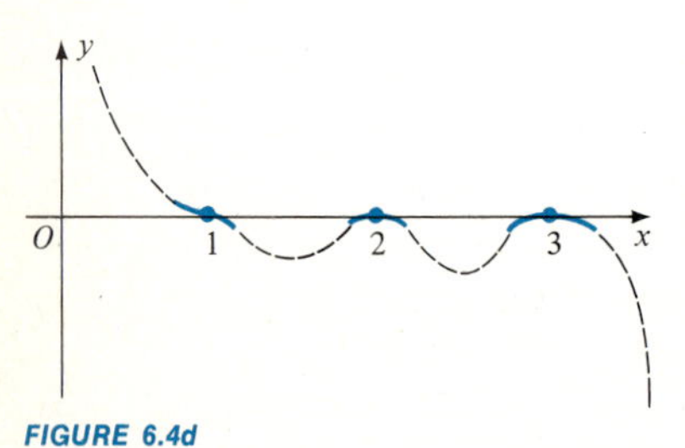

FIGURE 6.4d

EXAMPLE 2

Make a schematic graph of $P(x) = x^5(x+1)^2(x-1)$ and use the graph to find the solution set of the inequality $P(x) > 0$.

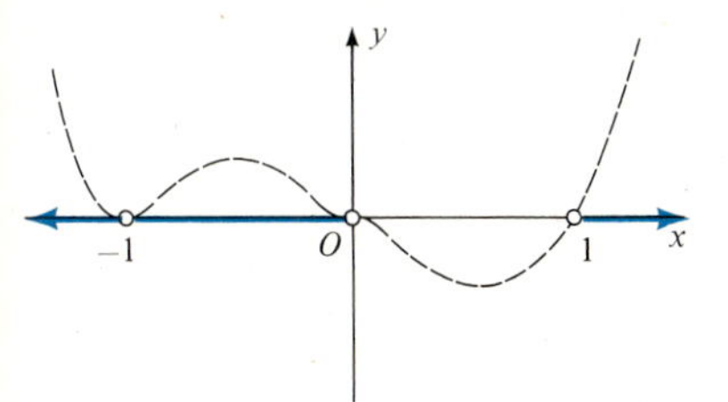

FIGURE 6.4e

SOLUTION:

We see that $P(x) = 0$ for $x = -1, 0$, and 1, and that $P(x)$ is positive for values of x greater than 1. The factor $x + 1$ has the exponent 2, so that the graph is tangent to the x-axis at $(-1, 0)$. The factor x has the exponent 5; hence, the graph crosses the axis at $(0, 0)$, somewhat like a cubic. The factor $x - 1$ has exponent 1; therefore, the graph crosses the x-axis at $(1, 0)$, like a straight line. The schematic graph is shown in Figure 6.4e.

Because $P(x) > 0$ wherever the graph lies above the x-axis, we can easily read off the solution of the inequality. The intervals where x satisfies the inequality are indicated by the heavily drawn portions of the x-axis. Thus, the solution consists of all values of x such that

$$x < -1 \quad \text{or} \quad -1 < x < 0 \quad \text{or} \quad x > 1$$

RATIONAL FUNCTIONS (OTHER THAN POLYNOMIALS)

Making a schematic graph of a rational function that is not simply a polynomial is not much more difficult than what we have done for the polynomial graphs. As for the polynomials, it is shown in calculus that the graph of a rational function is a continuous curve except where the denominator of the fraction is zero.

Let us look first at some simple examples. Figure 6.4f shows the graphs of the functions

I. $f(x) = \dfrac{1}{x-a}$ II. $f(x) = \dfrac{1}{(x-a)^2}$ III. $f(x) = \dfrac{1}{(x-a)^3}$

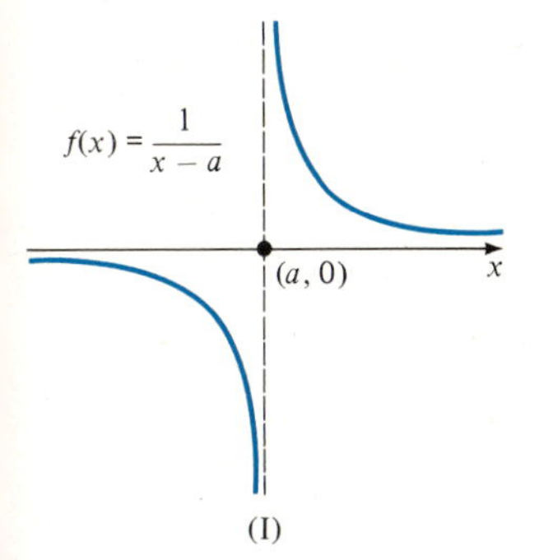

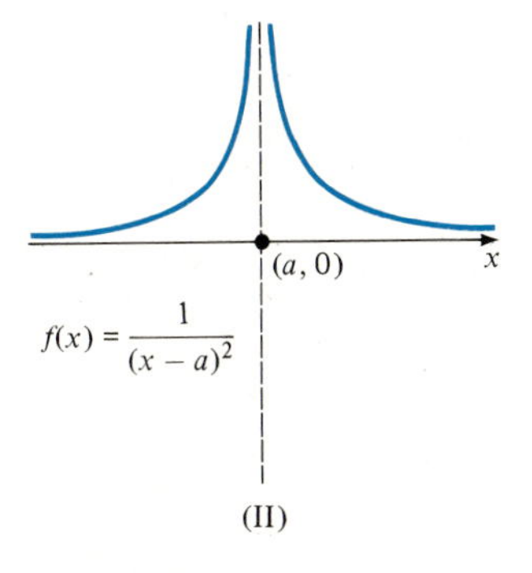

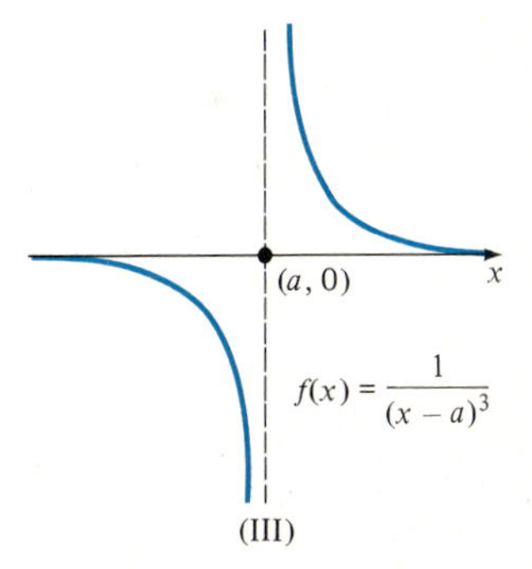

FIGURE 6.4f

First, we note that none of the three functions is defined for $x = a$. However, we need to discuss the behavior of the functions for values x near $x = a$. The accompanying table of values (Table 6.4a) gives a good indication that the values of these three functions grow larger and larger in absolute value as x gets closer and closer to $x = a$. In fact, we can make these absolute values as large as we wish by taking x close enough to a. Geometrically, this means that the curve recedes farther and farther away from the x-axis as $x - a$ gets closer

TABLE 6.4a

$x - a$	$f(x) = (x - a)^{-1}$	$f(x) = (x - a)^{-2}$	$f(x) = (x - a)^{-3}$
1	1	1	1
0.5	2	4	8
0.1	10	100	1000
0.01	100	10^4	10^6
0.001	1000	10^6	10^9
...	...	...	...
−1	−1	1	−1
−0.5	−2	4	−8
−0.1	−10	100	−1000
−0.01	−100	10^4	-10^6
−0.001	−1000	10^6	-10^9
...	...	...	...

and closer to zero. The graph does not intersect the vertical line $x = a$ but approaches it ever more closely as x "approaches" the value a. The line $x = a$, which the curve approaches in this manner, is called an *asymptote.* Notice in Figures 6.4f(I) and (III), *where the exponent in the denominator is odd, that the graph changes from below the x-axis to above the axis as $x - a$ changes from negative to positive values.* In Figure 6.4f(II), *where the exponent is even, the graph remains on one side of the axis.*

The preceding discussion can be extended to more general rational functions. Suppose that $R(x)$ is a rational function

$$R(x) = \frac{P(x)}{Q(x)}$$

where $P(x)$ and $Q(x)$ are polynomials with no common factor. The graph of this rational function will have a vertical asymptote for each real value of x where the denominator $Q(x) = 0$ and will have a point in common with the x-axis for each real value where the numerator $P(x) = 0$. At each real value of x where either $P(x) = 0$ or $Q(x) = 0$, *the graph changes from one side of the x-axis to the other if the exponent of the corresponding factor is odd; the graph remains on one side of the x-axis if the exponent is even.* A simple example will make this clear.

EXAMPLE 3

Make a schematic graph of the rational function defined by

$$R(x) = \frac{(x-1)^2}{x(x-2)}$$

SOLUTION:

From the preceding discussion, we see that the graph has the lines $x = 0$ and $x = 2$ as vertical asymptotes. Because $R(x) = 0$ for $x = 1$, the graph has the point $(1, 0)$ in common with the x-axis; here, the graph is tangent to the x-axis because the factor $x - 1$ occurs to an even power. There is a change of sign at $x = 0$ and at $x = 2$ because both x and $x - 2$ have odd exponents. For values of x greater than 2, $R(x)$ is positive, so that the graph lies above the x-axis to the right of the asymptote $x = 2$. Therefore, the graph is below the axis for values of x between 1 and 2 and also between 0 and 1. The graph is above the x-axis for negative values of x. The schematic graph appears in Figure 6.4g.

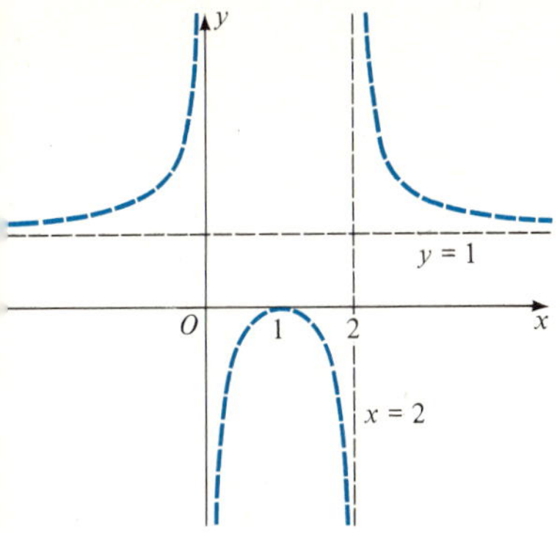

FIGURE 6.4g

It is sometimes important to know the behavior of a rational function and its graph for large absolute values of x. Consider the rational function in Example 3,

$$R(x) = \frac{(x-1)^2}{x(x-2)} = \frac{x^2 - 2x + 1}{x(x-2)}$$

If we divide numerator and denominator by x^2, the highest power of x that occurs, we obtain

$$R(x) = \frac{1 - (2/x) + (1/x^2)}{(1 - 2/x)}$$

For large absolute values of x, the terms $-2/x$ and $1/x^2$ in the numerator and $-2/x$ in the denominator are small in absolute value, so that the value of $R(x)$ is close to 1. In fact, by taking the absolute value of x large enough, we can make the value of $R(x)$ as close to 1 as we please. For instance, $R(10) = 1.0125$, $R(100) \approx 1.000102$, $R(1000) \approx 1.000001$, $R(-10) \approx 1.008333$, $R(-100) \approx 1.000098$, $R(-1000) \approx 1.000001$. Geometrically, these results mean that the line $y = 1$ is a horizontal asymptote to the graph of $R(x)$. Figure 6.4g shows this behavior.

We shall omit any formal proof, but considerations such as those of the preceding paragraph should make Theorem 6.4a plausible to you.

▼ THEOREM 6.4a

As x increases without bound in absolute value, the rational function defined by

$$R(x) = \frac{a_nx^n + a_{n-1}x^{n-1} + \cdots + a_1x + a_0}{b_mx^m + a_{m-1}x^{m-1} + \cdots + b_1x + b_0} \qquad a_n \neq 0,\ b_m \neq 0$$

I. increases without bound in absolute value if $n > m$;

II. approaches zero in value, and the graph has the x-axis as an asymptote if $n < m$;

III. approaches a_n/b_m in value, and the graph has the horizontal line $y = a_n/b_m$ as an asymptote if $n = m$.

EXAMPLE 4

Find all the vertical and horizontal asymptotes and make a schematic graph of the function defined by

$$F(x) = \frac{3x^2}{(x+2)(x^2+1)}$$

SOLUTION:

1. The only real value for which the denominator is zero is $x = -2$. Hence, the line $x = -2$ is the only vertical asymptote.
2. Because the degree of the numerator is less than the degree of the denominator, Part II of Theorem 6.4a applies. Thus, the x-axis is a horizontal asymptote to the graph.
3. The numerator of $F(x)$ is zero for $x = 0$, and because the factor x is to an even power, the graph is tangent to the x-axis at $(0, 0)$.
4. The graph changes from one side of the x-axis to the other at $x = -2$ because the factor $x + 2$ occurs to an odd power.
5. For positive values of x, $F(x)$ is positive, so that the graph is above the x-axis.

The schematic graph that appears in Figure 6.4h puts all the preceding information together.

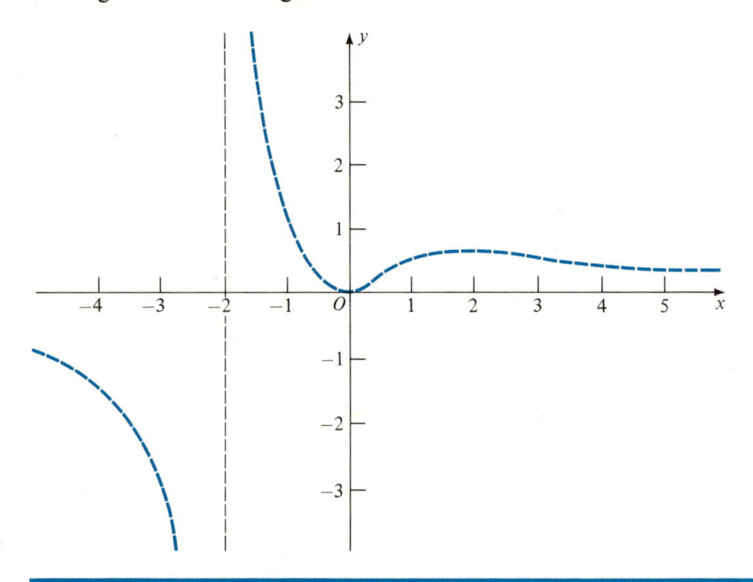

FIGURE 6.4h

As a final example, we return to the lens equation at the beginning of this section. Recall that we had the following formula for the image distance y in terms of the object distance x and the focal length f of the lens.

$$y = \frac{fx}{x-f} \tag{2}$$

EXAMPLE 5

Suppose that the lens of Equation 2 has a focal length of 3 centimeters. Make a graph of the corresponding function.

SOLUTION:

The function is defined by the equation

$$y = \frac{3x}{x-3}$$

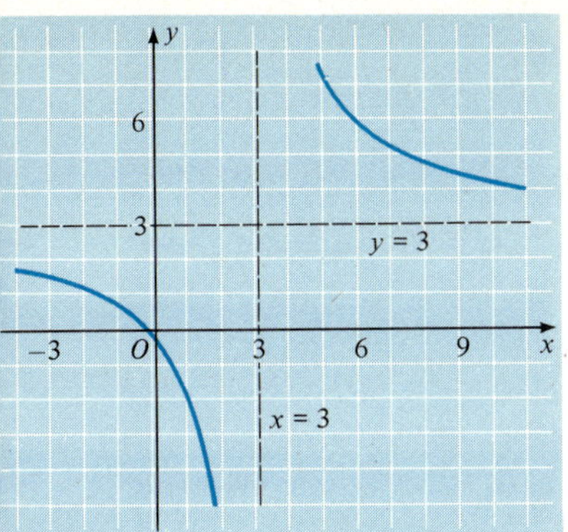

FIGURE 6.4i

The graph has a vertical asymptote at $x = 3$, the value for which the denominator, $x - 3$, is zero. Referring to Theorem 6.4a, Part III, we see that $a_n = 3$ and $b_m = 1$, so that $a_n/b_m = 3$, and the line $y = 3$ is a horizontal asymptote. Because both x and $x - 3$ occur with exponent 1 in the fraction, there is a change of sign at the values $x = 0$ and $x = 3$. Figure 6.4i shows the required graph.

Notice that y gets closer and closer to 3 as x gets larger and larger. This means that the image gets closer and closer to the focal point as the object is moved farther and farther away from the lens. In order to have a real image (such as would be needed on a photographic film or on a screen), y must be positive, and so x must be greater than 3. If x is between 0 and 3, y is negative. This result also has a physical interpretation. The image is now on the same side of the lens as the object, which is exactly the case if the lens is being used as a magnifying glass. For instance, if the object is 1 millimeter long and is held at a distance of 2 centimeters from the center of the lens, then $x = 2$ and

$$y = \frac{3 \cdot 2}{2 - 3} = -6$$

which means that the image distance is 6 centimeters to the right of the lens (on the same side as the object), and we get what is called a "virtual" image. It is shown in physics that the magnification is the ratio of the image distance to the object distance; for our example, this is the ratio of 6 to 2, that is, 3 to 1. If you were looking through the lens, the object would appear to be 3 millimeters long and at a distance of 6 centimeters from the lens.

EXERCISE 6.4

In Problems 1–20, make a schematic graph of the given function.

1 $P(x) = (x - 3)(x + 2)(x + 6)$
2 $P(x) = (x - 2)(x + 2)(x + 4)$
3 $P(z) = z(z + 3)(z - 4)$
4 $P(z) = z(2 - z)(z + 5)$
5 $P(w) = (w + 1)(w - 4)^2$
6 $P(w) = (w - 2)^2(w + 3)$
7 $P(x) = (x - 3)^2(x + 3)^2$
8 $P(x) = x^3(x + 4)^2$
9 $P(x) = 3(x - 1)^4(3x + 5)$
10 $P(x) = -6x^4(x - 3)^3$
11 $P(x) = x^2(x - 3)^2(x + 1)(x - 5)$
12 $P(x) = x^3(x - 1)^2(x - 3)^3(x + 5)$
13 $f(x) = \dfrac{1}{x + 2}$
14 $f(x) = \dfrac{-6}{x - 3}$
15 $g(x) = \dfrac{x + 1}{x - 1}$
16 $g(x) = \dfrac{(x + 1)^2}{1 - x}$
17 $p(x) = \dfrac{x + 3}{x^2 - 4}$
18 $p(x) = \dfrac{x^2}{9 - x^2}$
19 $F(x) = \dfrac{(x - 2)^3}{x(x^2 - 16)}$
20 $F(t) = \dfrac{t^3(t + 1)}{(t - 1)^2(t + 2)}$

In Problems 21–28, find the solution of the stated inequality for the function defined in the indicated problem.

21 $P(x) > 0$; Problem 3
22 $P(x) \le 0$; Problem 4
23 $P(x) \ge 0$; Problem 11
24 $P(x) \le 0$; Problem 12
25 $p(x) \ge 0$; Problem 17
26 $p(x) \le 0$; Problem 18
27 $F(x) < 0$; Problem 19
28 $F(t) < 0$; Problem 20

29 **a** If a lens has a focal length of 20 centimeters, for what object distances will there be a real image? (See Equation 2 of this section. Remember x is positive and the image will be real if y is positive.)
b At what distance from the lens should an object be placed so that the object and image distances are equal?
c, d Repeat the preceding questions for the focal length f.

30 If the degree of the numerator of a rational fraction exceeds the degree of the denominator by 1, then the graph of the corresponding function has an *oblique* asymptote that can be found by division. For example, if

$$f(x) = \frac{x^2}{x - 1}$$

then we can divide to write

$$f(x) = x + 1 + \frac{1}{x - 1}$$

For large absolute values of x, the fraction $1/(x - 1)$ is near zero, so that the value of $f(x)$ is close to the value of $x + 1$. This means that $y = x + 1$ is an asymptote to the graph. Make schematic graphs of the functions

$$f(x) = \frac{1}{x - 1}, \quad f(x) = \frac{x}{x - 1}, \quad \text{and } f(x) = \frac{x^2}{x - 1}$$

to see the effect of multiplying by x.

31 Find the asymptotes to the graph of

$$f(x) = \frac{2x^3 + x^2}{x^2 - 1}$$

Hint: See Problem 30.

32 Find the asymptotes to the graph of

$$f(x) = \frac{x^3}{(x + 1)(x + 2)}$$

Hint: See Problem 30.

C **33** For the equation

$$y = \frac{2x^2}{(x + 1)(x - 10)}$$

complete the accompanying table. Your results should reinforce your knowledge that $y = 2$ is a horizontal asymptote to the graph.

x	50	100	500	1000	5000	10,000
y						

C **34** At what points does the graph of the rational function in Problem 33 cross the line $y = 1$? Give answers to the nearest hundredth.

USING YOUR KNOWLEDGE 6.4

1 If two electrical resistances, r_1 and r_2, are connected in parallel, the equivalent single resistance, R, is given by the equation

$$\frac{1}{R} = \frac{1}{r_1} + \frac{1}{r_2}$$

Solve this equation for R. You should obtain

$$R = \frac{r_1 r_2}{r_1 + r_2}$$

Now let $r_1 = 10$ ohms and $r_2 = x$ ohms. Then, R is the rational function defined by

$$R(x) = \frac{10x}{10 + x}$$

Make a graph of this function for $x \geq 0$. How would you interpret the fact that the graph is asymptotic to the line $R = 10$?

The average rate of change of a function from one point to another is a basic idea in calculus. If the function is defined by $y = f(x)$ and x changes from x_1 to x_2, then the change in x is denoted by $\Delta x = x_2 - x_1$, and the corresponding change in y by $\Delta y = y_2 - y_1$. (Δ is the Greek letter, capital delta.) The average rate of change of y with respect to x is defined by the ratio

$$\frac{\Delta y}{\Delta x} = \frac{y_2 - y_1}{x_2 - x_1}$$

For example, if $y = x^2$, $x_1 = 2$, and $x_2 = 4$, then $y_1 = {x_1}^2 = 4$, and $y_2 = {x_2}^2 = 16$. Thus, the average rate of change of y with respect to x over the interval from $x = 2$ to $x = 4$ is

$$\frac{\Delta y}{\Delta x} = \frac{16 - 4}{4 - 2} = \frac{12}{2} = 6$$

This result means that, over the interval from $x = 2$ to $x = 4$, the function increases at the average rate of 6 units of y per unit of x.

2 A sewage treatment plant discharges its treated sewage through a pipe that extends into a large lake. If the concentration, in grams per liter, of the sewage in the lake x meters from the end of the pipe is

$$C(x) = \frac{0.5}{(x + 1)^2}$$

a What is the concentration 1 meter from the end of the pipe? **b** What is the concentration 4 meters from the end of the pipe? **c** What is the average rate of change of the concentration as we go from 1 meter to 4 meters from the end of the pipe?

3 In the early research in quantitative learning theory, it was assumed that a person could successfully perform $N(x)$ tasks in a given time after doing x practice tasks, where

$$N(x) = \frac{a(x + b)}{x + c}$$

and a, b, and c are appropriately selected numbers. If $a = 100$, $b = 1$, and $c = 29$, at what average rate does N change as x changes from 1 to 11?

6.5 VARIATION

DIRECT VARIATION

It is shown in physics that there is a simple relationship between a force, F, and the acceleration, a, that the force produces. This relationship is given by

$$a = kF$$

where k is a positive constant. In the language of science and technology, we say that "a is directly proportional to F" or that "a varies directly as F." The meaning of the statement that one variable *varies directly as* another is defined as in Definition 6.5a. The statements

▼ **DEFINITION 6.5a**

A variable y varies directly as a variable x if

$$y = kx$$

where k is a nonzero constant.

y varies with x,
y varies directly with x, and
y is proportional to x

are all equivalent and are translated as

$$y = kx$$

The constant k is called the *constant of proportionality* (or the constant of variation).

As another example, the formula for the perimeter of a square, $p = 4s$, shows that p varies directly as the length of the side of the square. In this case, 4 is the constant of proportionality.

Now, let us consider the formula $A = \pi r^2$ for the area of a circle of radius r. Here, A varies as the *square* or *second power* of the radius r, and the constant of proportionality is π. To take account of this type of variation, we generalize the preceding definition in Definition 6.5b. For example, the area A of a square varies directly as the *second power* of the length of its side s. The constant of proportionality here is 1, because $A = s^2$. Similarly, because the volume V of a sphere of radius r is

$$V = \tfrac{4}{3}\pi r^3$$

we say that the volume of a sphere is directly proportional to the cube of the radius. Here the constant of proportionality is $\frac{4}{3}\pi$.

▼ DEFINITION 6.5b

A variable y varies directly as the nth power of a variable x if

$$y = kx^n$$

where $n > 0$ and k is a nonzero constant.

EXAMPLE 1

The tension T on a spring varies directly as the distance s that the spring is stretched.

a. Write the equation of variation.

b. Find the constant of proportionality if it is known that it takes a 50-pound force to stretch the spring 2 inches.

SOLUTION:

a. By Definition 6.5a, the equation is

$$T = ks$$

b. We substitute $T = 50$ and $s = 2$ into the preceding equation to obtain

$$50 = k \cdot 2$$

so that

$$k = 25 \quad \text{and} \quad T = 25s$$

EXAMPLE 2

The kinetic energy KE of an object (the energy possessed by the object by virtue of its motion) is proportional to the square of its velocity v.

a. Write the equation of variation.

SOLUTION:

a. By Definition 6.5b, the equation is

$$KE = kv^2$$

b. Compare the kinetic energy of an automobile traveling at 15 mph to that of the same automobile traveling at 60 mph.

b. The kinetic energy, say KE_1, of an automobile traveling at 15 mph is

$$KE_1 = k(15)^2$$

and the kinetic energy, say KE_2, of the same automobile traveling at 60 mph is

$$KE_2 = k(60)^2$$

Thus,

$$\frac{KE_2}{KE_1} = \frac{k(60)^2}{k(15)^2} = \left(\frac{60}{15}\right)^2 = 4^2 = 16$$

so that the kinetic energy of an automobile traveling at 60 mph is 16 times that of the same automobile traveling at 15 mph. In case of a collision, this energy has to be dissipated somehow, which is something to think about when driving your car.

INVERSE VARIATION

Not all variations in mathematics and science are direct. For example, Robert J. Ringer, author of the best-selling book, *Winning Through Intimidation*, claims that the results R that a person obtains are *inversely proportional to* the degree i to which the person is intimidated. Here, the phrase, "*inversely proportional to*" means that the results R decrease as the degree to which the person is intimidated increases. Similarly, the statement that the acceleration a of an object is inversely proportional to its mass m, means that a decreases as m increases. The statement *y is inversely proportional to x* has the same meaning as the statement *y varies inversely as x*, which is defined in Definition 6.5c.

As in the case of direct variation, we can generalize Definition 6.5c as in 6.5d.

▼ DEFINITION 6.5c

A variable y varies inversely as a variable x if

$$y = \frac{k}{x}$$

where k is a nonzero constant.

▼ DEFINITION 6.5d

A variable y varies inversely as the nth power of a variable x if

$$y = \frac{k}{x^n}$$

where $n > 0$ and k is a nonzero constant.

EXAMPLE 3

Psychologists theorize that the time required to do a given task usually decreases with practice.

a. Some psychologists believe that the time T (minutes) required to do a task in the xth trial is inversely proportional to the number x of trials. Write the equation of variation.

b. Theorists in the area of fast learning believe that the time T required to do a task in the xth trial is inversely proportional to the *square* of x. Write the equation of variation and find the constant of proportionality if it is known that a mechanic needs 10 minutes to rotate the tires on a car in the second trial.

SOLUTION:

a. Because T is inversely proportional to x, the equation is

$$T = \frac{k}{x}$$

b. Here T is inversely proportional to the square of x. Thus,

$$T = \frac{k}{x^2}$$

We were also given that $T = 10$ when $x = 2$, which gives

$$10 = \frac{k}{2^2}$$

or

$$k = 40 \quad \text{and} \quad T = \frac{40}{x^2}$$

JOINT VARIATION

The variations we have discussed so far involve only two variables. There can be variations involving additional variables. For example, the area A of a triangle *varies jointly* as the length b of the base and the height h. Thus, we can write

$A = kbh$ $\quad$ k a nonzero constant

Of course, we know that $k = \frac{1}{2}$ in this case. We define *joint variation* in Definition 6.5e. As in the preceding cases, we can also generalize this definition. (See Definition 6.5f.)

▼ DEFINITION 6.5e

A variable z varies jointly as the variables x and y if

$$z = kxy$$

where k is a nonzero constant.

▼ DEFINITION 6.5f

A variable z varies jointly as the mth power of x and the nth power of y if

$$z = kx^m y^n$$

where $m, n > 0$ and k is a nonzero constant.

EXAMPLE 4

The lifting force P exerted by the atmosphere on the wings of an airplane varies jointly as the wing area A in square feet and the square of the plane's speed v in miles per hour (mph). Suppose that the lift is 1200 pounds for a wing area of 100 square feet and a speed of 75 mph.

a. Write the equation of variation.

b. Find the constant of proportionality.

SOLUTION:

a. In accordance with Definition 6.5f, the equation is

$$P = kAv^2$$

b. We substitute the given values, $P = 1200$, $A = 100$, and $v = 75$ into the preceding equation to obtain

$$1200 = (k)(100)(75)^2$$

and

$$k = \frac{4}{1875}$$

Hence,

$$P = \frac{4Av^2}{1875}$$

Sometimes the word "joint" is omitted in a statement of variation, and the word "and" usually indicates a product. Example 5 illustrates many of the preceding ideas.

EXAMPLE 5

Newton's Law of Gravitational Attraction states that the force F with which two particles of mass m_1 and m_2, respectively, attract each other varies directly as the product of the masses and inversely as the square of the distance r between them. Write the equation of variation for this law.

SOLUTION:

The required equation is

$$F = \frac{km_1m_2}{r^2}$$

EXERCISE 6.5

In Problems 1–10, write the corresponding equation of variation, using k for the constant of proportionality.

1 The pressure P exerted by a fluid at a given point varies directly as the depth d below the surface of the fluid.

2 The distance d traveled by an object moving at a constant rate is proportional to the time t.

3 The area A of a circle varies directly as the square of the radius r.

4 If the voltage is constant in an electric circuit, the current I varies inversely as the resistance R of the circuit.

5 For a wire of fixed length, the electrical resistance R varies inversely as the square of the diameter d.

6 The momentum p of a moving body is directly proportional to its mass m and its velocity v.

7 The power P in an electric circuit varies jointly as the resistance R and the square of the current I.

8 The strength S of a horizontal beam of rectangular cross section and of length L varies jointly as the breadth b and the square of the depth d and inversely as the length L.

9 The vibrational frequency f of a stretched string is inversely proportional to its length L, directly proportional to the square root of its tension T, and inversely proportional to the square root of its linear density m.

10 The velocity v of sound in a given medium is directly proportional to the square root of the elasticity E of the medium and is inversely proportional to the square root of the density d of the medium.

11 The weight W of a human brain is directly proportional to the body weight B. If a 120-pound person has a brain weighing 3 pounds:
a Find the constant of proportionality.
b Determine the weight of the brain of a person weighing 200 pounds.

12 The amount of oil A used by a ship traveling at a constant speed varies jointly as the distance s and the square of the speed v. If the ship uses 500 barrels of oil in traveling 200 miles at 20 mph:
a Find the constant of proportionality.
b Determine how many barrels of oil the ship would use in traveling 250 miles at 16 mph.

13 The weight of a model dam varies directly as the cube of its height. What would be the weight of a model dam 64 inches high if a similar model 12 inches high weighed 243 pounds?

14 The amount of sediment that a stream will carry is directly proportional to the sixth power of its speed. How much more sediment will a stream carry if its speed is increased by 50 percent?

15 If the temperature of a gas remains constant, the pressure P varies inversely as the volume V.
a Write the corresponding equation of variation.
b A gas initially at a pressure of 16 pounds per square inch and with a volume of 500 cubic feet is compressed to a volume of 25 cubic feet while the temperature is held constant. Find the final pressure.

(*Note:* The pressure is measured in pounds per square *inch*, but the volume is in cubic feet.)

16 The horsepower HP that a rotating shaft can safely transmit varies jointly as the cube of its diameter d and the number R of revolutions it makes per minute. If a 2-inch shaft rotating at 1200 revolutions per minute can safely transmit 288 HP:
a Find the constant of proportionality.
b Find the number of horsepower that a 1.5-inch shaft can safely transmit at 1800 revolutions per minute.

17 The force of attraction F between two spheres varies directly as the product of their masses m_1 and m_2 and inversely as the square of the distance d between their centers.
a Write the corresponding equation of variation.
b If it is known that the force of attraction between two spheres is 360 dynes when the distance between their centers is d, find the force if the distance is tripled.

18 The force with which the earth attracts an object above its surface varies inversely as the square of the distance of the object from the center of the earth. How much will a meteorite that weighs 72 pounds on the surface of the earth weigh when it is 1980 miles above the earth's surface? (The radius of the earth is about 3960 miles.)

19 The illumination I in foot-candles (ft-c) upon a wall varies directly as the intensity i in candlepower (cp) of the source of light and inversely as the square of the distance d from the light. If the illumination is 5 ft-c at a distance of 10 feet from a light of 500 cp, what is the illumination at a distance of 15 feet from a light of 300 cp?

20 The strength S of a horizontal beam of rectangular cross section and of length L when supported at both ends varies jointly as the breadth b and the square of the depth d and inversely as the length L. A 2-by-4-inch beam, 8 feet long and resting on the 2-inch side, will safely support 600 pounds. What is the safe load when the beam is resting on the 4-inch side?

C **21** In Problem 18, find the weight of the meteorite when it is 200 miles above the earth's surface. Give your answer to the nearest tenth of a pound.

C **22** The current I (amperes) in a certain electric circuit is inversely proportional to the resistance R (ohms). If the current is 5 amperes when the resistance is 22 ohms, find the current when the resistance is increased to 28 ohms. Give your answer to the nearest tenth of an ampere.

SELF-TEST

1 Find, in standard form, an equation of the line with slope -4 and passing through $(5, -1)$.

2 Find the slope-intercept equation of the line that passes through $(5, 2)$ and $(6, -4)$.

3 The equation of a line is $3y+9x=5$. Find:
a the slope of the line;
b the y-intercept of the line.

4 Find, in standard form, an equation of the line that passes through $(-1, 2)$ and is parallel to the line $2y = 4x + 7$.

5 Find, in standard form, an equation of the line that passes through $(3, -4)$ and is perpendicular to the line $2y = 4x + 7$.

6 a Graph the parabola $y = -x^2 - x + 2$.
b Use your graph to solve the inequality $-x^2 - x + 2 > 0$.
c How many real roots does the quadratic equation corresponding to the parabola in (a) have?

7 An open trough is made by bending a long piece of sheet metal, 24 inches wide, into a rectangular shape. Find the largest possible cross-sectional area for this trough.

8 Identify and sketch the graph of $4x^2 + 16y^2 = 144$.

9 Identify and sketch the graph of $9y^2 - 4x^2 = 36$.

10 Identify and sketch the graph of $4x^2 + 4y^2 = 9$.

11 Identify and sketch the graph of $y^2 + 2y - 1 = 4x$.

12 Make a schematic graph of the polynomial function $P(x) = -2(x-1)^3(x+1)^2(x-2)$.

13 If $P(x)$ is the polynomial in Problem 12, solve the inequality $P(x) \geq 0$.

14 Find the vertical asymptotes to the graph of the equation

$$R(x) = \frac{2x^3}{(x-1)^2(x+2)}$$

15 Find the horizontal asymptote to the graph in Problem 14.

16 Make a schematic graph of the function R in Problem 14.

17 The pressure P of a gas enclosed in a container of fixed volume varies directly with the absolute temperature T of the gas. If the pressure on the walls of the container is 5 pounds per square inch when the absolute temperature is 460°, find the formula for P in terms of T.

18 The distance that an object falls in t seconds is directly proportional to the square of t. If the distance fallen at the end of 2 seconds is 64 feet, find the formula for the distance d in terms of t.

19 The pressure P exerted by a gas held at constant temperature is inversely proportional to the volume V. If the pressure is 5 pounds per square inch when the volume is 10 cubic inches, find the pressure when the volume is 15 cubic inches.

20 According to Newton's second law of motion, the acceleration a of a body is directly proportional to the force F acting on the body and inversely proportional to the mass m of the body. Write the equation of variation for this law.

CHAPTER 7

EXPONENTIAL AND LOGARITHMIC FUNCTIONS

7.1 EXPONENTIAL FUNCTIONS

Figure 7.1a shows the per capita *G*ross *N*ational *P*roduct (GNP) for several nations. The per capita GNP is the total output of goods and services measured in dollars per person. If we let x be

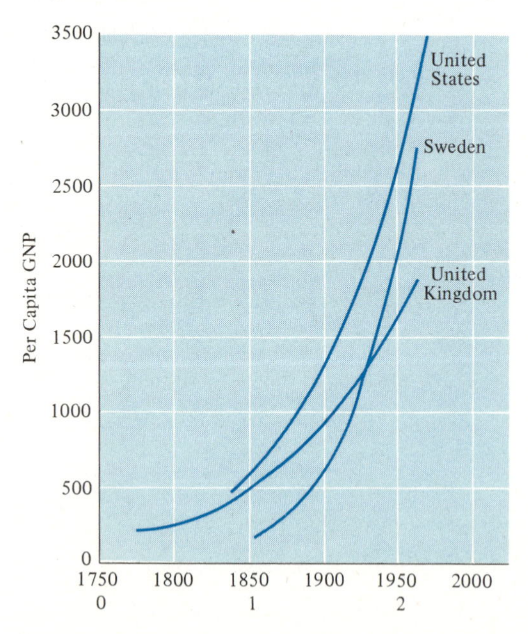

FIGURE 7.1a

the number of centuries, starting from 1750, so that $x = 1$ corresponds to 1850, $x = 2$ to 1950, and so on, and we let $100y$ dollars be the per capita GNP, so that $y = 1$ corresponds to \$100, $y = 2$ to \$200, and so on, then the equation

$$y = 5^x$$

closely approximates the per capita GNP for the United States. For example, for $x = 1$ (that is, in 1850), we have $y = 5^1$, which corresponds to \$500, and for $x = 2$ (that is, in 1950), we have $y = 5^2 = 25$, which corresponds to \$2500. Can you predict the per capita GNP for 1980? Because 1980 would correspond to $x = 2.3$, the preceding equation would give $y = 5^{2.3}$. We shall see later how to calculate a decimal approximation for $5^{2.3}$. (Of course, if you have a hand calculator with a y^x key on it, you can easily find $5^{2.3} = 40.5$, approximately.)

Because 2.3 is the same as $\frac{23}{10}$, we have no difficulty in interpreting $5^{2.3}$; this means $5^{23/10}$ or $(\sqrt[10]{5})^{23}$. We can interpret 5^x, where x is a rational number, in the same way, although we still have the problem of calculating a decimal value for 5^x. However, there are many places in mathematics where we need a meaning for 5^x with x an irrational number. For example, what does $5^{\sqrt{2}}$ mean, and, more generally, how can we define b^x when x is an irrational number? We shall not try to give a complete answer to this question, but we can provide an intuitive discussion of how to define b^x for irrational x.

We note first that if $b > 1$ and r is a positive rational number, then $b^r > 1$. Hence, if x and y are rational numbers with $x > y$, then $x - y$ is a positive rational number, so that $b^{x-y} > 1$. Multiplying both sides by the positive number b^y, we obtain $b^x > b^y$. Thus, we have Theorem 7.1a, which is illustrated in the following discussion.

▼ THEOREM 7.1a

If x and y are rational numbers with $x > y$, then for $b > 1$,

$$b^x > b^y$$

We know that $\sqrt{2}$ can be expressed as a decimal and that we can calculate as many decimal places as we please. For example, it can be shown that $\sqrt{2}$ is between 1.4142135 and 1.4142136. Now consider Table 7.1a, which has been constructed with a small, hand

TABLE 7.1a

$5^{1.4} = 9.5182697$	$5^{1.5} = 11.18034$
$5^{1.41} = 9.6726997$	$5^{1.42} = 9.8296353$
$5^{1.414} = 9.7351710$	$5^{1.415} = 9.7508518$
$5^{1.4142} = 9.7383052$	$5^{1.4143} = 9.7398726$
$5^{1.41421} = 9.7384619$	$5^{1.41422} = 9.7386186$
$5^{1.414213} = 9.7385089$	$5^{1.414214} = 9.7384933$
$5^{1.4142135} = 9.7385168$	$5^{1.4142136} = 9.7385183$

calculator having a y^x key on it. Note that the numbers in the table behave as predicted by Theorem 7.1a. In the left-hand column, as the exponents increase toward $\sqrt{2}$, the numbers also increase; in the right-hand column, as the exponents decrease toward $\sqrt{2}$, the numbers also decrease. Furthermore, the number in the left column is always less than the corresponding number in the right column, but the numbers in the two columns are getting closer and closer together; in the last line of the table, they agree to five decimal places.

In view of the preceding discussion, it should be plausible to you that there is exactly one number, say s, which lies between the number in the right column and the corresponding number in the left column no matter how far out we go in the decimal expansion of $\sqrt{2}$. Accordingly, we define

$$5^{\sqrt{2}} = s$$

where we know that $9.7385168 < s < 9.7385183$.

By following a procedure similar to the preceding one, we can define b^x for $b > 1$ and x any irrational number. If $b = 1$, then $b^x = 1$ for all real x, and we shall not be interested in this trivial case. If $0 < b < 1$, then we must modify Theorem 7.1a to state that b^r, where r is a rational number, decreases with increasing r. With this modification, the same type of argument can be used to define b^x, where x is an irrational number and $0 < b < 1$. If $b \le 0$, then b^x is not always a real number. For instance, if $b = 0$, then b^x is meaningless for $x \le 0$, and if $b < 0$, then $b^{1/2}$ is not a real number. For this reason, we require that the base be a positive number and, to eliminate the trivial case, that the base not be 1. With these restrictions, we see that if b is any admissible base, then b^x is defined for every real value of x and describes a function, which is named in Definition 7.1a.

▼ DEFINITION 7.1a

If $b > 0$ and $b \neq 1$, then the *exponential function* with base b is defined by $f(x) = b^x$, where the domain of f is the set of all real numbers.

EXAMPLE 1

Graph the given exponential functions on the same coordinate axes.

a. $f(x) = 2^x$
b. $f(x) = (\frac{1}{2})^x$
c. $f(x) = 3^x$
d. $f(x) = (\frac{1}{3})^x$

SOLUTION:

We first make a table with convenient values for x and then find the corresponding values for $f(x)$ as shown in Table 7.1b. The

TABLE 7.1b

x	-2	-1	0	1	2
$f(x) = 2^x$	$\frac{1}{4}$	$\frac{1}{2}$	1	2	4
$f(x) = (\frac{1}{2})^x$	4	2	1	$\frac{1}{2}$	$\frac{1}{4}$
$f(x) = 3^x$	$\frac{1}{9}$	$\frac{1}{3}$	1	3	9
$f(x) = (\frac{1}{3})^x$	9	3	1	$\frac{1}{3}$	$\frac{1}{9}$

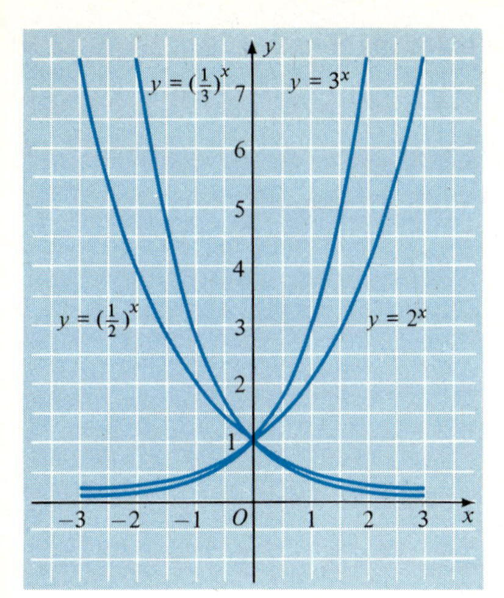

FIGURE 7.1b

graphs of the four functions are shown in Figure 7.1b. We have made use of the fact (proved in calculus) that the graphs are smooth curves.

Notice in Figure 7.1b that, as we proceed from left to right, the graphs of 2^x and 3^x are steadily rising, while those of $(\frac{1}{2})^x$ and $(\frac{1}{3})^x$ are steadily falling. This is the graphical reflection of the fact (also proved in calculus) that b^x defines a function that increases with increasing values of x if $b > 1$ and that decreases with increasing values of x if $0 < b < 1$. You should also notice that $(\frac{1}{2})^x = 2^{-x}$ and $(\frac{1}{3})^x = 3^{-x}$. Thus, the graphs of $(\frac{1}{2})^x$ and 2^x are symmetric to each other with respect to the y-axis and similarly for the graphs of $(\frac{1}{3})^x$ and 3^x. Notice that (0, 1) is the only point common to all four graphs

APPLICATIONS OF THE EXPONENTIAL FUNCTION

Many of the functions that occur in science and mathematics are exponential. The number of bacteria in a colony, under certain circumstances, grows exponentially; the decay of radioactive materials occurs exponentially; and economic growth can be represented by an exponential function. In order to work with these functions, we must know some of their properties. The most important property (one that is proved in advanced mathematics) is that the exponential function obeys all the laws of exponents. Another important property is given in Theorem 7.1b, which is a direct consequence of the steadily increasing (for $b > 1$) and the steadily decreasing (for $0 < b < 1$) character of the function defined by b^x. This theorem is used in almost all applications of the exponential function. Notice the important role it plays in Example 2.

▼ THEOREM 7.1b

The exponential function defined by

$f(x) = b^x \qquad b > 0, b \neq 1$

is a one-to-one function.

EXAMPLE 2

The half-life of strontium-90 is 25 years; that is, given any quantity of strontium-90, one half of it will be left 25 years hence. If we initially have q_0 milligrams of strontium-90, the quantity remaining at the end of t years is given by the formula

$$q = q_0 2^{-kt}$$

find the value of k.

SOLUTION:

We know that at the end of 25 years, the amount of strontium-90 remaining is $q_0/2$, Therefore,

$$\frac{q_0}{2} = q_0 2^{-25k}$$

Dividing by q_0 and writing $\frac{1}{2}$ as 2^{-1}, we obtain

$$2^{-1} = 2^{-25k}$$

Because the exponential function is one-to-one, the exponents on both sides must be equal, so that

$$-1 = -25k \quad \text{and} \quad k = \tfrac{1}{25} = 0.04$$

Example 2 shows that the formula for the amount of strontium-90 remaining at the end of t years is

$$q = q_0 2^{-0.04t}$$

For instance, if the initial amount of strontium-90 is 100 milligrams, so that $q_0 = 100$, then at the end of 10 years, that is, for $t = 10$, the amount remaining is $(100)(2^{-0.4})$ milligrams. With a calculator having a y^x key, we can easily evaluate the factor $2^{-0.4}$. The procedure is as follows:

Algebraic logic

[2] [y^x] [.4] [+/−] [=] $\approx .757858$

RPN logic

[2] [ENTER] [.4] [CHS] [y^x] $\approx .757858$

(Answers have been rounded to six decimal places.) At the end of 10 years, there would be about (100)(.757858) or approximately 75.8 milligrams of strontium-90 left.

EXPONENTIAL EQUATIONS

We can sometimes solve exponential equations (equations with the unknown in the exponent) by reducing both sides to the same base, as in Example 3.

EXAMPLE 3

Solve the equation $2^x = 4^{x-1} \cdot 8^{1-2x}$.

SOLUTION:

Since $4 = 2^2$ and $8 = 2^3$, the given equation can be rewritten as

$$2^x = (2^2)^{x-1}(2^3)^{1-2x}$$

or

$$2^x = 2^{2x-2} \cdot 2^{3-6x}$$

or

$$2^x = 2^{-4x+1}$$

Because the exponential function is one-to-one, the exponents on both sides must be equal. Thus,

$$x = -4x + 1 \quad \text{or} \quad x = \tfrac{1}{5}$$

(If you have a calculator with a y^x key, you can check this result in the given equation.)

Example 3 shows that we can solve an exponential equation if we can do two things: (1) bring the equation into the form $b^{f(x)} = b^{g(x)}$, and (2) solve the equation $f(x) = g(x)$. When these two steps cannot be carried out, an approximation procedure can be used to solve the equation. (See Using Your Knowledge 7.1.)

EXERCISE 7.1

In Problems 1–14, graph the given function.

1 $f(x) = 6^x$
2 $f(x) = (\frac{1}{6})^x$
3 $y = (\frac{1}{5})^x$
4 $y = 5^x$
5 $g(x) = 2^{-x}$
6 $F(x) = 3^{-x}$
7 $y = 2^{2x}$
8 $y = 3^{2x}$
9 $y = -2^x$
10 $y = -3^x$
11 $f(x) = 2^{x-1}$
12 $G(x) = 3^{x+1}$
13 $H(x) = (0.2)^{-x}$
14 $F(x) = -(0.2)^x$

In Problems 15–22, solve the given equation.

15 $3^x = 3^{2x+1}$
16 $5^{x-1} = 1$
17 $5^{-x} = 25$
18 $2^{x(x-1)} = 4$
19 $3^{x(x+4)} = 3^{-4}$
20 $3^x = 9^{x+1} \cdot 27^{1-2x}$
21 $5^{2x+1} = 25^x \cdot 5^{3x}$
22 $4^{x+2} = 2^{-2x} \cdot 8^{x-1}$

23 The half-life of lead-210 is approximately 22 years. If there are q_0 milligrams of lead-210 initially, the quantity q remaining at the end of t years is given by $q = q_0 2^{-kt}$. Find the value of k.

24 Repeat Problem 23 for radium, whose half-life is 1690 years.

25 Repeat Problem 23 for carbon-14, whose half-life is approximately 5600 years.

26 A radioactive substance decays so that the quantity q remaining at the end of t years is given by $q = q_0 2^{-kt}$. If the half-life of the substance is T years, show that

$$q = q_0 2^{-t/T}$$

27 In Problem 26, suppose that $T = 10$. Use your calculator to find what percent of q_0 remains at the end of:
a 1 year **b** 2 years **c** 5 years **d** 15 years **e** 45 years

28 Return to Figure 7.1a and read the year that the per capita GNP for the United States reached:
a \$2000 **b** \$3000
Use your calculator to check your answers in the equation

$$y = 100 \cdot 5^x$$

Recall that $x = 1$ corresponds to 1850, $x = 2$ to 1950, and so on.

USING YOUR KNOWLEDGE 7.1

As you learned in the preceding section, the method used in Example 3 for solving an exponential equation works only in rather simple cases. It depends first of all on your being able to reduce the equation to the form $b^{f(x)} = b^{g(x)}$. This reduction cannot be done in any convenient way for an equation such as

$$3^x = 2^x + 2$$

and approximation techniques are needed to solve this equation.

Figure 7.1c shows the graphs of $y = 3^x$ and $y = 2^x + 2$. The two graphs intersect at just the one point where x is approximately 1.4. This means that the equation $3^x = 2^x + 2$ has one real root and that root is close to 1.4. We can find a closer approximation by using a calculator with a y^x key as follows:

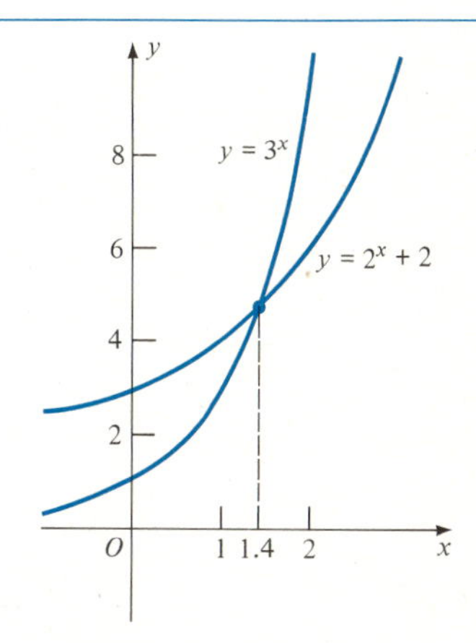

FIGURE 7.1c

We first write the equation with all terms on the left

$$3^x - 2^x - 2 = 0$$

and let $F(x) = 3^x - 2^x - 2$. Next we use the calculator to find $F(1.4)$. Thus,

Algebraic logic

[3] [y^x] [1.4] [−] [2] [y^x] [1.4] [−] [2] [=] ≈ 0.0165

RPN logic

[3] [ENTER] [1.4] [y^x] [2] [ENTER] [1.4] [y^x] [−] [2] [−] ≈ 0.0165

In the same way, we also calculate $F(1.3) \approx -0.291$. The change in sign from $F(1.3) \approx -0.291$ to $F(1.4) \approx 0.0165$ shows that the root is between 1.3 and 1.4. [Keep in mind that we are looking for the value of x for which $F(x) = 0$.] Because it appears that the root is much closer to 1.4 than to 1.3, we calculate $F(1.39) \approx -0.0161$. Because $F(1.40) \approx +0.0165$, we estimate that the root is about half way between 1.39 and 1.40. Then we calculate $F(1.395) \approx +0.00015$. This result, together with the value $F(1.39) \approx -0.0161$ indicates that the root is slightly less than 1.395. Hence, we calculate $F(1.3945) \approx -0.0015$, which shows that the root is between 1.3945 and 1.395. Therefore, correct to three decimal places, the root is 1.395.

The preceding ideas can be used to approximate the real roots of any equation in one variable. Approximate the roots of the following equations to three decimal places.

[C] **1** $5^x = 2^x + 100$ [C] **2** $5^{x+1} = 3 \cdot 2^x + 5$

7.2 LOGARITHMIC FUNCTIONS

Mitosis is a process in which a single cell or bacterium divides and forms two daughter cells. Each daughter cell then doubles in size and divides, and so on. If we assume that these cells divide every hour and that no cells die, then the number of cells present as a function of the time t is shown in Table 7.2a. You can check that a formula for this function is

$$N = 2^t$$

where t is the time (hours) and N is the number of cells at the end of t hours. This formula enables us to calculate the number of cells at any time t, but we are often interested in the inverse question: How long does it take for there to be a given number of cells? For example, how long does it take for there to be 2000 cells pre-

TABLE 7.2a

Time, t (hours)	0	1	2	3	4	5	6	7	8 . . .
Number, N, of cells	1	2	4	8	16	32	64	128	256 . . .

sent? To answer this question, we have to solve the equation $2000 = 2^t$ for t. This can be done using the procedure presented in Using Your Knowledge 7.1. Another method will be developed in this section.

THE LOGARITHM

In order to make it convenient to deal with exponential equations of this kind, we introduce the notion of a *logarithm*. The numbers in the first line of Table 7.2a are called the *logarithms to the base* 2 of the corresponding numbers in the second line. In the formula $N = 2^t$, the t is the exponent applied to the base 2 to yield the number N. The language of logarithms says that *the logarithm of N to the base* 2 *is t*. For instance, because $8 = 2^3$, the logarithm of 8 to the base 2 is 3. For brevity, we write the last statement as $\log_2 8 = 3$. With this notation and the information in the table, we can see that

$\log_2 16 = 4$ because $2^4 = 16$;
$\log_2 2 = 1$ because $2^1 = 2$; and
$\log_2 256 = 8$ because $2^8 = 256$.

Furthermore, we can solve the equation $2000 = 2^t$ by writing

$$t = \log_2 2000$$

However, we must learn more about logarithms before we can express t in a more familiar decimal form.

▼ DEFINITION 7.2a

Let $b > 0$ and $b \neq 1$. If y is any positive number, then

$$x = \log_b y \quad \text{if and only if} \quad b^x = y$$

(For $\log_b y$, read "the logarithm of y to the base b.")

The preceding ideas can be generalized to the function defined by $y = b^x$ as in Definition 7.2a. You should be sure to memorize the very important statement:

$$x = \log_b y \quad \textit{if and only if} \quad b^x = y \tag{1}$$

EXAMPLE 1

Write the given logarithmic equation in exponential form.

a. $\log_5 25 = 2$
b. $\log_3 \frac{1}{9} = -2$
c. $\log_{10} 1000 = 3$

SOLUTION:

a. Using Equation 1, we see that the equivalent of $\log_5 25 = 2$ is

$$5^2 = 25$$

b. Again by Equation 1, the equivalent of $\log_3 (\frac{1}{9}) = -2$ is

$$3^{-2} = \frac{1}{9}$$

c. Similarly, the equivalent of $\log_{10} 1000 = 3$ is

$$10^3 = 1000$$

EXAMPLE 2

Solve the equation $\log_3(x+1)=2$.

SOLUTION:

By Equation 1, the equation $\log_3(x + 1) = 2$ is equivalent to

$$3^2 = x + 1$$

Thus, we obtain

$$x = 8$$

▼ DEFINITION 7.2b

For $b > 0$ and $b \neq 1$, the function g defined by

$$g(x) = \log_b x \qquad x > 0$$

is called the *logarithm function* with base b.

According to Equation 1, if we solve the equation $y = b^x$ for x, we obtain $x = \log_b y$. Hence, if we follow the procedure of Section 5.4 and interchange the x and the y in the last equation to write $y = \log_b x$, we arrive at the fact that the function defined by $g(x) = \log_b x$ and the function defined by $f(x) = b^x$ are inverses of each other. The function g is the subject of Definition 7.2b. Because the domain of the exponential function is the set of all real numbers and the range is the set of all positive numbers, the logarithm function has the set of all positive numbers for its domain and the set of all real numbers for its range. Note carefully that the logarithm function is defined for positive numbers *only*. Figure 7.2a shows the graphs of $f(x) = 2^x$ and $g(x) = \log_2 x$. Notice that the graphs of these two inverse functions have the usual symmetry with respect to the line $y = x$. (See Section 5.4.)

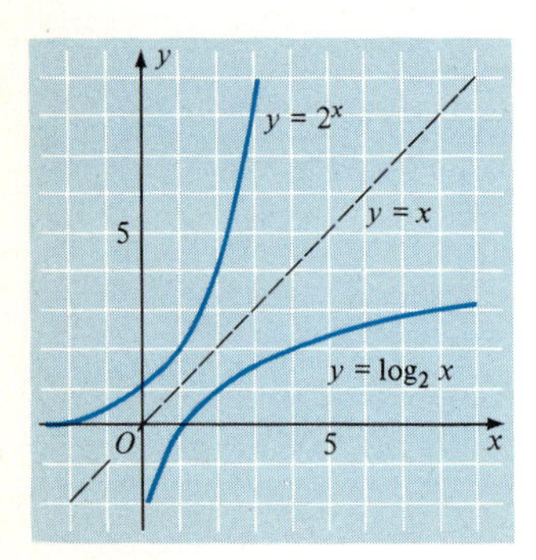

FIGURE 7.2a

PROPERTIES OF THE LOGARITHM

▼ THEOREM 7.2a

I. $b^{\log_b M} = M \qquad M > 0$

II. $\log_b b^M = M$ **for any real number M**

III. $\log_b 1 = 0$

Some of the important properties of the logarithm function are given in Theorem 7.2a. As the proofs show, these properties follow as direct consequences of the properties of the exponential function.

To prove Part I, we let $x = \log_b M$ for $M > 0$. Then by Equation 1,

$$M = b^x = b^{\log_b M}$$

Thus,

$$b^{\log_b M} = M$$

as was to be shown. To prove Part II, we let $x = \log_b b^M$. Then, again by Equation 1,

$$b^M = b^x$$

Because the exponential function is one-to-one, it follows that $x = M$, that is,

$$\log_b b^M = M$$

as was to be proved. Similarly, for Part III, we let $x = \log_b 1$. Then, by Equation 1,

$$b^x = 1.$$

But, we know that $b^0 = 1$, so that it follows from the one-to-one character of the exponential function that $x = 0$, that is,

$$\log_b 1 = 0$$

As illustrations of the theorem, we have

$$10^{\log_{10} 2} = 2, \quad \log_2 2^5 = 5, \quad \text{and } \log_{10} 1 = 0$$

EXAMPLE 3

Find the numerical value of $\log_{27} 243$.

SOLUTION:

First, we know that $\log_{27} 243 = y$ if and only if $27^y = 243$. Because $27 = 3^3$ and $243 = 3^5$, we may rewrite the last equation as

$$(3^3)^y = 3^5$$

or

$$3^{3y} = 3^5$$

which shows that

$$y = \tfrac{5}{3}$$

Therefore, $\log_{27} 243 = \frac{5}{3}$.

You must not forget that logarithms are simply exponents and that the properties of logarithms derive directly from those of exponents. The basic laws of logarithms, which are stated in Theorem 7.2b are essentially restatements of the laws of exponents. We shall prove Parts I and III and leave the other parts for the problems.

▼ THEOREM 7.2b

If M and N are positive numbers and b is any admissible base, then

I. $\log_b MN = \log_b M + \log_b N$

II. $\log_b \dfrac{M}{N} = \log_b M - \log_b N$

III. $\log_b M^k = k \log_b M$ **k any real number**

IV. $\log_b b = 1$

To prove Part I, let $x = \log_b M$ and $y = \log_b N$. Then,

$$b^x = M \quad \text{and} \quad b^y = N$$

Thus,

$$MN = b^x b^y = b^{x+y}$$

so that

$$\log_b MN = x + y = \log_b M + \log_b N$$

To prove Part III, let $x = \log_b M$. Then, as before,

$$b^x = M$$

and

$$M^k = (b^x)^k = b^{kx}$$

Thus,

$$\log_b M^k = kx = k \log_b M$$

EXAMPLE 4

When 10 is the base of the logarithm, it is customary to omit it and to write simply $\log N$. From a table of logarithms, we can find that $\log 2 = 0.301$, $\log 3 = 0.477$, and $\log 5 = 0.699$, all correct to three decimal places. Use these numbers and the properties of logarithms to find the value of $\log(\sqrt{18}/5)$.

SOLUTION:

$\log \frac{\sqrt{18}}{5} = \log \sqrt{18} - \log 5$	Theorem 7.2b, Part II
$= \log(2 \cdot 3^2)^{1/2} - \log 5$	
$= \frac{1}{2} \log(2 \cdot 3^2) - \log 5$	Theorem 7.2b, Part III
$= \frac{1}{2}(\log 2 + 2 \log 3) - \log 5$	Theorem 7.2b, Parts II and III
$= \frac{1}{2} \log 2 + \log 3 - \log 5$	
$= 0.151 + 0.477 - 0.699$	
$= -0.071$	

EXAMPLE 5

Solve:

a. $\log_2(x-2) + \log_2(x-3) = 1$

b. $\log_3(x-2) - \log_3 2 = \log_3(3x+1) + \log_3 \frac{1}{20}$

SOLUTION:

a. By Theorem 7.2b, Part I, the given equation may be rewritten

$$\log_2(x-2)(x-3) = 1$$

Therefore, by Equation 1,

$$(x-2)(x-3) = 2^1$$

or

$$x^2 - 5x + 6 = 2$$
$$x^2 - 5x + 4 = 0$$
$$(x-1)(x-4) = 0$$

and

$$x = 1 \quad \text{or} \quad x = 4$$

CHECK:

For $x = 1$, $x - 2 = -1$, which is not in the domain of the logarithm function. Therefore, $x = 1$ is not a solution of the given equation. For $x = 4$, $\log_2(x-2) + \log_2(x-3) = \log_2 2 + \log_2 1 = 1 + 0 = 1$, which checks. Thus, the solution is $x = 4$.

SOLUTION:

b. By Theorem 7.2b, Parts I and II, the given equation may be rewritten as

$$\log_3\left(\frac{x-2}{2}\right) = \log_3\left(\frac{3x+1}{20}\right)$$

Hence,

$$\frac{x-2}{2} = \frac{3x+1}{20}$$

or

$$10x - 20 = 3x + 1$$
$$7x = 21$$
$$x = 3$$

You should verify that $x = 3$ does check in the given equation and is thus the desired solution.

Notice the procedure in Example 5. We combine the logarithms, using Theorem 7.2b as needed, and try to write the equation in the form

$$\log_b A = N$$

so that

$$A = b^N$$

or in the form

$$\log_b A = \log_b C$$

so that

$$A = C$$

In either case, we obtain an equation that is free of logarithms and that we may attack by the methods we have already learned. Of course, this procedure requires all the logarithms to have the same base. An equation as simple-looking as

$$\log_2 x + \log_3(x + 1) = 1$$

cannot be solved except by approximation techniques.

EXAMPLE 6

Combine the logarithms and write the result as the logarithm of a single expression.

$3 \log \sqrt{a} - \log b^2 - 3 \log \sqrt[4]{a} + 5 \log \sqrt{b}$

SOLUTION:

$$\begin{aligned}
&3 \log \sqrt{a} - \log b^2 - 3 \log \sqrt[4]{a} + 5 \log \sqrt{b} \\
&\quad = 3 \log a^{1/2} - \log b^2 - 3 \log a^{1/4} + 5 \log b^{1/2} \\
&\quad = \tfrac{3}{2} \log a - 2 \log b - \tfrac{3}{4} \log a + \tfrac{5}{2} \log b \\
&\quad = \tfrac{3}{4} \log a + \tfrac{1}{2} \log b \\
&\quad = \log a^{3/4} + \log b^{1/2} \\
&\quad = \log(a^{3/4}b^{1/2}) \\
&\quad = \log(a^3b^2)^{1/4} \\
&\quad = \log \sqrt[4]{a^3b^2}
\end{aligned}$$

EXAMPLE 7

Show that for $x > y > 0$,

$$\log_b \sqrt{\frac{x^3(x+y)}{x-y}} = \frac{3}{2} \log_b x$$

SOLUTION:

We first write

$$\sqrt{\frac{x^3(x + y)}{x - y}} = \left[\frac{x^3(x + y)}{x - y}\right]^{1/2}$$

$+\frac{1}{2}\log_b(x+y)$

$-\frac{1}{2}\log(x-y)$

Then,

$$\log_b\sqrt{\frac{x^3(x+y)}{x-y}}$$

$$= \tfrac{1}{2}\log_b\frac{x^3(x+y)}{x-y} \qquad \text{Theorem 7.2b, Part III}$$

$$= \tfrac{1}{2}[\log_b x^3 + \log_b(x+y) - \log_b(x-y)] \qquad \text{Theorem 7.2b, Parts I and II}$$

$$= \tfrac{1}{2}[3\log_b x + \log_b(x+y) - \log_b(x-y)] \qquad \text{Theorem 7.2b, Part III}$$

$$= \tfrac{3}{2}\log_b x + \tfrac{1}{2}\log_b(x+y) - \tfrac{1}{2}\log_b(x-y)$$

which was to be shown.

EXERCISE 7.2

In Problems 1–6, write the given equation in exponential form.

1 $\log_9 729 = 3$ **2** $\log_{81} 27 = \frac{3}{4}$ **3** $\log_2 \frac{1}{256} = -8$

4 $\log_{10} 1 = 0$ **5** $\log_{10} 300 = 2.4771$ **6** $\log_5 \frac{1}{125} = -3$

In Problems 7–12, write the given equation in logarithmic form.

7 $2^7 = 128$ **8** $81^{1/2} = 9$ **9** $64^{1/6} = 2$

10 $10^{0.4771} = 3$ **11** $10^{-3} = 0.001$ **12** $M = a^{-4}$

In Problems 13–20, find the numerical value of each logarithm.

13 $\log_2 256$ **14** $\log_2 64$ **15** $\log_{27} 81$

16 $\log_{16} 64$ **17** $\log_{10} 1000$ **18** $\log_{10} 0.0001$

19 $\log_3 \sqrt[5]{3}$ **20** $\log_5 \sqrt[6]{5}$

In Problems 21–40, solve the given equation.

21 $\log_3 x = 5$ **22** $\log_{10} y = -5$ **23** $\log_x 64 = 2$

24 $\log_x 2 = \frac{1}{6}$ **25** $\log_y 0.1 = -\frac{1}{2}$ **26** $\log_y 128 = 7$

27 $\log_3(x-2) = 2$ **28** $\log_{10} x^2 = 2$ **29** $\log_3(x-1) - \log_3(x-3) = 1$

30 $\log_2(3+x) - \log_2(7-x) = 2$ **31** $\log_5 3x = \log_5 3 - \log_5 7$ **32** $\log_5 5x = \log_5 5 - \log_5 9$

33 $\log_6(x+1) = \log_6 8 - \log_6 2$ **34** $\log_6(3x+1) = \log_6 15 - \log_6 3$ **35** $\log_2(x+2) + \log_2(x+6) = 5$

36 $\log_3(x+10) + \log_3(x+4) = 3$ **37** $\log_3(x-2) + \log_3(x-3) = 2\log_3\sqrt{2}$ **38** $\log_4(x-2) + \log_4(x-1) = 2\log_4\sqrt{6}$

39 $\log_2(x^2+4x+7) = 2$ **40** $\log_2(x^2+4x+3) = 3$

In Problems 41–46, the base is omitted and is understood to be 10. Use $\log 2 = 0.301$, $\log 3 = 0.477$, and $\log 5 = 0.699$ to find the value of the given logarithm.

41 $\log 36$ **42** $\log\sqrt{75}$ **43** $\log\frac{24}{5}$

44 $\log\frac{81}{64}$ **45** $\log\sqrt[5]{7.5}$ **46** $\sqrt{\frac{128}{1250}}$

In Problems 47–50, combine the logarithms and write the result as the logarithm of a single number or expression. The base is understood to be 10.

47 $\log \frac{26}{7} - \log \frac{15}{63} + \log \frac{5}{26}$

48 $\log 9 - \log 8 - \log \sqrt{75} + \log \sqrt{\frac{25}{27}}$

49 $\log b^3 + \log 2 - \log \sqrt{b} + \log \sqrt{b^3/2}$

50 $\log a - \frac{1}{6} \log b - \frac{1}{2} \log a + \frac{1}{3} \log b$

In Problems 51–56, use the properties of logarithms to transform the left side of the given equation into the right side.

51 $\log_b \sqrt{\dfrac{x^2 - 16}{(x+4)^2}} = \dfrac{1}{2} \log_b(x-4) - \dfrac{1}{2} \log_b(x+4),\ x > 4$

52 $\log_b \left[\dfrac{(x-3)^{7/3}}{x^2(x^2-9)}\right] = \dfrac{4}{3} \log_b(x-3) - \log_b(x+3) - 2 \log_b x,\ x > 3$

53 $\log_b \left(y - \dfrac{1}{y}\right)^3 = 3 \log_b(y-1) + 3 \log_b(y+1) - 3 \log_b y,\ y > 1$

54 $\log_2 \left(\dfrac{z^2 \cdot 2^z}{2^{z^2}}\right) = z - z^2 + 2 \log_2 z,\ z > 0$

55 $\log_{10} \left(\dfrac{x^3 \cdot 10^{x^3}}{10^x}\right) = x^3 - x + 3 \log_{10} x,\ x > 0$

56 $\log_b \left(\dfrac{\sqrt{x} - 1}{\sqrt{x} + 1}\right) = 2 \log_b(\sqrt{x} - 1) - \log_b(x-1),\ x > 1$

57 Prove Theorem 7.2b, Part II.

58 Prove Theorem 7.2b, Part IV.

59 Assume $M, N, b > 0$ and $b \neq 1$. Prove that $\log_b M = \log_b N$ if and only if $M = N$.

60 Show that the functions defined by $f(x) = b^x$ and $g(x) = \log_b x$ satisfy Definition 5.4b, which proves that $g = f^{-1}$.

Hint: You must show that $f(g(x)) = x$ for every $x > 0$ and that $g(f(x)) = x$ for all real values of x.

USING YOUR KNOWLEDGE 7.2

The acidity of an aqueous (water) solution of a chemical is determined by the concentration of hydronium ions H_3O^+ in moles per liter. This concentration is usually symbolized by $[H_3O^+]$. In chemistry, the "pH" of a solution is defined by the equation

$$\text{pH} = -\log[H_3O^+]$$

where log means $\log_{10}$. For pure water at 25°C,

$$[H_3O^+] = 1.0 \times 10^{-7}$$

so that

$$\begin{aligned} \text{pH} &= -\log(1.0 \times 10^7) \\ &= -\log 1.0 - \log 10^{-7} \\ &= 0 + 7 \\ &= 7 \end{aligned}$$

(Keep in mind that log 1 = 0, and by the definition of a logarithm, $\log_b b^x = x$.) Because the pH of water is 7, it is usually agreed that a *neutral* solution has a pH of 7; *acidic* solutions have pH's less than 7; and *basic* solutions have pH's greater than 7.

If your calculator has a *log* key, you can use it to find the pH of a solution if you know the corresponding value of $[H_3O^+]$. For example, if $[H_3O^+] = 6.3 \times 10^{-4}$ (the value for apples), then

$$\begin{aligned} pH &= -\log(6.3 \times 10^{-4}) \\ &= -\log 6.3 - (-4) \\ &= 4 - \log 6.3 \end{aligned}$$

This pH can be computed on your calculator as follows:

Algebraic logic

[4] [−] [6.3] [log] [=] ≈ 3.201

RPN logic

[4] [ENTER] [6.3] [log] [−] ≈ 3.201

Thus, the pH is 3.2. (pH's are usually given to one decimal place.)

You can also use your calculator to find the value of $[H_3O^+]$ if the pH is given. If the pH is 3.2, the procedure is as follows:

Algebraic logic

[3.2] [+/−] [INV] [log] ≈ .00063096

RPN logic

[10] [ENTER] [3.2] [CHS] [y^x] ≈ .00063096

Thus, $[H_3O^+] = .00063 = 6.3 \times 10^{-4}$. (If an answer is very large or very small, the calculator will give it in scientific notation.) Notice that the RPN Logic procedure amounts to solving the equation

$$pH = -\log[H_3O^+]$$

by first writing it in the exponential form

$$[H_3O^+] = 10^{-pH}$$

Use your calculator to complete the entries in Table 7.2b.

TABLE 7.2b

	$[H_3O^+]$	pH
C 1. Sauerkraut	3.2×10^{-4}	
C 2. Tomatoes		4.2
C 3. Milk	4.0×10^{-7}	
C 4. Human blood		7.4
C 5. Seawater	5.0×10^{-9}	
C 6. Carrots		5.0

7.3 COMMON LOGARITHMS[1]

In Torrance, California, the noise limit for motorists has been set at 82 decibels [db(A)]. For a sound of intensity I, the loudness L, measured in decibels, is given by

$$L = 10 \log_{10}(I/I_0) \quad (1)$$

where I_0 is the minimum sound intensity detectable by the human

[1] This section should be omitted if students are using scientific calculators.

ear. For example, the sound of a riveting machine 30 feet away is 10^{10} times as intense as I_0. Hence, its loudness is

$$\begin{aligned} L &= 10 \log_{10}(10^{10} I_0/I_0) \\ &= 10 \log_{10}(10^{10}) \\ &= 10(10) \qquad \text{By Theorem 7.2a, Part II} \\ &= 100 \text{ decibels} \end{aligned}$$

Many applications of logarithms, in addition to the preceding one, use the base 10. When the base is 10, the logarithms are called *common logarithms*, and in writing them the base is omitted. Thus, $\log M$ is understood to mean $\log_{10} M$. Throughout this text, if no base appears with the abbreviation "log," you should assume that the base is 10.

CHARACTERISTIC AND MANTISSA

Common logarithms are especially helpful in long and tedious computations, although their importance for this purpose has diminished greatly in recent years because of the availability of hand-held calculators. The convenience of common logarithms results from the fact that any positive number can be written in *scientific notation* as a number between 1 and 10 times a power of 10. Thus, if x is any positive number, then

$$x = m \cdot 10^n \qquad \text{where } 1 \le m < 10 \text{ and } n \text{ is an integer} \tag{2}$$

For example, $235.6 = 2.356 \times 10^2$ and $0.002356 = 2.356 \times 10^{-3}$. Consequently for any positive number x, we have

$$\log x = \log(m \cdot 10^n) = \log m + \log 10^n = \log m + n$$

The result is

$$\log x = n + \log m \qquad \text{where } n \text{ is an integer and } 1 \le m < 10 \tag{3}$$

Because we can read n by inspection of the number x, Equation 3 tells us that to obtain the logarithm of any positive number, we need know only the logarithm of the corresponding number between 1 and 10. The integer n is called the *characteristic*, and the number $\log m$ is called the *mantissa* of $\log x$. It is important to know that the mantissa of a logarithm is always a number between 0 and 1. Thus, if $\log m = y$, then $m = 10^y$, and because $1 \le m < 10$, it follows that $0 \le y < 1$, as we stated.

EXAMPLE 1

Find the mantissa and the characteristic of:

a. $\log 31.2$
b. $\log 4.31$
c. $\log 0.0013$

SOLUTION:

a.
$$\begin{aligned} \log 31.2 &= \log(3.12 \cdot 10) \\ &= \log 3.12 + \log 10 \\ &= \log 3.12 + 1 \end{aligned}$$

Thus, the mantissa of $\log 31.2$ is $\log 3.12$, and its characteristic is 1.

b. $\log 4.31 = \log(4.31 \cdot 10^0)$

Thus, the mantissa of log 4.31 is log 4.31, and the characteristic is 0.

c. $\log 0.0013 = \log(1.3 \cdot 10^{-3})$

Thus, the mantissa of log 0.0013 is log 1.3, and its characteristic is -3.

In Example 1, the mantissas of the given logarithms were left in the form log x. Tables that give the approximate values of log m for m between 1 and 10 are available. The table of mantissas in the Appendix (Table A2) is a four-place table, part of which is reproduced in Table 7.3a. Notice that all the mantissas are numbers between 0 and 1, as we predicted. This is a very important fact, as you will see in Example 2b. To find log 3.12 in this table, we first locate the row headed 3.1 and then move across to the column headed 2. Thus,

$\log 3.12 = 0.4942$

TABLE 7.3a

x	0	1	2	3	4	5	6	7	8	9
3.0	.4771	.4786	.4800	.4814	.4829	.4843	.4857	.4871	.4886	.4900
3.1	.4914	.4928	.4942	.4955	.4969	.4983	.4997	.5011	.5024	.5038
3.2	.5051	.5065	.5079	.5092	.5105	.5119	.5132	.5145	.5159	.5172
3.3	.5185	.5198	.5211	.5224	.5237	.5250	.5263	.5276	.5289	.5302
3.4	.5315	.5328	.5340	.5353	.5366	.5378	.5391	.5403	.5416	.5428
3.5	.5441	.5453	.5465	.5478	.5490	.5502	.5514	.5527	.5539	.5551
3.6	.5563	.5575	.5587	.5599	.5611	.5623	.5635	.5647	.5658	.5670
3.7	.5682	.5694	.5705	.5717	.5729	.5740	.5752	.5763	.5775	.5786
3.8	.5798	.5809	.5821	.5832	.5843	.5855	.5866	.5877	.5888	.5899
3.9	.5911	.5922	.5933	.5944	.5955	.5966	.5977	.5988	.5999	.6010

EXAMPLE 2

Find:

a. log 328

b. log 0.000338

SOLUTION:

a.
$$\begin{aligned}\log 328 &= \log 3.28 \cdot 10^2\\ &= \log 3.28 + \log 10^2\\ &= \log 3.28 + 2 \log 10^2\\ &= \log 3.28 + 2\\ &= 0.5159 + 2\\ &= 2.5159\end{aligned}$$

b.
$$\begin{aligned}\log 0.000338 &= \log 3.38 \cdot 10^{-4}\\ &= \log 3.38 + \log 10^{-4}\end{aligned}$$

$$= \log 3.38 + (-4)\log 10$$
$$= \log 3.38 + (-4)$$
$$= 0.5289 + (-4)$$

In Example 2b, we did not add the mantissa and characteristic, because the latter is negative. If we had, we would have obtained

$$\log 0.000338 = -3.4711$$
$$= -3 - 0.44711$$

where the decimal part is *negative* and thus not convenient to use with tables, which list only positive decimals. It is customary to write

$$\log 0.000338 = 0.5289 + (-4)$$

as

$$\log 0.000338 = 6.5289 - 10$$

Notice that 6 has been added and subtracted so that only the form and not the value of the logarithm has been changed. Of course, we could also write

$$\log 0.000338 = 5.5289 - 9 \quad \text{or} \quad \log 0.000338 = 40.5289 - 44$$

but the conventional way is to use the "-10" notation and write

$$\log 0.000338 = 6.5289 - 10$$

ANTILOGARITHMS

We are now ready to discuss reversing the process of finding the logarithm of a number. For instance, in Example 2a, we found that $\log 328 = 2.5159$. If we wished to find x when $\log x = 2.5159$, we could "reverse" the process; that is, given the number y, we now wish to find a number x so that $y = \log x$. The number x is called the antilogarithm of y and is denoted by antilog y. Hence,

antilog $y = x$ if and only if $\log x = y$

Thus, the antilog of 2.5159, that is, the solution of $\log x = 2.5159$, can be found by reversing the process of finding logarithms as follows:

$$\log x = 2.5159$$
$$= 0.5159 + 2$$
$$= \log 3.28 + 2$$
$$= \log 3.28 + \log 10^2$$
$$= \log 3.28 \cdot 10^2$$
$$= \log 328$$

Because from Table 7.3a $\log 3.28 = 0.5159$

Thus,

$x = 328$

EXAMPLE 3

Find:

a. antilog 4.5237

b. antilog 7.5416 − 10

SOLUTION:

a.
$$\begin{aligned} \log x &= 4.5237 \\ &= 0.5237 + 4 \\ &= \log 3.34 + \log 10^4 \\ &= \log 3.34 \cdot 10^4 \\ &= \log 33{,}400 \end{aligned}$$

Thus,

$x = 33{,}400$

b.
$$\begin{aligned} \log x &= 7.5416 - 10 \\ &= 0.5416 - 3 \\ &= \log 3.48 + \log 10^{-3} \\ &= \log 3.48 \cdot 10^{-3} \\ &= \log 0.00348 \end{aligned}$$

Thus,

$x = 0.00348$

INTERPOLATION

You may have noticed that Table 7.3a or Appendix Table A2 gives the logarithms of three-digit numbers only. Thus, using these tables we are unable to find log 2.724 or read the antilog of 0.6944 directly. We can, however, find approximations to these numbers by using a procedure called *linear interpolation.* This procedure is illustrated in Example 4.

EXAMPLE 4

Find log 2.724.

SOLUTION:

From Table A2 in the Appendix, we find that

$\log 2.72 = 0.4346 \quad \text{and} \quad \log 2.73 = 0.4362$

Now,

$\log 2.72 < \log 2.724 < \log 2.73$

or

$0.4346 < \log 2.724 < 0.4362$

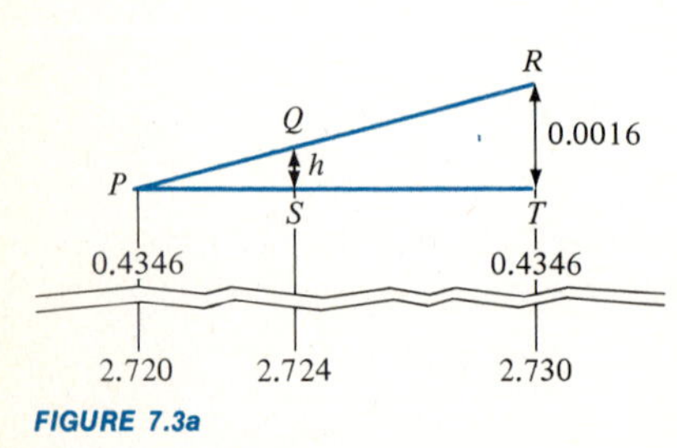

FIGURE 7.3a

Because a very short section of the logarithm curve is used, we obtain a sufficiently good approximation by replacing the curve by a straight line, as shown in Figure 7.3a. From the figure, you can see that

$\log 2.724 = 0.4346 + SQ = 0.4346 + h$

Because PQS and PRT are similar right triangles, we have

$$\frac{h}{0.0016} = \frac{0.004}{0.010}$$

which shows that the assumption of a straight line is equivalent to the assumption that the differences in the logarithms are proportional to the differences in the numbers. We now solve for h to obtain

$$h = \frac{0.004}{0.010}(0.0016) = 0.00064$$

which we round off to 0.0006. Thus,

$$\begin{aligned}\log 2.724 &= 0.4346 + h \\ &= 0.4346 + 0.0006 \\ &= 0.4352\end{aligned}$$

To shorten the procedure, we may write

$$0.010\left\{\begin{array}{l} 0.004\left\{\begin{array}{l}2.72 \\ 2.724\end{array}\right. \\ \quad\quad 2.73\end{array}\right. \quad \begin{array}{c}\textit{Number} \\ \\ \\ \\ \end{array} \quad \left.\begin{array}{l}\left.\begin{array}{c}0.4346 \\ ?\end{array}\right\} h \\ 0.4362\end{array}\right\} \mathbf{0.0016}$$

		Number	Log		
0.010	0.004	2.72	0.4346	h	**0.0016**
		2.724	?		
		2.73	0.4362		

Then write

$$\frac{\mathbf{h}}{\mathbf{0.0016}} = \frac{0.004}{0.010}$$

and solve as before.

EXAMPLE 5

Find antilog 0.6994.

SOLUTION:

From Table A2, we find log 5.00 = 0.6990 and log 5.01 = 0.6998, so that we know that antilog 0.6994 is between 5.00 and 5.01. We write

		Number	Log		
0.01	h	5.00	0.6990	0.0004	0.0008
		x	0.6994		
		5.01	0.6998		

$$\frac{h}{0.01} = \frac{0.0004}{0.0008}$$

Thus,

$$h = \frac{0.0004}{0.0008}(0.01) = 0.005$$

and

$$x = 5.00 + h = 5.005$$

that is

$$\text{antilog } 0.6994 = 5.005$$

Of course, if you wanted to find the antilogarithm of $7.6994 - 10$, you would first find antilog 0.6994, as in Example 5. Then, because the characteristic is $7 - 10$, or -3, you would multiply the 5.005 by 10^{-3} to obtain

$$\text{antilog}(7.6994 - 10) = 0.005005$$

Furthermore, if you wanted to find antilog(-2.3006), you would first change the negative logarithm, -2.3006, to the "-10" form to get a positive mantissa, so that you could use the table. Thus, you could add and subtract 10 to write

$$-2.3006 = (10 - 2.3006) - 10 = 7.6994 - 10$$

Now, you can see that

$$\begin{aligned}\text{antilog}(-2.3006) &= \text{antilog}(7.6994 - 10)\\ &= 0.005005.\end{aligned}$$

Some people prefer to add and subtract the smallest positive integer that will give a positive mantissa. For example,

$$-2.3006 = (3 - 2.3006) - 3 = 0.6994 - 3$$

The important idea is to get a *positive decimal* part to refer to the table.

EXERCISE 7.3

In Problems 1–10, use Appendix Table A2 to find the value of the given logarithm.

1 a log 1 **b** log 10 **c** log 100

2 a log 2 **b** log 20 **c** log 200

3 a log 3 **b** log 30 **c** log 300

4 a log 2.18 **b** log 21.8 **c** log 218

5 a log 0.31 **b** log 0.031 **c** log 0.00031

6 a log 6.81 **b** log 0.681 **c** log 0.0681

7 a log 42 **b** log 420 **c** log 4,200

8 a log 113 **b** log 1,130 **c** log 11,300

9 a log 0.00123 **b** log 0.000123 **c** log 0.0000123

10 a log 0.000321 **b** log 0.0000321 **c** log 0.00000321

In Problems 11–18, find the antilogarithm of the given number.

11 1.2672

12 3.5729

13 0.6928

14 7.7016 – 10

15 8.7839 – 10

16 4.8960 – 10

17 2.9991 – 10

18 3.7118 – 10

In Problems 19–24, use interpolation to find the indicated logarithm.

19 log 4.715
20 log 7.273
21 log 5.757
22 log 988.4
23 log 0.7729
24 log 0.1093

In Problems 25–32, use interpolation to find the indicated antilogarithm.

25 antilog 0.3520
26 antilog 1.6560
27 antilog 0.4010
28 antilog(9.0017 − 10)
29 antilog(8.5870 − 10)
30 antilog(6.9502 − 10)
31 antilog(−3.7826)
32 antilog(−0.7142)

In Problems 33–38, use Equation 1 at the beginning of this section to solve the problem. Recall that I_0 represents the minimum sound intensity that can be detected by the human ear.

33 The noise made by a jet plane at a distance of 100 feet is 10^{14} times as intense as I_0. How many decibels is that?

34 City traffic noise can be 10^9 times as intense as I_0. How many decibels is that?

35 The loudest noise created in a laboratory came from a steel and concrete horn and was 10^{21} times as intense as I_0. How many decibels is that?

36 The noise in a typing room is 10^7 times as intense as I_0. How many decibels is that?

37 The noise in the average home is 10^4 times as intense as I_0. How many decibels is that?

38 If the noise in the home becomes 1.5×10^4 times as intense as I_0, about how many decibels is that?

USING YOUR KNOWLEDGE 7.3

Let us return to Equation 1 at the beginning of Section 7.3 and solve this equation for the ratio I/I_0 to obtain the result

$$I/I_0 = \text{antilog}(L/10)$$

This is a handy formula for finding the ratio of the intensity of a given sound to the minimum intensity the human ear can hear. For example, if the loudness, L, is 82 decibels, then

$$I/I_0 = \text{antilog } 8.2$$

If you wish to use a table, then you must write

$$\text{antilog } 8.2 = 10^8 \times \text{antilog } 0.2$$

Appendix Table A2 gives antilog 0.2 ≈ 1.58, so that

$$I/I_0 = 1.58 \times 10^8$$

If you have a scientific calculator, you can use the same procedure as that described in Using Your Knowledge 7.2.

Find I/I_0 for the given values of L.

1 $L = 77$ decibels
2 $L = 65$ decibels
3 $L = 45$ decibels
4 $L = 28$ decibels

7.4 COMPUTATION WITH LOGARITHMS[2]

How much interest would you earn in 10 years on a \$20,000 certificate of deposit paying 6 percent compounded quarterly? If your answer is \$12,000, you are wrong. The correct answer is \$16,280.37. Compounding the interest makes the big difference. In order to check this answer, you need to know that if the sum of P dollars is invested at the rate r compounded annually, the amount A accumulated at the end of t years is given by

$$A = P(1 + r)^t \tag{1}$$

This amount includes the principal P and the accumulated interest; the interest alone would be obtained by subtracting P. If the interest is compounded n times per year, then the rate per period is r/n, and in t years, the number of periods is nt. Thus, the corresponding formula is

$$A = P\left(1 + \frac{r}{n}\right)^{nt} \tag{2}$$

In our problem $P = 20{,}000$, $r = 0.06$, $n = 4$, and $t = 10$. Hence, the amount at the end of 10 years is

$$\begin{aligned} A &= 20{,}000\left(1 + \frac{0.06}{4}\right)^{40} \\ &= 20{,}000(1.015)^{40} \end{aligned}$$

Using logarithms, we obtain

$$\begin{aligned} \log A &= \log[20{,}000(1.015)^{40}] \\ &= \log 20{,}000 + 40 \log 1.015 \\ &= 4.3010 + 0.2580 \\ &= 4.5590 \end{aligned}$$

Thus,

$$A = \text{antilog } 4.5590 \approx 36{,}230$$

Therefore, the interest earned is approximately

$$\$36{,}230 - \$20{,}000 = \$16{,}230$$

The discrepancy (we said that the interest is \$16,280.37) is caused by our using a four-place table, whereas the bank used either a more accurate table or else a computer.

---

[2] Most of this section should be omitted if students are using scientific calculators. We suggest that you go directly to Theorem 7.4a, the Change of Base Theorem.

If you have a calculator with a y^x key, then you can avoid the use of tables entirely. The calculation is as follows:

Algebraic logic

[20,000] [×] [1.015] [y^x] [40] [−] [20,000] [=] $\approx$ 16,280.37

RPN logic

[1.015] [ENTER] [40] [y^x] [20,000] [×] [20,000] [−] $\approx$ 16,280.37

(Both answers have been rounded off to the nearest cent.)

Without either a calculator or an electronic computer, you can see that logarithms furnish the only practical way to carry out the computation.

LOGARITHMIC COMPUTATIONS

In the remainder of this section we shall illustrate the use of logarithms and the theorems we have studied to perform certain arithmetical operations. You should be sure that you have mastered Theorems 7.2a and 7.2b before proceeding further.

EXAMPLE 1

Use logarithms to approximate $N = (0.653)(92.9)(214)$.

SOLUTION:

We first use Theorem 7.2b and then the table to write

$$\begin{aligned}\log N &= \log 0.653 + \log 92.9 + \log 214\\ &= (9.8149 - 10) + 1.9680 + 2.3304\\ &= 4.1133\end{aligned}$$

Now, by interpolation in the table, we find the approximate value $N = 12{,}980$

EXAMPLE 2

Use logarithms to calculate

$$N = \frac{(0.301)(6.25)}{(93.5)(0.05)}$$

SOLUTION:

By Theorem 7.2b, we can write

$$\begin{aligned}\log N &= \log 0.301 + \log 6.25 - \log 93.5 - \log 0.05\\ &= (9.4786 - 10) + (0.7959) - (1.9708) - (8.6990 - 10)\\ &= -0.3953\\ &= 9.6047 - 10\end{aligned}$$

Now, by interpolation in the table, we find the approximate value

$N = 0.4025$

EXAMPLE 3

Use logarithms to find $N = \sqrt{0.11200}$.

SOLUTION:

By Theorem 7.2b, we have

$$\log N = \log(0.11200)^{1/2}$$
$$= \tfrac{1}{2} \log 0.11200$$
$$= \tfrac{1}{2}(9.0492 - 10)$$
$$= \tfrac{1}{2}(19.0452 - 20)$$

In this step we added and subtracted 10 so that the division by 2 would give a result in the "-10" form.

$$= 9.5246 - 10$$

Now, by interpolation in the table, we obtain the approximate value

$$N = 0.3347$$

EXAMPLE 4

Use logarithms to find an approximate value of $N = 2^{-1.4}$.

SOLUTION:

Using Theorem 7.2b, we have

$$\log N = -1.4 \log 2$$
$$= -(1.4)(0.3010)$$
$$= -0.4214$$
$$= 9.5786 - 10$$

Here we added and subtracted 10 to get a *positive* decimal part so that we could use the table.

The table gives the approximate value

$$N = 0.3790$$

CHANGE OF BASE

In calculus, it is quite important to be able to change from one system of logarithms to another. Suppose that we have a system with base a and another system with base b, and we want to express $\log_b N$ in terms of $\log_a N$. Then, we let $\log_b N = x$, so that, by the definition of a logarithm,

$$N = b^x$$

We now take the logarithm to base a of both sides to obtain

$$\log_a N = \log_a b^x$$

or

$$\log_a N = x \log_a b$$

Solving for x, we find

$$x = \frac{\log_a N}{\log_a b}$$

Thus,

$$\log_b N = \frac{\log_a N}{\log_a b}$$

▼ **THEOREM 7.4a**

Change of base theorem

If a and b are positive numbers different from 1, then for any positive number, N,

$$\log_b N = \frac{\log_a N}{\log_a b}$$

This important result is stated in Theorem 7.4a. Taking $N = a$ and using the fact that $\log_a a = 1$, we obtain the interesting equation

$$\log_b a = \frac{1}{\log_a b}$$

EXAMPLE 5

Find $\log_6 18$.

SOLUTION:

By the Change of Base Theorem,

$$\log_6 18 = \frac{\log 18}{\log 6}$$

If we use tables to calculate the decimal answer, then we find

$$\frac{\log 18}{\log 6} = \frac{1.2553}{0.7782} = x$$

and

$$\begin{aligned}\log x &= \log \frac{1.2553}{0.7782} \\ &= \log 1.2553 - \log 0.7782 \\ &= 0.0988 - (0.8911 - 1) \\ &= 0.2077\end{aligned}$$

The table gives

$$x = \text{antilog } 0.2077 = 1.613$$

Thus,

$$\log_6 18 \approx 1.613$$

If we use a scientific calculator to compute the decimal value, then the procedure is as follows:

Algebraic logic

[18] [log] [÷] [6] [log] [=] ≈ 1.61315

RPN logic

[18] [log] [6] [log] [÷] ≈ 1.61315

(Both answers have been rounded off to five decimal places.)

EXERCISE 7.4

In Problems 1–18, use logarithms to calculate an approximate value for the given expression. In calculations involving negative numbers, first determine the sign of the result and then use logarithms of the absolute values. (Answers in the Appendix are given to four digits. Answers obtained with four-place logarithms may be slightly different.)

1 $(7.28)(0.0004)$

2 $(-0.0026)(-4.57)$

3 $(-4.02)(0.0071)(1.36)$

4 $(0.112)(0.066)(-8.4513)$

5 $\dfrac{(3.67)(0.487)}{(0.0005)(121.1)}$

6 $\dfrac{(0.0912)(7560)}{(0.163)(37.45)}$

7 $\dfrac{(60.13)(422)}{(319)(-2.84)}$

8 $\dfrac{(52{,}100)(0.0034)}{(14{,}600)(81{,}430)}$

9 $(2.35)^3$

10 $(3.41)^3$

11 $\sqrt{51.8}$

12 $\sqrt{0.132}$

13 $\sqrt[3]{-119}$

14 $\sqrt[3]{0.008}$

15 $(0.4231)^{0.3}$

16 $(0.21516)^{3/7}$

17 $5^{-1.2}$

18 $4^{-0.32}$

In Problems 19–24, evaluate the given logarithm.

19 $\log_3 14$

20 $\log_8 9$

21 $\log_5 17$

22 $\log_6 15$

23 $\log_3 21$

24 $\log_4 6$

25 If the volume stays constant, then the ratio of the pressure P to the absolute temperature T of a gas remains constant. Thus, we have

$$\frac{P}{T} = \frac{P_1}{T_1}$$

If, at an absolute temperature of 463°, the pressure is 5.30 pounds per square inch, find the pressure when the temperature is raised to 563°.

26 The lifting force F (in pounds) exerted by the atmosphere on the wings of an airplane is

$$F = \frac{4AV^2}{1875}$$

where A is the wing area in square feet, and V is the plane's speed in miles per hour (mph). Find F if the wing area is 100 square feet and the speed of the plane is 75 mph.

In Problems 27–30, find the approximate amount accumulated at the end of the given time if interest is compounded as stated.

	Principal	*Rate*	*Time*	*Compounded*
27	\$100	6%	10 years	annually
28	\$1000	6%	20 years	semiannually
29	\$1000	8%	30 years	quarterly
30	\$5000	9%	15 years	monthly

USING YOUR KNOWLEDGE 7.4

An equation of the type $a^x = b$ can always be solved for x by first taking the logarithm of both sides. For example, the equation

$$2^x = 3$$

becomes

$$x \log 2 = \log 3$$

so that

$$x = \frac{\log 3}{\log 2}$$

Using the table of common logarithms, we obtain

$$x = \frac{0.4771}{0.3010} = 1.585$$

Of course, you can use logarithms to do this last division, or, if you have just a four-function calculator, you can do the division on it. If you have a scientific calculator, you can compute the value of x by the same procedure that we used in Example 5 of this section.

1 If interest is compounded monthly, how long will it take for money to double if the annual rate is:
a 6 percent **b** 8 percent **c** 12 percent

Hint: Use Formula (2).

2 The present population of the world is about 4×10^9 and is increasing at the rate of 2 percent per year. How long will it take for the population to reach 6×10^9?

Hint: If the initial population is P_0 and the rate of increase is r (expressed as a decimal), then the population at the end of t years is $P = P_0(1 + r)^t$.

3 The land area of the earth is about 1.4×10^{14} square meters. Under the conditions of Problem 2, how long would it be before there were just 2 square meters per person?

4 According to Example 2 of Section 7.1, if we started with 100 milligrams of strontium-90, then in t years there would be $q = (100)(2^{-0.04t})$ milligrams left. In how many years would there be 75 milligrams left?

7.5 EXPONENTIAL EQUATIONS

Figure 7.5a shows the probability that a person will have an accident while driving a car if the person has a given blood alcohol level (BAC). The formula relating the probability P percent of having an accident to the blood alcohol level b percent is

$$P = e^{kb} \tag{1}$$

where k is a certain constant and the base is the number e (approximately 2.71828), a number of great importance in physics and mathematics. When appropriate numbers are substituted for P and b in Equation 1, the result is an *exponential equation* in k, that is, an equation in which the unknown occurs in an exponent. We shall have more to say about Equation 1 shortly.

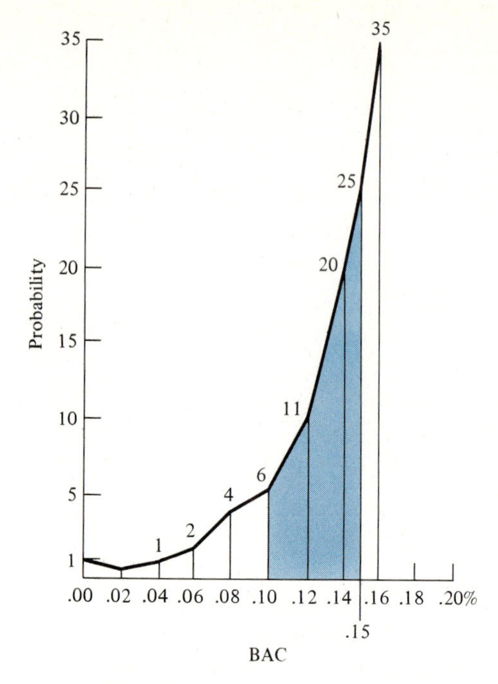

FIGURE 7.5a

NATURAL LOGARITHMS

The number e in Equation 1 is the base of the system of *natural logarithms*, which is used throughout calculus. We shall use the abbreviation *ln* for $\log_e$, that is,

$$\ln x = \log_e x$$

(The notation *ln x* is usually read "ell en of *x*.") Table A3 in the Appendix is a table of natural logarithms of numbers from 1.00 to 10.09. If you have a scientific calculator, it has a key labeled ln x (or LN). To obtain the natural logarithm of a number, you key in the number, then press the ln x key, and the display will show the logarithm. For example, the table gives $\ln 2 = 0.6931$; a calculator might show $\ln 2 = .69314718$.

Figure 7.5a shows that when the BAC is 0.15 percent, the probability of having an accident is 25 percent; that is, $P = 25$ when $b = 0.15$. If we substitute these values into Equation 1, we obtain

$$25 = e^{0.15k} \tag{2}$$

an exponential equation with k as the unknown. Can we find k? The answer is yes. We take the natural logarithm of both sides of Equation 2 and solve the resulting equation as follows:

$$\ln 25 = 0.15k$$

(The definition of a logarithm tells us that $\ln e^{0.15k} = 0.15k$.)

$$k = \frac{\ln 25}{0.15} \tag{3}$$

If you have a scientific calculator, you can find the numerator by keying in the 25 and then pressing the ln x key (or the LN key). The calculation is completed by dividing by 0.15. If you do not have a scientific calculator, you can turn to Table A3 to obtain

$$\begin{aligned} \ln 25 &= \ln(2.5 \times 10) \\ &= \ln 2.5 + \ln 10 \\ &= 0.9163 + 2.3026 \\ &= 3.2189 \end{aligned}$$

Thus, from Equation 3,

$$k = \frac{3.2189}{0.15} = 21.5$$

(We have rounded off the answer to the nearest tenth.) Now, Equation 1 can be written as

$$P = e^{21.5b} \tag{4}$$

EXAMPLE 1

At what blood alcohol level will the probability of having an accident while driving be 50 percent?

SOLUTION:

To answer this question, we substitute $P = 50$ into Equation 4 and then solve for b by taking the natural logarithm of both sides as follows:

$$\begin{aligned} 50 &= e^{21.5b} \\ \ln 50 &= 21.5b \end{aligned}$$

so that

$$b = \frac{\ln 50}{21.5}$$

From Table A3, we find

$$\begin{aligned} \ln 50 &= \ln 5 + \ln 10 \\ &= 1.6094 + 2.3026 \\ &= 3.9120. \end{aligned}$$

Thus,

$$b = \frac{3.9120}{21.5} \approx 0.18$$

The procedure for finding b with a scientific calculator is as follows:

Algebraic logic

[50] [ln x] [÷] [21.5] [=] ≈ 0.182 (Rounded off)

RPN logic

[50] [LN] [21.5] [÷] ≈ 0.182 (Rounded off)

Thus, when a person's BAC is about 0.18 percent, the probability that the person has an accident while driving is 50 percent.

A bank claims that passbook savings compounded daily at 5 percent per year yield a 5.13 percent rate of return. This means that if you deposited \$100 at the beginning of the year, you would receive \$105.13 at the end of the year. Do you think you would get much more money if the interest were compounded *continuously* rather than just daily? Continuous compounding means that if P_0 dollars are invested at the rate r, then the accumulated amount $P(t)$ at the end of t years is given by

$$P(t) = P_0 e^{rt} \tag{5}$$

For $t = 1$, $r = 5\% = 0.05$, and $P_0 = 100$, we have

$$P(1) = 100e^{(0.05)(1)} = 100e^{0.05} \tag{6}$$

A scientific calculator with an e^x key gives this value very quickly. The procedure is as follows:

Algebraic logic

[100] [×] [.05] [e^x] [=] ≈ 105.127

RPN logic

[100] [ENTER] [.05] [e^x] [×] ≈ 105.127

Note: If your calculator does not have an e^x key but does have an ln x and an inv key, then you can find e^x by keying in the value of x and then pressing the inv and the ln x keys in that order. (Some calculators may require a different procedure. Refer to your instruction manual.)

To evaluate $P(1)$ from Equation 6 without a calculator, you can use Table A4 in the Appendix to find values of e^x. From this table, we find $e^{0.05} = 1.0513$, which, upon substitution into Equation 6, yields

$$P(1) = 105.13$$

Thus, if you invest only \$100 for 1 year, there is no difference between daily and continuous compounding at the nominal rate of 5 percent. A difference would show up only if a considerably larger sum were deposited, because to six decimal places,

$$e^{0.05} = 1.051271 \quad \text{and} \quad \left(1 + \frac{0.05}{365}\right)^{365} = 1.051267$$

It is shown in calculus, that e^r is the number that is approached more and more closely as t is taken larger and larger in the expression

$$\left(1 + \frac{r}{t}\right)^t$$

EXPONENTIAL EQUATIONS

In Example 1, we solved the exponential equation

$$50 = e^{21.5b}$$

Exponential equations occur in a variety of problems and can sometimes be solved as in the next examples.

EXAMPLE 2

If P_0 dollars are invested at 5 percent per year compounded continuously. How long will it take for the money to double?

SOLUTION:

By Formula 5, the amount at the end of t years is

$$P(t) = P_0 e^{0.05t}$$

We are asked to find t such that $P(t) = 2P_0$. Hence, we must solve for t in the equation

$$2P_0 = P_0 e^{0.05t}$$

Dividing by P_0, we obtain the exponential equation

$$2 = e^{0.05t}$$

which we can solve by taking the natural logarithm of both sides. Thus,

$$\ln 2 = 0.05t$$

or

$$t = \frac{\ln 2}{0.05} \approx 13.86$$

(This result can be obtained by using either the tables or a calculator.) We have found that the amount doubles in about 13.86 years.

The study of radioactive materials is another area where exponential equations occur. Nuclear plants produce radioactive wastes, some of which are safely stored, whereas some, such as krypton-85, are released as gas into the atmosphere. It is known that radioactive elements decay exponentially with time. For example, krypton-85 has a decay rate of 6.44 percent a year. This means that if N_0 grams of krypton-85 are present initially, then the number of grams remaining at the end of t years is given by

$$N(t) = N_0 e^{-0.0644t} \tag{7}$$

EXAMPLE 3

a. Use Equation 7 to find the half-life of krypton-85; that is, find the time it takes for $N(t)$ to equal $\frac{1}{2}N_0$.

b. Find how long it takes for 90 percent of the krypton-85 to decay.

SOLUTION:

a. We have to solve the equation

$$\tfrac{1}{2}N_0 = N_0 e^{-0.0644t}$$

or

$$2^{-1} = e^{-0.0644t}$$

Therefore, we take the natural logarithm of both sides to obtain

$$-\ln 2 = -0.0644t$$

and

$$t = \frac{\ln 2}{0.0644} \approx 10.76$$

Thus, the half-life of krypton-85 is about 10.76 years.

b. Here, because 10 percent of the krypton-85 remains, we have to solve the equation

$$\tfrac{1}{10}N_0 = N_0 e^{-0644t}$$

or

$$10^{-1} = e^{-0.0644t}$$

As before, we take the natural logarithm of both sides to obtain

$$-\ln 10 = -0.644t$$

and

$$t = \frac{\ln 10}{0.0644} \approx 35.75$$

Thus, it takes about 35.75 years for 90 percent of the krypton-85 to decay.

In the preceding examples, the base was e, so that natural logarithms were the most convenient to use. However, it is possible to use common logarithms to solve certain exponential equations. The method is illustrated in Example 4.

EXAMPLE 4

Solve the equation $4^{2x-1} = 7^x$.

SOLUTION:

We take the logarithm of both sides to obtain

$$\log 4^{2x-1} = \log 7^x$$
$$(2x - 1)\log 4 = x \log 7$$
$$2x \log 4 - \log 4 = x \log 7$$
$$x(2 \log 4 - \log 7) = \log 4$$
$$x(\log 16 - \log 7) = \log 4$$
$$x = \frac{\log 4}{\log 16 - \log 7}$$

Using the table of common logarithms, we obtain

$$x = \frac{0.6021}{1.2041 - 0.8451} \approx 1.677$$

To evaluate x with a calculator, we can use the procedure:

Algebraic logic

[16] [log] [−] [7] [log] [=] [STO] [4] [log] [÷] [RCL] [=] ≈ 1.67694

RPN logic

[16] [LOG] [7] [LOG] [−] [4] [LOG] [÷] [1/x] ≈ 1.67694

EXERCISE 7.5

In Problems 1–6, use Equation 4 to find the blood alcohol level at which the probability of having an accident while driving is:

1 55 percent
2 60 percent
3 65 percent
4 70 percent
5 75 percent
6 80 percent

In Problems 7–10, find the amount of time necessary for money to double if invested at the given annual rate compounded continuously.

7 6 percent
8 8 percent
9 10 percent
10 12 percent

In Problems 11–16, find the half-life of the given substance.

11 Niobium-94, decay rate 0.0035 percent per year
12 Hydrogen-3, decay rate 5.59 percent per year
13 Cobalt-60, decay rate 13.08 percent per year
14 Strontium-90, decay rate 2.50 percent per year
15 Radium-88, decay rate 0.0436 percent per year
16 Thorium-232, decay rate 4.99×10^{-9} percent per year

In Problems 17–24, solve the given equation.

17 $5^{2x} = 7$
18 $3^{-2x} = 6$
19 $4^{3x-2} = 5$
20 $6^{4x-3} = 9$
21 $2^{3-x} = 6^{1-x}$
22 $3^{2-x} = 4^x$
23 $3^{x+1} = 5^{x-1}$
24 $4^{3+x} = 5^{x+1}$

In Problems 25–28, simplify the given expression.

25 $\ln xe^{-x}$

26 $\ln e^{-x}/x$

27 $\ln \dfrac{e^x}{e^x + 1}$

28 $\ln \sqrt{x}e^{-x^3/2}$

29 The world population P in 1974 was approximately 3.9 billion, and the annual growth rate was 2 percent. If we assume continuous growth, then $P = 3.9e^{0.02t}$, where t is the time in years after 1974.
a Assume no change in the growth rate and find the population in 1984.
b How many years after 1974 would it be before the population doubled?

30 In 1978, the world population was about 4 billion, and the growth rate was 2 percent per year. Assume that the growth is continuous and that the rate remains unchanged (see Problem 29). Find the number of years after 1978 it would take for the world population to reach 5 billion.

31 Under ideal conditions, the number of bacteria present in a culture at the end of t hours is $N(t) = 1000e^{kt}$. If there were 40,000 bacteria present at the end of 1 hour, how many bacteria would be present at the end of 3 hours?

32 Suppose that because of inflation, prices rise at the continuous rate of 12 percent per year. How long will it take for the price of steak to increase by 50%?

Hint: $A = A_0e^{rt}$

33 The radioactive isotope carbon-14 decays at the rate of 0.01204 percent per year. The amount of carbon-14 in a living plant or animal stays constant at the concentration of carbon-14 in the atmosphere, but when the plant or animal dies, the carbon-14 immediately begins to diminish at the stated decay rate. A piece of wood from a tree that was buried in the ash of a volcanic eruption of Mt. Mazama in southern Oregon was analyzed, and the fraction of undecomposed carbon-14 was found to be 0.451. About how long ago did the volcanic eruption occur?

34 **a** A piece of bone found in an archeological dig was analyzed and found to contain 5 percent of its original amount of carbon-14. Estimate the age of the bone.

Note: See Problem 33.

b If 50 percent of the original amount of carbon-14 were found in the bone, what would you estimate as its age?

35 It was found that when a laboratory animal was given a single injection of a certain drug, the amount of the drug in the animal's body decreased at the continuous rate of 5 percent per hour. If the original injection was 5 milliliters of the drug, how much would remain in the animal's body after 24 hours?

36 If a certain radioactive substance decays from 50 milligrams to 5 milligrams in 20 seconds, what is the half-life of the substance?

37 The half-life of uranium-238 is 4.5 billion years. Find the annual rate of decay of uranium-238.

38 A person can invest P dollars in an insured savings account that pays 5 percent compounded quarterly, or he can put P dollars into a more risky investment that pays 9 percent compounded monthly. How long would it take for the more risky investment to grow twice as large as the safe one?

Hint: Use Formula 2, Section 7.4.

USING YOUR KNOWLEDGE 7.5

In the theory of learning, it is assumed that a person's performance of a certain task improves with practice but has an upper limit that the person cannot exceed. For example, the number of words per minute that a person can type is given by an equation of the form

$$n = N(1 - e^{-kw})$$

where n is the number of words per minute the person can type after w weeks of practice and N is the upper limit that n cannot exceed. The k is a constant that has to be determined experimentally.

Suppose that a person can type 50 words per minute after 4 weeks of practice and 70 words per minute after 8 weeks of practice. Can we find the N and the k in the preceding equation for that person? Yes. Here is how it is done.

Substitute the known data into the equation to obtain

$$50 = N(1 - e^{-4k})$$

and

$$70 = N(1 - e^{-8k})$$

Because $e^{-8k} = (e^{-4k})^2$, let $x = e^{-4k}$, so that $e^{-8k} = x^2$. The two equations become

$$50 = N(1 - x) \quad \text{and} \quad 70 = N(1 - x^2)$$

which we solve for N to obtain

$$N = \frac{50}{1 - x} \quad \text{and} \quad N = \frac{70}{1 - x^2}$$

Therefore,

$$\frac{50}{1 - x} = \frac{70}{1 - x^2}$$

or, upon simplification,

$$5x^2 - 7x + 2 = 0$$

From this equation, we obtain $x = 1$ or $x = \frac{2}{5} = 0.4$. The first x-value makes no sense in the problem, because it gives $n = 0$ for all values of w. The desired solution is $x = 0.4$. You should be able to complete the calculation to obtain

$$e^{-4k} = 0.4$$
$$k \approx 0.23$$
$$N \approx 83$$

According to the theory, this person would never be able to type more than 83 words per minute with any amount of practice.

1 In the preceding problem, how many words per minute would the person be able to type after 10 weeks practice?

2 Suppose that after 4 weeks, a second person can type 50 words per minute and after 8 weeks, 80 words per minute. Find k and N for this person. How many weeks of practice would it take for this person to be able to type 100 words per minute?

SELF-TEST

1 Make a graph of each of the functions:
a $f(x) = 4^x$ **b** $g(x) = 4^{-x}$
c $h(x) = \log_4 x$

2 a The half-life of plutonium is about 23,100 years. If the initial amount of plutonium is q_0, then the amount q remaining at the end of t years can be written

$$q = q_0 2^{-kt}$$

Find the value of k.
b Use the value you found in (a) to determine how much plutonium would be left at the end of 46,200 years if the initial amount was 100 milligrams. (You should be able to do this without logarithms or calculator.)

3 a Write $\log_6 36 = 2$ in exponential form. **b** Write $32^{2/5} = 4$ in logarithmic form.

4 Find the numerical value of:
a $\log_3 81$ **b** $\log_4 8$

5 Solve the equation $\log_2(x + 3) + \log_2(x + 2) = 1$.

6 Solve $\log_2(x + 1) - \log_2 3 = \log_2(2x + 1) - \log_2 \frac{1}{5}$.

7 Simplify:
a $\log_7 7^{1.6}$ **b** $\log_b \sqrt[3]{x^4}$

8 Combine the logarithms and write as the logarithm of a single expression:

$$\log a^2 - \log 2 - \tfrac{1}{2} \log a + \log \sqrt{2}$$

9 Give the characteristic of each logarithm:
a log 53.72 **b** log 96,200
c log 0.00372 **d** log 0.5671

10 Find the mantissa of each logarithm:
a log 4.21 **b** log 89.6
c log 0.0231

11 Find each of the following logarithms:
a log 46.23 **b** log 0.09637

12 Find each antilogarithm:
a antilog 2.6340
b antilog(8.6649 − 10)

13 Use logarithms to calculate $\dfrac{(0.12)(3.14)}{(8.01)(0.03)}$

14 Use logarithms to find $\sqrt[5]{0.0296}$.

15 Use logarithms to find $3^{-2.2}$.

16 Find $\log_4 17$.

17 Solve the equation $3^{2x-1} = 5$.

18 Cesium-137 decays at the rate of 2.3 percent per year, so that if we have N_0 grams initially, then at the end of t years, the number N of grams remaining is $N = N_0 e^{-0.023t}$. Find the half-life of cesium-137.

19 The half-life of carbon-14 is about 5760 years. If the quantity q of carbon-14 is given in terms of the initial quantity q_0 by $q = q_0 e^{-kt}$, find the value of k.

20 If P dollars are invested at an annual rate of 8 percent compounded continuously, find how long it takes to double.

CHAPTER 8

THE CIRCULAR AND THE TRIGONOMETRIC FUNCTIONS

8.1 THE SINE AND COSINE FUNCTIONS

Figure 8.1a shows a portion of a cardiogram. Notice the repetitive character of the graph; this type of repetition is characteristic of the graphs of the periodic phenomena that occur all around and within us. The alternation of day and night, the phases of the moon, various kinds of wave motion, and the beating of our hearts are all examples of periodic phenomena.

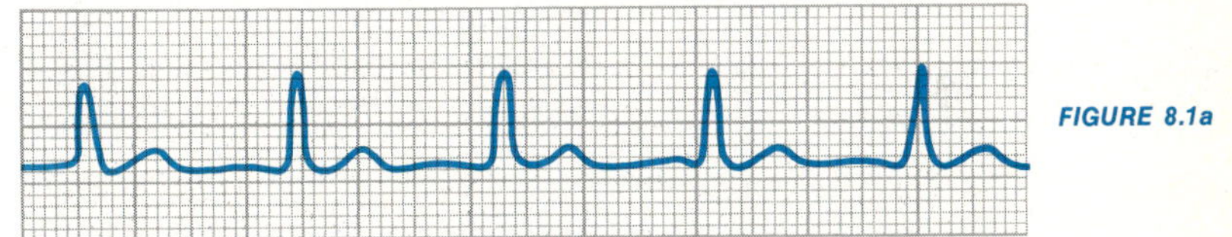

FIGURE 8.1a

There are certain simple functions that are important in the study of periodic phenomena, and this chapter is concerned with these functions. We start with a circle of radius 1 (a unit circle) with its center at the origin. This circle, whose equation is

$$x^2 + y^2 = 1$$

crosses the x-axis at the two points $(-1, 0)$ and $(1, 0)$, as shown in Figure 8.1b. The point $(1, 0)$ will be of special interest because we select it as the initial point from which to measure distances along the circumference of the circle.

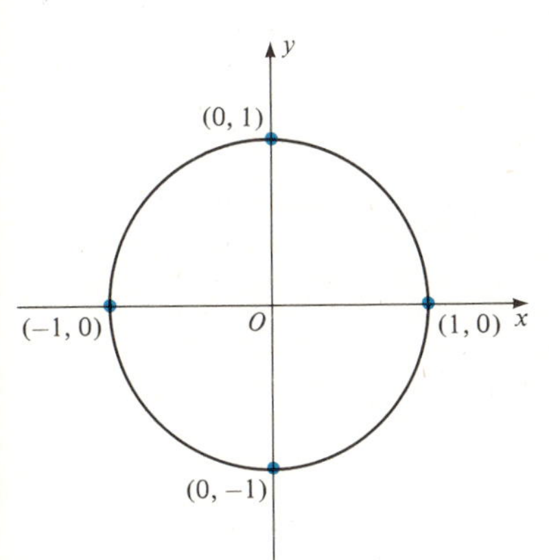

FIGURE 8.1b
The unit circle $x^2 + y^2 = 1$.

POINTS ON THE UNIT CIRCLE

If we are given any real number t, we start at the point $(1, 0)$ and measure a distance t units along the circle. By common agreement, we go in the *counterclockwise* direction if t is *positive* and in

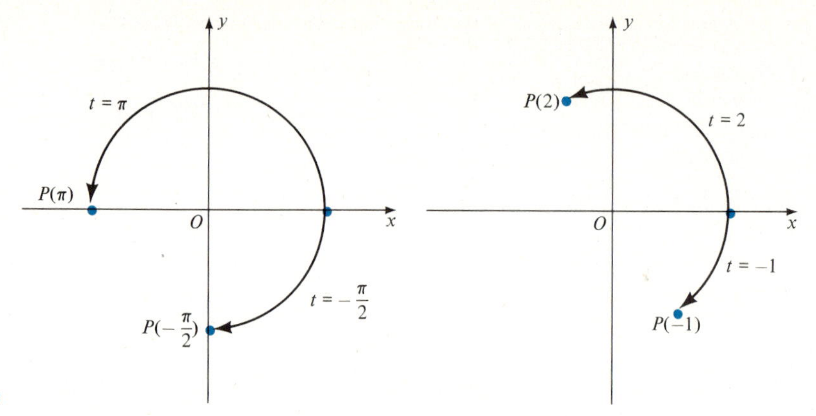

FIGURE 8.1c

the *clockwise* direction if t is *negative*. For each value of t, this procedure locates a unique point, say $P(t)$, on the circle. Figure 8.1c shows four points located in this way. (Recall that the entire length of the circumference of the unit circle is 2π.)

Since the length of the circumference of the circle is 2π, the numbers t and $t + 2\pi$ determine the same point (see Figure 8.1d); that is,

$$P(t + 2\pi) = P(t)$$

More generally, for any integer k, it is clear that

$$P(t + 2k\pi) = P(t)$$

This periodic behavior of the point P as it depends on t is extremely important for the remainder of this discussion. Note that exactly one point corresponds to a given value of t, but infinitely many values of t correspond to a given point.

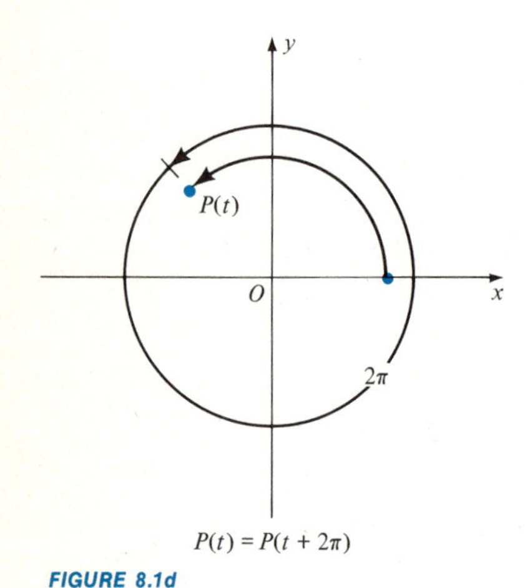

FIGURE 8.1d

THE SINE AND THE COSINE

Because each real value of t determines a unique point $P(t)$ on the unit circle, the rectangular coordinates of the point may be regarded as uniquely determined by the value of t. Thus, x and y are both functions of t that are defined for all real values of t. Furthermore, $x^2 + y^2 = 1$, so that

$$-1 \le x \le 1 \quad \text{and} \quad -1 \le y \le 1$$

To summarize, we say that x and y are both functions of t with the set of all real numbers for their domain and the set of all real numbers between -1 and 1, inclusive, for their range. These special functions are defined formally in Definition 8.1a. It is customary to denote the cosine and sine functions, respectively, by writing

$$x = \cos t \quad \text{and} \quad y = \sin t$$

Figure 8.1e shows the geometric representation of $\cos t$ and $\sin t$.

▼ DEFINITION 8.1a

Let $P(t)$ be the point that is t units along the unit circle from the point (1, 0), and let the rectangular coordinates of $P(t)$ be (x, y). Then, x *is a function of t called the cosine function,* and y *is a function of t called the sine function.*

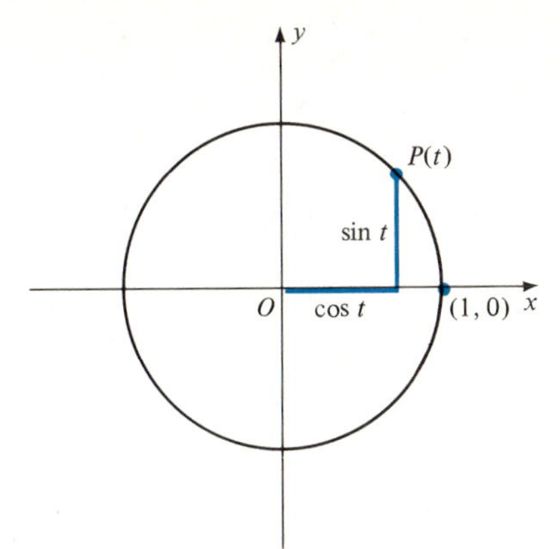

FIGURE 8.1e

TABLE 8.1a

II	I
$x = \cos t < 0$ $y = \sin t > 0$	$x = \cos t > 0$ $y = \sin t > 0$
III	IV
$x = \cos t < 0$ $y = \sin t < 0$	$x = \cos t > 0$ $y = \sin t < 0$

Because $x = \cos t$ and $y = \sin t$ are the coordinates of a point on the unit circle for a given value of t, they are positive or negative depending on the quadrant in which the point $P(t)$ lies. Table 8.1a summarizes the signs associated with the two functions for each of the four quadrants.

Many properties of the sine and cosine functions are readily deduced from their definitions. Because $(\cos t, \sin t)$ is a point on the unit circle, these coordinates must satisfy the equation of the circle. This fact results in Theorem 8.1a. Equation 1 is a basic identity, an equation that is true for all real values of t.[1] Notice that we have used the common convention of writing $\cos^2 t$ for $(\cos t)^2$ and $\sin^2 t$ for $(\sin t)^2$. Because $\cos^2 t$ and $\sin^2 t$ are *nonnegative* real numbers whose sum is 1, we must have

▼ THEOREM 8.1a

For all real values of t

$$\cos^2 t + \sin^2 t = 1 \qquad (1)$$

$$0 \le \cos^2 t \le 1 \quad \text{and} \quad 0 \le \sin^2 t \le 1$$

Hence, the following inequalities are valid for all real values of t.

$$-1 \le \cos t \le 1 \quad \text{and} \quad -1 \le \sin t \le 1$$

From Theorem 8.1a and Table 8.1a, we can deduce the following relationships:

$$\cos t = \sqrt{1 - \sin^2 t} \qquad \text{for } P(t) \text{ in Q I or Q IV}$$
$$\cos t = -\sqrt{1 - \sin^2 t} \qquad \text{for } P(t) \text{ in Q II or Q III}$$
$$\sin t = \sqrt{1 - \cos^2 t} \qquad \text{for } P(t) \text{ in Q I or Q II}$$
$$\sin t = -\sqrt{1 - \cos^2 t} \qquad \text{for } P(t) \text{ in Q III or Q IV}$$

[1] In general, an identity is an equation that is true for all values of the variables for which the members of the equation are defined.

Thus, if we are given the value of $\sin t$ or $\cos t$ and the quadrant in which the point $P(t)$ lies, then we can find the value of the other function.

EXAMPLE 1

If $\sin t = 0.6$ and $\pi/2 < t < \pi$, find $\cos t$.

SOLUTION:

For $\pi/2 < t < \pi$, the point $P(t)$ will lie in the second quadrant. Therefore, $\cos t$ is negative, and we have

$$\begin{aligned} \cos t &= -\sqrt{1 - \sin^2 t} \\ &= -\sqrt{1 - (0.6)^2} \\ &= -\sqrt{1 - 0.36} \\ &= -\sqrt{0.64} \\ &= -0.8 \end{aligned}$$

PERIODIC FUNCTIONS

The periodic character of the point $P(t)$ with respect to the number t implies a similar behavior for the coordinates of P. In order to give a precise description of this behavior, we first define a periodic function. (See Definition 8.1b.)

▼ DEFINITION 8.1b

If there is a nonzero number k, such that a nonconstant function f satisfies the equation

$$f(t + k) = f(t) \tag{2}$$

for all real values of t, then f is called a *periodic* function. The smallest positive value of k for which Equation 2 is true is called the *period* of f.

Because x and y are the rectangular coordinates of the point $P(t)$, it follows from the fact that $P(t + 2\pi) = P(t)$, that

$$\cos(t + 2\pi) = \cos t \quad \text{and} \quad \sin(t + 2\pi) = \sin t$$

Thus, the sine and cosine functions are two simple examples of periodic functions; they both have the period 2π. Of course, if m is any integer, then

$$\cos(t + 2m\pi) = \cos t \quad \text{and} \quad \sin(t + 2m\pi) = \sin t$$

For example,

$$\cos(t - 4\pi) = \cos t \quad \text{and} \quad \sin(t + 10\pi) = \sin t.$$

The periodicity of the sine and cosine functions is one of the reasons for their many applications in problems dealing with periodic phenomena. Such problems arise in the study of respiration, blood flow, alternating electric currents, and mechanical vibrations, as well as in many other important areas.

EXAMPLE 2

Find the smallest positive value of t for which it is true that:

a. $\sin t = \sin 9\pi/4$

b. $\cos t = \cos 15$

SOLUTION:

a. Since $9\pi/4 = 2\pi + \pi/4$, we have

$$\begin{aligned} \sin \frac{9\pi}{4} &= \sin\left(2\pi + \frac{\pi}{4}\right) \\ &= \sin \frac{\pi}{4} \end{aligned}$$

Therefore,

$$t = \frac{\pi}{4}$$

b. Because $\pi = 3.142$, we find that $4\pi < 15 < 5\pi$. Hence, we may write

$$\cos 15 = \cos(15 - 4\pi)$$

so that

$$t = 15 - 4\pi \approx 2.43$$

SOME PROPERTIES OF THE SINE AND COSINE

In Figure 8.1f, the directed line segment OA represents $\cos t$ or $\cos(t + 2m\pi)$, and the directed segment AP represents $\sin t$ or $\sin(t + 2m\pi)$, where m is any integer. Evidently, as t varies from 0 to $\pi/2$, $\cos t$ varies from 1 to 0, and $\sin t$ varies from 0 to 1. The symmetry of the circle enables us to find the values of $\cos t$ and $\sin t$ if their values are known only for $0 \le t \le \pi/2$. For instance, because the points $P(t)$ and $P(-t)$ are symmetric to each other with respect to the x-axis (see Figure 8.1g), they must have the same x-coordinates, and their y-coordinates must be negatives of each other. This observation gives us Theorem 8.1b. This theorem can be used to express the sine and cosine of a negative number in terms of the sine and cosine of the corresponding positive number, as Example 3 illustrates.

▼ THEOREM 8.1b

$\cos(-t) = \cos t$

and

$\sin(-t) = -\sin t$

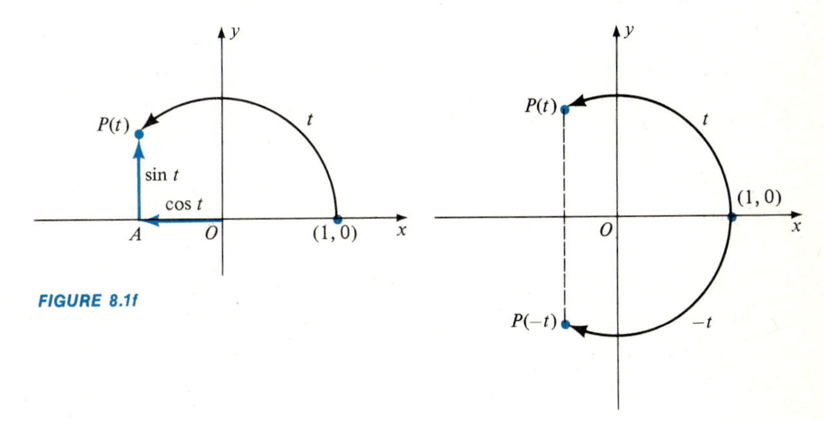

FIGURE 8.1f

FIGURE 8.1g

EXAMPLE 3

Express $\cos(-\pi/3)$ and $\sin(-\pi/3)$ in terms of functions of the positive number $\pi/3$.

SOLUTION:

By Theorem 8.1b, we can immediately write

$$\cos\left(-\frac{\pi}{3}\right) = \cos\frac{\pi}{3} \quad \text{and} \quad \sin\left(-\frac{\pi}{3}\right) = -\sin\frac{\pi}{3}$$

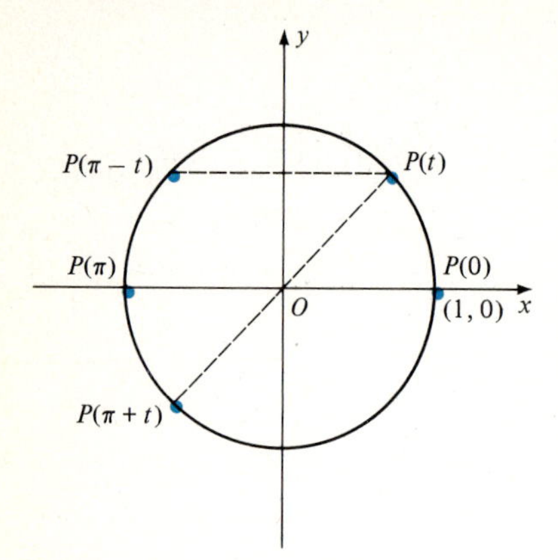

FIGURE 8.1h

Figure 8.1h illustrates the fact that the points $P(t)$ and $P(\pi - t)$ are symmetric to each other with respect to the y-axis. Thus, these points have the same y-coordinates, and their x-coordinates are negatives of each other. These considerations give us Theorem 8.1c.

Figure 8.1h also illustrates the fact that the points $P(t)$ and $P(\pi + t)$ are symmetric to each other with respect to the origin. If the coordinates of $P(t)$ are $(\cos t, \sin t)$, then the coordinates of $P(\pi + t)$ are $(-\cos t, -\sin t)$. Thus, we have Theorem 8.1d.

▼ THEOREM 8.1c

$\cos(\pi - t) = -\cos t$

and

$\sin(\pi - t) = \sin t$

▼ THEOREM 8.1d

$\cos(\pi + t) = -\cos t$

and

$\sin(\pi + t) = -\sin t$

The formulas in the preceding three theorems along with the periodicity of the two functions enable us to reduce the sine and cosine of a number outside the interval $0 \le t \le \pi/2$ to the sine and cosine of a number in that interval. As we shall see later, this is often the key to being able to use a table of sines and cosines. You really do not need to memorize these theorems; a quick sketch (or even a mental picture) will give you the results of the theorem.

EXAMPLE 4

Express $\cos 5\pi/3$ in terms of $\cos t$, where t is between 0 and $\pi/2$

SOLUTION:

Because $5\pi/3 = 2\pi - \pi/3$, we use the periodicity to write

$$\cos \frac{5\pi}{3} = \cos\left(2\pi - \frac{\pi}{3}\right) = \cos\left(-\frac{\pi}{3}\right)$$

Then, by Theorem 8.1b,

$$\cos\left(-\frac{\pi}{3}\right) = \cos \frac{\pi}{3}$$

EXAMPLE 5

Express $\sin(-17\pi/5)$ in terms of $\sin t$, where t is between 0 and $\pi/2$.

SOLUTION:

$$\sin\left(-\frac{17\pi}{5}\right) = -\sin \frac{17\pi}{5} \qquad \text{Theorem 8.1b}$$

$$= -\sin\left(2\pi + \frac{7\pi}{5}\right)$$

$$= -\sin \frac{7\pi}{5} \qquad \text{Periodicity}$$

$$= -\sin\left(\pi + \frac{2\pi}{5}\right)$$

$$= \sin \frac{2\pi}{5} \qquad \text{Theorem 8.1d}$$

EXAMPLE 6

Express $\sin(-15)$ in terms of $\sin t$, where $0 \le t \le \pi/2$ ($\pi/2 \approx 1.57$).

SOLUTION:

Using the approximate value 3.142 for π, we find that

$4\pi < 15 < 5\pi$

Furthermore, $15 - 4\pi \approx 2.43$, which is outside the desired interval, but $5\pi - 15 \approx 0.708$, which is between 0 and $\pi/2$. Thus, we may write

$$\begin{aligned} \sin(-15) &= \sin(4\pi - 15) && \text{Periodicity} \\ &= -\sin[\pi + (4\pi - 15)] && \text{Theorem 8.1d} \\ &= -\sin(5\pi - 15) \\ &\approx -\sin 0.708 \end{aligned}$$

EXERCISE 8.1

In Problems 1–10, the sine or cosine of t is given along with the quadrant of $P(t)$. Find the other function in each case.

1 $\sin t = 1/2$, $P(t)$ in Q I

2 $\sin t = \sqrt{3}/2$, $P(t)$ in Q I

3 $\cos t = \sqrt{3}/2$, $P(t)$ in Q IV

4 $\cos t = -\frac{1}{2}$, $P(t)$ in Q III

5 $\cos t = -\frac{1}{3}$, $P(t)$ in Q II

6 $\sin t = -\frac{2}{5}$, $P(t)$ in Q IV

7 $\sin t = -1/\sqrt{2}$, $P(t)$ in Q III

8 $\cos t = -\frac{1}{10}$, $P(t)$ in Q II

9 $\cos t = \frac{1}{2}$, $P(t)$ in Q IV

10 $\sin t = \frac{1}{4}$, $P(t)$ in Q II

11 Show that $\sin(3000\pi + t) = \sin t$.

12 Show that $\cos(1313\pi + t) = -\cos t$.

13 Show that $\sin(1316\pi - t) = -\sin t$.

14 Show that $\cos(1316\pi - t) = \cos t$

In Problems 15–30, express the given function in terms of the same function of a number between 0 and $\pi/2$. Leave answers in terms of π. For example, $\sin 2 = \sin(\pi - 2)$.

15 $\cos\left(-\frac{\pi}{4}\right)$

16 $\sin\left(-\frac{\pi}{3}\right)$

17 $\sin\left(\pi + \frac{\pi}{3}\right)$

18 $\cos\left(-\pi + \frac{\pi}{3}\right)$

19 $\cos\left(\pi - \frac{\pi}{6}\right)$

20 $\sin\left(\pi - \frac{\pi}{4}\right)$

21 $\sin \frac{7\pi}{4}$

22 $\cos \frac{5\pi}{4}$

23 $\cos \frac{13\pi}{3}$

24 $\sin \frac{17\pi}{6}$

25 $\cos\left(-\frac{19\pi}{5}\right)$

26 $\sin\left(-\frac{27\pi}{8}\right)$

27 $\sin 12$

28 $\cos 8$

29 $\cos(-7)$

30 $\sin(-20)$

31 The point $P(\pi/4)$ bisects the first quadrant arc of the unit circle. Show that this implies that $\sin \pi/4 = \cos \pi/4 = 1/\sqrt{2}$ or $\sqrt{2}/2$.

32 Use the result of Problem 25 and any of the theorems of this section to find the sine and cosine of:

a $\frac{3\pi}{4}$ **b** $\frac{5\pi}{4}$ **c** $\frac{7\pi}{4}$ **d** $-\frac{27\pi}{4}$ **e** $-\frac{127\pi}{4}$

33 Show that $\sin 2 \ne \sin 1$.

Hint: Look at the points $P(1)$ and $P(2)$ on the unit circle, and show that if $\sin 2 = \sin 1$, then $\pi = 3$.

[2] Problem 31 is needed for Using Your Knowledge 8.1.

34 Show that $\sin 3 \neq \sin 1$. (See the hint in Problem 33).

35 If $\sin s = \sin t$, what must be the relationship between the numbers s and t?

36 Suppose that $\sin s = \frac{1}{2}$ and $\cos s = -\sqrt{3}/2$, $0 \leq s \leq 2\pi$. Is this information enough to determine a unique value of s?

37 If $\sin s = -\frac{1}{2}$, $0 \leq s \leq 2\pi$, what other information do you need to determine a unique value of s?

38 If $\sin t = \frac{7}{5} + \cos t$, what values (if any) may $\cos t$ have?

Hint: Use Theorem 8.1a.

39 Suppose that f and g are two periodic functions with periods p and q, respectively. If p/q is a rational number, show that the function $f + g$ is periodic.

Hint: Because p/q is a rational number, we must have $p/q = (ac)/(bc)$, where a and b are relatively prime integers and c is a factor (not necessarily rational) common to p and q. Now show that f and g both have the period abc.

40 The function $\sin 3t$ has the period given by $3t = 2\pi$, that is, $2\pi/3$. Similarly, the function $\sin 4t$ has the period given by $4t = 2\pi$, that is, $2\pi/4$ or $\pi/2$. Is the function given by $\sin 3t + \sin 4t$ periodic? If so, what is its period?

Hint: See Problem 39.

USING YOUR KNOWLEDGE 8.1

In Problem 31 of the preceding exercise, we found that $\sin \pi/4 = \cos \pi/4 = \sqrt{2}/2$ by using the fact that $P(\pi/4)$ bisects the first quadrant arc of the unit circle. Thus, the point $P(\pi/4)$ must be on the line $y = x$ as well as on the unit circle $x^2 + y^2 = 1$. From these two equations, we see that the coordinates of $P(\pi/4)$ must be $(\sqrt{2}/2, \sqrt{2}/2)$.

Now let's look at the point $P(\pi/3)$ in Figure 8.1i. The point $P(\pi/3)$ is two thirds of the way along the circle from the point $(1, 0)$ to the point $P(\pi/2)$. Hence, the arc from $(1, 0)$ to $P(\pi/3)$ is one sixth of the *entire* circle, and the straight-line distance between these two points must be 1, the radius of the circle. (You should recall that the side of a regular hexagon inscribed in a circle is equal to the radius of the circle.) We see now that triangle OAP in the figure is an equilateral triangle, and if we draw the line PM perpendicular to OA, then M bisects the segment OA. This means that the length of OM is $\frac{1}{2}$. Now, you know that the length of OP is 1, and OMP is a right triangle. You should be able to show that the length of MP is $\sqrt{3}/2$, so that the coordinates of $P(\pi/3)$ are $(\frac{1}{2}, \sqrt{3}/2)$.

1 What is the value of: **a** $\cos \dfrac{\pi}{3}$ **b** $\sin \dfrac{\pi}{3}$

2 Find the value of: **a** $\cos \dfrac{\pi}{6}$ **b** $\sin \dfrac{\pi}{6}$

Hint: The point $P(\pi/6)$ is symmetric to the point $P(\pi/3)$ with respect to the line $y = x$, so that you should be able to read off the coordinates of $P(\pi/6)$ from those of $P(\pi/3)$.

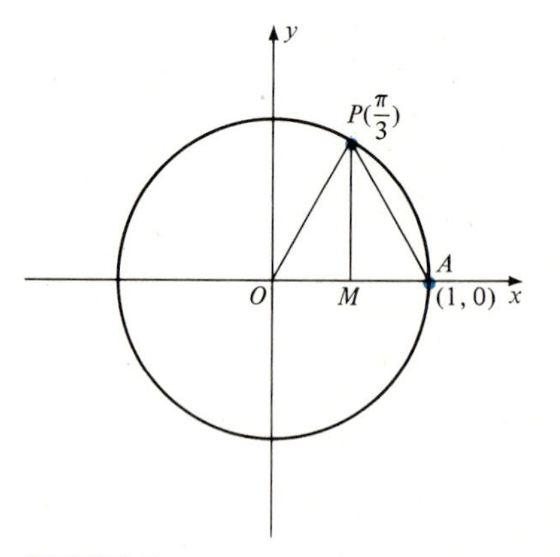

FIGURE 8.1i

3 Use the results of Problem 1 to find the sine and cosine of:

a $\dfrac{2\pi}{3}$ **b** $\dfrac{5\pi}{3}$ **c** $-\dfrac{11\pi}{3}$

4 Use the results of Problem 2 to obtain the sine and cosine of:

a $\dfrac{5\pi}{6}$ **b** $\dfrac{7\pi}{6}$ **c** $-\dfrac{17\pi}{6}$

8.2 ANGLES AND THEIR MEASURE

We made no mention of angles in Section 8.1, even though the sine and cosine functions were originally defined for angles and were used mostly for the measurement of triangles. This application of these functions is the reason for the name *trigonometric* functions. (The word *trigonometry* was derived from two Greek words meaning "triangle measurement.") As we shall see later, there is a close relationship between the circular functions of numbers that we defined in Section 8.1 and the trigonometric functions of angles that are usually defined in elementary trigonometry.

When you studied geometry, you probably were taught that an angle is the union of two rays with a common end point such as $\angle BAC$ in Figure 8.2a. The rays are called the *sides* of the angle, and the common point is called the *vertex* of the angle. Thus, in Figure 8.2a, AB and AC are the sides, and A is the vertex of the angle. Quite often, an angle is named by a single letter as $\angle A$, or by a small Greek letter α (alpha), β (beta), γ (gamma), θ (theta), ϕ (phi), or ψ (psi).

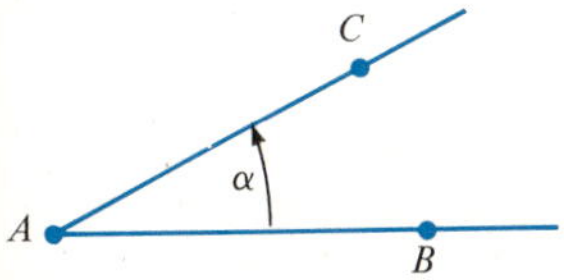

FIGURE 8.2a

For many of the applications of trigonometry, we need a somewhat more general notion of an angle; we want to take account of the amount of rotation used to go from one side to the other side of the angle. In accordance with this idea, we regard one side of the angle as the *initial* side and the other as the *terminal* side, and we think of the angle as consisting not only of two rays with a common point but also of an associated measure of the amount of rotation *from the initial to the terminal side.*

Figure 8.2b shows three different angles α_1, α_2, α_3, all with the same initial and terminal sides. Such angles are called *coterminal* angles. With an appropriate system of measurement, we can distinguish two different coterminal angles from each other.

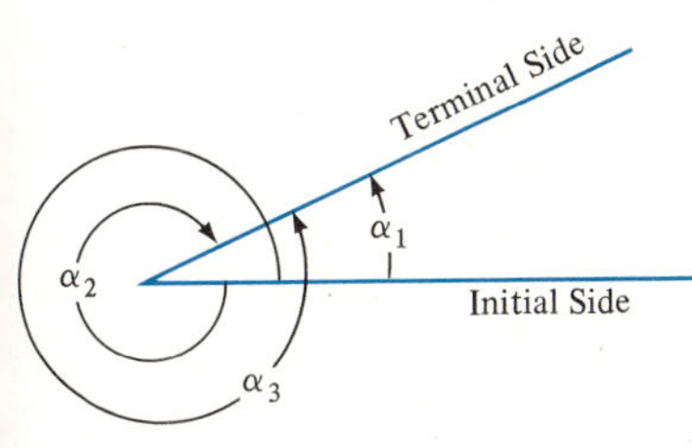

FIGURE 8.2b

There are two commonly used systems for measuring angles, the *degree* system and the *radian* system. The degree system, probably the more familiar of the two, is employed by surveyors and navigators. In this system, we imagine the angle placed with its vertex at the center of a circle whose circumference has been divided into 360 equal parts. We first consider angles for which the amount of rotation from initial to terminal side is not greater than one half of one complete revolution. Such an angle that intercepts an arc equal to exactly one of these parts has a measure of one degree (written 1°). Other angles are measured by the number of parts of the circumference they intercept. We may indicate the degree measure of an angle α by $m^\circ(\alpha)$, which is read "the degree measure of alpha". For instance, an angle that intercepts one fourth of the circumference of the circle has a measure of 90°; if the angle intercepts

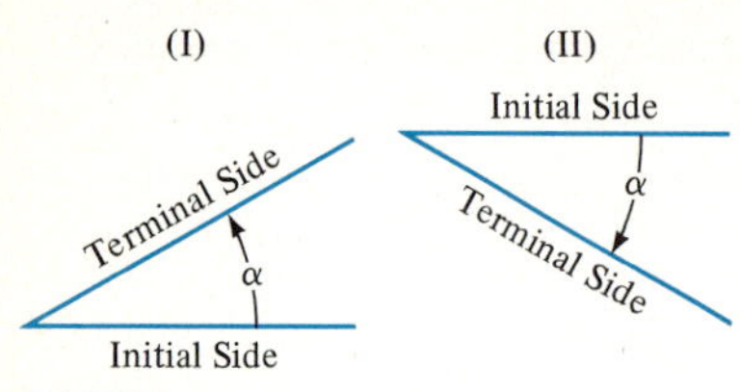

FIGURE 8.2c

(I) Positive rotation, $m^\circ(\alpha) < 0$;
(II) Negative rotation, $m^\circ(\alpha) < 0$.

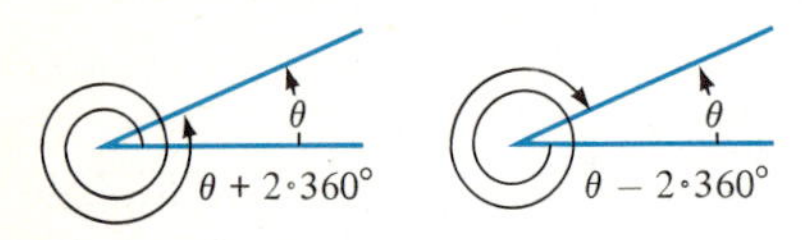

FIGURE 8.2d

one sixth of the circumference, its measure is 60°. Angles that measure between 0° and 90° are called *acute* angles.

By taking rotations into account, we can set up a one-to-one correspondence between the set of all possible angles and the set of all possible degree measures k°, where k is a real number. We agree to count the measure as *positive* if the rotation from the initial to the terminal side of the angle is *counterclockwise*; the measure is *negative* if the rotation is *clockwise*. (See Figure 8.2c.)

With the preceding conventions, we see that two coterminal angles will always have degree measures that differ by an integer multiple of 360°, as Figure 8.2d illustrates. Notice that we often use the name of the angle, such as θ, for the measure of the angle. This should cause no confusion because the context always makes clear which meaning is intended.

EXAMPLE 1

Sketch an angle of:

a. 930°

b. −930°

SOLUTION:

a.

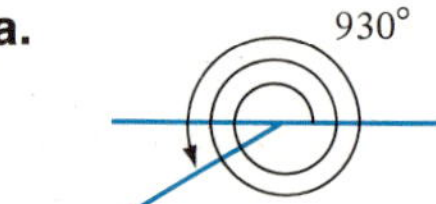

b.

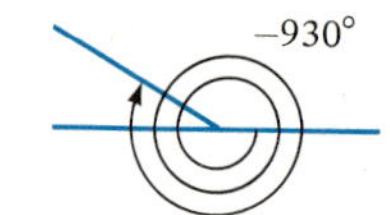

▼ DEFINITION 8.2a

If an acute angle, regarded as a central angle in a unit circle, intercepts an arc of *unit* length, the angle is a *unit angle*. The unit angle is called a *radian*.

The second method of measuring angles arises from the desire to have a natural way of associating an angle θ with a real number t. As we saw in Section 8.1, each real number t determines a unique point $P(t)$ on the unit circle. In particular, $t = 1$ determines the point $P(1)$, which in turn determines the acute angle AOP in Figure 8.2e. This is the angle that is selected as the natural unit angle. (See Definition 8.2a.)

You should note that one of the main reasons for the use of radian measure is that it simplifies many operations in calculus and more advanced mathematics. When working with angles measured in radians, it is customary to write $\theta = t$ rather than $\theta = t$ radians.

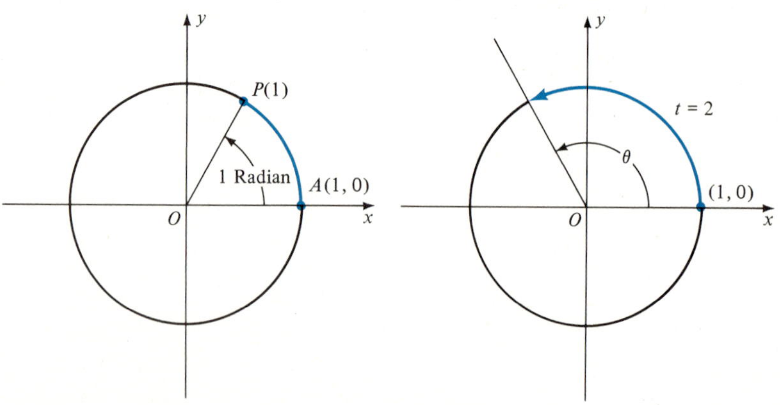

FIGURE 8.2e

FIGURE 8.2f

For instance, if the measure of θ is 2 radians, we usually write $\theta = 2$ instead of $\theta = 2$ radians. The notation $\theta = 2$ means that the angle is generated by rotating the terminal side of θ counterclockwise until its intersection with the unit circle has traveled 2 units. See Figure 8.2f. It must be strongly emphasized that *whenever no unit is indicated for the measure of an angle, radian measure is understood.*

Definition 8.2a allows us to associate an angle of t radians with each real number t. If $0 \leq t \leq \pi$, the angle will be the usual geometric angle between 0° and 180°. For other values of t, we use the notion of angles generated by the plane rotation of a ray about its end point from an initial to a terminal position. It is important to notice that rotations of θ radians and $(\theta + 2m\pi)$ radians (m an integer) from the same initial position result in coterminal angles, as shown in Figure 8.2g.

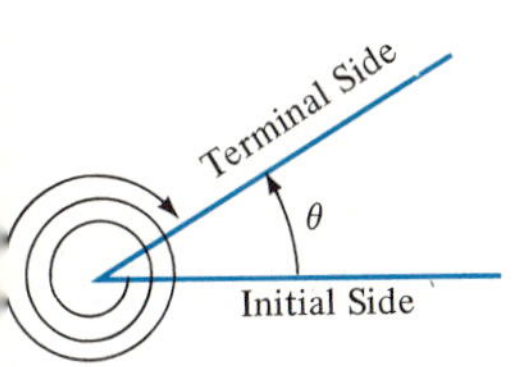

FIGURE 8.2g

It is sometimes necessary to change the measure of an angle from one system to the other. We can do this by noting that in both systems, the angle may be regarded as a central angle of a circle. Because the measures of the angle in the two systems are proportional to the number of units in the circumference of the circle, we see that *the number of degrees in the angle is to* 360 *as the number of radians is to* 2π. Hence, to convert from one system to the other, we need only the equation

$$360^\circ = 2\pi \text{ radians}$$

or, upon division by 2,

$$180^\circ = \pi \text{ radians}$$

From this equation, we obtain the results

$$1^\circ = \frac{\pi}{180} \text{ radians} \quad \text{and} \quad 1 \text{ radian} = \left(\frac{180}{\pi}\right)^\circ$$

These results are easy to remember if you realize that a ray generates an angle of 180° or π radians in one half of a complete revolution. They tell us that to convert from degrees to radians, we multiply by $\pi/180$, and to convert from radians to degrees, we multiply by $180/\pi$. The arithmetic involved in these conversions is conveniently done on a hand-held calculator.

EXAMPLE 2

Convert 2 radians to degrees.

SOLUTION:

$$2 \text{ radians} = (2)\left(\frac{180}{\pi}\right)^\circ = 114.6^\circ$$

EXAMPLE 3

Convert $3\pi/5$ radians to degrees.

SOLUTION:

$$\frac{3\pi}{5} \text{ radians} = \left(\frac{3\pi}{5}\right)\left(\frac{180}{\pi}\right)^\circ = 108^\circ$$

EXAMPLE 4

Convert 105° to radians, leaving the answer in terms of π.

SOLUTION:

$$105^\circ = (105)\left(\frac{\pi}{180}\right) \text{ radians} = \frac{7\pi}{12} \text{ radians}$$

Radian measure is used almost exclusively in advanced mathematics and in many areas of science, and the word *radians* is usually omitted. Thus, an angle of 2.5 means 2.5 radians, *not* degrees. If degrees are intended, the degree symbol *must* be used.

There is a third system for measuring angles, the *centesimal* system, in which the unit angle, called a *grad*, is one one-hundredth of a right angle, that is 0.9°. We shall not use this system.

In engineering and in other areas of applied science, angles are frequently given in degrees and decimal fractions of a degree. This convenient system is the one that we shall use when working with degree measure. Land surveys and astronomical observations often give angles in the sexagesimal system—degrees (°), minutes (′), and seconds (″). Just as in the measurement of time, where 1 hour is 60 minutes and 1 minute is 60 seconds, 1 degree is 60 minutes, and 1 minute is 60 seconds. If an angle is given in degrees, minutes, and seconds, it might be necessary to convert to decimal form in order to use a calculator or tables. For instance, to convert

$$31^\circ 52' 21''$$

to the decimal system, we have to change the following expression to decimal form.

$$31 + \tfrac{52}{60} + \tfrac{21}{3600}$$

Such computations are conveniently done on a calculator. If your calculator uses algebraic logic, the procedure is as follows:

Algebraic logic

[31] [+] [52] [÷] [60] [+] [21] [÷] [3600] [=] 31.8725

If your calculator uses RPN logic, then it is handier to rewrite the expression in the form

$$31 + \frac{(52)(60) + 21}{3600}$$

and then use the procedure:

RPN logic

[52] [ENTER] [60] [×] [21] [+] [3600] [÷] [31] [+] 31.8725

Thus,

$$31^\circ 52' 21'' = 31.8725^\circ$$

EXERCISE 8.2

In Problems 1–10, sketch an angle with the given degree measure and convert from degree to radian measure. Leave answers in terms of π.

1 135°
2 295°
3 -275°
4 -115°
5 330°
6 215°
7 660°
8 750°
9 -900°
10 -450°

In Problems 11–20, sketch an angle with the given radian measure and convert from radian to degree measure. Give answers to the nearest tenth of a degree.

11 $\frac{7\pi}{4}$
12 $\frac{3\pi}{2}$
13 $\frac{5\pi}{6}$
14 $\frac{2\pi}{5}$
15 $-\frac{3\pi}{8}$
16 $-\frac{\pi}{12}$
17 5
18 0.1
19 -2
20 -6

21 Draw an acute angle θ and a coterminal angle $\theta - 2\pi$.

22 Draw an acute angle α and a coterminal angle $\alpha + 720^\circ$

23 Draw an angle θ with measure between $\pi/2$ and π and a coterminal angle $\theta + 4\pi$.

24 Draw an acute angle β and an angle $\beta + 3\pi$ with the same initial side as β. What can you say about the terminal sides of the two angles?

In Problems 25–28, convert to degrees. Give answers to three decimal places.

25 $96^\circ 25' 32''$
C 26 $47^\circ 18' 39''$
C 27 $-142^\circ 14' 52''$
28 $-35^\circ 46' 27''$

In Problems 29 and 30, convert to radians. Give answers to three decimal places.

29 $85^\circ 37' 46''$
C 30 $-36^\circ 48' 22''$

USING YOUR KNOWLEDGE 8.2

When a wheel rotates about its axis at a constant speed, the number of radians through which a fixed radius on the wheel turns per unit of time is called the angular velocity of the wheel. The Greek letter ω (omega) is often used to denote the angular velocity. If a wheel of radius r units rolls without slipping along a straight path, then the velocity of the center of the wheel is given by the formula

$$v = r\omega \tag{1}$$

Can you see why?

Suppose that a wheel is 6 inches in diameter and has an angular velocity of 10 radians per second. If the wheel rolls without slipping along a straight path, how far does it go in 1 minute? We can answer this question quite easily using Formula 1. Because the radius of the wheel is 3 inches or $\frac{1}{4}$ foot, the velocity is

$$v = (\tfrac{1}{4})(10) = 2.5 \text{ feet per second}$$

Hence, in 1 minute the wheel goes

$(2.5)(60) = 150$ feet

Try to use the preceding ideas to answer the following questions.

1 A wheel 1 foot in diameter travels at a speed of 10 feet per second along a straight path. What is the angular velocity of the wheel?

2 A wheel 2 feet in diameter rolls without slipping and with an angular velocity of 60 radians per second along a straight path. How far does the wheel go in 1 minute?

3 The wheel (with its tire) on a small automobile is 2 feet in diameter. When the car is traveling 50mph, what is the angular velocity of the wheel?

4 A bicycle wheel 32 inches in diameter is turning through 11 radians per second while the bicycle is traveling along a straight road. Find the speed of the bicycle in miles per hour.

8.3 THE TRIGONOMETRIC COSINE AND SINE FUNCTIONS

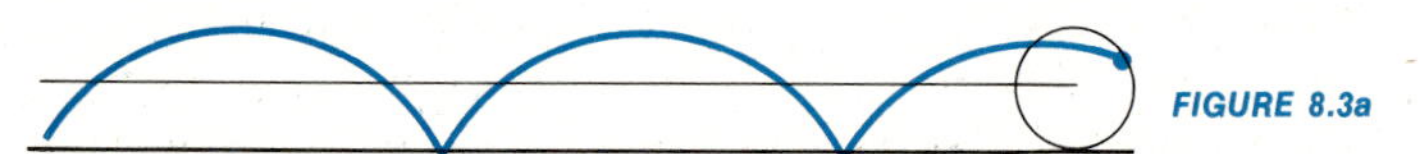

FIGURE 8.3a

Figure 8.3a shows the curve that is traced by a point on the rim of a wheel as the wheel rolls without slipping along a straight line. This curve is called a *cycloid*. If the straight line is chosen as the x-axis in a rectangular coordinate system and the origin is at the extreme left, where the tracing point is on the straight line, then it can be shown that the x- and y-coordinates of any point on the cycloid are given by the formulas

$$x = a(t - \sin t), \qquad y = a(1 - \cos t)$$

(*The derivation of these formulas can be found in most books on analytic geometry.*)

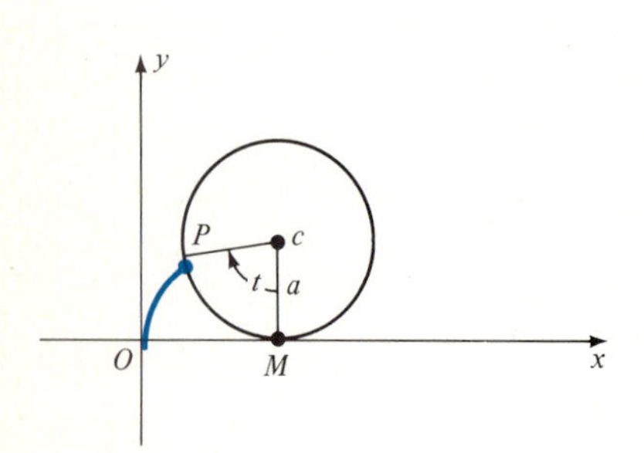

FIGURE 8.3b

where a is the radius of the wheel and t is a real number. The number t has the following interpretation: It is the number of radians in the angle between a radius (CM) of the wheel drawn to the point of contact with the straight line and a second radius (CP) drawn to the tracing point, as shown in Figure 8.3b. This interpretation leads us to define the cosine and sine functions of an angle, the functions that are called the *trigonometric cosine* and *sine functions*. As we noted in Section 8.2, there is a close relationship between the circular and the trigonometric functions. We shall explore this relationship in the following discussions.

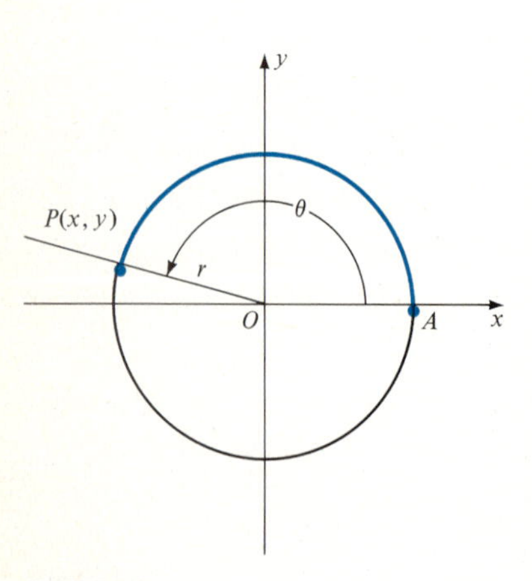

FIGURE 8.3c

An angle that is placed with its vertex at the origin and its initial line along the positive x-axis of a rectangular coordinate system is said to be in *standard position*. The angle θ in Figure 8.3c is in standard position.

To define the trigonometric functions of an angle θ, we first place the angle in standard position and then select a point $P(x, y)$

▼ **DEFINITION 8.3a**

Let the angle θ be in standard position and let $P(x, y)$ be a point on the terminal side of θ. Denote the distance OP by r. Then, the cosine of θ is defined to be x/r, and the sine of θ is defined to be y/r; that is,

$$\cos\theta = \frac{x}{r} \quad \text{and} \quad \sin\theta = \frac{y}{r}$$

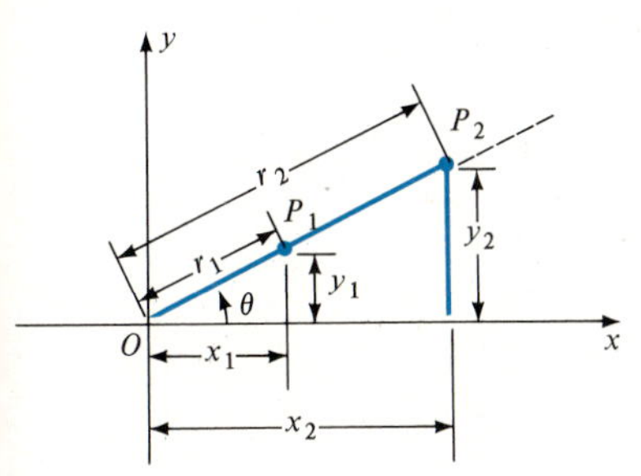

FIGURE 8.3d

on the terminal side, as in Figure 8.3c. The distance from the origin to the point P is denoted by r, a positive number. We then define the functions cosine and sine of θ in terms of x, y, and r. (See Definition 8.3a.) We can see by the similar triangles in Figure 8.3d that it makes no difference where on the terminal side of θ the point P is chosen; the ratios y/r and x/r remain constant for all such points.

If the point P is one unit from the origin, then $r = 1$ in Definition 8.3a, so that $\cos\theta = x$ and $\sin\theta = y$. Because for $r = 1$, the number of units of length in the arc AP in Figure 8.3c is exactly the number of radians in the angle θ, we see that the cosine and sine of θ agree in value with the circular cosine and sine as given by Definition 8.1a. This has a most important consequence:

If t is any real number, then $\cos t$ and $\sin t$ may be regarded as the corresponding functions of an angle of measure t radians.

SOME SPECIAL NUMBERS AND ANGLES

We shall now apply Definition 8.3a to find the sine and cosine in some important special cases. First, we look at the angles that terminate on one of the axes; these are called *quadrantal* angles. Figure 8.3e shows the quadrantal angles between 0° and 360°, that is, between 0 and 2π radians. For these angles, we can read off the values of the sine and the cosine functions, because $r = 1$ and the x- and y-values appear in the figure. Table 8.3a lists the desired function values.

The calculation of the sine and cosine for numbers or angles in general is usually carried out by calculus methods. However, there are a few special angles for which simple geometric ideas enable us to find the exact values. Some of these angles are 30°, 45°, and 60°, corresponding to the numbers $\pi/6$, $\pi/4$, and $\pi/3$, respectively. Figure 8.3f shows a 30° angle in standard position. For convenience, the point P on the terminal side of the angle has been taken at a dis-

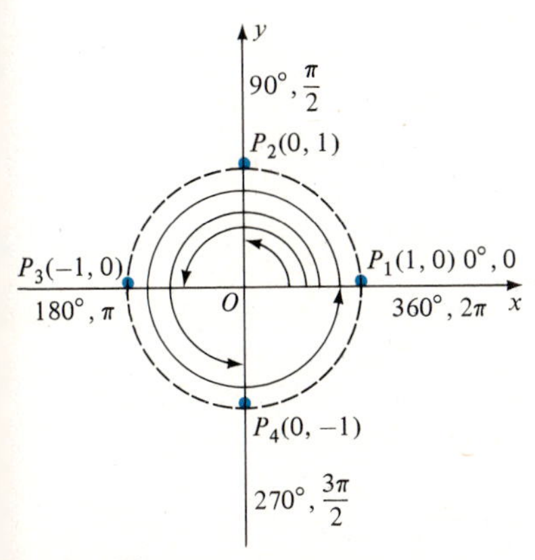

FIGURE 8.3e

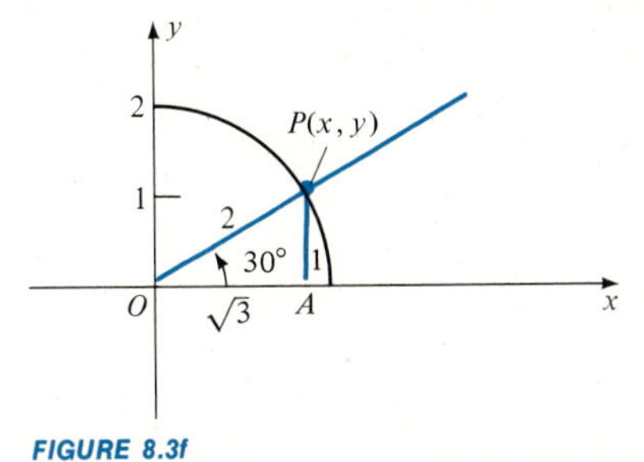

FIGURE 8.3f

TABLE 8.3a

Angle θ (degrees)	Angle θ (radians)	$\cos\theta$	$\sin\theta$
0°	0	1	0
90°	$\pi/2$	0	1
180°	π	−1	0
270°	$3\pi/2$	0	−1

tance of 2 units from the origin. We know from geometry that in a 30-60–degree right triangle, the side opposite the 30° angle is one half the hypotenuse. Thus, with $r = 2$, it follows that $y = 1$. Because $x^2 + y^2 = r^2$, we have $x^2 + 1 = 4$, or $x^2 = 3$ and $x = \sqrt{3}$. We may now use Definition 8.3a to obtain

$$\cos 30^\circ = \cos\frac{\pi}{6} = \frac{\sqrt{3}}{2} \qquad \sin 30^\circ = \sin\frac{\pi}{6} = \frac{1}{2}$$

By using the same fact from geometry, you should verify the numbers given in Figure 8.3g for a 60° angle and obtain the results

$$\cos 60^\circ = \cos\frac{\pi}{3} = \frac{1}{2} \qquad \sin 60^\circ = \sin\frac{\pi}{3} = \frac{\sqrt{3}}{2}$$

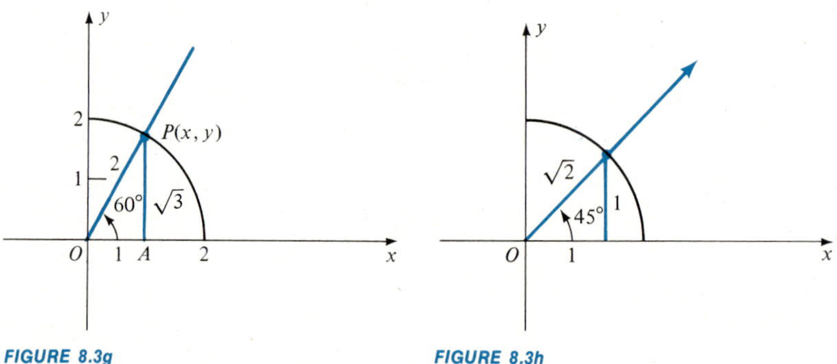

FIGURE 8.3g FIGURE 8.3h

Figure 8.3h shows a 45° angle in standard position. The corresponding right triangle is isosceles, so that we may select the point (1, 1) as the point P on the terminal side. Because $r^2 = x^2 + y^2$, we have $r^2 = 1 + 1$ and $r = \sqrt{2}$. Again, we use Definition 8.3a to obtain

$$\cos 45^\circ = \cos\frac{\pi}{4} = \frac{1}{\sqrt{2}} \qquad \sin 45^\circ = \sin\frac{\pi}{4} = \frac{1}{\sqrt{2}}$$

TABLE 8.3b

Angle θ (degrees)	Angle θ (radians)	$\cos\theta$	$\sin\theta$
30°	$\pi/6$	$\sqrt{3}/2$	$1/2$
45°	$\pi/4$	$1/\sqrt{2}$	$1/\sqrt{2}$
60°	$\pi/3$	$1/2$	$\sqrt{3}/2$

The preceding results are summarized in Table 8.3b. The 30°, 45°, and 60° angles are used so often that it would be wise either to memorize Table 8.3b or to learn how to reconstruct the entries quickly.

REFERENCE ANGLES

Although the angles listed in Table 8.3b are all first quadrant angles, the entries in the table can be used to find the functions of many other angles that can be related to these first-quadrant angles. A simple way of doing this is by means of *reference* angles. Figure 8.3i shows the reference angle θ_r for angles in each of the four quadrants. Notice carefully that the reference angle θ_r in each case is the positive acute angle between the terminal side of the angle θ

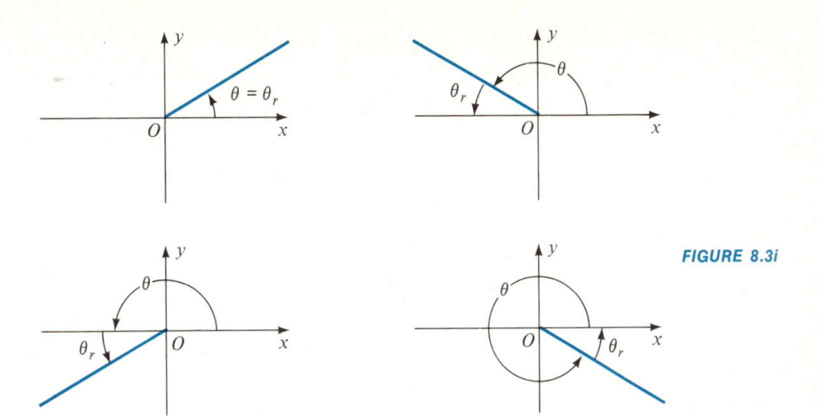

FIGURE 8.3i

and the x-axis. For instance, the reference angle of an angle of 147° is the acute angle 180° − 147°, or 33°; the reference angle of an angle of 213° is the acute angle 213° − 180°, or 33° again. The importance of the reference angle stems from the fact that $\cos\theta$ and $\sin\theta$ can differ from $\cos\theta_r$ and $\sin\theta_r$, respectively, only in algebraic sign. The signs of $\sin\theta$ and $\cos\theta$ are determined by the quadrant in which the terminal side of θ lies, according to Table 8.1a. The next examples illustrate the use of reference angles.

EXAMPLE 1

Use the reference angle to find cos 150° and sin 150°.

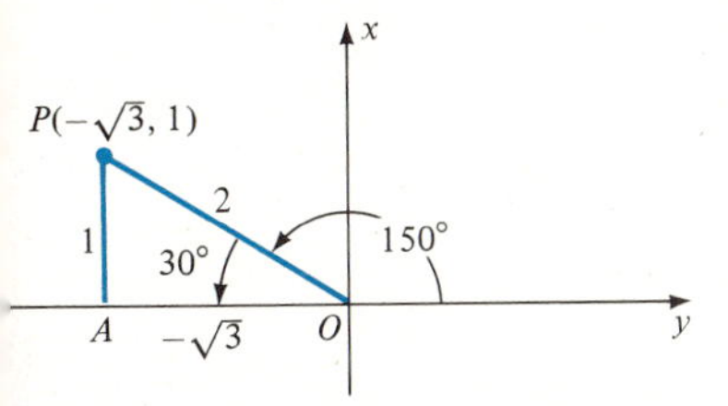

FIGURE 8.3j

SOLUTION:

Because 150° is a second-quadrant angle, the reference angle is $\theta_r = 180° - 150° = 30°$ (see Figure 8.3j). The cosine of a second-quadrant angle is negative, and the sine is positive, so that

$$\cos 150° = -\cos 30° = -\frac{\sqrt{3}}{2}$$

$$\sin 150° = \sin 30° = \frac{1}{2}$$

EXAMPLE 2

Find the cosine and the sine of 870°.

SOLUTION:

We note first that $360° = 2\pi$, so that we may subtract any integer multiple of 360° without altering the value of the functions. Thus,

$$\begin{aligned}\cos 870° &= \cos(870° - 2\cdot 360°)\\ &= \cos 150°\\ &= -\frac{\sqrt{3}}{2} \qquad \text{By Example 1}\end{aligned}$$

Similarly, we find

$$\sin 870° = \sin 150° = \tfrac{1}{2} \qquad \text{By Example 1}$$

EXAMPLE 3

Find $\sin 7\pi/4$.

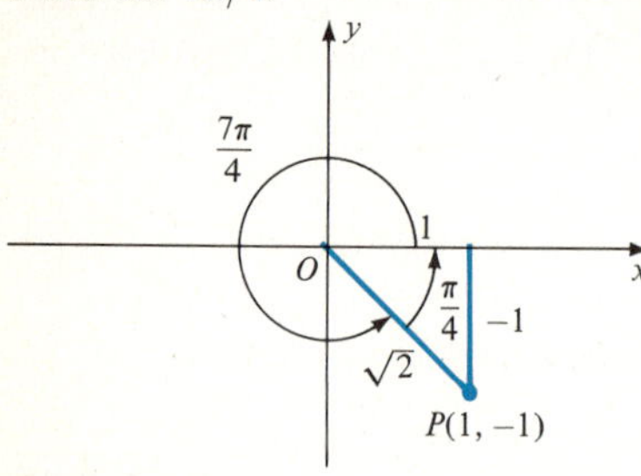

FIGURE 8.3k

SOLUTION:

Because $7\pi/4$ is the radian measure of a fourth-quadrant angle, the reference angle is $2\pi - 7\pi/4 = \pi/4$ (see Figure 8.3k). The sine of a fourth-quadrant angle is negative, so that

$$\sin\frac{7\pi}{4} = -\sin\frac{\pi}{4} = -\frac{1}{\sqrt{2}}$$

EXAMPLE 4

Find $\sin(-45\pi/4)$.

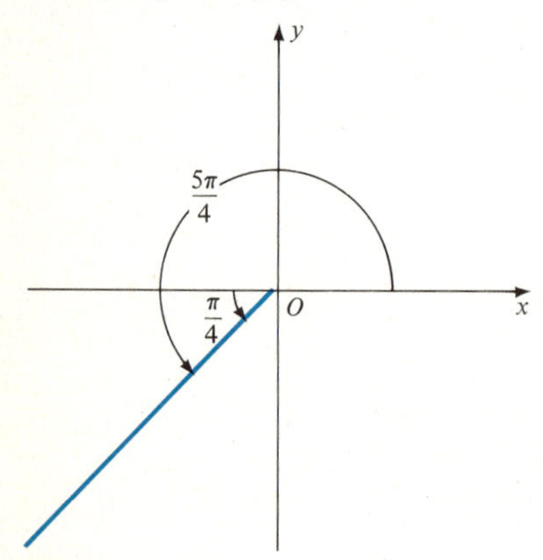

FIGURE 8.3l

SOLUTION:

$$\begin{aligned}\sin\left(-\frac{45\pi}{4}\right) &= -\sin\frac{45\pi}{4} && \text{Theorem 8.1b}\\ &= -\sin\left(\frac{45\pi}{4} - 10\pi\right) && \text{Periodicity of the sine}\\ &= -\sin\frac{5\pi}{4}\end{aligned}$$

Because $5\pi/4$ is a third-quadrant angle, the reference angle is $5\pi/4 - \pi$, or $\pi/4$. See Figure 8.3l. Thus,

$$\begin{aligned}\sin\left(\frac{-45\pi}{4}\right) &= -\sin\frac{5\pi}{4}\\ &= -\left(-\sin\frac{\pi}{4}\right)\\ &= \frac{1}{\sqrt{2}}\end{aligned}$$

EXAMPLE 5

Find $\cos 27\pi/2$.

SOLUTION:

$$\begin{aligned}\cos\frac{27\pi}{2} &= \cos\left(12\pi + \frac{3\pi}{2}\right)\\ &= \cos\frac{3\pi}{2} && \text{Periodicity of the cosine}\\ &= 0 && \text{Table 8.3a}\end{aligned}$$

EXERCISE 8.3

In each of Problems 1–6, find the cosine and sine of θ.

1 The point (3, 4) is on the terminal side of θ.

2 The point $(-12, 5)$ is on the terminal side of θ.

3 The point $(1, -2)$ is on the terminal side of θ.

4 The point $(15, -8)$ is on the terminal side of θ.

5 A point P on the terminal side of θ is 13 units from the origin, and the x-coordinate of P is -5.

6 A point P on the terminal side of θ is 17 units from the origin, and the y-coordinate of P is -15.

In Problems 7–22, find the cosine and sine of the given number.

7 23π

8 32π

9 100π

10 -101π

11 $\frac{9\pi}{2}$

12 $-\frac{7\pi}{2}$

13 $-\frac{21\pi}{2}$

14 $\frac{33\pi}{2}$

15 $\frac{23\pi}{6}$

16 $-\frac{15\pi}{4}$

17 $-\frac{31\pi}{3}$

18 $\frac{14\pi}{3}$

19 $\frac{59\pi}{6}$

20 $-\frac{41\pi}{6}$

21 $\frac{19\pi}{4}$

22 $-\frac{43\pi}{4}$

In Problems 23–38, find the cosine and sine of the given angle.

23 120°

24 -120°

25 315°

26 -315°

27 300°

28 210°

29 -450°

30 1200°

31 -600°

32 -1500°

33 990°

34 -1170°

35 765°

36 945°

37 -240°

38 930°

In Problems 39–50, find the reference angle θ_r and write the cosine and sine of the given angle in terms of functions of the reference angle.

39 125°

40 100°

41 250°

42 200°

43 305°

44 290°

45 375°

46 460°

47 -460°

48 -312°

49 562°

50 -1000°

In Problems 51–58, take π to be approximately 3.14, and find the reference number (the reference angle in radians). Then write the cosine and sine of the given number in terms of functions of the reference number.

51 2

52 4

53 5

54 6

55 10

56 12

57 -5

58 -10

59 For what values of t, $0 \leq t \leq 2\pi$, will $\cos t + \sin t = 0$?

60 Show that there are only two values of t between 0 and 2π for which $\sin t = 3 \cos t$.

Hint: $\cos^2 t + \sin^2 t = 1$.

61 Find $\cos t$ and $\sin t$ if $0 \leq t \leq 2\pi$ and $\sin t = 2 \cos t$.

62 For what angles between 0° and 180° is $\sin t > \cos t$?

Hint: This is equivalent to asking for what points on the upper half of the unit circle is the y coordinate greater than the x coordinate.

63 For what values of t between $-\pi/2$ and $\pi/2$ is $\sin t < \cos t$?

64 The function $\cos t$ has a period of 2π. Hence, the function $\cos 2t$ has a period of π.
a Explain.
b Is the function $f(t) = \cos t + \cos 2t$ periodic? If so, what is the period?

USING YOUR KNOWLEDGE 8.3

Let θ be a central angle in a circle of radius r. It follows from the definition of radian measure that the ratio of the radian measure of θ to 2π radians is equal to the ratio of the length of the intercepted arc to the entire circumference of the circle. Thus, we have the equation

$$\frac{\theta}{2\pi} = \frac{s}{2\pi r}$$

where s is the length of the arc. We can solve this equation to obtain the result

$$s = r\theta$$

which says that the length of the intercepted arc equals the length of the radius times the radian measure of the central angle. For example, a central angle of 2 radians in a circle of radius 3 feet intercepts an arc of length $(3)(2) = 6$ feet. The equation $s = r\theta$ has many interesting applications among which are the following problems.

[C] **1** The latitude of the city of Boulder, Colorado, is 40° north. If the radius of the earth is about 6400 kilometers, how far is it from Boulder to the North Pole?

Hint: This means that the angle at the center of the earth from Boulder to the North Pole is 50°. Keep in mind that in the equation $s = r\theta$, the θ is in radians.

[C] **2** Eratosthenes is the name of one of the famous ancient Greeks. Among other things, he was interested in astronomy, geography, and mathematics. He made a fairly accurate estimate of the radius of the earth back about 230 B.C. On a certain summer day, the sun was directly overhead at a city called Syene. At the same time, in the city of Alexandria, Eratosthenes found the sun to be inclined at an angle of 7°12′ to the vertical. The distance between the two cities is about 500 miles. Use this information along with Figure 8.3m to find an approximate value for the radius of the earth.

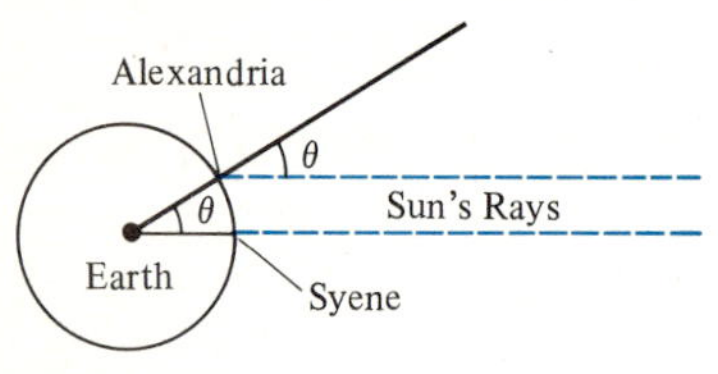

FIGURE 8.3m

8.4 OTHER TRIGONOMETRIC FUNCTIONS AND INVERSES

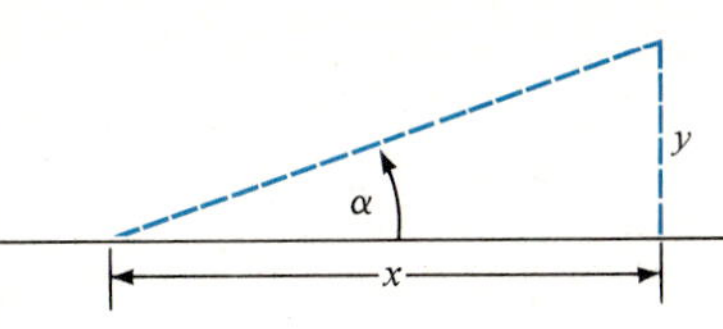

FIGURE 8.4a

Figure 8.4a shows the flight path of an airplane taking off from an airport. Suppose that the flight path is inclined at an angle α to the runway and that the horizontal distance x is known, what is the plane's altitude y? This type of problem arises quite frequently, and often the height y is inaccessible to direct measurement. It is possible for us to calculate y by using the sine and cosine of α as follows: From $\sin \alpha = y/r$, we have

$$y = r \sin \alpha \tag{1}$$

and from $\cos \alpha = x/r$, we find

$$r = \frac{x}{\cos \alpha} \tag{2}$$

If we substitute the value of r from Equation 2 into Equation 1, we obtain

$$\boldsymbol{y = \frac{x}{\cos \alpha} \sin \alpha = x \frac{\sin \alpha}{\cos \alpha}} \tag{3}$$

Because a large number of applications involve the ratio of $\sin \alpha$ to $\cos \alpha$, as in Equation 3, it is most convenient to regard this ratio as a new function of the angle. This function is called the *tangent function* and is denoted by *tan* α. With this notation, Equation 3 can be briefly written

$$y = x \tan \alpha$$

▼ DEFINITION 8.4a

1. The *tangent* function is defined by

$$\tan t = \frac{\sin t}{\cos t}$$

2. The *cotangent* function is defined by

$$\cot t = \frac{\cos t}{\sin t}$$

3. The *secant* function is defined by

$$\sec t = \frac{1}{\cos t}$$

4. The *cosecant* function is defined by

$$\csc t = \frac{1}{\sin t}$$

Definition 8.4a formalizes the preceding discussion and also introduces three additional functions that are important in many applications, both in mathematics and in other areas. In this definition, t may be regarded simply as a real number or, if it suits us, as the number of radians in an angle. If we wish to use degree measure, then we replace the t by the degree measure of the angle. For example, the equation

$$\tan 30^\circ = \frac{\sin 30^\circ}{\cos 30^\circ}$$

is equivalent to the equation

$$\tan \frac{\pi}{6} = \frac{\sin \pi/6}{\cos \pi/6}$$

If properties of the trigonometric functions of the angle are desired, then we place the angle in standard position with its vertex at the origin, its initial side along the positive x-axis, and its terminal side

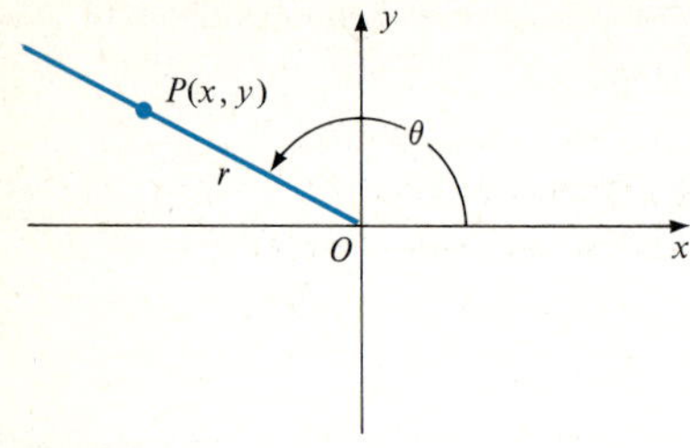

FIGURE 8.4b

passing through the point $P(t)$ on the unit circle and determined by the value of t, where t is the radian measure of the angle.

As a consequence of this last observation, we can now restate the definitions of the tangent, cotangent, secant, and cosecant functions of an angle directly in terms of the coordinates of an arbitrary point $P(x, y)$ on the terminal side of the angle and the distance r of the point P from the origin. (See Figure 8.4b.)

You can verify that Definition 8.4b is equivalent to Definition 8.4a by first referring to Definition 8.3a. For instance,

$$\tan\theta = \frac{\sin\theta}{\cos\theta} = \frac{y/r}{x/r} = \frac{y}{x}$$

EXAMPLE 1

Find tan 240°, cot 240°, sec 240°, and csc 240°.

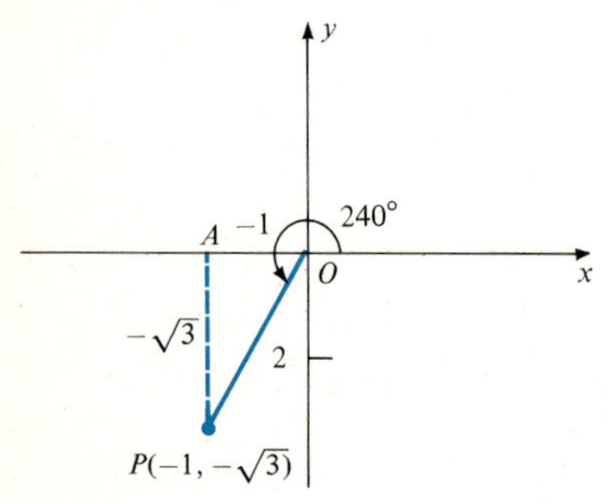

FIGURE 8.4c

SOLUTION:

Figure 8.4c shows an angle of 240° in standard position with a point P two units from the origin on the terminal side of the angle. From the geometry of the figure, we see that the coordinates of P must be $(-1, -\sqrt{3})$. Hence, we can use Definition 8.4b to read off the desired values:

$$\tan 240^\circ = \frac{-\sqrt{3}}{-1} = \sqrt{3} \qquad \cot 240^\circ = \frac{-1}{-\sqrt{3}} = \frac{1}{\sqrt{3}}$$

$$\sec 240^\circ = \frac{2}{-1} = -2 \qquad \csc 240^\circ = \frac{2}{-\sqrt{3}} = -\frac{2}{\sqrt{3}}$$

▼ DEFINITION 8.4b

(Definition 8.4a restated)

$$\tan\theta = \frac{y}{x} \qquad \cot\theta = \frac{x}{y}$$

$$\sec\theta = \frac{r}{x} \qquad \csc\theta = \frac{r}{y}$$

It is important to know the signs of the trigonometric functions for angles in the various quadrants. These signs follow from the definitions of the functions and are summarized in Table 8.4a, where all values of functions not explicitly shown are negative. (Table 8.4a includes the information given in Table 8.1a for the cosine and sine functions).

Table 8.4b gives some of the values of the six trigonometric functions. The results in this table can be extended to corresponding angles in the other three quadrants by using reference angles along with Table 8.4a. You should verify the entries in Table 8.4b by drawing suitable diagrams and using Definitions 8.3a and 8.4b. Notice carefully that certain entries are simply asterisks. These indicate that the value of the corresponding function is *not* defined. For instance, tan 90° is not defined because sin 90° = 1 and cos 90° = 0. Consequently, tan 90° = $\frac{1}{0}$, which is undefined.

You should memorize Table 8.4a or else be able to reconstruct the entries quickly, because they are used repeatedly in the work in trigonometry. A good way to remember the first five entries in the sine column is to think of them as

TABLE 8.4a

II	I
sin, csc +	all +
III	IV
tan, cot +	cos, sec +

TABLE 8.4b

Degrees	Radians	$\sin\theta$	$\cos\theta$	$\tan\theta$	$\cot\theta$	$\sec\theta$	$\csc\theta$
0°	0	0	1	0	***	1	***
30°	$\pi/6$	1/2	$\sqrt{3}/2$	$1/\sqrt{3}$	$\sqrt{3}$	$2/\sqrt{3}$	2
45°	$\pi/4$	$\sqrt{2}/2$	$\sqrt{2}/2$	1	1	$\sqrt{2}$	$\sqrt{2}$
60°	$\pi/3$	$\sqrt{3}/2$	1/2	$\sqrt{3}$	$1/\sqrt{3}$	2	$2/\sqrt{3}$
90°	$\pi/2$	1	0	***	0	***	1
180°	π	0	−1	0	***	−1	***
270°	$3\pi/2$	−1	0	***	0	***	−1
360°	2π	0	1	0	***	1	***

$$\frac{\sqrt{0}}{2}, \frac{\sqrt{1}}{2}, \frac{\sqrt{2}}{2}, \frac{\sqrt{3}}{2}, \frac{\sqrt{4}}{2}$$

The first five entries in the cosine column are the same numbers in reverse order. The entries in the other columns can be obtained by using the definitions of the other functions in terms of the sine and the cosine.

EXAMPLE 2

Find $\tan 11\pi/6$, $\cot 11\pi/6$, $\sec 11\pi/6$, and $\csc 11\pi/6$.

SOLUTION:

Regarding $11\pi/6$ as the radian measure of an angle, we find the reference angle to be $2\pi - 11\pi/6 = \pi/6$. Since $11\pi/6$ radians is a fourth-quadrant angle, we have

$$\tan\frac{11\pi}{6} = -\tan\frac{\pi}{6} = \frac{-1}{\sqrt{3}}$$

$$\cot\frac{11\pi}{6} = -\cot\frac{\pi}{6} = -\sqrt{3}$$

$$\sec\frac{11\pi}{6} = \sec\frac{\pi}{6} = \frac{2}{\sqrt{3}}$$

$$\csc\frac{11\pi}{6} = -\csc\frac{\pi}{6} = -2$$

EXAMPLE 3

Find $\tan 930°$ and $\sec 930°$.

SOLUTION:

Because $930° = 2 \cdot 360° + 210°$, we may write

$$\tan 930° = \tan 210°$$

$$\sec 930° = \sec 210°$$

The reference angle for 210° is $210° - 180° = 30°$, and 210° is a third-quadrant angle. Hence,

$$\tan 930^\circ = \tan 30^\circ = \frac{1}{\sqrt{3}}$$

$$\sec 930^\circ = -\sec 30^\circ = \frac{-2}{\sqrt{3}}$$

Notice that we have assumed the almost obvious fact that all the functions we have defined in this section are periodic with period 2π radians or 360° just as the cosine and sine functions are. This periodicity follows because increasing or decreasing an angle by an integer multiple of 2π leaves the terminal side of the angle unchanged.

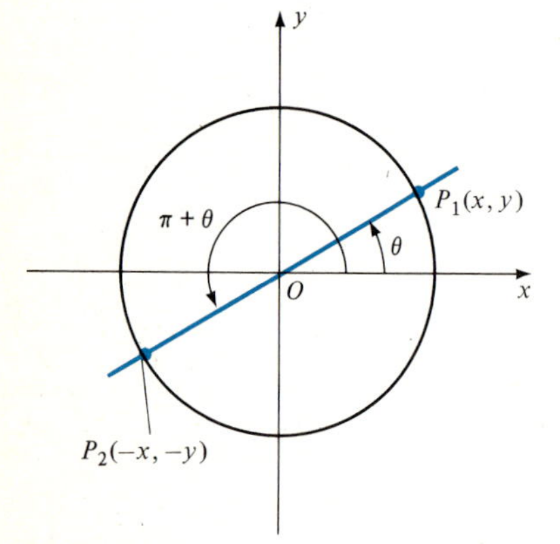

FIGURE 8.4d

Although the tangent and cotangent functions are periodic with period 2π, this is not their basic period as it is for the other four functions that we have defined. To see this, consider an angle θ and the angle $\theta + \pi$. As Figure 8.4d illustrates, these two angles always have their terminal sides in a straight line but on opposite sides of the origin. As a consequence of this fact, if the unit circle cuts the terminal side of angle θ in a point with corrdinates (x, y), then it cuts the terminal side of the angle $\theta + \pi$ in a point with coordinates $(-x, -y)$. Thus, we have Theorem 8.4a. The equations in this theorem express the fact that *the tangent and cotangent functions are periodic with period* π, which is the basic period for both functions.

EXAMPLE 4

Show that $\tan 1000^\circ = \tan 100^\circ$

SOLUTION:

Since $180^\circ = \pi$ radians, we know from Theorem 8.4a that the tangent function is periodic with period 180°. Furthermore,

$$1000^\circ = 5 \cdot 180^\circ + 100^\circ$$

Hence, the periodicity of the tangent function shows that

$$\tan 1000^\circ = \tan 100^\circ$$

▼ THEOREM 8.4a

$$\tan(\theta + \pi) = \tan\theta$$

and

$$\cot(\theta + \pi) = \cot\theta$$

INVERSES

There are many problems in trigonometry that require us to find an angle (or a number) when a value of one of its trigonometric functions is given. This is the inverse of the problem we have considered thus far. For example, if we are given

$$\sin\theta = \tfrac{1}{2}$$

what is the value of θ? We can see immediately that 30° (or $\pi/6$) is a possible answer, but 150° (or $5\pi/6$) and, more generally, $30^\circ + m \cdot 360^\circ$ and $150^\circ + m \cdot 360^\circ$, where m is any integer, are also possible answers. In order to specify a unique answer, we must know the quadrant of θ as well as its domain. If we restrict θ to be a second-quadrant angle between 0° and 360°, then the naswer 150°

is unique. For the present, we shall consider only the values obtainable from Table 8.4b and in the interval from 0° to 360° (or 0 to 2π). The next examples illustrate the ideas involved.

EXAMPLE 5

Find the angle θ if $\cos\theta = -\sqrt{3}/2$, $0° \le \theta \le 360°$, and $\tan\theta > 0$.

SOLUTION:

Because the given cosine is negative and the tangent is positive, the angle θ must be in the third quadrant. Thus, the cosine of the reference angle is $\sqrt{3}/2$, so that the reference angle is 30° and $\theta = 180° + 30°$ or 210°.

EXAMPLE 6

Find the number t between 0 and 2π such that $\cot t = -1/\sqrt{3}$ and $\sin t > 0$.

SOLUTION:

Since the cotangent is negative and the sine is positive, the number t must correspond to a second-quadrant angle, that is, $\pi/2 < t < \pi$. Because the cotangent of the reference number is $1/\sqrt{3}$, the reference number is $\pi/3$, and $t = \pi - \pi/3 = 2\pi/3$.

EXAMPLE 7

Find θ if $\sec\theta = -\sqrt{2}$ and $0° \le \theta \le 360°$.

SOLUTION:

Because the quadrant of θ is not specified, there are two solutions, one in the second quadrant and one in the third quadrant (the two quadrants where the secant is negative). Thus, the secant of the reference angle is $\sqrt{2}$; the reference angle is 45°; and

$$\theta = 180° - 45° = 135° \quad \text{or} \quad \theta = 180° + 45° = 225°$$

EXERCISE 8.4

In Problems 1–12, find the values of the six trigonometric functions of the given angle. Try to do this without referring to tables.

1 300° **2** $-270°$ **3** 240°

4 $-600°$ **5** $-765°$ **6** 1290°

7 $-540°$ **8** 780° **9** $-930°$

10 1020° **11** 495° **12** $-870°$

In Problems 12–24, find the values of the six circular functions of the given numbers. Try to do this without referring to tables.

13 $\frac{7\pi}{6}$ **14** $\frac{4\pi}{3}$ **15** $\frac{7\pi}{4}$

16 $-\frac{\pi}{6}$ **17** $-\frac{15\pi}{4}$ **18** $\frac{29\pi}{3}$

19 $-\frac{31\pi}{6}$ **20** $-\frac{17\pi}{3}$ **21** $\frac{19\pi}{4}$

22 $-\frac{23\pi}{6}$ **23** $\frac{57\pi}{2}$ **24** 1001π

In Problems 25–44, find the values of θ, $0^\circ \le \theta \le 360^\circ$.

25 $\sin\theta = -\frac{1}{2}$, θ in Q III

26 $\sin\theta = \frac{1}{2}$, θ in Q II

27 $\cos\theta = \frac{1}{2}$, θ in Q IV

28 $\cos\theta = -\frac{1}{2}$, θ in Q II

29 $\tan\theta = \sqrt{3}$, θ in Q III

30 $\tan\theta = -\sqrt{3}$, θ in Q IV

31 $\sec\theta = -\dfrac{2}{\sqrt{3}}$, θ in Q II

32 $\csc\theta = -2$, θ in Q III

33 $\cot\theta = 1$, $\sin\theta < 0$

34 $\cot\theta = -\sqrt{3}$, $\cos\theta < 0$

35 $\sin\theta = -\dfrac{\sqrt{3}}{2}$, $\tan\theta > 0$

36 $\cos\theta = \dfrac{\sqrt{2}}{2}$, $\csc\theta > 0$

37 $\sin\theta = -\dfrac{\sqrt{3}}{2}$

38 $\cos\theta = -\dfrac{\sqrt{2}}{2}$

39 $\tan\theta = 1$

40 $\cot\theta = 0$

41 $\sin\theta = 0$

42 $\cos\theta = 1$

43 $\sec\theta$ undefined

44 $\cot\theta$ undefined

USING YOUR KNOWLEDGE 8.4

The pair of functions, sine and cosine, are called cofunctions, as are the pair, tangent and cotangent, and the pair, secant and cosecant. If f denotes one of a pair of cofunctions then it is true that

$$f(t) = \operatorname{cof}\left(\frac{\pi}{2} - t\right) \tag{1}$$

For example, you might have noticed that $\pi/6 = \pi/2 - \pi/3$ and $\sin(\pi/6) = \cos(\pi/3)$.

1 Prove the general result.

Hint: The points $P(t)$ and $P(\pi/2 - t)$ on the unit circle are symmetrically located with respect to the line $y = x$, so that if the coordinates of one of the points are (a, b), then those of the other are (b, a).

Equation 1 can be used to express the value of a given function in terms of a function of a number between 0 and $\pi/4$ or of an angle between 0° and 45° if the problem is stated in degrees. Here are two examples:

a $\displaystyle \sin\frac{5\pi}{3} = \sin\left(2\pi - \frac{\pi}{3}\right) = -\sin\frac{\pi}{3} = -\sin\left(\frac{\pi}{2} - \frac{\pi}{6}\right) = -\cos\frac{\pi}{6}$

b $\tan 260^\circ = \tan(180^\circ + 80^\circ) = \tan 80^\circ = \tan(90^\circ - 10^\circ) = \cot 10^\circ$

Reduce each of the following to a function of a number between 0 and $\pi/4$ or to an angle between 0° and 45° if the problem is stated in degrees.

2 $\sec\dfrac{3\pi}{7}$ **3** $\cos\dfrac{5\pi}{8}$ **4** $\sin\left(-\dfrac{5\pi}{9}\right)$ **5** $\tan\dfrac{7\pi}{12}$ **6** $\csc\left(-\dfrac{4\pi}{7}\right)$ **7** $\sin 72^\circ$ **8** $\cot 108^\circ$

9 $\tan 265^\circ$ **10** $\cos(-435^\circ)$

8.5 BASIC IDENTITIES AND TRIGONOMETRIC EQUATIONS

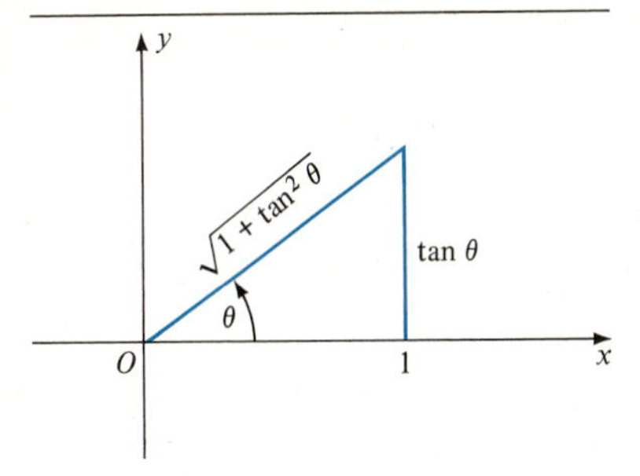

FIGURE 8.5a

There is a type of scientific calculator that is internally programmed to find $\tan \theta$ and then to find $\sin \theta$ and $\cos \theta$ by means of formulas expressing them in terms of $\tan \theta$. (We shall discuss the use of calculators for finding values of the trigonometric functions in Section 8.6.) The right triangle in Figure 8.5a shows schematically what the required formulas are. Thus, if θ is a first-quadrant angle, then

$$\sin \theta = \frac{\tan \theta}{\sqrt{1 + \tan^2 \theta}} \qquad \cos \theta = \frac{1}{\sqrt{1 + \tan^2 \theta}}$$

One way to obtain this formula for $\sin \theta$ is to use the definition of the tangent

$$\tan \theta = \frac{\sin \theta}{\cos \theta}$$

and the basic identity

$$\cos^2 \theta + \sin^2 \theta = 1$$

(from Theorem 8.1a). This identity gives

$$\cos \theta = \pm\sqrt{1 - \sin^2 \theta}$$

which we substitute into the definition of the tangent to obtain

$$\tan \theta = \frac{\sin \theta}{\pm\sqrt{1 - \sin^2 \theta}}$$

an equation that we can solve for $\sin \theta$. By squaring both sides, we obtain

$$\tan^2 \theta = \frac{\sin^2 \theta}{1 - \sin^2 \theta}$$

so that

$$\tan^2 \theta(1 - \sin^2 \theta) = \sin^2 \theta$$

or

$$\tan^2 \theta - \tan^2 \theta \sin^2 \theta = \sin^2 \theta$$

and

$$\tan^2 \theta = \sin^2 \theta + \tan^2 \theta \sin^2 \theta$$

We can now factor the right side to obtain

$$\tan^2 \theta = \sin^2 \theta(1 + \tan^2 \theta)$$

Thus,

$$\sin^2 \theta = \frac{\tan^2 \theta}{1 + \tan^2 \theta}$$

and

$$\sin \theta = \pm\frac{|\tan \theta|}{\sqrt{1 + \tan^2 \theta}}$$

where the plus sign or the minus sign must be chosen to correspond to the quadrant of θ (*plus* if θ is in the first or the second quadrant where $\sin \theta$ is positive, and *minus* if θ is in the third or the fourth quadrant where $\sin \theta$ is negative). If the procedure for solving the equation

$$\tan \theta = \frac{\sin \theta}{\pm\sqrt{1 - \sin^2 \theta}}$$

bothers you, let $\boldsymbol{T = \tan \theta}$ and $\boldsymbol{S = \sin \theta}$ The equation becomes the algebraic equation

$$T = \frac{S}{\pm\sqrt{1 - S^2}}$$

which we wish to solve for S. You should obtain the formula for $\cos \theta$ by using the same ideas.

The preceding problem illustrates the use of a basic identity and the solution of a *trigonometric* equation, an equation in which the unknown is involved in trigonometric functions. This section is concerned with other basic identities and their use in simplifying trigonometric expressions and in proving identities as well as in the solution of trigonometric equations.

BASIC IDENTITIES

A number of basic identities can be obtained from the definitions of the trigonometric functions. These identities are useful not only in changing one trigonometric expression into another but also in investigating essential properties of the functions. One of these identities was given in Theorem 8.1a and is repeated here for your convenience. By dividing this identity by $\cos^2 t$ and by $\sin^2 t$, we immediately obtain two new identities as stated in Theorems 8.5a and 8.5b. Just as $\cos^2 t$ stands for $(\cos t)^2$, $\tan^2 t$ stands for $(\tan t)^2$, and similarly for the other functions.

▼ THEOREM 8.1a

$$\cos^2 t + \sin^2 t = 1$$

▼ THEOREM 8.5a

$$1 + \tan^2 t = \sec^2 t$$

▼ THEOREM 8.5b

$$1 + \cot^2 t = \csc^2 t$$

From Theorem 8.1a, we see that the cosine and sine functions must obey the inequalities

$$-1 \leq \cos t \leq 1 \quad \text{and} \quad -1 \leq \sin t \leq 1$$

Similarly, Theorems 8.5a and 8.5b show that

$$|\sec t| \geq 1 \quad \text{and} \quad |\csc t| \geq 1$$

The identities in the preceding theorems do not restrict the values of the tangent and cotangent functions. In fact, the definitions of these functions show that they may assume all real values. Another way of stating these conclusions is to say that the sine and cosine functions have as range all real numbers between -1 and 1, inclusive; the secant and cosecant functions have as range all real numbers whose absolute values are greater than or equal to 1; and the tangent and cotangent functions have all real numbers for their range.

PROVING IDENTITIES

The following examples show how the basic identities can be used to prove other identities.

EXAMPLE 1

Prove the identity $\cos^2 \theta - \sin^2 \theta = 1 - 2 \sin^2 \theta$.

SOLUTION:

We shall prove this identity by changing the left side into the right side. From the basic identity $\cos^2 \theta + \sin^2 \theta = 1$, we find

$$\cos^2 \theta = 1 - \sin^2 \theta$$

which we substitute into the left side of the proposed identity to obtain

$$\begin{aligned} \cos^2 \theta - \sin^2 \theta &= (1 - \sin^2 \theta) - \sin^2 \theta \\ &= 1 - 2 \sin^2 \theta, \end{aligned}$$

which is the required right side. This completes the proof.

EXAMPLE 2

Prove the identity $\sec A = \cos A + \sin A \tan A$, $\cos A \neq 0$.

SOLUTION:

In this problem, we shall prove the identity by changing the right side into the left side. Using the definition of the tangent, we have

$$\begin{aligned} \cos A + \sin A \tan A &= \cos A + \sin A \cdot \frac{\sin A}{\cos A} \\ &= \cos A + \frac{\sin^2 A}{\cos A} \\ &= \frac{\cos^2 A + \sin^2 A}{\cos A} \\ &= \frac{1}{\cos A} \\ &= \sec A \end{aligned}$$

This proves the identity. Note that the condition $\cos A \neq 0$ assures us that both members of the identity are defined.

EXAMPLE 3

Prove the identity

$$\frac{1+\cos\theta}{\sin\theta} = \frac{\sin\theta}{1-\cos\theta} \qquad \sin\theta \neq 0$$

SOLUTION:

By changing the fractions into equivalent fractions with a common denominator, we can compare the fractions by comparing their numerators. Thus, we try to prove the identity by changing both sides into the same expression. The LCD of the two fractions is $\sin\theta(1-\cos\theta)$. Hence, the left side

$$\begin{aligned}\frac{1+\cos\theta}{\sin\theta} &= \frac{(1+\cos\theta)(1-\cos\theta)}{\sin\theta(1-\cos\theta)}\\ &= \frac{1-\cos^2\theta}{\sin\theta(1-\cos\theta)}\\ &= \frac{\sin^2\theta}{\sin\theta(1-\cos\theta)}\end{aligned}$$

and the right side

$$\begin{aligned}\frac{\sin\theta}{1-\cos\theta} &= \frac{\sin\theta\cdot\sin\theta}{\sin\theta(1-\cos\theta)}\\ &= \frac{\sin^2\theta}{\sin\theta(1-\cos\theta)}\end{aligned}$$

Because both sides are equal to the same expression, they are equal to each other.

The preceding examples show the three ways in which you can prove an identity: (1) by changing the left side into the right side, (2) by changing the right side into the left side, (3) by changing both sides into the same expression. Most often we use method 1 or 2, changing what seems to be the more complicated side into the other side. If neither succeeds, try the third method. The next examples will give you other hints. The important idea to keep in mind is that you do not work with the proposed identity as if it were a true equation. The two sides must be handled independently.

EXAMPLE 4

Prove that, if $\sin t\cos t \neq 0$, then

$$\frac{\sin t\cot t + \cos t}{\cot t} = 2\sin t$$

SOLUTION:

We shall prove the proposed identity by transforming the left side into the right side. Thus,

$$\begin{aligned}\frac{\sin t\cot t+\cos t}{\cot t} &= \frac{\sin t\cot t}{\cot t} + \frac{\cos t}{\cot t}\\ &= \sin t + \cos t \div \cot t\\ &= \sin t + \cos t \div \frac{\cos t}{\sin t}\end{aligned}$$

$$= \sin t + (\cos t)\frac{\sin t}{\cos t}$$

$$= \sin t + \sin t$$

$$= 2 \sin t$$

(The restriction $\sin t \cos t \neq 0$ was made to avoid division by zero.)

EXAMPLE 5

Prove that if $\sin \theta \neq 0$, then

$$\csc \theta + \cot \theta = \frac{\sin \theta}{1 - \cos \theta}$$

SOLUTION:

We start by expressing the left side in terms of $\sin \theta$ and $\cos \theta$.

$$\csc \theta + \cot \theta = \frac{1}{\sin \theta} + \frac{\cos \theta}{\sin \theta} = \frac{1 + \cos \theta}{\sin \theta}$$

It is a good idea to multiply the numerator and denominator of the last fraction by $1 - \cos \theta$, because this is the denominator of the right side of the proposed identity. This multiplication gives

$$\csc \theta + \cot \theta = \frac{(1 + \cos \theta)(1 - \cos \theta)}{\sin \theta(1 - \cos \theta)}$$

$$= \frac{1 - \cos^2 \theta}{\sin \theta(1 - \cos \theta)}$$

$$= \frac{\sin^2 \theta}{\sin \theta(1 - \cos \theta)}$$

$$= \frac{\sin \theta}{1 - \cos \theta}$$

Of course, we could have avoided the last few steps by using the identity that we proved in Example 3; namely,

$$\frac{1 + \cos \theta}{\sin \theta} = \frac{\sin \theta}{1 - \cos \theta}$$

In order to prove that a given equation is *not* an identity, we need to show that the equation is false for at least one of the permissible values of the variable. For instance, we can show that

$$\tan^2 t + \sec^2 t = 1$$

is not an identity by substituting the value $t = \pi/4$. This gives $1^2 + (\sqrt{2})^2 = 1$ or $1 + 2 = 1$, which is *false*. Therefore, the equation is *not* an identity. In fact, the equation is true for the values $t = m\pi$, $m = 0, \pm 1, \pm 2, \ldots$ and not for any other values.

SOLVING TRIGONOMETRIC EQUATIONS

As we have noted, equations in which the unknown is involved in trigonometric functions is called a *trigonometric equation*. The next examples illustrate the solution of such equations.

EXAMPLE 6

Find all the solutions of the equation:
$\sin t = \cos t$, $0 \leq t \leq 2\pi$.

SOLUTION:

We divide by $\cos t$ to obtain the equation

$$\frac{\sin t}{\cos t} = 1$$

that is,

$$\tan t = 1$$

From this equation, it follows that $t = \pi/4$ or $5\pi/4$. Since we divided by $\cos t$ in obtaining this solution, we must check that $\cos t \neq 0$. If $\cos t = 0$, then $t = \pi/2$ or $3\pi/2$, and $\sin t = 1$ or -1, so that the given equation is not satisfied if $\cos t = 0$. Thus, we have lost no solutions by performing this division, and the only solutions are $\pi/4$ and $5\pi/4$.

EXAMPLE 7

Find all the solutions of the equation $2 \cos^2 t - \sin t - 1 = 0$ if $0 \leq t \leq 2\pi$.

SOLUTION:

We first use the identity $\cos^2 t = 1 - \sin^2 t$ to write the first term of the equation in terms of $\sin t$. Thus, we obtain

$$2(1 - \sin^2 t) - \sin t - 1 = 0$$
$$2 - 2\sin^2 t - \sin t - 1 = 0$$
$$2\sin^2 t + \sin t - 1 = 0$$

which is a quadratic equation in $\sin t$. This suggests setting $\sin t$ equal to s to simplify the appearance of the equation, giving

$$2s^2 + s - 1 = 0$$

or

$$(2s - 1)(s + 1) = 0$$

The roots of this equation are $s = \frac{1}{2}$ and $s = -1$. It follows that $\sin t = \frac{1}{2}$, so that $t = \pi/6$ or $5\pi/6$, or $\sin t = -1$, so that $t = 3\pi/2$. Thus, the desired solutions are $\pi/6$, $5\pi/6$, and $3\pi/2$. You should check these in the original equation.

EXAMPLE 8

Find all the solutions of the equation $\tan^2 t + \sec^2 t = 7$.

SOLUTION:

From the basic identities, we have $\sec^2 t = 1 + \tan^2 t$, so that we may rewrite the equation entirely in terms of the tangent.

$$\tan^2 t + (1 + \tan^2 t) = 7$$

or

$$2\tan^2 t = 6$$

Hence,

$$\tan t = \pm\sqrt{3}$$

The numbers with smallest absolute value that satisfy the last equation are $\pm\pi/3$, and, because the period of the tangent is π, any other solution differs from one or the other of these by an integer multiple of π. Hence, we may write all the solutions in the form

$$t = m\pi \pm \frac{\pi}{3} \qquad m = 0, \pm 1, \pm 2, \pm 3, \ldots$$

The preceding examples illustrate the use of the basic identities to transform an equation containing more than one of the trigonometric functions into an equation in only one of these functions. If the transformed equation can be solved for that function of the unknown, then the unknown itself can be found by using tables or a calculator.

EXAMPLE 9

Find the values of t, $0 < t < \pi/2$, such that $\cot t > 3 \tan t$.

SOLUTION:

For $0 < t < \pi/2$, $\tan t$ is positive. Therefore, we may divide by $\tan t$ to obtain the inequality

$$\frac{\cot t}{\tan t} > 3$$

Since

$$\frac{\cot t}{\tan t} = (\cot t)\frac{1}{\tan t} = (\cot t)(\cot t) = \cot^2 t$$

the last inequality is the same as

$$\cot^2 t > 3$$

Because $\cot t$ is positive for $0 < t < \pi/2$, we may extract the positive square root of both sides to obtain

$$\cot t > \sqrt{3}$$

Using the fact that $\cot \pi/6 = \sqrt{3}$, we make a sketch as shown in Figure 8.5b. Evidently, if t is between 0 and $\pi/6$, then

$$x > \frac{\sqrt{3}}{2} \quad \text{and} \quad y < \frac{1}{2}$$

Because $\cot t = x/y$, this will mean that $\cot t > \sqrt{3}$. Therefore,

$$\cot t > 3 \tan t \qquad \text{for } 0 < t < \frac{\pi}{6}$$

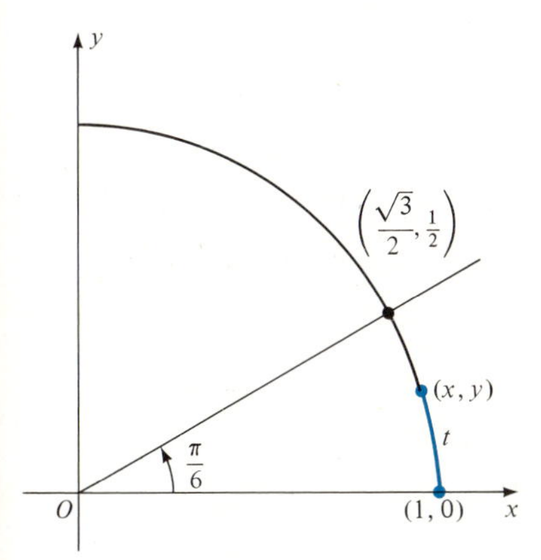

FIGURE 8.5b

EXERCISE 8.5

In Problems 1–20, prove that the given equation is an identity.

1 $\cos\theta = \sin\theta\cot\theta$

2 $\sin\theta = \cos\theta\tan\theta$

3 $\sec\theta\cot\theta = \csc\theta$

4 $\sin\theta\tan\theta = \sin^2\theta\sec\theta$

5 $\cos^2\theta - \sin^2\theta = 2\cos^2\theta - 1$

6 $\tan^2\theta + \sec^2\theta = 2\tan^2\theta + 1$

7 $\cot^2\theta + \csc^2\theta = 2\csc^2\theta - 1$

8 $1 + \sin^2\theta\sec^2\theta = \sec^2\theta$

9 $\tan t + \sec t = \dfrac{\cos t}{1 - \sin t}$

10 $\csc^2\theta - \cos^2\theta = 1 + \cos^2\theta\cot^2\theta$

11 $\tan\alpha + \cot\alpha = \sec\alpha\csc\alpha$

12 $\sec^2 t + \csc^2 t = \sec^2 t\csc^2 t$

13 $\dfrac{\cos\theta}{1 - \sin\theta} - \dfrac{1 - \sin\theta}{\cos\theta} = 2\tan\theta$

14 $\dfrac{1}{1 + \sin s} + \dfrac{1}{1 - \sin s} = 2\sec^2 s$

15 $\dfrac{\sin A + \sin B}{\sin A - \sin B} = \dfrac{\csc B + \csc A}{\csc B - \csc A}$

16 $\dfrac{1 - \sec^2 t}{1 - \csc^2 t} = \tan^4 t$

17 $\csc^4 A - \cot^4 A = (\sin^2 A + 2\cos^2 A)\csc^2 A$

18 $(\tan t + \cot t)^2 = \sec^2 t\csc^2 t$

19 $\dfrac{1 + \cos\theta}{1 - \cos\theta} = (\csc\theta + \cot\theta)^2$

20 $\cos^4 A + \cos^2 A\sin^2 A + \sin^4 A = 1 - \cos^2 A\sin^2 A$

In Problems 21–26, prove that the given equation is *not* an identity.

21 $\sin 3t = 3\sin t$

22 $\cos 2t = 2\cos t$

23 $\tan\dfrac{t}{2} = \dfrac{1}{2}\tan t$

24 $\cot^2 t + \csc^2 t = 1$

25 $\cos 2t = 1 + \sin^2 t$

26 $1 - \cos t = \sin 2t$

In Problems 27–40, find all the solutions of the given equation in the interval from 0 to 2π, inclusive.

27 $2\cos x = 1$

28 $\sqrt{3}\cot x = 1$

29 $2\sin y - 1 = 0$

30 $\tan y - 1 = 0$

31 $2\sin^2 t - 1 = 0$

32 $4\cos^2 t - 3 = 0$

33 $\sin t + \cos t = 0$

34 $\sin t + \sqrt{3}\cos t = 0$

35 $\sqrt{3}\sin\theta - \sec\theta\cos^2\theta = 0$

36 $\sec^2 x - \tan x = 1$

37 $3\sec^2 t = 4\tan^2 t$

38 $2\sin^2 x + 3\sin x = 2$

39 $2\sin^2\theta = 1 - \cos\theta$

40 $\cos^2 x = 1 - \sin x$

In Problems 41–46, find all the solutions of the given equation.

41 $3\sin y = 2\cos^2 y$

42 $\cos t = 2\sqrt{3}\sin^2 t$

43 $4\cot\theta = \sqrt{3}\csc^2\theta$

44 $\sec x = -\sqrt{2}\tan^2 x$

45 $2\cos^2 t - 5\sin t + 1 = 0$

46 $2\sin^2 t + \cos t + 1 = 0$

In Problems 47–52, find the values of t, $0 \le t \le \pi/2$, such that the given inequality is true.

47 $\tan t > 1$

48 $\sin t < \frac{1}{2}$

49 $\sec t > \csc t$

50 $\sec t < 2\tan t$

51 $\tan t < \cot t$

52 $\sin t > \cos t$

USING YOUR KNOWLEDGE 8.5

Figure 8.5c shows a pencil in a glass of water. The pencil appears to be bent because of the *refraction* of light. Refraction occurs because light travels more slowly in a given medium than in a less dense medium. If a light ray passes from a medium in which the velocity of light is v_1 into a second medium in which the velocity is v_2 then the ray is bent as indicated in Figure 8.5d. It is shown in physics that the angle of incidence α is related to the angle of refraction β by the equation

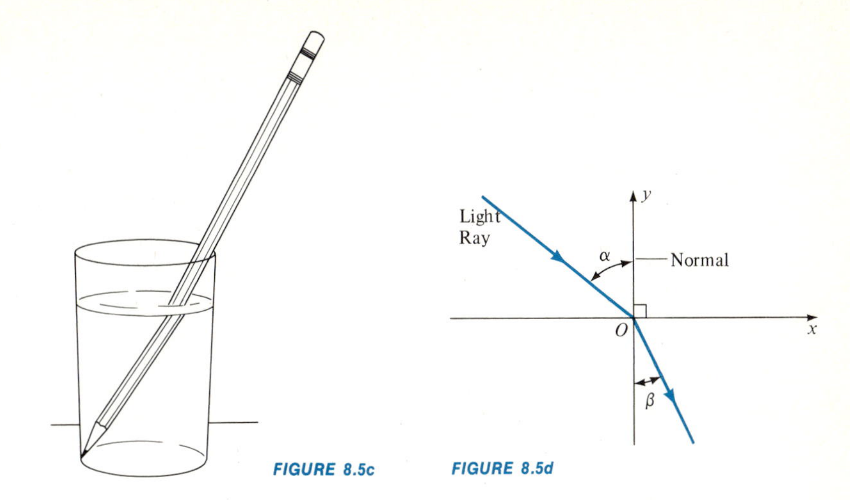

FIGURE 8.5c FIGURE 8.5d

$$\frac{\sin \alpha}{\sin \beta} = \frac{v_1}{v_2}$$

This equation is known as Snell's Law of Refraction. If the first medium is a vacuum, then the ratio v_1/v_2 is called the *index of refraction* of the second medium and is often denoted by the Greek letter μ (mu). Thus, we may write

$$\mu = \frac{\sin \alpha}{\sin \beta}$$

where the first medium is a vacuum. The index of refraction of air is about 1.0003, so that the index of refraction of a medium other than air can be found fairly accurately by considering a light ray passing from air into the second medium.

The index of refraction of water is about $\frac{4}{3}$. This means that a ray of light passing from air into water would be bent so that the angle of incidence α is related to the angle of refraction β by the equation

$$\frac{\sin \alpha}{\sin \beta} = \frac{4}{3}$$

Thus, $\sin \alpha = \frac{4}{3} \sin \beta$, which means that α is greater than β, the ray of light is bent toward the normal, and the angle of refraction is less than the angle of incidence. If one of the angles is known, we can find the other angle by using Snell's Law. For example, if $\sin \alpha = \frac{2}{3}$, then we have

$$\frac{\frac{2}{3}}{\sin \beta} = \frac{4}{3} \qquad \text{and} \qquad \sin \beta = (\tfrac{3}{4})(\tfrac{2}{3}) = \tfrac{1}{2}$$

so that $\beta = 30^\circ$. In the next section, we shall see how to find α from tables or by calculator; the approximate value is 42°. Our result shows that the ray of light is bent about 12° toward the normal.

1 If $\sin \alpha = \frac{2}{3}\sqrt{2}$ ($\alpha \approx 70.5^\circ$), find how much a ray of light is bent when it goes from air into water.
2 The index of refraction of benzene is about 1.5. If $\sin \alpha = \frac{3}{4}$ ($\alpha \approx 48.6^\circ$), find how much a ray of light is bent when it goes from air into benzene.
3 If a ray of light passes from air into water, is it possible for the angle of refraction to be 60°? Explain.
4 If a ray of light passes from air into benzene, is it possible for the angle of refraction to be 45°? Explain. See Problem 2.

8.6 TABLES OF THE CIRCULAR AND THE TRIGONOMETRIC FUNCTIONS[4]

During the Renaissance there was a very great interest in astrology, which increased the demand for accurate astronomical tables. Because astrological predictions needed precise future locations of the planets and the calculations necessary to obtain these locations involve the trigonometric functions, accurate trigonometric tables were required. Consequently, much effort was spent during this period improving the existing tables, which had not been changed very much since the time of Ptolemy, who lived around 150 A.D. Incidentally, the study of trigonometry goes back to the Greeks, about 200 B.C.

Most of the problems that arise in the application of the circular and the trigonometric functions require that we know values of these functions for numbers and angles other than the few simple ones that are given in Table 8.4b or that can be obtained from this table by using reference numbers or angles. We shall not be concerned here with the methods of calculating such tables of values but shall only remark that if you take a standard course in calculus, you will learn about such methods in connection with the subject of infinite series. In this book, we have included two tables, one for the circular functions of numbers (or angles in radian measure)—Appendix Table A5—and the second for the trigonometric functions of angles in degrees—Appendix Table A6.

Table A5 gives the values of the circular functions sine, tangent, cotangent, and cosine for numbers from 0 to 1.60 at intervals of 0.01. Because $\pi/2 \approx 1.57$, this table along with the use of reference numbers will furnish values of the functions of numbers (or angles in radians) in all four quadrants. To read the value of a function of a given number between 0 and 1.60, we go down the first column, which is headed *Rad.*, until we come to the given number; then we go across until we come to the column headed with the name of the desired function, where we read the value of the function. Table 8.6a reproduces a portion of Table A5. To find tan 0.53, for example, we

TABLE 8.6a *Values of the Circular Functions*

Rad.	Sin	Tan	Cot	Cos
0.50	0.47943	0.54630	1.8305	0.87758
0.51	0.48818	0.55936	1.7878	0.87374
0.52	0.49688	0.57256	1.7465	0.86782
0.53	0.50553	0.58592	1.7067	0.86281
0.54	0.51414	0.59943	1.6683	0.85771

[4] If students have scientific calculators available, this section, except for the Using Your Knowledge 8.6, may be omitted.

go down the first column until we come to 0.53, then across to the column headed *tan*, where we read 0.58592. Similarly, to find cos 0.52, we go down the first column to 0.52 and across to the column headed *cos*, where we read 0.86782.

For numbers between 0 and 1.60 but with more than two decimal places, we use *linear interpolation* just as we did for the table of logarithms. The following examples illustrate the procedure, which you may compare with that in Section 7.3. You should examine Table A5 carefully before studying these examples.

EXAMPLE 1

Find sin 1.083.

SOLUTION:

Since 1.083 is between 1.08 and 1.09, we assume that sin 1.083 is 0.3 of the way from sin 1.08 to sin 1.09; that is, we assume that the difference in the value of the function is proportional to the difference in the numbers. The accompanying diagram is essentially self-explanatory. Notice that the decimal point in the value of the sine is disregarded during the interpolation procedure.

Difference		Radians	Sin	Difference	
0.01	0.003	1.08	0.88196	x	467
		1.083	?		
		1.09	0.88663		

Thus,

$$\frac{x}{467} = \frac{0.003}{0.01} \quad \text{or} \quad x = (0.3)(467) = 140$$

Because the sine is increasing, we add 140 in the last three places of sin 1.08 to obtain sin 1.083 = 0.88336. (A more accurate value is 0.883368, which shows that the linear interpolation in this example yields a result that is in error by less than one unit in the fifth decimal place.)

For numbers that are outside the interval from 0 to 1.60, we can use reference numbers as illustrated by Example 2.

EXAMPLE 2

Find cos 2.

SOLUTION:

We first recognize that 2 is between $\pi/2$ and π, so that the reference number t_r is $\pi - 2$. If we use the approximate value 3.1416 for π, then $t_r \simeq 1.1416$. We can now interpolate in Table A5. We read cos 1.14 = 0.41759 and cos 1.15 = 0.40849, giving a tabular difference of 910 in the last three places. The corresponding proportion is

$$\frac{x}{910} = \frac{0.0016}{0.01}$$

so that $x = 0.16 \cdot 910 = 146$. Because the cosine is decreasing, we must subtract 146 in the last three places of cos 1.14 to obtain

$$\cos 1.1416 = 0.41613$$

However, an angle of 2 radians is a second-quadrant angle, making cos 2 negative. Consequently, we have $\cos 2 = -0.41613$.

Table A6 in the Appendix gives the trigonometric functions sine, tangent, cotangent, and cosine for angles measured in degrees from 0° to 90° at intervals of 0.1°. This table makes use of the fact that was obtained in Using Your Knowledge 8.4, namely, that a function of t is equal to the cofunction of $\pi/2 - t$ or, equivalently, that a trigonometric function of θ in degrees is equal to the cofunction of $90° - \theta$. Thus, Table A6 goes in one direction from 0° to 45° with the names of the functions at the top of each column corresponding to angles in the leftmost column, and then in the opposite direction from 45° to 90° with the names of the functions at the bottom of each column corresponding to angles in the rightmost column. To avoid gross errors, you must be sure to keep the construction of the table in mind. Except for this extra precaution, the table is read in the same way as the table of circular functions. Interpolation is also carried out in the same manner.

EXAMPLE 3

Find tan 36.47°.

SOLUTION:

For this problem, we must interpolate between 36.4° and 36.5°. The procedure is as follows:

$$0.1\left\{\begin{array}{l} 0.07\left\{\begin{array}{l} 36.4° \\ 36.47° \end{array}\right. \\ \phantom{0.07\{}36.5° \end{array}\right. \qquad \begin{array}{l} 0.73726 \\ ? \\ 0.73996 \end{array}\left.\begin{array}{l} \left.\begin{array}{l} \\ \end{array}\right\} x \\ \end{array}\right\} 270$$

Difference		Degrees	Tan	Difference	
0.1	0.07	36.4°	0.73726	x	270
		36.47°	?		
		36.5°	0.73996		

Thus,

$$\frac{x}{270} = \frac{0.07}{0.1} \quad \text{or} \quad x = 0.7 \cdot 270 = 189$$

Because the tangent is increasing, we add 189 in the last three places of tan 36.4° to obtain tan 36.47° = 0.73915.

EXAMPLE 4

Find cos(−510.52°).

SOLUTION:

First, we know that cos(−510.52°) = cos 510.52°. Furthermore, by the periodicity of the cosine, we have that cos 510.52° = cos(510.52° − 360°) = cos 150.52°. The reference angle is therefore $\theta_r = 180° - 150.52°$, or 29.48°. To find cos 29.48°, we must interpolate between 29.4° and 29.5°. The procedure is as follows:

Difference		Degrees	Cos	Difference	
0.1	0.08	29.4°	0.87121	x	85
		29.48°	?		
		29.5°	0.87036		

Therefore,

$$\frac{x}{85} = \frac{0.08}{0.1} \quad \text{or} \quad x = 0.8 \cdot 85 = 68$$

The cosine function is decreasing; hence, we subtract 68 in the last two places of cos 29.4° to obtain cos 29.48° = 0.87053. Because 150.52° is a second-quadrant angle, the cosine is negative. Thus, our final result is

$$\cos(-510.52°) = -0.87053$$

Having considered the problem of finding the value of a function when its argument, a number or an angle, is given, we must now turn to the inverse problem of finding the number or the angle when a circular or a trigonometric function value is given. If the function value occurs exactly as given, then it is only necessary to read the table to find the corresponding number or angle. If the exact value does not occur in the table, then interpolation is needed to obtain the desired number or angle. The following examples illustrate the procedure, which is much like that used for the logarithm tables.

EXAMPLE 5

Find t if $\sin t = 0.42910$, $0 < t < \pi/2$.

SOLUTION:

We find from Table A5 that 0.42910 is between 0.42594 = sin 0.44 and 0.43497 = sin 0.45. Accordingly, we interpolate between these two entries.

Difference		Radians	Sin	Difference	
0.01	x	0.44	0.42594	316	903
		?	0.42910		
		0.45	0.43497		

Thus,

$$\frac{x}{0.01} = \frac{316}{903} \quad \text{or} \quad x = 0.01 \cdot \frac{316}{903} = 0.0035$$

and

$$t = 0.4435$$

EXAMPLE 6

Find t if $\cot t = -0.25000$, $\pi/2 < t < \pi$.

SOLUTION:

We first find the reference number t_r such that $\cot t_r = 0.25000$. The table shows that 0.25000 is between $0.25619 = \cot 1.32$ and $0.24556 = \cot 1.33$. Hence, we interpolate between these two entries.

$$\begin{array}{cccc} \text{Difference} & \text{Radians} & \text{Cot} & \text{Difference} \\ 0.01\left\{\begin{array}{l} x\left\{\begin{array}{l} \\ \\ \end{array}\right. \\ \\ \end{array}\right. & \begin{array}{c} 1.32 \\ ? \\ 1.33 \end{array} & \begin{array}{c} 0.25619 \\ 0.25000 \\ 0.24556 \end{array} & \left.\begin{array}{l} \left.\begin{array}{l} \\ \\ \end{array}\right\} 619 \\ \\ \end{array}\right\} 1063 \end{array}$$

Thus, we have

$$\frac{x}{0.01} = \frac{619}{1063} \quad \text{or} \quad x = 0.01 \cdot \frac{619}{1063} = 0.0058$$

and

$$t_r = 1.3258$$

Therefore,

$$t = \pi - t_r = 3.1416 - 1.3258 = 1.8158$$

EXAMPLE 7

If $\cos \theta = -\frac{1}{3}$, $90^\circ < \theta < 180^\circ$, find θ.

SOLUTION:

Again, we first find the reference angle θ_r by using Table A6. We are given $\cos \theta = -\frac{1}{3} \simeq -0.33333$, so that $\cos \theta_r = 0.33333$, which lies between $\cos 70.5^\circ = 0.33381$ and $\cos 70.6^\circ = 0.33216$. The tabular difference is 165, and the difference between $\cos \theta_r$ and $\cos 70.5^\circ$ is 48. The interpolating equation is

$$\frac{x}{0.1} = \frac{48}{165}$$

which gives $x = 0.029$. Hence, the reference angle is $\theta_r = 70.529^\circ$, and $\theta = 180^\circ - \theta_r = 109.471^\circ$.

EXERCISE 8.6

In Problems 1–12, read the function value from the appropriate table.

1 $\sin 0.53$ **2** $\cos 0.86$ **3** $\tan 1.44$

4 $\cot 1.35$ **5** $\cot 0.27$ **6** $\sin 0.64$

7 $\cot 38.5^\circ$ **8** $\sin 47.2^\circ$ **9** $\tan 67.4^\circ$

10 $\cos 75.3^\circ$ **11** $\tan 50.7^\circ$ **12** $\cos 36.8^\circ$

In some cases it is more convenient to change from radians to degrees and use Table A6 rather than to interpolate in Table A5. Use this procedure in Problems 13–18.

13 $\tan \frac{5\pi}{12}$ **14** $\sin \frac{3\pi}{8}$ **15** $\cos \frac{7\pi}{24}$

16 $\cot \frac{13\pi}{12}$ **17** $\tan \frac{2\pi}{5}$ **18** $\sin \frac{7\pi}{5}$

In Problems 19–30, use interpolation as necessary in the tables to find the indicated function value.

19 $\cos 0.532$ **20** $\sin 0.864$ **21** $\cot 1.435$

22 $\tan 0.373$ **23** $\sin 0.281$ **24** $\tan 1.094$

25 $\sin 30.24^\circ$ **26** $\cos 46.15^\circ$ **27** $\tan 50.26^\circ$

28 $\cos 61.33^\circ$ **29** $\sin 28.37^\circ$ **30** $\cot 81.24^\circ$

In Problems 31–40, find the number t to two decimal places for the given function value.

31 $\tan t = 0.17166, 0 \le t \le \pi$ **32** $\sin t = \frac{1}{5}, 0 \le t \le \frac{\pi}{2}$ **33** $\cos t = -\frac{3}{4}, 0 \le t \le \pi$

34 $\cot t = 2, \pi \le t \le 2\pi$ **35** $\tan t = 1.4573, 0 \le t \le \pi$ **36** $\cot t = -2.3746, -\frac{\pi}{2} \le t \le \frac{\pi}{2}$

37 $\sin t = -\frac{2}{3}, 5\pi \le t \le \frac{11\pi}{2}$ **38** $\cos t = 0.41371, 7\pi \le t \le 8\pi$ **39** $\tan t = 1.3826, 3\pi \le t \le 4\pi$

40 $\cot t = -0.98072, -\frac{\pi}{2} \le t \le \frac{\pi}{2}$

In Problems 41–50, find the angle θ to the nearest hundredth of a degree for the given function value.

41 $\sin \theta = 0.53921, 0^\circ \le \theta \le 90^\circ$ **42** $\tan \theta = -1.0917, 0^\circ \le \theta \le 180^\circ$ **43** $\cos \theta = 0.13001, 180^\circ \le \theta \le 360^\circ$

44 $\cot \theta = 2.4951, 180^\circ \le \theta \le 360^\circ$ **45** $\cos \theta = -0.70066, 0^\circ \le \theta \le 180^\circ$ **46** $\sin \theta = -0.60000, -90^\circ \le \theta \le 90^\circ$

47 $\tan \theta = 0.78834, 0^\circ \le \theta \le 180^\circ$ **48** $\cot \theta = -2.5000, 180^\circ \le \theta \le 360^\circ$ **49** $\cos \theta = -0.68711, 0^\circ \le \theta \le 180^\circ$

50 $\sin \theta = 0.92329, 0^\circ \le \theta \le 90^\circ$

In Problems 51–56, find all the solutions of the given equation in the stated interval. Obtain answers to two decimal places.

51 $2 \sin^2 t = 1, 0 \le t \le \pi$ **52** $\tan^2 t = 2, 0 \le t \le \pi$ **53** $2 \sin^2 t - 7 \sin t + 5 = 0, 0 \le t \le \frac{\pi}{2}$

54 $3 \tan^2 t + \tan t - 2 = 0, 0 \le t \le \pi$ **55** $2 \cos^2 t - 4 \sin t = 1, 0 \le t \le \pi$ **56** $\sec^2 t - 2 \tan t = 5, 0 \le t \le \pi$

57 Suppose that the horizontal distance x from a point A to the base of a tree is 50 feet and that the angle of elevation α to the top of the tree is 52.3°. Find the height y of the tree.

Hint: If the angle α is placed in standard position, then y/x is one of the trigonometric functions of α.

58 Repeat Problem 57 if the angle α is 62.4°

USING YOUR KNOWLEDGE 8.6

Many of the small hand-held calculators are equipped with trigonometric functions. If you have one of these, then it is an easy matter to obtain values of the sine, cosine, and tangent functions by simply pressing the proper keys. Values of the cosecant, secant, and cotangent functions can be obtained by taking reciprocals of the sine, cosine, and tangent, respectively. Some calculators are equipped to use degree, radian, or even centesimal measure; you simply have to put the calculator into the proper mode. Consult the instruction book for your calculator.

For the inverse problem, finding the number or the angle when the function value is given, the calculator has a key marked *inv* or *arc*. To find θ when $\sin \theta$ is given, you key in the value of $\sin \theta$, and then press the *inv* or *arc* key and the *sin* key in that order. For instance, to find θ when you are given $\sin \theta = 0.56931$, key in .56931, press the *inv* or *arc* key, and then press the *sin* key. In the degree mode, the display will show 34.7021240, and in the radian mode, 0.60566632. (Your calculator might give a different number of decimal places.) If your calculator does not have the radian mode, you will have to convert from degrees to radians by multiplying by π and dividing by 180. Of course, if the answer is to be in degrees, then no conversion is needed.

If the given function is the cosecant, the secant, or the cotangent, then you must first find the reciprocal and use the corresponding reciprocal function according to Definition 8.4a.

$$\frac{1}{\csc\theta} = \sin\theta, \quad \frac{1}{\sec\theta} = \cos\theta, \quad \text{and} \quad \frac{1}{\cot\theta} = \tan\theta$$

Thus, to find θ when you are given $\cot\theta = 0.56537$, key in the 0.56537, then press, in the stated order, the reciprocal key ($1/x$), the *arc* or *inv* key, and the *tan* key. Your calculator will read 60.517485 (degrees).

It is extremely important to keep in mind that for *positive* values of the functions, the calculator gives the corresponding angle between 0° and 90° or, in radians, the number between 0 and $\pi/2$. For *negative* values, the calculator gives angles between $-90°$ and 0° for the inverse of the sine or the tangent, and angles between 90° and 180° for the inverse of the cosine. In radian mode, the numbers corresponding to these angles are given. You may use reference angles as in Section 8.3, but the calculator will not do this for you automatically.

As an illustration, suppose you want to solve the equation

$$7\tan^2 t - 31\tan t - 19 = 0$$

for values of t between 0 and 2π radians. You can solve this quadratic equation for $\tan t$ and find t all in one procedure:

Algebraic logic (radian mode)

[31] [x^2] [+] [4] [×] [7] [×] [19] [=] [$\sqrt{x}$] [STO] [31] [+] [RCL] [=] [÷] [14]
[=] [inv] [tan] 1.3724051 (Copy this number.)
[+] [π] [=] 4.5139978 (Copy this number.)
[31] [−] [RCL] [=] [÷] [14] [=] [inv] [tan] (Display shows negative.)
[+] [π] [=] 2.6420811 (Copy this number.)
[+] [π] [=] 5.7836737 (Copy this number.)

The numbers you copied off are the answers, which you may round off to fewer decimal places if you wish. For example, 1.372, 2.642, 4.514, and 5.784 are correct to three decimal places. Notice that you used the periodicity of the tangent function when you added π.

If your calculator uses RPN logic and does not have a radian mode, the procedure is modified:

RPN logic (no radian mode)

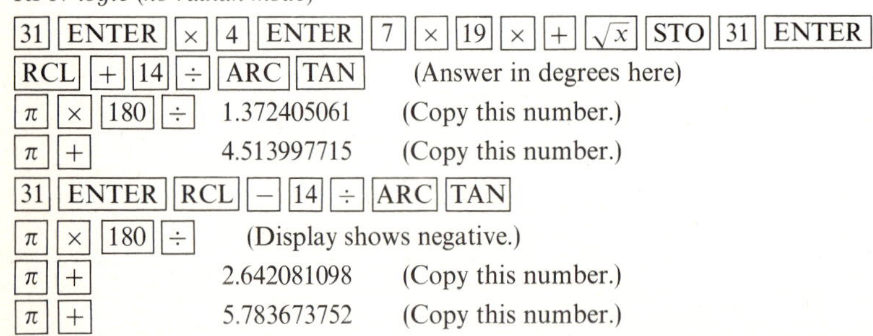

Note: You multiplied by π and divided by 180 to convert degrees to radians. You also used the periodicity of the tangent function when you added π.

C Use your calculator to do the following problems. In Problems 1–30, find the function value correct to five decimal places. (Be sure that your calculator is in the correct mode, degrees or radians.)

1 $\sin 0.532$ **2** $\cos 0.532$ **3** $\tan 0.532$ **4** $\cos 0.373$ **5** $\sin 0.281$ **6** $\tan 1.094$
7 $\cot 1.435$ **8** $\sec 1.257$ **9** $\csc 0.357$ **10** $\sin 3.76$ **11** $\cos 2.39$ **12** $\tan 1.72$
13 $\sin 30.24^\circ$ **14** $\cos 46.15^\circ$ **15** $\tan 50.26^\circ$ **16** $\sec 61.33^\circ$ **17** $\csc 28.37^\circ$ **18** $\cot 81.24^\circ$
19 $\sin 201.56^\circ$ **20** $\cos 137.81^\circ$ **21** $\tan 302.71^\circ$ **22** $\csc 125.73^\circ$ **23** $\sec 342.55^\circ$
24 $\cot 212.63^\circ$ **25** $\sin \frac{3\pi}{16}$ **26** $\cos \frac{7\pi}{24}$ **27** $\tan \frac{2\pi}{25}$ **28** $\cot \frac{13\pi}{24}$ **29** $\sec \frac{5\pi}{48}$ **30** $\csc \frac{37\pi}{96}$

In Problems 31–40, find the number t correct to two decimal places.

31 $\tan t = 0.17166, 0 \le t \le \pi$ **32** $\sin t = 1/5, 0 \le t \le \frac{\pi}{2}$ **33** $\cos t = -\frac{3}{4}, 0 \le t \le \pi$

34 $\cot t = 2, \pi \le t \le 2\pi$ **35** $\sec t = 1.4573, 0 \le t \le \pi$ **36** $\csc t = -2.3746, -\frac{\pi}{2} \le t \le \frac{\pi}{2}$

37 $\sin t = -\frac{2}{3}, 5\pi \le t \le \frac{11\pi}{2}$ **38** $\cos t = 0.41371, 7\pi \le t \le 8\pi$ **39** $\tan t = 1.3826, 3\pi \le t \le 4\pi$

40 $\cot t = -0.98072, -\frac{\pi}{2} \le t \le \frac{\pi}{2}$

In Problems 41–50, find the angle θ to the nearest hundredth of a degree.

41 $\sin \theta = 0.53921, 0^\circ \le \theta \le 90^\circ$ **42** $\tan \theta = -1.0917, 0^\circ \le \theta \le 180^\circ$
43 $\cos \theta = 0.13001, 180^\circ \le \theta \le 360^\circ$ **44** $\cot \theta = 2.4951, 180^\circ \le \theta \le 360^\circ$
45 $\cos \theta = -0.70066, 0^\circ \le \theta \le 180^\circ$ **46** $\sin \theta = -0.60000, -90^\circ \le \theta \le 90^\circ$
47 $\tan \theta = 0.78834, 0^\circ \le \theta \le 180^\circ$ **48** $\cot \theta = -2.5000, 180^\circ \le \theta \le 360^\circ$
49 $\sec \theta = -1.6871, 0^\circ \le \theta \le 180^\circ$ **50** $\csc \theta = 1.9232, 0^\circ \le \theta \le 90^\circ$

In Problems 51–60, find all the solutions of the given equation in the stated interval. Obtain answers to two decimal places.

51 $2\sin^2 t = 1, 0 \le t \le \pi$ **52** $\tan^2 t = 2, 0 \le t \le \pi$ **53** $2\sin^2 t - 7\sin t + 5 = 0, 0 \le t \le \frac{\pi}{2}$
54 $3\tan^2 t + \tan t - 2 = 0, 0 \le t \le \pi$ **55** $2\cos^2 t - 4\sin t = 1, 0 \le t \le \pi$
56 $\sec^2 t - 2\tan t = 5, 0 \le t \le \pi$ **57** $\sin^2 x = 3\cos x - 1, 0 \le \theta \le 2\pi$
58 $\sec^2 x + 5\tan x + 4 = 0, 0 \le \theta \le 2\pi$ **59** $2\cos x + \sin x + 1 = 0, 0 \le \theta \le 2\pi$
60 $\csc^2 x + 8\cot x + 12 = 0, 0 \le \theta \le 2\pi$

Now that we can use a calculator to find values of trigonometric functions for given angles and angles for given function values, we can do some more interesting problems involving Snell's Law of Refraction

$$\frac{\sin \alpha}{\sin \beta} = \mu$$

which was discussed in Using Your Knowledge 8.5. The critical angle of incidence for a given medium is the angle that corresponds to an angle of refraction of 90°. Light rays originating within the medium and striking the surface of the medium with angles of incidence greater than the critical angle are totally reflected; they do not leave the medium at all. For example, the index of refraction μ of water with respect to air is about $\frac{4}{3}$, so that the index of refraction of air with respect to water is $\frac{3}{4}$. Thus, if α is the angle of incidence of a ray originating in the water and β is the angle of refraction in the air, then

$$\frac{\sin \alpha}{\sin \beta} = \frac{3}{4} \qquad \text{and} \qquad \sin \beta = \tfrac{4}{3} \sin \alpha$$

If $\sin \alpha = \frac{3}{4}$, then $\sin \beta = 1$ and $\beta = 90°$. If $\sin \alpha > \frac{3}{4}$, then $\sin \beta > 1$, which is impossible. For $\sin \alpha = \frac{3}{4}$, your calculator gives $\alpha = 48.59°$, the critical angle for water. Rays of light originating in the water and striking the surface with angles of incidence greater than 48.59° are totally reflected from the surface back into the water; they do not enter the air at all.

61 Find the critical angle for benzene, for which the index of refraction with respect to air is about 1.5.

62 Find the critical angle for diamond, for which the index of refraction with respect to air is about 2.42. This critical angle is important to gem cutters, who wish to make a gem diamond as brilliant as possible. The bottom faces of the stone are cut at angles such that light rays inside the stone are totally reflected. Thus, rays of light entering the stone and striking the bottom surfaces are totally reflected internally and finally emerge through the upper surfaces of the diamond. This is what gives the cut stone such brilliance.

SELF-TEST

1 Find the period of:
a $\sin 5x$ **b** $\cos \pi x$

2 Without using your calculator, determine whether the following are positive or negative:
a $\cos 10$ **b** $\sin 932°$
c $\tan(-765°)$ **d** $\sec(-185°)$
e $\tan(-12)$

3 a If $\sin t = 1/\sqrt{2}$ and $\pi/2 < t < \pi$, what is the value of $\cos t$?
b What is the smallest positive value of t for which $\sin(-11\pi/6) = \sin t$?

4 Find the smallest positive value of t for which $\cos(34\pi/5) = \cos t$.

5 a What possible values can the difference in the degree measures of two coterminal angles have?
b Answer the same question for radian measure.

6 Sketch an angle of:
a 250° **b** $\frac{5\pi}{4}$

7 Convert:
a 210° to radians
b 3 radians to degrees

8 a If $(-2, 1)$ is a point on the terminal side of an angle θ, find $\sin \theta$, $\cos \theta$, and $\tan \theta$.
b If a point P on the terminal side of an angle θ is 5 units from the the origin, and the x-coordinate of P is 4, find $\sin \theta$, $\cos \theta$, and $\tan \theta$.

9 Give the value of each:
a $\sin 720°$ **b** $\cos 960°$ **c** $\tan 540°$

10 Give the value of each:
a $\cos\left(-\frac{11\pi}{4}\right)$ **b** $\sin \frac{11\pi}{6}$
c $\tan \frac{25\pi}{3}$

11 a What is the reference angle of 1226°?
b Express the sine, cosine, and tangent of 1226° in terms of the reference angle in (a).

12 Which of the following are undefined?
a $\tan 180°$ **b** $\sec \frac{\pi}{2}$ **c** $\csc \frac{3\pi}{2}$
d $\cot 360°$ **e** $\sin 2\pi$ **f** $\cos \frac{27\pi}{2}$

13 Give the exact value of each of the following:
a $\tan 210^\circ$ **b** $\sec 210^\circ$ **c** $\csc 240^\circ$
d $\sin 900^\circ$ **e** $\cos(-900^\circ)$
f $\cot(-900^\circ)$

14 Give the exact value of each of the following:

a $\tan \frac{7\pi}{3}$ **b** $\sec \frac{7\pi}{3}$ **c** $\csc\left(-\frac{11\pi}{6}\right)$

d $\cot\left(-\frac{17\pi}{6}\right)$ **e** $\sin\left(-\frac{7\pi}{6}\right)$

f $\cos\left(\frac{13\pi}{3}\right)$

15 For which of the following are there real values of t?
a $\sin t = 0.95$ **b** $\cos t = -1.35$
c $\tan t = 1000$ **d** $\csc t = 0.5$
e $\cot t = -0.2$ **f** $\sec t = -1.7$

16 Prove that
$\tan t + \cot t = \sec t \csc t$.

17 Find all the values of t for which $\cos^2 t = \frac{1}{4}$.

18 For $0 \le t \le 2\pi$, find the values of t for which $3 \csc^2 t = 4 \cot^2 t$.

19 Find t for each of the following:
a $\cos t = -0.76484$, $\pi < t < 2\pi$.
b $\sec t = 1.1357$, $\pi < t < 2\pi$.

20 Find θ for each of the following:
a $\tan \theta = -0.75$, $0^\circ < \theta < 180^\circ$.
b $\sin \theta = -0.53241$, $90^\circ < \theta < 270^\circ$.

CHAPTER 9

PROPERTIES OF THE TRIGONOMETRIC FUNCTIONS

9.1 GRAPHS OF THE SINE AND COSINE FUNCTIONS

In this section, we shall study the graphs of the sine and cosine functions, which are used in analyzing complex waveforms such as the cardiogram depicted in Figure 8.1a. In most of the applications of the graphs of the sine and cosine functions, we are concerned with the circular functions of numbers. In order to use the customary variables for graphing, we write

$$y = \sin x \quad \text{and} \quad y = \cos x$$

where x is a pure number (that may be interpreted as the radian measure of an angle at our convenience).

Of course, we can use tables or calculators to obtain points on the graph of the sine function; however, the accompanying table of values (Table 9.1a), by now quite familiar to you, will suffice to make a fairly accurate sketch. We confine ourselves to the interval from

TABLE 9.1a

x	0	$\pi/6$	$\pi/4$	$\pi/3$	$\pi/2$	π	$3\pi/2$
$y = \sin x$	0	$1/2$	$1/\sqrt{2}$	$\sqrt{3}/2$	1	0	-1
x (approx.)	0	0.52	0.79	1.05	1.57	3.14	4.71
y (approx.)	0	0.5	0.71	0.87	1	0	-1

$x = 0$ to $x = 2\pi$, because other sections of the graph are easily drawn by using the periodicity of the function. Reference numbers and the signs of the function in the various quadrants make it possible for us to manage with a table of values from $x = 0$ to $x = \pi/2$ only along with the values for $x = \pi$ and $x = 3\pi/2$.

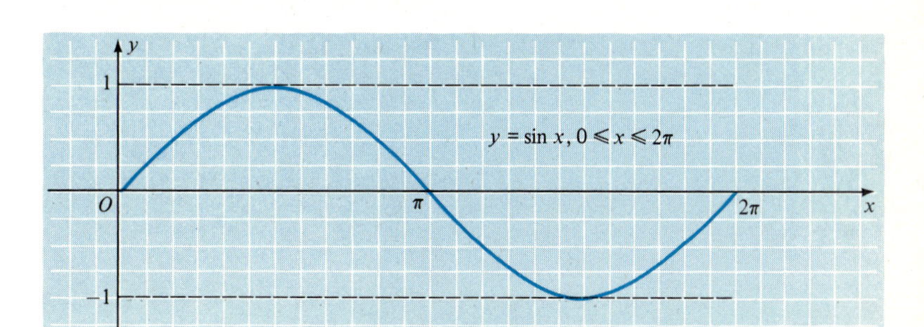

FIGURE 9.1a

Figure 9.1a shows the points obtained from Table 9.1a along with the corresponding points for values of x between $\pi/2$ and 2π. It is a little more convenient to select the units along the x-axis in terms of π, as is indicated in the figure. We have made use of the fact, proved in calculus, that the graph is a continuous, smooth curve.

You should notice carefully the salient features of the sine curve for the interval from $x = 0$ to $x = 2\pi$. This covers one period and gives one *cycle* of the sine function.

SALIENT FEATURES OF ONE CYCLE OF THE SINE CURVE

1. **The end points of this cycle are $(0, 0)$ and $(2\pi, 0)$.**
2. **The midpoint of the cycle is $(\pi, 0)$.**
3. **The quarter points of the cycle are $(\pi/2, 1)$ and $(3\pi/2, -1)$.**
4. **The curve rises from $(0, 0)$ to the first quarter point, falls to the midpoint, continues falling to the next quarter point, and then rises to the end point $(2\pi, 0)$.**

Figure 9.1b shows how these results follow from the circle definition of the sine function. It is worthwhile having this summary in mind

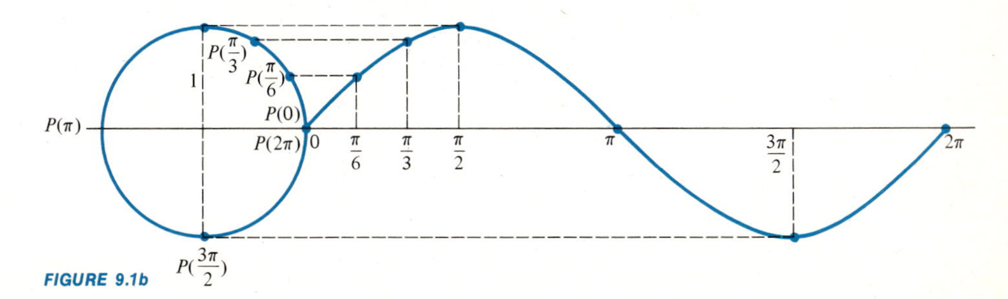

FIGURE 9.1b

because it enables you to make a fair sketch of the sine curve by plotting just the five points mentioned.

It is now an easy matter to see how the entire sine curve goes; we need use only the periodicity of the sine function to obtain the graph shown in Figure 9.1c from that of Figure 9.1a or 9.1b.

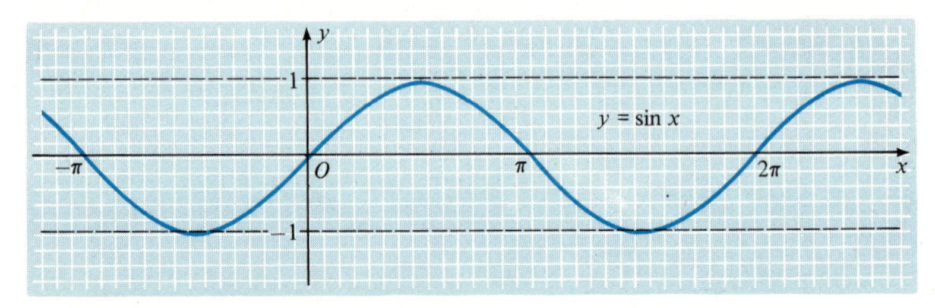

FIGURE 9.1c

The graph of the cosine function on the interval from 0 to 2π can be obtained by graphing points in the same way as for the sine function. You should verify that this procedure will result in the curve shown in Figure 9.1d. This figure shows one cycle of the cosine curve, and we can summarize the salient features just as we did for the sine function.

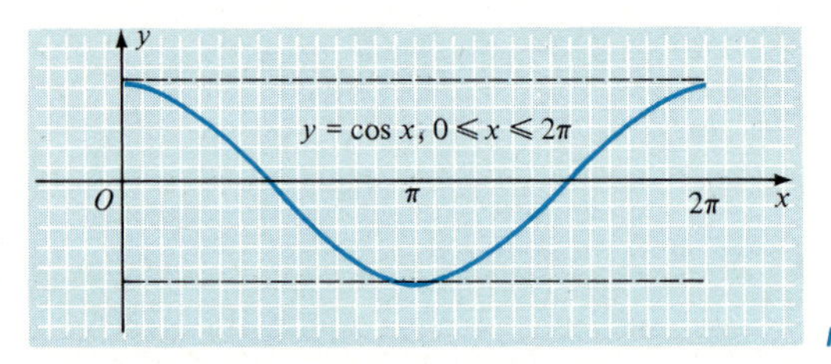

FIGURE 9.1d

SALIENT FEATURES OF ONE CYCLE OF THE COSINE CURVE

1. **The end points of this cycle are $(0, 1)$ and $(2\pi, 1)$.**
2. **The midpoint of the cycle is $(\pi, -1)$.**
3. **The quarter points of the cycle are $(\pi/2, 0)$ and $(3\pi/2, 0)$.**
4. **The curve falls from $(0, 1)$ to the first quarter point, continues to fall to the midpoint, rises to the next quarter point, and continues to rise to the end point $(2\pi, 1)$.**

Just as for the sine curve, you can make a fair sketch of the cosine curve by plotting only the five points named in the preceding summary. By using the periodicity of the cosine function, we obtain the graph in Figure 9.1e, which shows how the entire cosine curve goes.

Graphs that have the general shape of the sine curve in Figure 9.1c or of the cosine curve in Figure 9.1e are called *sine waves* or *sinusoids*. The graph over one period is called a *cycle* of the sinusoid, and one half the vertical distance from the lowest to the highest points on the curve is called the *amplitude*. Both the sine and cosine

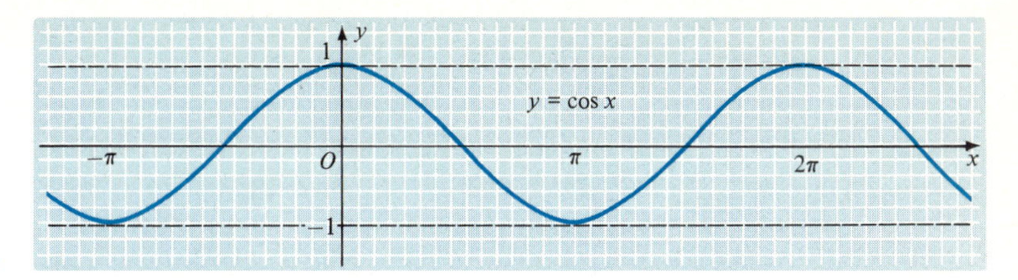

FIGURE 9.1e

curves given by the equations $y = \sin x$ and $y = \cos x$, respectively, have amplitude 1.

The graphs of equations of the form

$$y = A \sin(mx + b) \quad \text{and} \quad y = A \cos(mx + b)$$

where A, m, and b are constants with $A \neq 0$ and $m \neq 0$, are always sinusoids. For various values of A, m, and b, we obtain sine waves of various amplitudes and periods and variously placed with respect to the origin. In order to analyze such graphs, we consider the effect of simple modifications of the equation $y = \sin x$.

First, we consider the effect of multiplication of the sine function by the constant A. For each value of x, the value of $A \sin x$ is A times the y-value for $y = \sin x$, so that the amplitude of $y = A \sin x$ is $|A|$ times that of $y = \sin x$, which has amplitude 1. Hence, the amplitude of $y = A \sin x$ is $|A|$. Clearly, the graph of $y = A \cos x$ is similarly modified from that of $y = \cos x$.

EXAMPLE 1

Graph $y = 4 \sin x$, $-2\pi \leq x \leq 2\pi$.

SOLUTION:

We first sketch the graph of $y = \sin x$, $0 \leq x \leq 2\pi$, as a basis. Then, because the amplitude of $y = 4 \sin x$ is 4, we can sketch the graph we want by making each ordinate 4 times that of the corresponding ordinate of $y = \sin x$. Next, we can extend the graph over the required interval by using the periodicity of the sine function. The completed graph is shown in Figure 9.1f.

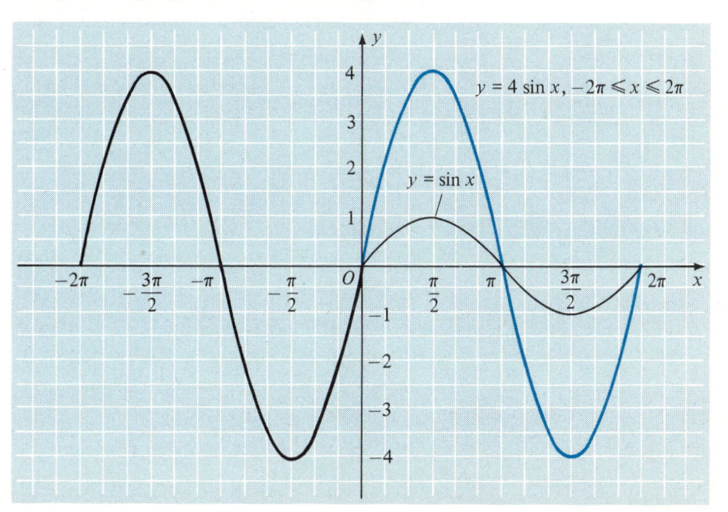

FIGURE 9.1f

We next consider the way in which the graph of

$$y = \sin mx \qquad m \neq 0$$

differs from that of the graph of $y = \sin x$. Clearly, the y-values in both cases range from -1 to 1, inclusive, so that both graphs have amplitude 1. Because one cycle of $y = \sin x$ is described as x varies from 0 to 2π, one cycle of $y = \sin mx$ will be described as mx varies from 0 to 2π, that is, as x varies from 0 to $2\pi/m$. Thus, the period of $y = \sin mx$ is $2\pi/m$. (We have assumed that m is a positive number, as it almost always is. However, if m is negative, then the period is properly written $2\pi/|m|$.) We see, then, that the graph of $y = \sin mx$, $m > 0$, is a sine wave with amplitude 1 and period $2\pi/m$, the distance along the x-axis in which the curve goes through one complete cycle.

EXAMPLE 2

Graph $y = \cos \frac{1}{2}x$, $-2\pi \leq x \leq 4\pi$.

SOLUTION:

We first sketch the graph of $y = \cos x$, $0 \leq x \leq 2\pi$, as a basis. Then, since the period of $y = \cos \frac{1}{2}x$ is

$$\frac{2\pi}{\frac{1}{2}} = 4\pi$$

we next sketch one cycle of $y = \cos \frac{1}{2}x$ over the interval $0 \leq x \leq 4\pi$. Finally, we use the periodicity of the cosine function to extend the graph over the required interval. The completed graph is shown in Figure 9.1g. The colored portion of the curve is the basic cycle

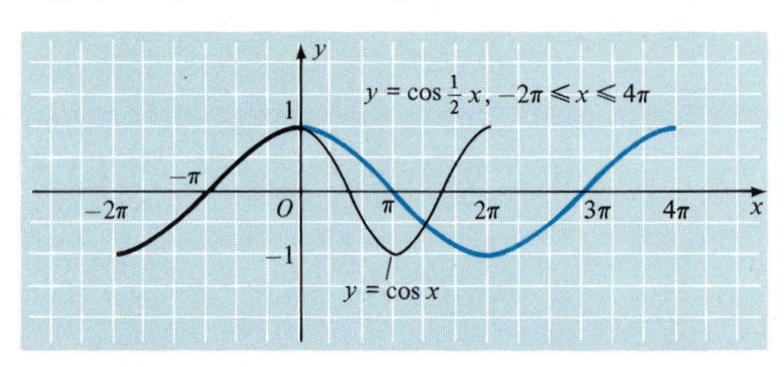

FIGURE 9.1g

extending over the interval $0 \leq x \leq 4\pi$. Note that you can easily sketch one cycle by dividing the interval from 0 to 4π into fourths and graphing the points where the function is -1, 0, or 1.

EXAMPLE 3

Graph $y = -3 \sin 2x$, $-\pi \leq x \leq 2\pi$.

SOLUTION:

We sketch the graph of $y = \sin x$, $0 \leq x \leq 2\pi$, as a basis. Because $A = -3$, the amplitude of the wave is $|A| = 3$. Furthermore, $m = 2$, so that the period is $2\pi/2 = \pi$. The graph of $y = -3 \sin 2x$ goes through two cycles in the distance that $y = \sin x$ goes through one cycle. Hence, we sketch the graph of $y = -3 \sin 2x$ over the interval $0 \leq x \leq \pi$, that is, through one cycle. This portion of the curve is shown in color in Figure 9.1h. The remainder of the graph is constructed by using the periodicity of the sine function.

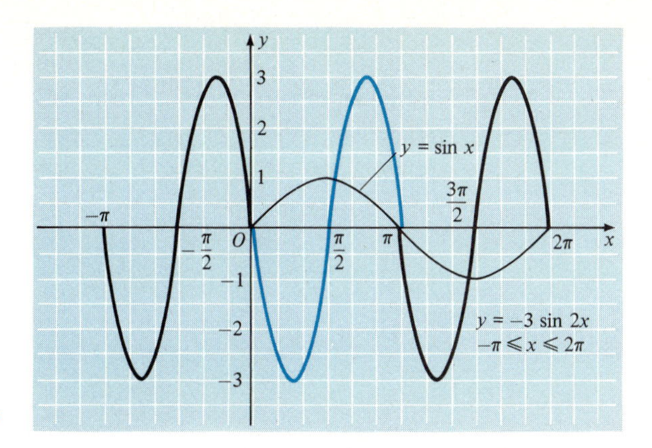

FIGURE 9.1h

You should note that the graph of $y = -f(x)$ can be obtained by turning the graph of $y = f(x)$ over the x-axis, so that the two graphs are symmetric to each other with respect to the x-axis. Thus, the graph of $y = -3 \sin 2x$ in Example 3 could be obtained in this way from the graph of $y = 3 \sin 2x$.

As a final step in our analysis, we examine the relationship between the graphs of $y = \sin(x + b)$ and $y = \sin x$. Of course, both curves have an amplitude of 1 and a period of 2π. Let us suppose that b is a positive number. If we select any value of x and write $x = x_1 - b$, then

$$y = \sin(x + b) = \sin(x_1 - b + b) = \sin x_1$$

This means that if we start at any point where $x = x_1$ on the graph of $y = \sin x$ and move b units to the *left*, we arrive at the corresponding point (same y-value) on the graph of $y = \sin(x + b)$. In Figure 9.1i, several points on the graph of $y = \sin x$ and the corresponding points on the graph of $y = \sin(x + b)$ are shown. Thus, if $b > 0$, the graph of $y = \sin(x + b)$ is simply the graph of $y = \sin x$ shifted to the left b units. In many applications, x represents the number of units of time, so that a shift of b units to the left means that it takes that much less time to reach the corresponding point. For this reason, the curve $y = \sin(x + b)$ is said to *lead* the curve

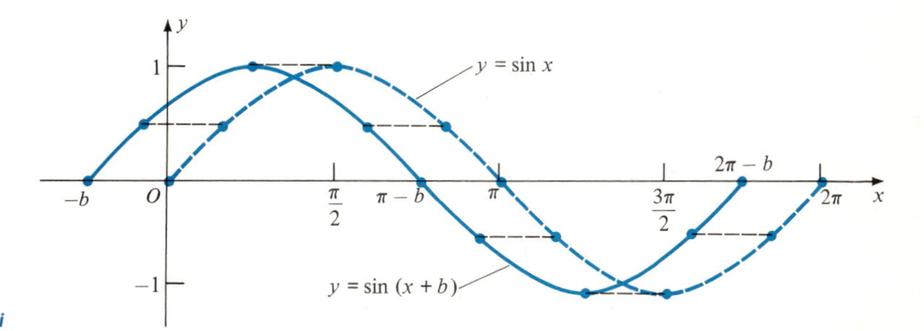

FIGURE 9.1i

$y = \sin x$. Similarly, if b is negative, the curve $y = \sin(x + b)$ is shifted to the right $|b|$ units from the curve $y = \sin x$ and is said to *lag* the curve $y = \sin x$. In both cases, the number $|b|$ is called the *phase shift* of the wave.

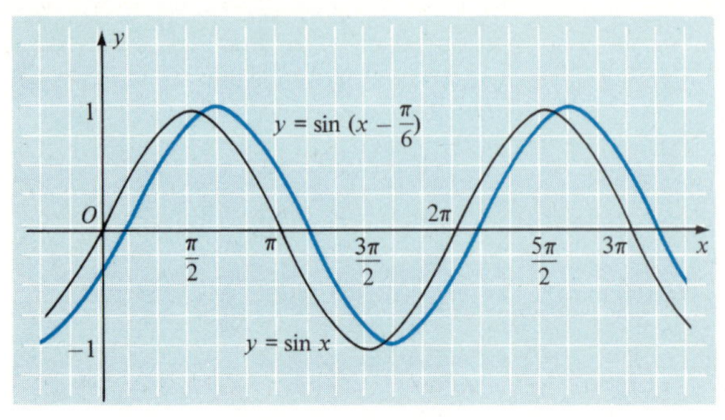

FIGURE 9.1j

Figure 9.1j shows the graphs of $y = \sin x$ and $y = \sin(x - \pi/6)$. As you can see, the figure displays the fact that the curves themselves are identical but are differently placed with respect to the origin. The graph of $y = \sin(x - \pi/6)$ can be obtained by shifting the graph of $y = \sin x$ to the right $\pi/6$ units. Thus, the phase shift is $\pi/6$, and the graph of $y = \sin(x - \pi/6)$ lags that of $y = \sin x$ by $\pi/6$ units.

Just as we were able to construct the graph of $y = \sin mx$ from that of $y = \sin x$, we can construct the graph of $y = \sin m(x + c)$ from that of $y = \sin(x + c)$. In both cases, the multiplication by m changes the period from 2π to $2\pi/m$. Hence, to analyze an equation of the form $y = A \sin(mx + b)$ or $y = A \cos(mx + b)$, we first rewrite the $mx + b$ in the form $m(x + b/m)$. This idea is used in Example 4, which illustrates the effect that the three constants A, m, and b have on the graph of $y = A \sin(mx + b)$ or $y = A \cos(mx + b)$.

EXAMPLE 4

Sketch the graph of $y = 2 \cos(2x - \pi/3)$.

SOLUTION:

We see at once that the amplitude of this wave is 2, because the constant $A = 2$. To use the information about the constants m and b, we first rewrite the equation in the form

$$y = 2 \cos 2\left(x - \frac{\pi}{6}\right)$$

Now, we can immediately read off the following facts about the graph:

1. It is a sinusoid of amplitude 2.
2. It has the period $2\pi/2 = \pi$.
3. It lags the graph of $y = 2 \cos 2x$ by $\pi/6$ units.

Using these facts, we can quickly sketch the graph shown in Figure 9.1k.

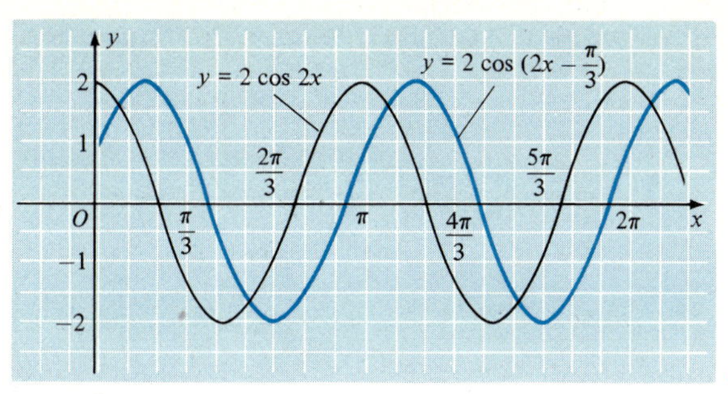

FIGURE 9.1k

EXAMPLE 5

Sketch the graph of $y = \frac{1}{4} \sin \pi x$.

SOLUTION:

By inspection of the equation, we can read off the following facts about the graph:

1. It is a sinusoid of amplitude $\frac{1}{4}$.
2. It has the period $2\pi/\pi = 2$.

We take advantage of the fact that the period is 2 by scaling the x-axis in integer units, which simplifies sketching the graph in Figure 9.1l.

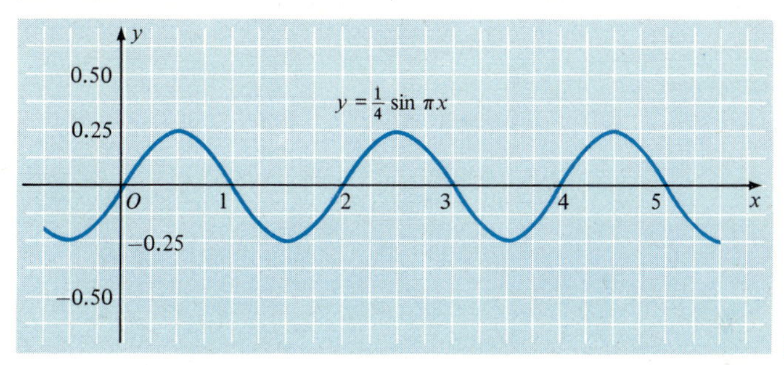

FIGURE 9.1l

EXERCISE 9.1

In Problems 1–18, sketch the graph of the given equation over the interval $-2\pi \le x \le 2\pi$.

1 $y = 2 \cos x$

2 $y = 4 \sin x$

3 $y = \frac{3}{2} \sin x$

4 $y = -\frac{1}{2} \cos x$

5 $y = -2 \sin x$

6 $y = \frac{5}{4} \cos x$

7 $y = \cos 2x$

8 $y = \sin 4x$

9 $y = \sin \frac{2}{3}x$

10 $y = 2 \cos \frac{1}{2}x$

11 $y = -\frac{3}{2} \sin 2x$

12 $y = -2 \cos 3x$

13 $y = \cos\left(x + \frac{\pi}{2}\right)$

14 $y = \sin\left(x - \frac{\pi}{4}\right)$

15 $y = 2 \cos\left(x - \frac{\pi}{2}\right)$

16 $y = 2 \sin\left(x + \frac{\pi}{3}\right)$

17 $y = 2 \cos\left(2x - \frac{5\pi}{3}\right)$

18 $y = \frac{7}{4} \sin\left(3x - \frac{3\pi}{4}\right)$

In Problems 19–24, sketch the first two cycles of the graph starting at $x = 0$ and going to the right of the y-axis.

19 $y = 3 \cos \pi x$

20 $y = -\frac{1}{2} \sin \frac{\pi x}{4}$

21 $y = \sin \frac{\pi x}{3}$

22 $y = -\cos \dfrac{\pi x}{6}$

23 $y = 2 \cos\left(\dfrac{\pi x}{2} + \dfrac{\pi}{6}\right)$

24 $y = -\sin\left(\pi x - \dfrac{\pi}{4}\right)$

In Problems 25–30, inspect the graph of the indicated preceding problem and find the values of x for which $y = 0$ (the zeros of the function) on the specified interval.

25 Problem 8

26 Problem 9

27 Problem 14

28 Problem 15

29 Problem 21

30 Problem 24

31 At the beginning of Section 8.3, the equations of the cycloid were given. Take $a = 2$, so that the equations become

$$x = 2(t - \sin t) \qquad y = 2(1 - \cos t)$$

Make a three-column table of values with headings t, x, and y for $0 \leq t \leq \pi$ at intervals of $\pi/8$, that is, for $t = 0, \pi/8, 2\pi/8, 3\pi/8$, etc. The first lines of your table should look like this:

t	x	y
0	0	0
$\pi/8$	0.02	0.15
$2\pi/8$	0.16	0.59

Plot the points (x, y) that you get from this table and draw a smooth curve through these points. You should end up with half of the first arch of the cycloid, and you can sketch the rest by symmetry.

USING YOUR KNOWLEDGE 9.1

The graph of an equation of the form

$$y = f(x) + g(x)$$

can often be easily constructed by the method of *addition of ordinates*. We shall demonstrate this method by construction of the graph of

$$y = \sin x + \cos 2x \qquad 0 \leq x \leq 2\pi$$

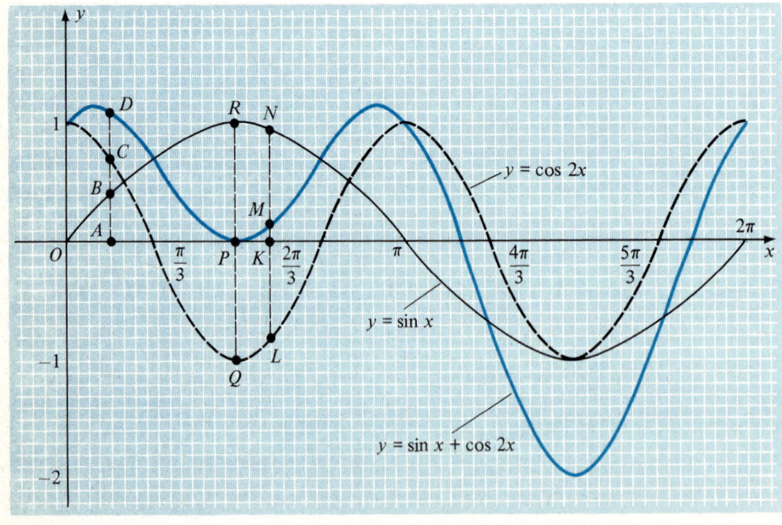

FIGURE 9.1m

Figure 9.1m shows the graph of $y = \sin x$ as a solid black curve and the graph of $y = \cos 2x$ as a dashed black curve, drawn on the same coordinate axes. The colored curve is the graph of $y = \sin x + \cos 2x$. To obtain points on this final curve, we added corresponding ordinates of the two component curves.

There is an important caution to be observed here. These ordinates are directed line segments, and the direction must be taken into account. For instance, the point D is obtained by adding the directed segments $\overrightarrow{AB}$ and $\overrightarrow{AC}$; that is

$$\overrightarrow{AD} = \overrightarrow{AB} + \overrightarrow{AC}$$

(The half-arrows above the letter designation are to emphasize that the segment is directed from the first to the second named point. Thus, $\overrightarrow{AC}$ signifies the segment directed from A to C.) Similarly, the point P is found by adding the directed segments $\overrightarrow{PQ}$ and $\overrightarrow{PR}$, which are equal in length but opposite in sign; the point N is obtained by adding $\overrightarrow{KL}$ and $\overrightarrow{KM}$ and noting that these are oppositely directed with the positive segment $\overrightarrow{KM}$ longer than the negative segment $\overrightarrow{KL}$. You can check some points if you wish by using the tables or a calculator. For example, at $x = \pi/3$,

$$\begin{aligned} y &= \sin\frac{\pi}{3} + \cos\frac{2\pi}{3} \\ &= \frac{\sqrt{3}}{2} + \left(-\frac{1}{2}\right) \\ &\simeq 0.37 \end{aligned}$$

Use addition of ordinates to sketch the graphs of the following equations from $x = 0$ to $x = 2\pi$.

1 $y = \sin x + \cos x$ **2** $y = \cos x - \sin x$ **3** $y = \sin 2x + \cos x$ **4** $y = \sin 2x - 2\cos x$
5 $y = 2\cos x - \frac{1}{2}\sin 2x$ **6** $y = \sin x - \frac{1}{4}\cos 4x$ **7** $y = x - 2\sin x$ **8** $y = x + \cos x$
9 $y = \frac{1}{2}x - \cos 2x$ **10** $y = x - 2 + \sin 2x$

9.2 GRAPHS OF OTHER CIRCULAR FUNCTIONS

One of the details involved in solving certain heat transfer problems is to find the roots of an equation of the form

$$\tan x = \frac{h}{x}$$

where h is a positive constant. There is no formula method for obtaining the solutions of such equations, and various methods of approximating the roots are used. One very simple method consists of graphing $y = \tan x$ and $y = h/x$ on the same coordinate axes and reading off the x-coordinates of the points of intersection. Can you see why these values of x are solutions of the original equation?

In order to carry out the graphical method just described, it is, of course, necessary to be able to graph the equation $y = \tan x$. This problem, along with that of graphing the cotangent, secant, and cosecant functions is easily handled by making use of the definitions of these functions in terms of the sine and cosine functions:

$$\tan x = \frac{\sin x}{\cos x} \qquad \cot x = \frac{\cos x}{\sin x} \qquad \sec x = \frac{1}{\cos x} \qquad \csc x = \frac{1}{\sin x}$$

Let us discuss the equation $y = \tan x$ first. The general behavior of the graph can be obtained by referring to the graphs of the sine and cosine functions. Because $\sin 0 = 0$ and $\cos 0 = 1$, $\tan 0 = 0$. As x increases from 0 to $\pi/2$, $\sin x$ increases from 0 to 1, and $\cos x$ decreases from 1 to 0. Thus, $\tan x$ increases as x increases; $\tan x$ reaches the value 1 for $x = \pi/4$, $\sqrt{3}$ for $x = \pi/3$, over 57 for $x = 89\pi/180 \simeq 1.5533$, more than 1,255 for $x = 1.57$, and continues to increase without bound as x gets closer and closer to $\pi/2$. (A calculator gives 1.5707963 for the value of $\pi/2$, and tan 1.5707 is over 10,000; tan 1.57079 is over 150,000; tan 1.570796 is over 3,000,000; etc.) This discussion shows that the graph of $y = \tan x$ approaches the line $x = \pi/2$ more and more closely as x itself gets closer and closer to $\pi/2$. A line that is approached by a graph in this fashion is called an *asymptote* to the graph. Thus, the line $x = \pi/2$ is a vertical asymptote to the graph of $y = \tan x$. We use this last fact, plot a few points and draw a smooth curve through them to obtain the graph over the interval 0 to $\pi/2$ as shown in Figure 9.2a. The remainder of the graph is obtained by first noting that

$$\tan(-x) = \frac{\sin(-x)}{\cos(-x)} = \frac{-\sin x}{\cos x} = -\tan x$$

This means that the graph is symmetric with respect to the origin; that is, every line through the origin that cuts the graph, cuts it in two points on opposite sides of and equidistant from the origin. This symmetry allows us to sketch the graph from $-\pi/2$ to 0 as shown in Figure 9.2a. Once we have the graph over the interval $-\pi/2$ to $\pi/2$, we can use the periodicity of the tangent function to

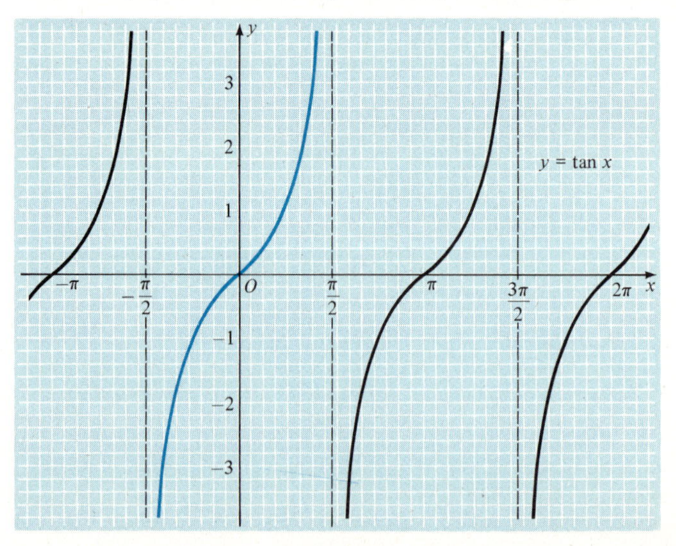

FIGURE 9.2a

draw the rest of the graph. (Recall that the period of the tangent function is π.) Notice that the lines $x = \pm\pi/2$, $\pm 3\pi/2$, $\pm 5\pi/2$, and so on, are all asymptotes to the graph. This fact, along with the symmetry, should enable you to draw a sketch of the tangent function quickly with very little point plotting.

For many applications, the most important portion of the tangent curve is the portion between $x = -\pi/2$ and $x = \pi/2$. As we noted earlier, this is the interval that is used by hand-held calculators to give the inverses of values of the tangent function. As in the case of the sine and cosine functions, you should note the salient features of one cycle of the tangent function. We shall consider the cycle to be over the interval from $-\pi/2$ to $\pi/2$. (See the portion in color, Figure 9.2a.)

SALIENT FEATURES OF ONE CYCLE OF THE TANGENT CURVE

1. **The curve has a vertical asymptote at each end of the cycle that extends from $x = -\pi/2$ to $x = \pi/2$.**
2. **The midpoint of the cycle is (0, 0).**
3. **The quarter points are $(-\pi/4, -1)$ and $(\pi/4, 1)$.**
4. **The curve is asymptotic to the negative portion of $x = -\pi/2$, rises steadily, and is asymptotic to the positive portion of $x = \pi/2$.**

The definition of the cotangent shows that it is the reciprocal of the tangent. Thus, the graph of $y = \cot x$ can be obtained from that of $y = \tan x$. At the points where $\tan x = 0$, $\cot x$ is not defined, and at the points where $\tan x$ is not defined, $\cot x = 0$. The vertical lines at odd multiples of $\pi/2$ are asymptotes of the graph of the tangent function; those at multiples of π are asymptotes of the graph of the cotangent function. The graph appears in Figure 9.2b.

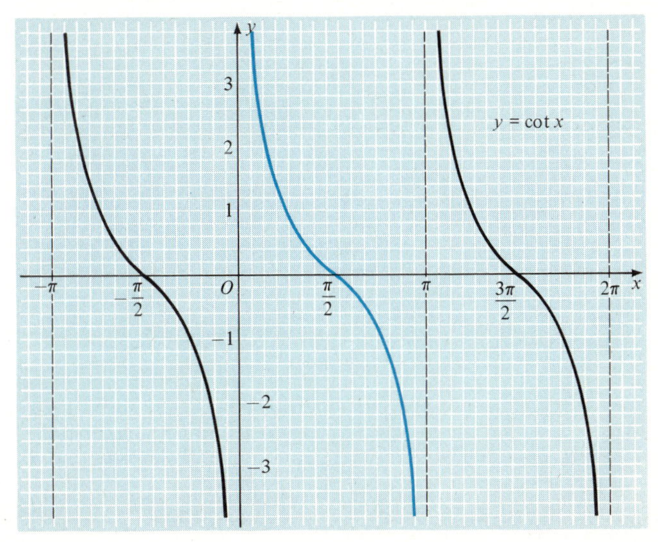

FIGURE 9.2b

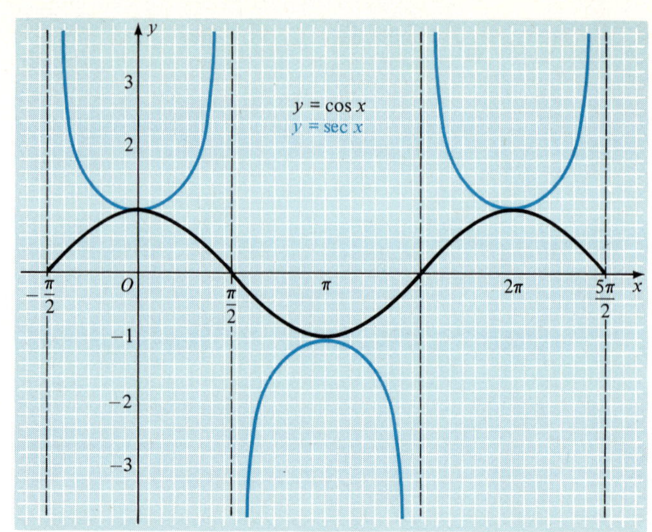

FIGURE 9.2c

The graphs of $y = \sec x$ and $y = \csc x$ can be obtained in a similar way. Figure 9.2c shows the graphs of the cosine function and its reciprocal, the secant function. It is proved in calculus that these graphs are tangent to each other at multiples of π, the points where both functions have absolute value 1. The vertical lines at odd multiples of $\pi/2$ are asymptotes to the graph of $y = \sec x$.

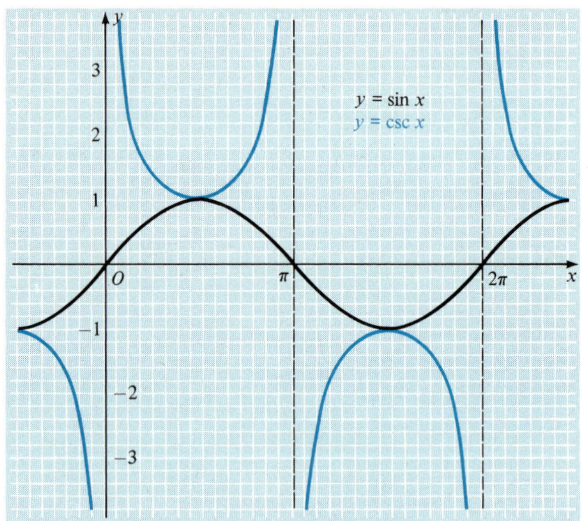

FIGURE 9.2d

Figure 9.2d shows the graphs of the sine function and its reciprocal, the cosecant function. These graphs are tangent to each other at odd multiples of $\pi/2$, where both functions have the value 1 or -1. The vertical lines at multiples of π are asymptotes to the graph of $y = \csc x$.

If we replace x by $ax + b$ and multiply the functions in the preceding discussion by a constant A, we obtain effects corresponding to those we saw in the case of the sine and cosine functions. The next two examples are illustrations of what can happen.

EXAMPLE 1

Sketch the graph of $y = -\tan x/2$ over the integral from $-\pi$ to π.

SOLUTION:

Because the period of $\tan x/2$ is $\pi/\frac{1}{2} = 2\pi$, we are being asked to graph one cycle of the curve.

1. The curve has the vertical asymptotes $x = -\pi$ and $x = \pi$.
2. The midpoint of the cycle is $(0, 0)$.
3. The quarter points are $(-\pi/2, 1)$ and $(\pi/2, -1)$.
4. The curve is asymptotic to the positive portion of the line $x = -\pi$, falls steadily, and is asymptotic to the negative portion of the line $x = \pi$.

The graph is shown in Figure 9.2e. If you compare this with the graph of $y = \tan x$, you will see that the effect of the minus sign is to turn the graph over the x-axis.

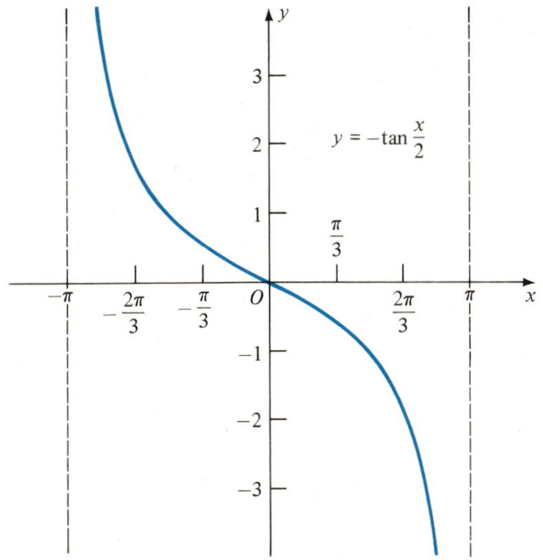

FIGURE 9.2e

EXAMPLE 2

Graph $y = \tan(x/2 - \pi/4)$ over the interval from $-\pi/2$ to $3\pi/2$.

SOLUTION:

Because the period of $\tan x$ is π, the period of

$$\tan\left(\frac{x}{2} - \frac{\pi}{4}\right) = \tan\frac{1}{2}\left(x - \frac{\pi}{2}\right)$$

is 2π, just as in Example 1. Hence, we are to sketch the graph over one period. At the ends of the interval, where $x = -\pi/2$ and $x = 3\pi/2$, we find

$$\frac{x}{2} - \frac{\pi}{4} = -\frac{\pi}{2} \quad \text{and} \quad \frac{x}{2} - \frac{\pi}{4} = \frac{\pi}{2}$$

respectively. Thus, the graph has vertical asymptotes at the ends of the interval. For $x = 0$, we obtain $y = \tan(-\pi/4) = -1$, and for $x = \pi$, $y = \tan(\pi/4) = 1$. For $x = \pi/2$, $y = \tan 0 = 0$. With these three points and the asymptotes, we can make a fair sketch of the graph. See Figure 9.2f.

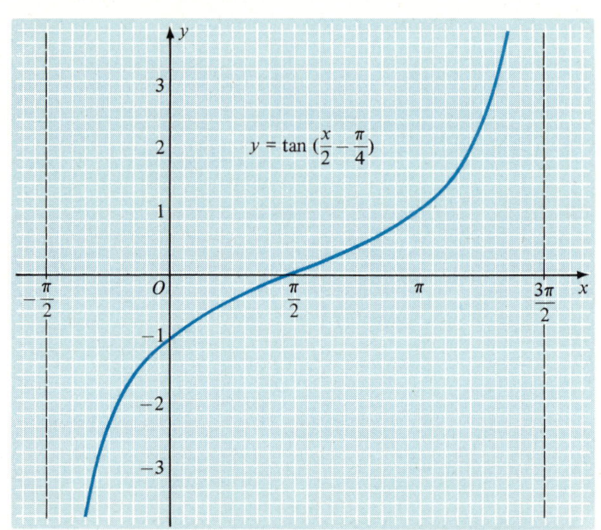

FIGURE 9.2f

EXAMPLE 3

Graph $y = 2 \sec(2x - \pi/3)$ over the interval from 0 to 2π.

SOLUTION:

We first sketch the graph of $y = 2 \cos(2x - \pi/3)$ to use as a reference. This graph was constructed in Example 4 of Section 9.1 (see Figure 9.1k). The period of the cosine and of the secant here is $2\pi/2$ or π. The graph of the secant will be tangent to that of the cosine at the points where $\cos(2x - \pi/3) = \pm 1$, that is, where $2x - \pi/3 = n\pi$, n an integer. This gives $x = n\pi/2 + \pi/6$, and the values between 0 and 2π are $\pi/6$, $2\pi/3$, $7\pi/6$, and $5\pi/3$. The vertical asymptotes of the secant graph will occur at the points where $\cos(2x - \pi/3) = 0$, that is, where $2x - \pi/3 = (2n + 1)\pi/2$, n an integer. This gives $x = (2n + 1)(\pi/4) + \pi/6$, and the values in the interval of interest are $5\pi/12$ for $n = 0$, $11\pi/12$ for $n = 1$, $17\pi/12$ for $n = 2$, and $23\pi/12$ for $n = 3$. The graph is shown in Figure 9.2g.

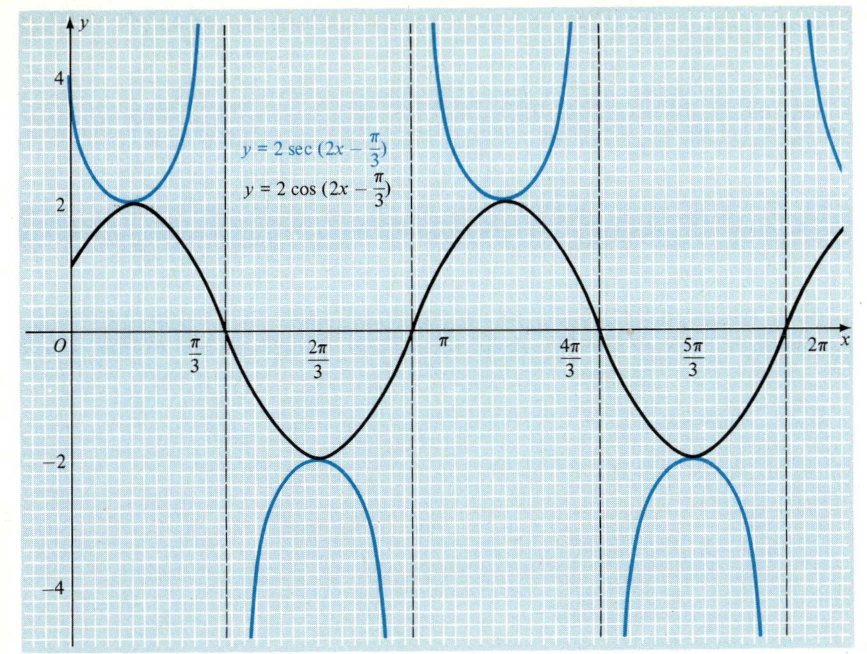

FIGURE 9.2g

EXERCISE 9.2

In Problems 1–10, make a graph of the given equation over the interval $0 \le x \le 2\pi$.

1 $y = \tan 2x$

2 $y = \sec \frac{x}{2}$

3 $y = \cot \frac{x}{2}$

4 $y = \csc 2x$

5 $y = 2\tan \frac{x}{2}$

6 $y = -3\sec 2x$

7 $y = -\frac{1}{2}\csc \frac{x}{2}$

8 $y = -\frac{1}{3}\cot 3x$

9 $y = \tan 4x$

10 $y = -\sec 4x$

In Problems 11–18, make a graph of the given equation over the indicated interval.

11 $y = \tan\left(x + \frac{\pi}{4}\right), -\frac{\pi}{4} \le x \le \frac{7\pi}{4}$

12 $y = \cot\left(x - \frac{\pi}{4}\right), -\frac{\pi}{4} \le x \le \frac{7\pi}{4}$

13 $y = \tan\left(x - \frac{\pi}{3}\right), -\frac{\pi}{6} \le x \le \frac{5\pi}{6}$

14 $y = \cot\left(\frac{\pi}{6} - x\right), -\frac{\pi}{3} \le x \le \frac{2\pi}{3}$

15 $y = \sec\left(x - \frac{\pi}{4}\right), -\frac{\pi}{4} \le x \le \frac{7\pi}{4}$

16 $y = \csc\left(\frac{\pi}{6} - x\right), 0 \le x \le 2\pi$

17 $y = \csc\left(2x + \frac{\pi}{3}\right), 0 \le x \le \pi$

18 $y = \sec\left(2x - \frac{\pi}{3}\right), 0 \le x \le \pi$

19 Make a graph of the function described by the following equations for $-\pi/2 < x < 3\pi/2$: $y = \sec x - \tan x$ for $x \ne \pi/2$ and $y = 0$ for $x = \pi/2$.

Note: Although $\sec x - \tan x$ is not defined for $x = \pi/2$, it is proved in calculus that for x near this value, $\sec x - \tan x$ is near zero.

20 Graph the function described by the following equations for $-\pi < x < \pi$: $y = \csc x - \cot x$ for $x \neq 0$ and $y = 0$ for $x = 0$. See the note in Problem 19. A similar situation exists here.

21 Find the value, correct to one decimal place, of the smallest positive root of the equation $\tan x = 1/x$.

Hint: Make a graph of $y = \tan x$ and $y = 1/x$ on the same set of axes and read the value of x at the point of intersection. You should use graph paper with 10 divisions to the inch and let 1 inch represent 1 unit on the x-axis.

22 Do the same as in Problem 21 for the equation $\tan x = 2/x$.

USING YOUR KNOWLEDGE 9.2

At the beginning of Section 9.2, the question of solving equations of the type $\tan x = h/x$ was raised, and in Problems 21 and 22 of the preceding exercise, you were asked to use a graphical method to find an approximation of a root of such an equation. Although the graphical method is quite useful, it is also quite limited in its accuracy. For instance, if you had to find, correct to three decimal places, the smallest positive root of $\tan x = 1/(2x)$, the graphical method would be too inefficient to use. However, this method does give us a good starting point for a much better method of approximation.

In Figure 9.2h, you see the graphs of $y = \tan x$ and $y = 1/(2x)$. The point of intersection appears to be where x is slightly less than 0.7. Because we are looking for a nonzero root, let us multiply by $2x$ and write the equation in the form

$$2x \tan x - 1 = 0$$

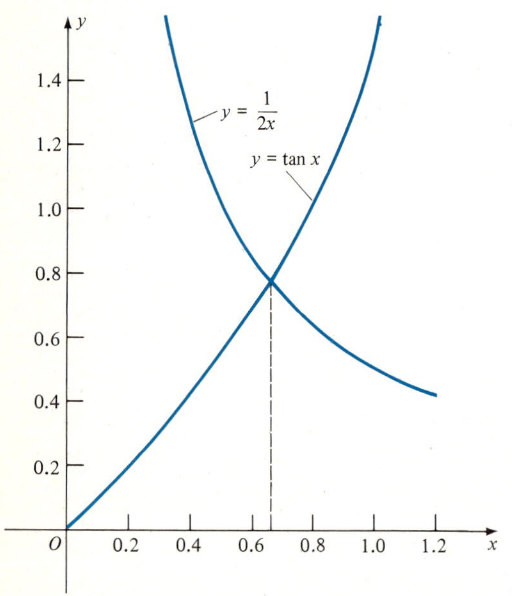

FIGURE 9.2h

For convenience, we shall put $f(x) = 2x \tan x - 1$. It can be shown by calculus methods that the graph of f for $0 < x < \pi/2$ is a continuous curve. We shall accept and use this continuity here. Now, use your calculator, *radian mode*, to evaluate $f(0.6)$. You should get about -0.18. Next, evaluate $f(0.7)$ to get about $+0.18$. Since the graph of $f(x)$ is continuous, it must cross the x-axis somewhere between $x = 0.6$ and $x = 0.7$. (This is because the graph is *below* the axis at $x = 0.6$ and *above* the axis at $x = 0.7$.) Furthermore, from the fact that $f(0.6) \approx -0.18$ and $f(0.7) \approx +0.18$, it seems that $x = 0.65$ would be a good guess for the root.

Now, use your calculator to evaluate $f(0.65)$. You should get about -0.012. This is negative, which shows that the root is greater than 0.65. (There is a change of sign in the value of $f(x)$ from $x = 0.65$ to $x = 0.7$.) Next, evaluate $f(0.66)$ to get about $+0.024$. Now you know that the root is between 0.65 and 0.66. From the values of $f(x)$ at these two points, we guess the root is about 0.653. Use your calculator to evaluate $f(0.653)$. You should get about -0.0010, showing that the root is greater than 0.653. Next, evaluate $f(0.654)$ to get about 0.0026, which indicates the root is between 0.653 and 0.654. All you need do now is determine which of these two numbers is the root correct to three decimal places. This determination is easy. Just evaluate $f(0.6535)$ to get about $+0.00082$, showing that the root is between 0.653 and 0.6535, so that 0.653 is correct to three decimal places.

Although it takes many words to describe the process we have used, it is actually quite simple, as Table 9.2a shows. All it takes is a little intelligent guesswork and a hand calculator.

TABLE 9.2a

	x	$f(x)$	
Root here →	0.6 0.7	-0.18 $+0.18$	$\frac{18}{36} = 0.5$
Root here →	0.65 0.66	-0.012 $+0.024$	$\frac{12}{36} \approx 0.3$
Root here →	0.653 0.654	-0.0010 $+0.0026$	
	0.6535	$+0.00082$	← Shows answer is 0.653.

Use your calculator to find, correct to three decimal places, the smallest positive root of the given equation.

C **1** $\tan x = \dfrac{1}{x}$ C **2** $\tan x = \dfrac{2}{x}$ C **3** $x \sin x = 0.5$ C **4** $x \cos x = 0.5$

9.3 THE ADDITION FORMULAS

A common error made by beginning students in trigonometry stems from their failure to realize that

$$\cos(s + t) \neq \cos s + \cos t$$

It is very easy to verify this inequality by substituting simple values for the variables. Thus, $s = t = 0$ gives $\cos 0 = 1$ on the left and $\cos 0 + \cos 0 = 2$ on the right; or $s = \pi/6$, $t = \pi/3$ gives $\cos \pi/2 = 0$ on the left and $\cos \pi/6 + \cos \pi/3 = \sqrt{3}/2 + \frac{1}{2} = (\sqrt{3} + 1)/2 \neq 0$ on the right.

In this section, we shall obtain the formulas for $\cos(s + t)$, $\cos(s - t)$, $\sin(s + t)$, and $\sin(s - t)$. These formulas are among the most important formulas in trigonometry, and they are surprisingly

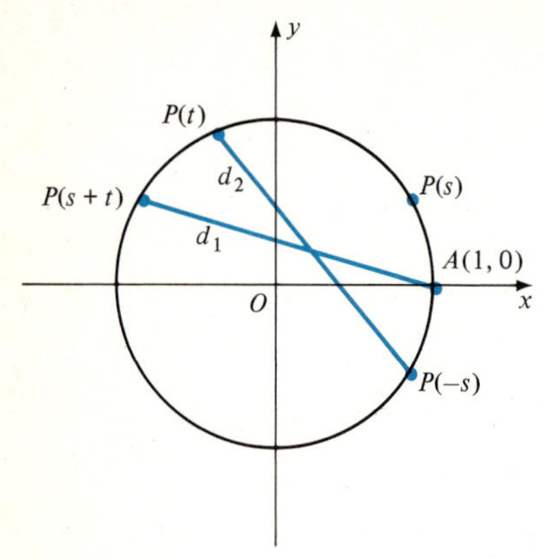

FIGURE 9.3a

easy to derive. The basic tool that is needed is the formula that we obtained in Section 5.1 for the distance between two points (x_1, y_1) and (x_2, y_2) in the plane:

$$d = \sqrt{(x_2 - x_1)^2 + (y_2 - y_1)^2}$$

We shall first derive the formula for $\cos(s + t)$. Figure 9.3a shows the four points $P(s)$, $P(t)$, $P(-s)$, and $P(s + t)$. You should recall that these points are found by measuring off, on the circumference of the unit circle, the appropriate arc lengths starting from the point $A(1, 0)$. We list the coordinates of the four points for use in the distance formula.

$$\begin{aligned} P(s) &= (\cos s, \sin s) \\ P(t) &= (\cos t, \sin t) \\ P(-s) &= [\cos(-s), \sin(-s)] = (\cos s, -\sin s) \\ P(s + t) &= [\cos(s + t), \sin(s + t)] \end{aligned}$$

Because the arc length from A to the point $P(s + t)$ is $s + t$, the arc length from $P(-s)$ to A is s, and the arc length from A to $P(t)$ is t, we see that the arc length from A to $P(s + t)$ is equal to the arc length from $P(-s)$ to $P(t)$. Therefore, the straight line distance, say d_1, from A to $P(s + t)$ is equal to the straight line distance, say d_2, from $P(-s)$ to $P(t)$. Hence, the squares of these distances are equal, and we have

$$d_1^2 = d_2^2$$

By using the distance formula and the basic identity

$$\cos^2 \theta + \sin^2 \theta = 1$$

we obtain

$$\begin{aligned} d_1^2 &= [\cos(s + t) - 1]^2 + [\sin(s + t) - 0]^2 \\ &= \cos^2(s + t) - 2\cos(s + t) + 1 + \sin^2(s + t) \\ &= 2 - 2\cos(s + t) \end{aligned}$$

$$\begin{aligned} d_2^2 &= [\cos t - \cos(-s)]^2 + [\sin t - \sin(-s)]^2 \\ &= (\cos t - \cos s)^2 + (\sin t + \sin s)^2 \\ &= \cos^2 t - 2\cos s \cos t + \cos^2 s + \sin^2 t + 2\sin s \sin t + \sin^2 s \\ &= 2 - 2(\cos s \cos t - \sin s \sin t) \end{aligned}$$

Because $d_1^2 = d_2^2$, we obtain

$$2 - 2\cos(s + t) = 2 - 2(\cos s \cos t - \sin s \sin t)$$

Subtracting 2 from both sides and dividing the result by -2, we obtain the desired formula as stated in Theorem 9.3a. Notice that the preceding derivation does not depend on any particular location

▼ THEOREM 9.3a

For any real numbers, s and t,

$$\cos(s + t) = \cos s \cos t - \sin s \sin t$$

of $P(s)$ or $P(t)$. Also since $P(s + 2k\pi) = P(s)$ for any integer k, and likewise for $P(t)$, s and t may be any real numbers.

EXAMPLE 1

Verify that Theorem 9.3a correctly gives $\cos(\pi/6 + \pi/3) = \cos \pi/2$.

SOLUTION:

By Theorem 9.3a,

$$\cos\left(\frac{\pi}{6} + \frac{\pi}{3}\right) = \cos\frac{\pi}{6}\cos\frac{\pi}{3} - \sin\frac{\pi}{6}\sin\frac{\pi}{3}$$

$$= \left(\frac{\sqrt{3}}{2}\right)\left(\frac{1}{2}\right) - \left(\frac{1}{2}\right)\left(\frac{\sqrt{3}}{2}\right) = 0 = \cos\frac{\pi}{2}$$

Because the numbers s and t may be regarded as the number of radians in two given angles, Theorem 9.3a can be restated as follows:

For any two angles, α and β,

$$\cos(\alpha + \beta) = \cos\alpha\cos\beta - \sin\alpha\sin\beta$$

This formula may be used to find the exact value of the cosine of an angle that can be written as the sum of two of the simple angles in Table 8.4b. Example 2 illustrates this idea.

EXAMPLE 2

Find the exact value of $\cos 75°$ by using the fact that $75° = 30° + 45°$.

SOLUTION:

By the preceding formula,

$$\cos 75° = \cos(30° + 45°)$$

$$= \cos 30° \cos 45° - \sin 30° \sin 45°$$

$$= \left(\frac{\sqrt{3}}{2}\right)\left(\frac{\sqrt{2}}{2}\right) - \left(\frac{1}{2}\right)\left(\frac{\sqrt{2}}{2}\right)$$

Note that $\cos 45° = \frac{1}{\sqrt{2}} = \frac{\sqrt{2}}{2}$.

$$= \frac{\sqrt{6} - \sqrt{2}}{4}$$

By writing $s - t$ as $s + (-t)$, we can use Theorem 9.3a to obtain the formula for the cosine of the difference of two numbers. Thus,

$$\cos(s-t) = \cos[s+(-t)]$$

$$= \cos s \cos(-t) - \sin s \sin(-t)$$

Recall that $\cos(-t) = \cos t$ and $\sin(-t) = -\sin t$.

$$= \cos s \cos t + \sin s \sin t$$

We restate this result in Theorem 9.3b.

▼ THEOREM 9.3b

For any real numbers, s and t,

$$\cos(s - t) = \cos s \cos t + \sin s \sin t$$

EXAMPLE 3

Use Theorem 9.3b to find the exact value of $\cos 15^\circ$.

SOLUTION:

Because s and t may be regarded as the radian measures of angles α and β, we know that

$$\cos(\alpha - \beta) = \cos\alpha\cos\beta + \sin\alpha\sin\beta$$

Hence, we take $15^\circ = 45^\circ - 30^\circ$, and write

$$\begin{aligned}\cos 15^\circ &= \cos(45^\circ - 30^\circ)\\ &= \cos 45^\circ \cos 30^\circ + \sin 45^\circ \sin 30^\circ\\ &= \left(\frac{\sqrt{2}}{2}\right)\left(\frac{\sqrt{3}}{2}\right) + \left(\frac{\sqrt{2}}{2}\right)\left(\frac{1}{2}\right)\\ &= \frac{\sqrt{6}+\sqrt{2}}{4}\end{aligned}$$

We can now describe a good method of obtaining one of the results that was briefly discussed in Using Your Knowledge 8.4, namely, that $\cos(\pi/2 - t) = \sin t$. All we need do is expand the left side of this equation by the preceding formula and then use $\sin \pi/2 = 1$, $\cos \pi/2 = 0$ to obtain Theorem 9.3c. Details are left to you.

▼ **THEOREM 9.3c**

For any real number, t,

$$\cos\left(\frac{\pi}{2} - t\right) = \sin t$$

If we replace t by $\pi/2 - t$ in Theorem 9.3c, we obtain the corresponding result for the sine function. (See Theorem 9.3d.) You should carry through the details.

▼ **THEOREM 9.3d**

For any real number, t,

$$\sin\left(\frac{\pi}{2} - t\right) = \cos t$$

Because the secant and cosecant functions are the reciprocals of the cosine and sine functions, respectively, Theorems 9.3c and 9.3d make it obvious that

$$\sec\left(\frac{\pi}{2} - t\right) = \csc t \quad \text{and} \quad \csc\left(\frac{\pi}{2} - t\right) = \sec t$$

By expressing the tangent and the cotangent in terms of the sine and the cosine, you can verify that

$$\tan\left(\frac{\pi}{2} - t\right) = \cot t \quad \text{and} \quad \cot\left(\frac{\pi}{2} - t\right) = \tan t$$

It has already been mentioned that the pairs of functions, sine and cosine, tangent and cotangent, secant and cosecant, are called *cofunctions* of each other. With this terminology, we can summarize all of the preceding results, including Theorems 9.3c and 9.3d, in one simple theorem, Theorem 9.3e. This theorem is the reason why the table of trigonometric functions of angles is read in one direction for angles from 0° to 45° and in the opposite direction for angles from 45° to 90°.

▼ **THEOREM 9.3e**

Let f and cof denote any pair of cofunctions. Then for any real number, t,

$$f\left(\frac{\pi}{2} - t\right) = cof(t)$$

The formula for $\sin(s + t)$ is another important consequence of Theorem 9.3e. Thus, we have

$$\begin{aligned}\sin(s+t) &= \cos\left[\frac{\pi}{2}-(s+t)\right]\\ &= \cos\left[\left(\frac{\pi}{2}-s\right)-t\right]\\ &= \cos\left(\frac{\pi}{2}-s\right)\cos t+\sin\left(\frac{\pi}{2}-s\right)\sin t\\ &= \sin s\cos t+\cos s\sin t\end{aligned}$$

This result is restated in Theorem 9.3f.

▼ THEOREM 9.3f

For any real numbers, s and t,

$$\sin(s+t) = \sin s\cos t+\cos s\sin t$$

EXAMPLE 4

Use Theorem 9.3f to find $\sin 75^\circ$.

SOLUTION:

Because $75^\circ = 45^\circ + 30^\circ$, we know that

$$\begin{aligned}\sin 75^\circ &= \sin(45^\circ+30^\circ)\\ &= \sin 45^\circ\cos 30^\circ+\cos 45^\circ\sin 30^\circ\\ &= \left(\frac{\sqrt{2}}{2}\right)\left(\frac{\sqrt{3}}{2}\right)+\left(\frac{\sqrt{2}}{2}\right)\left(\frac{1}{2}\right)\\ &= \frac{\sqrt{6}+\sqrt{2}}{4}\end{aligned}$$

Notice that $75^\circ + 15^\circ = 90^\circ$, so that our answer is in agreement with the answer to Example 3.

The replacement of t by $-t$ in the formula for $\sin(s+t)$ immediately yields the formula for $\sin(s-t)$. (See Theorem 9.3g.)

▼ THEOREM 9.3g

For any real numbers, s and t,

$$\sin(s-t) = \sin s\cos t-\cos s\sin t$$

The formulas in Theorems 9.3a, 9.3b, 9.3f, and 9.3g are usually called the addition and subtraction formulas for the circular (or the trigonometric) functions. We strongly advise you to memorize these formulas, which we have collected together here for convenience and for comparison to make them easier to remember.

THE ADDITION AND SUBTRACTION FORMULAS

$$\begin{aligned}\cos(s+t) &= \cos s\cos t-\sin s\sin t\\ \cos(s-t) &= \cos s\cos t+\sin s\sin t\\ \sin(s+t) &= \sin s\cos t+\cos s\sin t\\ \sin(s-t) &= \sin s\cos t-\cos s\sin t\end{aligned}$$

EXAMPLE 5

Expand $\sin(45^\circ - \theta)$ to express it in terms of functions of θ alone.

SOLUTION:

By the subtraction formula for the sine function,

$$\sin(45^\circ-\theta) = \sin 45^\circ\cos\theta-\cos 45^\circ\sin\theta$$

$$= \frac{\sqrt{2}}{2} \cos \theta - \frac{\sqrt{2}}{2} \sin \theta$$

$$= \frac{\sqrt{2}}{2} (\cos \theta - \sin \theta)$$

The addition and subtraction formulas for the sine and cosine functions enable us to obtain corresponding formulas for the tangent function. Thus,

$$\tan(s + t) = \frac{\sin(s + t)}{\cos(s + t)} = \frac{\sin s \cos t + \cos s \sin t}{\cos s \cos t - \sin s \sin t}$$

By dividing numerator and denominator of the last fraction by $\cos s \cos t$, we obtain the result given in Theorem 9.3h.

By replacing t by $-t$ in the preceding formula, you can obtain Theorem 9.3i.

▼ THEOREM 9.3h

For all real numbers, s and t, for which $\tan s$, $\tan t$, and $\tan(s + t)$ are defined,

$$\tan(s + t) = \frac{\tan s + \tan t}{1 - \tan s \tan t}$$

▼ THEOREM 9.3i

For all real numbers, s and t, for which $\tan s$, $\tan t$, and $\tan(s - t)$ are defined

$$\tan(s - t) = \frac{\tan s - \tan t}{1 + \tan s \tan t}$$

EXAMPLE 6

Find the exact value of:

a. $\tan 75^\circ$

b. $\tan 15^\circ$

SOLUTION:

a. We use Theorem 9.3h as follows:

$$\tan 75^\circ = \tan(45^\circ + 30^\circ)$$

$$= \frac{\tan 45^\circ + \tan 30^\circ}{1 - \tan 45^\circ \tan 30^\circ}$$

$$= \frac{1 + 1/\sqrt{3}}{1 - (1)(1/\sqrt{3})}$$

$$= \frac{\sqrt{3} + 1}{\sqrt{3} - 1} \cdot \frac{\sqrt{3} + 1}{\sqrt{3} + 1}$$

$$= \frac{4 + 2\sqrt{3}}{2}$$

$$= 2 + \sqrt{3}$$

b. Theorem 9.3i gives

$$\tan 15^\circ = \tan(45^\circ - 30^\circ)$$

$$= \frac{\tan 45^\circ - \tan 30^\circ}{1 + \tan 45^\circ \tan 30^\circ}$$

$$= \frac{1 - 1/\sqrt{3}}{1 + (1)(1/\sqrt{3})}$$

$$= 2 - \sqrt{3}$$

You should verify that the value found in Example 6a agrees with that calculated by using the results of Example 4 and Example 2, and dividing to obtain tan 75°.

EXAMPLE 7

If α is a second-quadrant angle such that $\tan \alpha = -\frac{3}{4}$ and β is a fourth-quadrant angle such that $\cos \beta = \frac{3}{5}$, find $\sin(\alpha + \beta)$.

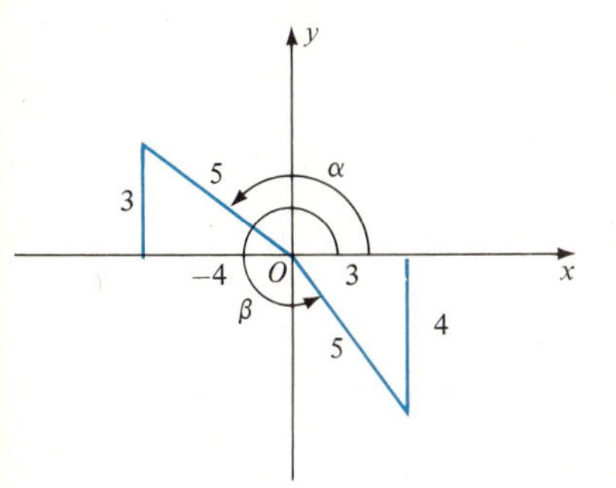

FIGURE 9.3b

SOLUTION:

By sketching α as shown in Figure 9.3b and using the geometry of the figure, we find

$$\sin \alpha = \tfrac{3}{5} \quad \text{and} \quad \cos \alpha = -\tfrac{4}{5}$$

Similarly, we sketch β and use geometry to obtain $\sin \beta = -\frac{4}{5}$ because $\cos \beta = \frac{3}{5}$ was given. Thus,

$$\begin{aligned}\sin(\alpha + \beta) &= \sin \alpha \cos \beta + \cos \alpha \sin \beta \\ &= (\tfrac{3}{5})(\tfrac{3}{5}) + (-\tfrac{4}{5})(-\tfrac{4}{5}) \\ &= \tfrac{9}{25} + \tfrac{16}{25} \\ &= 1\end{aligned}$$

Note that α and β might differ from the angles in the figures by integral multiples of 360°; this would not alter the values we found. The answer shows that $\alpha + \beta = 90^\circ + k \cdot 360^\circ$, where k is an integer that depends on the actual sizes of α and β.

EXAMPLE 8

Find all the solutions of the equation $\sin(x + \pi/4) = \sqrt{2} \sin x$ in the interval $0 \le x \le 2\pi$.

SOLUTION:

We use the formula for $\sin(s + t)$ to express the left side of the equation in terms of $\sin x$ and $\cos x$. This gives

$$\sin x \cos \frac{\pi}{4} + \cos x \sin \frac{\pi}{4} = \sqrt{2} \sin x$$

or

$$\frac{1}{\sqrt{2}} \sin x + \frac{1}{\sqrt{2}} \cos x = \sqrt{2} \sin x$$

and

$$\begin{aligned}\sin x + \cos x &= 2 \sin x \\ \cos x &= \sin x\end{aligned}$$

Because $\cos x = 0$ obviously does not satisfy this equation, we may divide by $\cos x$ to obtain

$$\tan x = 1$$

so that $x = \pi/4$ or $x = 5\pi/4$. *You should check these answers in the original equation.*

EXERCISE 9.3

In Problems 1–10, use the addition formulas to verify the given identity.

1 $\cos(180^\circ - \theta) = -\cos\theta$ **2** $\cos(270^\circ - A) = -\sin A$ **3** $\sin(90^\circ + A) = \cos A$

4 $\cos\left(\frac{\pi}{2} + B\right) = -\sin B$ **5** $\tan(3\pi - x) = -\tan x$ **6** $\tan(\theta - \pi) = \tan\theta$

7 $\sin(270^\circ + C) = -\cos C$ **8** $\cos(450^\circ - A) = \sin A$

9 $\sin\left(2m\pi + \frac{\pi}{2} + \theta\right) = \cos\theta$, m an integer **10** $\cos\left(2m\pi + \frac{\pi}{2} + \theta\right) = -\sin\theta$, m an integer

In Problems 11–22, use the addition formulas to find the exact function value.

11 $\sin 105^\circ$ **12** $\cos(-75^\circ)$ **13** $\tan 255^\circ$ **14** $\cos(-15^\circ)$ **15** $\tan 165^\circ$ **16** $\sin 285^\circ$

17 $\cos(-105^\circ)$ **18** $\sin(-75^\circ)$ **19** $\tan(-105^\circ)$ **20** $\cos(-195^\circ)$ **21** $\sin 195^\circ$ **22** $\tan 105^\circ$

In Problems 23–28, express in terms of functions of θ alone.

23 $\sin\left(\frac{\pi}{6} + \theta\right)$ **24** $\cos(\theta - 30^\circ)$ **25** $\tan(\theta - 45^\circ)$ **26** $\cos\left(\frac{\pi}{3} - \theta\right)$ **27** $\sin(210^\circ + \theta)$ **28** $\tan(\theta + 225^\circ)$

In Problems 29–32, make sketches of the angles A and B, and use the addition formulas to find the required function values.

29 $\tan A = \frac{4}{3}$, A in Q III, $\cot B = -\frac{5}{12}$, B in Q IV. Find $\sin(A - B)$.

30 $\cos A = \frac{3}{5}$, A in Q IV, $\csc B = \frac{17}{15}$, B in Q II. Find $\tan(A + B)$.

31 $\sin A = -\frac{12}{13}$, A in Q III, $\sec B = \frac{13}{12}$, B in Q IV. Find $\cos(A + B)$.

32 $\tan A = \frac{20}{21}$, A in Q I, $\sec B = \frac{5}{4}$, B in Q IV. Find $\tan(A - B)$.

In Problems 33–40, prove the given identity.

33 $\cos A = \cos(A + B)\cos B + \sin(A + B)\sin B$

Hint: $A = (A + B) - B$.

34 $\sin A = \sin(A - B)\cos B + \cos(A - B)\sin B$

Note: See Problem 33.

35 $\sin\left(\theta + \frac{\pi}{3}\right) - \cos\left(\theta - \frac{\pi}{6}\right) = 0$ **36** $\sec(A + B) = \dfrac{\sec A \sec B}{1 - \tan A \tan B}$

37 $\sin A + \sin B = 2\sin\dfrac{A + B}{2}\cos\dfrac{A - B}{2}$

Hint: Let $s = (A + B)/2$ and $t = (A - B)/2$. Then solve for A and B in terms of s and t.

38 $\sin A - \sin B = 2\cos\dfrac{A + B}{2}\sin\dfrac{A - B}{2}$

Note: See Problem 37.

39 $\cos A + \cos B = 2\cos\frac{A+B}{2}\cos\frac{A-B}{2}$

Note: See Problem 37.

40 $\cos A - \cos B = -2\sin\frac{A+B}{2}\sin\frac{A-B}{2}$

Note: See Problem 37.

41 Find a formula for $\cot(s + t)$ in terms of $\cot s$ and $\cot t$.

42 Find a formula for $\cot(s - t)$ in terms of $\cot s$ and $\cot t$.

In Problems 43–46, solve the given equation for all values of x such that $0 \leq x \leq 2\pi$.

43 $\sin\left(x + \frac{\pi}{2}\right) = \sin x$

44 $\sin\left(x - \frac{\pi}{3}\right) = \cos\left(x + \frac{\pi}{6}\right)$

45 $\sin\left(x + \frac{\pi}{4}\right) = 2\cos\left(x - \frac{\pi}{4}\right)$

46 $\tan\left(x + \frac{\pi}{4}\right) = 1 + \tan x$

47 A picture 5 feet high hangs on a wall so that the bottom edge of the picture is 4 feet above the floor. A camera, whose lens is 5 feet above the floor is pointed at the picture from a point x feet away from the wall. (See Figure 9.3c.) Show that the angle subtended by the picture at the camera lens is given by the formula

$$\tan\theta = \frac{5x}{x^2 - 4}$$

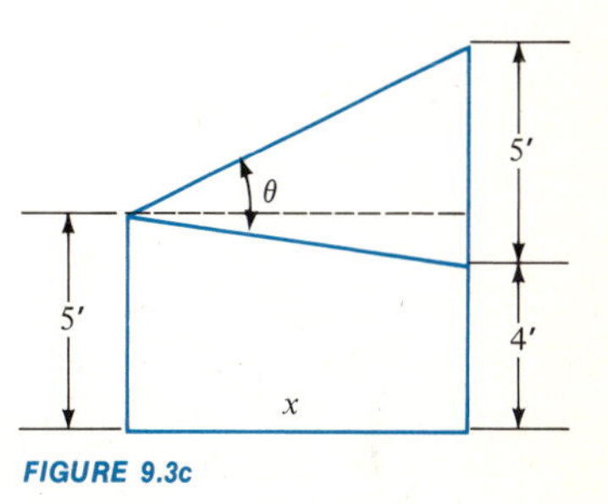

FIGURE 9.3c

48 In Problem 47, suppose the camera lens is 3 feet above the floor. Find the formula for $\tan\theta$.

USING YOUR KNOWLEDGE 9.3

In many areas of science and engineering, we have some type of system, which may be electrical or mechanical, into which there is an input that is expressed in the form

$$A \sin kt$$

where A and k are constants. Frequently, the system responds with an output that is expressed in the form

$$a \sin kt + b \cos kt$$

where a and b are constants that depend not only on A and k but also on the system.

In order to compare the output with the input (we like to know what we got for what we paid), it is desirable, if possible, to write the output in the form

$$B \sin(kt + \theta)$$

For then we could compare the output amplitude, B, with the input amplitude, A, and we could look at the phase angle, θ, to determine by how much the output lags or leads the input. Certain ideas in physics make it plausible that the output is expressible in the desired form, and we can show mathematically that this is possible.

Suppose that

$$a \sin kt + b \cos kt = B \sin(kt + \theta)$$

is an identity. We expand the right side to obtain the equation

$$a \sin kt + b \cos kt = B \sin kt \cos\theta + B \cos kt \sin\theta$$

If this equation is to be an identity, then it must be true for $t = 0$, which, upon substitution, yields the equation

$$b = B \sin \theta \tag{1}$$

Similarly, for $t = \pi/(2k)$ or $kt = \pi/2$, we obtain a second equation

$$a = B \cos \theta \tag{2}$$

Squaring and adding the last two equations, we find that

$$\begin{aligned} a^2 + b^2 &= B^2 \cos^2 \theta + B^2 \sin^2 \theta \\ &= B^2(\cos^2 \theta + \sin^2 \theta) \\ &= B^2 \end{aligned}$$

Consequently,

$$B = \pm\sqrt{a^2 + b^2}$$

Because it is simpler to have a positive amplitude, we choose the plus sign. Then, solving Equations 1 and 2 for the sine and cosine, we obtain

$$\sin \theta = \frac{b}{B} \quad \text{and} \quad \cos \theta = \frac{a}{B} \qquad B = \sqrt{a^2 + b^2} \tag{3}$$

For convenience, we restrict θ to the interval $-\pi < \theta < \pi$, which results in the smallest absolute value for θ. Because Equations 3 give us both the sine and the cosine, the quadrant of θ is determined.

Now let us look at an example. Suppose that we wish to express

$$\sin t + \sqrt{3} \cos t$$

in the form $B \sin(t + \theta)$. Then, from Equations 3, we find

$$B = \sqrt{1^2 + (\sqrt{3})^2} = \sqrt{4} = 2 \qquad \sin \theta = \frac{\sqrt{3}}{2} \qquad \cos \theta = \frac{1}{2}$$

Thus, $\theta = \pi/3$, and we have shown that

$$\sin t + \sqrt{3} \cos t = 2 \sin\left(t + \frac{\pi}{3}\right)$$

By expanding the right-hand side, you can show that this equation is actually an identity.

If you wanted to graph the equation $y = \sin t + \sqrt{3} \cos t$, there would be a great advantage in putting it into the form $y = 2 \sin(t + \pi/3)$. From this last form, you see at once that the graph is that of a sinusoid, that the maximum and minimum values of y are 2 and -2, occurring at points where $t + \pi/3$ is an odd multiple of $\pi/2$, and that $y = 0$ when $t + \pi/3$ is an integral multiple of π. These facts are not as obvious from the original form of the equation.

Try to use the preceding ideas to write each of the following in the form $B \sin(t + \theta)$.

1 $\sin t + \cos t$
2 $\sin t - \cos t$
3 $\sqrt{3} \sin t - \cos t$
4 $3 \sin t + 4 \cos t$ *Hint:* Use tables or a calculator to find θ.

9.4 THE DOUBLE- AND HALF-ANGLE FORMULAS

One of the important uses of the addition formulas developed in Section 9.3 is in the derivation of formulas for trigonometric functions of multiples of a given angle. For example, by setting s and t both equal to θ in the formula for $\sin(s + t)$, we obtain

$$\begin{aligned}\sin 2\theta &= \sin(\theta + \theta)\\ &= \sin\theta\cos\theta + \cos\theta\sin\theta\\ &= 2\sin\theta\cos\theta\end{aligned}$$

▼ THEOREM 9.4a

$\sin 2\theta = 2\sin\theta\cos\theta$

▼ THEOREM 9.4b

$$\begin{aligned}\cos 2\theta &= \cos^2\theta - \sin^2\theta\\ &= 2\cos^2\theta - 1\\ &= 1 - 2\sin^2\theta\end{aligned}$$

This proves Theorem 9.4a. A similar procedure, using the formula for $\cos(s + t)$, results in Theorem 9.4b. The details of the proof are left to you. Note that the last two forms of the formula for $\cos 2\theta$ are obtained by using $\sin^2\theta = 1 - \cos^2\theta$ and $\cos^2\theta = 1 - \sin^2\theta$, respectively.

The formulas in Theorems 9.4a and 9.4b are usually called the *double-angle* formulas. Their application is illustrated in the following examples.

EXAMPLE 1

If $\sin\theta = \frac{4}{5}$ and θ is a second-quadrant angle, find $\sin 2\theta$ and $\cos 2\theta$.

SOLUTION:

Because θ is a second-quadrant angle,

$$\cos\theta = -\sqrt{1 - \sin^2\theta} = -\sqrt{1 - (\tfrac{4}{5})^2} = -\sqrt{1 - \tfrac{16}{25}} = -\tfrac{3}{5}$$

Thus,

$$\sin 2\theta = 2\sin\theta\cos\theta = 2(\tfrac{4}{5})(-\tfrac{3}{5}) = -\tfrac{24}{25}$$

and

$$\cos 2\theta = \cos^2\theta - \sin^2\theta = (-\tfrac{3}{5})^2 - (\tfrac{4}{5})^2 = -\tfrac{7}{25}$$

EXAMPLE 2

Express $\cos 3\theta$ in terms of $\cos\theta$.

SOLUTION:

$$\begin{aligned}\cos 3\theta &= \cos(2\theta + \theta)\\ &= \cos 2\theta\cos\theta - \sin 2\theta\sin\theta && \text{Addition formula for the cosine}\\ &= (2\cos^2\theta - 1)\cos\theta - 2\sin\theta\cos\theta\sin\theta && \text{Double-angle formulas}\\ &= 2\cos^3\theta - \cos\theta - 2\sin^2\theta\cos\theta\\ &= 2\cos^3\theta - \cos\theta - 2(1 - \cos^2\theta)\cos\theta && \sin^2\theta = 1 - \cos^2\theta\\ &= 4\cos^3\theta - 3\cos\theta\end{aligned}$$

We have seen that, in general, $\cos 2\theta \neq 2\cos\theta$. In Example 3, we shall find two angles for which the equality does hold.

EXAMPLE 3

For what values of θ between 0° and 360° is it true that $\cos 2\theta = 2 \cos \theta$?

SOLUTION:

We wish to solve the equation

$$\cos 2\theta = 2 \cos \theta$$

By the double-angle formula for the cosine, an equivalent equation is

$$2 \cos^2 \theta - 1 = 2 \cos \theta$$

Thus, we have the quadratic equation in $\cos \theta$

$$2 \cos^2 \theta - 2 \cos \theta - 1 = 0$$

which we can solve by the quadratic formula to obtain

$$\cos \theta = \frac{2 \pm \sqrt{12}}{4} = \frac{1 \pm \sqrt{3}}{2}$$

We discard the plus sign because it gives a value greater than 1 and we know that $|\cos \theta| \leq 1$. The minus sign gives $\cos \theta = -0.36603$, and the table gives us the reference angle $\theta_r = 68.53°$. Since $\cos \theta$ is negative, θ must be a second- or a third-quadrant angle. Thus,

$$\theta = 180° - \theta_r = 111.47°$$

or

$$\theta = 180° + \theta_r = 248.53°$$

If you use a calculator, you can evaluate $(1 - \sqrt{3})/2$ and then press the *inv* key (or *arc* key) and the *cos* key. The calculator will read 111.47 (degree mode). To find the second solution, you can subtract 180 and change sign to obtain 68.53 (the number of degrees in the reference angle). Add 180 and you have the second solution. With a little practice on the calculator, you will be able to get both answers in less time than it takes just to do the interpolation in the table.

CHECK:

For $\theta = 111.47°$, $\cos 2\theta = \cos 222.94° = -0.73206$, and $2 \cos \theta = 2 \cos 111.47° = 2(-0.36603) = -0.73206$. Thus, $\cos 2\theta = 2 \cos \theta$, as required. The check for the second solution is left for you to do.

▼ THEOREM 9.4c

For every real number, t, such that $\tan t$ and $\tan 2t$ are both defined,

$$\tan 2t = \frac{2 \tan t}{1 - \tan^2 t}$$

Another double-angle formula that is frequently encountered is that for the tangent function. (See Theorem 9.4c.) The result is easily obtained from the formula for $\tan(s + t)$, and the details are left for you to carry out.

EXAMPLE 4

Show that $\tan(t + \pi/4) = \sec 2t + \tan 2t$.

SOLUTION:

By the addition formula for the tangent function,

$$\tan\left(t+\frac{\pi}{4}\right)=\frac{\tan t+\tan \pi/4}{1-\tan t\tan \pi/4}=\frac{\tan t+1}{1-\tan t}$$

The denominator of the formula for $\tan 2t$ in Theorem 9.4c is $1-\tan^2 t$, which suggests that we multiply numerator and denominator of the last fraction by $1+\tan t$ to obtain

$$\tan\left(t+\frac{\pi}{4}\right)=\frac{(1+\tan t)^2}{1-\tan^2 t}=\frac{1+\tan^2 t+2\tan t}{1-\tan^2 t}$$

We now use the basic identity $1+\tan^2 t=\sec^2 t$ to obtain

$$\tan\left(t+\frac{\pi}{4}\right)=\frac{\sec^2 t}{1-\tan^2 t}+\frac{2\tan t}{1-\tan^2 t}$$

By Theorem 9.4c, the second fraction on the right side is exactly $\tan 2t$. In the first fraction, let us change to sines and cosines. This gives

$$\begin{aligned}\frac{\sec^2 t}{1-\tan^2 t}&=\frac{1/\cos^2 t}{1-(\sin^2 t)/(\cos^2 t)}\\&=\frac{1/\cos^2 t}{(\cos^2 t-\sin^2 t)/\cos^2 t}\\&=\frac{1}{\cos^2 t-\sin^2 t}\\&=\frac{1}{\cos 2t}\\&=\sec 2t\end{aligned}$$

This completes the problem and proves the identity

$$\tan\left(t+\frac{\pi}{4}\right)=\sec 2t+\tan 2t$$

HALF-ANGLE FORMULAS

▼ THEOREM 9.4d

$$\cos\frac{\theta}{2}=\pm\sqrt{\frac{1+\cos\theta}{2}}$$

where the sign is to be chosen to agree with the sign of $\cos \theta/2$.

▼ THEOREM 9.4e

$$\sin\frac{\theta}{2}=\pm\sqrt{\frac{1-\cos\theta}{2}}$$

where the sign is to be chosen to agree with the sign of $\sin \theta/2$.

From the second two forms of the formula for $\cos 2\theta$ in Theorem 9.4b, we can obtain additional useful formulas, known as the *half-angle* formulas. We start with the identity

$$\cos 2\alpha=2\cos^2\alpha-1$$

and solve for $\cos^2\alpha$ to obtain

$$\cos^2\alpha=\frac{1+\cos 2\alpha}{2}$$

If we now let $2\alpha=\theta$, that is, $\alpha=\theta/2$, then we obtain the result stated in Theorem 9.4d. Theorem 9.4e is obtained in the same way by using the formula $\cos 2\alpha=1-2\sin^2\alpha$. You should carry out the details for yourself.

EXAMPLE 5

Use the preceding formulas to find the exact value of:

a. $\cos \frac{\pi}{8}$

b. $\sin \frac{\pi}{8}$

SOLUTION:

a. Because $\pi/8 = \frac{1}{2}(\pi/4)$, we let θ be $\pi/4$ in the formula for $\cos \theta/2$. Thus,

$$\cos \frac{\pi}{8} = \sqrt{\frac{1 + \cos(\pi/4)}{2}} = \sqrt{\frac{1 + \sqrt{2}/2}{2}} = \sqrt{\frac{2 + \sqrt{2}}{4}} = \frac{1}{2}\sqrt{2 + \sqrt{2}}$$

b. Except for the sign in the numerator under the radical, the calculation of $\sin \theta/2$ duplicates that of $\cos \theta/2$. The result is

$$\sin \frac{\pi}{8} = \frac{1}{2}\sqrt{2 - \sqrt{2}}$$

▼ THEOREM 9.4f

$$\tan \frac{\theta}{2} = \pm\sqrt{\frac{1 - \cos \theta}{1 + \cos \theta}} = \frac{1 - \cos \theta}{\sin \theta} = \frac{\sin \theta}{1 + \cos \theta}$$

where the sign in the first form is to be chosen to agree with that of $\tan \theta/2$.

Because

$$\tan \frac{\theta}{2} = \frac{\sin \theta/2}{\cos \theta/2}$$

Theorem 9.4f follows at once from the formulas for the sine and the cosine of the half-angle. Derivation of the alternate forms is left to the exercises.

The next examples illustrate further the use of the preceding formulas.

EXAMPLE 6

If $\tan A = \frac{3}{4}$ and A is a third-quadrant angle, find $\tan A/2$.

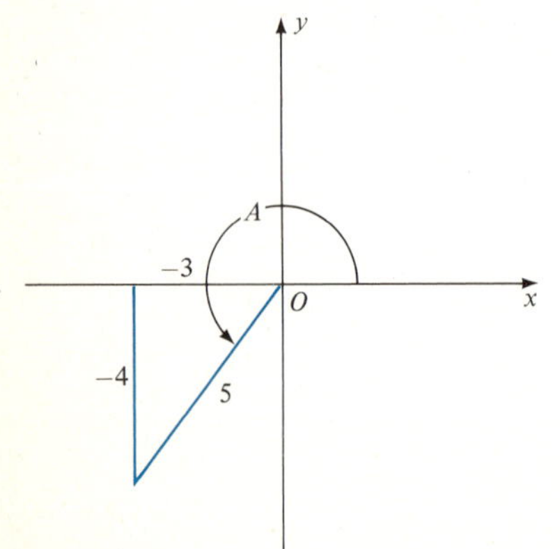

FIGURE 9.4a

SOLUTION:

A quick sketch (see Figure 9.4a) shows that $\sin A = -\frac{3}{5}$ and $\cos A = -\frac{4}{5}$. Hence, using the second form given in Theorem 9.4f, we obtain

$$\tan \frac{A}{2} = \frac{1 - (-\frac{4}{5})}{-\frac{3}{5}} = \frac{\frac{9}{5}}{-\frac{3}{5}} = -3$$

EXAMPLE 7

Solve the equation $\sin 2x + \cos x = 0$ for $0 \leq x \leq 2\pi$.

SOLUTION:

By using the identity $\sin 2x = 2 \sin x \cos x$, we obtain the equivalent equation

$$2 \sin x \cos x + \cos x = 0$$

which factors into

$$\cos x(2 \sin x + 1) = 0$$

Thus,

$$\cos x = 0 \quad \text{or} \quad \sin x = -\tfrac{1}{2}$$

giving the values of x: $\pi/2$, $7\pi/6$, $3\pi/2$, and $11\pi/6$. You should check these results.

EXAMPLE 8

Find all the solutions of the equation $\cos 4t + \cos 2t + 1 = 0$.

SOLUTION:

Because $\cos 4t = 2 \cos^2 2t - 1$, we may rewrite the equation as $2 \cos^2 2t - 1 + \cos 2t + 1 = 0$

or

$$2 \cos^2 2t + \cos 2t = 0$$

This equation factors into

$$\cos 2t(2 \cos 2t + 1) = 0$$

Consequently, either $\cos 2t = 0$ or $2 \cos 2t + 1 = 0$. If $\cos 2t = 0$, then $2t = 2m\pi \pm \pi/2$, $m = 0, \pm 1, \pm 2, \pm 3, \ldots$, so that

$$t = m\pi \pm \frac{\pi}{4} \qquad m = 0, \pm 1, \pm 2, \pm 3, \ldots$$

If $2 \cos 2t + 1 = 0$, then $\cos 2t = -\frac{1}{2}$, and

$$2t = 2m\pi \pm \frac{2\pi}{3} \qquad m = 0, \pm 1, \pm 2, \pm 3, \ldots$$

so that

$$t = m\pi \pm \frac{\pi}{3} \qquad m = 0, \pm 1, \pm 2, \pm 3, \ldots$$

Hence, all the solutions of the given equation are

$$t = m\pi \pm \frac{\pi}{4} \quad \text{or} \quad t = m\pi \pm \frac{\pi}{3} \qquad m = 0, \pm 1, \pm 2, \pm 3, \ldots$$

EXERCISE 9.4

In Problems 1–4, find $\cos 2\theta$ and $\sin 2\theta$ for θ in the specified quadrant.

1 $\sin \theta = -\frac{3}{5}$, Q IV

2 $\tan \theta = \frac{5}{12}$, Q III

3 $\cot \theta = \frac{8}{15}$, Q I

4 $\cos \theta = -\frac{7}{25}$, Q II

In Problems 5–14, find the exact function value by using the theorems of this section.

5 $\cos \frac{\pi}{12}$ **6** $\sin \frac{5\pi}{12}$ **7** $\tan \frac{3\pi}{8}$ **8** $\cos \frac{7\pi}{8}$ **9** $\sin \frac{11\pi}{12}$ **10** $\tan \frac{7\pi}{12}$

11 $\cos 15^\circ$ **12** $\sin 75^\circ$ **13** $\sin 157.5^\circ$

14 $\tan 112.5^\circ$ **15** Prove Theorem 9.4b. **16** Prove Theorem 9.4c.

17 Prove Theorem 9.4e.

18 Derive the second two forms of the formula in Theorem 9.4f.

Hint: Rationalize the numerator (denominator) of the right side of the first form. Be sure to show that the plus–minus sign is not needed in these two forms.

In Problems 19–28, prove the given identity.

19 $\tan B + \cot B = 2 \csc 2B$ **20** $\cos^4 t - \sin^4 t = \cos 2t$ **21** $\tan \frac{x}{2} + \cot \frac{x}{2} = 2 \csc x$

22 $\sin^4 \frac{A}{2} + \cos^4 \frac{A}{2} = 1 - \frac{1}{2} \sin^2 A$ **23** $\frac{1 - \tan^2 C}{1 + \tan^2 C} = \cos 2C$ **24** $\cos 4x = 1 - 8 \sin^2 x \cos^2 x$

25 $\cot x - \tan x = 2 \cot 2x$ **26** $\frac{1 + \sin \theta - \cos \theta}{1 + \sin \theta + \cos \theta} = \tan \frac{\theta}{2}$ **27** $1 - \cos t = 2 \sin^2 \frac{t}{2}$

28 $\csc 2A + \cot 2A = \cot A$

29 Find a formula for $\sin 3t$ in terms of $\sin t$.

30 Express $\sin^4 x$ in terms of cosines of multiples of x, all to the first power.

Hint: $\sin^2 x = \frac{1}{2}(1 - \cos 2x)$ from Theorem 9.4e.

31 Express $\cos^4 x$ in terms of cosines of multiples of x, all to the first power. See Problem 30.

32 If A and B are the acute angles of a right triangle, show that $\sin(A - B) = \cos 2B$.

Hint: Use the fact that $A = 90^\circ - B$.

In Problems 33–42, find all the solutions of the given equation in the interval $0 \leq x \leq 2\pi$.

33 $2 \cos 2x + 1 = 0$ **34** $2 \sin 3x = 1$ **35** $4 \cos^2 2x = 3$

36 $\sin^2 3x = 3 \cos^2 3x$ **37** $\cos 2x + \sin 2x + 1 = 0$ **38** $\cos 2x - \cos x = 0$

39 $\tan \frac{x}{2} = \sin x$ **40** $\cot \frac{x}{2} + \sin x = 0$ **41** $\csc 2x + \sec x = 0$

42 $\tan \frac{x}{2} + \cot \frac{x}{2} = 4$

In Problems 43–52, find all the solutions of the given equation.

43 $4 \sin^2 2x = 3$ **44** $\cos^2 3x = 3 \sin^2 3x$ **45** $\sin 2x = \sin x$

46 $\tan^2 4x = 1$ **47** $\tan 2x + \tan x = 0$ **48** $\cos 2x = 2 \sin^2 x$

49 $\cos^2 \frac{x}{2} - \sin^2 \frac{x}{2} = \cos^2 x$ **50** $2 \sin^2 \frac{x}{2} = \cos x$ **51** $4 \cos^2 \frac{x}{2} = 2 + \sqrt{3}$

52 $2 \cos^2 2x - \sin 2x - 1 = 0$

USING YOUR KNOWLEDGE 9.4

We have already mentioned that the study of trigonometry dates back to the ancient Greeks. Sometime between 200 and 100 B.C., a Greek astronomer named Hipparchus developed a table of "chords." Hipparchus was familiar with the work of the Babylonians, and he adopted their system of dividing the circle into 360 equal parts, thus being responsible for introducing into trigonometry the degree system of measuring angles, which we use today.

Although the work of Hipparchus is lost, the greatest of the ancient Greek astronomers, Ptolemy, in about 150 A.D., noted the work of Hipparchus and constructed a table of "chords" himself. We do not know how Hipparchus constructed his table, but we do know all the interesting details of Ptolemy's method.

By this time, you are probably wondering what a table of "chords" is, and you can find out by referring to Figure 9.4b. By taking half the angle θ, we see that

$$\sin\frac{\theta}{2} = \frac{\frac{1}{2}\text{ chord }\theta}{R}$$

where "chord θ" is the length of the chord that subtends the central angle θ. Because chord $\theta = 2R\sin\theta/2$, a table of chords is equivalent to a table of sines. Ptolemy used a circle of radius 60, but we shall simplify matters and use a unit circle; the ideas are all the same. With $R = 1$, we now have

$$\sin\frac{\theta}{2} = \frac{1}{2}\text{ chord }\theta$$

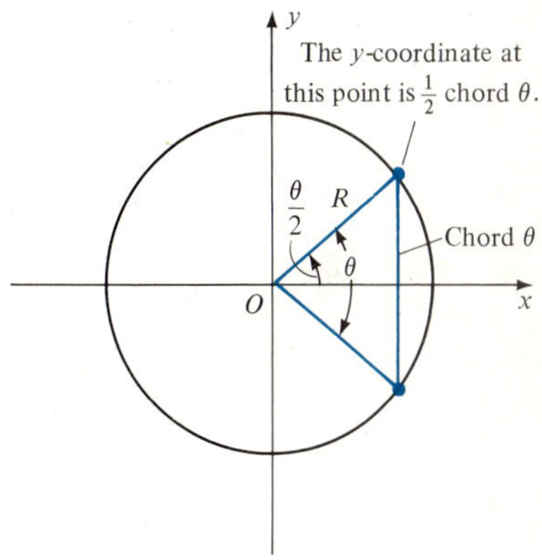

FIGURE 9.4b

The ancient Greek mathematicians were extremely interested in the problem of constructing regular polygons (equal sides and equal angles), and Ptolemy had figured out how to construct a regular pentagon (five-sided polygon) inscribed in a circle. From his construction, he was able to calculate the exact length of a side of the pentagon in terms of the radius of the circle. Because one side of such a pentagon subtends an angle of 72° $(=\frac{1}{5}\cdot 360^\circ)$, the number of units in the length of the side gives chord 72°, and half of this number is sin 36°. Thus, Ptolemy's calculation of the length of the side of the pentagon is equivalent to finding the exact value of sin 36°. We shall not bother you with the details of the calculation but give you the result

$$\sin 36^\circ = \tfrac{1}{4}\sqrt{10 - 2\sqrt{5}}$$

Knowing the sine and cosine of 30°, we can use the difference formula to find $\sin(36^\circ - 30^\circ)$, that is, sin 6°. Ptolemy had a geometrical method for calculating chord $(72^\circ - 60^\circ)$, which is exactly 2 sin 6°. From chord θ, Ptolemy could also calculate chord $\theta/2$, so that he was able to calculate the equivalent of 2 sin 3°, and then 2 sin 1.5°, and 2 sin 0.75°. By interpolating between these last two, he found a very accurate value for chord 1°, and, from this chord, 0.5°. He also had a geometrical method for calculating chord $(\theta_1 + \theta_2)$ from chord θ_1 and chord θ_2, which then enabled him to construct a table of chords from 0° to 180° at intervals of 0.5°. This table was so accurate, that it became the standard table used for hundreds of years.

As you can see from the preceding discussion, Ptolemy had essentially the equivalents of our addition and half-angle formulas for the sine function. Now here are a couple of things for you to try.

1 See if you can verify that the exact value of $\sin 6^\circ$ is $\frac{1}{8}(\sqrt{30 - 6\sqrt{5}} - \sqrt{5} - 1)$

2 Using Table A6, interpolate between $\sin 0.7^\circ$ and $\sin 0.8^\circ$ to find $\sin 0.75^\circ$. Then, interpolate between your answer and the tabular value of $\sin 1.5^\circ$ to find $\sin 1^\circ$. Compare your answer with the tabular value of $\sin 1^\circ$. You might be surprised at the very close agreement. This suggests that a good many entries in a table of sines can be calculated by interpolation rather than by direct evaluation of the sine.

3 There is another way, more modern than Ptolemy's method, of finding the exact value of $\sin 36^\circ$. Because $5 \cdot 36^\circ = 180^\circ$ and $\sin 180^\circ = 0$, we know that $\theta = 36^\circ$ must be one of the solutions of the equation $\sin 5\theta = 0$. By using the addition formulas and the double-angle formulas along with the basic identity $\cos^2\theta = 1 - \sin^2\theta$, it is fairly easy to obtain the formula

$$\sin 5\theta = 16 \sin^5\theta - 20 \sin^3\theta + 5 \sin\theta$$

Solve the equation $\sin 5\theta = 0$ by first using this formula and thus verify the exact value given in the text for $\sin 36^\circ$.

SELF-TEST

1 a Give the period and amplitude of the curve $y = -3 \sin 2x$.
b What is the phase shift of $y = 4 \sin(x - \pi/4)$?
c What is the phase shift of $y = 4 \sin(2x - \pi/4)$?

2 Graph the equation $y = -3 \sin 2x$, $0 \le x \le 2\pi$.

3 Graph the equation $y = \frac{1}{2} \cos \pi x/2$, $0 \le x \le 4$.

4 Sketch the graph of $y = 4 \sin(x - \pi/4)$, $0 \le x \le 2\pi$.

5 Sketch the graph of $y = \frac{1}{2} \sin(2x - \pi/4)$, $0 \le x \le 2\pi$.

6 Sketch the graph of $y = \tan 4x$, $0 \le x \le \pi$.

7 Sketch the graph of $y = \cot 2x$, $0 \le x \le \pi$.

8 Sketch the graph of $y = \sec x/2$, $0 \le x \le 4\pi$.

9 Sketch the graph of $y = \tan(\pi/3 - x)$, $-\pi/6 \le x \le 5\pi/6$.

10 a Use the formula for $\cos(s + t)$ to find the exact value of $\cos 105^\circ$.
b Find the exact value of $\tan 195^\circ$.

11 a Express $\cos(30^\circ + \theta)$ in terms of $\sin\theta$ and $\cos\theta$.
b Express $\tan(45^\circ + x)$ in terms of $\tan x$.

12 Express in terms of an acute angle less than 45°:
a $\cos 63^\circ$ **b** $\tan 100^\circ$

13 If $\sin A = \frac{2}{3}$ and A is a second-quadrant angle, and $\sin B = \frac{1}{3}$ and B is a first-quadrant angle, find $\sin(A + B)$.

14 Find all the solutions of the equation $2 \sin(x - \pi/4) = \cos(x + \pi/4)$, $0 \le x \le 2\pi$.

15 If $\cos\theta = \frac{1}{3}$ and θ is a fourth-quadrant angle, find the exact value of $\sin 2\theta$.

16 Find a formula for $\cos 4\theta$ in terms of $\cos\theta$.

17 If $\cos\theta = -\frac{2}{9}$ and θ is a third-quadrant angle, find the exact value of $\cos\theta/2$, where $\pi < \theta < 3\pi/2$.

18 Find all the solutions of $\sqrt{3} \cos t/2 = \sin t$, $0 \le t \le 2\pi$.

19 Find all the solutions of $\sin 2t + \sin t = 0$.

20 Prove the identity $\csc 2x - \cot 2x = \tan x$.

CHAPTER 10

SOLUTION OF TRIANGLES

10.1 RIGHT TRIANGLES

Suppose we want to know the height of the cliff in Figure 10.1a. We would measure the horizontal distance QS from the foot of the cliff to a point S and the angle of elevation (angle above the horizontal) of the line of sight from S to a point P at the edge of the cliff and vertically above the point Q in the figure. Thus, we would know one side and the adjacent acute angle of a right triangle, and some simple trigonometry would enable us to calculate the distance QP, the height of the cliff.

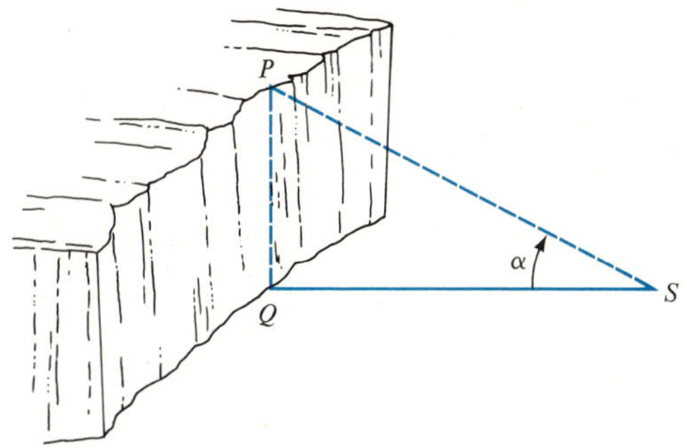

FIGURE 10.1a

RIGHT-TRIANGLE DEFINITIONS OF THE TRIGONOMETRIC FUNCTIONS

In order to simplify the trigonometry of the right triangle, we first restate the definitions of the trigonometric functions of the acute angles in terms of the lengths of the sides of the triangle. Con-

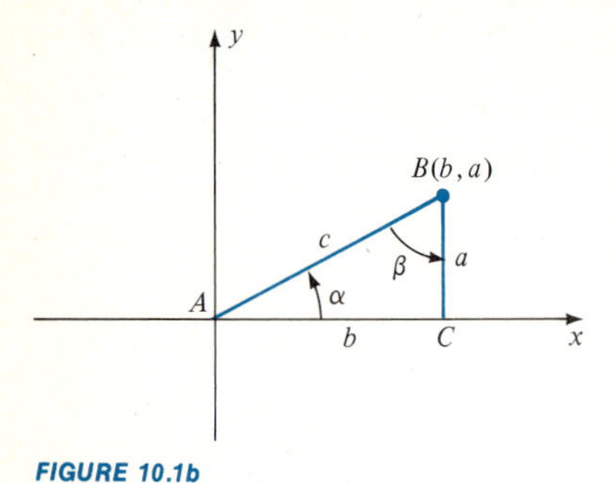

FIGURE 10.1b

sider the right triangle ABC with right angle at C in Figure 10.1b. Let the angle BAC be denoted by α and the angle ABC by β; and let the length of the side opposite α be denoted by a, the length of the side opposite β by b, and the length of the hypotenuse by c. Figure 10.1b shows the triangle placed with α in the standard position, so that the coordinates of the point B are (b, a) and the distance of B from the origin is c. Thus, by the definitions of the trigonometric functions,

$$\sin \alpha = \frac{a}{c} = \frac{\text{opposite side}}{\text{hypotenuse}}$$

$$\cos \alpha = \frac{b}{c} = \frac{\text{adjacent side}}{\text{hypotenuse}}$$

$$\tan \alpha = \frac{a}{b} = \frac{\text{opposite side}}{\text{adjacent side}}$$

It should be understood that, where we have written opposite side, adjacent side, and hypotenuse, the respective lengths of these sides must be used. The definitions of the reciprocal functions—cosecant, secant, and cotangent—lead to similar statements in terms of the sides of the triangle.

Clearly, either acute angle of the triangle may be put into standard position, so that the definitions of the trigonometric functions in terms of opposite side, adjacent side, and hypotenuse are quite independent of the position of the right triangle. Hence, we have Definition 10.1a.

▼ **DEFINITION 10.1a**

Let θ be an acute angle of a right triangle. Then

$$\sin \theta = \frac{\text{opposite side}}{\text{hypotenuse}}$$

$$\cos \theta = \frac{\text{adjacent side}}{\text{hypotenuse}}$$

$$\tan \theta = \frac{\text{opposite side}}{\text{adjacent side}}$$

$$\csc \theta = \frac{\text{hypotenuse}}{\text{opposite side}}$$

$$\sec \theta = \frac{\text{hypotenuse}}{\text{adjacent side}}$$

$$\cot \theta = \frac{\text{adjacent side}}{\text{opposite side}}$$

For instance, in Figure 10.1a,

$$\sin \alpha = \frac{PQ}{PS}, \quad \cos \alpha = \frac{QS}{PS}, \quad \text{and } \tan \alpha = \frac{PQ}{QS}$$

Also, in Figure 10.1b, by using Definition 10.1a, we can give the trigonometric functions of β without having to put β into standard position. Thus,

$$\sin \beta = \frac{b}{c}, \quad \cos \beta = \frac{a}{c}, \quad \text{and } \tan \beta = \frac{b}{a}$$

(Note that for the angle β, b is the opposite side and a is the adjacent side.)

SOLUTION OF RIGHT TRIANGLES

In dealing with the trigonometry of triangles, we regard a triangle as "solved" if we know all the sides and the angles. We can always solve a right triangle if we know two sides or if we know a side and one of the acute angles. The following examples illustrate the procedure. We use the lettering of Figure 10.1b as standard.

EXAMPLE 1

Solve the right triangle in which $a = 3$ and $b = 5$.

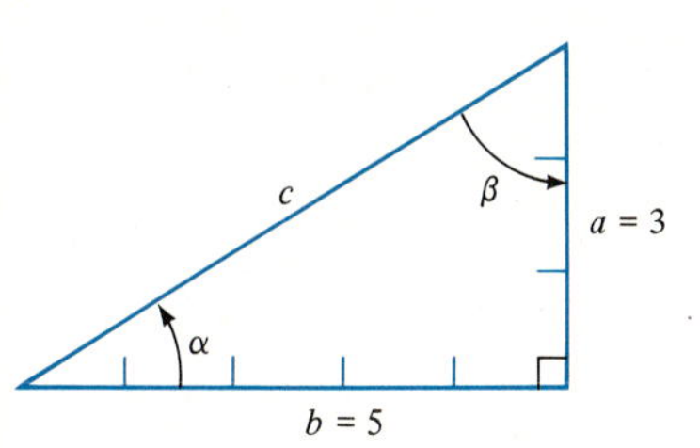

FIGURE 10.1c

SOLUTION:

We sketch the triangle (Figure 10.1c) and use Definition 10.1a to write

$$\tan \alpha = \frac{a}{b} = \frac{3}{5} = 0.6$$

Then, from tables or calculator, we obtain $\alpha = 30.96°$. Since the sum of the angles of any triangle is 180°, we must have $\alpha + \beta = 90°$. Thus, $\beta = 90° - \alpha = 59.04°$. There are now two ways of finding c: We can use the Pythagorean Theorem, $c^2 = a^2 + b^2$, so that

$$c = \sqrt{3^2 + 5^2} = \sqrt{34} \approx 5.831$$

or we can use

$$\sin \alpha = \frac{a}{c}$$

which we solve for c to obtain

$$c = \frac{a}{\sin \alpha} = \frac{3}{0.51444} \approx 5.832$$

(The discrepancy in the last digit is due to the fact that the value of α is only approximate.)

With a calculator, you can do this problem in one schedule of operations using the formulas

$$\tan \alpha = \frac{a}{b} \qquad \beta = 90° - \alpha \qquad c = \frac{a}{\sin \alpha}$$

Of course, in this problem, you do not need a calculator to divide a by b, but we shall indicate this division on the calculator to show you what to do if a long division were required at this stage:

Algebraic logic

1.	[3] [÷] [5] [=] [inv] [tan]	30.963757	(Copy for answer.)
2.	[STO] [90] [−] [RCL] [=]	59.036243	(Copy for answer.)
3.	[3] [÷] [RCL] [sin] [=]	5.8309519	(Copy for answer.)

RPN logic

1.	[3] [ENTER] [5] [÷] [ARC] [TAN]	30.96375653	(Copy for answer.)
2.	[STO] [90] [ENTER] [RCL] [−]	59.03624347	(Copy for answer.)
3.	[3] [ENTER] [RCL] [SIN] [÷]	5.830951895	(Copy for answer.)

Rounding off the copied numbers to four digits gives

$$\alpha = 30.96° \qquad \beta = 59.04° \qquad c = 5.831$$

EXAMPLE 2

Solve the right triangle in which $a = 8$ and $\alpha = 39.3°$.

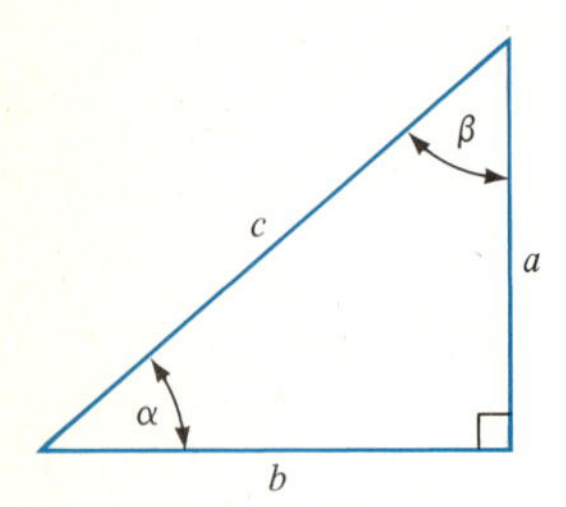

FIGURE 10.1d

SOLUTION:

We sketch the triangle (Figure 10.1d) and use the fact that $\alpha + \beta = 90°$ to find $\beta = 90° - 39.3° = 50.7°$. Next, we solve the equation $\cot \alpha = b/a$ for b to obtain

$$b = a \cot \alpha = (8)(1.2218) = 9.774$$

To find c, we use the same equation as in Example 1,

$$c = \frac{a}{\sin \alpha} = \frac{8}{0.63338} = 12.63$$

The calculator solution goes as follows:

Algebraic logic

1. [39.3] [STO] [90] [−] [RCL] [=] 50.7
2. [8] [÷] [RCL] [tan] [=] ≈ 9.774
3. [8] [÷] [RCL] [sin] [=] ≈ 12.63

RPN logic

1. [39.3] [STO] [90] [ENTER] [RCL] [−] 50.7
2. [8] [ENTER] [RCL] [TAN] [÷] ≈ 9.774
3. [8] [ENTER] [RCL] [SIN] [÷] ≈ 12.63

With your calculator, you can perform an easy check using the Pythagorean Theorem:

$$a^2 + b^2 = (8)^2 + (9.774)^2 = 159.5$$
$$c^2 = (12.63)^2 = 159.5$$

Because these results agree to four digits, they are acceptable.

EXAMPLE 3

In Figure 10.1a, suppose that the distance QS is 1000 feet and the angle α is 42.7°. Find the height of the cliff.

SOLUTION:

We use the equation $\tan \alpha = PQ/QS$ and solve for PQ to obtain

$$PQ = QS \tan \alpha = (1000)(\tan 42.7°) \approx 922.77$$

Thus, the top of the cliff is about 923 feet above ground level.

Some problems, as in Example 4, involve more than one right triangle for their solution.

EXAMPLE 4

A vertical flagpole stands at one corner of the top of a building that is 112 feet high. At a point P in the same horizontal plane as the base of the building, the angle of elevation of the line of

SOLUTION:

As usual, the first thing to do is make a sketch showing the given data and what is to be found. (See Figure 10.1e.) In the sketch, we let α be the angle of elevation of the line to the bottom of the pole and β be the angle of elevation of the line to the top of the pole. We also let x be the height of the pole, h be the height of the building, and d be the distance from the point P to the base of the building.

sight to the bottom of the pole is 33.7°, and the angle of elevation of the line of sight to the top of the pole is 37.5°. Find the height of the flagpole.

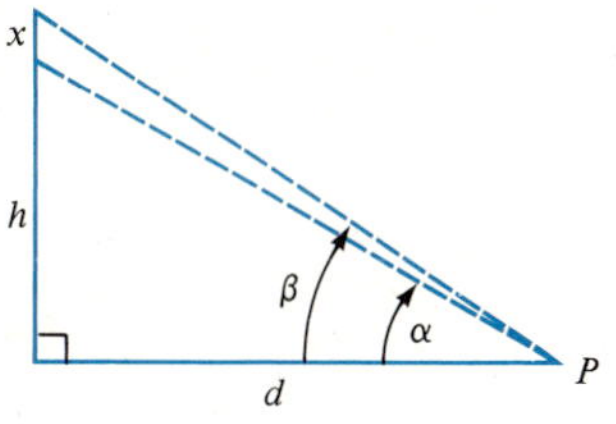

FIGURE 10.1e

Since both x and d are unknown, we shall need two equations involving them. From the smaller right triangle, we have

$$\frac{d}{h} = \cot \alpha$$

and from the larger triangle,

$$\frac{d}{x + h} = \cot \beta$$

We next solve the two equations for d to obtain

$$d = h \cot \alpha \quad \text{and} \quad d = (x + h) \cot \beta$$

Thus, we obtain the equation for x:

$$(x + h) \cot \beta = h \cot \alpha$$

or

$$x + h = h \frac{\cot \alpha}{\cot \beta}$$

and

$$x = h\left(\frac{\cot \alpha}{\cot \beta} - 1\right)$$

Now, we are ready to substitute in the data and compute the value of x. We find

$$x = 112\left(\frac{\cot 33.7^\circ}{\cot 37.5^\circ} - 1\right) \approx 16.86$$

Thus, the flagpole is about 16.9 feet high.

Note: The 16.86 was obtained with a calculator. If you use the table of trigonometric functions and four-place logarithms, your answer may be slightly different.

You should pay particular attention to the procedure that was used in Example 4 and familiarize yourself with using letters in place of the numerical values. If possible, get a formula for the final answer before you substitute in any of the numbers. This is the best and the easiest procedure because it avoids all unnecessary calculations and allows the best possible control over the accuracy of the results of arithmetic done with approximate measurements—and it also helps you learn some algebra.

EXERCISE 10.1

Throughout this exercise, unless it is otherwise indicated, find lengths to the nearest tenth of a unit and angles to the nearest tenth of a degree.

In Problems 1–10, two parts of a standard right triangle are given. Solve the triangle.

1 $a = 12, b = 9$

2 $a = 12, b = 24$

3 $a = 20, \alpha = 52^\circ$

4 $a = 15, \beta = 35^\circ$

5 $a = 5.8, \alpha = 42.6^\circ$

6 $b = 6.3, \beta = 62.5^\circ$

7 $c = 18, \alpha = 25^\circ$

8 $c = 25, \beta = 16.3^\circ$

9 $a = 6, c = 15$

10 $a = 12, c = 30$

11 From a point on the ground 100 meters from the foot of a building, the angle of elevation of the top of the building is 16.7°. How high is the building?

12 Repeat Problem 11 if the angle of elevation is 63.4°.

13 From the top of a lighthouse 150 meters high, the angle of depression (angle below the horizontal) of a ship at sea is 4.2°. How many kilometers is it from the point at sea level directly below the observer to the ship?

14 From the top of a lighthouse 450 feet high, the angle of depression of a ship at sea is 4.2°. How many miles is it from the point at sea level directly below the observer to the ship?

15 The area of a parallelogram is given by the formula $A = bh$, where b is the length of the base and h is the altitude (height). Two adjacent sides of a parallelogram are 16 and 20 inches long, and the included angle measures 50°. Find the area to the nearest tenth of a square inch.

16 Repeat Problem 15 if the sides are 32 and 36 inches long and the included angle is 125°.

17 The base of an isosceles triangle is 12 inches long, and the equal sides are each 20 inches long. Find the angle opposite the base.

18 The equal angles of an isosceles triangle each measures 50°, and each of the equal sides is 25 inches long. Find the length of the third side.

19 In a circle of radius 5 inches, a chord of length 6 inches is drawn. At one end of the chord, a line tangent to the circle is drawn and, at the other end, a line through the center of the circle. At what distance from the center of the circle does the tangent line meet the line through the center?

Hint: Let θ be the central angle subtended by the chord and use the double-angle formula $\cos\theta = 1 - 2\sin^2(\theta/2)$. You should be able to get the exact length without the use of tables or calculator.

20 Find the exact length of one side of a regular octagon inscribed in a circle of radius 10 inches.

21 A rectangle is 30 feet long and 25 feet wide. Find the smaller angle between a diagonal of the rectangle and one of the sides.

22 One of the diagonals of a parallelogram is 8 inches long and makes an angle of 32.6° with the base of the parallelogram. If the area of the parallelogram is 22 square inches, what are the lengths of the sides? (See Problem 15.)

23 In Figure 10.1f, an observer at P sees the moon when it is on the horizon at the same time that a second observer at Q also sees the moon. Both observers note the position of the moon relative to the stars and, by comparing their data, calculate that the angle α is 0.95°. The radius of the earth is 3963 miles. How far is the moon from the center of the earth?

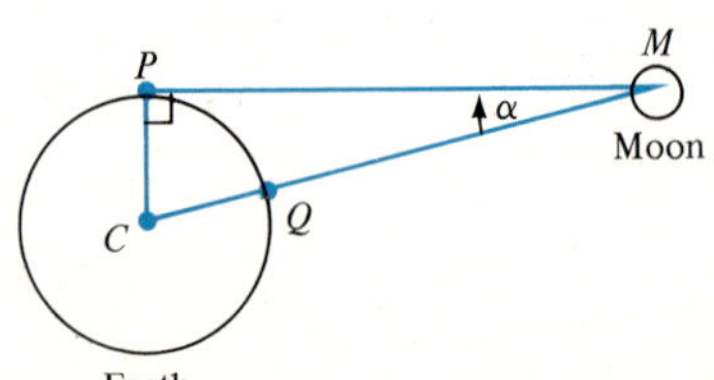

FIGURE 10.1f

24 A regular decagon (ten-sided polygon) is inscribed in a circle of radius R. Show that the area of the decagon is $5R^2 \sin 36^\circ$.

25 To find the width of a canyon, a surveyor sights straight across the canyon from a point Q on her side to a point P on the opposite side. She finds that this line of sight has a bearing of N 50.2° E. (This means that she must rotate her scope 50.2° from due north toward the east.) She then goes 200 meters in a direction N 39.8° W and reaches a point R. At R she again sights point P and finds it to be due east. Find the width of the canyon.

Hint: First draw a diagram and identify the right triangle that is involved.

26 To find the width of a north-south–running river, a surveyor locates points A and B on an east-west line on opposite sides of the river and then goes due north 125 yards from A to a point C. The surveyor then finds the bearing of CB to be S 63.4° E. Find the width of the river. (See Problem 25.)

27 In Figure 10.1g, show that

$$x = \frac{d \sin \alpha \sin \beta}{\sin(\alpha - \beta)}$$

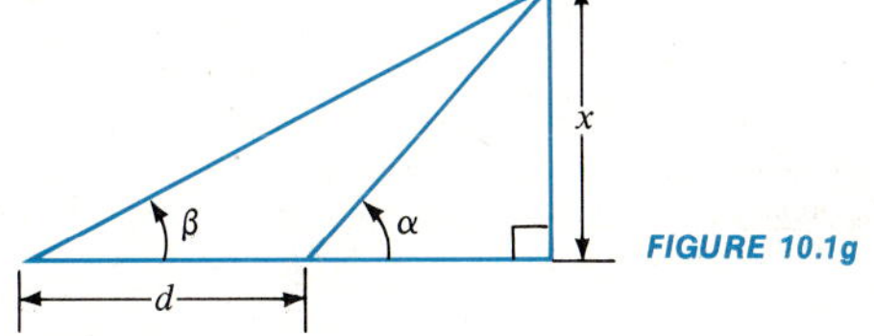

FIGURE 10.1g

Hint: For the right side, compare $a^2 + b^2$ with $(a + b)^2$. For the left side, note that $\alpha + \beta = 90°$,

$(\sin \alpha + \sin \beta)^2 = (\sin \alpha + \cos \alpha)^2 = 1 + 2 \sin \alpha \cos \alpha = 1 + \sin 2\alpha.$

Hence, $0 < (\sin \alpha + \sin \beta)^2 \leq 2$

28 Show that the inequality

$$\frac{a + b}{\sqrt{2}} \leq c < a + b$$

is true for every right triangle.

USING YOUR KNOWLEDGE 10.1

C **1** At an instant when the sun was directly overhead, an airplane and its shadow were sighted from the top of a tower 152 feet high. The angle of elevation of the plane was 68.2°, and the angle of depression of the shadow was 8.5°. Find the height of the plane to the nearest 10 feet.

C **2** A boy whose eye level is 5 feet 6 inches from the floor stands 10 feet away from a wall, holds his camera up to his eye, and sights it at a mural painted on the wall. The bottom edge of the mural is 4 feet above the floor, and the top of the mural is 8 feet above the floor. Assume these numbers are exact and find, to the nearest tenth of a degree, the angle that the mural subtends in the camera lens.

C **3** The Empire State Building in New York City is about 1250 feet high. Suppose that from the top of this building, the angle of depression of the top of a second building is 56.3° and the angle of depression of the bottom of the second building is 68.2°. What would be the height of the second building?

C **4** The highest structure in the United States is the KTHI-TV mast in Fargo, North Dakota. The mast is 2063 feet high. Suppose that a building is built on the same level as the mast and is such that the angle of elevation of the top of the mast from the top of the building is 76.05° and the angle of depression of the foot of the mast is 5.71°. What would have to be the height of the building and how far away from the mast would it be? (Answer to the nearest foot.)

10.2 COMPUTING WITH APPROXIMATE NUMBERS

When computing with approximate numbers, we must always be on guard against the tendency to present results with a false appearance of accuracy. In this section, we shall simply state some practical rules that are usually followed by people experienced in computing with approximate numbers. There are certain problems in science and technology that demand a detailed analysis of the possible errors (not mistakes) caused by working with approximate numbers, but we shall not be concerned with such analysis here.

Numbers that are the direct result of measurement are always approximate. Their accuracy is limited to the accuracy of the measuring instruments. For instance, if we measure a length correct to the nearest tenth of a millimeter and report 50.2 millimeters, we mean that the actual ("true") length is between 50.15 and 50.25 millimeters. A closer measurement might narrow the interval in which the "true" length lies but can never give us the exact "true" length. As you have already seen, approximate numbers also occur when we make decimal approximations of irrational numbers (such as $\pi \approx 3.14$). Most of the entries in the logarithm and trigonometric tables are approximate.

ROUNDING OFF

There are many instances in which we must "round off" a given number. We frequently round off tabular values if we intend to use fewer digits than the table presents, and we always round off in obtaining a decimal approximation of an irrational number. Let us use the word *residue* for the number represented by the discarded digits (the dropped digits). Then we can state the rules for rounding off quite simply.

RULES FOR ROUNDING OFF

1. **If the residue is less than half a unit in the last retained place, simply drop it.**
2. **If the residue is half a unit or more in the last retained place, add one unit in that place and drop the residue.**

EXAMPLE 1

Round off the value $\pi \approx 3.1415926$ to:

a. three decimal places

b. five decimal places

SOLUTION:

a. Because 0.0005926 is more than half a unit in the third decimal place, we round off by adding one unit in the third place. Thus, the desired result is 3.142.

b. Since 0.0000026 is less than half a unit in the fifth decimal place, we round off to 3.14159.

ADDITION AND SUBTRACTION OF APPROXIMATE NUMBERS

A practical rule to be observed when adding or subtracting approximate numbers is as follows:

RULE FOR ADDING OR SUBTRACTING APPROXIMATE NUMBERS

The result of an addition or subtraction of approximate numbers must not be given to more decimal places than are possessed by that one of the numbers with the *fewest* decimal places.

Usually the result is stated to the maximum number of places compatible with this rule.

EXAMPLE 2

Find the value of $21.262 + 23.75 - 39.6$ if these are all approximate numbers.

SOLUTION:

$$\begin{array}{r} 21.262 \\ +\ 23.75 \\ \hline 45.012 \\ -\ 39.6 \\ \hline 5.412 \end{array}$$

The least precise of the given numbers is 39.6, with just one decimal place. Therefore, the answer correctly rounded off is 5.4.

SIGNIFICANT DIGITS

▼ DEFINITION 10.2a

Significant digits of a number

1. **The digits 1, 2, 3, 4, 5, 6, 7, 8, and 9 are always significant.**
2. **The digit 0 is significant if it is preceded and followed by other significant digits.**
3. **The digit 0 is never significant when its only function is to place the decimal point.**

Before stating the rule for multiplication and division of approximate numbers, we must define the *significant digits* of a number. (See Definition 10.2a.) For example, according to the definition, all six digits in each of the numbers 1.41421 and 1.73205 are significant. The number 0.00721 has only three significant digits, 7, 2, and 1; the zeros in this number do nothing but place the decimal point.

It is generally agreed that, when an approximate number has a decimal point, *final* zeros are taken to be significant. For instance, 7.20 is regarded as having three significant digits. Notice that the position of the decimal point has nothing to do with the number of significant digits.

There is one ambiguous case left: For example, how many significant digits does the number 73,200 have? This question cannot be answered without more information about the final zeros. We can eliminate the ambiguity by using scientific notation. Thus, we would write 73,200 in the form

7.32×10^4 if only **three** digits are significant;
7.320×10^4, if **four** digits are significant;
7.3200×10^4 if all **five** digits are significant.

MULTIPLICATION AND DIVISION OF APPROXIMATE NUMBERS

We are now ready to state the rule for multiplication and division of approximate numbers.

RULE FOR MULTIPLYING OR DIVIDING APPROXIMATE NUMBERS

The result of a multiplication or division of approximate numbers must never be given with more significant digits than are possessed by that one of the numbers with the *fewest* significant digits.

EXAMPLE 3

Find the product of the approximate numbers 0.00227 and 35.63.

SOLUTION:

By direct multiplication, we obtain 0.0808801, which we round off to three significant digits, 0.0809.

EXAMPLE 4

Express $180/\pi$ correct to two decimal places.

SOLUTION:

Because the value of π is a little more than 3, we see that four significant digits are required in the result. To be safe, we use a five-digit approximation to π, and then round off to two decimal places. Thus,

$$\frac{180}{\pi} \approx \frac{180}{3.1416} \approx 57.2956 \approx 57.30$$

Of course, this is approximately the number of degrees in one radian. (Do not overlook the fact that the 180 is exact.)

ACCURACY OF SOLUTIONS OF TRIANGLES

In solving triangles, we use a table of relationships between the accuracy of measurement and computation of sides and angles (Table 10.2a). The rules for the accuracy to be stated in the results are usually derived by calculus methods. For instance, if the data included two sides with *three* significant digits and an angle correct only to the nearest degree, then the other side would be given to only *two* significant digits and the remaining angles to the nearest degree. In general, results should not be stated with more accuracy than the least accurate of the given data.

TABLE 10.2a

Sides	Angles
Two significant digits	Nearest degree
Three significant digits	Nearest tenth of a degree
Four significant digits	Nearest hundredth of a degree

EXAMPLE 5

In a right triangle, the hypotenuse is measured to be 10.352 meters long and one of the acute angles to be 37.4°. Find the side opposite this angle.

SOLUTION:

We have given $c = 10.352$ meters and $\alpha = 37.4°$. Hence,

$$\begin{aligned} a = c \sin \alpha &= 10.352 \sin 37.4° \\ &= 6.2876 \qquad \text{By tables or calculator} \end{aligned}$$

Since the angle is given to the nearest tenth of a degree, we must round off the answer to three significant digits. Thus, the side is approximately 6.29 meters long. Notice that the lesser accuracy is in the given angle, which therefore governs the accuracy of the final result.

Most people who have to compute with approximate data carry one or more extra digits through the intermediate calculations and then round off the final results. This is the procedure that we followed in Example 5 and that is followed in the remainder of this text.

EXERCISE 10.2

In Problems 1–6, the given numbers are all approximate, correct to the given number of decimal places. Perform the indicated operations and give the answer correctly rounded off.

1 $6.09 + 5.4 + 849.0 + 7.01$

2 $92.33 + 4.785 + 3.5 + 7.13$

3 $1.962 + 0.3427 + 2.400 + 3.29$

4 $0.003 + 1.4729 + 0.051 + 0.126$

5 $0.081 + 2.315 - 0.0969$

6 $4.321 + 0.120 - 3.72$

In Problems 7–14, the given numbers are all approximate, correct to the stated number of significant digits. Perform the indicated operations and give the answer correctly rounded off and in scientific notation.

7 $(2000.)(8.00)$

8 $(700.0)(0.40)$

9 $(456)(0.00500)$

10 $(50{,}000.)(0.062)$

11 $48.00 \div 0.120$

12 $11.2 \div 2.24$

13 $480.0 \div 0.12$

14 $0.04560 \div 0.0024$

Problems 15–26 should be done on a calculator. If you do not have a calculator, you may skip the computation but try to figure out how many significant digits should be given in the answer. Again, all the data numbers are approximate.

15 $(6.81)(10.32) - 5.273$

16 $(0.36)(0.472) - 4.00$

17 $\dfrac{(5.62)(3.81)}{6.83}$

18 $\dfrac{(4.371)(2.637)(25.2)}{(62.0)(0.971)}$

19 $(3.26)(0.953)^2$

20 $(5.72)^2(1.50) - 48.2$

21 $\dfrac{1}{\sqrt{6.32} + \sqrt{3.71}}$

22 $\dfrac{1}{\sqrt{6.3218} - \sqrt{6.3198}}$

Hint: Rationalize the denominator.

23 $\dfrac{81.2 \sin 43.5^\circ}{95.7}$

24 $\dfrac{32 \sin 29.7^\circ}{126}$

25 $\dfrac{5.623 \sin 65.74^\circ}{\sin 57.32^\circ}$

26 $\dfrac{4.32 \sin 27.25^\circ}{\sin 32.76^\circ}$

In Problems 27–36, two parts of a right triangle are given. Solve the triangle. Assume the data are approximate numbers and give answers consistent with Table 10.2a.

27 $a = 36.9, \alpha = 38.1^\circ$

28 $a = 102, \beta = 43^\circ$

29 $c = 597.2, \alpha = 19.83^\circ$

30 $c = 32.8, \beta = 25.6^\circ$

31 $a = 135, b = 217$

32 $a = 48.72, b = 32.5$

33 $a = 75, c = 105$

34 $b = 7.36, c = 9.514$

35 $c = 45.63, \alpha = 30.1^\circ$

36 $a = 8.0, \beta = 63.91^\circ$

USING YOUR KNOWLEDGE 10.2

We can check the reasonableness of the rules given in the preceding section by means of a small amount of calculation. Thus, consider the product (4.23)(57.8), where the numbers are both approximate. Because 4.23 represents a number between 4.225 and 4.235 and 57.8 represents a number between 57.75 and 57.85, we see that

$$(4.225)(57.75) \le (4.23)(57.8) < (4.235)(57.85) \qquad \text{or} \qquad 243.99375 \le 244.494 < 244.99475$$

This shows that 244 is a reasonable result for the product of the two approximate numbers. Notice that there is only a slight doubt about the accuracy of the third digit, but that no more than three digits in the answer can be justified.

Similarly, we have for the quotient 57.8/4.23

$$\frac{57.75}{4.235} < \frac{57.8}{4.23} < \frac{57.85}{4.225} \qquad \text{or} \qquad 13.636364 < 13.664303 < 13.692308$$

Hence, we see that 13.7 is a reasonable result for the quotient. Notice that we selected the possible errors in the approximate numbers so that the range of the answer would be as large as possible. For instance, in the case of the quotient, the lower bound is obtained by using the smallest possible numerator and the largest possible denominator, whereas the upper bound is obtained by using the largest possible numerator and the smallest possible denominator.

Addition and subtraction problems can be analyzed in the same way as multiplication and division problems. Use the preceding ideas to analyze the accuracy of the result in each of the following problems. All of the numbers are to be taken as approximate.

C **1** (37.2)(2.56) C **2** (48.23)(5.79) C **3** 37.2 ÷ 2.56 C **4** 48.23 ÷ 5.79 C **5** 52.3 sin 63.7°
C **6** 4.92 cos 43.5° **7** 37.2 + 2.56 **8** 48.23 + 5.79 **9** 37.2 − 2.56 **10** 48 23 − 5.79

10.3 OBLIQUE TRIANGLES, CASE I: GIVEN ONE SIDE AND TWO ANGLES

In Figure 10.3a, Q is a point in the horizontal plane and vertically below the point P on the cliff. To find the height PQ of the cliff, the horizontal distance d and the angles α and β are measured, and, from these data, the height is to be calculated. Although this problem can be solved by using the two right triangles PQR and PQS, it is much simpler to use the oblique triangle PRS and one of the right triangles. If you think about it for a moment, you will realize that if PR were known, then it would be a simple matter to find PQ, because

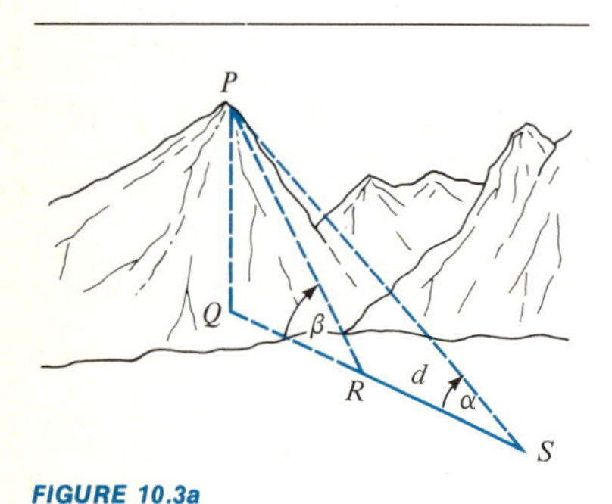

FIGURE 10.3a

$$PQ = PR \sin \beta$$

AREA OF A TRIANGLE

To be able to find PQ from the given data in an efficient manner, we shall develop an important relationship among the sides and the angles of any triangle. However, we need a simple preliminary result first. We know from geometry that the area of any triangle is one half the product of its base and its altitude (height). Consider the two triangles in Figure 10.3b, where we have used the usual convention of denoting the angles by the Greek letters correspond-

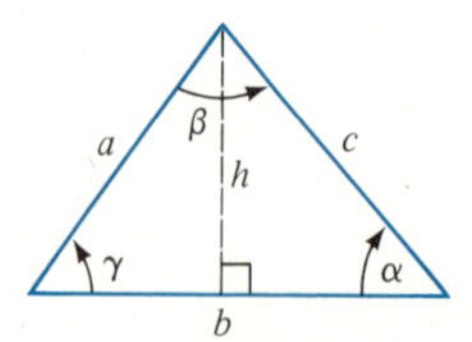

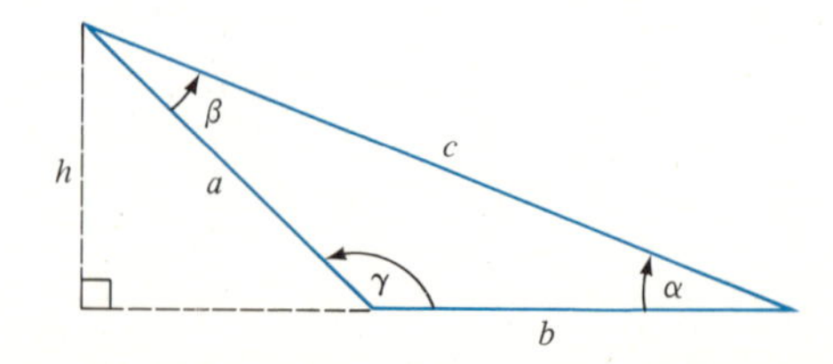

FIGURE 10.3b

ing to the English letters on the opposite sides. In both triangles, $h = a \sin \gamma$. (In the triangle on the right, $h = a \sin(180° - \gamma) = a \sin \gamma$.) Consequently, the formula for the area is

$A = \frac{1}{2}ab \sin \gamma$ Substitute for h in the formula $A = \frac{1}{2}bh$.

Since we are free to select any side of the triangle as base and use the corresponding altitude to that side, we may state the general result in Theorem 10.3a.

▼ THEOREM 10.3a

The area of any triangle is one half the product of any two sides and the sine of the included angle.

Using the standard lettering for the sides and angles, we can write three formulas for the area:

$$A = \frac{1}{2}ab \sin \gamma \quad (1)$$
$$A = \frac{1}{2}bc \sin \alpha \quad (2)$$
$$A = \frac{1}{2}ac \sin \beta \quad (3)$$

EXAMPLE 1

Find the area of the oblique triangle in which $b = 5$ inches, $c = 12$ inches, and $\alpha = 30°$.

SOLUTION:

By Formula 2, we have

$$A = \frac{1}{2}(5)(12) \sin 30° = \frac{1}{2}(5)(12)(\frac{1}{2}) = 15$$

Thus, the area is 15 square inches.

THE LAW OF SINES

Because the area of a given triangle is a unique number, all three of the formulas for A must give the same number. Thus, from Equations 1 and 2, we obtain

$$\frac{1}{2}ab \sin \gamma = \frac{1}{2}bc \sin \alpha$$

which, upon division by $\frac{1}{2}b \sin \alpha \sin \gamma$, gives

$$\frac{a}{\sin \alpha} = \frac{c}{\sin \gamma}$$

Using Equations 2 and 3 in a similar way, we find

$$\frac{b}{\sin \beta} = \frac{a}{\sin \alpha}$$

▼ THEOREM 10.3 b

The law of sines

The sides of any triangle are proportional to the sines of the opposite angles; that is,

$$\frac{a}{\sin \alpha} = \frac{b}{\sin \beta} = \frac{c}{\sin \gamma}$$

We now combine the last two results in Theorem 10.3b. You should satisfy yourself that this theorem holds for right triangles. Thus, if $\gamma = 90°$, then $\sin \gamma = 1$, and the equations reduce to

$$\frac{a}{\sin \alpha} = c \quad \text{or} \quad \sin \alpha = \frac{a}{c}$$

and

$$\frac{b}{\sin \beta} = c \quad \text{or} \quad \sin \beta = \frac{b}{c}$$

in agreement with our previous definitions for right triangles.

SOLUTION OF SAA PROBLEMS

The Law of Sines is a handy tool for solving an oblique triangle when one side and two angles are given. We shall call this problem Case I and denote it by SAA (*s*ide, *a*ngle, *a*ngle). The remainder of this section is devoted to Case I problems.

We first return to Figure 10.3a, noting that $\angle PRS = 180° - \beta$, so that

$$\angle RPS + \alpha + (180° - \beta) = 180° \qquad \text{The sum of the angles of any triangle is } 180°.$$

which gives

$$\angle RPS = \beta - \alpha$$[1]

Next, we see that in $\triangle PRS$ the angle RPS is opposite the side RS and angle α is opposite side PR. Hence, by the Law of Sines,

$$\frac{PR}{\sin \alpha} = \frac{d}{\sin(\beta - \alpha)}$$

or

$$PR = \frac{d \sin \alpha}{\sin(\beta - \alpha)} \qquad \text{Multiply by } \sin \alpha$$

Now,

$$PQ = PR \sin \beta$$

$$= \frac{d \sin \alpha \sin \beta}{\sin(\beta - \alpha)} \qquad \text{Because } PR = \frac{d \sin \alpha}{\sin(\beta - \alpha)}$$

(If you did Problem 27, Exercise 10.1, you will appreciate the ease of this method of solution.)

EXAMPLE 2

Solve the triangle in which $c =$ 25.2 inches, $\alpha = 23.5°$, and $\beta =$ 33.3°.

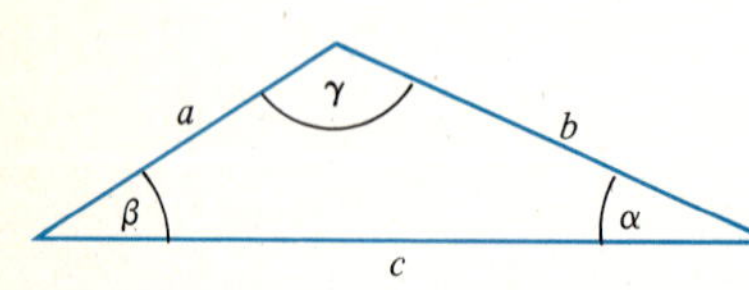

FIGURE 10.3c

SOLUTION:

First we make a schematic sketch of the triangle. (See Figure 10.3c.) Then, because $\alpha + \beta + \gamma = 180°$,

$$\gamma = 180° - (\alpha + \beta)$$
$$= 123.2°$$

Next, from the Law of Sines,

$$a = \frac{c \sin \alpha}{\sin \gamma} = \frac{25.2 \sin 23.5°}{\sin 123.2°} = 12.01$$

and

$$b = \frac{c \sin \beta}{\sin \gamma} = \frac{25.2 \sin 33.3°}{\sin 123.2°} = 16.53$$

---

[1] If you recall some more geometry, you can obtain this result from the theorem that says that an exterior angle of a triangle is the sum of the two remote interior angles.

The calculations can be done on a calculator or by using the table of trigonometric functions and logarithms.[2] The calculator solution is as follows:

Algebraic logic

[25.2] [÷] [123.2] [sin] [=] [STO] [×] [23.5] [sin] [=] ≈ 12.009
[RCL] [×] [33.3] [sin] [=] ≈ 16.534

RPN logic

[25.2] [ENTER] [123.2] [SIN] [÷] [STO] [23.5] [SIN] [×] ≈ 12.009
[RCL] [33.3] [SIN] [×] ≈ 16.534

Thus, the answers, correctly rounded off, are $\gamma = 123.2°$, $a = 12.0$ inches, and $b = 16.5$ inches.

EXAMPLE 3

The Guinness book of World Records lists among the world's notable sculptures a tall figure that was found in 1968 on a hill above Tarapacár, Chile. One way of finding the height of such a figure without taking it down or otherwise damaging it is illustrated by the following (fictitious) problem: Suppose that the hill slopes upward at an angle of 31.9° and, at a point on the hill 350 feet down from the foot of the figure, the angle of elevation of the top of the figure is 60.0°. Find the height of the figure.

SOLUTION:

We first make a sketch (Figure 10.3d) to show the given data. Because the hill slopes upward at an angle of 31.9°, $\angle BAD$ is 31.9°, so that $\alpha = 60.0° - 31.9° = 28.1°$. Also, triangle ADC is a right triangle with right angle at D. Therefore, $\gamma = 90° - \angle CAD = 30.0°$. We can now use the Law of Sines, which tells us that

$$\frac{BC}{\sin \alpha} = \frac{AB}{\sin \gamma}$$

that is,

$$\frac{x}{\sin 28.1°} = \frac{350}{\sin 30°}$$

Hence,

$$x = \frac{350 \sin 28.1°}{\sin 30°}$$

$$= 329.7$$

so that the figure is about 330 feet tall.

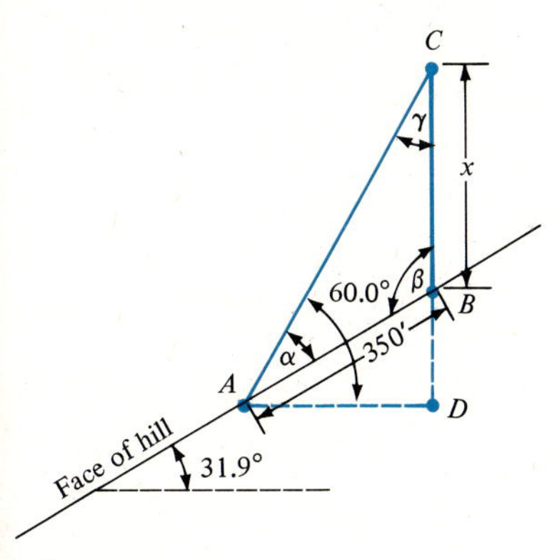

FIGURE 10.3d

Surveyors usually specify directions relative to due north or due south. For instance, the notation N 32.5° E means that the direction is obtained by turning 32.5° toward the east from due north. Similarly, S 45° W means the direction that is obtained by turning 45° toward the west from due south.

[2] There are tables that give the logarithms of the trigonometric functions, but with the advent of electronic calculators, such tables are seldom used.

EXAMPLE 4

A surveyor wishes to find the distance between two points A and B that are separated by a deep hole. From A, he finds the bearing of B to be N 26.40° E. He then proceeds to a point C, 1000 meters due west from A, and finds the bearing of CB to be N 32.50° E. Find the distance AB to the nearest meter.

SOLUTION:

We first sketch triangle ABC to show the given data. (See Figure 10.3e.) The figure makes it clear that

$$\alpha = 90^\circ + 26.4^\circ = 116.4^\circ$$

$$\gamma = 90^\circ - 32.5^\circ = 57.5^\circ$$

Therefore,

$$\beta = 180^\circ - (116.4^\circ + 57.5^\circ) = 6.1^\circ \qquad \alpha + \beta + \gamma = 180^\circ$$

Now, we use the Law of Sines to write

$$\frac{c}{\sin \gamma} = \frac{b}{\sin \beta}$$

from which we obtain

$$c = \frac{b \sin \gamma}{\sin \beta} = \frac{1000 \sin 57.5^\circ}{\sin 6.1^\circ} = 7936.8$$

Hence, AB is 7937 meters, to the nearest meter.

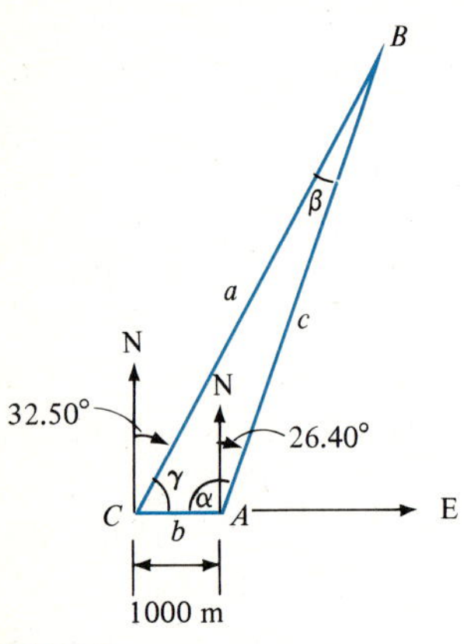

FIGURE 10.3e

EXERCISE 10.3

In Problems 1–10, find the area of the triangle.

1 $a = 10, b = 60, \gamma = 60^\circ$
2 $b = 20, c = 10\sqrt{2}, \alpha = 135^\circ$
3 $a = 7, c = 8, \beta = 30^\circ$
4 $a = 12, b = 9, \gamma = 120^\circ$
5 $b = 6, c = 10, \alpha = 45^\circ$
6 $a = 8, b = 20, \gamma = 150^\circ$
7 $b = 31.5, c = 52.6, \alpha = 49.3^\circ$
8 $a = 101.3, b = 92.7, \gamma = 105.6^\circ$
9 $a = 85.2, c = 90.3, \beta = 112.8^\circ$
10 $b = 25.7, c = 40.8, \alpha = 36.4^\circ$

11 Find the area of the parallelogram whose sides are 15 and 18 inches long and one of whose angles is 32.5°.

12 Find the area of the parallelogram whose sides are 27.3 and 40.5 inches long and one of whose angles is 110.8°.

In Problems 13–18, solve the triangle. Assume that the data are exact and give the sides to three significant digits.

13 $a = 5, \beta = 50^\circ, \gamma = 60^\circ$
14 $a = 12, \alpha = 35^\circ, \beta = 105^\circ$
15 $b = 43, \alpha = 65^\circ, \gamma = 70^\circ$
16 $b = 50, \alpha = 95^\circ, \beta = 60^\circ$
17 $c = 110, \alpha = 85^\circ, \beta = 48^\circ$
18 $c = 96, \beta = 47^\circ, \gamma = 75^\circ$

In Problems 19–26, solve the triangle. Follow the rules for approximate data.

19 $a = 80.2, \beta = 56.7^\circ, \gamma = 61.3^\circ$
20 $b = 51.7, \alpha = 38.5^\circ, \beta = 16.4^\circ$
21 $c = 110.3, \alpha = 81.7^\circ, \gamma = 52.8^\circ$
22 $a = 392.5, \alpha = 112.9^\circ, \beta = 46.1^\circ$
23 $b = 86.7, \beta = 93.5^\circ, \gamma = 61.5^\circ$
24 $a = 2187, \alpha = 35.9^\circ, \beta = 53.7^\circ$
25 $c = 32.96, \beta = 47.12^\circ, \gamma = 108.31^\circ$
26 $c = 95.73, \alpha = 105.62^\circ, \gamma = 38.97^\circ$

27 The base of a parallelogram is 18.0 centimeters long; the smaller angle between the base and an adjacent side is 32.8°; and the shorter diagonal forms an angle of 38.5° with the base. Find the length of the adjacent side.

28 The base of a parallelogram is 30.6 inches long; the larger angle between the base and an adjacent side is 110.7°; and the longer diagonal forms an angle of 28.4° with the base. Find the length of the adjacent side.

29 In a triangle, α, β, and c are given. Find a formula for the area in the terms of these three parts. Start with the formula

$$A = \tfrac{1}{2}bc \sin \alpha$$

and use the Law of Sines to express b in terms of α, β, and c. You should end up with the formula

$$A = \frac{c^2 \sin \alpha \sin \beta}{2 \sin(\alpha + \beta)}$$

30 From the Law of Sines, obtain the equation

$$\frac{a}{b} = \frac{\sin \alpha}{\sin \beta}$$

Then subtract 1 from both sides to get

$$\frac{a - b}{b} = \frac{\sin \alpha - \sin \beta}{\sin \beta}$$

Use a similar procedure to get

$$\frac{a + b}{b} = \frac{\sin \alpha + \sin \beta}{\sin \beta}$$

Now combine the last two equations to obtain

$$\frac{a - b}{a + b} = \frac{\sin \alpha - \sin \beta}{\sin \alpha + \sin \beta}$$

Next, use the formulas obtained in Problems 37 and 38, Exercise 9.3 to arrive at the result

$$\frac{a - b}{a + b} = \frac{\tan \frac{1}{2}(\alpha - \beta)}{\tan \frac{1}{2}(\alpha + \beta)}$$

This is one form of the Law of Tangents, a formula that is sometimes used to solve triangles with logarithms. The other two forms can be written by a cyclic interchange of the parts of the triangle. Write these forms.

31 The face of a hill is inclined at an angle of 15.1° to the horizontal. A large, flat slab of rock has its base 1200 feet up the hill and is inclined at an angle of 36.6° to the horizontal. At the bottom of the hill, the angle of elevation of the top of the rock is 25.3°. Assume that all measurements are in the same vertical plane and find the length of the slab.

32 An airplane flies directly over two ground stations that are 500 meters apart on the same level plain. An observer at A finds the angle of elevation of the plane to be 37.6° at the same instant that an observer at B finds the angle of elevation to be 39.8°. If the plane is flying in the direction from A to B and has just passed over B, what is the plane's altitude? (Assume all measurements to be in the same vertical plane.)

33 Do Problem 32 if the plane has not yet passed over B.

34 The navigator of a plane flying at an altitude of 10,000 feet finds the angle of depression of a ground station ahead to be 10.7°. One minute later, he finds the angle of depression of the same station to be 20.7°. If the station is still straight ahead, what is the ground speed of the plane (the speed relative to the ground)?

35 In checking a triangular piece of land, a surveyor finds side AB to be 106.2 meters long. From A, the bearing of C is N 15.87° E, and the bearing of B is N 48.32° E. From B, the bearing of C is N 32.51° W. Find the area of this piece of land.

36 A surveyor finds the side AB of a triangular piece of land to be 1524 feet long. The surveyor also finds the bearing of AB to be S 31.2° E, that of AC to be S 25.7° W, and that of BC to be S 78.3° W. If an acre is 43,560 square feet, how many acres are in this piece of land?

USING YOUR KNOWLEDGE 10.3

An interesting application of the Law of Sines occurs in the solution of a problem that is often met in road construction, the problem of laying out a circular curve that passes through a given point and is tangent to two given intersecting lines. In Figure 10.3f, the x-axis has been taken as one of the tangent lines; the other tangent line is the line OS. The angle β between the two tangent lines and the location of the point P through which the circle is to pass are known. We need to find R, the radius of the circle, and to locate C, the center of the circle.

Because the location of P is given, we know the angle α and the distance y from the x-axis to P. Let C be the center of the desired circle, and draw a line from C to the point of tangency, Q, on the x-axis. On CQ, mark the point D so that $QD = y$. Thus, $CP = CQ = R$, and $CD = R - y$. We also know that the line OC bisects the angle between the two tangents, so that angle $QOC = \frac{1}{2}\beta$. Since Q is the point of tangency, the line CQ is perpendicular to the x-axis, and OCQ is a right triangle.

We now let angle PCQ be denoted by θ. Then, we see from the geometry of the diagram that

$$\begin{aligned}\angle COP &= \tfrac{1}{2}\beta - \alpha \\ \angle OCP &= 90^\circ - \tfrac{1}{2}\beta - \theta \\ \angle OPC &= 180^\circ - (90^\circ - \tfrac{1}{2}\beta - \theta) - (\tfrac{1}{2}\beta - \alpha) \\ &= 90^\circ + \alpha + \theta\end{aligned}$$

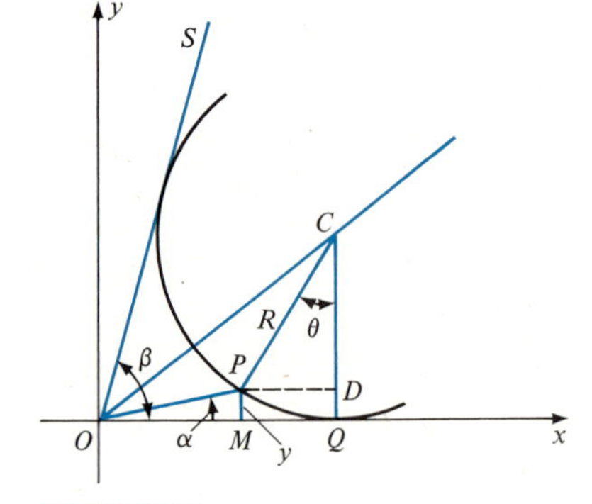

FIGURE 10.3f

From triangle OCQ, we have

$$\frac{QC}{OC} = \sin \tfrac{1}{2}\beta \qquad \text{or} \qquad OC = \frac{QC}{\sin \frac{1}{2}\beta} = \frac{R}{\sin \frac{1}{2}\beta}$$

Now, we apply the Law of Sines to triangle OPC to obtain

$$\frac{OC}{PC} = \frac{\sin \angle OPC}{\sin \angle COP} \qquad \text{or} \qquad \frac{R}{R \sin \frac{1}{2}\beta} = \frac{\sin(90^\circ + \alpha + \theta)}{\sin(\frac{1}{2}\beta - \alpha)}$$

$$\text{or} \qquad \frac{1}{\sin \frac{1}{2}\beta} = \frac{\cos(\alpha + \theta)}{\sin(\frac{1}{2}\beta - \alpha)}$$

In this equation, α and β are both known, so we can solve for θ. First, we find

$$\cos(\alpha + \theta) = \frac{\sin(\frac{1}{2}\beta - \alpha)}{\sin \frac{1}{2}\beta}$$

from which we can find $\alpha + \theta$, and, by subtracting α, we have θ.

Finally, we go to the right triangle PCD, where θ is now known and where $PC = R$. We write

$$\cos \theta = \frac{CD}{PC} = \frac{R - y}{R} = 1 - \frac{y}{R}$$

from which we find

$$\frac{y}{R} = 1 - \cos \theta \qquad \text{and} \qquad R = \frac{y}{1 - \cos \theta}$$

Because y is known, we can now calculate R and OC.

As an example, suppose that the angle β between the tangents is 60°, that $\alpha = 10°$ and $y = 40$ feet. (For illustrative purposes, all the data are assumed to be exact.) To solve the problem, we need the three equations

$$\cos(\alpha + \theta) = \frac{\sin(\frac{1}{2}\beta - \alpha)}{\sin \frac{1}{2}\beta}, \qquad R = \frac{y}{1 - \cos\theta}, \qquad \text{and} \qquad OC = \frac{R}{\sin\frac{1}{2}\beta}$$

Using the given data, we have $\frac{1}{2}\beta = 30°$ and $\frac{1}{2}\beta - \alpha = 20°$. The first equation gives

$$\cos(10° + \theta) = \frac{\sin 20°}{\sin 30°}$$

from which we obtain

$$10° + \theta = 46.84° \quad \text{and} \quad \theta = 36.84°$$

From the second equation,

$$R = \frac{40}{1 - \cos 36.84°}$$

$$= 200.31$$

The third equation gives

$$OC = \frac{200.31}{\sin 30°} = 400.62$$

Thus, the center of the circle is on the bisector about 400.6 feet from the intersection of the tangent lines, and the radius of the circle is about 200.3 feet.

Use the preceding ideas to solve the following problems.

1 Suppose that two roads intersect at an angle of 160° and that they are to be joined by a circular curve whose center line passes through a point P such that $\alpha = 10°$ and $y = 30$ feet. Find the radius and locate the center C of the center line of the curve. See Figure 10.3g. (Assume the given data are exact.)

2 Repeat Problem 1 if the roads intersect at an angle of 80° and $\alpha = 25°$, $y = 60$ feet.

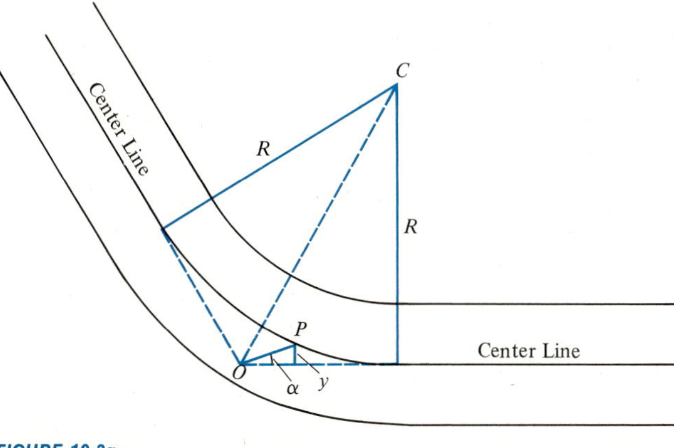

FIGURE 10.3g

10.4 OBLIQUE TRIANGLES, CASE II: GIVEN TWO SIDES AND THE INCLUDED ANGLE

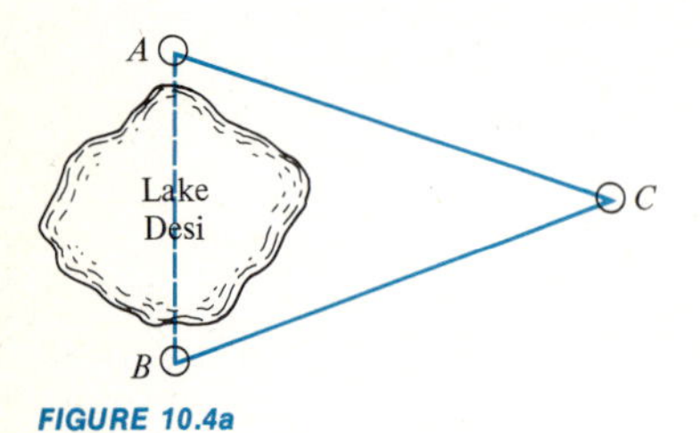

FIGURE 10.4a

Figure 10.4a shows a commonly occurring problem, that of finding the distance between two points A and B separated by an obstruction that prevents direct measurement. In this case, A and B are two points on opposite sides of a lake. Both points can be reached from a third point, C, so that the lengths of AC and BC as well as the measure of $\angle ACB$ can be found. The Law of Sines will not handle this problem because we do not know any angle opposite one of the known sides.

This problem is a Case II problem: Two sides and the included angle are known. We shall abbreviate this information by SAS (*s*ide, *a*ngle, *s*ide). To solve a Case II problem, it is convenient to have a relationship among the sides and angles other than the Law of Sines. We shall derive such a relationship next.

THE LAW OF COSINES

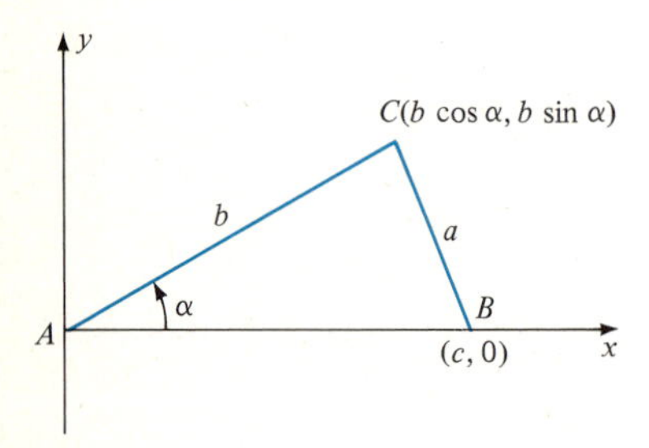

FIGURE 10.4b

Let triangle ABC (Figure 10.4b) be placed so that angle α is in standard position. Then, the coordinates of B are $(c, 0)$, and the coordinates of C are $(b \cos \alpha, b \sin \alpha)$. We use the distance formula

$$d = \sqrt{(x_2 - x_1)^2 + (y_2 - y_1)^2}$$

to obtain an expression for a^2, the square of the distance from B to C. Thus,

$$\begin{aligned} a^2 &= (b \cos \alpha - c)^2 + (b \sin \alpha - 0)^2 \\ &= b^2 \cos^2 \alpha - 2bc \cos \alpha + c^2 + b^2 \sin^2 \alpha \\ &= b^2(\cos^2 \alpha + \sin^2 \alpha) + c^2 - 2bc \cos \alpha \end{aligned}$$

or

$$a^2 = b^2 + c^2 - 2bc \cos \alpha$$

Note that it does not matter whether α is an acute angle or not; the coordinates of C will always be $(b \cos \alpha, b \sin \alpha)$.

Obviously, we may put any one of the angles of the triangle in the standard position, so that the same procedure will give us a formula for the square of the side opposite that angle. The result is Theorem 10.4a.

It is interesting to note that if $\gamma = 90°$, the last of the equations in Theorem 10.4a reduces to $c^2 = a^2 + b^2$. Thus, the Law of Cosines is truly a generalization of the famous Pythagorean Theorem.

▼ THEOREM 10.4a

The law of cosines

In any triangle, the square of a side equals the sum of the squares of the other two sides diminished by twice their product times the cosine of the included angle. In symbols:

$$a^2 = b^2 + c^2 - 2bc \cos \alpha$$
$$b^2 = a^2 + c^2 - 2ac \cos \beta$$
$$c^2 = a^2 + b^2 - 2ab \cos \gamma$$

SOLUTION OF SAS PROBLEMS

EXAMPLE 1

In Figure 10.4a, let the length of AC be 500 feet, that of BC be 600 feet (both to the nearest foot), and let $\angle ABC = 60.0°$. Find the distance from A to B.

SOLUTION:

In our usual notation, we have given $b = 500$ feet, $a = 600$ feet, and $\gamma = 60.0°$. Using the third equation of Theorem 10.4a, we obtain

$$\begin{aligned} c^2 &= a^2 + b^2 - 2ab\cos\gamma \\ &= (600)^2 + (500)^2 - 2(600)(500)\cos 60° \\ &= 360{,}000 + 250{,}000 - 300{,}000 \\ &= 310{,}000 \end{aligned}$$

Therefore,

$$c = \sqrt{310{,}000} = 556.776$$

and the distance from A to B is approximately 557 feet.

EXAMPLE 2

Solve the triangle in which $a = 25.0$, $b = 32.0$, and $\gamma = 46.3°$.

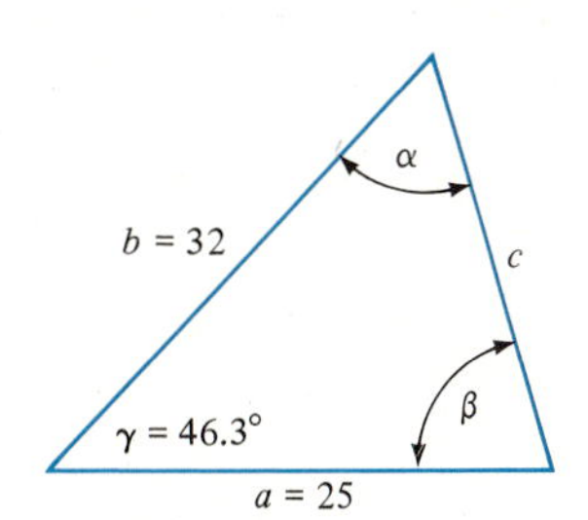

FIGURE 10.4c

SOLUTION:

We first make a sketch showing the notation. (See Figure 10.4c.) The side c can be found by using the Law of Cosines.

$$\begin{aligned} c^2 &= a^2 + b^2 - 2ab\cos\gamma \\ &= (25)^2 + (32)^2 - 2(25)(32)\cos 46.3° \end{aligned}$$

The calculator procedure for finding c is as follows:

Algebraic logic

[25] [x^2] [+] [32] [x^2] [−] [2] [×] [25] [×] [32] [×] [46.3] [cos] [=] [$\sqrt{x}$] ≈ 23.315

RPN logic

[25] [ENTER] [×] [32] [ENTER] [×] [+] [2] [ENTER] [25] [×] [32] [×] [46.3] [COS] [×] [−] [$\sqrt{x}$] ≈ 23.315

Although it is possible to use the Law of Cosines to find the angles α and β, it is much more convenient to use the Law of Sines. From the data, we see that $a < b$, so that we know that $\alpha < \beta$. Hence, α must be an acute angle. (A triangle cannot have more than one angle $\geq 90°$.) Thus,

$$\sin\alpha = \frac{a\sin\gamma}{c} = \frac{25\sin 46.3°}{23.315}$$

and

$$\alpha = 50.82°$$

Similarly, because $\alpha + \gamma > 90°$, β must be an acute angle. Hence, from

$$\sin\beta = \frac{b\sin\gamma}{c} = \frac{32\sin 46.3°}{23.315}$$

we find

$$\beta = 82.88°$$

As a check, we add the three angles:

$$50.82° + 82.88° + 46.3° = 180°$$

Note that we chose to calculate both angles by the Law of Sines so that an easy check on the result would be available. For this reason, our procedure is better than that of finding one angle by the Law of Sines and subtracting the sum of this angle and the given angle from 180° to obtain the third angle.

For final answers, we give $c \approx 23.3$, $\alpha \approx 50.8°$, $\beta \approx 82.9°$.

THE LAW OF TANGENTS

If you have a calculator, the Law of Cosines is convenient for Case II (SAS) problems, but if you want to use logarithms rather than a calculator, the Law of Cosines should be replaced by the Law of Tangents. The derivation of this law was outlined in detail in Problem 30, Exercise 10.3, and we shall only state the corresponding theorem here. (See Theorem 10.4b.)

Suppose we have an SAS problem in which a, b, and γ are given. Then from $\alpha + \beta + \gamma = 180°$, we immediately obtain $\alpha + \beta = 180° - \gamma$, so that the only unknown quantity in the preceding equation is $\alpha - \beta$. We can solve the equation in Theorem 10.4b for $\tan \frac{1}{2}(\alpha - \beta)$ and obtain a formula that is quite convenient for logarithmic calculation. The procedure is illustrated in Example 3.

▼ THEOREM 10.4b

The law of tangents

In any triangle, the ratio of the difference to the sum of any pair of sides equals the ratio of the tangent of half the difference to the tangent of half the sum of the corresponding opposite angles. A representative relationship is

$$\frac{a-b}{a+b} = \frac{\tan(\alpha-\beta)/2}{\tan(\alpha+\beta)/2}$$

EXAMPLE 3

Use the Law of Tangents to solve the triangle in Example 2.

SOLUTION:

Since we were given $a = 25.0$, $b = 32.0$, and $\gamma = 46.3°$, we see that $b > a$, and we therefore use Theorem 10.4b to write

$$\frac{b-a}{b+a} = \frac{\tan \frac{1}{2}(\beta - \alpha)}{\tan \frac{1}{2}(\beta + \alpha)}$$

We solve this equation for $\tan \frac{1}{2}(\beta - \alpha)$ to obtain

$$\tan \tfrac{1}{2}(\beta - \alpha) = \frac{b-a}{b+a} \tan \tfrac{1}{2}(\beta + \alpha)$$

From the given data, we find

$$b - a = 32 - 25 = 7$$

$$b + a = 32 + 25 = 57$$

$$\tfrac{1}{2}(\beta + \alpha) = \tfrac{1}{2}(180° - \gamma) = \tfrac{1}{2}(180° - 46.3°) = 66.85°$$

Thus,

$$\tan \tfrac{1}{2}(\beta - \alpha) = \tfrac{7}{57} \tan 66.85°$$

$$= \tfrac{7}{57}(2.3388)$$
$$= 0.28722$$

(The preceding calculation can be done with logarithms.) Tables now give $\frac{1}{2}(\beta - \alpha) = 16.03°$. Next, β and α are found from the two equations

$$\tfrac{1}{2}(\beta + \alpha) = 66.85°$$
$$\tfrac{1}{2}(\beta - \alpha) = 16.03°$$

$\beta = 82.88°$ By adding

$\alpha = 50.82°$ By subtracting

(Notice that the Law of Tangents automatically gives the correct quadrants for the angles being found.) Knowing the angles, we can now use the Law of Sines to find c. Thus,

$$c = \frac{a \sin \gamma}{\sin \alpha} = \frac{25 \sin 46.3°}{\sin 50.82°} = 23.317$$

The final answers are $c \approx 23.3$, $\alpha \approx 50.8°$, and $\beta \approx 82.9°$, as before.

As you can see from the last examples, the Law of Sines and the Law of Tangents are much more suitable for logarithmic calculation than is the Law of Cosines. Some people even prefer the Law of Tangents for a calculator solution.

EXAMPLE 4

A surveyor, starting at a point A, paced off a distance of 600 feet in the direction N 72.0° E. He then turned and paced off 750 feet in the direction N 35.0° W. Assume that the distances are correct to the nearest foot, and locate his final position relative to his starting point.

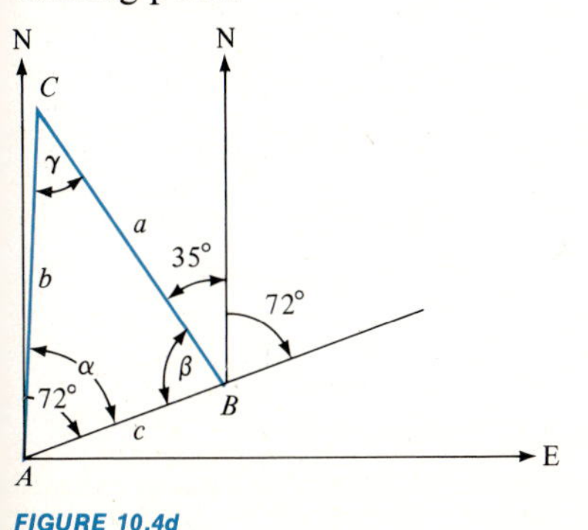

FIGURE 10.4d

SOLUTION:

Method 1 (Law of Cosines) We first make a sketch to show the given data. (See Figure 10.4d.) From the sketch, we see that $\beta = 180° - 35° - 72°$, or 73°. Thus, in triangle ABC, we have given $a = 750$, $c = 600$, and $\beta = 73°$. We use the Law of Cosines to write

$$b^2 = a^2 + c^2 - 2ac \cos \beta$$
$$= (750)^2 + (600)^2 - 2(750)(600) \cos 73°$$
$$\approx 659{,}365$$
$$b \approx 812.0$$

Next, we use the Law of Sines to obtain

$$\sin \alpha = \frac{a \sin \beta}{b} = \frac{750 \sin 73°}{812} \approx 0.88329$$

$$\alpha \approx 62.0°$$

The sketch shows that the bearing of AC can be obtained by subtracting α from 72°. Thus, the surveyor is about 812 feet in the direction N 10° E from his starting point.

Method 2 (Law of Tangents) Because a, c, and β are known, we write the Law of Tangents in the form

$$\frac{a-c}{a+c} = \frac{\tan \frac{1}{2}(\alpha - \gamma)}{\tan \frac{1}{2}(\alpha + \gamma)}$$

and solve for $\tan \frac{1}{2}(\alpha - \gamma)$ to obtain

$$\tan \tfrac{1}{2}(\alpha - \gamma) = \frac{a-c}{a+c} \tan \tfrac{1}{2}(\alpha + \gamma)$$

From the given data, we find

$$a - c = 150 \qquad a + c = 1350 \qquad \alpha + \gamma = 180^\circ - 73^\circ = 107^\circ$$

Hence,

$$\tan \tfrac{1}{2}(\alpha - \gamma) = \tfrac{150}{1350} \tan 53.5^\circ = 0.15016$$
$$\tfrac{1}{2}(\alpha - \gamma) = 8.54^\circ$$

From this result and $\frac{1}{2}(\alpha + \gamma) = 53.5^\circ$, we have $\alpha = 62.04^\circ$ and $\gamma = 44.96^\circ$. Next, the Law of Sines gives

$$b = \frac{a \sin \beta}{\sin \alpha} = \frac{750 \sin 73^\circ}{\sin 62.04^\circ} = 812.01$$

As in the first method, we arrive at the result that the surveyor is about 812 feet in the direction N 10° E from his starting point.

EXERCISE 10.4

In Problems 1–6, find the exact length of the third side of the triangle. Give your answers in radical form.

1 $a = 4$, $b = 3\sqrt{3}$, $\gamma = 30^\circ$
2 $a = 10$, $b = 8\sqrt{2}$, $\gamma = 45^\circ$
3 $b = 6$, $c = 5$, $\alpha = 60^\circ$
4 $a = 5$, $c = 8$, $\beta = 120^\circ$
5 $a = 5\sqrt{2}$, $b = 6$, $\gamma = 135^\circ$
6 $b = 5\sqrt{3}$, $c = 6$, $\alpha = 150^\circ$

In Problems 7–12, solve the triangle in the indicated preceding problem.

7 Problem 1
8 Problem 2
9 Problem 3
10 Problem 4
11 Problem 5
12 Problem 6

In Problems 13–20, solve the triangle. If you have a calculator available, use the Law of Cosines, otherwise use the Law of Tangents. Use the rules for approximate data.

13 $a = 17$, $b = 23$, $\gamma = 32^\circ$
14 $b = 50$, $c = 75$, $\alpha = 47^\circ$
15 $a = 35$, $c = 53$, $\beta = 106^\circ$
16 $a = 43$, $b = 105$, $\gamma = 160^\circ$
17 $a = 37.9$, $b = 46.3$, $\gamma = 20.7^\circ$
18 $b = 101$, $c = 98.3$, $\alpha = 136.2^\circ$
19 $b = 126.3$, $c = 137.9$, $\alpha = 49.25^\circ$
20 $a = 37.36$, $b = 52.98$, $\gamma = 126.32^\circ$

21 Use the Law of Cosines to find the exact length of one side of a regular dodecagon (12-sided polygon) inscribed in a circle of radius a.

22 Repeat Problem 21 for a regular octagon.

23 Two adjacent sides of a parallelogram are 27.3 and 36.5 feet long, respectively, and the included angle is 48.3°. Find the lengths of the two diagonals.

24 Repeat Problem 23 if the included angle is 126.5°.

25 A girl scout walked 200 yards in the direction S 35° E and then walked 100 yards in the direction S 48° W. Find her location relative to her starting point.

26 A surveyor goes a distance of 1 mile in the direction N 25° W from a point *A* and then 4 miles N 45° E. Find the surveyor's location relative to the point *A*.

27 An artillery gunner at *A* knows that his target at *C* is 5 kilometers in the direction N 38° W from his base at *B*. He also knows that the base is 7 kilometers in the direction N 48° E from him. In what direction and how far is the target from him?

28 A horizontal shaft is bored for a distance of 560 feet from the entrance to a mine straight into a mountain side. A turn to the right through an angle of 138° is then made, and the shaft is continued for another 210 feet. It is then decided to bore a second shaft from the entrance to the end of the first shaft. In what direction should the second shaft be bored?

USING YOUR KNOWLEDGE 10.4

1 Here is a problem that you can solve by using your knowledge of the Law of Cosines: The accompanying diagram represents the earth and the moon. The moon is directly overhead at the point *A*, and $\angle ACB$ is 60°. How much closer is the moon to the point *A* than to the point *B*? The radius of the earth is about 3960 miles, and the distance of the moon from the center of the earth is about 239,000 miles.

2 In Problem 1, suppose that arc *AB* is 3000 miles (about the distance from Florida to California). How much closer is the moon to point *A* than to point *B*?

Hint: $\angle ACB$ is about $\frac{3000}{3960}$ radian.

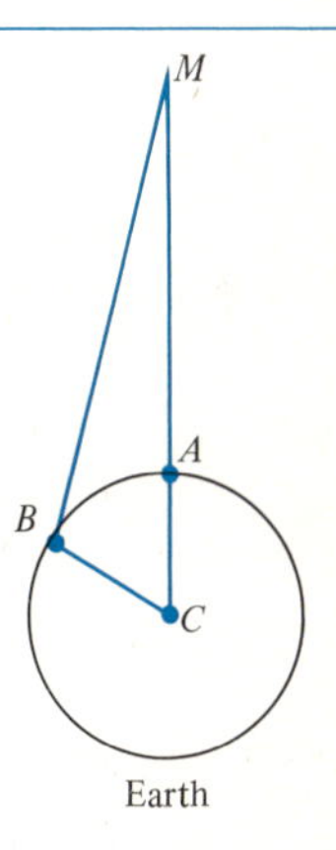

10.5 OBLIQUE TRIANGLES, CASE III: GIVEN THE THREE SIDES

Figure 10.5a shows a truss, a sturdy type of braced support that is used a great deal in building and bridge construction. Suppose that the truss is intended for a roof, part of which is slanting (*BC* and *ED*) and part flat (*CD*). If *AB* is 9.00 feet long, *AC* is 7.75 feet long, and *BC* is 5.50 feet long, what is the slope of the slanting portion of the roof? Recall that the slope is the ratio of the vertical rise to the corresponding horizontal run, and you will realize that we are asking for the tangent of $\angle ABC$.

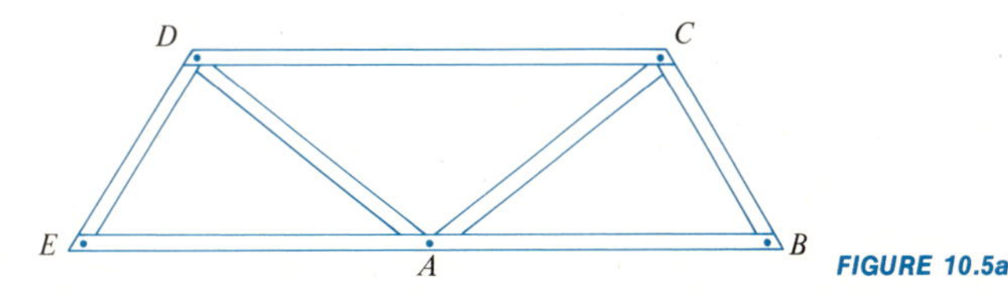

FIGURE 10.5a

SOLUTION OF SSS PROBLEMS

Note that we have posed a problem in which the sides of a triangle are given, and we wish to find one of the angles. This is a Case III (SSS) problem that can be solved by the Law of Cosines. With our usual notation, we have given a, b, and c, and we wish to find β. Hence, we write the Law of Cosines formula that involves β,

$$b^2 = a^2 + c^2 - 2ac \cos \beta$$

and solve for $\cos \beta$ to obtain

$$\cos \beta = \frac{a^2 + c^2 - b^2}{2ac}$$

This is a convenient formula if you have a calculator, but it is not well suited for logarithmic calculation. However, we can remedy this by a slight modification. If we add 1 to both sides, we have

$$\begin{aligned} 1 + \cos \beta &= 1 + \frac{a^2 + c^2 - b^2}{2ac} \\ &= \frac{a^2 + 2ac + c^2 - b^2}{2ac} \\ &= \frac{(a + c)^2 - b^2}{2ac} \end{aligned}$$

so that we can factor the numerator as the difference of two squares. Thus,

$$1 + \cos \beta = \frac{(a + c - b)(a + c + b)}{2ac}$$

which is much more suitable for logarithmic calculation. Corresponding formulas are easily written for the other two angles.

EXAMPLE 1

Find the slope of BC using the dimensions of the truss given at the beginning of this section.

SOLUTION:

We are given $a = 5.50$ feet, $b = 7.75$ feet, $c = 9.00$ feet, and are to find $\tan \beta$. Thus, for a solution by calculator, we use the formula

$$\cos \beta = \frac{a^2 + c^2 - b^2}{2ac}$$

to obtain

$$\begin{aligned} \cos \beta &= \frac{(5.50)^2 + (9.00)^2 - (7.75)^2}{2(5.50)(9.00)} \\ &= 0.51704545 \end{aligned}$$

and

$$\beta = 58.87^\circ \qquad \tan \beta = 1.66$$

For a logarithmic solution, we use the formula we derived for $1 + \cos\beta$:

$$1 + \cos\beta = \frac{(a + c - b)(a + c + b)}{2ac}$$

$$= \frac{(5.50 + 9.00 - 7.75)(5.50 + 9.00 + 7.75)}{2(5.50)(9.00)}$$

$$= \frac{(6.75)(22.25)}{99}$$

The use of four-place logarithms gives the approximate value 1.518 for the last fraction, and subtracting 1, we find

$$\cos\beta = 0.518$$

Interpolating in the table, we find $\beta = 58.8°$, and $\tan\beta = 1.65$. The discrepancy between the answers is caused by the relative inaccuracy of four-place logarithms.

EXAMPLE 2

Solve the triangle in which $a = 4$, $b = 5$, $c = 8$.

SOLUTION:

We use the Law of Cosines formulas solved for the cosines of the respective angles. Thus, we have

$$\cos\alpha = \frac{b^2 + c^2 - a^2}{2bc} = \frac{25 + 64 - 16}{80} = 0.9125$$

$$\alpha = 24.15°$$

$$\cos\beta = \frac{a^2 + c^2 - b^2}{2ac} = \frac{16 + 64 - 25}{64} = 0.859375$$

$$\beta = 30.75°$$

$$\cos\gamma = \frac{a^2 + b^2 - c^2}{2ab} = \frac{16 + 25 - 64}{40} = -0.575$$

$$\gamma = 125.10°$$

As a check we add the three angles:

$$24.15° + 30.75° + 125.10° = 180.00°$$

EXAMPLE 3

A surveyor walked 195 meters due north from a point A to a point B. At B, he turned and walked 365 meters in a southwesterly direction to a point C, which he found to be 296 meters from A. Find the bearings of C

SOLUTION:

We first make a sketch to show the given data. (See Figure 10.5b.) We have given $a = 365$ meters, $b = 296$ meters, and $c = 195$ meters. In order to answer the question, we must find α and β. We use the Law of Cosines as follows:

$$\cos\alpha = \frac{b^2 + c^2 - a^2}{2bc} = \frac{(296)^2 + (195)^2 - (365)^2}{2(296)(195)}$$

with respect to A and B; that is, find the direction from A to C and the direction from B to C.

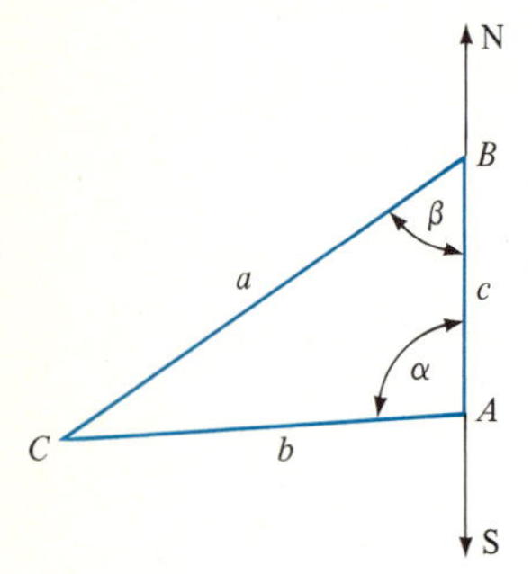

FIGURE 10.5b

$$= -0.0656964657 \quad \text{and} \quad \alpha = 93.77° \qquad \text{By calculator}$$

$$\cos\beta = \frac{a^2 + c^2 - b^2}{2ac} = \frac{(365)^2 + (195)^2 - (296)^2}{2(365)(195)}$$

$$= 0.5875237092 \quad \text{and} \quad \beta = 54.02° \qquad \text{By calculator}$$

From the figure, we see that to get the bearing of AC we must subtract α from 180°. Thus, the bearing of AC is S 86.2° W. The bearing of BC is S 54.0° W. (Answers have been rounded to the nearest tenth of a degree.)

Additional formulas more suitable for logarithmic calculation as well as for direct calculation of the area from the three sides are developed in Problems 19–23 of Exercise 10.5.

EXERCISE 10.5

In Problems 1–4, solve the triangle. Assume the data are exact and find the angles to the nearest hundredth of a degree.

1 $a = 3, b = 4, c = 6$
2 $a = 8, b = 10, c = 15$
3 $a = 30, b = 20, c = 25$
4 $a = 12, b = 13, c = 7$

In Problems 5–10, solve the triangle. Use the rules for approximate data.

5 $a = 13.2, b = 15.4, c = 17.7$
6 $a = 21.5, b = 36.9, c = 20.3$
7 $a = 57.1, b = 93.7, c = 80.4$
8 $a = 101, b = 95.2, c = 48.3$
9 $a = 375.4, b = 481.2, c = 156.8$
10 $a = 287.1, b = 305.2, c = 356.3$

11 A girl scout walked 100 yards due south from a point A to a point B. She then turned and walked 90 yards to a point C, where she found herself to be 50 yards from A. What is the bearing of C from A?

12 A surveyor paced off 110 meters due east from a point A to a point B. She then turned and went 200 meters in a northwesterly direction to a point C, which she found to be 150 meters from A. What is the bearing of C from A?

13 An artillery gunner at a point G knows that he is 1500 meters N 45° W from his base at B. He also knows that his target at T is in an easterly direction from him and is 2000 meters from G and 1800 meters from B. In what direction must he shoot to hit his target?

14 A hunter goes 5 miles in the direction N 30° E from a point A to a point B. He then goes 4 miles in a southeasterly direction to a point C, which is 7 miles from A. In what direction must he go to get back to A?

15 Find the area of a parallelogram with its adjacent sides 6 inches and 4 inches long, respectively, if the shorter diagonal is 3 inches long. Assume the data are exact.

16 Find the area of a parallelogram with its adjacent sides 23.5 and 32.6 centimeters long, respectively, if one of the diagonals is 36.1 centimeters long.

17 Two adjacent sides of a plane quadrilateral are 50.00 and 100.00 inches long. The angle between these two sides is 60.00°. The other two sides are 60.00 and 80.00 inches long, with the 60.00-inch side adjacent to the 50.00-inch side. Find the area of the quadrilateral.

18 Two adjacent sides of a plane quadrilateral are 25 and 30 feet long. The angle between these two sides is 45°. The other two sides are 40 and 20 feet long, with the 20-foot side adjacent to the 30-foot side. Find the area of the quadrilateral.

19 In this section, we obtained the formula

$$1 + \cos\beta = \frac{(a + c - b)(a + b + c)}{2ac}$$

Use the same idea to obtain the formula

$$1 - \cos\beta = \frac{(a + b - c)(b + c - a)}{2ac}$$

20 We know that the area of a triangle is given by the formula

$A = \frac{1}{2}ac \sin\beta$ and $\sin\beta = \sqrt{1 - \cos^2\beta} = \sqrt{(1 - \cos\beta)(1 + \cos\beta)}$

Let $s = \frac{1}{2}(a + b + c)$, the semi-perimeter of the triangle, and use the results in Problem 19 to show that

$A = \sqrt{s(s - a)(s - b)(s - c)}$

This neat result is known as Heron's Formula. Heron (about 75 A.D.) gave a simple proof of the formula, but it is quite probable that the formula itself goes back to Archimedes, about 300 years earlier.

21 Let r be the radius of the circle inscribed in the triangle ABC, and let s be the semi-perimeter as in Problem 20. Show that the area of the triangle is given by $A = rs$. Then use the result of Problem 20 to show that the radius of the circle is given by

$$r = \sqrt{\frac{(s - a)(s - b)(s - c)}{s}}$$

22 Use the results of Problems 19 and 21 to show that

$$\tan\tfrac{1}{2}\beta = \frac{r}{s - b}$$

It follows by symmetry that

$$\tan\tfrac{1}{2}\alpha = \frac{r}{s - a} \quad \text{and} \quad \tan\tfrac{1}{2}\gamma = \frac{r}{s - c}$$

These formulas are quite convenient for solving Case III (SSS) problems using logarithms.

23 Use the formulas obtained in Problems 21 and 22 to solve the triangle in Problem 9.

24 Repeat Problem 23 for the triangle in Problem 10.

25 Show that the radius R of the circle circumscribed about the triangle ABC is given by

$$R = \frac{a}{2\sin\alpha} = \frac{b}{2\sin\beta} = \frac{c}{2\sin\gamma}$$

Hint: It will help you to recall that a central angle is measured by the intercepted arc, and an inscribed angle is measured by one half the intercepted arc. Make a sketch showing the radii drawn to the vertices and the perpendiculars from the center of the circle to the sides of the triangle.

USING YOUR KNOWLEDGE 10.5

1 In Problem 22 of Exercise 10.5, we found that

$$\tan\tfrac{1}{2}\alpha = \frac{r}{s - a}$$

where r is the radius of the inscribed circle and s is the semi-perimeter of the triangle. Use the formula

$\sin\alpha = 2\sin\tfrac{1}{2}\alpha\cos\tfrac{1}{2}\alpha$

to obtain the formula

$$\sin\alpha = \frac{2rs}{bc}$$

2 Use the result of Problem 25 of Exercise 10.5 and the preceding result to show that the radius R of the circumscribed circle and the radius r of the inscribed circle satisfy the equation

$$Rr = \frac{abc}{4s}$$

3 Given the triangle with $a = 10$, $b = 12$, $c = 14$, find r by using the formula obtained in Problem 21 of Exercise 10.5. Then use the result of the preceding problem to find R.

4 Use the answer to Problem 3 and the result of Problem 25 of Exercise 10.5 to show that $\sin\alpha = 2\sqrt{6}/7$ for the triangle of Problem 3.

10.6 OBLIQUE TRIANGLES, CASE IV: GIVEN TWO SIDES AND THE ANGLE OPPOSITE ONE OF THEM

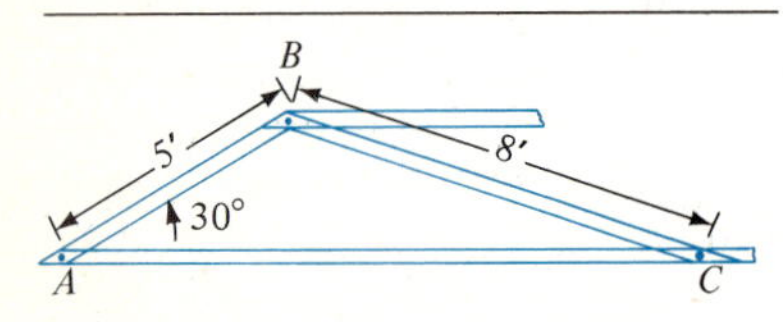

FIGURE 10.6a
Trusses used in the arch of a bridge

Figure 10.6a shows a portion of a truss that a carpenter is building. The short, slanting member, AB, is 5 feet long and makes an angle of 30° with the horizontal member. If the member BC is 8 feet long, how long is AC? (Assume the data are exact.)

SOLUTION OF SAA PROBLEMS

In the preceding problem, we are given two sides and the angle opposite one of them. We classify this as a Case IV (SSA) problem, and we can handle it by using the Law of Sines, as Example 1 shows.

EXAMPLE 1
Find the length of AC in the truss of Figure 10.6a.

SOLUTION:
We have given $a = 8$ feet, $c = 5$ feet, and $\alpha = 30°$. Hence, we can find γ by using the Law of Sines. Because $c < a$, we know that $\gamma < \alpha$. Thus,

$$\sin\gamma = \frac{c\sin\alpha}{a} = \frac{5\sin 30°}{8} = 0.3125$$

$$\gamma = 18.2°$$

Then, since $\alpha + \beta + \gamma = 180°$, we find

$$\beta = 180° - 30° - 18.2° = 131.8°$$

Another application of the Law of Sines gives

$$b = \frac{a\sin\beta}{\sin\alpha} = \frac{8\sin 131.8°}{\sin 30°} = 11.9$$

Thus, AC is about 11.9 feet long.

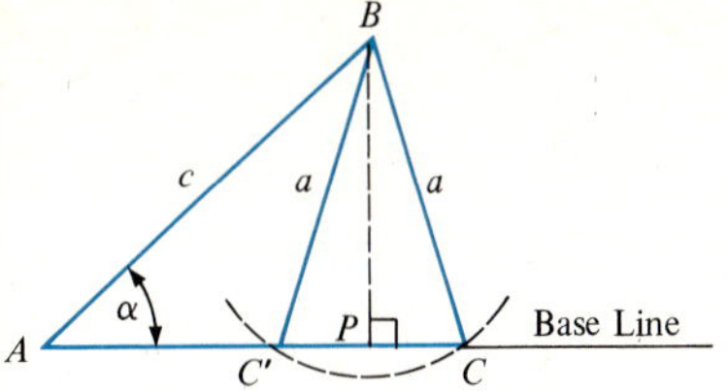

FIGURE 10.6b

Case IV problems are subject to a peculiarity not found in the other three cases. Suppose that the given angle, say α, is acute, as indicated in Figure 10.6b. Let P be the foot of the perpendicular from B to the base line, so that h is the altitude (height) of the triangle. Then $h = c \sin \alpha$. We can see now that if

1. $a < h$, no triangle is possible;
2. $a = h$, one triangle, a right triangle, is possible;
3. $h < a < c$, then a circle of radius a centered at B will cut the base line twice, once between A and P and once to the right of P, so that two triangles are possible. (ABC and ABC' in Figure 10.6b);
4. $a \geq c$, there is one triangle possible.

If $\alpha \geq 90°$ and if

5. $a \leq c$, no triangle is possible;
6. $a > c$, there is one triangle possible.

The preceding discussion makes clear why Case IV is commonly called the "ambiguous" case. The more troublesome type of problem occurs when the given angle is acute. The best way to handle this situation is to draw a base line and sketch in the given angle and the adjacent side as in Figure 10.6b. Then compare the side opposite the angle with the height of the triangle.

EXAMPLE 2

Solve the triangle in which $a = 12.6$, $b = 8.5$, and $\beta = 47.2°$.

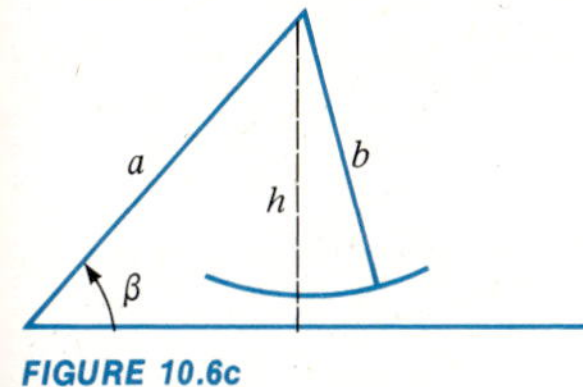

FIGURE 10.6c

SOLUTION:

First we sketch a base line showing the angle β and the adjacent side a, as in Figure 10.6c. For this problem, the height of the triangle (if there is one), is given by

$$\begin{aligned} h &= a \sin \beta \\ &= 12.6 \sin 47.2° \\ &= 9.24 \end{aligned}$$

Because $b = 8.5$, $b < h$, and no triangle is possible.

EXAMPLE 3

Solve the triangle in which $c = 12.6$, $a = 10.8$, and $\alpha = 47.2°$.

SOLUTION:

We make the same type of sketch as in Example 2. (See Figure 10.6d.) Again,

$$h = c \sin \alpha = 9.24$$

Because $h < a < c$, two triangles are possible, CAB and $C'AB$, as shown in the figure. For triangle CAB, we have

$$\begin{aligned} \sin \gamma &= \frac{c \sin \alpha}{a} \\ &= \frac{12.6 \sin 47.2°}{10.8} = 0.85602 \end{aligned}$$

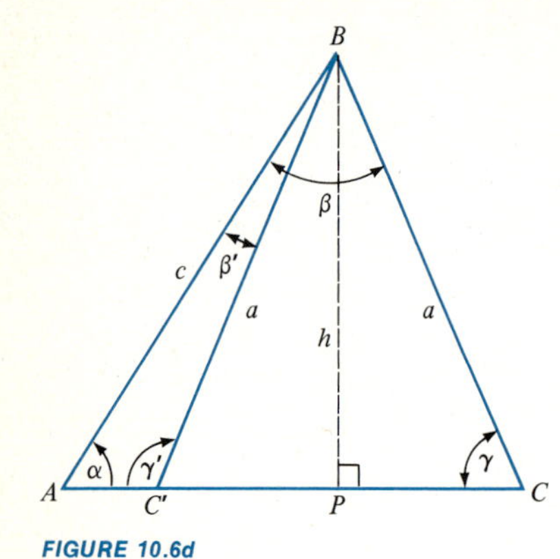

FIGURE 10.6d

$$\gamma = 58.87^\circ$$
$$\beta = 180^\circ - (\gamma + \alpha) = 73.93^\circ$$
$$b = \frac{a \sin \beta}{\sin \alpha}$$
$$= \frac{10.8 \sin 73.93^\circ}{\sin 47.2^\circ} = 14.14$$

Thus, in triangle CAB, $\gamma \approx 58.9^\circ$, $\beta \approx 73.9^\circ$, $b \approx 14.1$.

To solve triangle $C'AB$, we first note that triangle CBC' is isosceles, so that angles BCC' and $BC'C$ are equal. Therefore,

$$\gamma' = 180^\circ - \gamma = 121.13^\circ$$

and

$$\beta' = 180^\circ - (\gamma' + \alpha) = 11.67^\circ$$

Hence,

$$b' = \frac{a \sin \beta'}{\sin \alpha}$$
$$= \frac{10.8 \sin 11.67^\circ}{\sin 47.2^\circ}$$
$$= 2.977$$

Thus, in triangle $C'AB$, $\gamma' \approx 121.1^\circ$, $\beta' \approx 11.7^\circ$, $b' \approx 2.98$.

EXAMPLE 4

Solve the triangle in which $a = 12.6$, $b = 15.7$, and $\beta = 47.2^\circ$.

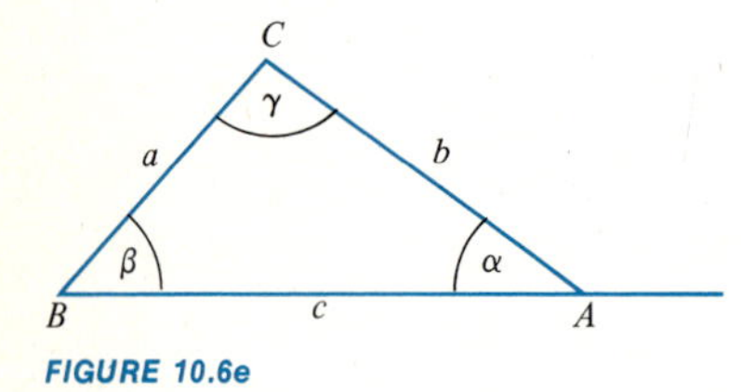

FIGURE 10.6e

SOLUTION:

Again we make a sketch to show the base line, the angle β, and the side a, as in Figure 10.6e. Because $b > a$, there is only one triangle that will include the angle β, and we can solve this triangle by using the Law of Sines. Thus,

$$\sin \alpha = \frac{a \sin \beta}{b}$$
$$= \frac{12.6 \sin 47.2^\circ}{15.7} = 0.58885$$
$$\alpha = 36.08^\circ$$
$$\gamma = 180^\circ - (\alpha + \beta) = 96.72^\circ$$

and

$$c = \frac{b \sin \gamma}{\sin \beta}$$
$$= \frac{15.7 \sin 96.72^\circ}{\sin 47.2^\circ}$$

$= 21.25$

Hence, $\alpha \approx 36.1°$, $\gamma \approx 96.7°$, and $c \approx 21.3$.

EXAMPLE 5

Solve the triangle in which $c = 10.8$, $a = 12.9$, and $\alpha = 125.2°$.

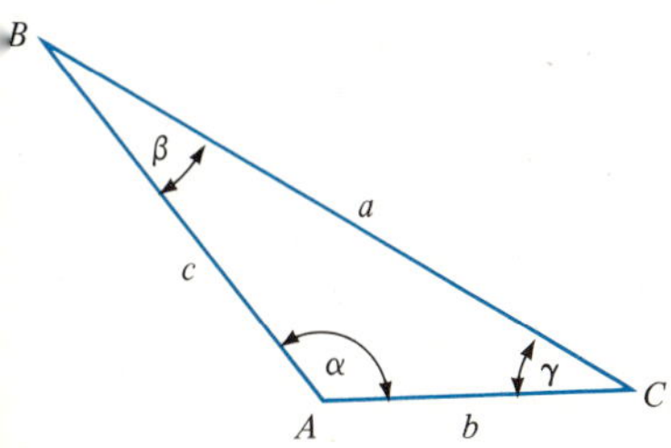

FIGURE 10.6f

SOLUTION:

A quick sketch (Figure 10.6f) shows that, because $a > c$, there is only one triangle that can include the given angle. Again, a solution is simply obtained by using the Law of Sines. We first find γ as follows:

$$\sin \gamma = \frac{c \sin \alpha}{a}$$

$$= \frac{10.8 \sin 125.2°}{12.9} = 0.68412$$

$$\gamma = 43.17°$$

Then we have

$$\beta = 180° - (\alpha + \beta) = 11.63°$$

and

$$b = \frac{a \sin \beta}{\sin \alpha}$$

$$= \frac{12.9 \sin 11.63°}{\sin 125.2°} = 3.182$$

Thus, $\beta \approx 11.6°$, $\gamma \approx 43.2°$, and $b \approx 3.18$.

EXERCISE 10.6

In Problems 1–10, solve the triangle. If no triangle is possible, state why. Assume that the given data are exact.

1 $a = 4, b = 8, \alpha = 30°$
2 $b = 8, c = 12, \beta = 45°$
3 $a = 4, c = 6, \alpha = 30°$
4 $a = 10, c = 9, \gamma = 60°$
5 $b = 12, c = 5, \gamma = 30°$
6 $b = 20, c = 9, \gamma = 30°$
7 $a = 6, c = 8, \alpha = 140°$
8 $b = 12, c = 15, \beta = 135°$
9 $a = 8, c = 6, \alpha = 160°$
10 $b = 15, c = 12, \beta = 135°$

In Problems 11–14, solve the triangle. Use the rules for approximate data.

11 $a = 21.7, c = 16.5, \gamma = 38.2°$
12 $b = 36.5, c = 35.0, \gamma = 65.3°$
13 $b = 127.4, c = 138.5, \gamma = 52.70°$
14 $a = 328.9, b = 237.4, \alpha = 146.50°$

15 A carpenter is building a triangular truss (see the accompanying diagram). He nails a 4-foot member at an angle of 45° to the horizontal member and makes the member opposite the 45° angle 6 feet long. Find the slope of the 6-foot side when the truss is standing vertically.

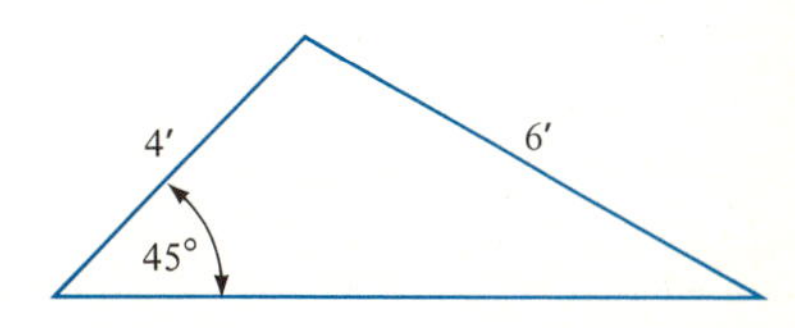

16 In building a triangular truss, a carpenter fastened a 5-foot member at an angle of 15° to the horizontal member (see the accompanying diagram). He then fastened a 6-foot member from the first to the horizontal member. Find the slope of the 6-foot member when the truss is standing vertically.

17 In a parallelogram, two adjacent sides are 8 inches and 10 inches long, respectively. The diagonal joining these sides makes an angle of 32° with the 10-inch side. Find the area of the parallelogram.

18 Repeat Problem 17 if the 32° angle is between the diagonal and the 8-inch side.

19 A hunter leaves a straight, east–west road at a point A and goes 2 miles in the direction N 58.4° W. He then turns and walks 4 miles in a southeasterly direction, arriving back on the road at a point B. How far from A is B?

20 A man in a jeep is driving in fairly flat country along a straight, north–south road. He leaves the road at a point A and drives 25 miles in the direction N 40° W to a point B, where he discovers that he forgot to fill his gas tank and has only enough gas for 20 more miles. The nearest gas station is at a point G, 4.5 miles north of A. If the man is willing to walk along the road (he might hitch a ride) but not cross-country, in what direction should he drive to minimize his walking distance to the gas station?

USING YOUR KNOWLEDGE 10.6

If you have a calculator available, you will find it interesting to tackle the ambiguous case in a different way. Suppose that a, b, and α are given. Then, by the Law of Cosines,

$$a^2 = b^2 + c^2 - 2bc \cos \alpha$$

This is a quadratic equation in the unknown c, and we rewrite it in standard form:

$$c^2 - (2b \cos \alpha)c + (b^2 - a^2) = 0$$

The quadratic formula gives

$$\begin{aligned} c &= \frac{2b \cos \alpha \pm \sqrt{4b^2 \cos^2 \alpha - 4b^2 + 4a^2}}{2} \\ &= b \cos \alpha \pm \sqrt{a^2 - b^2(1 - \cos^2 \alpha)} \\ &= b \cos \alpha \pm \sqrt{a^2 - b^2 \sin^2 \alpha} \end{aligned}$$

From this solution, we can obtain all the information that we detailed in Section 10.6 concerning the ambiguous case. For instance, if $\alpha < 90°$ and

$$b \sin \alpha < a < b \qquad \text{then} \qquad 0 < \sqrt{a^2 - b^2 \sin^2 \alpha} < b \cos \alpha$$

(Do you see why?) Thus, there are two real positive solutions of the quadratic equation corresponding to there being two triangles possible in this case.

Notice that there is nothing extra to remember; the Law of Cosines and the solution of the quadratic equation for the third side of the triangle tell the whole story. For example, if we return to the truss problem of Section 10.6, we had given $a = 8$, $c = 5$, and $\alpha = 30°$. Then, by the Law of Cosines,

$$a^2 = b^2 + c^2 - 2bc \cos \alpha \qquad \text{or} \qquad 64 = b^2 + 25 - 5\sqrt{3}b$$

which we rearrange to read

$$b^2 - 5\sqrt{3}b - 39 = 0$$

The quadratic formula gives

$$b = \frac{5\sqrt{3} \pm \sqrt{75 + 156}}{2} = \frac{8.66 \pm 15.20}{2}$$

We discard the minus sign because b must be positive and obtain $b \approx 11.9$, in agreement with our previous result. Note that only one triangle is possible here.

As another example, suppose $a = 5$, $b = 2.75$, $\beta = 36°$. Then we have

$$b^2 = a^2 + c^2 - 2ac \cos \beta \qquad \text{or} \qquad (2.75)^2 = 5^2 + c^2 - 10c \cos 36°$$

which simplifies to

$$c^2 - 8.0902c + 17.4375 = 0$$

This equation gives

$$c = \frac{8.0902 \pm \sqrt{65.451 - 69.75}}{2}$$

Since the quantity under the radical sign is negative, there are no real values of c, and no triangle is possible. Note that this information is more easily obtained by comparing b with $a \sin \beta$.

The following problems are selected from Exercise 10.6. Do not solve the triangle completely, but do find the missing side(s). If no triangle is possible, state why. Use the Law of Cosines.

1 Problem 1 **2** Problem 3 **3** Problem 5 **4** Problem 7 **5** Problem 9 **6** Problem 11

SELF-TEST

1 Give the indicated trigonometric function as a ratio of lengths in the accompanying figure. For instance,

$$\sin \angle EFC = \frac{EC}{EF}$$

a $\sin \angle AFE$ **b** $\cos \angle AFB$
c $\tan \angle DAE$ **d** $\sec \angle EAF$

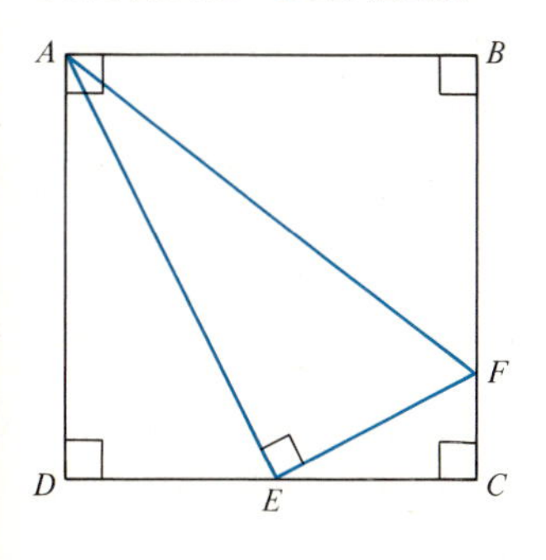

2 In the following problems, the triangle is the standard right triangle. Give the required angles to the nearest tenth of a degree and the required lengths to the nearest tenth of a unit.

a If $a = 5$ and $b = 10$, find α.
b If $a = 5$ and $\alpha = 31.2°$, find c.
c If $b = 6$ and $\alpha = 47.8°$, find a.
d If $b = 6$ and $c = 20$, find α.

3 In the accompanying figure, with right angles indicated by the small squares, d, α, and β are known. Find x and y in terms of the known parts.

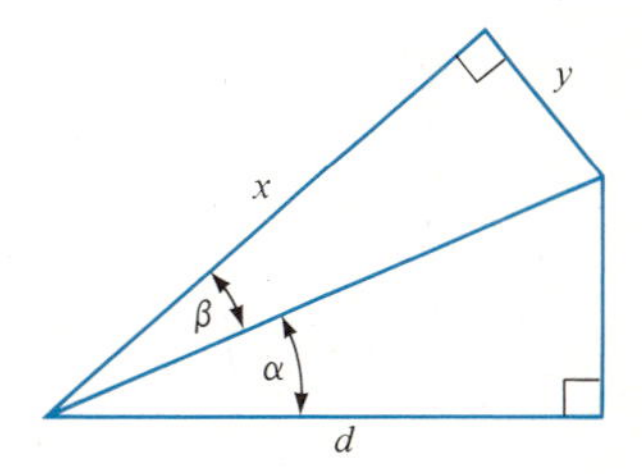

4 Evaluate each of the following if the numbers are all approximate.

a $52.375 + 69.37 - 48.2$

b $\dfrac{(36.7)(41.35)}{0.0028}$

5 In the following problems, take all measurements as approximate numbers.

a In a right triangle, $a = 5375.6$ meters and $\alpha = 27.8°$. Find c.
b In a right triangle, $b = 36.3$ inches and $\alpha = 32.35°$. Find a.

6 In the following problems, take all measurements as approximate numbers.

a In a right triangle, $a = 42.3$ feet and $c = 65.78$ feet. Find b.
b In the triangle of (a), find α.

Problems 7–13 do not stress computation. You should be able to obtain the exact solutions of the numerical problems by using the functions of the simple angles.

7 In an oblique triangle, $a = 10$, $c = 6\sqrt{2}$, and $\beta = 45°$.
a Find the area of the triangle.
b Find b.

8 a In an oblique triangle, $\alpha = 30°$, $\beta = 105°$, and $c = 15$. Find a.
b For the data in the accompanying figure, find x.

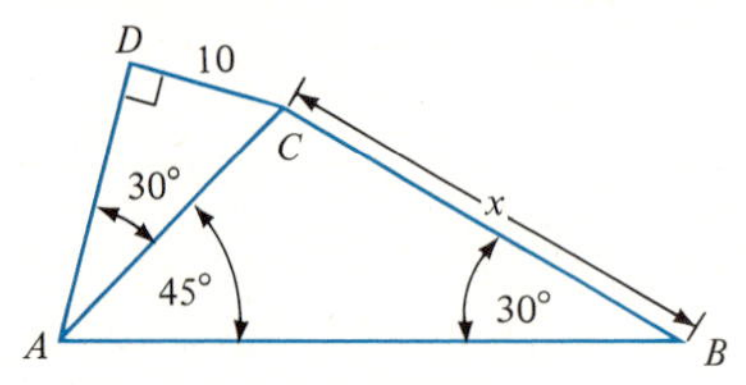

9 A surveyor found the side AB of a triangular tract of land to be 100 meters long. She also found the bearing of AB to be N 80° E, the bearing of AC to be N 50° E, and the bearing of BC to be N 5° E. Find the exact length of BC.

10 In the accompanying figure, we are given that C, B, and D are on a straight line. The angle at C is 90°, and b, α, and θ are known. Express x in terms of the known parts.

11 In triangle ABC, the length of AB is 5 inches; that of AC is 4 inches; and angle BAC is 60°.
a Find the length of BC.
b Find $\sin \angle ABC$.
c Find $\sin \angle ACB$.

12 In a triangle, $a = \sqrt{3} + 1$, $b = \sqrt{3} - 1$, and $\gamma = 60°$. Use the Law of Tangents to find:
a $\tan \frac{1}{2}(\alpha - \beta)$ **b** α and β

13 A ship sails 10 miles in the direction N 60° E and then sails 5 miles due north. Find the distance of the ship from its starting point.

14 Two adjacent sides of a parallelogram are 10 feet and 12 feet long, respectively, and the diagonal joining their ends is 13 feet long. Find the angle between the two sides to the nearest tenth of a degree.

15 The accompanying figure shows a roof truss. Find the slope of the 15-foot side.

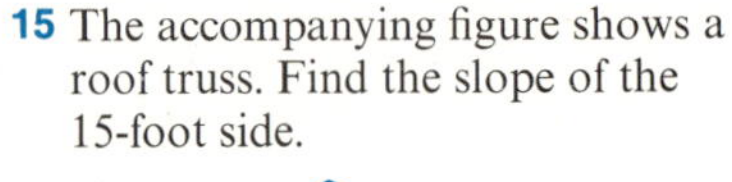

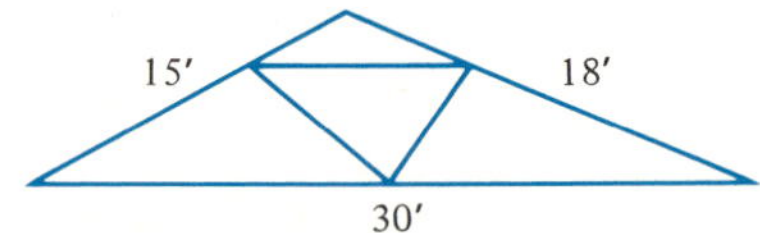

16 A surveyor goes 2 miles due south from a point A to a point B. From B, he goes 3 miles in a southwesterly direction to a point C. If C is 4 miles from A, find the bearing of C with respect to A.

17 Find the number of possible triangles for the given data:
a $a = 4$, $c = 5$, $\alpha = 60°$
b $a = 4.7$, $b = 5$, $\alpha = 60°$
c $a = 7$, $c = 6$, $\alpha = 30°$

18 Solve the triangle for which $a = 6$, $b = 5$, $\alpha = 150°$. Find the angles to the nearest tenth of a degree and the side to three significant digits.

19 Solve the triangle in which $a = 31.47$ meters, $\beta = 72.61°$, and $\gamma = 47.83°$.

20 Find the area (to the nearest square foot) of the quadrilateral shown in the accompanying figure.

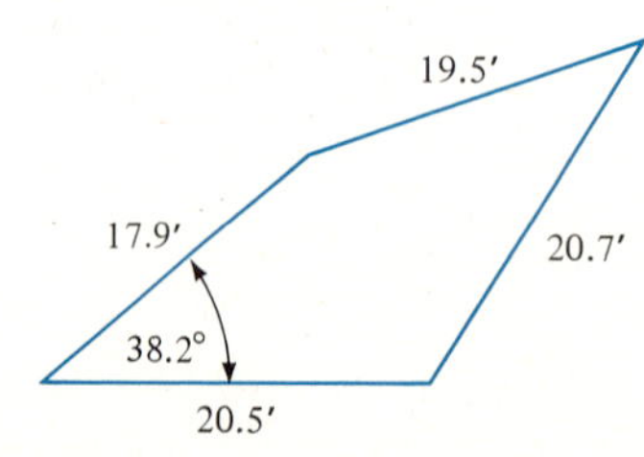

CHAPTER 11

INVERSE TRIGONOMETRIC FUNCTIONS, POLAR COORDINATES, AND VECTORS

11.1 INVERSE TRIGONOMETRIC FUNCTIONS

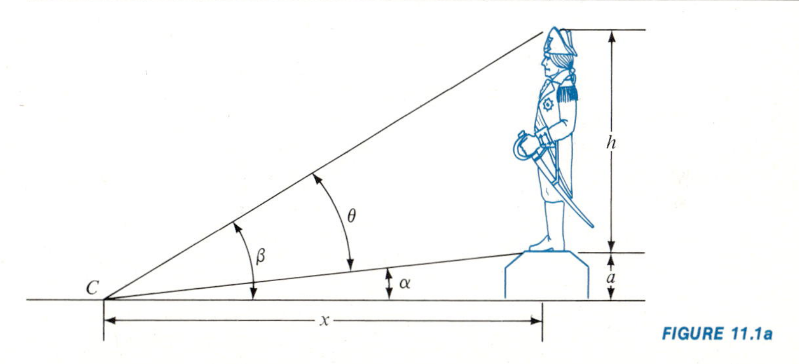

FIGURE 11.1a

Figure 11.1a shows a statue of height h standing on a pedestal of height a. The angle θ is the angle that the statue subtends at the point C, which is at a horizontal distance x from the foot of the statue. There is a standard problem in calculus that is equivalent to the question: At what distance x is the angle θ as large as possible? We shall not try to solve this problem here (the answer is $\sqrt{a(a+h)}$), but the first step is to express θ in terms of x. From the illustration, we see that $\theta = \beta - \alpha$ and

$$\tan \beta = \frac{a+h}{x} \qquad \tan \alpha = \frac{a}{x}$$

We would now like to solve these two equations for β and α, respectively, so that we could then substitute into the equation $\theta = \beta - \alpha$ to obtain the desired expression. We shall return to this problem later in this section.

There are other problems that require solving equations such as $y = \sin x$ and $y = \cos x$ for x in terms of y. In order to handle problems of this type, we introduce the *inverse* trigonometric functions. As you saw in Section 5.4, the inverse of a function f is obtained by solving the equation $y = f(x)$ for x and then interchanging the x and the y. Although the result is always a relation, it is a function only if the given function f is one-to-one.

THE INVERSE SINE FUNCTION

We consider the sine function first. As you can easily see by setting y equal to $\frac{1}{2}$, for example, in the equation $y = \sin x$, the sine is not a one-to-one function. For $y = \frac{1}{2}$, we have $x = \pi/6 + 2m\pi$ or $x = 5\pi/6 + 2m\pi$, where m is any integer. Any value of y between -1 and 1, inclusive, will also give infinitely many values of x. Geometrically, this means that any horizontal line between $y = -1$ and $y = 1$, inclusive, intersects the graph of $y = \sin x$ in infinitely many points. Consequently, the inverse of the sine function is *not* a function.

We can avoid the difficulty by restricting the domain of $y = \sin x$ so that the restricted function is one-to-one. To make this function as simple and useful as possible, we choose the set of customarily tabulated values as part of the domain. Thus, for values of y between 0 and 1, inclusive, we restrict x to the interval $0 \leq x \leq \pi/2$. For values of y such that $-1 \leq y < 0$, it turns out that the simplest choice is that which makes the graph a continuous curve. As you can see by referring to Figure 11.1b, this requires that for negative values of y, we restrict x to the interval $-\pi/2 \leq x < 0$. Accordingly, we define the *restricted* sine function, which we shall designate by capitalizing the *s* and writing $y = \text{Sin}\, x$, as in Definition 11.1a.

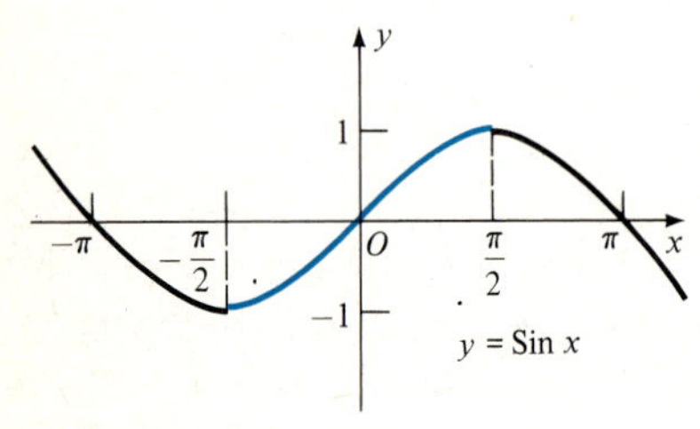

FIGURE 11.1b

▼ DEFINITION 11.1a

The *restricted sine function* is the set of pairs

$$\left\{(x, \sin x) \middle| -\frac{\pi}{2} \leq x \leq \frac{\pi}{2}\right\}$$

This function is designated by writing $y = \text{Sin}\, x$.

The portion of the graph in color in Figure 11.1b is the graph of $y = \text{Sin}\, x$. Notice that this graph displays the one-to-one character of the restricted sine function. If we substitute $y = \frac{1}{2}$ in the equation $y = \text{Sin}\, x$, then we obtain the single value $x = \pi/6$; for $y = -\frac{1}{2}$, we obtain $x = -\pi/6$. (Note that $5\pi/6$ is *incorrect* because it is not in the interval from $-\pi/2$ to $\pi/2$.) For each value of y between -1 and 1, inclusive, we have exactly one value of x. Thus, the inverse of the function defined by $y = \text{Sin}\, x$ is also a function, as is indicated by Definition 11.1b. Notice that the symbol for the function is capitalized to emphasize the restriction on the y-values.

▼ DEFINITION 11.1b

The *inverse sine function* is the set of pairs

$$\left\{(x, y) \middle| x = \sin y, -\frac{\pi}{2} \leq y \leq \frac{\pi}{2}\right\}$$

This function is symbolized by writing $y = \text{Sin}^{-1} x$ (or $y = \text{Arcsin}\, x$).

According to Definition 11.1b, if we write $y = \text{Sin}^{-1} x$, then $x = \sin y$ *and* $-\pi/2 \leq y \leq \pi/2$. For example,

$\text{Sin}^{-1} \frac{1}{2} = \pi/6$, because $\pi/6$ is the only number between $-\pi/2$ and $\pi/2$, such that the sine of that number has the value $\frac{1}{2}$.

$\text{Sin}^{-1}(-\frac{1}{2}) = -\pi/6$, because $-\pi/6$ is the only number between $-\pi/2$ and $\pi/2$, such that the sine of that number has the value $-\frac{1}{2}$.

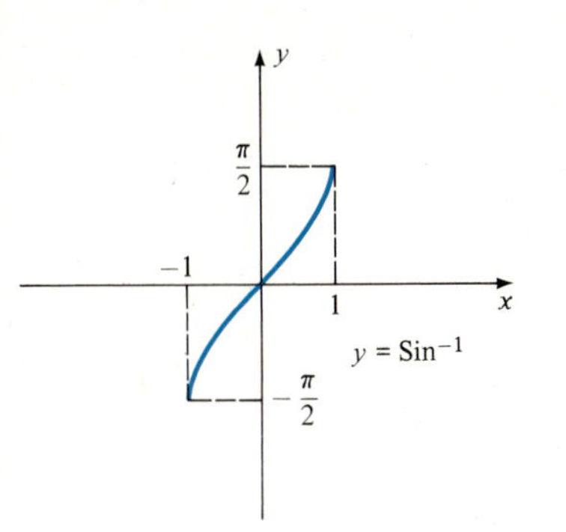

FIGURE 11.1c

Notice the agreement between these remarks and the graph of the inverse sine function that appears in Figure 11.1c. You should memorize both the definition of the inverse sine function and the graph of this function.

If we write $f(x) = \text{Sin } x$ and $g(x) = \text{Sin}^{-1} x$, then it follows from the definitions of these functions that

$$\text{for } -1 \leq x \leq 1, f(g(x)) = \text{Sin}(\text{Sin}^{-1} x) = x$$

and

$$\text{for } -\frac{\pi}{2} \leq x \leq \frac{\pi}{2}, g(f(x)) = \text{Sin}^{-1}(\text{Sin } x) = x$$

so that f and g satisfy the definition of inverse functions given in Section 5.4. Details are left for the exercises.

▼ THEOREM 11.1a

$$\text{Sin}^{-1}(-x) = -\text{Sin}^{-1} x \qquad -1 \leq x \leq 1$$

Theorem 11.1a follows at once from the definition of the inverse sine function and reflects the symmetry of the graph in Figure 11.1c. The detailed proof of this theorem is left to you.

You should note another important property of the inverse sine function that is displayed by Figure 11.1c: $\text{Sin}^{-1} x$ *increases* from $-\pi/2$ to $\pi/2$ as x *increases* from -1 to 1.

EXAMPLE 1

Find:

a. $\text{Sin}^{-1} 0.23770$

b. $\text{Sin}^{-1}(-0.23770)$

SOLUTION:

a. If we write $y = \text{Sin}^{-1} 0.23770$, then $\sin y = 0.23770$ and $0 \leq y \leq \pi/2$. From Table A5, we read $y = 0.24$. Thus, $\text{Sin}^{-1} 0.23770 = 0.24$.

b. By Theorem 11.1a, $\text{Sin}^{-1}(-0.23770) = -\text{Sin}^{-1} 0.23770$, so that we may use the result of (a) to obtain

$$\text{Sin}^{-1}(-0.23770) = -0.24$$

Example 1 illustrates the simple procedure to be used in finding the value of the inverse sine function when the argument is a negative number. We use the tables to find the number whose sine is the absolute value of the given argument and put a minus sign in front of the result. If you have a scientific calculator, the procedure is even simpler, because these calculators are programmed to give the values of the inverse trigonometric functions that agree with the definitions used in this text. For example, to find $\text{Sin}^{-1}(-0.23770)$, you should be sure to put your calculator in the radian mode. Then key in .2377, press the +/− (or CHS) key to change the sign, and then press the INV (or ARC) key, and the sin key, in that order. The calculator will read approximately -0.2399973, which you can round off to -0.24000 to agree with the number of significant digits in the given sine.

Example 2 shows an easy geometric procedure that you can use to find trigonometric functions of a number given as an inverse sine.

EXAMPLE 2

Find the sine, cosine, and tangent of $\mathrm{Sin}^{-1}(-\frac{3}{5})$.

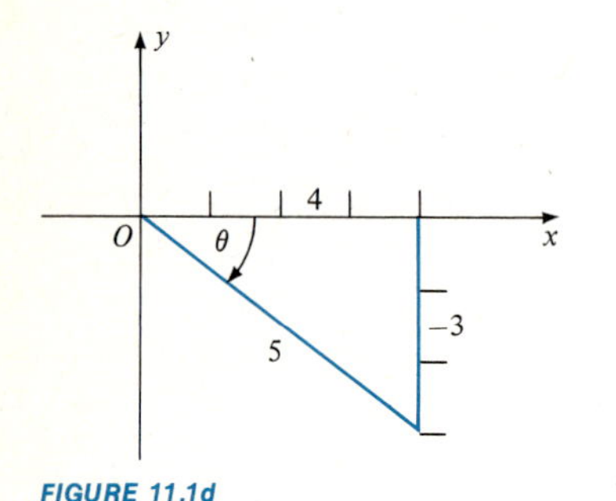

FIGURE 11.1d

SOLUTION:

Let $\theta = \mathrm{Sin}^{-1}(-\frac{3}{5})$. Then $\sin\theta = -\frac{3}{5}$ and $-\pi/2 < \theta < 0$. Now, we can interpret θ as the radian measure of an angle, sketch the angle as in Figure 11.1d, and read off the required values from the sketch:

$$\sin[\mathrm{Sin}^{-1}(-\tfrac{3}{5})] = -\tfrac{3}{5}$$
$$\cos[\mathrm{Sin}^{-1}(-\tfrac{3}{5})] = \tfrac{4}{5}$$
$$\tan[\mathrm{Sin}^{-1}(-\tfrac{3}{5})] = -\tfrac{3}{4}$$

The first part of Example 2 suggests the identity $\sin(\mathrm{Sin}^{-1} x) = x$, $-1 \le x \le 1$. If you let $\theta = \mathrm{Sin}^{-1} x$, then $\sin\theta = x$ by the definition of $\mathrm{Sin}^{-1} x$. Thus,

$$\sin(\mathrm{Sin}^{-1} x) = \sin\theta = x \qquad -1 \le x \le 1$$

Notice that we have written *sin* not *Sin*. Now see Example 3.

EXAMPLE 3

Is it true that $\mathrm{Sin}^{-1}(\sin x) = x$ for all real values of x?

SOLUTION:

Let $y = \mathrm{Sin}^{-1}(\sin x)$. Then, y is the *unique* number such that

$$\sin y = \sin x \quad \textit{and} \quad -\frac{\pi}{2} \le y \le \frac{\pi}{2}$$

Thus, if $-\pi/2 \le x \le \pi/2$, then $y = x$; that is, $\mathrm{Sin}^{-1}(\sin x) = x$. However, if x is outside this interval, then $y \ne x$, that is, $\mathrm{Sin}^{-1}(\sin x) \ne x$. For instance, if $x = 3\pi/2$, then $\sin x = -1$, and

$$\mathrm{Sin}^{-1}\left(\sin\frac{3\pi}{2}\right) = \mathrm{Sin}^{-1}(-1) = -\frac{\pi}{2} \ne \frac{3\pi}{2}$$

THE INVERSE COSINE FUNCTION

We arrive at a definition of the inverse cosine function in a manner similar to that for the inverse sine function. Again, we know that the cosine is not a one-to-one function, so that we first define a *restricted* cosine function. Of course, we choose the domain to include the usually tabulated values, and we try to make the graph a continuous curve. Thus, if $y = \cos x$, then for $0 \le y \le 1$, we choose x such that $0 \le x \le \pi/2$. Then, to make the graph a continuous curve, we must choose $\pi/2 < x \le \pi$ for $-1 \le y < 0$. These restrictions give us the portion of the graph that is shown in color in Figure 11.1e and lead to Definition 11.1c. Note the capitalization to remind you of the restriction on the domain. As you can see from the graph in Figure 11.1e, the restricted cosine function is one-to-one, so that the inverse is also a function. We then have Definition 11.1d.

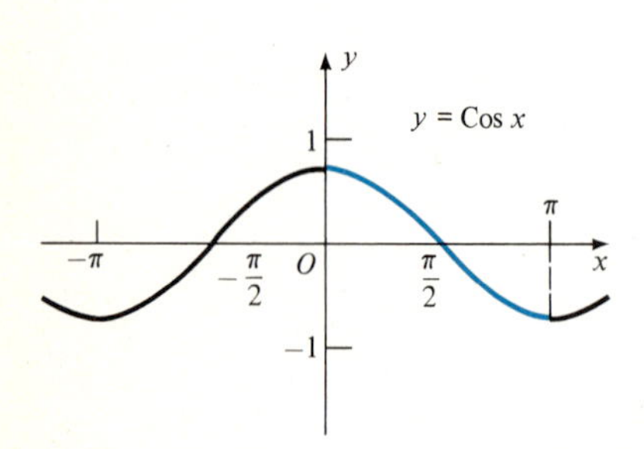

FIGURE 11.1e

As in the case of the inverse sine function, the symbol is capitalized to emphasize the restriction on the y-values. The graph of

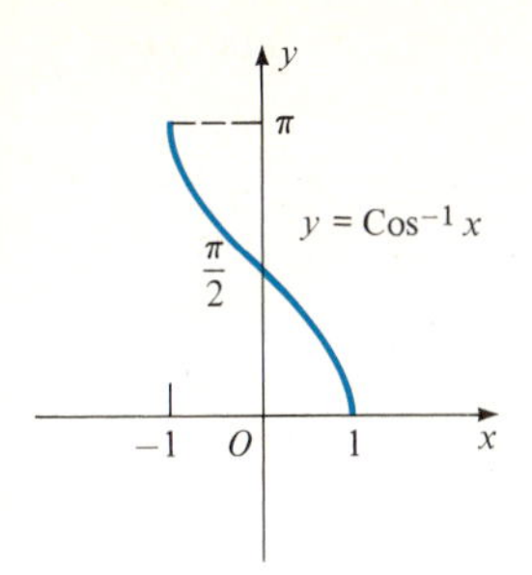

FIGURE 11.1f

▼ **DEFINITION 11.1c**

The *restricted cosine function* is the set of pairs

$$\{(x, \cos x) \mid 0 \le x \le \pi\}$$

This function is symbolized by $y = \operatorname{Cos} x$.

▼ **DEFINITION 11.1d**

The *inverse cosine function* is the set of pairs

$$\{(x, y) \mid x = \cos y,\ 0 \le y \le \pi\}$$

This function is designated by $y = \operatorname{Cos}^{-1} x$ (or $y = \operatorname{Arccos} x$).

$y = \operatorname{Cos}^{-1} x$ is shown in Figure 11.1f. The domain of this function is the interval $-1 \le x \le 1$, and the range is the interval $0 \le y \le \pi$. Notice that $\operatorname{Cos}^{-1} x$ *decreases* from π to 0 as x *increases* from -1 to 1.

EXAMPLE 4

Find:

a. $\operatorname{Cos}^{-1} \dfrac{\sqrt{3}}{2}$

b. $\operatorname{Cos}^{-1}\left(-\dfrac{\sqrt{3}}{2}\right)$

SOLUTION:

a. Let $\theta = \operatorname{Cos}^{-1} \sqrt{3}/2$, so that $\cos \theta = \sqrt{3}/2$ and $0 \le \theta \le \pi$. Thus, $\theta = \pi/6$, and $\operatorname{Cos}^{-1} \sqrt{3}/2 = \pi/6$.

b. Let $\phi = \operatorname{Cos}^{-1}(-\sqrt{3}/2)$. Then $\cos \phi = -\sqrt{3}/2$ and $0 \le \phi \le \pi$. We see that ϕ must correspond to a second-quadrant angle, and we may use θ from part (a) as the reference number. Thus,

$$\phi = \pi - \theta = \frac{5\pi}{6} \quad \text{or} \quad \operatorname{Cos}^{-1}\left(-\frac{\sqrt{3}}{2}\right) = \frac{5\pi}{6}$$

▼ **THEOREM 11.1b**

$$\operatorname{Cos}^{-1}(-x) = \pi - \operatorname{Cos}^{-1} x \qquad -1 \le x \le 1$$

Part (b) of Example 4 is a special instance of Theorem 11.1b. The details of the proof are left to you.

EXAMPLE 5

Evaluate $\operatorname{Cos}^{-1}(-0.79608)$.

SOLUTION:

By Theorem 11.1b,

$$\operatorname{Cos}^{-1}(-0.79608) = \pi - \operatorname{Cos}^{-1} 0.79608$$

From the tables, we find $\operatorname{Cos}^{-1} 0.79608 = 0.650$. Thus,

$$\operatorname{Cos}^{-1}(-0.79608) = \pi - 0.650 \approx 2.492$$

The calculator procedure is as follows. (Be sure that your calculator is in the radian mode.)

Algebraic logic

[.79608] [+/−] [INV] [cos] ≈ 2.4916

RPN logic

[.79608] [CHS] [ARC] [COS] ≈ 2.4916

THE INVERSE TANGENT FUNCTION

Considerations similar to those used for the inverse sine and cosine functions lead to the following definitions for the restricted tangent function and its inverse. (See Definitions 11.1e and 11.1f.)

▼ DEFINITION 11.1e

The *restricted tangent function* is the set of pairs

$$\left\{(x, \tan x) \middle| -\frac{\pi}{2} < x < \frac{\pi}{2}\right\}$$

We symbolize this function by writing $y = \text{Tan}\, x$.

▼ DEFINITION 11.1f

The *inverse tangent function* is the set of pairs

$$\left\{(x, y) \middle| x = \tan y, -\frac{\pi}{2} < y < \frac{\pi}{2}\right\}$$

This function is symbolized by writing $y = \text{Tan}^{-1} x$ (or $y = \text{Arctan}\, x$).

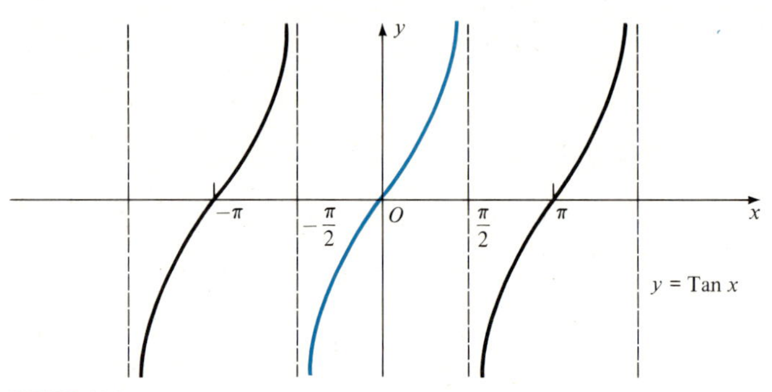

FIGURE 11.1g

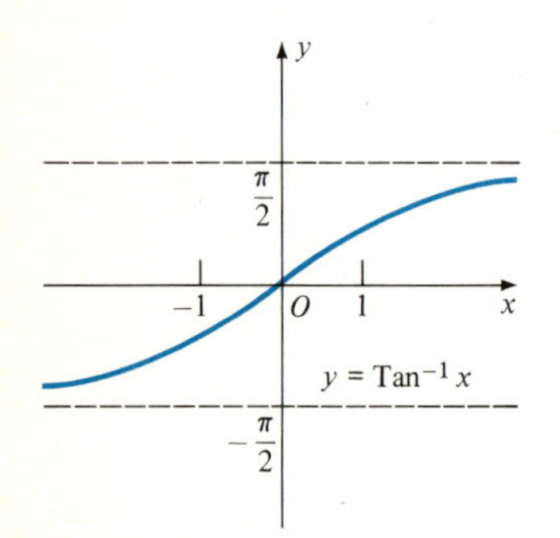

FIGURE 11.1h

Note the capitalization to remind you of the restriction on the domain. As you can see from Figure 11.1g, the restricted tangent function is one-to-one, so that its inverse is also a function.

The graph of $y = \text{Tan}^{-1} x$ appears in Figure 11.1h. Notice that the graph has the asymptotes $y = -\pi/2$ and $y = \pi/2$. The domain of the inverse tangent function is the set of all real numbers, and the range is the set $\{y | -\pi/2 < y < \pi/2\}$. Note that $\text{Tan}^{-1} x$ *increases* as *x increases*; as $|x|$ becomes larger and larger, $\text{Tan}^{-1} x$ gets closer and closer to $-\pi/2$ if x is negative and to $\pi/2$ if x is positive.

EXAMPLE 6

Find:

a. $\text{Tan}^{-1} \sqrt{3}$

b. $\text{Tan}^{-1}(-\sqrt{3})$

SOLUTION:

a. Let $\theta = \text{Tan}^{-1} \sqrt{3}$, so that $\tan \theta = \sqrt{3}$, and $-\pi/2 < \theta < \pi/2$. Therefore, $\theta = \pi/3$, and $\text{Tan}^{-1} \sqrt{3} = \pi/3$.

b. Let $\phi = \text{Tan}^{-1}(-\sqrt{3})$. Then $\tan \phi = -\sqrt{3}$, and $-\pi/2 < \phi < \pi/2$. Thus, $\phi = -\pi/3$, and $\text{Tan}^{-1}(-\sqrt{3}) = -\pi/3$.

▼ THEOREM 11.1c

$\text{Tan}^{-1}(-x) = -\text{Tan}^{-1} x$.

In general, as Example 6 indicates, we have Theorem 11.1c. Notice that the inverse tangent function and the inverse sine function both have the same kind of symmetry; they both satisfy the equation

$$f(-x) = -f(x)$$

so that their graphs are symmetric with respect to the origin. See Figures 11.1c and 11.1h.

EXAMPLE 7

Evaluate $\text{Tan}^{-1}(-2.1198)$.

SOLUTION:

From Table A5, we find $\text{Tan}^{-1}\, 2.1198 = 1.13$. Therefore, by Theorem 11.1c,

$$\text{Tan}^{-1}(-2.1198) = -1.13$$

Example 7 shows how you evaluate $\text{Tan}^{-1}\, x$ for a negative value of x. You find $\text{Tan}^{-1}\, x$ in the tables and simply place a minus sign in front of the answer. If you use a calculator, the same procedure as for $\text{Sin}^{-1}\, x$ and $\text{Cos}^{-1}\, x$ will give $\text{Tan}^{-1}(-2.1198) \approx -1.1300$.

EXAMPLE 8

Return to Figure 11.1a and express θ in terms of x.

SOLUTION:

We had $\theta = \beta - \alpha$, and

$$\tan \beta = \frac{a+h}{x} \qquad \tan \alpha = \frac{a}{x}$$

It is evident from the figure that both β and α are acute angles, so that we may solve the last two equations for β and α to obtain

$$\beta = \text{Tan}^{-1}\left(\frac{a+h}{x}\right) \quad \text{and} \quad \alpha = \text{Tan}^{-1}\left(\frac{a}{x}\right)$$

Therefore,

$$\theta = \text{Tan}^{-1}\left(\frac{a+h}{x}\right) - \text{Tan}^{-1}\left(\frac{a}{x}\right)$$

EXAMPLE 9

Express $\pi/4 + \text{Tan}^{-1}\, \frac{1}{3}$ as a single inverse tangent.

SOLUTION:

We can solve this problem easily by using the formula for the tangent of the sum of two numbers. First, we observe that $\text{Tan}^{-1}\, \frac{1}{3}$ is between 0 and $\pi/4$, so that $\pi/4 + \text{Tan}^{-1}\, \frac{1}{3}$ is between 0 and $\pi/2$. Then we let $\theta = \text{Tan}^{-1}\, \frac{1}{3}$ and write

$$\begin{aligned} \tan\left[\frac{\pi}{4} + \text{Tan}^{-1}\left(\frac{1}{3}\right)\right] &= \tan\left(\frac{\pi}{4} + \theta\right) \\ &= \frac{\tan \pi/4 + \tan \theta}{1 - \tan \pi/4 \tan \theta} \\ &= \frac{1 + \frac{1}{3}}{1 - \frac{1}{3}} = \frac{\frac{4}{3}}{\frac{2}{3}} = 2 \end{aligned}$$

Therefore,

$$\frac{\pi}{4} + \text{Tan}^{-1}\left(\frac{1}{3}\right) = \text{Tan}^{-1}\, 2$$

EXAMPLE 10

Simplify $\sin(2\ \mathrm{Sin}^{-1} x)$.

SOLUTION:

Let $\theta = \mathrm{Sin}^{-1} x$. Then

$$\sin(2\ \mathrm{Sin}^{-1} x) = \sin 2\theta = 2 \sin \theta \cos \theta$$

Because $\theta = \mathrm{Sin}^{-1} x$, we know that $\sin \theta = x$ and

$$\cos \theta = \sqrt{1 - \sin^2 \theta} = \sqrt{1 - x^2}$$

Thus,

$$\sin(2\ \mathrm{Sin}^{-1} x) = 2x\sqrt{1 - x^2}$$

EXAMPLE 11

Simplify the expression $\cos[\mathrm{Tan}^{-1} \frac{1}{7} - \mathrm{Tan}^{-1}(-\frac{3}{4})]$.

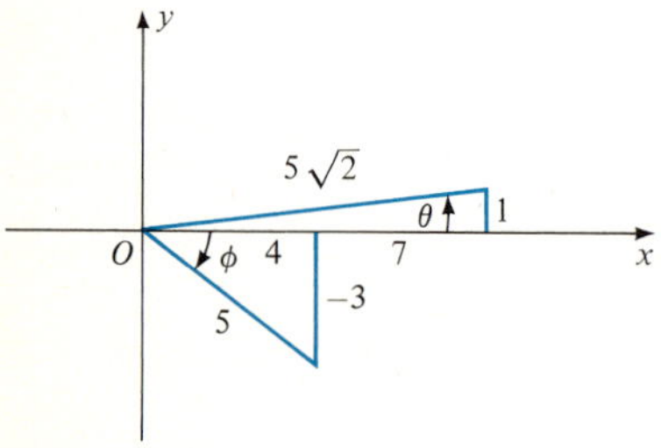

SOLUTION:

Let $\theta = \mathrm{Tan}^{-1} \frac{1}{7}$ and $\phi = \mathrm{Tan}^{-1}(-\frac{3}{4})$. We may interpret θ and ϕ as angles and make the sketch shown in Figure 11.1i. The figure allows us to read off the other trigonometric functions of θ and ϕ that we need. We can now write

FIGURE 11.1i

$$\cos[\mathrm{Tan}^{-1} \tfrac{1}{7} - \mathrm{Tan}^{-1}(-\tfrac{3}{4})] = \cos(\theta - \phi)$$

and use the formula

$$\cos(\theta - \phi) = \cos \theta \cos \phi + \sin \theta \sin \phi$$

Reading the sine and cosine values from the figure, we obtain

$$\cos(\theta - \phi) = \left(\frac{7}{5\sqrt{2}}\right)\left(\frac{4}{5}\right) + \left(\frac{1}{5\sqrt{2}}\right)\left(-\frac{3}{5}\right)$$

$$= \frac{28 - 3}{25\sqrt{2}} = \frac{1}{\sqrt{2}} = \frac{\sqrt{2}}{2}$$

Note: You might guess that, because $\cos(\theta - \phi) = \sqrt{2}/2$ in Example 11, $\theta - \phi = \pi/4$. You can verify this guess by finding $\tan(\theta - \phi)$, which turns out to be 1. Then, since

$$0 < \theta < \frac{\pi}{2} \quad \text{and} \quad -\frac{\pi}{2} < \phi < 0$$

and both the tangent and the cosine are positive, $\theta - \phi$ must be a first-quadrant angle. Thus, $\theta - \phi = \mathrm{Tan}^{-1} 1 = \pi/4$.

Because the inverse cotangent, cosecant, and secant functions are used much more rarely than the three functions that we have discussed, they are left to the problems. You should understand that the choice of the ranges of the inverse trigonometric functions is essentially arbitrary. The selections made here are the customary and most useful ones. We have already noted that hand-held calculators are internally programmed to give the values corresponding to the definitions presented in this text. We often call the functions chosen here the *principal-valued* inverse trigonometric functions to distinguish them from other choices that might be made.

EXERCISE 11.1

In Problems 1–6, evaluate the given expressions.

1 a $\text{Sin}^{-1}\dfrac{\sqrt{3}}{2}$ **b** $\text{Sin}^{-1}\left(-\dfrac{\sqrt{3}}{2}\right)$ **2 a** $\text{Cos}^{-1}\dfrac{1}{\sqrt{2}}$ **b** $\text{Cos}^{-1}\left(-\dfrac{1}{\sqrt{2}}\right)$ **3 a** $\text{Tan}^{-1}(-1)$ **b** $\text{Sin}^{-1}(-1)$

4 a $\text{Cos}^{-1}(-1)$ **b** $\text{Tan}^{-1} 0$ **5 a** $\text{Sin}^{-1}\left[\cos\left(-\dfrac{\pi}{4}\right)\right]$ **b** $\text{Cos}^{-1}\left[\cos\left(\dfrac{5\pi}{3}\right)\right]$

6 a $\text{Tan}^{-1}\left[\tan\left(\dfrac{5\pi}{4}\right)\right]$ **b** $\text{Sin}^{-1}\left[\tan\left(\dfrac{3\pi}{4}\right)\right]$

7 If the inverse cotangent function is the set of pairs

$$\{(x, y)\,|\,x = \cot y, 0 < y < \pi\}$$

draw the graph of the equation $y = \text{Cot}^{-1} x$.

8 If the inverse secant function is the set of pairs

$$\left\{(x, y)\,|\,x = \sec y, 0 \le y < \frac{\pi}{2} \text{ for } x \ge 1 \text{ and } -\pi \le y < -\frac{\pi}{2} \text{ for } x \le -1\right\}$$

draw the graph of the equation $y = \text{Sec}^{-1} x$.

9 If the inverse cosecant function is the set of pairs

$$\left\{(x, y)\,|\,x = \csc y, 0 < y \le \frac{\pi}{2} \text{ for } x > 1 \text{ and } -\pi < y \le -\frac{\pi}{2} \text{ for } x < -1\right\}$$

draw the graph of the equation $y = \text{Csc}^{-1} x$.

10 Show that $\cos(\text{Cos}^{-1} x) = x$ for $-1 \le x \le 1$.

11 Show that $\tan(\text{Tan}^{-1} x) = x$ for all real x.

12 Let $k(n) = (n^3 - 6n^2 + 17n)\pi/72$. Show that $\sin[k(n)] = \sin[\text{Sin}^{-1}(\sqrt{n}/2)] = \sqrt{n}/2$ for $n = 0, 1, 2, 3, 4$.

13 Is it true that $\text{Cos}^{-1}(-x) = \text{Cos}^{-1} x$? Explain.

14 Prove Theorem 11.1a. **15** Prove Theorem 11.1b. **16** Prove Theorem 11.1c.

17 Under what restriction on x is it true that $\text{Tan}^{-1}(\tan x) = x$?

18 Show that the solutions of the equation $\sin\theta = b$, $-1 \le b \le 1$, may be written $\theta = n\pi + (-1)^n \text{Sin}^{-1} b$, $n = 0, \pm 1, \pm 2, \pm 3, \ldots$.

19 Show that the solutions of the equation $\cos\theta = c$, $-1 \le c \le 1$, may be written $\theta = 2n\pi \pm \text{Cos}^{-1} c$, $n = 0, \pm 1, \pm 2, \pm 3, \ldots$.

20 Show that the solutions of the equation $\tan\theta = b$, b any real number, may be written $\theta = n\pi + \text{Tan}^{-1} b$, $n = 0, \pm 1, \pm 2, \pm 3, \ldots$.

21 Without using tables, show that $\text{Tan}^{-1}\frac{1}{2} + \text{Tan}^{-1}\frac{1}{3} = \pi/4$.

Hint: Let $\alpha = \text{Tan}^{-1}\frac{1}{2}$ and $\beta = \text{Tan}^{-1}\frac{1}{3}$. Then take the tangent of the sum.

22 Without using tables, show that $\text{Tan}^{-1}(2 + \sqrt{3}) - \text{Tan}^{-1}(2 - \sqrt{3}) = \pi/3$. See the hint in Problem 21.

23 Show that $\text{Sin}^{-1}\frac{3}{5} + \text{Sin}^{-1}\frac{4}{5} = \pi/2$.

Hint: If you draw a circle of radius 5 with center at the origin, the point (3, 4) lies on this circle.

24 Show that $\text{Cos}^{-1}\frac{3}{5} + \text{Cos}^{-1}\frac{5}{13} = \text{Cos}^{-1} -\frac{33}{65}$.

25 Consider the sum, $\text{Tan}^{-1} 2 + \text{Tan}^{-1} 3$. A student let $\alpha = \text{Tan}^{-1} 2$ and $\beta = \text{Tan}^{-1} 3$. Then, by using the formula for the tangent of a sum, this student obtained $\tan(\alpha + \beta) = -1$. Whereupon, the student wrote $\alpha + \beta = \text{Tan}^{-1}(-1) = -\pi/4$, which is obviously wrong. Can you see where the mistake is?

In Problems 26–33, simplify the given expression.

26 $\sin(2\,\text{Cos}^{-1}\frac{4}{5})$

27 $\cos(2\,\text{Sin}^{-1}\frac{5}{13})$

28 $\tan(\frac{1}{2}\,\text{Sin}^{-1}\frac{24}{25})$

29 $\sin(\frac{1}{2}\,\text{Tan}^{-1}\frac{5}{12})$

30 $\sin[\text{Sin}^{-1}\frac{1}{3} + \text{Cos}^{-1}\frac{2}{3})$

31 $\tan(\text{Tan}^{-1}\frac{1}{7} - \text{Tan}^{-1}\frac{1}{3})$

32 $\sin(\text{Tan}^{-1}\frac{3}{4} - \text{Sin}^{-1}\frac{4}{5})$

33 $\cos(\text{Cos}^{-1}x + \text{Sin}^{-1}x)$

34 Solve for x:

$$\text{Tan}^{-1}x + \text{Tan}^{-1}\left(\frac{3}{5}\right) = \frac{\pi}{4}$$

35 Solve for x:

$$\text{Sin}^{-1}x + \text{Sin}^{-1}\left(\frac{3}{5}\right) = \text{Tan}^{-1}\left(\frac{56}{33}\right)$$

36 For the accompanying figure show that $\phi = \text{Tan}^{-1}\left(\frac{a+h}{x} + \tan\alpha\right) - \text{Tan}^{-1}\left(\frac{a}{x} + \tan\alpha\right)$

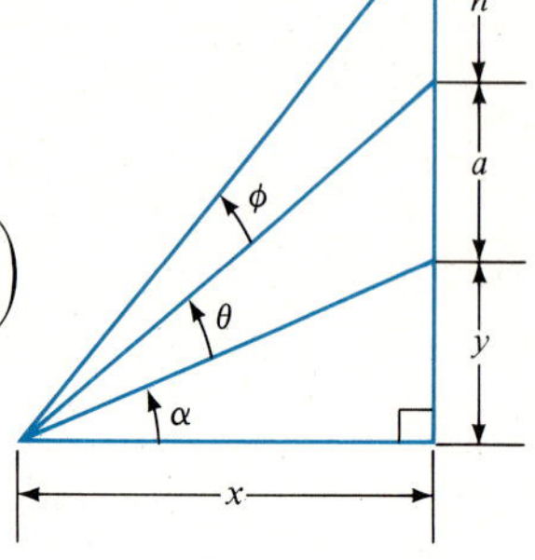

USING YOUR KNOWLEDGE 11.1

Let us return to the calculus problem described at the beginning of this section. In Example 8, we found that the angle θ is given by

$$\theta = \text{Tan}^{-1}\left(\frac{a+h}{x}\right) - \text{Tan}^{-1}\left(\frac{a}{x}\right) \qquad (1)$$

and we said that the maximum value of θ occurs for $x = \sqrt{a(a+h)}$. Although we do not have the tools here to obtain this value of x, we can show that this value does give the maximum value of θ. We proceed as follows. You should carry out the details.

1. Take the tangent of both sides of Equation 1 to obtain

$$\tan\theta = \frac{hx}{x^2 + a^2 + ah} \qquad \text{so that} \qquad \theta = \text{Tan}^{-1}\left(\frac{hx}{x^2 + a^2 + ah}\right)$$

2. We can see that θ will be as large as possible if $F(x) = hx/(x^2 + a^2 + ah)$ is as large as possible. Explain why.

3. Now put $x = \sqrt{a^2 + ah}$ to obtain

$$F(\sqrt{a^2 + ah}) = \frac{h}{2\sqrt{a^2 + ah}}$$

4. Because we claim that this is the maximum value of $F(x)$, we subtract $F(x)$ from $F(\sqrt{a^2 + ah})$ to get

$$F(\sqrt{a^2 + ah}) - F(x) = \frac{h(x - \sqrt{a^2 + ah})^2}{2\sqrt{a^2 + ah(x^2 + a^2 + ah)}}$$

5. The result of step 4 shows that $F(x)$ has its maximum value for $x = \sqrt{a^2 + ah}$. Explain why.

Hint: $F(\sqrt{a^2 + ah}) - F(x) \geq 0$ for all values of x.

11.2 POLAR COORDINATES

Various coordinate systems are used in the plane, and one of the most useful is the *polar coordinate* system. This system is a fairly natural one, in which a point is located by giving its *distance* from a fixed reference point, called the *pole*, and its *direction* with respect to a fixed axis, called the polar *axis*.

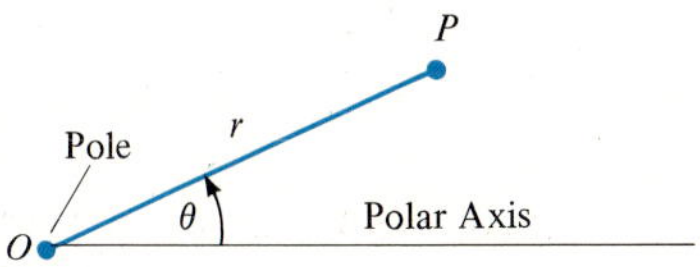

FIGURE 11.2a

Figure 11.2a shows a point plotted in polar coordinates. We use the letter r for the distance of P from the pole and the letter θ for the angle from the polar axis to the line OP. Thus, the point P is located relative to the pole (or origin), O, by the ordered pair, (r, θ), where r is the number of units in the length of OP and θ is the number of radians in the angle from the polar axis to the line OP. The two numbers, r and θ, are the *polar coordinates* of P.

The polar axis from which θ is measured is usually chosen as a horizontal ray originating at O. We frequently superimpose a rectangular coordinate system on the polar system, taking the origin at the pole and the polar axis as the positive x-axis. To be consistent with our previous agreement about angles, we measure positive values of θ in the counter-clockwise sense and negative values in the clockwise sense. We agree to allow negative values of r with the convention that such a value correspond to a distance measured in a direction exactly opposite that of the terminal ray of the angle θ. For example, in Figure 11.2b, several points are plotted in polar coordinates. These points are labeled in two ways, once with a positive value for r and once with a negative value for r. Of course, the value of θ may be increased or decreased by any integer multiple of 2π.

PRIMARY POLAR COORDINATES

Notice that we have now encountered a major difference between rectangular and polar coordinates. In a system of rectangular coordinates, there is a one-to-one correspondence between the set of all real number-pairs and the set of all points in the plane. This is not so for polar coordinates. Although it is true that a given pair of polar coordinates has a unique point corresponding to it, each point in the plane has infinitely many different pairs of polar coordinates. However, for a given point, other than the pole, there are just two *primary* distinct pairs of coordinates: one pair with *positive* r and the *smallest possible nonnegative* θ, say (r_1, θ_1), and the other with coordinates $(-r_1, \pi + \theta_1)$. (The coordinates shown in Figure 11.2b are the primary coordinates for each point.) All other pairs of coordinates of the given point are of the form

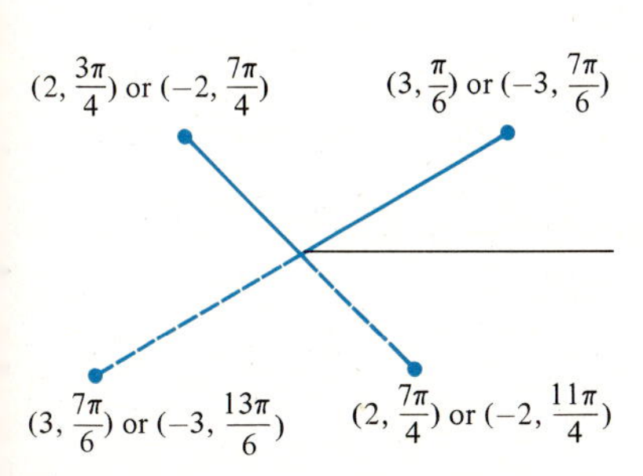

FIGURE 11.2b

$$(r_1, \theta_1 + 2m\pi) \text{ or } (-r_1, \pi + \theta_1 + 2m\pi)$$

where m is an integer. Because no direction is determined by a single point, the coordinates of the pole itself are $(0, \theta)$ where θ is entirely arbitrary. Although the multiplicity of coordinates for a specified point sometimes causes us some difficulty, we are more than compensated for this by gaining a greater versatility in describing curves in polar coordinates.

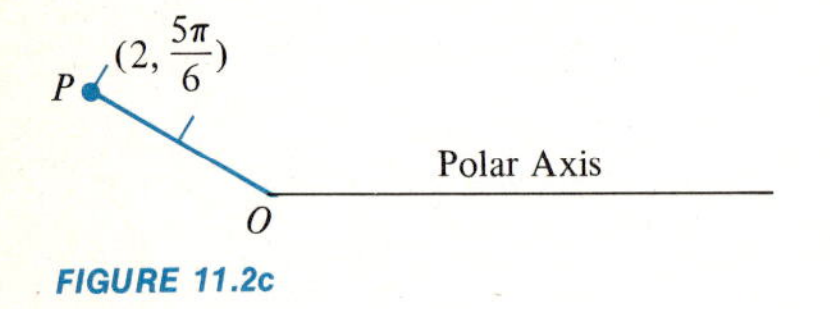

FIGURE 11.2c

Figure 11.2c shows the point P that corresponds to the coordinates $(2, 5\pi/6)$. The other primary pair of coordinates is $(-2, 11\pi/6)$. This point has the two basic sets of polar coordinates:

$$\left\{\left(2, \frac{5\pi}{6} + 2m\pi\right)\right\} \quad \text{and} \quad \left\{\left(-2, \frac{11\pi}{6} + 2m\pi\right)\right\}$$

where $m = 0, \pm 1, \pm 2, \ldots$.

EXAMPLE 1

Plot the point with coordinates $(-3, \pi/4)$. What are the two primary pairs of coordinates? Give the two basic sets of coordinates.

SOLUTION:

The point is plotted in Figure 11.2d. Recall that the primary coordinates are (r_1, θ_1) and $(-r_1, \pi + \theta_1)$, where r_1 is positive and θ_1 is the smallest nonnegative value possible for the given point. Thus, we must use $r = 3$, so that one primary pair of coordinates is $(3, 5\pi/4)$ and the other is $(-3, 9\pi/4)$. The two basic sets of coordinates are

$$\left(3, \frac{5\pi}{4} + 2m\pi\right) \quad \text{and} \quad \left(-3, \frac{9\pi}{4} + 2m\pi\right) \qquad \text{where } m = 0, \pm 1, \pm 2, \ldots.$$

Notice that the given coordinates $(-3, \pi/4)$ are obtained by taking $m = -1$ in the second basic set.

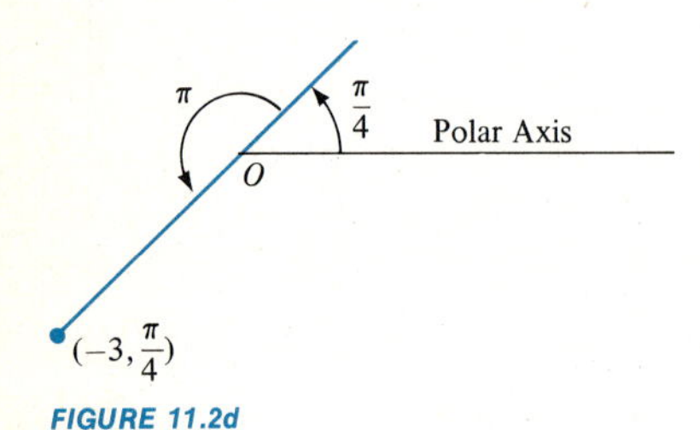

FIGURE 11.2d

Example 2 illustrates one of the minor difficulties that is caused by the multiplicity of polar coordinates for a given point.

EXAMPLE 2

The equation $r = \tan \theta - 4$ describes a certain curve in polar coordinates. Is the point with coordinates $(3, 5\pi/4)$ on this curve?

SOLUTION:

If we substitute $(3, 5\pi/4)$ into the equation, we come up with

$$3 = \tan \frac{5\pi}{4} - 4$$

or

$$3 = 1 - 4$$

which is false. Does this mean that the given point is not on the curve? No. If you will refer to Example 1, you will see that the given point has another primary set of coordinates, $(-3, 9\pi/4)$. If we substitute $(-3, 9\pi/4)$ into the given equation, we obtain

$$-3 = \tan \frac{9\pi}{4} - 4$$

or

$$-3 = 1 - 4$$

which is true. Thus, the pair $(-3, 9\pi/4)$ does satisfy the equation, and the point is on the curve.

THE POLAR COORDINATE LINES

In a rectangular coordinate system, a point (a, b) is located as the intersection of two "coordinate lines," $x = a$ and $y = b$. The coordinate lines are simply the lines obtained by setting x equal to c_1 and y equal to c_2 for various values of the constants c_1 and c_2. In polar coordinates, the coordinate lines are determined similarly by setting r equal to c_1 and θ equal to c_2, where c_1 and c_2 are constants. The graph of $r = c_1$ is a circle of radius c_1 with its center at the pole O. The graph of $\theta = c_2$ is a straight line passing through the pole and making an angle of c_2 radians with the polar axis. Thus, in polar coordinates, the coordinate lines consist of all the circles concentric at the pole and all the straight lines that pass through the pole. You are undoubtedly familiar with the type of ruled paper that is used for graphing in rectangular coordinates. Figure 11.2e shows a sheet of polar coordinate paper, ruled with the coordinate lines. This kind of paper makes it quite easy to plot points in polar coordinates.

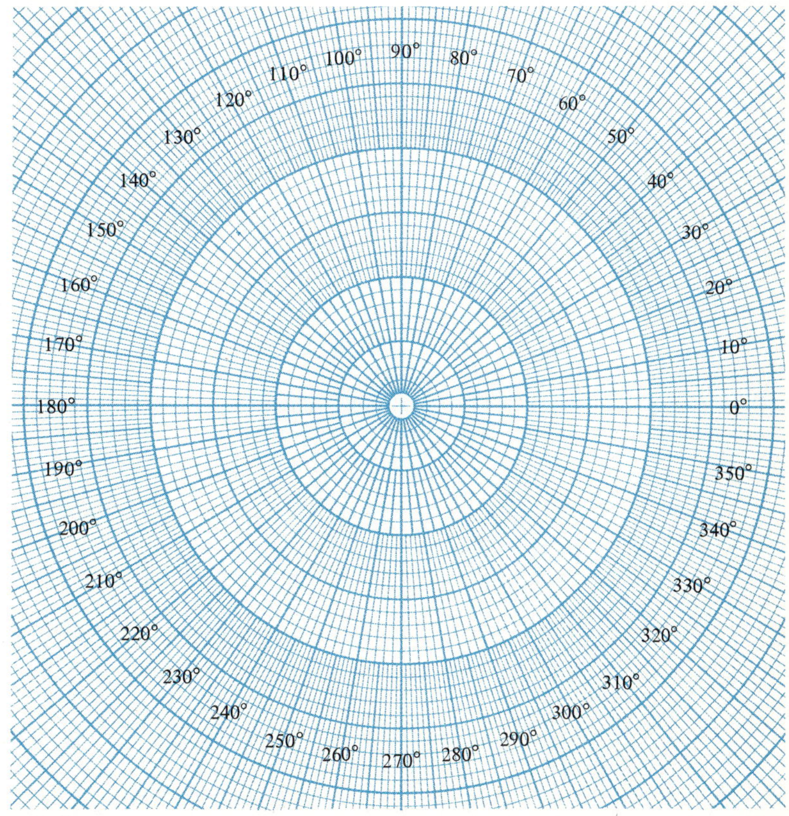

FIGURE 11.2e

THE DISTANCE FORMULA

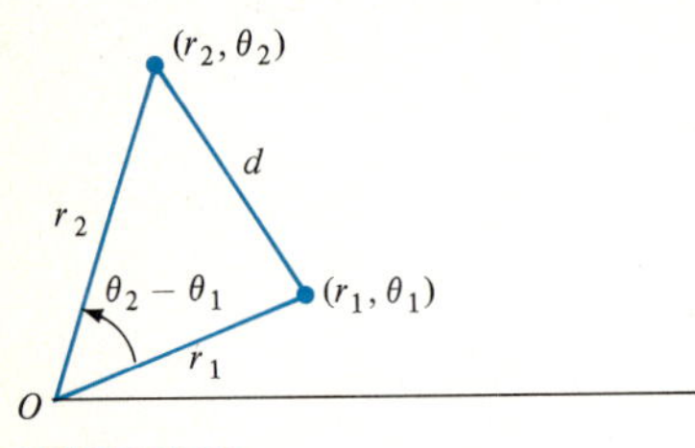

FIGURE 11.2f

The distance formula for polar coordinates is readily obtained by using the Law of Cosines for the triangle shown in Figure 11.2f. This gives the following expression for the distance d between the two points (r_1, θ_1) and (r_2, θ_2):

$$d = \sqrt{r_1{}^2 + r_2{}^2 - 2r_1r_2 \cos(\theta_2 - \theta_1)}$$

EXAMPLE 3

Find the distance between the points $(2, \pi/6)$ and $(3, 5\pi/6)$.

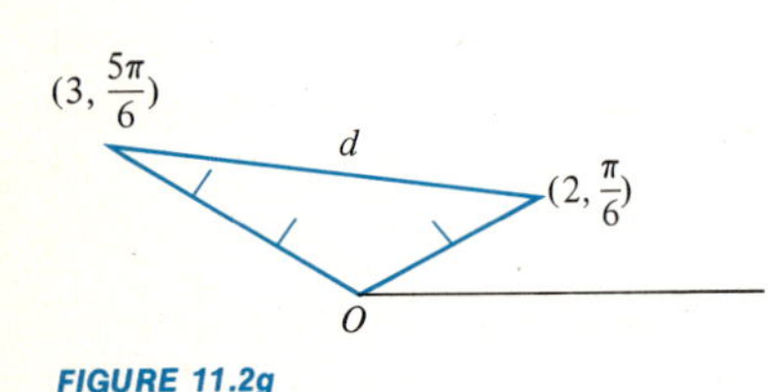

FIGURE 11.2g

SOLUTION:

The points are plotted in Figure 11.2g. Using the distance formula, we obtain

$$d = \sqrt{2^2 + 3^2 - 2(2)(3)\cos\left(\frac{5\pi}{6} - \frac{\pi}{6}\right)}$$

$$= \sqrt{4 + 9 - 12\cos\frac{2\pi}{3}}$$

$$= \sqrt{4 + 9 + 6}$$

$$= \sqrt{19} \approx 4.4$$

EXAMPLE 4

A circle has its center at the point $(2, \pi/2)$ and passes through the pole. (See Figure 11.2h.) Find a polar coordinate equation for this circle.

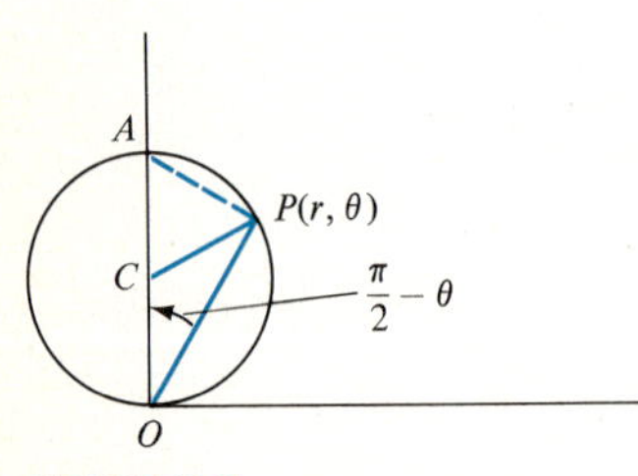

FIGURE 11.2h

SOLUTION:

As indicated in Figure 11.2h, we let a general point on the circle have coordinates (r, θ). Because the circle passes through the pole, its radius must be 2, the distance of the center from the pole. Consequently, we can equate the square of the distance from the center to the general point to the square of the radius. In this way, we obtain

$$r^2 + 2^2 - 2(r)(2)\cos\left(\frac{\pi}{2} - \theta\right) = 2^2$$

or

$$r^2 - 4r\sin\theta = 0$$

This equation factors into

$$r(r - 4\sin\theta) = 0$$

which is satisfied if $r = 0$ or if $r = 4\sin\theta$. Clearly, $r = 0$ cannot be the required equation because it describes only the pole. Thus, the required equation is $r = 4\sin\theta$; this can be verified by joining the point (r, θ) in Figure 11.2h to the point where the circle cuts the line $\theta = \pi/2$ and using the right triangle that is so formed.

RELATIONSHIP BETWEEN RECTANGULAR AND POLAR COORDINATES

With two different coordinate systems in the plane, it is often helpful to be able to go back and forth between the two. The equations that enable us to make such transformations of coordinates are not difficult to obtain. As indicated earlier, we superimpose the two coordinate systems so that the pole is at the origin and the polar axis coincides with the positive x-axis as shown in Figure 11.2i. From this figure, we see that

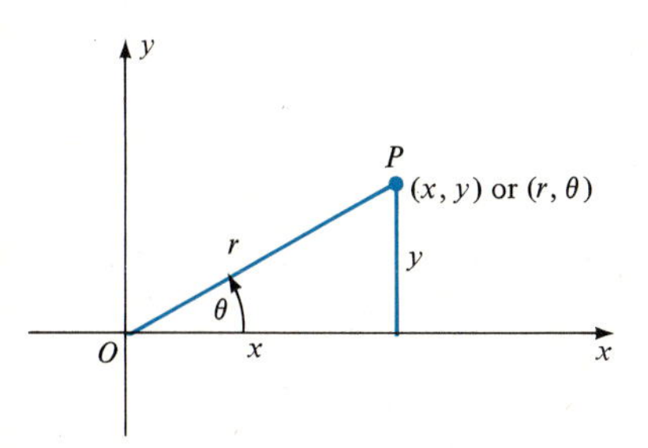

FIGURE 11.2i

$$\begin{aligned} x &= r \cos \theta \\ y &= r \sin \theta \end{aligned} \tag{1}$$

and

$$r^2 = x^2 + y^2 \qquad \sin \theta = \frac{x}{r} \qquad \cos \theta = \frac{y}{r} \tag{2}$$

With Equations 1, we can quickly find the rectangular coordinates of a point if its polar coordinates are given. For instance, the rectangular coordinates of the point with polar coordinates $(4, \pi/3)$ are

$$x = 4 \cos \frac{\pi}{3} = 2$$

$$y = 4 \sin \frac{\pi}{3} = 2\sqrt{3}$$

Similarly, to go in the opposite direction, from rectangular to polar coordinates, we use Equations 2. Thus, for the point (3, 4) in rectangular coordinates,

$$r = \sqrt{3^2 + 4^2} = 5, \sin \theta = \tfrac{3}{5}, \text{ and } \cos \theta = \tfrac{4}{5}$$

so that

$$\theta = \text{Tan}^{-1} \tfrac{3}{4}$$

However, for the point $(-3, -4)$,

$$r = 5, \sin \theta = -\tfrac{3}{5}, \text{ and } \cos \theta = -\tfrac{4}{5}$$

so that

$$\theta = \pi + \text{Tan}^{-1} \tfrac{3}{4}$$

Notice the use of both sine and cosine; this is to avoid the error of obtaining a wrong quadrant for the angle θ.

Using Equations 1 and 2, we can transform the equation of a curve from rectangular to polar coordinates or from polar to rectangular coordinates. The next two examples show the procedure to be used. Transformations of this kind are sometimes helpful in graphing an equation, as will be seen later.

EXAMPLE 5

Transform the equation $x^2 + y^2 = 4y$ into polar coordinates.

SOLUTION:

Using Equations 1 and 2, we have $x^2 + y^2 = r^2$ and $y = r \sin \theta$, which we substitute into the given equation to obtain

$$r^2 = 4r \sin \theta$$

Since $r \neq 0$ in general, we may divide both members by r to arrive at the result

$$r = 4 \sin \theta$$

Notice that this example provides another check on the answer to Example 4.

EXAMPLE 6

Transform the equation $r = 2 \sin 2\theta$ into rectangular coordinates.

SOLUTION:

We first express $\sin 2\theta$ in terms of $\cos \theta$ and $\sin \theta$ to obtain

$$r = 4 \cos \theta \sin \theta$$

Then, because $x = r \cos \theta$ and $y = r \sin \theta$, we multiply both sides by r^2 to obtain

$$r^3 = 4(r \cos \theta)(r \sin \theta)$$

Using Equations 1 and 2, we now have

$$\pm(x^2 + y^2)^{3/2} = 4xy$$

or, by squaring both sides,

$$(x^2 + y^2)^3 = 16x^2y^2$$

EXERCISE 11.2

In Problems 1–6, plot the given points in polar coordinates and give their rectangular coordinates.

1 a $\left(5, \frac{\pi}{6}\right)$ **b** $\left(-5, \frac{\pi}{6}\right)$ **c** $\left(5, -\frac{\pi}{6}\right)$ **2 a** $\left(1, \frac{3\pi}{2}\right)$ **b** $\left(1, -\frac{\pi}{2}\right)$ **c** $\left(-1, \frac{\pi}{2}\right)$ **3 a** $\left(2, \frac{\pi}{4}\right)$ **b** $\left(-2, -\frac{\pi}{4}\right)$ **c** $\left(2, \frac{3\pi}{4}\right)$

4 a $\left(-3, \frac{2\pi}{3}\right)$ **b** $\left(3, \frac{8\pi}{3}\right)$ **c** $\left(3, \frac{14\pi}{3}\right)$ **5 a** $\left(4, \frac{5\pi}{6}\right)$ **b** $\left(-4, \frac{11\pi}{6}\right)$ **c** $\left(4, -\frac{5\pi}{6}\right)$

6 a $\left(2, \frac{\pi}{3}\right)$ **b** $\left(2, -\frac{\pi}{3}\right)$ **c** $\left(-2, -\frac{\pi}{3}\right)$

In Problems 7–10, plot the given points in rectangular coordinates and find their two primary pairs of polar coordinates.

7 a $(1, -1)$ **b** $(-1, 1)$ **c** $(-1, -1)$ **8 a** $(1, \sqrt{3})$ **b** $(-1, \sqrt{3})$ **c** $(1, -\sqrt{3})$ **9 a** $(0, 2)$ **b** $(-2, 0)$ **c** $(0, -2)$
10 a $(\sqrt{3}, -1)$ **b** $(-\sqrt{3}, 1)$ **c** $(-\sqrt{3}, -1)$

In Problems 11–14, determine: (a) if the given pair of polar coordinates satisfies the equation $r = \cos \theta - 1$ and (b) if the point described by the given coordinates is on the curve.

11 $\left(\frac{3}{2}, \frac{\pi}{3}\right)$ **12** $\left(\frac{1}{2}, \frac{\pi}{3}\right)$ **13** $\left(-\frac{1}{2}, \frac{4\pi}{3}\right)$ **14** $\left(1, \frac{3\pi}{2}\right)$

In Problems 15–18, determine: (a) if the given pair of polar coordinates satisfies the equation $r = 2/(1 - \sin\theta)$ and (b) if the point described by the given coordinates is on the curve.

15 $\left(4, \frac{\pi}{6}\right)$ **16** $\left(2, \frac{\pi}{2}\right)$ **17** $\left(-\frac{4}{3}, \frac{\pi}{6}\right)$ **18** $\left(-8 - 4\sqrt{3}, -\frac{\pi}{3}\right)$

In Problems 19–22, two points are given in polar coordinates. Find the distance between them. Simplify, but leave answers in radical form if necessary.

19 $\left(2, \frac{2\pi}{3}\right)$ and $(4, \pi)$ **20** $\left(3, \frac{2\pi}{3}\right)$ and $\left(4, \frac{\pi}{3}\right)$ **21** $\left(3, \frac{2\pi}{3}\right)$ and $\left(-4, \frac{\pi}{3}\right)$

22 $\left(4, \frac{\pi}{4}\right)$ and $\left(2, \frac{11\pi}{12}\right)$

In Problems 23–26, transform the given equation into polar coordinates.

23 $x^2 + y^2 = 2y$ **24** $x^2 + y^2 = 2x - 4y$ **25** $y^2 = 2x$ **26** $x^2 + y^2 = \frac{y}{x}$

In Problems 27–30, transform the given equation into rectangular coordinates.

27 $r^2 = 4\cos 2\theta$ **28** $r = 2\sin\theta$ **29** $r^2 = \sin\theta$ **30** $r = \sin\theta + \cos\theta$

31 A circle has its center at $(2, \pi/3)$ and passes through the origin. Show that an equation of the circle is $r = 4\cos(\theta - \pi/3)$.

32 A circle of radius 3 has its center at $(2, 0)$. Show that an equation of the circle is $r^2 - 4r\cos\theta = 5$.

33 A circle has its center at $(5, \text{Tan}^{-1}\frac{3}{4})$ and passes through the origin. Show that an equation of the circle is $r = 8\cos\theta + 6\sin\theta$.

34 Show that the points $A(\sqrt{3}, 0)$, $B(3, \pi/2)$, and $C(3, \pi/6)$ are the vertices of a right triangle

USING YOUR KNOWLEDGE 11.2

The conic sections, which were defined in Section 6.3, can also be defined as follows: A conic section is a set of points such that the ratio of the distance of each point from a fixed point (called the *focus*) to its distance from a fixed line (called the *directrix*) is a constant e (called the *eccentricity*). The equation of a conic section comes out in an especially simple form if the focus is taken at the pole and the directrix is a line perpendicular to the polar axis, as in Figure 11.2j. Let the directrix be p units to the left of the origin and let $P(r, \theta)$ be a point on the graph. Then, by the definition,

$$\frac{OP}{DP} = e \tag{1}$$

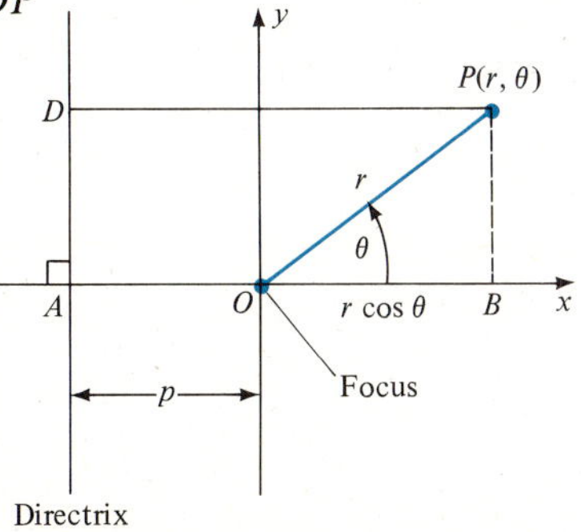

FIGURE 11.2j

1. We know that $OP = r$. Show that $DP = p + r\cos\theta$.
2. Substitute the results of Problem 1 in Equation 1 and solve the resulting equation for r. You should obtain

$$r = \frac{ep}{1 - e\cos\theta}$$

3. If $e = 1$, show that the definition agrees with that given for a parabola in Exercise 6.3, Problem 17.
4. If $e < 1$, show that the range of r is finite. (The curve is an ellipse.)
5. If $e > 1$, show that the range of r is *in*finite. (The curve is a hyperbola.)
6. The earth travels around the sun in an elliptical path with the sun at the focus and with an eccentricity of about 0.0167, so that the equation of the ellipse may be taken as

$$r = \frac{149.6 \times 10^6}{1 - 0.0167\cos\theta}$$

where distance is measured in kilometers. Find the approximate maximum and minimum distances of the earth from the sun.

11.3 GRAPHS OF POLAR COORDINATE CURVES

In this section, we shall study some basic techniques that are quite helpful in graphing curves in polar coordinates. We have already seen in Section 11.2 that the equation $r = a$, where a is a positive constant, describes a circle of radius a with center at the pole. If $a = 0$, the graph is just the pole itself, and if a is negative, the graph is again a circle with center at the pole and radius $|a|$. We also saw that the equation $\theta = b$, where b is a constant, describes a straight line through the pole and making an angle of b radians with the polar axis. These coordinate curves are the simplest of the curves that we can describe in polar coordinates.

When graphing curves, it is occasionally helpful to transform from polar to rectangular coordinates. For example, the equation

$$r = a\cos\theta + b\sin\theta$$

may be multiplied by r to obtain

$$r^2 = ar\cos\theta + br\sin\theta$$

which, in rectangular coordinates, is

$$x^2 + y^2 = ax + by$$

We recognize this as the equation of a circle that passes through the origin and cuts the x-axis at $(a, 0)$ and the y-axis at $(0, b)$. See Figure 11.3a, where a and b are assumed to be positive.

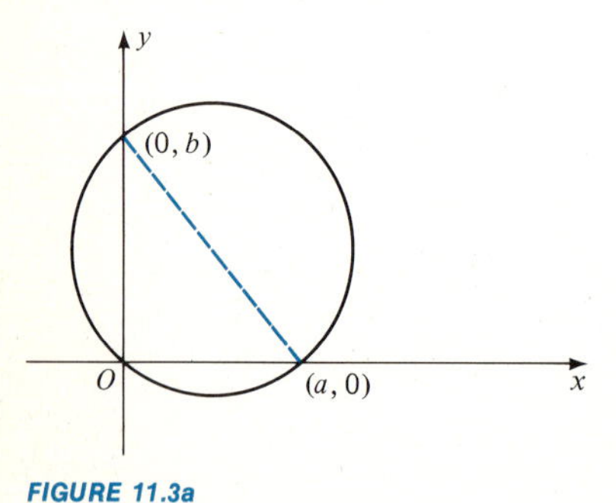

FIGURE 11.3a

Unfortunately, the information obtained by changing to rectangular coordinates will often not be very helpful, so that we must usually work directly with the polar coordinates. Frequently, quite simple considerations will furnish enough information on which to base the sketch of a curve. As before, it is convenient to regard the system of polar coordinates as superimposed on a system of rectangular coordinates with the pole at the origin and the polar axis along the positive x-axis. The points where the graph intersects the x- and y-axes can then be found by substituting the values $n\pi/2$, $n = 0, \pm 1, \pm 2, \ldots$, for θ into the equation and calculating the corresponding values of r. If trigonometric functions are involved in the equation, the variation in the values of these functions will be helpful in drawing the graph. These ideas will be illustrated in the examples to follow.

SYMMETRY

In graphing a curve in polar coordinates, it is often useful to determine if any simple symmetry exists. As in rectangular coordinates, the usual symmetries to be investigated are with respect to the axes and the origin. Suppose we wish to test an equation $r = f(\theta)$ to determine if its graph is symmetric with respect to the y-axis. Then, as is shown in Figure 11.3b, if $P(r, \theta)$ is a point on the graph and if the graph is symmetric to the y-axis, then there is a point $Q(r_1, \theta_1)$ on the graph located so that the y-axis is the perpendicular bisector of the line segment PQ. Conversely, if there is such a symmetric point Q corresponding to each point P on the graph, then the graph is symmetric to the y-axis. In terms of the coordinates of P, the only possible coordinates for Q are $(r, \pi - \theta + 2n\pi)$ or $(-r, -\theta + 2n\pi)$, where $n = 0, \pm 1, \pm 2, \ldots$. Thus, such symmetry exists if and only if the equation reduces to its original form when (r, θ) is replaced by (r_1, θ_1), where for some integer n, either

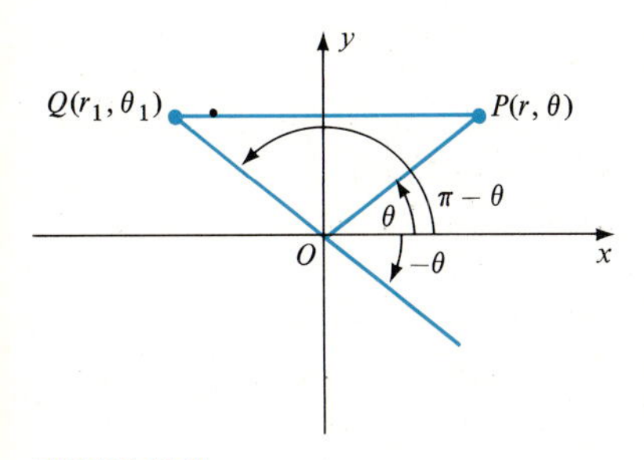

FIGURE 11.3b

$$r_1 = r \quad \text{and} \quad \theta_1 = \pi - \theta + 2n\pi$$

or

$$r_1 = -r \quad \text{and} \quad \theta_1 = -\theta + 2n\pi$$

This means that whenever the point (r, θ) is on the graph, the point (r_1, θ_1) is also on the graph, because the equation is satisfied by (r_1, θ_1) whenever it is satisfied by (r, θ).

EXAMPLE 1

Show that the graph of $r = a \sin 2\theta$ is symmetric with respect to the y-axis

SOLUTION:

If we use the first part of the preceding test and replace θ by $\pi - \theta + 2n\pi$, we obtain

$$\begin{aligned} r &= a \sin 2(\pi - \theta + 2n\pi) \\ &= a \sin(2\pi - 2\theta + 4n\pi) \\ &= a \sin(-2\theta) \end{aligned}$$

or

$$r = -a \sin 2\theta$$

which is not the original equation. But if we replace r by $-r$ and θ by $-\theta + 2n\pi$, we obtain

$$\begin{aligned} -r &= a \sin 2(-\theta + 2n\pi) \\ &= a \sin(4n\pi - 2\theta) \\ &= a \sin(-2\theta) \end{aligned}$$

or

$$r = a \sin 2\theta$$

which is the original equation. Hence, the graph is symmetric with respect to the y-axis. Note that both parts of the test must fail before we can state that the symmetry does not exist.

Exactly the same type of analysis yields tests for symmetry with respect to the x-axis and with respect to the origin. The details are left to you. Table 11.3a summarizes the two tests for each of the three symmetries.

TABLE 11.3a

Graph symmetric with respect to	**If there is an integer n, such that the equation reduces to the original equation upon replacement of (r, θ) by**
x-axis	$(r, -\theta + 2n\pi)$ or $(-r, \pi - \theta + 2n\pi)$*
y-axis	$(r, \pi - \theta + 2n\pi)$ or $(-r, -\theta + 2n\pi)$*
origin	$(r, \pi + \theta + 2n\pi)$ or $(-r, \theta + 2n\pi)$*

* At least one of the two tests must be satisfied in each case.

EXAMPLE 2

Test the equation $r^2 = \sin 2\theta$ for the three symmetries in Table 11.3a.

SOLUTION:

x-axis Substitute $(r, -\theta + 2n\pi)$ to obtain

$$r^2 = \sin(-2\theta + 4n\pi)$$

or

$$r^2 = -\sin 2\theta$$

which is not the original equation. Substitute $(-r, \pi - \theta + 2n\pi)$ to obtain

$$(-r)^2 = \sin(2\pi - 2\theta + 4n\pi)$$

or

$$r^2 = -\sin 2\theta$$

which is not the original equation. Therefore, the graph is not symmetric with respect to the x-axis.

y-axis Substitute $(r, \pi - \theta + 2n\pi)$ to obtain

$$r^2 = \sin(2\pi - 2\theta + 4n\pi)$$

or

$$r^2 = -\sin 2\theta$$

which is not the original equation. Substitute $(-r, -\theta + 2n\pi)$ to obtain

$$(-r)^2 = \sin(-2\theta + 4n\pi)$$

or

$$r^2 = -\sin 2\theta$$

which is not the original equation. Therefore, the graph is not symmetric with respect to the y-axis.

Origin Substitute $(r, \pi + \theta + 2n\pi)$ to obtain

$$r^2 = \sin(2\pi + 2\theta + 4n\pi)$$

or

$$r^2 = \sin 2\theta$$

which is the original equation. Thus, the graph is symmetric with respect to the origin.

Before we turn to sketching polar coordinate curves, we state one helpful result that is proved in calculus: For all the simple curves that we shall consider, if a curve passes through the origin, it does so in the directions given by the values of θ for which $r = 0$. For instance, the curve $r^2 = \sin 2\theta$, which was considered in Example 2, passes through the origin in the directions given by the solutions of the equation

$$\sin 2\theta = 0$$

that is,

$$2\theta = n\pi \qquad n = 0, \pm 1, \pm 2, \ldots$$

or

$$\theta = \tfrac{1}{2}n\pi \qquad n = 0, \pm 1, \pm 2, \ldots$$

The curve passes through the origin at angles of 0, $\pi/2$, π, and $3\pi/2$. (The other values will duplicate these four directions.)

EXAMPLE 3

Sketch the curve $r^2 = \sin 2\theta$.

SOLUTION:

1. We have just seen that this curve passes through the pole for $\theta = 0, \pi/2, \pi$, and $3\pi/2$. Because there are no other values

between 0 and 2π that make $r = 0$, the curve cuts the axes only at the origin. The curve passes through the origin in the directions given by the four preceding values of θ.

2. In Example 2, we found this curve to be symmetric to the origin but not to either the x- or the y-axis.
3. Because r^2 must be nonnegative, we see that the right side, $\sin 2\theta = 2 \sin \theta \cos \theta$, must also be nonnegative. Thus, for $r \neq 0$, $\sin \theta$ and $\cos \theta$ must have like signs, so that none of the graph can be in the second or fourth quadrants.
4. Because of the symmetry found in (2), it will suffice to calculate a few points for $0 \leq \theta \leq \pi/2$. First, however, let us note that as θ increases from 0 to $\pi/4$, r increases from 0 to 1, and as θ increases from $\pi/4$ to $\pi/2$, r decreases from 1 to 0. This follows, of course, from the variation in the sine function. The accompanying table gives a few points that will be adequate for sketching the curve. Figure 11.3c shows the required graph, which has been drawn by making use of all the information obtained in this discussion. This curve is called a *lemniscate*, which means "bow-shaped."

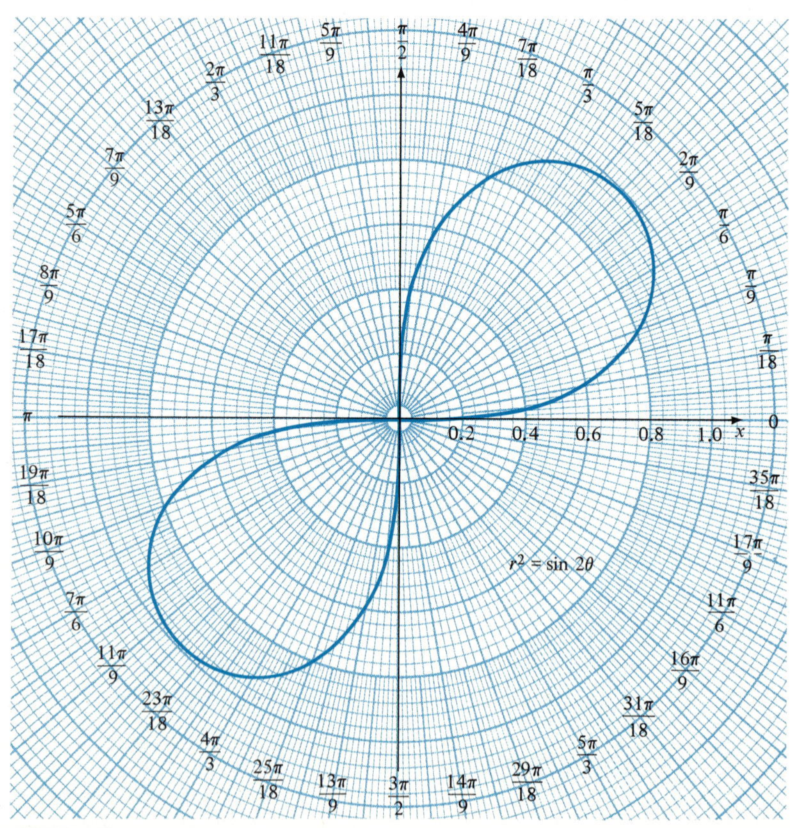

FIGURE 11.3c

θ	0	$\pi/12$	$\pi/6$	$\pi/4$	$\pi/3$	$5\pi/12$	$\pi/2$
r	0	0.71	0.93	1.00	0.93	0.71	0

As the solution of Example 3 indicates, the following procedure is helpful in making the graph of a polar coordinate equation.

Procedure for graphing polar coordinate equations

1. Determine where the curve cuts the x- and y-axes. Find the values of θ (if there are any) for which $r = 0$. These values give the directions in which the graph passes through the origin.
2. Determine any simple symmetry.
3. Find what values of θ (if any) cannot be used because they do not yield real values of r.
4. Note the variation in the values of r as θ increases through its permissible values. Then calculate and plot a few points. Make use of the symmetry to avoid calculating too many points. Draw a smooth curve through the plotted points, and draw in the remainder of the curve by using the symmetry.

EXAMPLE 4

Discuss and sketch the curve $r = 2(1 + \cos\theta)$.

SOLUTION:

1. Because the period of $\cos\theta$ is 2π, we may restrict θ to values between 0 and 2π. The curve cuts the axes at the points $(4, 0)$, $(2, \pi/2)$, $(0, \pi)$, and $(2, 3\pi/2)$, and $r = 0$ for $\theta = \pi$, only.
2. Substituting $(r, -\theta)$ for (r, θ), we obtain

   $$r = 2[1 + \cos(-\theta)]$$

   or

   $$r = 2(1 + \cos\theta)$$

   the original equation. Thus, the graph is symmetric with respect to the x-axis. You should show that the graph is not symmetric to the y-axis or the origin.
3. It is clear from the equation that all values of θ are permissible.
4. As θ increases from 0 to $\pi/2$, r decreases from 4 to 2. As θ increases from $\pi/2$ to π, r decreases from 2 to 0. As θ increases from π to $3\pi/2$, r increases from 0 to 2. As θ increases from $3\pi/2$ to 2π, r increases from 2 to 4.

See the accompanying table of values and the graph in Figure 11.3d. Because of its heart shape, this curve is called a *cardioid*.

θ	0	$\pi/6$	$\pi/4$	$\pi/3$	$\pi/2$	$2\pi/3$	$3\pi/4$	$5\pi/6$	π
r	4	3.73	3.41	3.00	2.00	1.00	0.59	0.27	0

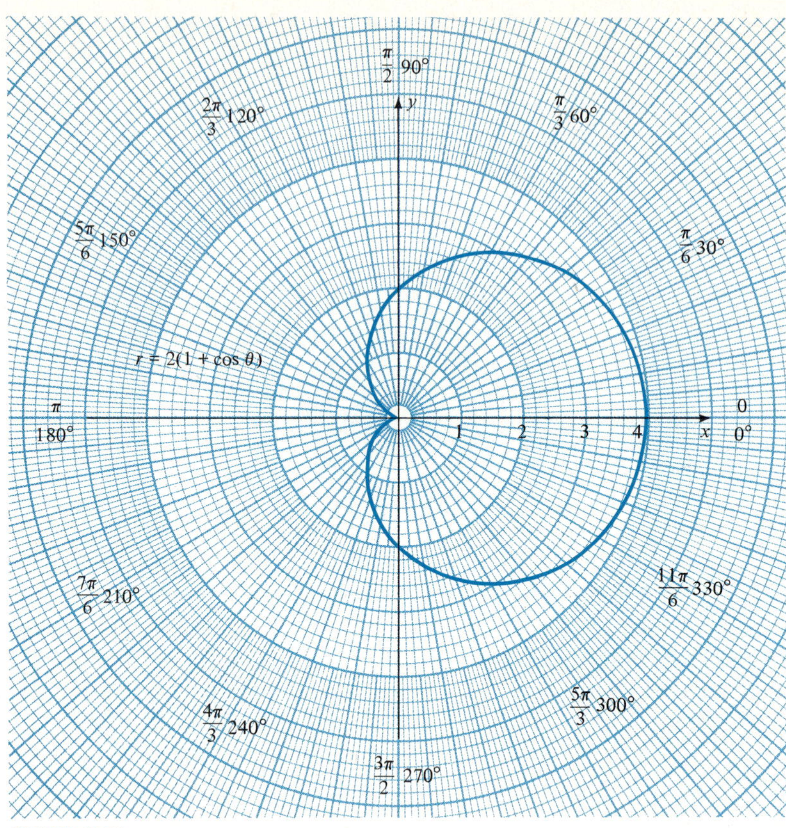

FIGURE 11.3d

EXAMPLE 5

Graph the equation $r = 5 \sin 3\theta$.

SOLUTION:

1. The curve crosses the axes at the points $(0, 0)$, $(-5, \pi/2)$, $(0, \pi)$, and $(5, 3\pi/2)$. Notice that the first and third of these pairs describe the same point, the origin. The second and last pairs also describe the same point, the point 5 units down on the y-axis. Since $\sin 3(\pi + \theta) = \sin(3\pi + 3\theta) = -\sin 3\theta$, the coordinates $(-r, \pi + \theta)$ satisfy the equation whenever (r, θ) do. Therefore, the entire curve is traced out as θ goes from 0 to π; it would be retraced if θ were to go from π to 2π. Hence, we may confine θ to the interval $0 \leq \theta \leq \pi$. If $r = 0$, then $\sin 3\theta = 0$, so that $3\theta = 0, \pi, 2\pi$, or 3π, and $\theta = 0, \pi/3, 2\pi/3$, or π. Thus, the curve passes through the origin for these values of θ.
2. If we replace (r, θ) by $(r, \pi - \theta)$, we obtain

$$r = 5 \sin(3\pi - 3\theta) \quad \text{or} \quad r = 5 \sin 3\theta$$

the original equation. This means that the graph is symmetric with respect to the y-axis. You should show that the graph is not symmetric with respect to the x-axis or the origin.

3. All values of θ are permissible.

4. As θ varies from 0 to $\pi/6$, r increases from 0 to 5. As θ varies from $\pi/6$ to $\pi/3$, r decreases from 5 to 0. As θ varies from $\pi/3$ to $\pi/2$, r decreases from 0 to -5. We need not go beyond $\pi/2$, because half the curve is traced and the remainder will be obtained by symmetry. The accompanying table of values gives the coordinates of points at intervals of 10° or $\pi/18$ radians as θ goes from 0 to $\pi/2$. The graph is shown in Figure 11.3e. This curve is called a *three-leaved rose curve.*

θ	0°	10°	20°	30°	40°	50°	60°	70°	80°	90°
r	0	2.50	4.33	5.00	4.33	2.50	0	−2.50	−4.33	−5.00

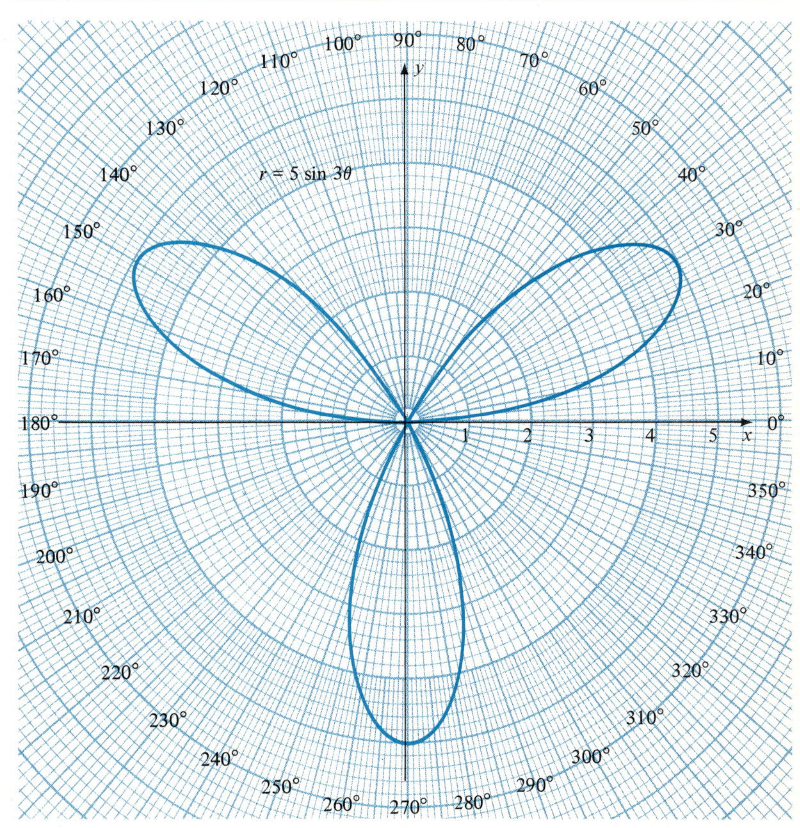

FIGURE 11.3e

EXAMPLE 6

Graph $r = 1 + 2\cos\theta$.

SOLUTION:

1. The points where the curve cuts the axes are $(3, 0)$, $(1, \pi/2)$ $(-1, 3\pi/2)$. For $r = 0$, we obtain $\cos\theta = -\frac{1}{2}$, so that $\theta = 2\pi/3$ and $\theta = 4\pi/3$ are the directions in which the curve passes through the origin.
2. By replacing (r, θ) by $(r, -\theta)$, we obtain

$$r = 1 + 2\cos(-\theta)$$
or

$$r = 1 + \cos 2\theta$$

the original equation. Hence, the curve is symmetric to the x-axis. You should show that no other simple symmetry exists.

3. All values of θ are permissible.
4. As θ increases from 0 to π, r decreases from 3 to -1. The accompanying table gives a few points that can be used to sketch the curve from $\theta = 0$ to $\theta = \pi$. The remainder of the graph can be constructed by using the symmetry found in (2). The graph appears in Figure 11.3f. This curve is called a *limaçon*, which means "snail-shaped."

θ	0	$\pi/6$	$\pi/4$	$\pi/3$	$\pi/2$	$2\pi/3$	$3\pi/4$	$5\pi/6$	π
r	3	2.73	2.41	2.00	1.00	0	-0.41	-0.73	-1.00

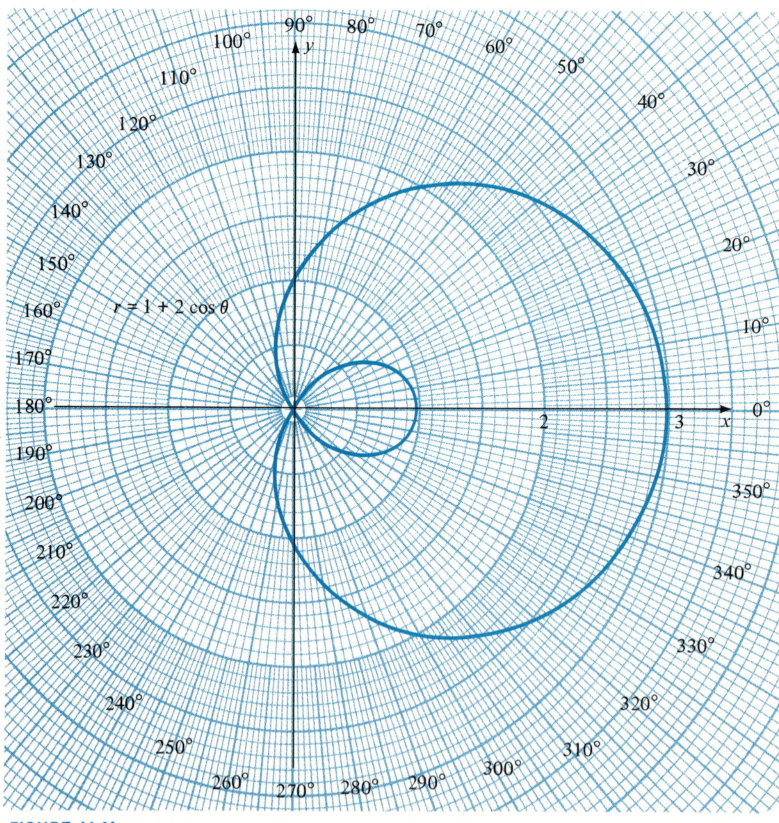

FIGURE 11.3f

EXERCISE 11.3

In Problems 1–8, transform the given equation into rectangular coordinates and sketch the curve.

1 $r \sin \theta = -2$ **2** $r \cos \theta = 3$ **3** $r = 5$ **4** $\theta = \frac{3\pi}{4}$ **5** $r = 4 \cos \theta$ **6** $r + 2 \sin \theta = 0$

7 $r = \cos \theta - \sin \theta$ **8** $r = \cos \theta + 2 \sin \theta$

In Problems 9–24, discuss the given equation and sketch its graph.

9 $r + 2 = \sin \theta$ **10** $r = 1 + 2 \sin \theta$ **11** $r = 3 - 6 \cos \theta$

12 $r = 2 + 3 \sin \theta$ **13** $r = 3 + 2 \cos \theta$ **14** $r = 3 \sin 2\theta$

15 $r = \sin 4\theta$ **16** $r = 2 \cos 3\theta$ **17** $r = 2(1 + \sin \theta)$

18 $r = 4(1 - \cos \theta)$ **19** $r^2 = \cos 2\theta$ **20** $r = \sin^2 \theta$

21 $r\theta = 1$ (hyperbolic spiral) **22** $r = \theta$ (spiral of Archimedes) **23** $r = e^\theta$ (logarithmic spiral)

24 $r = \ln \theta$

25 Transform the equation $(x^2 + y^2)^2 = x^3 - 3xy^2$ into polar coordinates. Then discuss and sketch the curve.

Hint: Your work will be made easier if you see the formula $\cos 3\theta = 4 \cos^3 \theta - 3 \cos \theta$.

26 Some people state the symmetry tests leaving out the term $2n\pi$, which we have included. Consider the equation $r = \sin(\theta/2 + \pi/4)$. Show that the graph of this equation is symmetric with respect to the origin. Note that the first part of the test in Table 11.3a fails and that the second part succeeds, but only for odd values of n. Thus, omitting the term $2n\pi$ would make the test fail completely. Show that this curve is symmetric to both axes and the origin. Sketch the curve.

Hint: Calculate points for θ between $-90°$ and $90°$ and then use the symmetry.

27 The points $(a/\sqrt{2}, 0)$ and $(a/\sqrt{2}, \pi)$ in polar coordinates are called the *foci* of the lemniscate $r^2 = a^2 \cos 2\theta$. Show that if (r, θ) is any point on the graph, then the product of the distances of (r, θ) from the two foci is $a^2/2$.

USING YOUR KNOWLEDGE 11.3

Certain problems in calculus call for finding the area of a region bounded by two curves given in polar coordinates. The first step in such a problem consists of finding the points of intersection of the two curves. We can do this part of the problem here. For instance, to find the points of intersection of $r = 2 \cos \theta$ and $r = 2(1 - \cos \theta)$, we proceed as follows:

1. Equate the two expressions for r to obtain the equation

   $$2 \cos \theta = 2(1 - \cos \theta) \quad \text{or} \quad \cos \theta = \tfrac{1}{2}$$

2. This gives $\theta = \frac{\pi}{3}$ or $\frac{5\pi}{3}$

3. Substitute these values into the equation $r = 2 \cos \theta$ to obtain $r = 2 \cos \pi/3 = 1$ and $r = 2 \cos 5\pi/3 = 1$. Thus, the curves intersect at $(1, \pi/3)$ and at $(1, 5\pi/3)$.
4. Because $r = 0$ at the origin but θ is not unique, we can check to see whether or not the two curves go through the origin. Thus, for $r = 0$, $2 \cos \theta = 0$ gives $\theta = \pi/2$ or $\theta = 3\pi/2$, and for $r = 0$, $2(1 - \cos \theta) = 0$ gives $\theta = 0$. Hence, the curves also intersect at the origin.

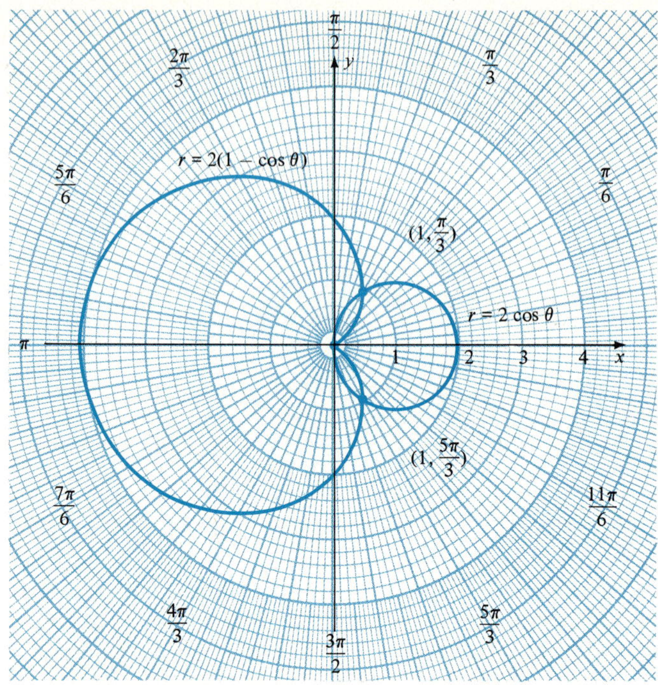

FIGURE 11.3g

It is a good idea to sketch the curves as a check on your work. (See Figure 11.3g.)

Find the points of intersection of the given curves:

1 $r = \sin\theta$ and $r = 1 - \sin\theta$ **2** $r = 2\cos 2\theta$ and $r = 1$ **3** $r = \cos 2\theta$ and $r = \cos\theta$
4 $r = 1 - \cos\theta$ and $r = 1 + \sin\theta$

11.4 VECTORS IN THE PLANE

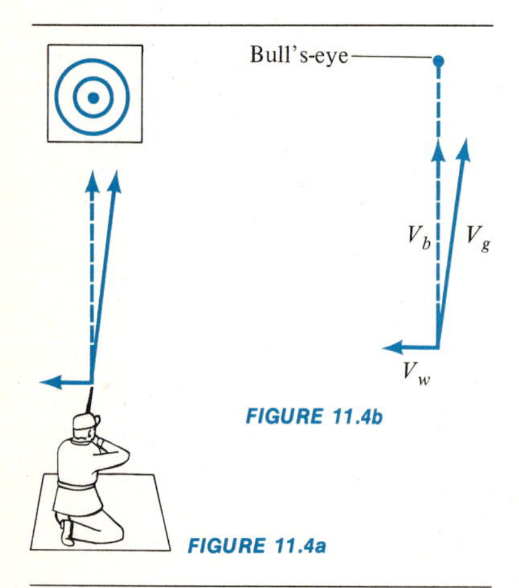

FIGURE 11.4b

FIGURE 11.4a

Figure 11.4a shows a marksman aiming his rifle at a target. Suppose that in still air, the bullet travels 3000 feet per second and that the target is 150 feet due north of the marksman. Also, suppose there is a wind blowing from east to west that imparts an east-west velocity of 5 mph to the bullet. Where should the marksman aim to hit the bull's-eye? Of course, we know that the gun must be aimed slightly east of the bull's-eye to compensate for the effect of the wind, but we want a more precise answer than this.

We can make a diagram as in Figure 11.4b to show what happens. The bullet acts as if it had two velocities, V_g imparted by the gun and V_w by the wind. The net result is that the bullet travels with a velocity V_b that is composed of these two velocities. In this section, we shall learn how such velocities are properly combined.

VECTOR QUANTITIES

Notice that we have been careful to use the word *velocity* rather than *speed*. A velocity is a quantity that must be described by giving its magnitude and its direction. The magnitude of the velocity is the speed. Thus, in the marksman's problem, the velocity imparted to the bullet by the wind is 5 mph due west. The speed is 5 mph, and the direction is due west. If the gun is aimed in a direction N x° E, that is, x degrees to the east of due north, then the velocity imparted by the gun is 3000 feet per second in the direction N x° E (read "north, x degrees east"). Both these velocities are known as *vector quantities*. As a preliminary definition, we can say that vector quantities are quantities that require both a magnitude and a direction for their description.

The fact that a vector quantity has both magnitude and direction suggests that it can be represented geometrically by a directed line segment such as OP in Figure 11.4c. The length of the segment is proportional to the magnitude, and the direction of the segment is the direction of the vector quantity. Directed line segments used for such a representation are frequently denoted by symbols such as **a**, **b**, **r**, and so on. (In this text, boldface type will be used to denote vector quantities. In writing by hand, it is customary to use an arrow over a letter or a wavy line under the letter, as $\vec{a}$ or $\underset{\sim}{a}$, for the same purpose.)

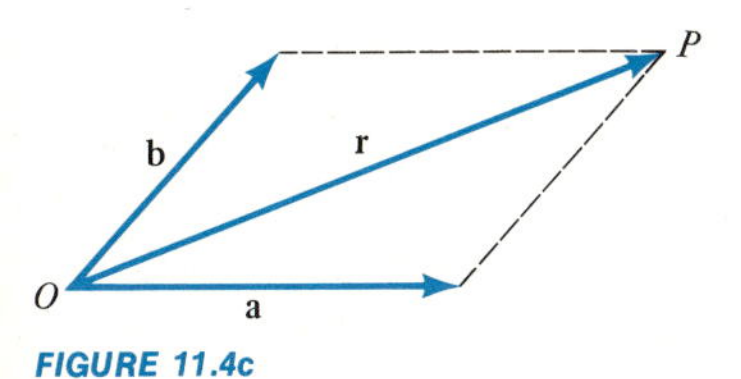

FIGURE 11.4c

ADDITION OF VECTOR QUANTITIES

Countless physical experiments have shown that two vector quantities **a** and **b**, acting at the same point, are equivalent to a single vector quantity known as the *resultant*, **r**. Furthermore, this resultant is correctly represented by the diagonal of the parallelogram having the arrows representing **a** and **b** as its sides. (See Figure 11.4c.) The resultant **r** is called the *sum* of **a** and **b**, and we write

$$\mathbf{r} = \mathbf{a} + \mathbf{b}$$

It is important to notice that this equation uses the plus sign in a different sense from that of ordinary addition. Here, the plus sign means that the quantities represented by **a** and **b** are combined by the *parallelogram law of composition*.

A large part of the preceding discussion is summarized in Definition 11.4a. Commonly occurring vector quantities are forces, displacements, and velocities. The directed line segments in Figure 11.4c are examples of vectors.

▼ DEFINITION 11.4a

***Vector quantities* are quantities that can be represented by directed line segments and that are added according to the parallelogram law. Directed line segments used in this manner are called *vectors*.**

VECTORS AND ORDERED PAIRS OF NUMBERS

Although we now have a geometric representation of vector quantities, we still need a convenient analytic representation for them. Our earlier work involving the location of points in a plane by means of ordered pairs of numbers suggests that it is possible to represent a vector by means of such a number pair. For example, the coordinates (4, 3) represent a certain point P, with respect to an origin O. (See Figure 11.4d.) This pair of numbers, with reference

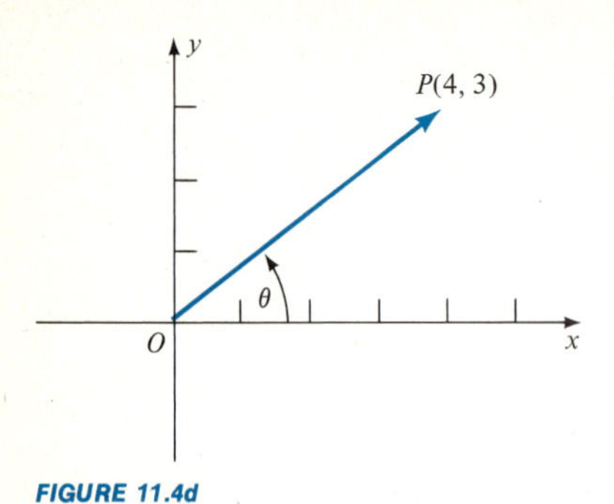

FIGURE 11.4d

to O, determines a direction and a magnitude, the direction being given by the angle θ and the magnitude by the length of the line segment from O to P. It is therefore reasonable to interpret an ordered pair (a_1, a_2), where a_1 and a_2 are not both zero, as a directed line segment extending *from* the origin *to* the point with coordinates (a_1, a_2).

In order to regard (a_1, a_2) as a vector, we require a condition for equality and a definition of addition of ordered pairs that is equivalent to the parallelogram law. Because distinct ordered pairs locate distinct points, Definition 11.4b is a sensible definition of equality.

▼ DEFINITION 11.4b

Equality of vectors

$(a_1, a_2) = (b_1, b_2)$ if and only if $a_1 = b_1$ and $a_2 = b_2$.

▼ DEFINITION 11.4c

Addition of vectors

If $\mathbf{a} = (a_1, a_2)$ and $\mathbf{b} = (b_1, b_2)$, then $\mathbf{a} + \mathbf{b} = (a_1 + b_1, a_2 + b_2)$

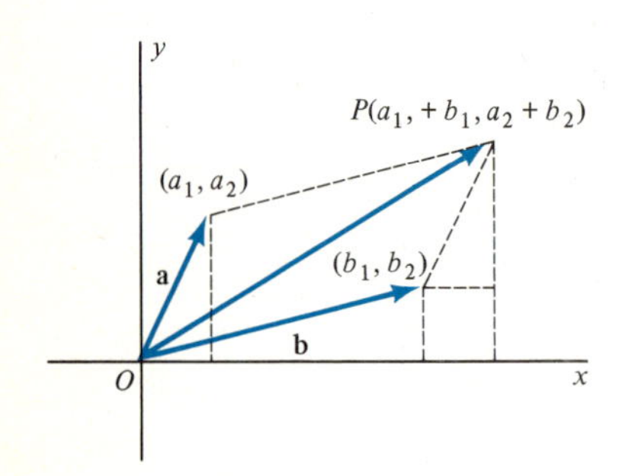

FIGURE 11.4e

Using simple geometry, we can see that the vertex P of the parallelogram in Figure 11.4e is represented by the ordered pair $(a_1 + b_1, a_2 + b_2)$, which suggests the operation of addition as expressed in Definition 11.4c. Thus, if $\mathbf{a} = (2, 4)$ and $\mathbf{b} = (-3, 2)$, then

$$\begin{aligned}\mathbf{a} + \mathbf{b} &= (2, 4) + (-3, 2)\\ &= (2 - 3, 4 + 2)\\ &= (-1, 6)\end{aligned}$$

The magnitude of a vector corresponds to the length of the vector and is defined in Definition 11.4d.
The vector $\mathbf{0} = (0, 0)$ is called the *zero vector*; it has magnitude zero but does not have a direction. The direction of a nonzero vector is specified by the angle θ that the vector makes with the positive x direction. (See Definition 11.4e.)

▼ DEFINITION 11.4d

Magnitude of a vector

If $\mathbf{a} = (a_1, a_2)$ is a vector, then the *magnitude* of a, denoted by $|\mathbf{a}|$, is the number

$$\sqrt{a_1^2 + a_2^2}$$

▼ DEFINITION 11.4e

Direction of a vector

The direction of $\mathbf{a} \neq \mathbf{0}$ is given by

$$\cos\theta = \frac{a_1}{|\mathbf{a}|} \qquad \sin\theta = \frac{a_2}{|\mathbf{a}|}$$

EXAMPLE 1

Find the magnitude and the direction of the vector (4, 3) shown in Figure 11.4d.

SOLUTION:

Let $\mathbf{a} = (4, 3)$. Then the magnitude of $\mathbf{a}$ is

$$|\mathbf{a}| = \sqrt{4^2 + 3^2} = \sqrt{25} = 5$$

Also, $\cos\theta = \frac{4}{5}$ and $\sin\theta = \frac{3}{5}$, so that $\theta = 36.87°$. Thus, the vector (4, 3) makes an angle of 36.87° with the positive x direction.

EXAMPLE 2

Find the magnitude and the direction of the resultant of the vectors $\mathbf{a} = (1, 2)$ and $\mathbf{b} = (-4, 4)$. Illustrate with a sketch.

SOLUTION:

By Definition 11.4c,

$$\begin{aligned}\mathbf{r} &= \mathbf{a} + \mathbf{b}\\ &= (1, 2) + (-4, 4)\\ &= (1 - 4, 2 + 4)\\ &= (-3, 6)\end{aligned}$$

Thus,

$$|\mathbf{r}| = \sqrt{(-3)^2 + 6^2} = \sqrt{45} = 3\sqrt{5} \approx 6.71$$

$$\cos\theta = \frac{-3}{3\sqrt{5}} = -\frac{1}{\sqrt{5}} \quad \text{and} \quad \sin\theta = \frac{6}{3\sqrt{5}} = \frac{2}{\sqrt{5}}$$

so that

$$\theta = 116.57°$$

Figure 11.4f illustrates the solution.

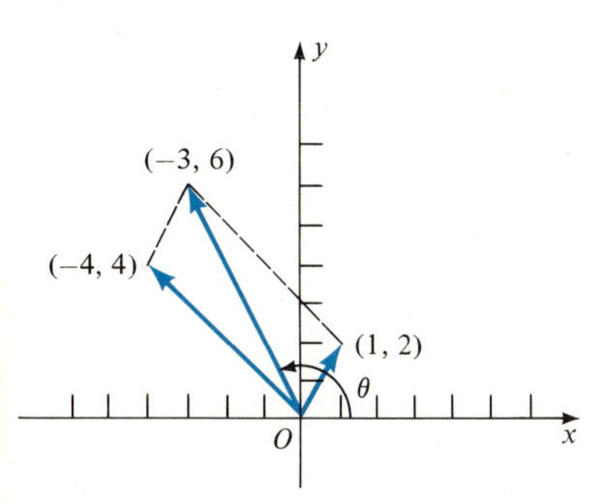

FIGURE 11.4f

PROPERTIES OF VECTORS

Because vectors can be represented by ordered pairs of real numbers, it follows that certain properties of vectors can be deduced from properties of real numbers. Theorems 11.4a, 11.4b, and 11.4c illustrate this idea.

▼ **THEOREM 11.4a**

Vector addition is commutative:

$\mathbf{a} + \mathbf{b} = \mathbf{b} + \mathbf{a}$

PROOF OF THEOREM 11.4a:

Let $\mathbf{a} = (a_1, a_2)$ and $\mathbf{b} = (b_1, b_2)$. Then

$$\mathbf{a} + \mathbf{b} = (a_1 + b_1, a_2 + b_2)$$
$$\mathbf{b} + \mathbf{a} = (b_1 + a_1, b_2 + a_2)$$

By the Commutative Law of Addition for real numbers,

$$a_1 + b_1 = b_1 + a_1 \quad \text{and} \quad a_2 + b_2 = b_2 + a_2$$

so that

$$\mathbf{a} + \mathbf{b} = \mathbf{b} + \mathbf{a}$$

as was to be shown.

▼ **THEOREM 11.4b**

Vector addition is associative:

$(\mathbf{a} + \mathbf{b}) + \mathbf{c} = \mathbf{a} + (\mathbf{b} + \mathbf{c})$

Theorem 11.4b is a simple consequence of the definition of addition of vectors and the Associative Law of Addition for real numbers. The proof of the theorem is left for the exercises. For example, if $\mathbf{a} = (1, 3)$, $\mathbf{b} = (-2, 5)$, and $\mathbf{c} = (4, -6)$, then

$$\mathbf{a} + \mathbf{b} = (-1, 8)$$

so that

$$(\mathbf{a} + \mathbf{b}) + \mathbf{c} = (-1, 8) + (4, -6) = (3, 2)$$

and

$$\mathbf{b} + \mathbf{c} = (2, -1)$$

so that

$$\mathbf{a} + (\mathbf{b} + \mathbf{c}) = (1, 3) + (2, -1) = (3, 2)$$

as before.

Suppose that $\mathbf{a} = (a_1, a_2)$. Then, by the definition of addition, we may write

$$2\mathbf{a} = 2(a_1, a_2) = (a_1, a_2) + (a_1, a_2) = (2a_1, 2a_2)$$

If this process is repeated a number of times, we arrive at the relationship $k(a_1, a_2) = (ka_1, ka_2)$, where k is a positive integer. This suggests the more general relationship given by Definition 11.4f. Note that the first part of this definition is equivalent to the statement

$$k\mathbf{a} = \mathbf{a}k$$

▼ DEFINITION 11.4f

If k is a real number, then

$$k(a_1, a_2) = (a_1, a_2)k = (ka_1, ka_2)$$

The real number k is often called a *scalar*, and the operation defined in Definition 11.4f is called *scalar multiplication.*

The proof of Theorem 11.4c follows directly from Definitions 11.4c and 11.4f. The details are left for the exercises.

▼ THEOREM 11.4c

If k is any real number, then

$$k(\mathbf{a} + \mathbf{b}) = k\mathbf{a} + k\mathbf{b}$$

It should be fairly obvious that multiplication of a vector by a positive number k does not change the direction of the vector, but it does multiply the magnitude by k. Multiplication of a vector by a negative number k always reverses the direction of the vector and multiplies the magnitude by $|k|$. Multiplication of a vector by zero always yields the zero vector.

SUBTRACTION OF VECTORS

▼ DEFINITION 11.4g

$$-\mathbf{a} = (-1)\mathbf{a}$$

▼ DEFINITION 11.4h

$$\mathbf{a} - \mathbf{b} = \mathbf{a} + (-\mathbf{b})$$

Vector subtraction may be accomplished by means of Definitions 11.4g and 11.4h. Thus, if $\mathbf{a} = (1, 2)$ and $\mathbf{b} = (-4, 4)$, then

$$\begin{aligned}\mathbf{a} - \mathbf{b} &= (1, 2) + (-1)(-4, 4) \\ &= (1, 2) + (4, -4) \\ &= (5, -2)\end{aligned}$$

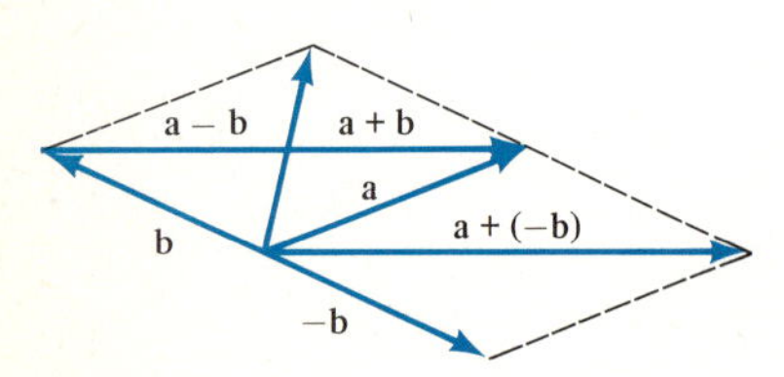

FIGURE 11.4g

For two vectors $\mathbf{a}$ and $\mathbf{b}$, the sum and difference are shown in Figure 11.4g. This figure illustrates an important fact about vectors. Because $\mathbf{a} - \mathbf{b}$ is a vector $\mathbf{x}$ such that $\mathbf{a} = \mathbf{b} + \mathbf{x}$, it may be represented in the diagram as the vector that extends from the tip of $\mathbf{b}$ to the tip of $\mathbf{a}$. But $\mathbf{x}$ is also the vector $\mathbf{a} + (-\mathbf{b})$, which is shown in the diagram. We know that $\mathbf{a} - \mathbf{b}$ and $\mathbf{a} + (-\mathbf{b})$ are the same vector, $\mathbf{x}$, yet they appear as two different line segments in the figure. Evidently, the location of the arrow in the diagram is not important; only the magnitude and the direction are the fundamental characteristics of a vector. To illustrate further, consider the vector $\mathbf{a} = (2, 1)$, which can be written as the sum or difference of two other vectors in infinitely many ways. Thus, $(2, 1) = (4, -1) - (2, -2) = (4, 4) - (2, 3)$, so that $\mathbf{a}$ may be presented geometrically as shown in Figure 11.4h. The important idea is that a given vector may be represented by an arrow of the correct length and direction but located anywhere in the plane, simply by writing it as the difference of two appropriate vectors.

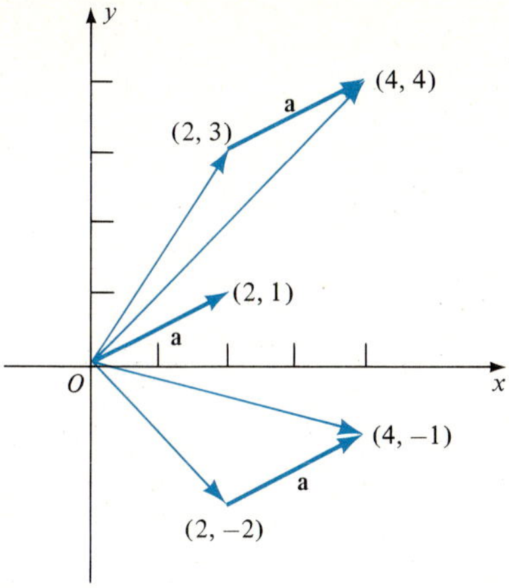

FIGURE 11.4h

THE* i *AND* j *UNIT VECTORS

By using the addition rule for vectors and Definition 11.4f, we can write any arbitrary vector $\mathbf{a}$ in a special form

$$\begin{aligned}\mathbf{a} = (a_1, a_2) &= (a_1, 0) + (0, a_2)\\ &= a_1(1, 0) + a_2(0, 1)\end{aligned}$$

This shows that every two-dimensional vector can be expressed as the sum of the two special vectors (1, 0) and (0, 1) each multiplied by a properly chosen scalar. It is customary to denote these vectors by

$$\mathbf{i} = (1, 0) \quad \text{and} \quad \mathbf{j} = (0, 1)$$

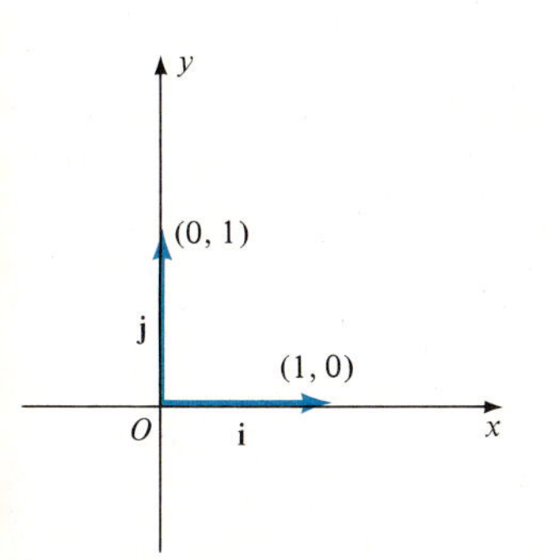

FIGURE 11.4i

and $\mathbf{i}$ and $\mathbf{j}$ may be represented as shown in Figure 11.4i. Because $\mathbf{i}$ and $\mathbf{j}$ each has a magnitude of one unit, they are called unit vectors. In terms of these two unit vectors, we may now write any vector $\mathbf{a} = (a_1, a_2)$ in the form

$$\mathbf{a} = a_1\mathbf{i} + a_2\mathbf{j}$$

The real numbers a_1 and a_2 are called the *rectangular components* of $\mathbf{a}$. The vectors $\mathbf{i}$ and $\mathbf{j}$ are called a *base set* or simply a *basis*, because every vector in the plane may be written in terms of these two, as we have just indicated.

EXAMPLE 3

If $\mathbf{a} = 3\mathbf{i} - 5\mathbf{j}$ and $\mathbf{b} = -\mathbf{i} + 2\mathbf{j}$, find:

a. $2\mathbf{a} + 3\mathbf{b}$
b. $|2\mathbf{a} + 3\mathbf{b}|$
c. $|2\mathbf{a}| + |3\mathbf{b}|$

SOLUTION:

a. $$\begin{aligned}2\mathbf{a} + 3\mathbf{b} &= 2(3\mathbf{i} - 5\mathbf{j}) + 3(-\mathbf{i} + 2\mathbf{j})\\ &= 6\mathbf{i} - 10\mathbf{j} - 3\mathbf{i} + 6\mathbf{j}\\ &= 3\mathbf{i} - 4\mathbf{j}\end{aligned}$$

b. $|2\mathbf{a} + 3\mathbf{b}| = \sqrt{3^2 + (-4)^2} = 5$

c. $$\begin{aligned}|2\mathbf{a}| + |3\mathbf{b}| &= |6\mathbf{i} - 10\mathbf{j}| + |-3\mathbf{i} + 6\mathbf{j}|\\ &= \sqrt{6^2 + (-10)^2} + \sqrt{(-3)^2 + 6^2}\\ &= 2\sqrt{34} + 3\sqrt{5}\end{aligned}$$

EXAMPLE 4

A vector **v** has a magnitude of 8 units and makes an angle of 60° with the positive x direction. Express **v** in terms of the **i**, **j** basis.

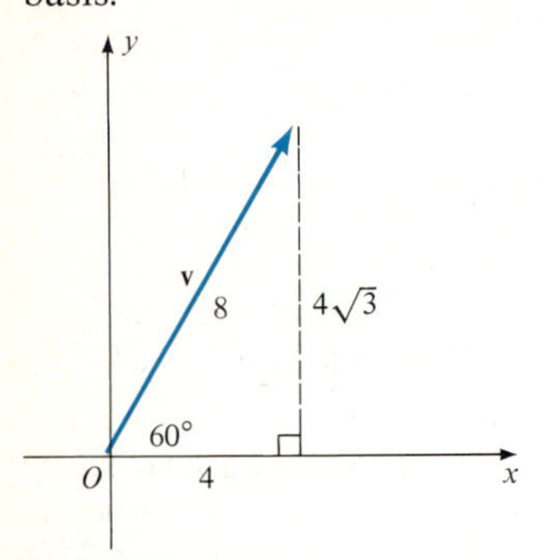

FIGURE 11.4j

SOLUTION:

From Figure 11.4j, we see that the horizontal component of **v** is 8 cos 60°, or 4, and the vertical component is 8 sin 60°, or $4\sqrt{3}$. Thus,

$$\mathbf{v} = 4\mathbf{i} + 4\sqrt{3}\mathbf{j}$$

EXAMPLE 5

A vector **a** has magnitude 10 and makes an angle of 30° with the positive **x** direction. A second vector **b** has magnitude 8 and makes an angle of 150° with the positive x direction. For these two vectors, find the resultant, its magnitude, and its direction.

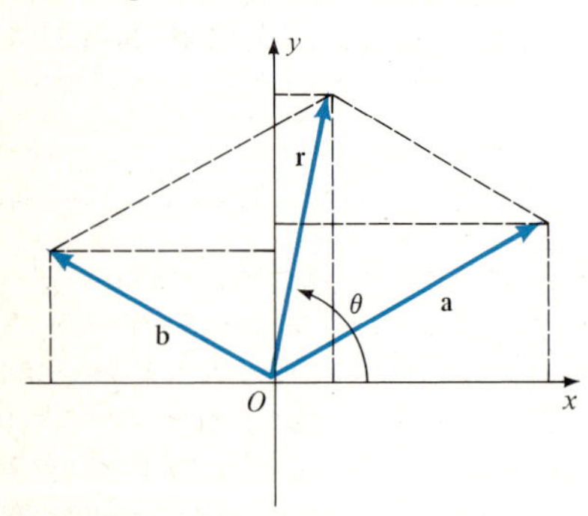

FIGURE 11.4k

SOLUTION:

We can do this problem by first expressing the two given vectors in terms of the **i**, **j** basis. Thus,

$$\mathbf{a} = (10 \cos 30°)\mathbf{i} + (10 \sin 30°)\mathbf{j} = 5\sqrt{3}\mathbf{i} + 5\mathbf{j}$$
$$\mathbf{b} = (8 \cos 150°)\mathbf{i} + (8 \sin 150°)\mathbf{j} = -4\sqrt{3}\mathbf{i} + 4\mathbf{j}$$

The resultant is therefore

$$\begin{aligned}\mathbf{r} &= \mathbf{a} + \mathbf{b}\\ &= (5\sqrt{3}\mathbf{i} + 5\mathbf{j}) + (-4\sqrt{3}\mathbf{i} + 4\mathbf{j})\\ &= \sqrt{3}\mathbf{i} + 9\mathbf{j}\end{aligned}$$

The magnitude of the resultant is

$$|\mathbf{r}| = \sqrt{(\sqrt{3})^2 + 9^2} = \sqrt{84} \approx 9.17$$

If θ is the angle that **r** makes with the positive x direction, then,

$$\theta = \text{Tan}^{-1}(9/\sqrt{3}) = \text{Tan}^{-1}(3\sqrt{3}) \approx 79.11°$$

You should study Figure 11.4k carefully. (Note that both $\sin\theta$ and $\cos\theta$ are positive, so that the principal inverse tangent may be used.)

EXERCISE 11.4

In Problems 1–10, $\mathbf{a} = (2, 4)$, $\mathbf{b} = (4, -3)$, and $\mathbf{c} = (-3, 2)$. Calculate the specified quantities.

1 a $\mathbf{a} + \mathbf{b}$ **b** $|\mathbf{a} + \mathbf{b}|$

2 a $\mathbf{b} + \mathbf{c}$ **b** $|\mathbf{b} + \mathbf{c}|$

3 a $\mathbf{b} - \mathbf{a}$ **b** $|\mathbf{b} - \mathbf{a}|$

4 a $\mathbf{a} - \mathbf{c}$ **b** $|\mathbf{a} - \mathbf{c}|$

5 $2\mathbf{a} + 3\mathbf{b}$

6 $3\mathbf{a} + 2\mathbf{c}$

7 $\mathbf{a} + \mathbf{b} - \mathbf{c}$

8 $\mathbf{a} + 2\mathbf{b} + 3\mathbf{c}$

9 $|3\mathbf{a} - \mathbf{b}|$

10 $|2\mathbf{a} - \mathbf{b} + \mathbf{c}|$

11 A vector **v** has its tail at the point $(-2, -3)$ and its head at $(3, 9)$. Express **v** in terms of the vectors $\mathbf{a} = (-2, -3)$, $\mathbf{b} = (3, 9)$. Is this the same as the vector $\mathbf{u} = (5, 12)$? Explain.

12 The vector $(8, 8)$ is the resultant of the vector $(3, 2)$ and another vector **b**. What is the magnitude of **b**?

13 The sides of a triangle represent three vectors placed head to tail. What must be the sum of the three vectors?

14 The sides of a closed polygon of n sides represent n vectors placed head to tail. What must be the sum of the n vectors?

In Problems 15–22, $\mathbf{a} = 2\mathbf{i} + 3\mathbf{j}$ and $\mathbf{b} = 4\mathbf{i} - \mathbf{j}$. Calculate the indicated quantities.

15 $\mathbf{a} + \mathbf{b}$

16 $\mathbf{a} - \mathbf{b}$

17 $2\mathbf{a} - 3\mathbf{b}$

18 $|\mathbf{a}|\,|\mathbf{b}|$

19 $|\mathbf{a} + \mathbf{b}|$

20 $|\mathbf{a}| + |\mathbf{b}|$

21 $|2\mathbf{a} - 3\mathbf{b}|$

22 $|2\mathbf{a}| - |3\mathbf{b}|$

In Problems 23–28, two vectors **a** and **b** are described by their magnitudes and the angles they make with the positive x direction. Find the resultant, its magnitude, and the angle it makes with the positive x direction.

23 **a**: 50, 60°; **b**: 40, 120°

24 **a**: 20, 60°; **b**: 40, −60°

25 **a**: 30, 45°; **b**: 40, 135°

26 **a**: 6, 30°; **b**: 8, 120°

27 **a**: 25, 37.2°; **b**: 30, 125.4°

28 **a**: 12.40, 75.6°; **b**: 8.20, −21.7°

29 Let $\mathbf{c} = -(\mathbf{a} + \mathbf{b})$. Then $\mathbf{a} + \mathbf{b} + \mathbf{c} = \mathbf{0}$, so that **a**, **b**, and **c** may be taken as three vectors drawn head to tail and forming a closed triangle. Use this idea to show that $|\mathbf{a} + \mathbf{b}| \le |\mathbf{a}| + |\mathbf{b}|$. When does the equality hold?

30 Refer to Problem 30 and show similarly that $|\mathbf{a} - \mathbf{b}| \ge ||\mathbf{a}| - |\mathbf{b}||$. When does the equality hold?

31 A unit vector is a vector of unit magnitude. Thus, a unit vector may be formed from any nonzero vector by dividing the vector by its own magnitude. The resulting unit vector has the same direction as the original vector. Find a unit vector having the same direction as the resultant of the vectors $\mathbf{a} = 2\mathbf{i} + 5\mathbf{j}$ and $\mathbf{b} = 3\mathbf{i} - \mathbf{j}$.

32 Prove Theorem 11.4b.

33 Prove Theorem 11.4c.

34 Let $\mathbf{a} = a_1\mathbf{i} + a_2\mathbf{j}$ and $\mathbf{b} = b_1\mathbf{i} + b_2\mathbf{j}$ be any two nonzero vectors. Show that **a** and **b** are perpendicular if and only if

$$a_1b_1 + a_2b_2 = 0$$

Hint: The vectors **a**, **b**, and $\mathbf{a} - \mathbf{b}$ form a right triangle with the right angle between **a** and **b**.

35 Use the result stated in Problem 34 to find a unit vector that is perpendicular to the vector $\mathbf{a} = 3\mathbf{i} - 2\mathbf{j}$.

36 A triangle is drawn with its sides parallel to the vectors $2\mathbf{i} + 5\mathbf{j}$, $3\mathbf{i} - 4\mathbf{j}$, and $15\mathbf{i} - 6\mathbf{j}$, respectively. Show that the triangle is a right triangle. (See Problem 34.)

USING YOUR KNOWLEDGE 11.4

In doing Problem 14, Exercise 11.4, you found that the sum of n vectors that form a closed polygon when placed head to tail is the zero vector. Use this knowledge to solve the following problems.

1 If the vectors $2\mathbf{i}$, $\mathbf{i} + \mathbf{j}$, and $\mathbf{i} - 3\mathbf{j}$ are placed head to tail, what vector will form the fourth side of a closed polygon?

2 If the vectors $x\mathbf{i} + y\mathbf{j}$, $(x - y)\mathbf{i} + (x + y)\mathbf{j}$, and $-4\mathbf{i} + 3\mathbf{j}$ form a closed triangle, find the values of x and y.

11.5 APPLICATIONS OF VECTORS

Suppose that a person can row a boat at the rate of 4 mph in still water. If that person attempts to row straight across a river, which is 100 yards wide and flows at the rate of 3 mph, where will the boat land on the opposite shore? What will be the actual speed of the boat?

To solve this problem, we first choose a set of axes with the origin at the starting point of the boat and the x-axis pointing straight across the river as shown in Figure 11.5a. Thus, the component of the boat's velocity due to rowing is represented by the vector $\mathbf{a} = 4\mathbf{i}$, and the component due to the flow of the river by the vector $\mathbf{b} = -3\mathbf{j}$. The resulting velocity of the boat is then represented by the sum of these two vectors; that is,

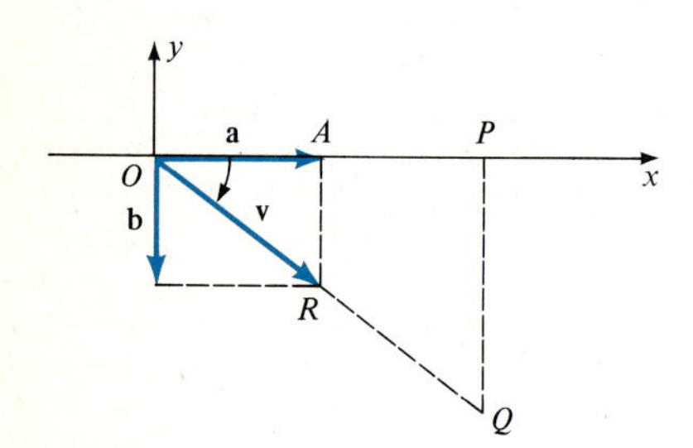

FIGURE 11.5a

$$\mathbf{v} = \mathbf{a} + \mathbf{b} = 4\mathbf{i} - 3\mathbf{j}$$

The actual speed of the boat is therefore $|\mathbf{v}| = \sqrt{4^2 + (-3)^2} = 5$ mph. Because $\tan\theta = -\frac{3}{4}$ (see Figure 11.5a), we can find the distance PQ from the relationship

$$\frac{PQ}{OP} = \tan\theta$$

Since $OP = 100$, we find

$$PQ = 100 \tan\theta = (100)(-\tfrac{3}{4}) = -75$$

Hence, the boat will land 75 yards below the x-axis, or 75 yards downstream.

The preceding problem and its solution is an illustration of the way in which vectors are used to solve problems involving velocities, displacements, or forces. Some of the simpler problems of this type are discussed in this section.

EXAMPLE 1

Solve the marksman's problem that was described at the beginning of Section 11.4.

SOLUTION:

In Figure 11.5b (not drawn to scale), T represents the bull's-eye, A the point at which the marksman must aim in order to allow for the wind, and O is the muzzle of the gun. If we let $\mathbf{w}$ represent the component of the bullet's velocity that is imparted by the wind, then $\mathbf{w} = \frac{22}{3}\mathbf{i}$ (5 mph is equivalent to $\frac{22}{3}$ feet per second). We let $\mathbf{g}$ be the component of the velocity imparted by the gun, so that $\mathbf{g}$ is a vector in the direction of OA in the figure. We write

$$\mathbf{g} = x\mathbf{i} + y\mathbf{j} \qquad \text{where } x^2 + y^2 = (3000)^2$$

(Recall that the gun imparts a speed of 3000 feet per second to the bullet.) If $\mathbf{v}$ represents the actual velocity of the bullet, a vector in the direction of OT in the figure, then

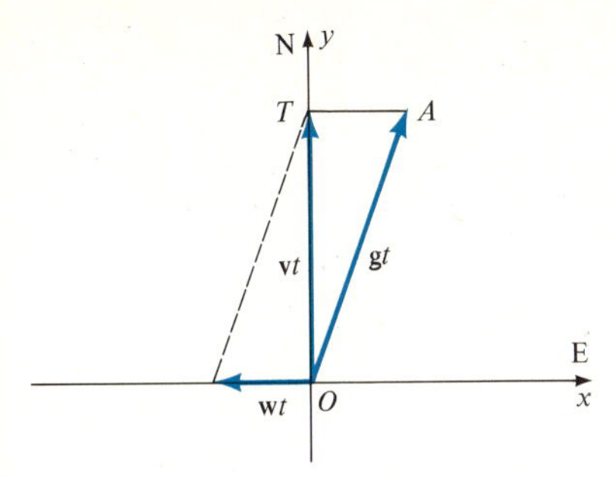

FIGURE 11.5b

$$\mathbf{v} = \mathbf{w} + \mathbf{g}$$

Now, suppose that it takes t seconds for the bullet to strike the target. Then $\mathbf{v}t = 150\mathbf{j}$, because the distance from O to T is 150 feet. Furthermore,

$$\mathbf{v}t = (\mathbf{w} + \mathbf{g})t = \mathbf{w}t + \mathbf{g}t$$

so that

$$150\mathbf{j} = -\tfrac{22}{3}t\mathbf{i} + (x\mathbf{i} + y\mathbf{j})t$$

or

$$0\mathbf{i} + 150\mathbf{j} = (-\tfrac{22}{3}t + xt)\mathbf{i} + yt\mathbf{j}$$

Using the definition of equality of vectors, we obtain the two equations:

$$0 = -\tfrac{22}{3}t + xt \quad \text{and} \quad 150 = yt$$

Since $t \neq 0$, the first of these equations gives $x = \frac{22}{3}$, and the second gives $t = 150/y$. Using the equation $x^2 + y^2 = (3000)^2$, we obtain

$$y = \sqrt{(3000)^2 - (\tfrac{22}{3})^2}$$

Referring to Figure 11.5b, we see that $\overrightarrow{AT} = xt\mathbf{i}$, and hence $|\overrightarrow{AT}| = xt$. Combining these results, we obtain

$$\begin{aligned} xt &= \left(\frac{22}{3}\right)\left(\frac{150}{y}\right) \\ &= \frac{1100}{\sqrt{(3000)^2 - (\frac{22}{3})^2}} \\ &\simeq 0.37 \text{ feet or } 4.4 \text{ inches} \end{aligned}$$

Hence, the marksman should aim at a point about 4.4 inches east of the center of the bull's-eye.

Example 2 is an illustration of the application of displacement vectors.

EXAMPLE 2

An explorer drives his jeep 10 miles in the northeast direction from a point O to point A. He then drives 20 miles in the northwest direction from A to a point B and, finally, 20 miles in the southwest direction from B to a point C. How far is C and in what direction is it from O?

SOLUTION:

First we make a sketch as in Figure 11.5c to show the explorer's route. We may regard the directed segments $\overrightarrow{OA}$, $\overrightarrow{AB}$, and $\overrightarrow{BC}$ corresponding to the path as displacement vectors $\mathbf{a}$, $\mathbf{b}$, and $\mathbf{c}$. Thus, the vector $\overrightarrow{OC} = \mathbf{d}$, which is the sum of $\mathbf{a}$, $\mathbf{b}$, and $\mathbf{c}$, gives the distance and direction of C from O. We find $\mathbf{d}$ by resolving the first three vectors into their rectangular components and then adding:

Since $\mathbf{a}$ is inclined at 45° to the x-axis,

$$\mathbf{a} = (10 \cos 45°)\mathbf{i} + (10 \cos 45°)\mathbf{j}$$

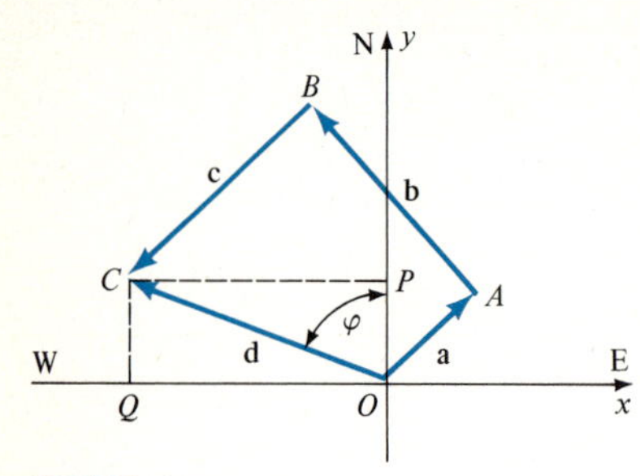

FIGURE 11.5c

$$= 5\sqrt{2}\mathbf{i} + 5\sqrt{2}\mathbf{j}$$

Similarly,

$$\mathbf{b} = (20 \cos 135^\circ)\mathbf{i} + (20 \sin 135^\circ)\mathbf{j}$$
$$= -10\sqrt{2}\mathbf{i} + 10\sqrt{2}\mathbf{j}$$

and

$$\mathbf{c} = (20 \cos 225^\circ)\mathbf{i} + (20 \cos 225^\circ)\mathbf{j}$$
$$= -10\sqrt{2}\mathbf{i} - 10\sqrt{2}\mathbf{j}$$

Thus,

$$\mathbf{d} = \mathbf{a} + \mathbf{b} + \mathbf{c}$$
$$= -15\sqrt{2}\mathbf{i} + 5\sqrt{2}\mathbf{j}$$

so that

$$|\mathbf{d}| = \sqrt{(15\sqrt{2})^2 + (5\sqrt{2})^2} = \sqrt{500} = 10\sqrt{5} \approx 22.4$$

From Figure 11.5c, we see that the angle ϕ is given by

$$\tan \phi = \frac{CP}{QC} = \frac{15\sqrt{2}}{5\sqrt{2}} = 3$$

Hence,

$$\phi \simeq 71.6^\circ$$

Therefore, the point C is about 22.4 miles N 71.6° W from O.

EXAMPLE 3

Two forces $\mathbf{F}_1$ and $\mathbf{F}_2$ are acting on an object at O. (See Figure 11.5d.) If $\mathbf{F}_1$ is 6 pounds in the direction N 60° E and $\mathbf{F}_2$ is 12 pounds in the direction N 60° W, what single additional force acting on the object would keep it from moving?

SOLUTION:

Because the required additional force must balance the two forces $\mathbf{F}_1$ and $\mathbf{F}_2$, this force, say $\mathbf{F}$, must be equal in magnitude and opposite in direction to the resultant of $\mathbf{F}_1$ and $\mathbf{F}_2$. If $\mathbf{r}$ denotes this resultant, then

$$\mathbf{r} = \mathbf{F}_1 + \mathbf{F}_2 \quad \text{and} \quad \mathbf{F} = -\mathbf{r}$$

From Figure 11.5d, we see that

$$\mathbf{F}_1 = (6 \cos 30^\circ)\mathbf{i} + (6 \sin 30^\circ)\mathbf{j}$$
$$= 3\sqrt{3}\mathbf{i} + 3\mathbf{j}$$

and

$$\mathbf{F}_2 = (12 \cos 150^\circ)\mathbf{i} + (12 \sin 150^\circ)\mathbf{j}$$
$$= -6\sqrt{3}\mathbf{i} + 6\mathbf{j}$$

Therefore,

$$\mathbf{r} = \mathbf{F}_1 + \mathbf{F}_2$$
$$= -3\sqrt{3}\mathbf{i} + 9\mathbf{j}$$

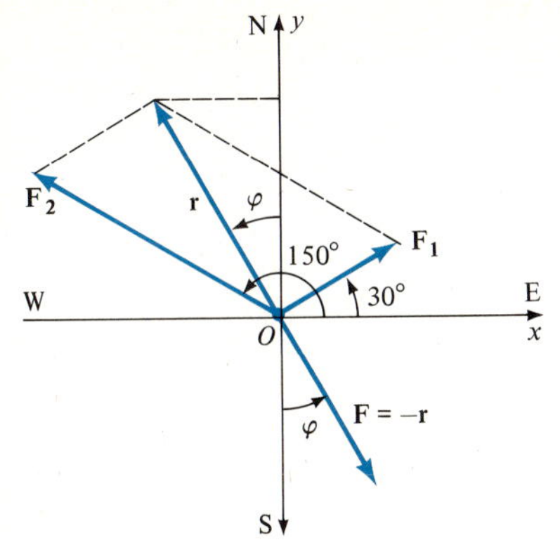

FIGURE 11.5d

$$|\mathbf{r}| = \sqrt{(-3\sqrt{3})^2 + 9^2} = \sqrt{108} = 6\sqrt{3} \approx 10.39$$

$$\tan\phi = \frac{3\sqrt{3}}{9} = \frac{1}{\sqrt{3}} \qquad \text{See the figure.}$$

$$\phi = 30^\circ$$

This means that the resultant is about 10.39 pounds in the direction N 30° W, so that the required force **F** is about 10.39 pounds in the direction S 30° E.

In Example 3, because $\mathbf{F} = -\mathbf{r}$, we see that $\mathbf{F} + \mathbf{F}_1 + \mathbf{F}_2 = \mathbf{0}$. This is an instance of the following Fundamental Law of Mechanics: *If a rigid mechanical system is at rest, then the vector sum of all the external forces acting on the system is the zero vector.* Example 4 illustrates another application of this law.

EXAMPLE 4

Figure 11.5e is a diagram showing a 7000-pound weight suspended from two cables. The coordinate system in the figure has been taken with the x-axis horizontal and the origin at the point where the two cables are fastened to the weight. If the right-hand cable makes an angle $\theta_1 = \text{Tan}^{-1}\frac{3}{4}$ with the horizontal and the left-hand cable makes an angle $\theta_2 = \text{Tan}^{-1}\frac{5}{12}$, as shown in the diagram, what is the tension (magnitude of the force) in each cable?

SOLUTION:

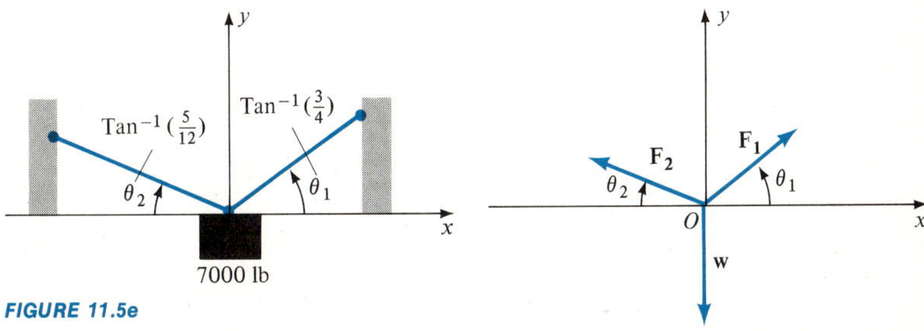

FIGURE 11.5e

FIGURE 11.5f

In Figure 11.5f, we use the same coordinate system as in Figure 11.5e to show the three forces that are involved. Here, $\mathbf{F}_1$ represents the force in the right-hand cable, $\mathbf{F}_2$ the force in the left-hand cable, and **w** the downward force due to the weight. By the Fundamental Law of Mechanics previously stated, the sum of these three vectors must be the zero vector; that is

$$\mathbf{F}_1 + \mathbf{F}_2 + \mathbf{w} = \mathbf{0}$$

We now resolve the vectors into their rectangular components and write them in the **i**, **j** form. Because $\tan\theta_1 = \frac{3}{4}$, the horizontal and vertical components of $\mathbf{F}_1$ must be in the ratio of 4 to 3, so that we may write

$$\mathbf{F}_1 = 4k\mathbf{i} + 3k\mathbf{j} \qquad \text{where } k \text{ is a constant}$$

Similarly,

$$\mathbf{F}_2 = -12m\mathbf{i} + 5m\mathbf{j} \qquad \text{where } m \text{ is a constant}$$

and

$$\mathbf{w} = -7000\mathbf{j}$$

From $\mathbf{F}_1 + \mathbf{F}_2 + \mathbf{w} = \mathbf{0}$, we obtain

$$(4k - 12m)\mathbf{i} + (3k + 5m - 7000)\mathbf{j} = \mathbf{0} = 0\mathbf{i} + 0\mathbf{j}$$

Consequently, we must have

$$4k - 12m = 0 \quad \text{and} \quad 3k + 5m - 7000 = 0$$

It follows that

$$k = 3m \quad \text{and} \quad 3(3m) + 5m = 7000$$

so that

$$m = 500 \quad \text{and} \quad k = 1500$$

Thus,

$$\mathbf{F}_1 = 6000\mathbf{i} + 4500\mathbf{j}$$

and

$$\mathbf{F}_2 = -6000\mathbf{i} + 2500\mathbf{j}.$$

The tension in the first cable is

$$|\mathbf{F}_1| = \sqrt{(6000)^2 + (4500)^2} = 7500 \text{ pounds}$$

and in the second cable is

$$|\mathbf{F}_2| = \sqrt{(6000)^2 + (2500)^2} = 6500 \text{ pounds}$$

EXERCISE 11.5

In Problems 1–6, find the resultant of the given set of coplanar forces.

1 10 pounds upward, 8 pounds to the right, 3 pounds downward

2 5 pounds in the northeast direction, 7 pounds in the northwest direction

3 3 pounds in the northeast direction, 6 pounds in the east direction, 5 pounds in the southeast direction

4 20 pounds acting upward along a line whose slope is $\frac{4}{3}$, 50 pounds acting upward along a line whose slope is $-\frac{3}{4}$

5 65 pounds acting downward along a line whose slope is $\frac{12}{5}$, 100 pounds acting downward along a line whose slope is $-\frac{4}{3}$, 85 pounds acting upward along a line whose slope is $\frac{15}{8}$.

6 100 pounds making an angle of 60° with the positive x-axis, 50 pounds making an angle of 120° with the positive x-axis, 80 pounds making an angle of 240° with the positive x-axis.

7 A plane is headed due west at right angles to a 50-mph wind blowing from the north. If the plane's speed with no wind is 700 mph, find the actual speed and direction of the flight.

8 A ship sailing due north meets an ocean current flowing from west to east at a speed of 4 knots (nautical miles per hour). If the ship's speed in still water is 25 knots, find the resultant speed and direction.

9 A person pushes a lawn mower with a force of 50 pounds acting in the direction of the handle. If the handle is held at a 30° angle with the horizontal, how many pounds are actually used in pushing the mower in the horizontal direction?

10 Referring to Problem 9, answer the question if the angle is 60°.

11 Two people are pushing a heavy box across a floor. One is pushing due east with a force of 60 pounds, and the other is pushing due north with a force of 50 pounds. Find the magnitude and the direction of the resultant force.

12 Repeat Problem 11 if the second person is pushing in the northeast direction.

13 If the wind is blowing toward the southeast at 40 mph, in what direction should a pilot head a plane that can fly 120 mph in still air in order to travel due east?

14 A 6000-pound weight is supported by two cables. The left-hand cable is inclined at an angle of 45° to the horizontal and the right-hand cable is inclined at an angle of 60° to the horizontal. Find the tension in each cable.

15 Repeat Problem 14 if the left-hand cable is inclined at 30° to the horizontal.

16 A boat can travel 9 mph in still water. In what direction should the boat head in order to land 7 miles downstream on the opposite side of a river that is $3\sqrt{3}$ miles wide and that flows from north to south at the rate of 6 mph?

17 A crane is holding a 3-ton beam above the ground. There is a 6-foot length of cable between the beam and the top of the crane.
a What force is required to push the beam 2 feet sideways?
b If the cable were lengthened an additional 12 feet, what force would be required to push the beam 2 feet sideways?

18 A boom 20 feet long is attached to the base of a vertical pole 20 feet high and is supported at its upper end by a cable passing through a pulley at the top of the pole. Find the tension in the cable when the boom is inclined at an angle of 30° to the horizontal and is supporting a weight of 5000 pounds from its upper end. Neglect the weight of the boom.

USING YOUR KNOWLEDGE 11.5

A car is stuck in the mud. The enterprising driver ties a rope to the car and to a tree 40 feet in front of the car. Suppose that the driver can pull the rope so tight that it will not stretch any more. What force can he apply to pull the car straight forward if he can displace the center of the rope one foot with a force of 100 pounds in the direction perpendicular to the original direction of the rope?

SELF-TEST

1 Find:
a $\text{Sin}^{-1}\dfrac{1}{\sqrt{2}}$ **b** $\text{Sin}^{-1}\left(-\dfrac{1}{\sqrt{2}}\right)$
c $\text{Sin}^{-1}\,0.81342$
d $\text{Sin}^{-1}\,(-0.81342)$

2 Find:
a $\text{Cos}^{-1}\,\frac{1}{2}$ **b** $\text{Cos}^{-1}\,(-\frac{1}{2})$
c $\text{Cos}^{-1}\,0.85252$
d $\text{Cos}^{-1}\,(-0.85252)$

3 Find:
a $\text{Tan}^{-1}\dfrac{1}{\sqrt{3}}$ **b** $\text{Tan}^{-1}\left(-\dfrac{1}{\sqrt{3}}\right)$
c $\text{Tan}^{-1}\,3.6021$
d $\text{Tan}^{-1}\,(-3.6021)$

4 a Find $\tan\,[\text{Sin}^{-1}\,(-\frac{2}{3})]$. **b** Is it always true that $\text{Sin}^{-1}\,(\sin x) = x$? Explain.

5 Write $\text{Tan}^{-1}\,x - \text{Tan}^{-1}\,y$ as a single inverse tangent. Assume $xy \geq 0$.

6 a What are the primary pairs of polar coordinates of the point $(-2, -\pi/3)$?
b Write the basic sets of coordinates of the point in (a).

7 a Determine which one of the points $(-1, 7\pi/6)$ and $(-3, 7\pi/6)$ is on the curve $r = 2(1 + \sin\theta)$.
b Find the distance between the two points (in polar coordinates) $(3, \pi/3)$ and $(5, 5\pi/3)$.

8 a A point has the rectangular coordinates $(-1, \sqrt{3})$. Find its primary polar coordinates.
b A point has the polar coordinates $(-2, 2\pi/3)$. Find its rectangular coordinates.

9 a Transform the equation $x^2 + y^2 = 2x$ into polar coordinates.
b Transform the equation $r^2 = \cos 2\theta$ into rectangular coordinates.

10 Discuss and sketch the graph of $r^2 = \cos 2\theta$.

11 Discuss and sketch the graph of $r = 3 - 2\cos\theta$.

12 Discuss and sketch the graph of $r = 1 - 2\sin\theta$.

13 If $\mathbf{a} = (3, 2)$ and $\mathbf{b} = (-7, 1)$, find:
a $\mathbf{a} + \mathbf{b}$ **b** $|\mathbf{a} + \mathbf{b}|$

14 Let $\mathbf{u} = (-1, 3)$ and $\mathbf{v} = (-2, 1)$.
a Find the magnitude and the direction of the resultant of $\mathbf{u}$ and $\mathbf{v}$.
b Find the magnitude and the direction of $\mathbf{u} - \mathbf{v}$.

15 Let $\mathbf{a} = 2\mathbf{i} + 3\mathbf{j}$ and $\mathbf{b} = -4\mathbf{i} - \mathbf{j}$. Find:
a $3\mathbf{a} + 2\mathbf{b}$ **b** $|\mathbf{a} + \mathbf{b}|$ **c** $|\mathbf{a}| + |\mathbf{b}|$

16 A motorboat can go 12 mph in still water. If the boat is headed straight across a river that is 180 yards wide and that flows at the rate of 5 mph, how far downstream will the boat land on the opposite shore?

17 A wind is blowing from the north at 50 mph. The pilot of a plane, whose speed in still air is 130 mph, wants to fly due east. In what direction should she head her plane?

18 A surveyor goes 100 meters northwest from a point O to a point A. He then goes 200 meters southwest from A to a point B and, finally, 200 meters southeast from B to a point C. Find the distance from O to C and the bearing of C with respect to O.

19 Two forces $\mathbf{F}_1$ and $\mathbf{F}_2$ act on an object. If $\mathbf{F}_1$ is 10 pounds due east and $\mathbf{F}_2$ is 5 pounds due north, what single additional force acting on the object would keep it from moving?

20 A 1000-pound weight is suspended from two cables. The right-hand cable is inclined at an angle of 30° to the horizontal, and the left-hand cable is inclined at 60° to the horizontal. Find the tension in each cable.

CHAPTER 12

SYSTEMS OF EQUATIONS

12.1 LINEAR SYSTEMS IN TWO VARIABLES

Figure 12.1a shows some Save-on rates that were available at the same time as National's introductory rate that was mentioned in Section 4.1. For example, the cost of renting a National car for 12 hours and driving x miles was y dollars, where

$$y = 0.13x + 8.50 \tag{1}$$

The corresponding cost for a Save-on Gremlin was y dollars, where

$$y = 0.10x + 10 \tag{2}$$

Equations 1 and 2 are examples of *linear equations* in two variables, x and y.

Of course, we may have equations in more than two variables. If we have to deal with a large number of variables, it is convenient to denote them all by the same letter, say x, but with different subscripts, such as $x_1, x_2, \ldots, x_n$. In general, an equation of the form

$$a_1x_1 + a_2x_2 + \cdots + a_nx_n = b \tag{3}$$

where the a's and the b are real numbers, is called a linear equation in the n variables $x_1, x_2, \ldots, x_n$. The expression on the left side of Equation 3 is a first-degree polynomial in the x's. For brevity, we can denote such a polynomial by the letter L. If $L_1, L_2, \ldots, L_n$ are n first-degree polynomials in the variables $x_1, x_2, \ldots, x_n$ and if $c_1, c_2, \ldots, c_n$ are n real numbers, then the equations

$SAVE-ON
RENT-A-CAR

FREE PICK-UP AND RETURN

SARASOTA
(813) 355-7704
6423 N. Tamiami Trail

TAMPA
(813) 879-1661
1902 N. Westmore Blvd.

TYPE	MILEAGE RATE			FLAT RATE	
	DAY	WEEK	MILE	DAY	WEEK
ECONOMY	$ 8.00	48.00	.10	$15.00	$79.00
GREMLIN	10.00	55.00	.10	17.00	89.00
COMPACT	11.00	60.00	.11	17.00	89.00

FIGURE 12.1a

$$L_1 = c_1$$
$$L_2 = c_2$$
$$\cdots$$
$$L_n = c_n$$

constitute a *system of linear equations* in the n variables. Although a system of equations may have more or fewer equations than variables, we shall restrict our attention to the case where the number of equations is the same as the number of variables.

To *solve* a system of equations means to find solutions that are common to all the equations in the system. For example, if we wish to find how many miles of driving will make the costs given by Equations 1 and 2 the same, then we must solve the system consisting of those two equations which we repeat here.

$$y = 0.13x + 8.50 \quad (1)$$
$$y = 0.10x + 10 \quad (2)$$

that is, we must find an ordered pair (x, y) that satisfies both equations.

THE SUBSTITUTION METHOD

One of the important ways of solving a system such as that consisting of Equations 1 and 2 is to substitute from one of the equations into the other to obtain an equation in one variable. This equation can be solved for its variable, and the resulting value can be substituted back into one of the original equations to find the value of the second variable. For obvious reasons, this method is called the *substitution method.*

We can solve the preceding system quite easily by using the substitution method. Thus, if we substitute $y = \mathbf{0.13x + 8.50}$ from Equation 1 into Equation 2, we obtain

$$\mathbf{0.13x + 8.50} = 0.10x + 10$$

or

$$0.03x = 1.50$$

and

$$x = 50$$

If we substitute $x = \mathbf{50}$ into Equation 1, we find

$$y = (0.13)(\mathbf{50}) + 8.50 = 15$$

Thus, the solution of the system is (50, 15), which means that driving a distance of 50 miles would have made the cost, \$15, the same for both cars.

EXAMPLE 1

Use the substitution method to solve the system:

$$3x + y = -5$$
$$x + 4y = 2$$

SOLUTION:

Solving the first equation for y, we obtain $y = -3x - 5$. Substituting this expression for y into the second equation, we obtain

$$x + 4(-3x - 5) = 2$$
$$-11x = 22$$
$$x = -2$$

By substituting $x = -2$ into the equation $y = -3x - 5$, we find $y = 1$. The solution $(-2, 1)$ can be checked by substitution into the original equations.

THE ELIMINATION METHOD

A second method of solving a system of equations is called the *elimination method*. In this method, we use addition or subtraction to transform the given system into an *equivalent* one in which one of the equations contains only one of the variables. Two systems are said to be *equivalent* if they have exactly the same solutions. To use the elimination method, we need to know what transformations are allowed. These transformations are listed in Theorem 12.1a. Although this theorem is stated for a system of two equations in two variables, a corresponding theorem is true for a system of n equations in n variables, $n = 3, 4, 5$, and so on.

Obviously, Systems I, II, and III in the statement of Theorem 12.1a are equivalent. The proof that Systems I and IV are equivalent is left for the exercises.

▼ THEOREM 12.1a

Let L_1 and L_2 be first-degree polynomials in x and y, and let k be any nonzero number. Then the system

I. $\begin{matrix} L_1 = c_1 \\ L_2 = c_2 \end{matrix}$

is equivalent to each of the following systems:

II. $\begin{cases} L_2 = c_2 \\ L_1 = c_1 \end{cases}$ **Interchanging the two equations**

III. $\begin{cases} kL_1 = kc_1 \\ L_2 = c_2 \end{cases}$ **Multiplying one equation by k**

IV. $\begin{cases} L_1 = c_1 \\ L_2 + kL_1 = c_2 + kc_1 \end{cases}$ **Adding k times one equation to the other**

EXAMPLE 2

Solve the system:

$$\left.\begin{matrix} 3x - 4y = 13 \\ 2x + 5y = 1 \end{matrix}\right\} \quad (4)$$

SOLUTION:

Suppose we decide to eliminate the variable x by addition. Then, we must transform the equations so that the coefficients of x will differ only in sign. Thus, we multiply the first equation by 2 and the second by -3 to obtain the system

$$\left.\begin{aligned} 6x - 8y &= 26 \\ -6x - 15y &= -3 \end{aligned}\right\} \quad (5)$$

Adding these two equations, we obtain

$$-23y = 23$$

so that

$$y = -1$$

Substituting $y = \mathbf{-1}$ into the first equation of System 4, we find

$$3x - 4(\mathbf{-1}) = 13 \quad \text{or} \quad x = 3$$

Thus, the solution of System 4 is (3, −1).

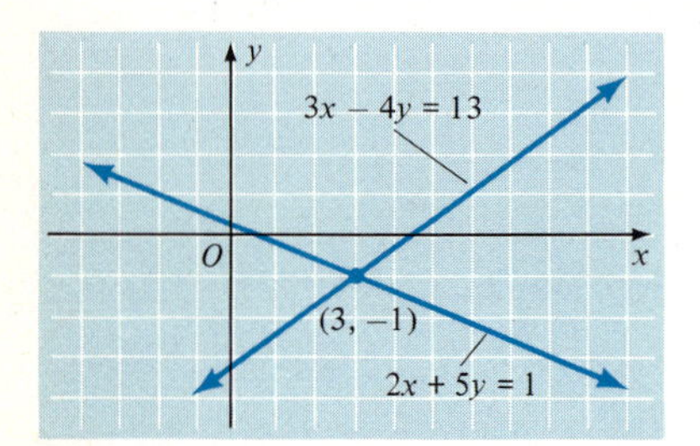

FIGURE 12.1b

Note: System 4 can also be solved by graphing the two equations on one set of axes and then reading the coordinates of the point of intersection. The graphs of the two lines in System 4 as well as the point of intersection (3, −1) are shown in Figure 12.1b.

EXAMPLE 3

Solve the system:

$$\left.\begin{aligned} 3x + 5y &= 2 \\ 6x + 10y &= 4 \end{aligned}\right\} \quad (6)$$

SOLUTION:

To eliminate the variable x by addition, we multiply the first equation by −2 and then add to obtain the system

$$\left.\begin{aligned} -6x - 10y &= -4 \\ 6x + 10y &= 4 \end{aligned}\right\} \quad (7)$$

Adding, we obtain

$$0 = 0$$

Thus, System 6 is equivalent to the system

$$\left.\begin{aligned} 3x + 5y &= 2 \\ 0 &= 0 \end{aligned}\right\} \quad (8)$$

Because the second equation in System 8 is true for all values of x and y, the solutions of the system are simply the solutions of the first equation. Of course, this system has infinitely many solutions, such as $(0, \frac{2}{5})$, $(\frac{2}{3}, 0)$, and $(-1, 1)$, to name a few. If we substitute $x = k$ into the first equation and solve for y, we obtain $y = (2 - 3k)/5$, so that the solutions of System 6 may be written

$$\left(k, \frac{2 - 3k}{5}\right) \qquad \text{where } k \text{ is any real number}$$

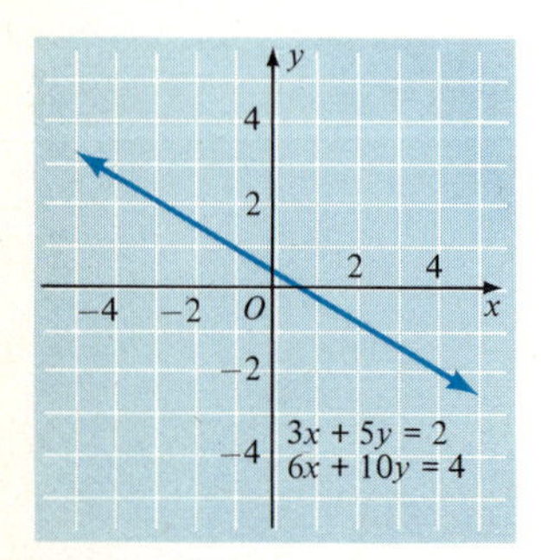

FIGURE 12.1c

The specific solutions given are for $k = 0, \frac{2}{3}$, and -1, respectively. If you look at System 6, you can see that the second equation is simply the first multiplied by 2. The two equations have the same graph, as shown in Figure 12.1c. We say that the lines represented by the two equations *coincide*.

EXAMPLE 4

Solve the system:

$$\left.\begin{aligned} 3x + 2y &= 6 \\ 6x + 4y &= -9 \end{aligned}\right\} \tag{9}$$

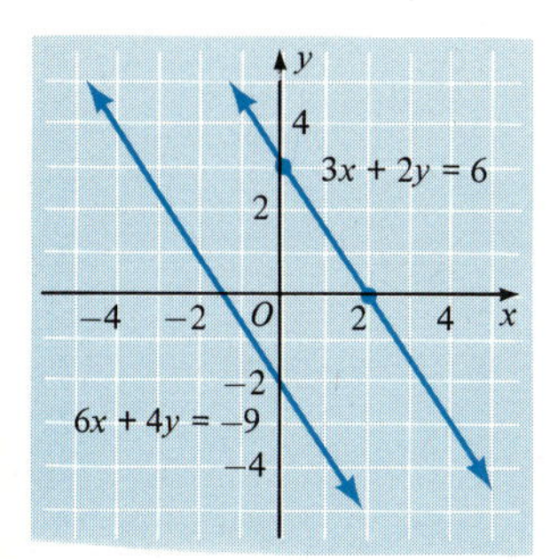

FIGURE 12.1d

SOLUTION:

To eliminate x by addition, we multiply the first equation by -2 and obtain

$$\left.\begin{aligned} -6x - 4y &= -12 \\ 6x + 4y &= -9 \end{aligned}\right\} \tag{10}$$

Adding, we obtain

$$0 = -21$$

Since this equation is *false* for all values of x and y, the given system has *no solution.* Note that if we write both equations in the slope-intercept form (see Section 6.1), we obtain the system

$$\left.\begin{aligned} y &= -\tfrac{3}{2}x + 3 \\ y &= -\tfrac{3}{2}x - \tfrac{9}{4} \end{aligned}\right\} \tag{11}$$

These equations show that the two lines have the same slope but different y-intercepts, so that they are parallel. The lines are shown in Figure 12.1d.

The three preceding examples show that there are three distinct possibilities for a system of two linear equations in two unknowns. You should keep these possibilities in mind when you are solving such a system. We list the possibilities here:

1. There is a *unique* solution, as in Example 2. Such a system is said to be *consistent* and *independent.* This corresponds to two lines *intersecting at one point.* (See Figure 12.1b.)
2. There are *infinitely many* solutions, as in Example 3. Such a system is called *dependent*; it corresponds to two lines being *coincident.* (See Figure 12.1c.)
3. There are *no* solutions, as in Example 4. Such a system is said to be *inconsistent*; it corresponds to *parallel* lines. (See Figure 12.1d.)

EXERCISE 12.1

In Problems 1–10, use the substitution method to solve the given system of equations and state if the system is dependent or inconsistent.

1 $x + y = 3$
$x - y = 1$

2 $3x + 2y = 19$
$2x + y = 11$

3 $y = -x + 8$
$3x - 4y = 3$

4 $x = -3y + 7$
$-x + 2y = -3$

5 $y + 2x = 4$
$2x + y = -2$

6 $-2x = -y + 1$
$-x - y = 5$

7 $-y + 2x = 10$
$3x + 4y = -7$

8 $2x - y = 25$
$5y = -x - 4$

9 $4x - 3y = 8$
$x - \tfrac{3}{4}y = 2$

10 $x + 2y - 1 = 0$
$3x - 2y - 19 = 0$

In Problems 11–22, solve the given system and state if the system is dependent or inconsistent.

11 $2x - 3y = 16$
$x - y = 7$

12 $3x - 2y = 8$
$-6x + 4y = 5$

13 $16x - 9y = -5$
$10x + 18y = -11$

14 $4x + 5y = 5$
$-10y - 4x = -7$

15 $-6x - 10y = 2$
$3x + 5y = 1$

16 $3(x + 2) = 2y$
$2(y + 5) = 7x$

17 $\frac{1}{5}x + \frac{2}{5}y = 1$
$\frac{1}{4}x - \frac{1}{3}y = -\frac{5}{12}$

18 $4(y - 4x) = 2(x + 5)$
$10(y - x) = 11y - 15x$

19 $-2x + 3y = 1$
$4x - 6y = 2$

20 $\frac{3}{2}x + \frac{5}{2}y = \frac{1}{2}$
$-\frac{3}{2}x - \frac{5}{2}y = \frac{1}{2}$

21 $\dfrac{2x}{5} = \dfrac{y}{2}$
$\dfrac{x}{3} = \dfrac{y}{3} + 1$

22 $2x = \frac{5}{2}y$
$\frac{3}{5}x - \frac{1}{4}y = 2$

In Problems 23–30, the given system is not linear. However, simplification of the equations or a change of variable leads to a linear system to be solved. Solve each system.

23 $\dfrac{x + y}{y - x} = 7$
$\dfrac{x + y + 1}{x + y - 1} = \dfrac{3}{4}$

24 $\dfrac{x - 1}{x + 1} = \dfrac{y - 5}{y - 3}$
$\dfrac{x + 2}{x - 2} = \dfrac{y - 5}{y - 7}$

25 $(x + 2)(y - 4) = xy$
$(x + 8)(y - 10) = xy$

26 $(3a + 2)(2b - 3) = 6ab$
$(4a + 5)(b - 5) = 4ab$

27 $\dfrac{4}{x} + \dfrac{12}{y} = 5$
$\dfrac{5}{x} - \dfrac{6}{y} = 1$
Hint: Let $u = 1/x$ and $v = 1/y$.

28 $\dfrac{2}{3x} + \dfrac{3}{4y} = \dfrac{7}{2}$
$\dfrac{1}{3x} - \dfrac{1}{2y} = 0$
Note: See the hint in Problem 27.

29 $\dfrac{2}{u - v} + \dfrac{3}{u + v} = 3$
$\dfrac{8}{u - v} - \dfrac{9}{u + v} = \dfrac{3}{2}$
Hint: Let $x = 1/(u - v)$, $y = 1/(u + v)$.

30 $\dfrac{2}{u - v} + \dfrac{1}{u + v} = \dfrac{2}{3}$
$\dfrac{5}{u - v} - \dfrac{6}{u + v} = \dfrac{1}{4}$
Note: See the hint in Problem 29.

In Problems 31–34, show that the given system has a solution by finding the solution of two of the equations and then showing that this solution satisfies the third equation.

31 $x - 5y = 27$
$2x + 3y = 2$
$5x + 8y = 3$

32 $9x + 4y = 10$
$5x - 2y = 6$
$3x + 14y = 2$

33 $(a - b)x - (a + b)y = 2a^2 - 2b^2$
$(a + b)x + (a - b)y = 4ab$
$x + y - 2b = 0$
a and b nonzero constants

34 $cx + dy = c^2 + d^2$
$cx - dy = c^2 - 2cd - d^2$
$x + 2y = 3c + d$
c and d constants

35 The height of the Empire State Building and its antenna is 1472 feet. The difference in height between the building and the antenna is 1028 feet. How tall is the antenna and how tall is the building?

36 It is estimated that in 1980, the average American created 8 pounds of waste per day. In 1969, the total collected (c) and uncollected (u) residential, commercial, and institutional wastes reached 250 million tons. If the collected wastes amounted to 70 million more tons than the uncollected wastes, how many tons of each were produced in in 1969?

37 In 1976, the beer industry produced 9060 million glass beer bottles. There are 8140 million more nonreturnables than returnables. How many million of each kind of beer bottle was produced in 1976?

38 The boiling point of water at sea level is 212°F (Fahrenheit) or 100°C (Celsius). The freezing point is 32°F or 0°C. If the equation for converting Celsius to Fahrenheit temperatures is of the form $F = aC + b$, find a and b.

39 If the numerator of a certain fraction is doubled and the denominator is increased by 8, the value of the fraction becomes $\frac{1}{4}$. If the numerator is increased by 7 and the denominator is tripled, the value becomes $\frac{5}{24}$. Find the fraction.

40 An airplane makes a 700-mile trip against a head wind in 2 hours and 20 minutes. The return trip takes 2 hours, the wind now being a tail wind. If the plane maintains constant speed with respect to still air and the speed of the wind remains constant, find the still-air speed of the plane and the speed of the wind.

41 A woman drives from her office to her apartment. When she drives at 50 mph, she arrives 4 minutes earlier than usual, and when she drives at 40 mph, she arrives 5 minutes later than usual. How far is her apartment from her office, and how long does it usually take her to drive the distance?

42 In Theorem 12.1a, prove that Systems I and IV are equivalent.

Hint: Write out the two systems in the form

$$\left.\begin{aligned} a_1x + b_1y &= c_1 \\ a_2x + b_2y &= c_2 \end{aligned}\right\} \tag{I}$$

$$\left.\begin{aligned} a_1x + b_1y &= c_1 \\ a_2x + b_2y + k(a_1x + b_1y) &= c_2 + kc_1 \end{aligned}\right\} \tag{IV}$$

Suppose that (u, v) is a solution of System I. Then,

$$a_1u + b_1v = c_1 \quad \text{and} \quad a_2u + b_2v = c_2$$

are both numerical identities. Use this to show that (u, v) satisfies System IV. On the other hand, suppose that (w, z) is a solution of IV. Write the numerical identities that this implies and use them to show that (w, z) is also a solution of I.

USING YOUR KNOWLEDGE 12.1

At the beginning of Section 5.4, we found a formula giving the temperature in terms of the number of chirps that a house cricket makes per minute. This formula is

$$F = \frac{n}{4} + 40 \tag{1}$$

where F is the temperature in degrees Fahrenheit and n is the number of chirps per minute that the cricket makes.

We also know how to tell the temperature from the speed with which a certain kind of ant crawls. The required formula is

$$C = 6d + 4 \tag{2}$$

where C is the temperature in degrees Celsius and d is the speed of the ant in centimeters per second.

The relationship between C and F is known to be

$$C = \frac{5}{9}(F - 32) \tag{3}$$

Do you see how to use a substitution method to find the relationship between d and n? Do the following problems to see how it goes.

1 Substitute Equations 1 and 2 into Equation 3. **2** Solve the equation you obtain in Problem 1 for d.
3 The result of Problem 2 tells you how fast the ant crawls in terms of how fast the cricket chirps. If the cricket is chirping 112 times per minute, how fast is the ant crawling?

12.2 SYSTEMS OF LINEAR EQUATIONS IN MORE THAN TWO VARIABLES

A florist has three types of orchids, which she classifies as large, medium, and small. She will make up a spray consisting of one large, two medium, and four small orchids for \$18, or a spray of two large and three small orchids for \$14, or a spray of two large, two medium, and five small orchids for \$24. What price is she setting on each type of orchid?

To answer this question, we first let x, y, and z dollars be the price of a large, a medium, and a small orchid, respectively. Then, for the \$18 spray, we have the equation

$$x + 2y + 4z = 18 \tag{1}$$

for the \$14 spray,

$$2x \qquad + 3z = 14 \tag{2}$$

and for the \$24 spray,

$$2x + 2y + 5z = 24 \tag{3}$$

Now, we have to solve the system of equations:

$$\left.\begin{aligned} x + 2y + 4z &= 18 \\ 2x \qquad + 3z &= 14 \\ 2x + 2y + 5z &= 24 \end{aligned}\right\} \tag{4}$$

Although System 4 which consists of three linear equations in three variables, can be solved by a substitution method similar to that used in Section 12.1, this method is quite awkward for larger systems unless one or more of the equations is extremely simple. For instance, an equation such as $y = 2x$ could be used immediately to reduce the number of equations to be handled.

THE ELIMINATION METHOD

The only important elementary method for solving systems of linear equations is the method of elimination by addition or subtraction. Except for having to handle more than two equations, this method proceeds essentially like that discussed in Section 12.1. We illustrate this by solving System 4:

$$\begin{aligned} x + 2y + 4z &= 18 \\ 2x \qquad + 3z &= 14 \\ 2x + 2y + 5z &= 24 \end{aligned}$$

Notice that the second equation involves only x and z. This suggests that we combine the other two equations to get another equation involving only x and z. Obviously, we can eliminate y between the first and third equations by subtracting the first equation from the third. The resulting equation is $x + z = 6$, and we have reduced the system to two equations in two variables:

$$\begin{aligned} 2x + 3z &= 14 \\ x + z &= 6 \end{aligned}$$

Now, we can multiply the second of these equations by 2 and subtract from the first to obtain

$$z = 2$$

Substituting $z = 2$ into the equation $x + z = 6$ gives us $x = 4$. Finally, we can substitute $x = 4$, $z = 2$ into the first equation of System 4 to obtain

$$4 + 2y + 8 = 18$$

or

$$2y = 6 \quad \text{and} \quad y = 3$$

Therefore, the solution of System 4 is (4, 3, 2), which means that the florist prices her orchids at $4 for a large, $3 for a medium, and $2 for a small bloom. You should check that these prices do give the florist's spray prices as stated.

Notice that the solution of System 4 was given as an ordered triple of numbers with x first, y second, and z third. It is customary to write solutions in this way.

EXAMPLE 1

Solve the system:

$$\left.\begin{aligned} 2x - y + 2z &= 3 \\ 2x + 2y - z &= 0 \\ -x + 2y + 2z &= -12 \end{aligned}\right\} \qquad (5)$$

SOLUTION:

We can eliminate x from two equations by first multiplying the first equation by -1 and adding the result to the second equation, and then by multiplying the third equation by 2 and adding the first equation to the result. This gives the equivalent system

$$\left.\begin{aligned} 2x - y + 2z &= 3 \\ 3y - 3z &= -3 \\ 3y + 6z &= -21 \end{aligned}\right\} \qquad (6)$$

We can then eliminate y from the third equation by multiplying the second equation in System 6 by -1 and adding the result to the third equation. This will give the system

$$\left.\begin{aligned} 2x - y + 2z &= 3 \\ 3y - 3z &= -3 \\ 9z &= -18 \end{aligned}\right\} \quad (7)$$

The solution of System 7 is now found by *back substitution.* The last equation gives $z = -2$, which can be substituted *back* into the second equation, yielding

$$3y - (3)(-2) = -3$$

or

$$y = -3$$

Next, we substitute $y = -3$, $z = -2$ *back* into the first equation, obtaining

$$2x - (-3) + (2)(-2) = 3$$

or

$$x = 2$$

Thus, the solution of System 5 is $(2, -3, -2)$. You should check this solution.

THE ECHELON FORM

The simple back substitution procedure that was used to solve System 7 makes use of the fact that the first variable, x, is missing from the second and third equations and that the second variable, y, is missing from the third equation. Such a system is said to be in *echelon* (or *triangular*) form. It is useful to know that every linear system can be converted into echelon form by using the transformations of Theorem 12.1a and that, as in the case of two equations in two variables, the system may be *consistent* and *independent* (have a *unique* solution), *dependent* (have *infinitely many* solutions), or *inconsistent* (have *no* solution).

EXAMPLE 2

Solve the following system by first transforming it into echelon form.

$$\left.\begin{aligned} x - 2y - 3z &= 2 \\ -3x + 5y + 4z &= 2 \\ x - 4y - 13z &= 4 \end{aligned}\right\} \quad (8)$$

SOLUTION:

We must first eliminate x from the second and third equations. This can be done by multiplying the first equation by 3 and adding the result to the second equation, and then by multiplying the first equation by -1 and adding the result to the third equation. This procedure gives

$$\left.\begin{aligned} x - 2y - 3z &= 2 \\ -y - 5z &= 8 \\ -2y - 10z &= 2 \end{aligned}\right\} \quad (9)$$

Next, we eliminate y from the third equation in System 9 by multiplying the second equation by -2 and adding the result to the third equation. This gives the desired echelon form:

$$\left.\begin{aligned} x - 2y - 3z &= 2 \\ - y - 5z &= 8 \\ 0 &= -14 \end{aligned}\right\} \qquad (10)$$

Since the last equation in System 10 is always *false*, the system is *inconsistent*; it has *no* solution.

HOMOGENEOUS SYSTEMS

In the systems we have discussed so far, at least one of the equations has had a nonzero constant term. A system in which *all* the constant terms are zeros is called a *homogeneous* system. A homogeneous system always has a solution, the *trivial* solution in which all the variables equal zero. Usually, when dealing with such systems, we are interested in finding whether or not there is a nontrivial (not all zeros) solution and if so, what the solution is. Such questions can be answered by changing the system to echelon form, as in Example 3.

EXAMPLE 3

Determine whether or not the following system has a nontrivial solution and if it has, find it.

$$\left.\begin{aligned} x + y + z &= 0 \\ 3x - y + z &= 0 \\ -5x + y - 2z &= 0 \end{aligned}\right\} \qquad (11)$$

SOLUTION:

We eliminate x from the second and third equations by multiplying the first equation by -3 and adding the result to the second equation, and then by multiplying the first equation by 5 and adding the result to the third equation. This gives the equivalent system

$$\left.\begin{aligned} x + y + z &= 0 \\ -4y - 2z &= 0 \\ 6y + 3z &= 0 \end{aligned}\right\} \qquad (12)$$

If we simplify the second equation of System 12 by dividing by -2 and the third equation by dividing by 3, we obtain the system

$$\left.\begin{aligned} x + y + z &= 0 \\ 2y + z &= 0 \\ 2y + z &= 0 \end{aligned}\right\} \qquad (13)$$

If we now subtract the second equation of System 13 from the third equation, we obtain the desired echelon form:

$$\left.\begin{aligned} x + y + z &= 0 \\ 2y + z &= 0 \\ 0 &= 0 \end{aligned}\right\} \qquad (14)$$

Because the third equation of System 14 is true for all values of x, y, and z, the system will be satisfied by any solution of the first two equations. If we solve the second equation for z, we obtain $z = -2y$, and if we then substitute this into the first equation, we obtain $x + y - 2y = 0$ or $x = y$. Thus, the system is equivalent to the two equations

$$x = y \quad \text{and} \quad z = -2y$$

where y is any real number. Therefore, we can let $y = k$, where k is any real number, and obtain $x = k$, $z = -2k$, so that the general solution is $(k, k, -2k)$. By taking $k = 1$, we have the solution $(1, 1, -2)$; for $k = -2$, we have the solution $(-2, -2, 4)$; for $k = \frac{1}{2}$, we have the solution $(\frac{1}{2}, \frac{1}{2}, -1)$; and so on. This means that System 11 is a *dependent* system with *infinitely many* solutions. You should check, without using any specific value for k, that $(k, k, -2k)$ satisfies System 11.

If we change the third equation of System 11 so that the system becomes

$$\left.\begin{aligned} x + y + z &= 0 \\ 3x - y + z &= 0 \\ -5x - y + z &= 0 \end{aligned}\right\} \tag{15}$$

then the first step of the reduction to echelon form is exactly the same as in Example 3. The result is

$$\left.\begin{aligned} x + y + z &= 0 \\ -4y - 2z &= 0 \\ 4y + 6z &= 0 \end{aligned}\right\} \tag{16}$$

If we add the second equation of System 16 to the third equation, the system is reduced to the echelon form

$$\left.\begin{aligned} x + y + z &= 0 \\ -4y - 2z &= 0 \\ 4z &= 0 \end{aligned}\right\} \tag{17}$$

The third equation of System 17 is satisfied by $z = 0$ only, and back substitution shows that $x = 0$ and $y = 0$ are the only values that satisfy the other two equations. Thus, reduction to echelon form shows that System 15 has only the trivial solution $(0, 0, 0)$.

WORD PROBLEMS

Example 4 illustrates the application of systems of equations to the solution of a word problem.

EXAMPLE 4

The sum of \$10 consists of nickels, dimes, and quarters. If there are 90 coins in all and there are twice as many nickels as quarters, how many of each coin are there?

SOLUTION:

Let n be the number of nickels, d the number of dimes, and q the number of quarters. The first statement of the problem translates into the equation

$$5n + 10d + 25q = 1000$$

where we have changed everything to cents. Dividing this equation by 5, we obtain

$$n + 2d + 5q = 200$$

Because there are 90 coins in all,

$$n + d + q = 90$$

and there are twice as many nickels as quarters, so that

$$n = 2q$$

Thus, we have the system

$$\left.\begin{aligned} n + 2d + 5q &= 200 \\ n + d + q &= 90 \\ n - 2q &= 0 \end{aligned}\right\} \qquad (18)$$

System 18 is solved by first substituting $n = 2q$ from the third equation into the other two equations to obtain

$$\begin{aligned} 2d + 7q &= 200 \\ d + 3q &= 90 \end{aligned}$$

If we multiply the second equation by 2 and subtract from the first equation, the result is

$$q = 20$$

Since $n = 2q$, we have $n = 40$, and by substitution into the equation $d + 3q = 90$, we find $d = 30$. Thus, there are 40 nickels, 30 dimes, and 20 quarters. The check is left to you.

EXERCISE 12.2

In Problems 1–30, solve the given system of equations.

1 $\begin{aligned} x + y - 2z &= 13 \\ x - 3y - z &= -3 \\ x - y + 4z &= -17 \end{aligned}$

2 $\begin{aligned} x + y + z &= 13 \\ x - 2y + 4z &= 10 \\ 3x + y - 3z &= 5 \end{aligned}$

3 $\begin{aligned} x + y + z &= -1 \\ 3x - y - 5z &= 13 \\ 5x + 3y + 2z &= 1 \end{aligned}$

4 $\begin{aligned} x + y + z &= 6 \\ x - y + 2z &= 5 \\ x - y - 3z &= 10 \end{aligned}$

5 $\begin{aligned} x - 2y - z &= 3 \\ 2x - 5y + z &= -1 \\ x - 2y - z &= -3 \end{aligned}$

6 $\begin{aligned} x - y + z &= 2 \\ x + y + z &= 4 \\ 2x + 2y - z &= -4 \end{aligned}$

7 $\begin{aligned} x - 3y + z &= 2 \\ x + 2y - z &= 1 \\ -7x + y + z &= 0 \end{aligned}$

8 $\begin{aligned} 3x + y + z &= 1 \\ x + 2y - z &= 1 \\ x + y + 2z &= -17 \end{aligned}$

9 $\begin{aligned} 2x + 3y + z &= 11 \\ 4x - 2y - z &= -1 \\ 6x - 2y - z &= -1 \end{aligned}$

10 $\begin{aligned} 2x + 4y + 3z &= 3 \\ 4x + 4y - 3z &= 2 \\ 10x - 8y - 9z &= 0 \end{aligned}$

11 $\begin{aligned} x + 2y + 2z &= 11 \\ 3x + 4y + z &= 14 \\ 2x + y + z &= 7 \end{aligned}$

12 $\begin{aligned} x + 3y + 4z &= 14 \\ x + 2y + z &= 7 \\ 2x + y + 2z &= 2 \end{aligned}$

13 $\begin{aligned} x - 3y + z &= 2 \\ x + 2y - z &= 1 \\ -7x + y + z &= -10 \end{aligned}$

14 $\begin{aligned} x + 4y + 3z &= 17 \\ 3x + 3y + z &= 16 \\ 2x + 2y + z &= 11 \end{aligned}$

15 $\begin{aligned} x + y - z &= 1 \\ z + x - y &= 3 \\ z - x + y &= 7 \end{aligned}$

16 $\begin{aligned} x - 2z - 2y &= 3 \\ 3z - 5y + 2x &= 1 \\ -2y + x - 2z &= -3 \end{aligned}$

17 $\begin{aligned} 2x + y + z &= 3 \\ 2y - x - z &= 3 \\ z + 3x + 4y &= 10 \end{aligned}$

18 $\begin{aligned} x + y + z &= 0 \\ -3x - 5y - 4z &= 0 \\ -5x - 3y - 4z &= 0 \end{aligned}$

19 $\begin{aligned} x - 3y - 4z &= -21 \\ 6z + 3x - 5y &= -20 \\ y + 4z + 5x &= -5 \end{aligned}$

20 $\begin{aligned} -x - 2y + 3z &= 0 \\ -x - y + 2z &= 0 \\ 2x - 3y + z &= 0 \end{aligned}$

21 $\begin{aligned} \frac{x}{2} + \frac{y}{2} - \frac{z}{3} &= 3 \\ \frac{x}{3} + \frac{y}{6} - \frac{z}{2} &= -5 \\ \frac{x}{6} - \frac{y}{3} + \frac{z}{6} &= 0 \end{aligned}$

22 $\begin{aligned} \frac{x}{3} + \frac{y}{4} - \frac{z}{3} &= 21 \\ \frac{x}{5} + \frac{y}{6} - \frac{z}{3} &= 0 \\ \frac{x}{10} + \frac{y}{3} - \frac{z}{6} &= 3 \end{aligned}$

23 $\begin{aligned} x - \frac{y+z}{3} &= 4 \\ y - \frac{x+z}{8} &= 5 \\ z - \frac{y-x}{2} &= -11 \end{aligned}$

24 $\begin{aligned} x - \frac{y+2}{5} &= z + 4 \\ y - \frac{z+4}{2} &= x - 6 \\ z - \frac{x-7}{3} &= y - 5 \end{aligned}$

25 $\begin{aligned} x + y &= 5 \\ y + z &= 3 \\ x + z &= 7 \end{aligned}$

26 $\begin{aligned} x + y + z &= 1 \\ x - 2y &= 0 \\ y - 2z &= 5 \end{aligned}$

27 $\begin{aligned} x + 2y - 4z &= 8 \\ 5x - 3z &= 2 \\ 2z - y &= -5 \end{aligned}$

28 $\begin{aligned} x + 2y - z &= -1 \\ 5x - 2z &= 1 \\ 3z - y &= 6 \end{aligned}$

29 $\begin{aligned} \frac{2}{x} + \frac{3}{y} + \frac{1}{z} &= 4 \\ \frac{3}{x} - \frac{5}{y} + \frac{2}{z} &= -5 \\ \frac{4}{x} - \frac{6}{y} + \frac{3}{z} &= 7 \end{aligned}$

Hint: Let $1/x = u$, $1/y = v$, and $1/z = w$.

30 $\begin{aligned} \frac{1}{x} + \frac{6}{y} + \frac{1}{z} &= 6 \\ \frac{2}{x} + \frac{3}{y} - \frac{2}{z} &= 8 \\ \frac{2}{x} + \frac{4}{z} &= 3 \end{aligned}$

Hint: Let $1/x = u$, $1/y = v$, and $1/z = w$.

31 Find a condition on a, b, and c so that the system

$$\begin{aligned} -4x + 3y &= a \\ 5x - 4y &= b \\ -3x + 2y &= c \end{aligned}$$

has a solution.

32 Show that the system

$$\begin{aligned} 3x + y + z &= -1 \\ x - y - z &= -5 \\ 2y - z &= -2 \\ 2x + 4z &= 6 \end{aligned}$$

does not have a solution. *Hint:* Solve the system consisting of the first three equations, and then show that the solution does not satisfy the last equation.

33 Find a value of k so that the system

$$\begin{aligned} 5x - y + 2z &= 2 \\ 3x + y - 3z &= 7 \\ x + 5y + z &= 5 \\ x + ky - z &= 9 \end{aligned}$$

has a solution. *Hint:* Solve the system consisting of the first three equations, and then substitute the values of x, y, and z into the last equation.

34 Repeat Problem 33 for the system:

$$\begin{aligned} 2x + 4z &= 6 \\ 3x + y + z &= -1 \\ 2y - z &= -2 \\ x - y + kz &= -5 \end{aligned}$$

35 On counting his cash, a man finds that he has $48 consisting of 294 coins, all dimes, quarters, and half-dollars. There are $3\frac{1}{2}$ times as many dimes as quarters. How many coins of each denomination are there?

36 Three pipes supply a water tank. The tank can be filled by pipes A and B running for 4 hours, or by pipes B and C running for 6 hours, or by pipes A and C running for 8 hours. What length of time is required for each pipe running alone to fill the tank?

37 Three contractors, A, B, and C, can do a construction job in 20 days. A and B together can do the work in 30 days, and B and C can do the work in 40 days. How many days would it take each contractor alone to do the work?

38 The points (0, 4), $(-12, -2)$, and $(-2, 6)$ lie on the parabola $y = ax^2 + bx + c$. Find a, b, and c.

Hint: The coordinates of each point must satisfy the equation.

39 The points given in Problem 38 lie on the parabola $x = ay^2 + by + c$. Find a, b, and c.

40 We learned earlier that the standard form of the equation of a circle is $(x - h)^2 + (y - k)^2 = r^2$. For some problems, it is convenient to multiply out the squares, collect terms, and write the equation in the form $x^2 + y^2 + Dx + Ey + F = 0$. Use this form to find the equation of the circle that passes through the three points given in Problem 38.

USING YOUR KNOWLEDGE 12.2

When the coefficients of a system of linear equations are not simple numbers as in the preceding problems, then an efficient, systematic procedure for carrying out the reduction to a special echelon form is desirable. A procedure that can be carried out with a four-function calculator is illustrated by the following solution of the system

$$\begin{aligned} 12x + 43y + 16z &= 286.7 \\ 13.2x - 10.5y + 3.8z &= 17.33 \\ 10x + 17y - 14z &= 2.5 \end{aligned}$$

1. To make the coefficients of x all ones, we divide each equation by the coefficient of x in that equation. We have rounded the remaining coefficients to three decimal places.

$$\begin{aligned} x + 3.583y + 1.333z &= 23.892 \\ x - 0.795y + 0.288z &= 1.313 \\ x + 1.700y - 1.400z &= 0.250 \end{aligned}$$

2. We subtract the second equation from the first and the third equation from the first to obtain the system

$$\begin{aligned} x + 3.583y + 1.333z &= 23.892 \\ 4.378y + 1.045z &= 22.579 \\ 1.883y + 2.733z &= 23.642 \end{aligned}$$

3. To make the coefficients of y in the second and third equation ones, we divide each equation by its y-coefficient. This produces the system

$$\begin{aligned} x + 3.583y + 1.333z &= 23.892 \\ y + 0.239z &= 5.157 \\ y + 1.451z &= 12.555 \end{aligned}$$

4. Next, we subtract the second equation from the third to obtain the echelon form:

$$\begin{aligned} x + 3.583y + 1.333z &= 23.892 \\ y + 0.239z &= 5.157 \\ 1.212z &= 7.398 \end{aligned}$$

5. We now solve the third equation for z, obtaining $z = 6.104$, which we substitute into the second equation to find $y = 3.698$. Then we substitute the y- and z-values into the first equation to obtain $x = 2.505$. Because of round-off errors that might have been introduced in the solution process, we round off the results to two decimal places and give the solution as (2.51, 3.70, 6.10).

A check in the original equations gives 286.82, 17.46, and 2.60 in place of the right-hand members 286.7, 17.33, and 2.5. Because the discrepancies are all less than 0.2 unit, the values of x, y, and z are probably correct to two decimal places. A better determination of the accuracy would require carrying more decimal places in the intermediate calculations.

Use your calculator to solve the following systems. Give answers to two decimal places.

C **1** $$\begin{aligned} 17x + 38y + 42z &= 44 \\ 15x - 33y - 23z &= 115 \\ 12x + 25y + 35z &= 73 \end{aligned}$$

C **2** $$\begin{aligned} 25x - 36y + 31z &= 137 \\ 40x + 27y - 53z &= 60 \\ 32x - 70y + 46z &= 115 \end{aligned}$$

12.3 MATRIX SOLUTION OF SYSTEMS OF EQUATIONS

The *Nine Chapters in the Mathematical Art* is a Chinese work that was written about 250 B.C. Chapter VIII of this work contains the following problem, which involves a system of linear equations in three unknowns.

> Three sheafs of good crop, two sheafs of mediocre crop, and one sheaf of bad crop are sold for 39 dou. Two sheafs of good, three mediocre, and one bad are sold for 34 dou; and one good, two mediocre, and three bad are sold for 26 dou.
>
> What is the price received for a sheaf of each of good crop, mediocre, and bad crop?

Using the notation of the preceding section, we would express this problem by the following system of equations

$$\begin{aligned} 3x + 2y + z &= 39 \\ 2x + 3y + z &= 34 \\ x + 2y + 3z &= 26 \end{aligned} \tag{1}$$

In the *Nine Chapters*, these conditions were given as *columns* in a rectangular array, which was read from right to left, like this

$$\begin{bmatrix} 1 & 2 & 3 \\ 2 & 3 & 2 \\ 3 & 1 & 1 \\ 26 & 34 & 39 \end{bmatrix}$$

We represent System 1 in a similar fashion, using *rows* rather than columns to correspond to the way we write the equations. Thus, we arrange the coefficients of the variables x, y, and z in a rectangular array enclosed in square brackets, but in our notation, each coefficient occupies the same position in the array as in the system of equations. Such a rectangular array of numbers is called a *matrix* (the plural is *matrices*). For example, the *matrix of coefficients* of System 1 is

$$A = \begin{bmatrix} 3 & 2 & 1 \\ 2 & 3 & 1 \\ 1 & 2 & 3 \end{bmatrix}$$

If we wish to include the constant terms appearing to the right of the equals signs, we write the *augmented matrix*, denoted by $A|D$

$$A|D = \left[\begin{array}{ccc|c} 3 & 2 & 1 & 39 \\ 2 & 3 & 1 & 34 \\ 1 & 2 & 3 & 26 \end{array}\right]$$

In general, the *coefficient matrix* corresponding to the system

$$\begin{aligned} a_1x + b_1y + c_1z &= d_1 \\ a_2x + b_2y + c_2z &= d_2 \\ a_3x + b_3y + c_3z &= d_3 \end{aligned} \qquad (2)$$

is

$$A = \begin{bmatrix} a_1 & b_1 & c_1 \\ a_2 & b_2 & c_2 \\ a_3 & b_3 & c_3 \end{bmatrix}$$ **and the augmented matrix is** $$A|D = \left[\begin{array}{ccc|c} a_1 & b_1 & c_1 & d_1 \\ a_2 & b_2 & c_2 & d_2 \\ a_3 & b_3 & c_3 & d_3 \end{array}\right]$$

We always interpret the matrix $A|D$ as the augmented matrix of a system of linear equations. For instance, if we are given the matrix

$$\left[\begin{array}{ccc|c} 2 & 1 & 3 & 5 \\ 6 & 9 & -2 & 7 \\ 8 & -1 & 2 & -3 \end{array}\right]$$

we understand it to be the augmented matrix of the system

$$\left.\begin{aligned} 2x + y + 3z &= 5 \\ 6x + 9y - 2z &= 7 \\ 8x - y + 2z &= -3 \end{aligned}\right\}$$

As we saw in Section 12.2, we can solve System 2 by first reducing it to echelon form by using the transformations corresponding to those of Theorem 12.1a. Of course, each new system obtained in the process has a corresponding augmented matrix. These matrices can be obtained in succession by using operations corresponding to the operations on the equations themselves. Because these operations change only the rows of the matrices, they are called *elementary row operations*, and the successive matrices are called *row-equivalent*. When two matrices, A and B, are row-equivalent, we shall write

$$A \sim B$$

▼ THEOREM 12.3a

The elementary row operations

Given the augmented matrix of a system of linear equations, each of the following row operations yields the augmented matrix of an equivalent system.

1. **Interchanging any two rows.**
2. **Multiplying each element in a row by a *nonzero* number.**
3. **Adding to each element of a row k times the corresponding element of another row, where k is any real number.**

Theorem 12.3a is the analog of Theorem 12.1a and describes the three elementary row operations. You should recognize these operations as corresponding exactly to the three operations on equations in Theorem 12.1a.

SOLUTION OF SYSTEMS USING MATRICES

We illustrate the use of Theorem 12.3a by solving the system of equations of Example 1 in Section 12.2. You should note the similarities between the operations on the matrices used in this solution and the transformations used to solve this system in Section 12.2.

EXAMPLE 1

Solve the system:

$$\left.\begin{aligned} 2x - y + 2z &= 3 \\ 2x + 2y - z &= 0 \\ -x + 2y + 2z &= -12 \end{aligned}\right\} \qquad (3)$$

SOLUTION:

The augmented matrix of System 3 is

$$\left[\begin{array}{rrr|r} 2 & -1 & 2 & 3 \\ 2 & 2 & -1 & 0 \\ -1 & 2 & 2 & -12 \end{array}\right]$$

Keep in mind that we want to obtain the augmented matrix of the echelon form of the given system. Consequently, we must get zeros in the first position of the second and third rows and in the second position of the third row. To get a zero for the first element of row II, we multiply row I by -1 and add the result to row II (Elementary Operation 3). This procedure is denoted by (II/II − I), which we read "row II is replaced by row II − row I." Thus, we have

$$\left[\begin{array}{rrr|r} 2 & -1 & 2 & 3 \\ 2 & 2 & -1 & 0 \\ -1 & 2 & 2 & -12 \end{array}\right] \xrightarrow{\text{(II/II} - \text{I)}} \left[\begin{array}{rrr|r} 2 & -1 & 2 & 3 \\ 0 & 3 & -3 & -3 \\ -1 & 2 & 2 & -12 \end{array}\right]$$

To get a zero for the first element of row III, we multiply row III by 2 and add row I to the result, so that row III is replaced by 2 row III + row I.

$$\begin{bmatrix} 2 & -1 & 2 & | & 3 \\ 0 & 3 & -3 & | & -3 \\ -1 & 2 & 2 & | & -12 \end{bmatrix} \xrightarrow{(\text{III}/2\text{III} + \text{I})} \begin{bmatrix} 2 & -1 & 2 & | & 3 \\ 0 & 3 & -3 & | & -3 \\ 0 & 3 & 6 & | & -21 \end{bmatrix}$$

To get a zero for the second element of row III, we multiply row II by -1 and add the result to row III. Thus,

$$\begin{bmatrix} 2 & -1 & 2 & | & 3 \\ 0 & 3 & -3 & | & -3 \\ 0 & 3 & 6 & | & -21 \end{bmatrix} \xrightarrow{(\text{III}/-\text{II} + \text{III})} \begin{bmatrix} 2 & -1 & 2 & | & 3 \\ 0 & 3 & -3 & | & -3 \\ 0 & 0 & 9 & | & -18 \end{bmatrix}$$

The last matrix corresponds to the system in echelon form that we obtained in Example 1, Section 12.2. As before, the solution obtained by back substitution is $(2, -3, -2)$.

EXAMPLE 2

Solve the system:

$$\left.\begin{aligned} 2x - y + 2z &= 3 \\ 2x + 2y - z &= 0 \\ 4x + y + z &= 5 \end{aligned}\right\} \quad (4)$$

SOLUTION:

The augmented matrix of this system is

$$\begin{bmatrix} 2 & -1 & 2 & | & 3 \\ 2 & 2 & -1 & | & 0 \\ 4 & 1 & 1 & | & 5 \end{bmatrix}$$

We transform this matrix into echelon form as follows:

$$\begin{bmatrix} 2 & -1 & 2 & | & 3 \\ 2 & 2 & -1 & | & 0 \\ 4 & 1 & 1 & | & 5 \end{bmatrix} \xrightarrow[(\text{III}/\text{III} - 2\text{I})]{(\text{II}/\text{II} - \text{I})} \begin{bmatrix} 2 & -1 & 2 & | & 3 \\ 0 & 3 & -3 & | & -3 \\ 0 & 3 & -3 & | & -1 \end{bmatrix}$$

$$\xrightarrow{(\text{III}/\text{III} - \text{II})} \begin{bmatrix} 2 & -1 & 2 & | & 3 \\ 0 & 3 & -3 & | & -3 \\ 0 & 0 & 0 & | & 2 \end{bmatrix}$$

The last row of the final matrix corresponds to the equation

$0x + 0y + 0z = 2$

which is always *false*. Thus, System 4 is *inconsistent*; it has *no* solution.

EXAMPLE 3

Solve the system:

$$\left.\begin{aligned} 2x - y + 2z &= 3 \\ 2x + 2y - z &= 0 \\ 4x + y + z &= 3 \end{aligned}\right\} \quad (5)$$

SOLUTION:

We transform the augmented matrix into echelon form as follows:

$$\begin{bmatrix} 2 & -1 & 2 & | & 3 \\ 2 & 2 & -1 & | & 0 \\ 4 & 1 & 1 & | & 3 \end{bmatrix} \xrightarrow[(\text{III}/\text{III} - 2\text{I})]{(\text{II}/\text{II} - \text{I})} \begin{bmatrix} 2 & -1 & 2 & | & 3 \\ 0 & 3 & -3 & | & -3 \\ 0 & 3 & -3 & | & -3 \end{bmatrix}$$

$$\xrightarrow{(\text{III}/\text{III} - \text{II})} \begin{bmatrix} 2 & -1 & 2 & | & 3 \\ 0 & 3 & -3 & | & -3 \\ 0 & 0 & 0 & | & 0 \end{bmatrix}$$

The last matrix corresponds to the system

$$\begin{aligned} 2x - y + 2z &= 3 \\ 3y - 3z &= -3 \\ 0x + 0y + 0z &= 0 \end{aligned}$$

The third of these equations is true for all values of x, y, and z. If we divide the second equation by 3, we obtain

$$y - z = -1$$

or

$$y = z - 1$$

If we let $z = k$ in this equation, we obtain $y = k - 1$. Substituting these values into the first equation gives

$$2x + 1 - k + 2k = 3$$

so that

$$2x = 2 - k$$

and

$$x = 1 - \tfrac{1}{2}k$$

Thus, if k is any real number, then $(1 - \frac{1}{2}k, k - 1, k)$ is a solution of System 5. The system is *dependent* and has *infinitely* many solutions. We can check this by substituting $x = 1 - \frac{1}{2}k$, $y = k - 1$, $z = k$ into 5. We obtain

$$\begin{aligned} 2(1 - \tfrac{1}{2}k) - (k - 1) + 2k &= 2 - k - k + 1 + 2k = 3 \\ 2(1 - \tfrac{1}{2}k) + 2(k - 1) - k &= 2 - k + 2k - 2 - k = 0 \\ 4(1 - \tfrac{1}{2}k) + (k - 1) + k &= 4 - 2k + k - 1 + k = 3 \end{aligned}$$

as required.

If you prefer to avoid the fractions, you may replace k by $2m$ and thus write the solution as $(1 - m, 2m - 1, 2m)$, where m is any real number. Particular solutions can be obtained by substituting particular values of m (or k). For instance, $m = 0$ gives the solution $(1, -1, 0)$; $m = 1$ gives the solution $(0, 1, 2)$; $m = -\frac{1}{4}$ gives $(\frac{5}{4}, -\frac{3}{2}, -\frac{1}{2})$; $m = \sqrt{2}$ gives $(1 - \sqrt{2}, 2\sqrt{2} - 1, 2\sqrt{2})$; and so on.

EXERCISE 12.3

For Problems 1–20, use matrices to solve Problems 1–20 in Exercise 12.2.

In Problems 21–26, solve the given system by using matrices. Be sure to supply zeros for any missing terms.

21 $\begin{aligned} 2x - y + 2z &= 5 \\ 2x + y - z &= -6 \\ 3x + 2z &= 3 \end{aligned}$

22 $\begin{aligned} 2x + 3y + 6z &= 17 \\ x + y + z &= 0 \\ 5x - 6y &= 8 \end{aligned}$

23 $\begin{aligned} x + y + z &= 3 \\ x - 2y + 3z &= 5 \\ 3x - 3y + 7z &= 14 \end{aligned}$

24 $x + 2z = 9$
$y - 2z = 5$
$x + 3y = -4$

25 $y + z = x - 7$
$x + 2z = 1 - 2y$
$x - 7 = 2y + 2z$

26 $x + z = 6 - y$
$x + 3y = 8$
$2x = 2y + 5 - 3z$

USING YOUR KNOWLEDGE 12.3

Of course, we can use matrices to solve systems involving more than three variables. We illustrate the procedure by solving the system

$$\begin{aligned} x_1 + x_2 + x_3 + x_4 &= -1 \\ x_1 + 2x_2 - x_3 - 2x_4 &= -3 \\ x_1 - 2x_2 + x_3 - 2x_4 &= 11 \\ x_1 - x_2 - x_3 + 4x_4 &= 0 \end{aligned}$$

First, we reduce the augmented matrix to echelon form:

$$\left[\begin{array}{cccc|c} 1 & 1 & 1 & 1 & -1 \\ 1 & 2 & -1 & -2 & -3 \\ 1 & -2 & 1 & -2 & 11 \\ 1 & -1 & -1 & 4 & 0 \end{array}\right] \xrightarrow[\text{(IV/IV - I)}]{\text{(II/II - I)} \atop \text{(III/III - I)}} \left[\begin{array}{cccc|c} 1 & 1 & 1 & 1 & -1 \\ 0 & 1 & -2 & -3 & -2 \\ 0 & -3 & 0 & -3 & 12 \\ 0 & -2 & -2 & 3 & 1 \end{array}\right]$$

$$\left[\begin{array}{cccc|c} 1 & 1 & 1 & 1 & -1 \\ 0 & 1 & -2 & -3 & -2 \\ 0 & -3 & 0 & -3 & 12 \\ 0 & -2 & -2 & 3 & 1 \end{array}\right] \xrightarrow[\text{(IV/2II + IV)}]{\text{(III/3II + III)}} \left[\begin{array}{cccc|c} 1 & 1 & 1 & 1 & -1 \\ 0 & 1 & -2 & -3 & -2 \\ 0 & 0 & -6 & -12 & 6 \\ 0 & 0 & -6 & -3 & -3 \end{array}\right]$$

$$\left[\begin{array}{cccc|c} 1 & 1 & 1 & 1 & -1 \\ 0 & 1 & -2 & -3 & -2 \\ 0 & 0 & -6 & -12 & 6 \\ 0 & 0 & -6 & -3 & -3 \end{array}\right] \xrightarrow{\text{(IV/IV - III)}} \left[\begin{array}{cccc|c} 1 & 1 & 1 & 1 & -1 \\ 0 & 1 & -2 & -3 & -2 \\ 0 & 0 & -6 & -12 & 6 \\ 0 & 0 & 0 & 9 & -9 \end{array}\right]$$

The last matrix corresponds to the system

$$\begin{aligned} x_1 + x_2 + x_3 + x_4 &= -1 \\ x_2 - 2x_3 - 3x_4 &= -2 \\ -6x_3 - 12x_4 &= 6 \\ 9x_4 &= -9 \end{aligned}$$

which is in echelon form and can easily be solved by back substitution to obtain in succession $x_4 = -1$, $x_3 = 1$, $x_2 = -3$, $x_1 = 2$. Thus, the solution of the given system is $(2, -3, 1, -1)$. You should check this result.

Use matrices to solve the following systems.

1
$$\begin{aligned} x_1 + x_2 + 2x_3 + 2x_4 &= 9 \\ 2x_1 + x_2 - 2x_3 - x_4 &= -8 \\ 4x_1 - 3x_2 + x_3 - 5x_4 &= 3 \\ 3x_1 - 2x_2 - 3x_3 + 4x_4 &= 6 \end{aligned}$$

2
$$\begin{aligned} x_1 + 2x_2 + x_3 + x_4 &= 0 \\ x_1 - 5x_2 - 3x_3 + 2x_4 &= 7 \\ 3x_1 + x_2 + 2x_3 - 2x_4 &= 7 \\ 2x_1 + 4x_2 + 2x_3 - x_4 &= 3 \end{aligned}$$

12.4 DETERMINANTS

The German mathematician Gottfried Wilhelm Leibniz was one of the originators of the theory of determinants. A *determinant* is a number associated with a square matrix, say A, and is denoted by **det** $\boldsymbol{A}$ (read "determinant of A"). [Other notations are $|A|$ and $\delta(A)$.]

We learned in Section 12.3 that a matrix is a rectangular array of numbers. A *row* of a matrix consists of the numbers placed in a *horizontal* line of the array; a *column* of a matrix consists of the numbers placed in a *vertical* line of the array. Thus, in the matrix

$$A = \begin{bmatrix} 2 & 5 & -9 \\ 4 & 7 & 13 \\ 3 & 8 & 6 \end{bmatrix}$$

the first row is composed of the numbers 2, 5, and -9, and the third row is 3, 8, and 6. Similarly, the first column consists of the numbers 2, 4, and 3, and the second column of the numbers 5, 7, and 8. Because the matrix A has three rows and three columns, it is called a "3 by 3" (3×3) matrix. If a matrix has m rows and n columns, it is said to be of *order* $m \times n$ (m by n). A matrix that has the same number, say n, of rows and columns is called a *square matrix of order* n. The matrix A in the preceding illustration is a square matrix of order 3; the matrix

$$B = \begin{bmatrix} 1 & 0 \\ -4 & 8 \end{bmatrix}$$

is a square matrix of order 2.

For convenience, a double subscript notation is used to identify the individual elements in each row and column. The symbol

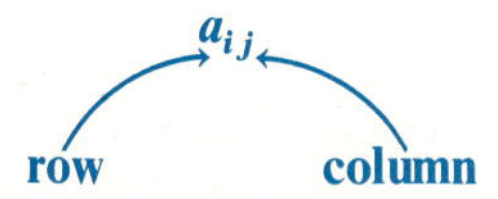

represents the element in the ith row and the jth column, so that the *first* subscript indicates the *row* and the *second* subscript indicates the *column*. For example, in the matrix A, a_{21} (read "a-two-one") is the element in the second row, first column, that is, $a_{21} = 4$, a_{23} is the element in the second row, third column, that is, $a_{23} = 13$. In the matrix B, $a_{21} = -4$ and $a_{12} = 0$.

We are now ready to consider the definition of the determinant of a square matrix. We first define the determinant of a matrix of order 1 and the determinant of a matrix of order 2. Then, we shall extend the definition to matrices of higher order.

▼ **DEFINITION 12.4a**

For a square matrix $A = [a_{11}]$ of order 1,

det $A = a_{11}$

Notice in Definition 12.4a that a matrix of order 1 consists of a

▼ **DEFINITION 12.4b**

For a square matrix of order 2,

$$A = \begin{bmatrix} a_{11} & a_{12} \\ a_{21} & a_{22} \end{bmatrix}$$

$$\det A = a_{11}a_{22} - a_{12}a_{21}$$

single element and the determinant of the matrix is simply the value of that element. Thus, if $A = [-3]$, then $\det A = -3$. According to Definition 12.4b, if

$$B = \begin{bmatrix} 1 & 0 \\ -4 & 8 \end{bmatrix}$$

so that $b_{11} = 1, b_{12} = 0, b_{21} = -4, b_{22} = 8$, then

$$\det B = b_{11}b_{22} - b_{12}b_{21} = (1)(8) - (0)(-4) = 8$$

As you will soon see, there is a simple scheme for obtaining the value of the determinant of a second-order matrix without memorizing the formula in Definition 12.4b. The determinant of a square matrix is usually written by replacing the brackets in the matrix notation by vertical lines. Thus, we symbolize $\det A$ by writing

$$\det A = \begin{vmatrix} a_{11} & a_{12} \\ a_{21} & a_{22} \end{vmatrix}$$

With this notation, we can indicate the evaluation of $\det A$ thus:

$$\det A = \begin{vmatrix} a_{11} & a_{12} \\ a_{21} & a_{22} \end{vmatrix} = a_{11}a_{22} - a_{12}a_{21}$$

As the diagram shows, we are simply forming products along the diagonals, preceding the product by a plus sign if we go down to the right and by a minus sign if we go down to the left.

EXAMPLE 1

Evaluate the determinants:

a. $\begin{vmatrix} -1 & 2 \\ 3 & -2 \end{vmatrix}$ **b.** $\begin{vmatrix} \frac{1}{2} & -5 \\ 2 & 8 \end{vmatrix}$

SOLUTION:

a. $\begin{vmatrix} -1 & 2 \\ 3 & -2 \end{vmatrix} = (-1)(-2) - (2)(3) = 2 - 6 = -4$

b. $\begin{vmatrix} \frac{1}{2} & -5 \\ 2 & 8 \end{vmatrix} = (\frac{1}{2})(8) - (-5)(2) = 4 - (-10) = 14$

THIRD-ORDER DETERMINANTS

▼ **DEFINITION 12.4c**

Let

$$\det A = \begin{vmatrix} a_{11} & a_{12} & a_{13} \\ a_{21} & a_{22} & a_{23} \\ a_{31} & a_{32} & a_{33} \end{vmatrix}$$

The *minor* M_{ij} of the element a_{ij} is the determinant obtained by deleting the ith row and the jth column of A. The *cofactor* of the element a_{ij} is defined by $A_{ij} = (-1)^{i+j}M_{ij}$

To define a third-order determinant, we first introduce the idea of a *minor* and of a *cofactor* of an element of a determinant. (See Definition 12.4c.) Note that to obtain M_{ij}, we delete the row and column of a_{ij}; the remaining elements of A form the determinant M_{ij}. To form A_{ij}, we attach the sign indicated by $(-1)^{i+j}$ to the minor M_{ij}. Notice that we have not yet defined the *value* of $\det A$, but we shall do so very shortly.

EXAMPLE 2

Let

$$\det A = \begin{vmatrix} 3 & 1 & 5 \\ 2 & -3 & 6 \\ 0 & -1 & -3 \end{vmatrix}$$

Find M_{11}, M_{32}, M_{21}, and the corresponding cofactors of det A.

SOLUTION:

To find M_{11}, we delete the first row and the first column of det A. Thus,

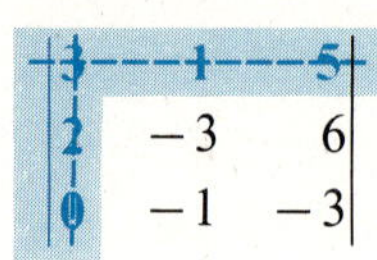

This gives

$$M_{11} = \begin{vmatrix} -3 & 6 \\ -1 & -3 \end{vmatrix} = (-3)(-3) - (6)(-1) = 15$$

so that

$$A_{11} = (-1)^{1+1}M_{11} = (1)(15) = 15$$

To find M_{32}, we delete the third row and the second column of det A to obtain

$$M_{32} = \begin{vmatrix} 3 & 5 \\ 2 & 6 \end{vmatrix} = (3)(6) - (5)(2) = 8$$

and

$$A_{32} = (-1)^{3+2}M_{32} = (-1)(8) = -8$$

To find M_{21}, we delete the second row and the first column of det A to obtain

$$M_{21} = \begin{vmatrix} 1 & 5 \\ -1 & -3 \end{vmatrix} = (1)(-3) - (5)(-1) = 2$$

and

$$A_{21} = (-1)^{2+1}M_{21} = (-1)(2) = -2$$

We are now ready to define the value of det A in terms of the first row elements and their cofactors. (See Definition 12.4d.)

▼ DEFINITION 12.4d

The number, det A, is the sum of the products formed by multiplying each element of the first row of A by its cofactor. Thus,

$$\det A = \begin{vmatrix} a_{11} & a_{12} & a_{13} \\ a_{21} & a_{22} & a_{23} \\ a_{31} & a_{32} & a_{33} \end{vmatrix}$$

$$= a_{11}A_{11} + a_{12}A_{12} + a_{13}A_{13}$$

$$= a_{11}M_{11} - a_{12}M_{12} + a_{13}M_{13}$$

EXAMPLE 3

Evaluate:

$$\det A = \begin{vmatrix} 2 & -4 & 3 \\ 3 & 0 & 5 \\ 1 & -1 & 2 \end{vmatrix}$$

SOLUTION:

By Definition 12.4d,

$$\begin{aligned} \det A &= a_{11}A_{11} + a_{12}A_{12} + a_{13}A_{13} \\ &= a_{11}M_{11} - a_{12}M_{12} + a_{13}M_{13} \\ &= (2)\begin{vmatrix} 0 & 5 \\ -1 & 2 \end{vmatrix} - (-4)\begin{vmatrix} 3 & 5 \\ 1 & 2 \end{vmatrix} + (3)\begin{vmatrix} 3 & 0 \\ 1 & -1 \end{vmatrix} \\ &= (2)(5) + (4)(1) + (3)(-3) \\ &= 5 \end{aligned}$$

HIGHER-ORDER DETERMINANTS

As you can see from Definition 12.4d, a third-order determinant is defined in terms of second-order determinants. A fourth-order determinant can be defined in the same way, in terms of third-order determinants. In general, an nth-order determinant can be defined in terms of determinants of order $n - 1$. Thus, if

$$\det A = \begin{vmatrix} a_{11} & a_{12} & \cdots & a_{1n} \\ a_{21} & a_{22} & \cdots & a_{2n} \\ \cdots & \cdots & \cdots & \cdots \\ a_{n1} & a_{n2} & \cdots & a_{n3} \end{vmatrix}$$

then the minor M_{ij} of the element a_{ij} is defined to be the determinant formed by deleting the ith row and the jth column of A. The cofactor A_{ij} of the element a_{ij} is defined by

$$\mathbf{A_{ij} = (-1)^{i+j}M_{ij}}$$

(Notice that the words are all exactly the same as those we used to define third-order determinants.) The value of det A is then defined by the formula:

$$\det A = a_{11}A_{11} + a_{12}A_{12} + \cdots + a_{1n}A_{1n} \tag{1}$$

Formula 1 gives what is called an expansion of det A by cofactors *along the first row*. It can be shown that the same value is obtained by using any row or column and the corresponding cofactors. For example, for a third-order determinant, we have the following expansion along the third row:

$$\begin{aligned} \det A &= a_{31}A_{31} + a_{32}A_{32} + a_{33}A_{33} \\ &= a_{31}M_{31} - a_{32}M_{32} + a_{33}M_{33} \end{aligned}$$

Note that corresponding to the *third* row, the first subscripts on the a, A, and M are all threes. Similarly, we have the following expansion along the second column

$$\begin{aligned} \det A &= a_{12}A_{12} + a_{22}A_{22} + a_{32}A_{32} \\ &= -a_{12}M_{12} + a_{22}M_{22} - a_{32}M_{32} \end{aligned}$$

Note this time that corresponding to the *second column*, the second subscripts on the a, A, and M are all twos.

EXAMPLE 4

Evaluate the determinant:

$$D = \begin{vmatrix} 1 & 0 & 2 & 0 \\ -2 & 3 & -1 & 0 \\ 0 & 4 & -3 & 2 \\ 1 & 2 & 0 & 1 \end{vmatrix}$$

SOLUTION:

We expand by cofactors along the fourth column because of the two zero elements in that column. (The first row would do as well.) We obtain

$$D = -0\begin{vmatrix} -2 & 3 & -1 \\ 0 & 4 & -3 \\ 1 & 2 & 0 \end{vmatrix} + 0\begin{vmatrix} 1 & 0 & 2 \\ 0 & 4 & -3 \\ 1 & 2 & 0 \end{vmatrix}$$

$$-2\begin{vmatrix} 1 & 0 & 2 \\ -2 & 3 & -1 \\ 1 & 2 & 0 \end{vmatrix} + 1\begin{vmatrix} 1 & 0 & 2 \\ -2 & 3 & -1 \\ 0 & 4 & -3 \end{vmatrix}$$

$$= -2(2 - 0 - 14) + (-5 - 0 - 16)$$

$$= 3$$

(We expanded the last two third-order determinants by cofactors along the first row.)

We can use the preceding ideas to obtain a theorem that will be helpful to us in the following section. Consider the second-order determinant with two identical rows:

$$\begin{vmatrix} a & b \\ a & b \end{vmatrix} = ab - ab = 0$$

Obviously, every second-order determinant with its two rows identical must have the value zero. Now suppose that two rows of a third-order determinant are identical. If we expand by minors along the remaining row, all the minors will be second-order determinants with two identical rows and thus with the value zero. It follows that the value of the third-order determinant must be zero. This reasoning can be extended step by step to a determinant of any order. The same argument applies (with rows replaced by columns) if there are two identical columns. Thus, we have Theorem 12.4a.

▼ **THEOREM 12.4a**

If two rows (or two columns) of a determinant are identical, then the value of the determinant is zero.

EXERCISE 12.4

In Problems 1–18, evaluate the given determinant.

1 $\begin{vmatrix} 3 & 1 \\ 5 & 2 \end{vmatrix}$

2 $\begin{vmatrix} 5 & 1 \\ 6 & 3 \end{vmatrix}$

3 $\begin{vmatrix} \frac{1}{8} & \frac{1}{6} \\ \frac{1}{4} & \frac{2}{3} \end{vmatrix}$

4 $\begin{vmatrix} -2 & -\frac{1}{2} \\ 3 & \frac{7}{4} \end{vmatrix}$

5 $\begin{vmatrix} \frac{3}{4} & \frac{5}{2} \\ \frac{1}{6} & -\frac{4}{3} \end{vmatrix}$

6 $\begin{vmatrix} -\frac{1}{8} & -\frac{1}{28} \\ \frac{3}{2} & \frac{3}{7} \end{vmatrix}$

7 $\begin{vmatrix} \frac{4}{5} & -\frac{1}{5} \\ \frac{1}{2} & -\frac{1}{2} \end{vmatrix}$

8 $\begin{vmatrix} \frac{3}{5} & \frac{1}{2} \\ -\frac{1}{4} & -\frac{1}{2} \end{vmatrix}$

9 $\begin{vmatrix} 2 & 1 & 3 \\ 1 & 2 & -1 \\ 3 & 1 & 5 \end{vmatrix}$

10 $\begin{vmatrix} 1 & 3 & 5 \\ 2 & 0 & 10 \\ -3 & 1 & -15 \end{vmatrix}$ **11** $\begin{vmatrix} 2 & 1 & 3 \\ 3 & 4 & 5 \\ 1 & 7 & 2 \end{vmatrix}$ **12** $\begin{vmatrix} 3 & 4 & 2 \\ 8 & 5 & 9 \\ 7 & 6 & 1 \end{vmatrix}$

13 $\begin{vmatrix} 8 & 7 & 6 \\ -2 & 3 & -1 \\ 3 & 0 & 4 \end{vmatrix}$ **14** $\begin{vmatrix} 1 & 6 & 7 \\ 3 & 0 & 3 \\ -5 & 4 & 1 \end{vmatrix}$ **15** $\begin{vmatrix} 1 & -1 & a \\ -1 & a & -1 \\ a & -1 & -1 \end{vmatrix}$

16 $\begin{vmatrix} a & 0 & a \\ b & a & b \\ 0 & b & a \end{vmatrix}$ **17** $\begin{vmatrix} 5 & 7 & -1 & 0 \\ 2 & 9 & 5 & 0 \\ 0 & 3 & 8 & -2 \\ 4 & 0 & -3 & 0 \end{vmatrix}$ **18** $\begin{vmatrix} a & 0 & 0 & 0 \\ b & c & 0 & 0 \\ d & e & f & 0 \\ g & h & i & j \end{vmatrix}$

In Problems 19–28, verify the given identity.

19 a $\begin{vmatrix} 1 & 2 \\ 3 & 4 \end{vmatrix} = -\begin{vmatrix} 2 & 1 \\ 4 & 3 \end{vmatrix}$ **b** $\begin{vmatrix} 1 & 2 \\ 3 & 4 \end{vmatrix} = -\begin{vmatrix} 3 & 4 \\ 1 & 2 \end{vmatrix}$

20 a $\begin{vmatrix} 1 & 2 & 3 \\ -4 & 5 & 0 \\ 1 & 2 & 3 \end{vmatrix} = 0$ **b** $\begin{vmatrix} 1 & 4 & 1 \\ 2 & 5 & 2 \\ 3 & 6 & 3 \end{vmatrix} = 0$

21 $\begin{vmatrix} a & b \\ c & d \end{vmatrix} = -\begin{vmatrix} b & a \\ d & c \end{vmatrix}$ **22** $\begin{vmatrix} a & b \\ c & d \end{vmatrix} = -\begin{vmatrix} c & d \\ a & b \end{vmatrix}$ **23** $\begin{vmatrix} a & b & c \\ d & e & f \\ a & b & c \end{vmatrix} = 0$

24 $\begin{vmatrix} a & a & d \\ b & b & c \\ c & c & f \end{vmatrix} = 0$ **25** $\begin{vmatrix} ka & kb \\ c & d \end{vmatrix} = k \cdot \begin{vmatrix} a & b \\ c & d \end{vmatrix}$ **26** $\begin{vmatrix} ma_1 & b_1 & c_1 \\ ma_2 & b_2 & c_2 \\ ma_3 & b_3 & c_3 \end{vmatrix} = m \cdot \begin{vmatrix} a_1 & b_1 & c_1 \\ a_2 & b_2 & c_2 \\ a_3 & b_3 & c_3 \end{vmatrix}$

27 $\begin{vmatrix} a_1 & b_1 & (c_1 + d_1) \\ a_2 & b_2 & (c_2 + d_2) \\ a_3 & b_3 & (c_3 + d_3) \end{vmatrix} = \begin{vmatrix} a_1 & b_1 & c_1 \\ a_2 & b_2 & c_2 \\ a_3 & b_3 & c_3 \end{vmatrix} + \begin{vmatrix} a_1 & b_1 & d_1 \\ a_2 & b_2 & d_2 \\ a_3 & b_3 & d_3 \end{vmatrix}$

Hint: Expand the left side by cofactors along the third column and use the fact that $(c_1 + d_1)A_{13} = c_1A_{13} + d_1A_{13}$, and so on.

28 $\begin{vmatrix} a_1 & b_1 & c_1 + ka_1 \\ a_2 & b_2 & c_2 + ka_2 \\ a_3 & b_3 & c_3 + ka_3 \end{vmatrix} = \begin{vmatrix} a_1 & b_1 & c_1 \\ a_2 & b_2 & c_2 \\ a_3 & b_3 & c_3 \end{vmatrix}$

Hint: First use the result in Problem 27.

In Problems 29–34, expand the determinant and solve the resulting equation.

29 $\begin{vmatrix} x & -1 \\ x & x \end{vmatrix} = 2$ **30** $\begin{vmatrix} 3x & -2 \\ 3 & 4 \end{vmatrix} = 18$ **31** $\begin{vmatrix} x & 1 & 4 \\ 3 & x & 2 \\ x & 1 & -3 \end{vmatrix} = -7$

32 $\begin{vmatrix} x & 1 & x \\ 3 & x & 2 \\ 4 & 1 & 3 \end{vmatrix} = -3$ **33** $\begin{vmatrix} 0 & 2 & x \\ 2 & -3 & 1 \\ 8 & -7 & x \end{vmatrix} = -2$ **34** $\begin{vmatrix} 0 & 2 & x \\ 8 & -7 & x \\ 2 & -3 & 1 \end{vmatrix} = 2$

USING YOUR KNOWLEDGE 12.4

Determinants can sometimes provide a neat way of writing a required condition or equation. We shall look at one simple instance of this type.

In Section 6.1, we studied the equation of a straight line in the plane. One of the problems in that section was to find an equation of a line passing through two given points. Suppose that the points are (x_1, y_1) and (x_2, y_2). Then, an equation of the line can be written in the form

$$\begin{vmatrix} x & y & 1 \\ x_1 & y_1 & 1 \\ x_2 & y_2 & 1 \end{vmatrix} = 0 \tag{1}$$

Here is the way to verify this fact. First, if we think of expanding the determinant by minors along the first row, we recognize at once that the coefficients of x and y will be constants, so that the resulting equation is linear. Next, we see that (x_1, y_1) and (x_2, y_2) both satisfy the equation, because if we substitute (x_1, y_1) or (x_2, y_2) for (x, y) in the first row of the determinant, the result is a determinant with two identical rows, which Theorem 12.4a tells us is zero. This shows that 1 is an equation of the required line.

As an illustration, we find an equation of the line through (1, 3) and (−5, −2). According to Equation 1, this equation is

$$\begin{vmatrix} x & y & 1 \\ 1 & 3 & 1 \\ -5 & -2 & 1 \end{vmatrix} = 0$$

or

$$5x - 6y + 13 = 0$$

(We expanded by minors along the first row.)

Use Equation 1 to find an equation of the line passing through the two given points.

1 (2, 7) and (0, 3) **2** (10, 12) and (−7, 1) **3** (−1, 4) and (8, 2) **4** (5, 0) and (0, −3)
5 $(a, 0)$ and $(0, b)$, $ab \neq 0$ **6** $(0, b)$ and $(-b/m, 0)$, $b \neq 0$

12.5 CRAMER'S RULE

Although the Scottish mathematician Colin Maclaurin has been credited with a result that he did not discover in calculus, an important rule that he did originate in algebra does not bear his name. This rule, known as Cramer's Rule, is used to solve a system of n linear equations in n unknowns. We shall justify the use of Cramer's Rule for a system of three equations in three unknowns; however, the method of proof can be extended to n linear equations in n unknowns.

We consider the system

$$\left.\begin{aligned} a_{11}x + a_{12}y + a_{13}z &= c_1 \\ a_{21}x + a_{22}y + a_{23}z &= c_2 \\ a_{31}x + a_{32}y + a_{33}z &= c_3 \end{aligned}\right\} \tag{1}$$

and we use the coefficients of x, y, and z to write the determinant

$$D = \begin{vmatrix} a_{11} & a_{12} & a_{13} \\ a_{21} & a_{22} & a_{23} \\ a_{31} & a_{32} & a_{33} \end{vmatrix}$$

The determinant D is called the determinant of System 1. We shall denote by D_x the determinant formed from D by replacing the coefficients of x by the corresponding c's; similarly, D_y and D_z will denote the determinants obtained from D by replacing the coefficients of y and z, respectively, by the corresponding c's. Thus,

$$D_x = \begin{vmatrix} c_1 & a_{12} & a_{13} \\ c_2 & a_{22} & a_{23} \\ c_3 & a_{32} & a_{33} \end{vmatrix} \qquad D_y = \begin{vmatrix} a_{11} & c_1 & a_{13} \\ a_{21} & c_2 & a_{23} \\ a_{31} & c_3 & a_{33} \end{vmatrix} \qquad D_z = \begin{vmatrix} a_{11} & a_{12} & c_1 \\ a_{21} & a_{22} & c_2 \\ a_{31} & a_{32} & c_3 \end{vmatrix}$$

We can now state Cramer's Rule for System 1 in terms of the preceding four determinants. (See Theorem 12.5a.) Before proving this theorem, we illustrate its use in Example 1.

▼ THEOREM 12.5a

Cramer's Rule

If $D \neq 0$, then System 1 has the unique solution

$$x = \frac{D_x}{D} \qquad y = \frac{D_y}{D} \qquad z = \frac{D_z}{D}$$

EXAMPLE 1

Solve the system:

$$\begin{aligned} x + y + 2z &= 7 \\ x - y - 3z &= -6 \\ 2x + 3y + z &= 4 \end{aligned}$$

SOLUTION:

In order to use Theorem 12.5a, we first have to calculate the determinant D of the system. If $D \neq 0$, then we have to calculate the other three determinants, D_x, D_y, and D_z. The determinant of the system is

$$D = \begin{vmatrix} 1 & 1 & 2 \\ 1 & -1 & -3 \\ 2 & 3 & 1 \end{vmatrix} = 11$$

Keep in mind that D_x is formed from D by replacing the first column (the x-coefficients) by the c's, and similarly for D_y and D_z. Thus,

$$D_x = \begin{vmatrix} 7 & 1 & 2 \\ -6 & -1 & -3 \\ 4 & 3 & 1 \end{vmatrix} = 22 \qquad D_y = \begin{vmatrix} 1 & 7 & 2 \\ 1 & -6 & -3 \\ 2 & 4 & 1 \end{vmatrix} = -11$$

$$D_z = \begin{vmatrix} 1 & 1 & 7 \\ 1 & -1 & -6 \\ 2 & 3 & 4 \end{vmatrix} = 33$$

Thus, the solution of the given system is

$$x = \frac{D_x}{D} = \frac{22}{11} = 2 \qquad y = \frac{D_y}{D} = -\frac{11}{11} = -1 \qquad z = \frac{D_z}{D} = \frac{33}{11} = 3$$

That $(2, -1, 3)$ is the solution can be checked by substitution into the original equations.

Although Theorem 12.5a is stated for a system of three equations in three unknowns, the method, as we noted, is quite general. Example 2 illustrates the use of Cramer's Rule for a system of two equations in two unknowns.

EXAMPLE 2

Solve the system:

$$2x + 3y = -4$$
$$5x - 7y = 19$$

SOLUTION:

For this system,

$$D = \begin{vmatrix} 2 & 3 \\ 5 & -7 \end{vmatrix} = -29 \qquad D_x = \begin{vmatrix} -4 & 3 \\ 19 & -7 \end{vmatrix} = -29$$

$$D_y = \begin{vmatrix} 2 & -4 \\ 5 & 19 \end{vmatrix} = 58$$

Therefore,

$$x = \frac{D_x}{D} = \frac{-29}{-29} = 1 \qquad y = \frac{D_y}{D} = \frac{58}{-29} = -2$$

The solution $(1, -2)$ can be checked in the given system.

The proof of Theorem 12.5a can be simplified by first making an easy observation. We know that the value of the determinant D is given by

$$D = a_{1k}A_{1k} + a_{2k}A_{2k} + a_{3k}A_{3k} \qquad k = 1 \text{ or } 2 \text{ or } 3 \tag{2}$$

(This is just the expansion of D by cofactors along the kth column.) Now, suppose that we replace the elements a_{1k}, a_{2k}, a_{3k} by the corresponding elements of a different column to obtain the sum

$$S = a_{1j}A_{1k} + a_{2j}A_{2k} + a_{3j}A_{3k} \qquad j \neq k$$

We can see that $S = 0$ because the effect of the replacement is the construction of a determinant with the jth and kth columns identical. Thus, we have the observation:

$$\text{If } j \neq k, \text{ then } a_{1j}A_{1k} + a_{2j}A_{2k} + a_{3j}A_{3k} = 0 \tag{3}$$

In words, Observation 3 says that the sum of the products of the elements of any column and the corresponding cofactors of a different column is zero. The same result holds using rows in place of columns.

We are now ready to give the proof of Theorem 12.5a. We return to System 1 and its determinant D. We multiply the first equation of the system by A_{11}, the cofactor of a_{11} in D; the second equation by A_{21}, the cofactor of a_{21}; and the third equation by A_{31}, the cofactor of a_{31}. Adding the resulting three equations yields

$$(a_{11}A_{11} + a_{21}A_{21} + a_{31}A_{31})x + (a_{12}A_{11} + a_{22}A_{21} + a_{32}A_{31})y$$
$$+ (a_{13}A_{11} + a_{23}A_{21} + a_{33}A_{31})z = c_1A_{11} + c_2A_{21} + c_3A_{31}$$

By referring to 2 and 3, we see that this equation is simply

$$Dx + 0 + 0 = c_1A_{11} + c_2A_{21} + c_3A_{31}$$

Furthermore, the right side is exactly the expansion of the determinant D_x by cofactors along the first column; hence, we have shown that

$$Dx = D_x$$

Using a similar procedure, we can obtain the equations $Dy = D_y$ and $Dz = D_z$. Since it was assumed that $D \neq 0$, it must be true that not all the cofactors of any column are zero. Thus, we have used only the permissible elementary operations on System 1 to form the system

$$Dx = D_x \qquad Dy = D_y \qquad Dz = D_z \tag{4}$$

Consequently, System 4 is equivalent to System 1. Obviously, if $D \neq 0$, System 4 and thus System 1 have the unique solution

$$x = \frac{D_x}{D} \qquad y = \frac{D_y}{D} \qquad z = \frac{D_z}{D}$$

This completes the proof of Theorem 12.5a.

If $D = 0$ and at least one of D_x, D_y, D_z is *not* zero, then Systems 1 and 4 are equivalent. In this case, there is no solution, and System 1 is *inconsistent*. If all four of D, D_x, D_y, D_z are zero, then System 4 might or might not be equivalent to System 1. It can be shown in this case that System 1 might or might not have solutions; the details are beyond the scope of our treatment.

EXERCISE 12.5

In Problems 1–24, find the determinant D of the given system. If $D \neq 0$, solve the system by Cramer's Rule. If $D = 0$, determine whether the system has infinitely many solutions or no solutions.

1 $\begin{aligned} 2x - 3y &= 16 \\ x - y &= 7 \end{aligned}$

2 $\begin{aligned} 3x - 2y &= 8 \\ -6x + 4y &= 5 \end{aligned}$

3 $\begin{aligned} 16x - 9y &= -5 \\ 10x + 18y &= -11 \end{aligned}$

4 $\begin{aligned} 4x + 5y &= 5 \\ -10x - 4y &= -7 \end{aligned}$

5 $\begin{aligned} -6x - 10y &= 2 \\ 3x + 5y &= 1 \end{aligned}$

6 $\begin{aligned} -2x + 3y &= 1 \\ 4x - 6y &= 2 \end{aligned}$

7 $\begin{aligned} 3(x + 2) &= 2y \\ 2(y + 5) &= 7x \end{aligned}$

8 $\begin{aligned} 4(y - 4x) &= 2(x + 5) \\ 10(y - x) &= 11y - 15x \end{aligned}$

9 $\begin{aligned} x + y - 2z &= 13 \\ x - 3y - z &= -3 \\ x - y + 4z &= -17 \end{aligned}$

10 $\begin{aligned} x + y + z &= 13 \\ x - 2y + 4z &= 10 \\ 3x + y - 3z &= 5 \end{aligned}$

11 $\begin{aligned} x + y + z &= -1 \\ 3x - y - 5z &= 13 \\ 5x + 3y + 2z &= 1 \end{aligned}$

12 $\begin{aligned} x + y + z &= 6 \\ x - y + 2z &= 5 \\ x - y - 3z &= 10 \end{aligned}$

13 $\begin{aligned} x - 2y - z &= 3 \\ 2x - 5y + z &= -1 \\ x - 2y - z &= -3 \end{aligned}$

14 $\begin{aligned} x - y + z &= 2 \\ x + y + z &= 4 \\ 2x + 2y - z &= -4 \end{aligned}$

15 $\begin{aligned} x - 3y + z &= 2 \\ x + 2y - z &= 1 \\ -7x + y + z &= 0 \end{aligned}$

16 $\begin{aligned} 3x + y + z &= 1 \\ x + 2y - z &= 1 \\ x + y + 2z &= -17 \end{aligned}$

17 $\begin{aligned} 2x - y + 2z &= 5 \\ 2x + y - z &= -6 \\ 3x + 2z &= 3 \end{aligned}$

18 $\begin{aligned} 2x + 3y + 6z &= 17 \\ x + y + z &= 0 \\ 5x - 6y &= 8 \end{aligned}$

19 $\begin{aligned} x + y + z &= 3 \\ x - 2y + 3z &= 5 \\ 3x - 3y + 7z &= 14 \end{aligned}$

20 $\begin{aligned} x + 2z &= 9 \\ y - 2z &= 5 \\ x + 3y &= -4 \end{aligned}$

21 $\begin{aligned} y + z &= x - 7 \\ x + 2z &= 1 - 2y \\ x - 7 &= 3y + 2z \end{aligned}$

22 $\begin{aligned} x + z &= 6 - y \\ x + 3y &= 8 \\ 2x &= 2y + 5 - z \end{aligned}$

23 $\begin{aligned} w + x + y + z &= 2 \\ 2w - x - 2y &= 9 \\ w - 2x + 2y &= 0 \\ w + 3x + y + z &= 0 \end{aligned}$

24 $\begin{aligned} w + x + 4y + z &= 0 \\ 2w - x + y - z &= -1 \\ 2w + 3y + z &= 0 \\ x + 2y - z &= 1 \end{aligned}$

25 For the following system, show that D, D_x, D_y, and D_z are all zero. Then, by reducing the system to echelon form, show that it has infinitely many solutions.

$$\begin{aligned} x + 2y - z &= 4 \\ 2x - y + z &= 6 \\ 5x + 5y - 2z &= 18 \end{aligned}$$

26 For the following system, show that D, D_x, D_y, and D_z are all zero. Then, by reducing the system to echelon form, show that it has no solution.

$$\begin{aligned} x + 2y - z &= 4 \\ 2x + 4y - 2z &= 12 \\ 3x + 6y - 3z &= 16 \end{aligned}$$

USING YOUR KNOWLEDGE 12.5

Have you ever tried to draw two intersecting lines and then a third line passing through the point of intersection? You have to be very precise to make your drawing look right. Perhaps this is a reflection of the fact that there is a very precise algebraic condition that has to be met if the system of linear equations

$$\left.\begin{aligned} a_1x + b_1y &= c_1 \\ a_2x + b_2y &= c_2 \\ a_3x + b_3y &= c_3 \end{aligned}\right\} \qquad (1)$$

has a solution. Suppose that the system consisting of the first two equations of System 1 has a nonzero determinant D. Then, the solution of these two equations is

$$x = \frac{D_x}{D} \qquad y = \frac{D_y}{D}$$

where

$$D = \begin{vmatrix} a_1 & b_1 \\ a_2 & b_2 \end{vmatrix} \qquad D_x = \begin{vmatrix} c_1 & b_1 \\ c_2 & b_2 \end{vmatrix} \qquad D_y = \begin{vmatrix} a_1 & c_1 \\ a_2 & c_2 \end{vmatrix}$$

If this solution also satisfies the third equation, then

$$\frac{a_3D_x}{D} + \frac{b_3D_y}{D} = c_3 \qquad \text{or} \qquad a_3D_x + b_3D_y = c_3D$$

that is,

$$a_3\begin{vmatrix} c_1 & b_1 \\ c_2 & b_2 \end{vmatrix} + b_3\begin{vmatrix} a_1 & c_1 \\ a_2 & c_2 \end{vmatrix} = c_3\begin{vmatrix} a_1 & b_1 \\ a_2 & b_2 \end{vmatrix}$$

Because

$$\begin{vmatrix} c_1 & b_1 \\ c_2 & b_2 \end{vmatrix} = -\begin{vmatrix} b_1 & c_1 \\ b_2 & c_2 \end{vmatrix}$$

we may change all the signs and write

$$a_3\begin{vmatrix} b_1 & c_1 \\ b_2 & c_2 \end{vmatrix} - b_3\begin{vmatrix} a_1 & c_1 \\ a_2 & c_2 \end{vmatrix} + c_3\begin{vmatrix} a_1 & b_1 \\ a_2 & b_2 \end{vmatrix} = 0$$

We can now recognize the left side of this equation as the expansion by minors along the third row of the determinant

$$\begin{vmatrix} a_1 & b_1 & c_1 \\ a_2 & b_2 & c_2 \\ a_3 & b_3 & c_3 \end{vmatrix}$$

of all the coefficients of System 1. Thus, we have arrived at the fact that, *if System 1 has a solution, then the determinant of all the coefficients must be zero.* Although we assumed that $D \neq 0$, it can be shown that the final result is valid without that assumption. Furthermore, the corresponding result for n linear equations in $n - 1$ unknowns can be obtained in the same way.

Unfortunately, the converse statement is *not* true. The coefficient determinant may be zero without the system having a solution. One example proves this: The system

$$3x + 3y = 0$$
$$3x + 3y = 1$$
$$3x + 3y = 2$$

is obviously inconsistent, but the coefficient determinant has two identical columns, so that it is zero. Thus, if the determinant is zero, the system might or might not have a solution.

Try to use the preceding ideas to do the following problems.

1 Determine for what value of k (if any) the following system has a solution.

Hint: You must not only make the determinant zero, but you must also check to see if the resulting system does have a solution.

$$2x + y = 1$$
$$3x + 2y = -1$$
$$7x + 3y = k$$

2 Determine for what values of m (if any) the following system has a solution.

$$x + 4y = 18$$
$$mx - y = 2$$
$$2x + my = 16$$

SELF-TEST

1 Use the substitution method to solve the system:

$$\begin{aligned} 11x - 3y &= 34 \\ y &= 5x - 14 \end{aligned}$$

2 Solve the system:

$$\begin{aligned} 3x + 4y &= -1 \\ 2x + 3y &= 1 \end{aligned}$$

3 Solve the system:

$$\begin{aligned} 2x - y &= -2 \\ -4x + 2y &= 4 \end{aligned}$$

4 Solve the system:

$$\begin{aligned} -3x + 4y &= 6 \\ 6x - 8y &= 9 \end{aligned}$$

5 Solve the system:

$$\frac{x-y}{x+y} = 4$$

$$\frac{x+y+8}{x+y+5} = 2$$

6 Reduce the following system to echelon form and solve.

$$\begin{aligned} x - y + 2z &= -3 \\ 3x + 2y + 3z &= 4 \\ 4x - y - z &= 3 \end{aligned}$$

7 Reduce the following homogeneous system to echelon form and find the nontrivial solutions (if there are any).

$$\begin{aligned} x + y + z &= 0 \\ 3x + y - z &= 0 \\ 4x + y - 2z &= 0 \end{aligned}$$

8 The sum of $1.00 consists of pennies, nickels, and dimes. There are 26 coins in all and three times as many pennies as nickels. How many coins of each denomination are there?

9 Solve the system:

$$\frac{4}{x} - \frac{9}{y} + \frac{2}{z} = 4$$

$$\frac{2}{x} - \frac{3}{y} + \frac{2}{z} = 1$$

$$\frac{5}{x} - \frac{3}{y} + \frac{7}{z} = 0$$

10 Use matrices to solve the system:

$$\begin{aligned} x + y - z &= -3 \\ x + 2y + z &= -10 \\ 5x + 7y - z &= -29 \end{aligned}$$

11 Use matrices to solve the system:

$$\begin{aligned} x + y + z &= -2 \\ 2x + 6y - z &= 3 \\ 3x + 2y + 3z &= -4 \end{aligned}$$

12 Use matrices to solve the system:

$$\begin{aligned} x - y - z &= 2 \\ 5x + 3y - 2z &= 6 \\ 7x + y - 4z &= 11 \end{aligned}$$

13 Evaluate the determinant:

$$\begin{vmatrix} 1 & 3 \\ -4 & -2 \end{vmatrix}$$

14 Evaluate the determinant:

$$\begin{vmatrix} 1 & 0 & -2 \\ 2 & 3 & -3 \\ 0 & -1 & 4 \end{vmatrix}$$

a by minors along the third row;
b by minors along the second column.

15 Evaluate the determinant:

$$\begin{vmatrix} 1 & 0 & 0 & 1 \\ 1 & 2 & 0 & 1 \\ 1 & 2 & 3 & 0 \\ 1 & 2 & 3 & 4 \end{vmatrix}$$

16 For what values of k is the following determinant zero?

$$\begin{vmatrix} 1 & 2 & k \\ 3 & k & 5 \\ 2 & 0 & 3 \end{vmatrix}$$

17 Solve the following system by Cramer's Rule:

$$\begin{aligned} 3x + 2y &= 1 \\ 4x - 5y &= -14 \end{aligned}$$

18 Evaluate D and D_x for the following system to show that it is inconsistent:

$$\begin{aligned} 2x + 3y + z &= 5 \\ x - 2y + 3z &= 4 \\ x + 12y - 7z &= 0 \end{aligned}$$

19 Use Cramer's Rule to solve the system:

$$\begin{aligned} x + y - 2z &= 7 \\ x + 2y + 2z &= 6 \\ 2x + 3y + 3z &= 8 \end{aligned}$$

20 Expand the determinant and solve the equation:

$$\begin{vmatrix} x & 2 & x \\ 2 & x & 1 \\ 3 & 2 & 2 \end{vmatrix} = -10$$

CHAPTER 13

COMPLEX NUMBERS AND THEORY OF EQUATIONS

13.1 COMPLEX NUMBERS

The equation $x^2 + 1 = 0$ is one of the most important equations in mathematics. If we try to solve this equation, we obtain $x^2 = -1$, which stops us because we know that there is no real number whose square is -1. We also know that the general quadratic equation $ax^2 + bx + c = 0$ has no real roots if the discriminant, $b^2 - 4ac$, is negative.

Most of the problems involving quadratic equations in the preceding chapters were deliberately set up to avoid equations that have no real-number solutions. However, in many of the applications of mathematics, expressions involving square roots of negative numbers arise and are given important practical interpretations. The designing of automobile suspension systems and the designing of the electrical circuits in a television receiver are just two of the practical problems in which these roots play a significant role.

In order to work with square roots of negative numbers, we first define a new number, denoted by the letter i, and having the property described in Definition 13.1a.

▼ **DEFINITION 13.1a**

The number i has the property that

$$i^2 = -1$$

We write $\sqrt{-1} = i$, and agree that the two square roots of -1 are i and $-i$.

If b is a positive number, we write

$$\sqrt{-b} = \sqrt{(-1)(b)} = (\sqrt{-1})(\sqrt{b}) = i\sqrt{b}$$

Thus, we can always represent the square root of a negative number as the product of a real number and the number i. For example,

$$\sqrt{-4} = \sqrt{(-1)(4)} = i\sqrt{4} = 2i$$

and

$\sqrt{-7} = \sqrt{(-1)(7)} = i\sqrt{7}$

We always agree to write the square root of a negative number as the product of i and the square root of a positive number as in these illustrations. *Incorrect results can be obtained if this agreement is not observed.*

EXAMPLE 1

Write the given expression as a product of i and a real number:

a. $\sqrt{-36}$

b. $\sqrt{-72}$

SOLUTION:

a. $\sqrt{-36} = \sqrt{(-1)(36)} = (\sqrt{-1})(\sqrt{36}) = 6i$

b. $\sqrt{-72} = \sqrt{(-1)(72)} = (\sqrt{-1})(\sqrt{72}) = 6\sqrt{2}i$

The number bi, where b is a real number, is called a *pure imaginary number*. (This is an unfortunate tradition, for the so-called imaginary numbers are no less "real" than the so-called real numbers; both kinds of numbers exist only in our minds.) We now define a more general set of numbers called the *complex numbers*. (See Definition 13.1b.) For example,

$$2 + 3i, \frac{1}{2} + \frac{\sqrt{3}}{2}i, \text{ and } 4.7 - 0.19i$$

are all complex numbers. In the complex number $a + bi$, the a is called the *real part*, and the b is called the *imaginary part*. In the three preceding numbers, the real parts are 2, $\frac{1}{2}$, and 4.7, respectively; the imaginary parts are 3, $\sqrt{3}/2$, and -0.19, respectively.

▼ DEFINITION 13.1b

The complex numbers are numbers of the form

$a + bi$

where a and b are real numbers.

COMPLEX-NUMBER ARITHMETIC

As a general guideline, we want the arithmetic of the complex numbers to be as nearly like that of the real numbers as possible. Consequently, we combine complex numbers just as if they were ordinary binomials, with the additional agreement that wherever i^2 occurs, it is to be replaced by -1. Before considering the arithmetic operations, we need the definition of equality of two complex numbers. For example, based on Definition 13.1c, if $a + bi = 2 - 3i$, then $a = 2$ and $b = -3$.

The sum and difference of two complex numbers are defined as in Definition 13.1d.

▼ DEFINITION 13.1c

Equality

If a, b, c, and d are real numbers, then $a + bi = c + di$ if and only if $a = c$ and $b = d$.

▼ DEFINITION 13.1d

If a, b, c, and d are real numbers, then

$(a + bi) + (c + di) = (a + c) + (b + d)i$

$(a + bi) - (c + di) = (a - c) + (b - d)i$

EXAMPLE 2

Write as complex numbers in the standard $a + bi$ form:

a. $(3 - 7i) + (2 + 9i)$

b. $(-5 + 2i) - (-8 - 6i)$

SOLUTION:

We use Definition 13.1d to obtain

a. $(3 - 7i) + (2 + 9i) = (3 + 2) + (-7 + 9)i = 5 + 2i$

b. $(-5 + 2i) - (-8 - 6i) = -5 + 2i + 8 + 6i = 3 + 8i$

To multiply two complex numbers $a + bi$ and $c + di$, we proceed as if they were ordinary binomials and then replace i^2 by -1. Thus,

$$\begin{aligned}(a + bi)(c + di) &= ac + adi + bci + bdi^2 \\ &= ac + adi + bci - bd \\ &= (ac - bd) + (ad + bc)i\end{aligned}$$

which is simply another complex number. We base the definition of the product on this result. (See Definition 13.1e.) Of course, you do not have to memorize this definition; simply multiply out and replace i^2 by -1, as we did in arriving at the definition.

▼ DEFINITION 13.1e

If a, b, c, and d are real numbers, then

$$(a + bi)(c + di) = (ac - bd) + (ad + bc)i$$

EXAMPLE 3

Write the following products in standard $a + bi$ form:

a. $(3 - 5i)(2 + 7i)$

b. $-4i(5 - 2i)$

SOLUTION:

a. $$\begin{aligned}(3 - 5i)(2 + 7i) &= 6 + 21i - 10i - 35i^2 \\ &= 6 + 11i + 35 \\ &= 41 + 11i\end{aligned}$$

b. $$\begin{aligned}-4i(5 - 2i) &= -20i + 8i^2 \\ &= -20i - 8 \\ &= -8 - 20i\end{aligned}$$

Before considering the quotient of two complex numbers, we introduce the idea of *conjugate* complex numbers. *Two complex numbers are conjugates of each other if they differ only in the signs of their imaginary parts.* The importance of this idea lies in the fact that the product of a complex number and its conjugate is a real number. For example, $2 + 5i$ and $2 - 5i$ are conjugates of each other; their product is $4 - 25i^2 = 4 + 25 = 29$. In general, the product of two conjugates $(a + bi)(a - bi) = a^2 + b^2$.

To divide one complex number by another, we multiply numerator and denominator by the conjugate of the denominator. This gives a real number for the denominator. Then we use the assumption that if A, B, and C are real numbers, then

$$\frac{A + Bi}{C} = \frac{A}{C} + \frac{B}{C}i$$

Thus, to write the quotient of $2 + 3i$ divided by $4 - i$, we proceed as follows:

$$\frac{2 + 3i}{4 - i} = \frac{(2 + 3i)(4 + i)}{(4 - i)(4 + i)} \qquad 4 + i \text{ is the conjugate of } 4 - i.$$

$$= \frac{8 + 2i + 12i + 3i^2}{16 + 1}$$

$$= \frac{8 + 14i - 3}{17}$$

$$= \frac{5 + 14i}{17}$$

$$= \frac{5}{17} + \frac{14}{17}i$$

In general,

$$\frac{a + bi}{c + di} = \frac{(a + bi)\mathbf{(c - di)}}{(c + di)\mathbf{(c - di)}}$$

$$= \frac{ac - adi + bci - bdi^2}{c^2 + d^2}$$

$$= \frac{ac + (bc - ad)i + bd}{c^2 + d^2}$$

$$= \frac{ac + bd}{c^2 + d^2} + \frac{bc - ad}{c^2 + d^2}i$$

Accordingly, we define the quotient to agree with this result. (See Definition 13.1f.)

▼ DEFINITION 13.1f

If a, b, c, and d are real numbers, and c and d are not both zero, then

$$\frac{a + bi}{c + di} = \frac{ac + bd}{c^2 + d^2} + \frac{bc - ad}{c^2 + d^2}i$$

Definition 13.1f gives us the quotient of two complex numbers as simply another complex number. Of course, you should *not* memorize the formula in this definition; it is far better to use the procedure of multiplying numerator and denominator by the conjugate of the denominator as we did in arriving at the definition.

EXAMPLE 4

Evaluate the fraction $(x + 1)/(x - 1)$ if $x = 2 + 3i$.

SOLUTION:

First, we substitute $x = 2 + 3i$ into the fraction to obtain

$$\frac{(2 + 3i) + 1}{(2 + 3i) - 1} = \frac{3 + 3i}{1 + 3i}$$

Next, we multiply numerator and denominator by $1 - 3i$, the conjugate of the denominator. Thus,

$$\frac{3 + 3i}{1 + 3i} = \frac{(3 + 3i)\mathbf{(1 - 3i)}}{(1 + 3i)\mathbf{(1 - 3i)}}$$

$$= \frac{3 - 9i + 3i - 9i^2}{1 + 9}$$

$$= \frac{3 - 6i + 9}{10}$$

$$= \frac{12}{10} - \frac{6}{10}i$$

$$= \frac{6}{5} - \frac{3}{5}i$$

EXAMPLE 5

Solve and check the equation $5x^2 - 2x + 1 = 0$.

SOLUTION:

In this equation, $a = 5$, $b = -2$, and $c = 1$. Thus, the quadratic formula gives

$$x = \frac{2 \pm \sqrt{(-2)^2 - 4(5)(1)}}{(2)(5)}$$

$$= \frac{2 \pm \sqrt{-16}}{10}$$

$$= \frac{2 \pm 4i}{10}$$

$$= \frac{1 \pm 2i}{5}$$

Therefore, the two solutions are $x = \frac{1}{5} + \frac{2}{5}i$ and $x = \frac{1}{5} - \frac{2}{5}i$.

CHECK:

For $x = \frac{1}{5} + \frac{2}{5}i$, the left side of the given equation becomes

$$\begin{aligned} 5(\tfrac{1}{5} + \tfrac{2}{5}i)^2 - 2(\tfrac{1}{5} + \tfrac{2}{5}i) + 1 &= 5(\tfrac{1}{25} + \tfrac{4}{25}i + \tfrac{4}{25}i^2) - \tfrac{2}{5} - \tfrac{4}{5}i + 1 \\ &= 5(-\tfrac{3}{25} + \tfrac{4}{25}i) - \tfrac{2}{5} - \tfrac{4}{5}i + 1 \\ &= -\tfrac{3}{5} + \tfrac{4}{5}i - \tfrac{2}{5} - \tfrac{4}{5}i + 1 \\ &= (-\tfrac{3}{5} - \tfrac{2}{5} + 1) + (\tfrac{4}{5} - \tfrac{4}{5})i \\ &= 0 + 0i \\ &= 0 \end{aligned}$$

as required. The verification for the other solution can be obtained by replacing i by $-i$ in the preceding check. The details are left for you.

We apply integral exponents to complex numbers with the same definitions as those we use for real numbers. For example, we write

$$i^3 = i^2 \cdot i = -i$$

$$i^4 = (i^2)^2 = (-1)^2 = 1$$

$$i^5 = i^4 \cdot i = 1 \cdot i = i$$

$$i^6 = i^4 \cdot i^2 = -1$$

$$i^7 = i^4 \cdot i^3 = -i$$

and so on. Similarly,

$$i^0 = 1$$

$$i^{-1} = \frac{1}{i} = \frac{1}{i} \cdot \frac{-i}{-i} = \frac{-i}{1} = -i$$

$$i^{-2} = \frac{1}{i^2} = \frac{1}{-1} = -1$$

and so on. By using the property $i^2 = -1$, we can reduce any integral power of i to one of the four numbers i, -1, $-i$, or 1.

EXAMPLE 6

Simplify:

a. i^{127}

b. i^{-127}

SOLUTION:

a. $i^{127} = (i^2)^{63} \cdot i = (-1)^{63} \cdot i = -1 \cdot i = -i$

b. $i^{-127} = i^{-128} \cdot i = (i^2)^{-64} \cdot i = (-1)^{-64} \cdot i = 1 \cdot i = i$

Note that part (b) may also be done by using the result of (a). Thus,

$$i^{-127} = \frac{1}{i^{127}} = \frac{1}{-i} = \frac{1}{-i} \cdot \frac{i}{i} = \frac{i}{-(i^2)} = i$$

EXAMPLE 7

If $P(x) = x^3$, evaluate $P(2 - 3i)$.

SOLUTION:

We use the fact that $(a + b)^3 = a^3 + 3a^2b + 3ab^2 + b^3$ to write

$$\begin{aligned} P(2 - 3i) &= (2 - 3i)^3 \\ &= (2)^3 + 3(2)^2(-3i) + 3(2)(-3i)^2 + (-3i)^3 \\ &= 8 - 36i + 54i^2 - 27i^3 \\ &= 8 - 36i - 54 + 27i \qquad i^2 = -1 \text{ and } i^3 = -i \\ &= -46 - 9i \end{aligned}$$

The preceding discussion indicates that the complex numbers give us a true extension of the real-number system of algebra. In fact, we may think of ordinary algebra as being that algebra that is based on the complex numbers; no further extension of the number system is needed to handle the problems of this basic algebra. For instance, not only have the complex numbers made it possible for us to extract square roots of any given real number, but they also have given us a more complete picture of the behavior of the quadratic equation in one unknown. Without the complex numbers, we would have to say that some quadratic equations, such as $x^2 + 1 = 0$, have no solutions, whereas others, such as $x^2 - 1 = 0$, do have solutions. It is much more satisfying to say that all quadratic equations have solutions in the set of complex numbers.

GRAPHICAL REPRESENTATION OF COMPLEX NUMBERS

Before considering additional properties of the complex numbers, we shall take a brief look at the graphical representation of these numbers. Although we cannot represent these numbers on the real-number line, we can represent them on a plane, because a complex number $a + bi$ is essentially characterized by the ordered pair of real numbers (a, b). Hence, we may represent $a + bi$ by the point (a, b) in a rectangular coordinate system in which the real part, a, of the number is measured along the x-axis and the imaginary part, b, is measured along the y-axis. We may also regard a

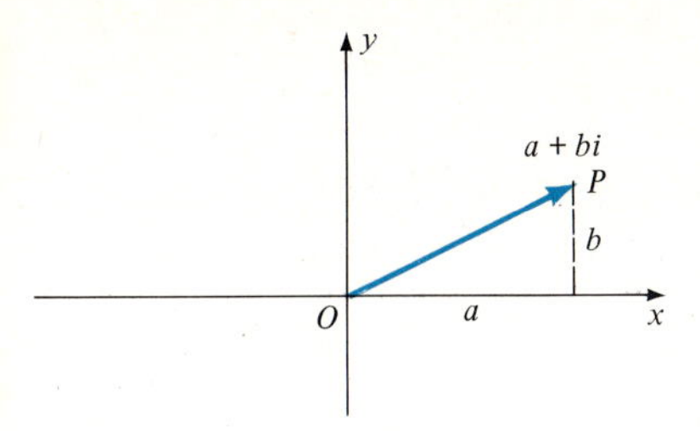

FIGURE 13.1a

and b as the horizontal and vertical components, respectively, of a *vector* (the directed line segment) from the origin to the point $P(a, b)$ and, consequently, take the vector $\overrightarrow{OP}$ (see Figure 13.1a) as a representation of the number $a + bi$. We may use either the point or the vector representation at our convenience.

The length or magnitude of the vector $\overrightarrow{OP}$ is $\sqrt{a^2 + b^2}$, the distance of the point $P(a, b)$ from the origin. This magnitude, which is always a nonnegative real number, is called the *absolute value* (or the *modulus*) of the complex number and is symbolized by

$$|a + bi|$$

As you can see, this is simply a generalization of the absolute value that we define for real numbers. In fact, if $b = 0$, then $|a + bi|$ reduces to $|a|$, the absolute value of the real number a.

EXAMPLE 8

Write in standard form and find the absolute value:

$$2 + i + \frac{1}{2 + i}$$

SOLUTION:

We first rewrite the second term by multiplying numerator and denominator by $2 - i$, the conjugate of the denominator:

$$2 + i + \frac{1}{2 + i} \cdot \frac{2 - i}{2 - i} = 2 + i + \frac{2 - i}{5}$$

Combining the two terms, we obtain

$$2 + i + \frac{1}{2 + i} = \frac{12 + 4i}{5} = \frac{12}{5} + \frac{4}{5}i$$

as the required standard form. The absolute value is

$$\begin{aligned}\left|2 + i + \frac{1}{2 + i}\right| &= \left|\frac{12}{5} + \frac{4}{5}i\right| \\ &= \sqrt{\left(\frac{12}{5}\right)^2 + \left(\frac{4}{5}\right)^2} \\ &= \frac{\sqrt{160}}{5} \\ &= \frac{4}{5}\sqrt{10}\end{aligned}$$

EXAMPLE 9

Show that the multiplication of a complex number by i leaves the absolute value unchanged but turns the vector representation through an angle of 90° counterclockwise.

SOLUTION:

Let the complex number be $a + bi$, so that the absolute value is

$$|a + bi| = \sqrt{a^2 + b^2}$$

Multiplying $a + bi$ by i, we obtain

$$i(a + bi) = ai + bi^2 = -b + ai$$

with absolute value

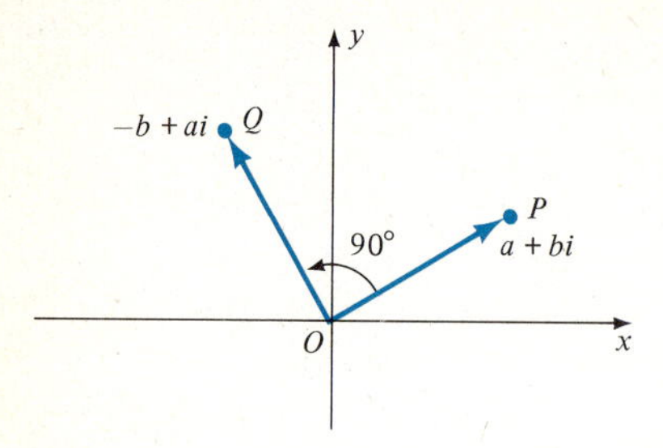

FIGURE 13.1b

$$|-b+ai| = \sqrt{(-b)^2+a^2} = \sqrt{a^2+b^2}$$

Thus, the absolute value is unchanged. From Figure 13.1b, we see that the slope of the vector $\overrightarrow{OP}$ is b/a and the slope of the vector $\overrightarrow{OQ} = i(a+bi) = -b+ai$ is $-a/b$. Because b/a and $-a/b$ are negative reciprocals of each other, the vectors $\overrightarrow{OP}$ and $\overrightarrow{OQ}$ are perpendicular. As the figure indicates, the vector $\overrightarrow{OQ}$ is obtained from $\overrightarrow{OP}$ by turning $\overrightarrow{OP}$ through 90° counterclockwise. (Note that multiplication by $-i$ would also leave the absolute value unchanged but would turn the vector 90° clockwise.)

In Example 9, we multiplied $a+bi$ by i, a number with absolute value 1, and found that

$$\begin{aligned}|i(a+bi)| &= (1)(\sqrt{a^2+b^2})\\ &= |i|\,|a+bi|\end{aligned}$$

This result is a special case of a more general one that we discuss next.

We now show that the absolute value of the product of two complex numbers is the product of their absolute values. We have

$$\begin{aligned}|(a+bi)(c+di)| &= |(ac-bd)+(ad+bc)i|\\ &= \sqrt{(ac-bd)^2+(ad+bc)^2}\\ &= \sqrt{a^2c^2+b^2d^2+a^2d^2+b^2c^2} \qquad \text{The cross product terms drop out.}\\ &= \sqrt{a^2(c^2+d^2)+b^2(c^2+d^2)}\\ &= \sqrt{(a^2+b^2)(c^2+d^2)}\\ &= \sqrt{a^2+b^2}\cdot\sqrt{c^2+d^2}\\ &= |a+bi|\cdot|c+di|\end{aligned}$$

This result enables us to prove Theorem 13.1a.

▼ THEOREM 13.1a

If the product of two complex numbers is zero, then at least one of the complex numbers is zero.

PROOF:
If a complex number is zero, then the absolute value is obviously also zero. Hence, if $(a+bi)(c+di)=0$, then $|(a+bi)(c+di)|=0$. As we have just seen,

$$|(a+bi)(c+di)| = |a+bi|\cdot|c+di| = \sqrt{a^2+b^2}\cdot\sqrt{c^2+d^2}$$

where both radicals are real numbers. Since the product of two real numbers cannot be zero unless at least one of the numbers is zero, we see that at least one of the radicals must be zero. If

$$\sqrt{a^2+b^2}=0$$

then

$$a^2+b^2=0$$

and

$$a^2 = -b^2$$

But a and b are real numbers, so that $a^2 = -b^2$ is possible only if $a = 0$ and $b = 0$, that is, if $a + bi = 0$.

If $a^2 + b^2 \neq 0$, then $c^2 + d^2 = 0$, and it would follow that $c + di = 0$. This concludes the proof of the theorem.

EXAMPLE 10

For what real values of x and y is the following equation true?

$$[x + y + (x - y + 2)i][x - 2y + (x + 2y - 4)i] = 0$$

SOLUTION:

By Theorem 13.1a, either

$$x + y + (x - y + 2)i = 0 \tag{1}$$

or

$$x - 2y + (x + 2y - 4)i = 0 \tag{2}$$

Writing $0 = 0 + 0i$ and using Definition 13.1c, we see that $a + bi = 0$ if and only if both $a = 0$ and $b = 0$. Thus, if Equation 1 is true, then

$$x + y = 0 \quad \text{and} \quad x - y + 2 = 0$$

This system of equations has the solution $x = -1$, $y = 1$. If Equation 2 is true, then

$$x - 2y = 0 \quad \text{and} \quad x + 2y - 4 = 0$$

a system that has the solution $x = 2$, $y = 1$. Hence, the given equation is true if $x = -1$ and $y = 1$ or if $x = 2$ and $y = 1$.

EXERCISE 13.1

In Problems 1–12, find in the standard $a + bi$ form: (a) the sum $w + z$, (b) the difference $w - z$, (c) the product wz, and (d) the quotient w/z.

1 $w = 3 + \sqrt{-4}$, $z = 5 - \sqrt{-9}$

2 $w = -2 - \sqrt{-25}$, $z = 4 + \sqrt{-16}$

3 $w = 3 + \sqrt{-50}$, $z = 7 + \sqrt{-2}$

4 $w = -4 + \sqrt{-20}$, $z = -3 + \sqrt{-5}$

5 $w = 3 + i$, $z = 2 + 3i$

6 $w = 2 + 3i$, $z = 4 - 5i$

7 $w = -3 + 9i$, $z = -2 + 10i$

8 $w = 4 + 3i$, $z = 2 - 3i$

9 $w = 2 + \sqrt{3}i$, $z = 3 - 3\sqrt{3}i$

10 $w = 3 + \sqrt{5}i$, $z = 4 + 2\sqrt{5}i$

11 $w = \dfrac{3}{2i}$, $z = 3 - 2i$

12 $w = -\dfrac{4}{3i}$, $z = \dfrac{2}{1 + i}$

In Problems 13–18, solve the given quadratic equation.

13 $x^2 + 6x + 10 = 0$

14 $x^2 + 4x + 5 = 0$

15 $2x^2 - 4x + 3 = 0$

16 $2x^2 + 5x + 4 = 0$

17 $3t^2 - 2t + 1 = 0$

18 $5s^2 + 6s + 5 = 0$

In Problems 19–24, simplify the given expression.

19 $i^6 + i^{10}$

20 $i^8 + i^{16}$

21 $(-i)^{14} - i^{22}$

22 $(-i)^{13} + i^{15}$

23 $i^{31} + i^{32} + i^{33}$

24 $-i^{21} + i^{22} + i^{23}$

In Problems 25–36, perform the indicated operations and find the absolute value of the resulting complex number.

25 $(1+i)^2(1-i)$

26 $(1-i)^3(1+i)$

27 $(\sqrt{3}-i)(\sqrt{2}+i)$

28 $(3-\sqrt{3}i)(\sqrt{3}+i)$

29 $\dfrac{4}{1+\sqrt{3}i}$

30 $\dfrac{12}{\sqrt{2}+i}$

31 $\dfrac{8-6i}{1-2i}$

32 $\dfrac{3+2i}{2-3i}$

33 $\dfrac{1-2i}{1+2i}+\dfrac{1+2i}{1-2i}$

34 $\dfrac{5-i}{5+i}-\dfrac{5+i}{5-i}$

35 $\left(\dfrac{4i}{1+i}\right)^3$

36 $\left(\dfrac{2i}{3-i}\right)^3$

37 If $P(x)=x^3$, evaluate $P(1+2i)$ and $P(1-2i)$.

38 If $P(x)=x^3$, evaluate $P(\sqrt{3}+i)$ and $P(\sqrt{3}-i)$.

39 If $P(x)=x^3+3x^2-16$, evaluate $P(2-i)$.

40 If $P(x)=x^4-x^2+5$, evaluate $P(1+i)$.

41 Show that $1-\sqrt{2}i$ is a solution of the equation

$$\frac{1}{x-2}-\frac{1}{x}=-\frac{2}{3}$$

42 Show that $-2+i$ is a solution of the equation

$$\frac{5}{x+5}-\frac{5}{x-1}=3$$

In Problems 43–48, find the real values of x and y for which the given equation is true.

43 $(x+y-1)+(x-y+3)i=0$

44 $(2x+y-4)+(x-3y-9)i=0$

45 $(6x-8y)-(5x-3y)i=29-15i$

46 $(5x+3y)-(2x-y)i=11-11i$

47 $[(2x+y)+(x-2y+5)i][(5x+3y)-(x+2y+7)i]=0$

48 $[(3x-4y+24)+(3x-2y)i][(x-2y)+(2x+y-15)i]=0$

49 Show that the absolute value of the ratio of two complex numbers is the ratio of their absolute values; that is

$$\left|\frac{a+bi}{c+di}\right|=\frac{|a+bi|}{|c+di|}$$

Hint: First multiply numerator and denominator of the given fraction by the conjugate of the denominator.

50 Show that $|(a+bi)+(c+di)|\le|a+bi|+|c+di|$.

Hint: Use the graphical interpretation and look at the triangle with its vertices at $(0,0)$, (a,b), and $(a+c, b+d)$.

In Problems 51–54, use your calculator to find the roots of the given equation correct to three decimal places. Keep in mind that if b^2-4ac is negative, then $\sqrt{b^2-4ac}=i\sqrt{4ac-b^2}$.

51 $37x^2+23x+52=0$

52 $19x^2-32x+46=0$

53 $23x^2-48x+103=0$

54 $43x^2+17x+29=0$

USING YOUR KNOWLEDGE 13.1

Bill and Joe found the old map with the message that is pictured here. Of course, they decided to go to the island to see if the treasure was still there. When they reached the island, they were able to find the stone markers but not the tree. Apparently, some long-ago storm had uprooted and blown the tree away. After a fruitless search, Joe was ready to give up and go home, but Bill, being somewhat mathematically inclined, said, "This looks like a nice little complex-number problem. Let me see what I can do with it." And, sure enough, after a short time, Bill announced that he knew where the treasure was and walked straight to the spot where the treasure was buried.

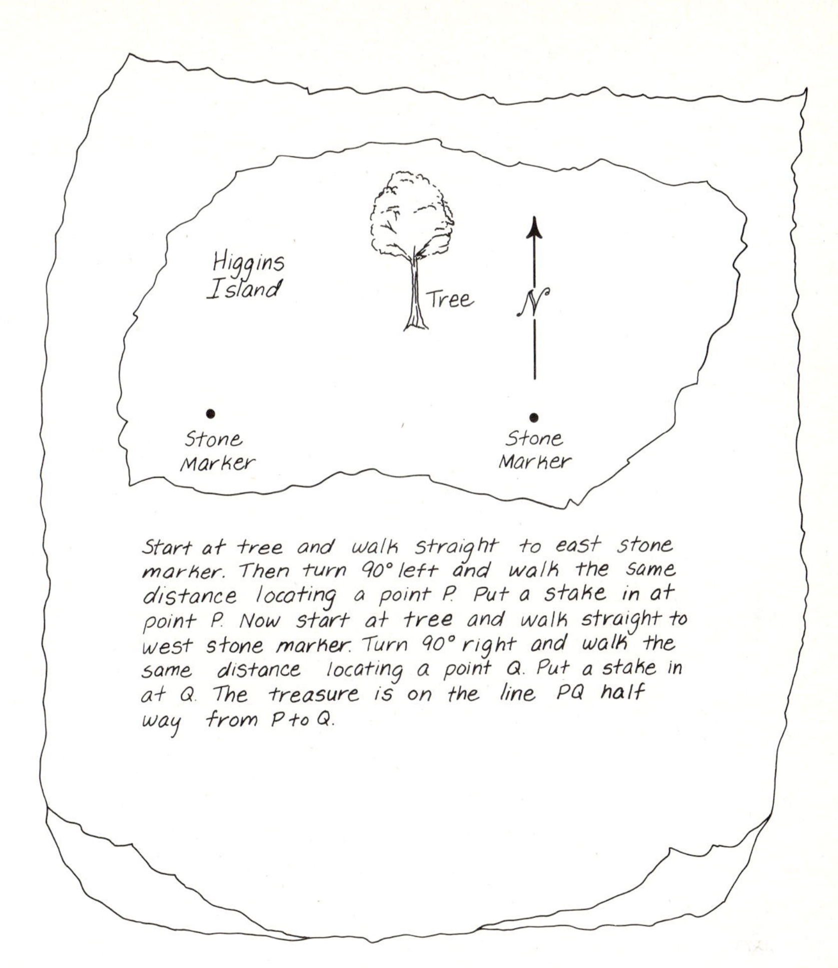

Would you like to stop reading here and see if you can duplicate Bill's feat? Otherwise, you can obtain the solution by following these steps:

1. Choose an x-axis through the stone markers, with the origin midway between them. The y-axis will then run due north and south. Mark the east stone marker $1 + 0i$ and the west stone marker $-1 + 0i$. Suppose that the tree was located at $a + ib$, where a and b are unknown.

2. The vector from the tree to the east stone marker is

$$(1 + 0i) - (a + ib) = (1 - a) - ib$$

Now, recalling Example 9 of this section, you see that you can form the vector from the marker to the point P by multiplying the preceding vector by i. Thus, the vector from the tree to the point P is

$$i[(1 - a) - ib] \quad \text{or} \quad b + (1 - a)i$$

3. Follow the same procedure as in step 2 to form the vector from the tree to the west stone marker and then the vector from the marker to the point Q. Remember that you want to turn clockwise this time.

4. Using the midpoint formula idea, take half the sum of the vectors found in steps 2 and 3 to locate the treasure. Do not be surprised if your answer is completely independent of the numbers a and b.

13.2 THE POLAR FORM OF A COMPLEX NUMBER[1]

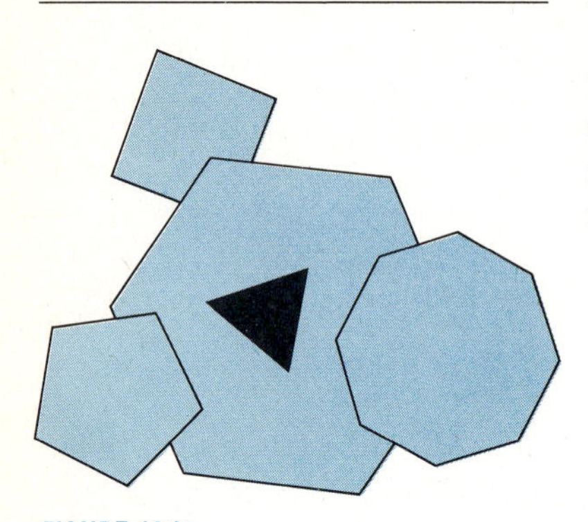

FIGURE 13.2a

Figure 13.2a is made up of five of the regular polygons. Such polygons are of interest in connection with complex numbers because, as we shall see in this section, when the roots of a complex number are displayed graphically, they fall on the vertices of a regular polygon.

RECTANGULAR AND POLAR FORMS

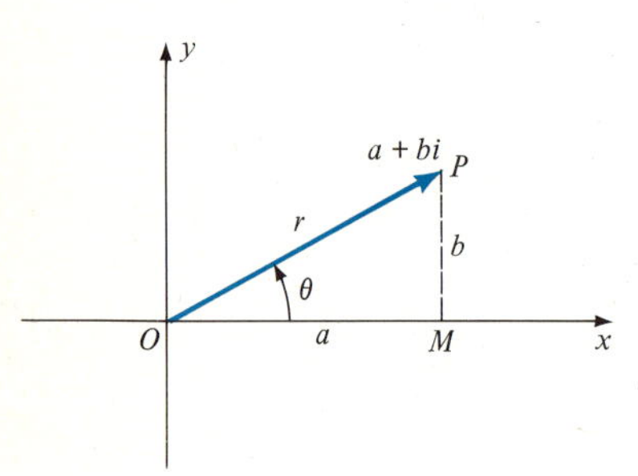

FIGURE 13.2b

In order to work with powers and roots of complex numbers in an efficient manner, we shall define the *trigonometric* or *polar form* of a complex number. The form $a + bi$ is known as the *rectangular form* of the number, because a and b may be taken as the rectangular components of the vector that represents the number. This graphical interpretation makes it easy to obtain the polar form. Figure 13.2b shows the vector representation of the number $a + bi$. The length or magnitude of the vector is labeled r, and we recognize at once that

$$r = \sqrt{a^2 + b^2} \tag{1}$$

is what we have called the *absolute value* of the complex number. Furthermore, the angle θ in the figure is given by

$$\tan \theta = \frac{b}{a} \tag{2}$$

We shall call θ the *angle of the complex number*, but just as the absolute value is often called the *modulus* of the number, the angle of the complex number is often called the *amplitude* or the *argument* of the number.

From the right triangle OMP in Figure 13.2b, we see that

$$a = r \cos \theta \quad \text{and} \quad b = r \sin \theta \tag{3}$$

Because we measure θ from the positive x direction, these two equations are correct regardless of the location of the point P in the plane. Using these equations, we may now write

$$a + bi = r \cos \theta + (r \sin \theta)i$$

[1] This section requires a knowledge of some basic analytic trigonometry.

or

$$a + bi = r(\cos\theta + i\sin\theta)$$

This new form, which displays the absolute value and the angle of the complex number is called the *polar form* of the number. As many authors do, we shall abbreviate the quantity in parentheses by *cis* θ; so that

$$\text{cis}\,\theta = \cos\theta + i\sin\theta$$

and we may write

$$a + bi = r\,\text{cis}\,\theta \tag{4}$$

As we know from trigonometry, the angle θ may be increased or decreased by an integral multiple of 360° without changing the value of either the sine or the cosine of the angle. Consequently, the polar form of a number may be written as

$$r\,\text{cis}(\theta + k\cdot 360^\circ) \tag{5}$$

where k may be zero or any positive or negative integer. This expression for a complex number is called the *complete polar form* of the number.

If a complex number is to be changed from rectangular to polar form, then Equations 1 and 2 may be used. When using Equation 2, we must be careful to determine the proper quadrant for the angle θ. The separate signs of a and b are important in this determination, as you can see from the graphical representation of the number. In order to be safe rather than sorry, it is a good idea to graph the number before changing from rectangular to polar form.

EXAMPLE 1

Write the number $1 - \sqrt{3}i$ in the specified form:

a. polar form

b. complete polar form

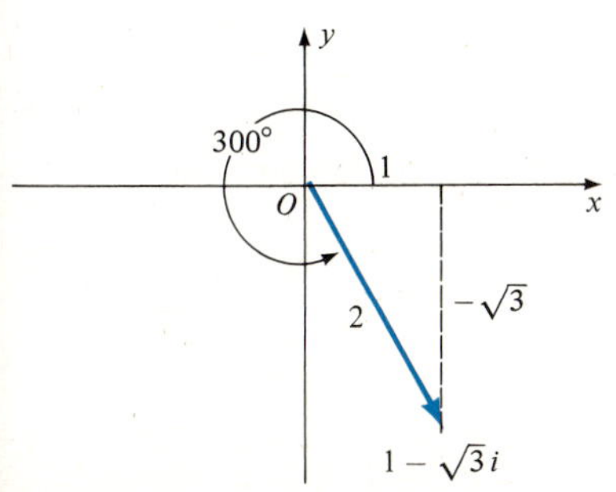

FIGURE 13.2c

SOLUTION:

a. The number $1 - \sqrt{3}i$ is shown graphically in Figure 13.2c. Either from the figure or from Equation 1, we obtain

$$r = \sqrt{(1)^2 + (-\sqrt{3})^2} = \sqrt{1+3} = 2$$

From Equation 2,

$$\tan\theta = \frac{-\sqrt{3}}{1}$$

and because θ must be a fourth-quadrant angle,

$$\theta = 300^\circ$$

Thus, $1 - \sqrt{3}i = 2\,\text{cis}\,300^\circ$.

b. The complete polar form is

$$2\,\text{cis}(300^\circ + k\cdot 360^\circ) \qquad k = 0, \pm 1, \pm 2, \pm 3, \ldots$$

Note that in the polar form, r is never negative and θ is always measured in the counterclockwise direction from the positive x direction. The number zero has absolute value zero, but the angle is ambiguous. We shall have no occasion to use a polar form for zero.

EXAMPLE 2
Write the number -8 in polar form.

SOLUTION:
From Figure 13.2d, we see that $\theta = 180°$ and $r = 8$. Hence,

$$-8 = 8 \text{ cis } 180°$$

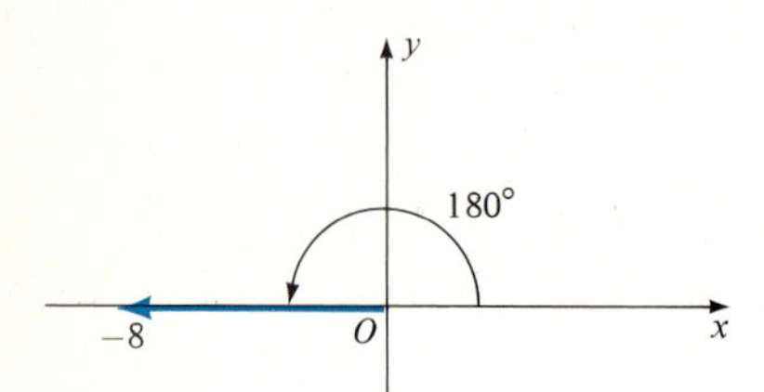

FIGURE 13.2d

EXAMPLE 3
Express the number $-2 + 3i$ in polar form.

SOLUTION:
The vector representing $-2 + 3i$ is shown in Figure 13.2e. The absolute value of the number is

$$r = \sqrt{(-2)^2 + (3)^2} = \sqrt{13}$$

and

$$\tan\theta = \frac{3}{-2} = -1.5$$

Because θ is a second-quadrant angle, $\theta = 123.7°$ (approximately). Thus,

$$-2 + 3i = \sqrt{13} \text{ cis } 123.7°$$

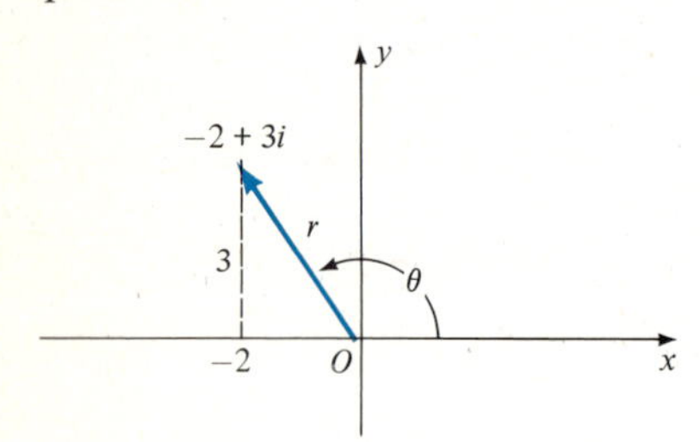

FIGURE 13.2e

EXAMPLE 4
Express the number 6 cis 225° in rectangular form.

SOLUTION:

$$\begin{aligned} 6 \text{ cis } 225° &= 6 \cos 225° + (6 \sin 225°)i \\ &= (6)\left(-\frac{\sqrt{2}}{2}\right) + (6)\left(-\frac{\sqrt{2}}{2}\right)i \\ &= -3\sqrt{2} - 3\sqrt{2}i \end{aligned}$$

MULTIPLICATION AND DIVISION

Multiplication and division are especially easy when the numbers are in polar form. If we have two numbers

$$a_1 + b_1 i = r_1 \text{ cis } \theta_1 \quad \text{and} \quad a_2 + b_2 i = r_2 \text{ cis } \theta_2$$

then direct multiplication gives

$$(r_1 \text{ cis } \theta_1)(r_2 \text{ cis } \theta_2) = r_1 r_2(\cos\theta_1 + i\sin\theta_1)(\cos\theta_2 + i\sin\theta_2)$$

$$= r_1 r_2[(\cos\theta_1 \cos\theta_2 - \sin\theta_1 \sin\theta_2) + i(\sin\theta_1 \cos\theta_2 + \cos\theta_1 \sin\theta_2)]$$
$$= r_1 r_2[\cos(\theta_1 + \theta_2) + i\sin(\theta_1 + \theta_2)]$$
$$= r_1 r_2 \operatorname{cis}(\theta_1 + \theta_2)$$

Notice that we have used the formulas for the cosine and sine of the sum of two angles.

We state the result just obtained in Theorem 13.2a.

▼ THEOREM 13.2a

The absolute value of the product of two numbers is the product of their absolute values, and the angle of the product is the sum of their angles.

$$(r_1 \operatorname{cis}\theta_1)(r_2 \operatorname{cis}\theta_2) = r_1 r_2 \operatorname{cis}(\theta_1 + \theta_2)$$

EXAMPLE 5

Evaluate (3 cis 40°)(2 cis 80°) and give the result in rectangular form.

SOLUTION:

By Theorem 13.2a,

$$\begin{aligned}(3 \operatorname{cis} 40^\circ)(2 \operatorname{cis} 80^\circ) &= (3)(2)\operatorname{cis}(40^\circ + 80^\circ)\\ &= 6 \operatorname{cis} 120^\circ\\ &= 6\cos 120^\circ + (6 \sin 120^\circ)i\\ &= (6)\left(-\frac{1}{2}\right) + (6)\left(\frac{\sqrt{3}}{2}\right)i\\ &= -3 + 3\sqrt{3}i\end{aligned}$$

We can obtain a useful result for the reciprocal of a number by using Theorem 13.2a as follows:

$$\begin{aligned}\operatorname{cis}\theta \operatorname{cis}(-\theta) &= \operatorname{cis}(\theta - \theta)\\ &= \operatorname{cis} 0\\ &= \cos 0 + i \sin 0\\ &= 1\end{aligned}$$

Hence, dividing by cis θ, we obtain

$$\operatorname{cis}(-\theta) = \frac{1}{\operatorname{cis}\theta}$$

and we have Theorem 13.2b.

▼ THEOREM 13.2b

$$\frac{1}{\operatorname{cis}\theta} = \operatorname{cis}(-\theta)$$

With this theorem, we can obtain an efficient method for dividing one complex number by another. Thus, for $r_2 \neq 0$,

$$\begin{aligned}\frac{r_1 \operatorname{cis}\theta_1}{r_2 \operatorname{cis}\theta_2} &= \frac{r_1}{r_2}\operatorname{cis}\theta_1 \frac{1}{\operatorname{cis}\theta_2}\\ &= \frac{r_1}{r_2}\operatorname{cis}\theta_1 \operatorname{cis}(-\theta_2) && \text{Theorem 13.2b}\\ &= \frac{r_1}{r_2}\operatorname{cis}(\theta_1 - \theta_2) && \text{Theorem 13.2a}\end{aligned}$$

We state this result in Theorem 13.2c.

▼ THEOREM 13.2c

The absolute value of the quotient of two complex numbers is the absolute value of the dividend divided by the absolute value of the divisor; the angle of the quotient is the angle of the dividend minus the angle of the divisor.

$$\frac{r_1 \operatorname{cis}\theta_1}{r_2 \operatorname{cis}\theta_2} = \frac{r_1}{r_2}\operatorname{cis}(\theta_1 - \theta_2) \qquad r_2 \neq 0$$

EXAMPLE 6

Express in simplified rectangular form:

$$\frac{4 \text{ cis } 130^\circ}{2 \text{ cis } 40^\circ}$$

SOLUTION:

By Theorem 13.2c,

$$\begin{aligned}\frac{4 \text{ cis } 130^\circ}{2 \text{ cis } 40^\circ} &= \frac{4}{2} \text{ cis}(130^\circ - 40^\circ)\\ &= 2 \text{ cis } 90^\circ\\ &= 2(\cos 90^\circ + i \sin 90^\circ)\\ &= 2i\end{aligned}$$

DE MOIVRE'S THEOREM

Theorem 13.2a leads to some interesting results for integral powers of a complex number. In this theorem, we set $r_1 = r_2 = r$ and $\theta_1 = \theta_2 = \theta$, and obtain

$$(r \text{ cis } \theta)^2 = r^2 \text{ cis } 2\theta$$

If we multiply both sides of this equation by r cis θ, we obtain

$$(r \text{ cis } \theta)^3 = r^3 \text{ cis } 3\theta$$

and we may continue in this manner to obtain the result in Theorem 13.2d for any positive integral exponent.

▼ THEOREM 13.2d

De Moivre's theorem

$$(r \text{ cis } \theta)^n = r^n \text{ cis } n\theta$$

By making use of Theorem 13.2b, we can verify that De Moivre's Theorem holds for negative integral exponents. We let $n = -m$, where m is a positive integer. Then,

$$(r \text{ cis } \theta)^{-m} = r^{-m}(\text{cis } \theta)^{-m} = r^{-m} \cdot \frac{1}{(\text{cis } \theta)^m}$$

Since m is a positive integer, Theorem 13.2d applies to give

$$r^{-m} \cdot \frac{1}{(\text{cis } \theta)^m} = r^{-m} \cdot \frac{1}{\text{cis } m\theta}$$

Now, Theorem 13.2b gives

$$r^{-m} \cdot \frac{1}{\text{cis } m\theta} = r^{-m} \text{ cis}(-m\theta)$$

which completes the verification.

EXAMPLE 7

Evaluate:

a. $(\sqrt{3} + i)^7$

b. $(1 - i)^{-5}$

SOLUTION:

a. We first change the number $\sqrt{3} + i$ into polar form to obtain $\sqrt{3} + i = 2 \text{ cis } 30^\circ$

Then

$$\begin{aligned}(\sqrt{3} + i)^7 &= 2^7 \text{ cis}(7 \cdot 30^\circ) = 2^7 \text{ cis } 210^\circ\\ &= 2^7 \cos 210^\circ + i2^7 \sin 210^\circ\\ &= 2^7\left(-\frac{\sqrt{3}}{2}\right) + i2^7\left(-\frac{1}{2}\right)\end{aligned}$$

$$= -64\sqrt{3} - 64i$$

b. Because $1 - i = \sqrt{2}\ \text{cis}\ 315°$, we find

$$\begin{aligned}(1 - i)^{-5} &= (\sqrt{2})^{-5}\ \text{cis}(-1575°)\\ &= (\sqrt{2})^{-5}\ \text{cis}(225° - 5 \cdot 360°)\\ &= (\sqrt{2})^{-5}\ \text{cis}\ 225°\\ &= \frac{1}{4\sqrt{2}}(\cos 225° + i \sin 225°)\\ &= \frac{1}{4\sqrt{2}}\left(-\frac{1}{\sqrt{2}} - \frac{1}{\sqrt{2}}i\right)\\ &= -\frac{1}{8} - \frac{1}{8}i\end{aligned}$$

ROOTS OF COMPLEX NUMBERS

Having disposed of integral powers of complex numbers, we now turn to roots of complex numbers. First, as for real numbers, we define an nth root of a complex number, $r\ \text{cis}\ \theta$, to be a complex number, $R\ \text{cis}\ \phi$, such that

$$(R\ \text{cis}\ \phi)^n = r\ \text{cis}\ \theta$$

Since n is a positive integer, we may apply De Moivre's Theorem to the left side of this equation to obtain

$$R^n\ \text{cis}\ n\phi = r\ \text{cis}\ \theta$$

or

$$R^n \cos n\phi + iR^n \sin n\phi = r \cos \theta + ir \sin \theta$$

For this equation to be valid, the real parts must be equal, and the imaginary parts must be equal. Thus,

$$R^n \cos n\phi = r \cos \theta \quad \text{and} \quad R^n \sin n\phi = r \sin \theta \tag{6}$$

By squaring and adding corresponding members of these two equations, we obtain

$$R^{2n}(\cos^2 n\phi + \sin^2 n\phi) = r^2(\cos^2 \theta + \sin^2 \theta)$$

or, because $\cos^2 A + \sin^2 A = 1$,

$$R^{2n} = r^2$$

Since R and r are both positive numbers, this equation gives

$$R = r^{1/n} \tag{7}$$

which is to be taken as the principal nth root of r. By substituting from Equation 7 into Equation 6 and dividing by r, we obtain

$$\cos n\phi = \cos\theta \quad \text{and} \quad \sin n\phi = \sin\theta$$

These equations are satisfied if and only if

$$n\phi = \theta + k \cdot 360° \qquad k = 0, \pm 1, \pm 2, \pm 3, \ldots$$

that is,

$$\phi = \frac{\theta}{n} + \frac{k \cdot 360°}{n} \qquad \text{where } k \text{ is any integer} \tag{8}$$

Using Equations 7 and 8, we see that

$$(r \text{ cis } \theta)^{1/n} = r^{1/n} \text{ cis}\left(\frac{\theta}{n} + \frac{k \cdot 360°}{n}\right) \qquad \text{where } k \text{ is any integer} \tag{9}$$

▼ THEOREM 13.2e

Every complex number r cis θ, $r \neq 0$, has exactly n distinct nth roots. These roots all have the same absolute value, the positive number $r^{1/n}$. The angles of these nth roots may be taken, respectively, as

$$\frac{\theta + k \cdot 360°}{n} \qquad k = 0, 1, 2, 3, \ldots, (n-1)$$

It appears from this formula that the values $k = 0, 1, 2, \ldots, (n-1)$ will give n different values of the angle on the right side, differing by integral multiples of $360°/n$. These n values will give different values for the nth root; however, any other value of k will give an angle that differs from one of these angles by an integral multiple of 360°. Because the sine and cosine functions are periodic with period 360°, such a value of k will not give an additional value of the nth root.

We summarize the preceding discussion in Theorem 13.2e.

EXAMPLE 8

Find the four fourth roots of $-8 + 8\sqrt{3}i$ and display them graphically.

SOLUTION:

We first write the given number in polar form:

$$-8 + 8\sqrt{3}i = 16 \text{ cis } 120°$$

Next, we find

$$16^{1/4} = 2, \frac{120°}{4} = 30°, \text{ and } \frac{360°}{4} = 90°$$

Hence, the required roots all have the absolute value 2, and the angles may be taken as

$$30°, 30° + 90°, 30° + 2 \cdot 90°, \text{ and } 30° + 3 \cdot 90°$$

respectively. (We have used $k = 0$, 1, 2, and 3.) The values of k and the four fourth roots are:

$k = 0$: $2 \text{ cis } 30° = \sqrt{3} + i$

$k = 1$: $2 \text{ cis } 120° = -1 + \sqrt{3}i$

$k = 2$: $2 \text{ cis } 210° = -\sqrt{3} - i$

$k = 3$: $2 \text{ cis } 300° = 1 - \sqrt{3}i$

(Notice that $k = 4$ would yield $2 \text{ cis } 390° = 2 \text{ cis } 30°$, a repetition of the result for $k = 0$.)

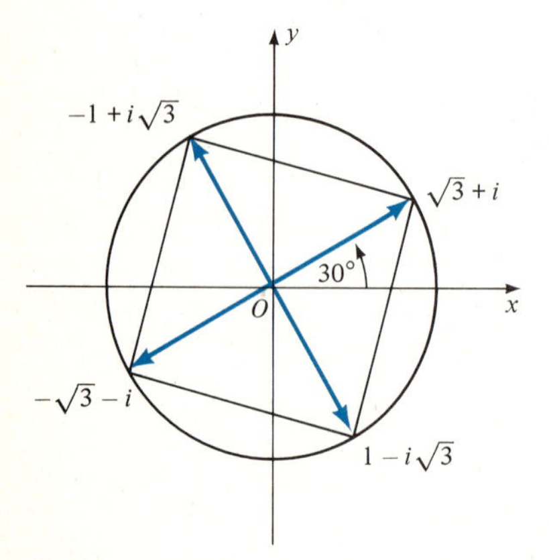

FIGURE 13.2f

The four roots are shown graphically in Figure 13.2f. The four points corresponding to the roots are the vertices of a square inscribed in a circle of radius 2 and with center at the origin.

From the preceding discussion, we can conclude that the points representing the nth roots of r cis θ, $r \neq 0$, are the vertices of a regular polygon of n sides inscribed in a circle of radius $r^{1/n}$, centered at the origin. The angle θ/n is the angle of that one of the roots that has the smallest nonnegative angle.

EXAMPLE 9

Find all the roots of the equation $x^3 + 64i = 0$. Express the roots in both polar and rectangular form.

SOLUTION:

We first write the equation in the form

$$x^3 = -64i$$

which shows that the problem is equivalent to that of finding the three cube roots of $-64i$. Hence, we first rewrite $-64i$ in complete polar form to obtain

$$x^3 = 64 \text{ cis}(270° + k \cdot 360°)$$

Then, the three cube roots are given by

$$\begin{aligned} x_k &= \sqrt[3]{64} \text{ cis } \left(\frac{270° + k \cdot 360°}{3}\right) \\ &= 4 \text{ cis}(90° + k \cdot 120°) \end{aligned}$$

with $k = 0$, 1, and 2. Thus,

$$x_0 = 4 \text{ cis } 90° = 4i$$

$$x_1 = 4 \text{ cis } 210° = 4\left(-\frac{\sqrt{3}}{2} - \frac{1}{2}i\right) = -2\sqrt{3} - 2i$$

$$x_2 = 4 \text{ cis } 330° = 4\left(\frac{\sqrt{3}}{2} - \frac{1}{2}i\right) = 2\sqrt{3} - 2i$$

The easiest way to check these answers is to multiply together the three factors corresponding to the roots:

$$\begin{aligned} &(x - 4i)(x + 2\sqrt{3} + 2i)(x - 2\sqrt{3} + 2i) \\ &\quad = (x - 4i)[(x + 2i)^2 - 12] \\ &\quad = (x - 4i)(x^2 + 4xi - 16) \\ &\quad = x^3 + 4x^2i - 16x - 4x^2i + 16x + 64i \\ &\quad = x^3 + 64i \end{aligned}$$

as required.

EXERCISE 13.2

In Problems 1–10, write the given complex number in the polar and in the complete polar form.

1 $-8 + 0i$
2 $18i$
3 $3\sqrt{2} - 3\sqrt{2}i$
4 $15 - 5i$
5 $10 - 10\sqrt{3}i$
6 $2\sqrt{6} - 2\sqrt{2}i$
7 $4 - 3i$
8 $-6 + 12i$
9 $-12 + 5i$
10 $-15 - 8i$

In Problems 11–20, write the given number in the rectangular form.

11 $6 \text{ cis } 60^\circ$
12 $18 \text{ cis } 135^\circ$
13 $17 \text{ cis } 270^\circ$
14 $41 \text{ cis } 90^\circ$
15 $2\sqrt{3} \text{ cis } 150^\circ$
16 $5\sqrt{2} \text{ cis } 225^\circ$
17 $4 \text{ cis } 70^\circ$
18 $10 \text{ cis } 140^\circ$
19 $8 \text{ cis } 345^\circ$
20 $20 \text{ cis } 235^\circ$

In Problems 21–34, perform the indicated operations and express the result in rectangular form.

21 $(2 \text{ cis } 18^\circ)(6 \text{ cis } 12^\circ)$
22 $(7 \text{ cis } 223^\circ)(5 \text{ cis } 227^\circ)$
23 $\dfrac{12 \text{ cis } 72^\circ}{3 \text{ cis } 42^\circ}$
24 $\dfrac{24 \text{ cis } 154^\circ}{6 \text{ cis } 64^\circ}$
25 $(2 \text{ cis } 30^\circ)^4$
26 $(4 \text{ cis } 10^\circ)^6$
27 $(\frac{1}{2} \text{ cis } 18^\circ)^{-5}$
28 $(\frac{1}{3} \text{ cis } 30^\circ)^{-6}$
29 $(\text{cis } 15^\circ)^{100}$
30 $(\text{cis } 60^\circ)^{50}$
31 $\left(\dfrac{\sqrt{2}}{2} - \dfrac{\sqrt{2}}{2}i\right)^{30}$
32 $\left(-\dfrac{1}{2} + \dfrac{\sqrt{3}}{2}i\right)^{40}$
33 $(\sqrt{3} + i)^5$
34 $(\sqrt{2} - i\sqrt{2})^9$

In Problems 35–48, find all the indicated roots in polar form. Represent the roots on a diagram as in Example 8 of this section.

35 $4 \text{ cis } 120^\circ$, square roots
36 $36 \text{ cis } 300^\circ$, square roots
37 $8 \text{ cis } 135^\circ$, cube roots
38 $27 \text{ cis } 180^\circ$, cube roots
39 $16 \text{ cis } 240^\circ$, fourth roots
40 $81 \text{ cis } 120^\circ$, fourth roots
41 $32 \text{ cis } 225^\circ$, fifth roots
42 $\text{cis } 180^\circ$, sixth roots
43 $-\dfrac{1}{\sqrt{2}} + \dfrac{1}{\sqrt{2}}i$, cube roots
44 -16, fourth roots
45 $-8 - 8\sqrt{3}i$, fourth roots
46 $-2 + 2\sqrt{3}i$, fourth roots
47 $16 - 16\sqrt{3}i$, fifth roots
48 $-1 - i$, fifth roots

In Problems 49–56, find all the roots of the given equation and express these roots in both polar and rectangular form.

49 $x^3 - 27 = 0$
50 $x^3 - 8i = 0$
51 $z^3 + 27i = 0$
52 $z^4 + 4 = 0$
53 $x^5 - 1 = 0$
54 $y^4 + 32 + 32\sqrt{3}i = 0$
55 $y^6 - 64 = 0$
56 $z^5 - 243i = 0$

USING YOUR KNOWLEDGE 13.2

Figure 13.2g shows a diagram of the simplest electric circuit that contains all the basic electrical components:

1. an electromotive force, E, measured in volts;
2. an inductance, L, measured in henrys;
3. a resistance, R, measured in ohms;
4. a capacitance, C, measured in farads.

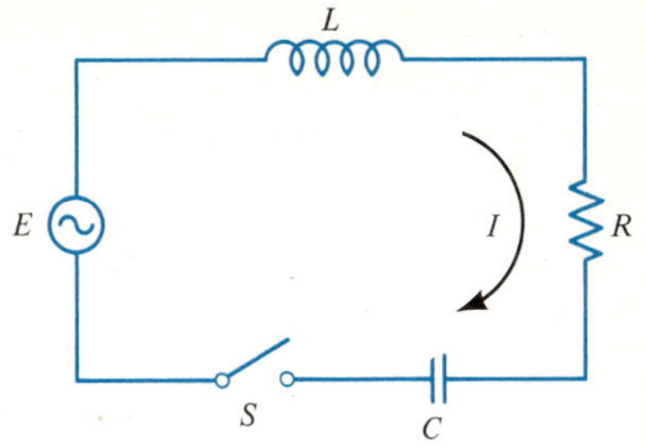

Suppose there is no current flowing and no charge on the capacitor at time $t = 0$, when switch S is closed. Current will flow for $t > 0$, and this current will depend on all the components of the circuit. Let

FIGURE 13.2g

$$E = E_0 \sin kt$$

Then, it is shown in electric circuit theory that if

$$\frac{R^2}{4L^2} - \frac{1}{LC} < 0$$

the steady-state current, I_s (measured in amperes), can be calculated as follows:

1. Write $F(s) = \dfrac{E_0 k}{Ls^2 + Rs + 1/C}$
2. Evaluate $F(ik)$ in the polar form $r \text{ cis } \theta$.
3. The formula for I_s is $I_s = r \cos(kt + \theta)$.

For example, if $L = 0.1$ henry, $R = 10$ ohms, $C = 0.002$ farad, and $E = 75 \sin 100t$, so that $E_0 = 75$ and $k = 100$, then

1. $F(s) = \dfrac{(75)(100)}{0.1s^2 + 10s + 500} = \dfrac{75{,}000}{s^2 + 100s + 5000}$
2. $F(ik) = F(100i) = \dfrac{75{,}000}{-10{,}000 + 10{,}000i + 5000} = \dfrac{15}{-1 + 2i} = \dfrac{15}{\sqrt{5} \text{ cis } \theta} = 3\sqrt{5} \text{ cis}(-\theta)$

   where $\cos \theta = -1/\sqrt{5}$ and $\sin \theta = 2/\sqrt{5}$, so that $\theta \approx 2.034$ (radians).
3. $I_s = 3\sqrt{5} \cos(100t - \theta) \approx 6.708 \cos(100t - 2.034)$

Calculate I_s for the circuit in Figure 13.2g if

C **1** $E = 120 \sin 400t$ C **2** $E = 120 \sin 120\pi t$

13.3 THE REMAINDER THEOREM AND SYNTHETIC DIVISION

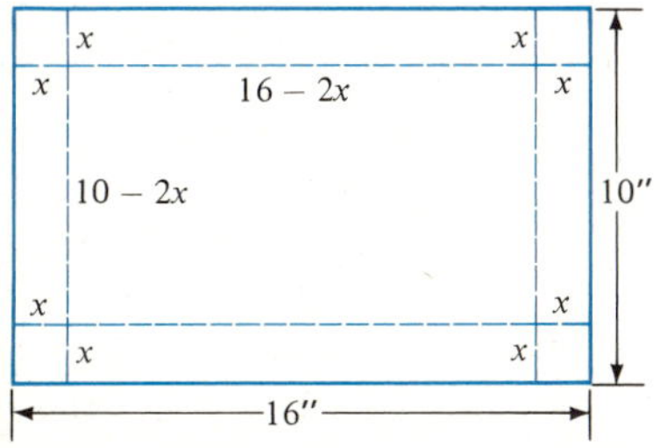

Figure 13.3a is a diagram of a rectangular sheet of tin, 10 inches wide by 16 inches long, that is to be made into a box by cutting equal squares from the four corners and bending up the sides. If the volume of the box is required to be 144 cubic inches, what must be the length of the sides of the cut-out squares?

As you can see from the diagram, if the sides of the squares are x inches long, then the base of the box will be $10 - 2x$ by $16 - 2x$ inches, and the height of the box will be x inches. Hence, the volume of the box, the height times the area of the base, will be

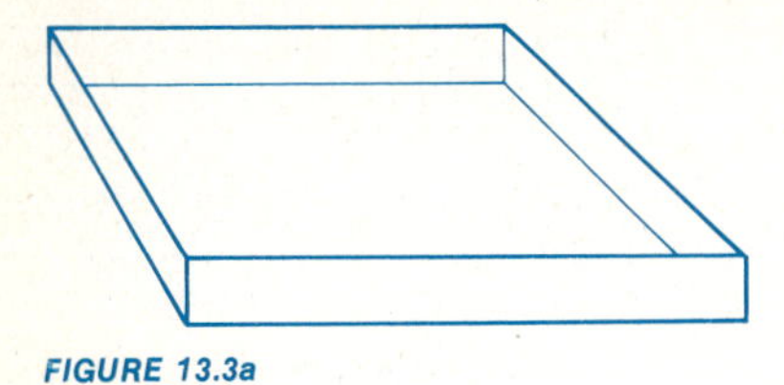

FIGURE 13.3a

$$x(10 - 2x)(16 - 2x) = 160x - 52x^2 + 4x^3$$

cubic inches. Because this volume is required to be 144 cubic inches, x must satisfy the equation

$$160x - 52x^2 + 4x^3 = 144$$

or, if we divide by 4 and rearrange the terms,

$$x^3 - 13x^2 + 40x - 36 = 0$$

POLYNOMIAL EQUATIONS

The last equation is an example of the type of equation that occurs in a great many of the applications of mathematics. The general form of such an equation is

$$a_nx^n + a_{n-1}x^{n-1} + \cdots + a_1x + a_0 = 0 \qquad a_n \neq 0 \tag{1}$$

where n is a positive integer and the a's are constants. The left side of Equation 1 is, of course, an nth-degree polynomial in x. Consequently, the equation is often called simply a *polynomial equation of the nth degree*. As before, we shall use the symbol $P(x)$ to denote a polynomial in x, so that for brevity, Equation 1 may be symbolized by

$$P(x) = 0 \tag{2}$$

For example, the equation in our box problem,

$$x^3 - 13x^2 + 40x - 36 = 0$$

is a cubic (third-degree) equation in x. The coefficients are $a_3 = 1$, $a_2 = -13$, $a_1 = 40$, $a_0 = -36$. Notice that, corresponding to the degree n, there are $n + 1$ coefficients; if powers of x are missing, they are supplied with zero coefficients.

EXAMPLE 1

Identify the degree and state the values of the coefficients of the equation $3x^5 + 4x^3 - 6x + 7 = 0$.

SOLUTION:

The degree is 5. The coefficients are $a_5 = 3$, $a_4 = 0$, $a_3 = 4$, $a_2 = 0$, $a_1 = -6$, and $a_0 = 7$. (Notice the zeros for the missing powers.)

We have previously studied the simple cases where $n = 1$ and $n = 2$ (linear and quadratic equations). Third- and fourth-degree equations are often called *cubic* and *quartic* equations, respectively.

One of the most important problems connected with polynomial equations is that of finding their solutions. We have seen that linear and quadratic equations can be solved by formula, so that we might optimistically hope that higher-degree equations could also be solved by formula. Unfortunately, they cannot. There are formulas for the solution of the general cubic and quartic equations, but they are too complicated to be of much practical use.

It has been proved that no algebraic formulas are possible for the general polynomial equation of degree five or higher. Nevertheless, there are important special cases that can be solved. If the equation has numerical coefficients, there are approximation methods for finding the solutions with sufficient accuracy for applications.

The process of approximating roots (or finding exact roots) of a polynomial equation requires the evaluation of the polynomial for specific values of x. We shall see that this evaluation is related to the division of a polynomial by a binomial of the form $x - c$, where c is a constant. If we divide a polynomial $P(x)$ by a binomial $x - c$ and carry out the division to the point where the remainder is a constant, say R, then we have found a polynomial $Q(x)$, the quotient, such that

$$P(x) = Q(x)(x - c) + R$$

Notice that this equation corresponds exactly to the statement we make for an arithmetic division: *The dividend is equal to the quotient times the divisor plus the remainder*. For example, if we divide $P(x) = 3x^3 + 5x^2 - 7x + 10$ by $x - 2$, we obtain the quotient $Q(x) = 3x^2 + 11x + 15$ and the remainder $R = 40$. (See the illustration that follows Theorem 13.3b below.) To check this division, you would show that

$$3x^3 + 5x^2 - 7x + 10 = (3x^2 + 11x + 15)(x - 2) + 40$$

that is, $P(x) = Q(x)(x - c) + R$. It is an important theorem that, given $P(x)$ and $x - c$, the polynomial $Q(x)$ and the constant R are unique. We state Theorem 13.3a without proof.

▼ THEOREM 13.3a

Given a polynomial $P(x)$ and a binomial $x - c$, where c is a constant, there exists a unique polynomial $Q(x)$ and a unique constant R, such that

$$P(x) = Q(x)(x - c) + R$$

If we substitute $x = c$ in the equation

$$P(x) = Q(x)(x - c) + R$$

we obtain

$$P(c) = Q(c)(c - c) + R$$

or, because $Q(c)$ is simply a number and $c - c = 0$,

$$P(c) = R$$

This result is the substance of the important Remainder Theorem, which is stated in Theorem 13.3b.

▼ THEOREM 13.3b

The remainder theorem

If a polynomial $P(x)$ is divided by a binomial $x - c$ and the division is carried to the point where the remainder does not involve x, then the remainder is the value of the polynomial when x is replaced by c. If R is the remainder, then

$$R = P(c)$$

To illustrate this theorem, let us divide the polynomial

$$P(x) = 3x^3 + 5x^2 - 7x + 10$$

by $x - 2$, using ordinary long division:

$$
\begin{array}{rrrrrrl}
 & & 3x^2 & + 11x & + 15 & & \text{(Quotient)} \\
\text{(Divisor)} & x-2 \,\big|\, & 3x^3 + & 5x^2 - & 7x + 10 & & \text{(Dividend)} \\
 & & 3x^3 - & 6x^2 & & & \\
\hline
 & & & 11x^2 & & & \\
 & & & 11x^2 - & 22x & & \\
\hline
 & & & & 15x & & \\
 & & & & 15x - & 30 & \\
\hline
 & & & & & +40 & \text{(Remainder)}
\end{array}
$$

Furthermore,

$$
\begin{aligned}
P(2) &= 3(2^3) + 5(2^2) - 7(2) + 10 \\
&= 24 + 20 - 14 + 10 = +40
\end{aligned}
$$

Thus, as we expected, $P(2)$ is the remainder when the polynomial is divided by $x - 2$.

SYNTHETIC DIVISION

In order to make the Remainder Theorem more useful, we need an efficient way of doing the division that it involves. If we rewrite the preceding long division and omit the powers of x, writing only the coefficients in their proper positions, we have the same division appearing next. Notice the repetition of the starred numbers in the first three columns.

$$
\begin{array}{rrrrrrl}
 & & 3^* & +11^* & +15^* & & \text{(Quotient)} \\
\text{(Divisor)} & 1-2 \,\big|\, & 3^* & +5 & -7 & +10 & \text{(Dividend)} \\
 & & 3^* & -6 & & & \\
\hline
 & & & +11^* & & & \\
 & & & +11^* & -22 & & \\
\hline
 & & & & +15^* & & \\
 & & & & +15^* & -30 & \\
\hline
 & & & & & +40 & \text{(Remainder)}
\end{array}
$$

We may take advantage of the fact that the first coefficient of the divisor is 1 by agreeing to keep this in mind and not writing it. Then, we can obtain the next simple form if we omit the unnecessary repetitions and write the preceding division in three lines:

$$
\begin{array}{r|rrrr}
-2 & 3 & +5 & -7 & +10 \\
 & & -6 & -22 & -30 \\
\hline
 & \mathbf{3} & \mathbf{+11} & \mathbf{+15} & \mathbf{+40}
\end{array}
$$

The boldface numbers are the important items in the result. The numbers 3, 11, and 15 are the coefficients of the quotient, and 40 is the remainder. It is convenient to make one more slight modification. To obtain the third line in the preceding form, we subtracted the second line from the first. We replace this subtraction with an addition by changing the sign of the indicated divisor; that is, we use $+2$ in place of -2, so that we have only to add at each step. The final form of the division is displayed next. This abbreviated form of division is called *synthetic division.*

$$\begin{array}{r|rrrr} 2 & 3 & +5 & -7 & +10 \\ & & +6 & +22 & +30 \\ \hline & 3 & +11 & +15 & +40 \end{array} \tag{3}$$

In summary, the process of synthetic division, which is used in dividing a polynomial $P(x)$ by a binomial $x - c$, may be carried out in the following steps.

Procedure for synthetic division

1. Write down the coefficients of the dividend, including signs, in order of descending powers of x. Be sure to supply a zero for each missing power.
2. If the divisor is $x - c$, write c for the indicated divisor. Next, write the first coefficient of the dividend in the first place in the third line.
3. Follow the arrows in 3. Each number at an arrowhead is obtained by multiplying the number at the tail by the indicated divisor. Except for the first coefficient, each number at the tail is obtained by adding the two numbers above it.
4. When the division is completed, the last number in the third line is the remainder. The other numbers in this line are the coefficients of the quotient in order of descending powers of x. Obviously, the highest power of x in the quotient is one less than that in the dividend.

EXAMPLE 2

Divide $2x^4 - 3x^2 + 5x - 7$ by $x + 3$.

SOLUTION:

The divisor in this example must be regarded as $x - (-3)$ to obtain the correct indicated divisor. The zero in the first line of the division is put in place of the missing x^3 term. You should follow through the steps of the procedure previously outlined to check the following division.

$$\begin{array}{r|rrrrr} -3 & 2 & 0 & -3 & +5 & -7 \\ & & -6 & +18 & -45 & +120 \\ \hline & 2 & -6 & +15 & -40 & +113 \end{array}$$

The quotient is thus $2x^3 - 6x^2 + 15x - 40$, and the remainder is 113.

EXAMPLE 3

If $P(x) = 2x^3 + 5x^2 - 8x + 6$, use synthetic division to find $P(-2)$.

SOLUTION:

By the Remainder Theorem, $P(-2)$ equals the remainder when $P(x)$ is divided by $x - (-2)$. Hence the synthetic division uses the indicated divisor -2. The division is as follows.

$$\begin{array}{r|rrrr} -2 & 2 & +5 & -8 & +6 \\ & & -4 & -2 & +20 \\ \hline & 2 & +1 & -10 & +26 \end{array}$$

Since the remainder is 26, $P(-2) = 26$.

If the remainder, when $P(x)$ is divided by $x - c$, is zero, then $x - c$ is an exact divisor of $P(x)$, that is, $x - c$ is a factor of the polynomial. Also, because the remainder is the value of $P(c)$, we see that $P(c) = 0$. If we write the Remainder Theorem, $P(x) = (x - c)Q(x) + R$, we see at once that $x - c$ is a factor of $P(x)$ if and only if $R = 0$. This important result is stated in Theorem 13.3c.

▼ THEOREM 13.3c

The factor theorem

The polynomial $P(x)$ has $x - c$ as a factor if and only if $P(c) = 0$.

The Factor Theorem is sometimes stated in the form: $x - c$ is a factor of the polynomial $P(x)$ if and only if c is a root (solution) of the equation $P(x) = 0$.

EXAMPLE 4

Use synthetic division to show that 2 is a root of the equation $P(x) = x^3 - 13x^2 + 40x - 36 = 0$, which we found in the box problem at the beginning of this section.

SOLUTION:

We evaluate $P(2)$ by synthetic division as follows.

$$\begin{array}{r|rrrr} 2 & 1 & -13 & +40 & -36 \\ & & +2 & -22 & +36 \\ \hline & 1 & -11 & +18 & 0 \end{array}$$

Because $P(2) = R = 0$, 2 is a root, as was to be shown.

The result of Example 4 shows that the box will have the required volume if the side of the cut-out squares is 2 inches. You should check that this is correct. Are any other answers possible? From the synthetic division in Example 4, we see that $x^3 - 13x^2 + 40x - 36 = (x - 2)(x^2 - 11x + 18)$. Thus, if there are any other solutions, they must come from solutions of the equation

$$x^2 - 11x + 18 = 0$$

This equation factors into

$$(x - 2)(x - 9) = 0$$

so that the roots are 2 and 9. The 2 is a repetition of our previous answer. The 9 is impossible because we cannot cut four 9-inch squares from a sheet of tin that is only 10 inches by 16 inches. Therefore, $x = 2$ is the only possible answer.

Because the remainder in the synthetic division is the value of $P(c)$, another name often used for this procedure is *synthetic substitution.* If you will refer to Using Your Knowledge 2.1, you will recognize the calculator procedure presented there as the synthetic substitution we have discussed here.

Since the theorems of this section have not assumed that the coefficients are real numbers, these theorems hold for imaginary coefficients as well. However, the great majority of applications involve polynomials with real coefficients, and we have limited ourselves to such "real" polynomials, but even in the case of real quadratic equations, we have seen that imaginary roots may occur. These remarks motivate Example 5.

EXAMPLE 5

Show that $1 + i\sqrt{2}$ is a root of the equation $x^3 + 2x^2 - 5x + 12 = 0$.

SOLUTION:

We evaluate $P(1 + i\sqrt{2})$ by synthetic division:

$$\begin{array}{r|rrrr} 1 + i\sqrt{2} & 1 & +2 & -5 & +12 \\ & & 1 + i\sqrt{2} & 1 + i4\sqrt{2} & -12 \\ \hline & 1 & 3 + i\sqrt{2} & -4 + i4\sqrt{2} & 0 \end{array}$$

Since the remainder is zero, $P(1 + i\sqrt{2}) = 0$, so that $1 + i\sqrt{2}$ is a root of the equation, as was to be shown.

EXERCISE 13.3

In Problems 1–4, state the degree and give the value of each coefficient of the given equation.

1 $5x^4 - 7x^3 + 20x - 3 = 0$

2 $\frac{3}{2}x^5 + \frac{1}{2}x^4 - 2x + 5 = 0$

3 $\sqrt{2}x^7 - 3x^5 + 3\sqrt{2}x - 4 = 0$

4 $-\frac{5}{2}x^6 + \frac{1}{2}x^4 - \frac{1}{2}\sqrt{3}x^3 - 7 = 0$

In Problems 5–16, use synthetic division to find the quotient and the remainder.

5 $(x^3 - 8x - 3) \div (x - 3)$

6 $(x^3 - 4x^2 - 25) \div (x - 5)$

7 $(x^3 + 4x^2 - 7x + 5) \div (x - 2)$

8 $(x^3 + 32x + 24) \div (x + 6)$

9 $(2y^4 - 3y^3 + y^2 - 3y - 6) \div (y - 2)$

10 $(y^5 - 4y^3 + 5y^2 - 5) \div (y + 1)$

11 $(x^4 + 4x^3 - 9x^2 - 8x + 14) \div (x - \sqrt{2})$

12 $(2z^4 + z^3 - 11z^2 - 3z + 15) \div (z + \sqrt{3})$

13 $(x^4 - bx^3 + 3b^2x^2 - 7b^3x - 6b^4) \div (x - 2b)$

14 $(2z^4 + 5cz^3 + 6c^2z^2 + 3c^3z - 10c^4) \div (z + 2c)$

15 $(2x^3 + 3x^2 + 2x + 3) \div (x - i) \quad (i^2 = -1)$

16 $(3x^3 - x^2 + 12x - 4) \div (x + 2i) \quad (i^2 = -1)$

In Problems 17–24, show by the Factor Theorem that the binomial (given first) is a factor of the given polynomial.

17 $x + 2;\ x^5 + 32$

18 $x + 1;\ x^7 + 1$

19 $x + 2;\ x^3 - 11x - 14$

20 $y - 4;\ 3y^3 - 10y^2 - 15y + 28$

21 $y + 1;\ y^5 + y^4 + 2y^3 + 5y^2 - 2y - 5$

22 $z - 1;\ 3z^6 - z^5 + 5z^4 - 3z - 4$

23 $z + i;\ z^6 + 1 \quad (i^2 = -1)$

24 $z - 2i;\ z^4 - 16 \quad (i^2 = -1)$

In Problems 25–30, find the indicated values by synthetic division.

25 $P(x) = 5x^3 - 6x^2 + x - 20;\ P(2)$ and $P(3)$

26 $P(x) = 3x^4 - 4x^3 + 10x - 11;\ P(-3)$ and $P(4)$

27 $P(x) = 4x^4 - 27x^3 + 13x - 21;\ P(6)$ and $P(-\frac{1}{2})$

28 $P(x) = 2x^5 + x^4 - 3x^3 + 85x^2 + 18;\ P(\frac{1}{2})$ and $P(-5)$

29 $P(x) = x^3 + 3x^2 - 2x + 1;\ P(2i)$ and $P(-i) \quad (i^2 = -1)$

30 $P(x) = 2x^3 - x^2 + 4x - 3;\ P(1 + i)$ and $P(3i) \quad (i^2 = -1)$

In Problems 31–36, use the given root to find the remaining roots of the given equation.

31 $5x^3 + 18x^2 - x + 28 = 0$; one root is -4

32 $7x^3 - 39x^2 - 26x + 48 = 0$; one root is 6

33 $x^3 + 6x^2 - 6x - 136 = 0$; one root is 4

34 $3x^3 + 13x^2 - 57x - 7 = 0$; one root is -7

35 $10x^3 - 23x^2 + 18x - 9 = 0$; one root is $\frac{3}{2}$

36 $12x^3 + 17x^2 - 6x - 8 = 0$; one root is $-\frac{2}{3}$

USING YOUR KNOWLEDGE 13.3

Use your calculator and the synthetic substitution procedure explained in Using Your Knowledge 2.1 to do the following problems.

C **1** Show that 5.182 is a root of the equation

$$x^3 + 1.818x^2 - 33.274x - 15.546 = 0$$

and find the other two roots, correct to three decimal places.

C **2** Show that -2.563 is a root of the equation

$$x^3 - 1.437x^2 - 4.252x + 15.378 = 0$$

and find the other two roots, correct to three decimal places.

13.4 POLYNOMIAL EQUATIONS

Carl Friedrich Gauss (1777–1855), one of the greatest mathematicians of all time, was the first to put complex numbers on a sound mathemtical basis and, in his doctoral thesis, gave the first proof of the Fundamental Theorem of Algebra. Gauss was only 21 or 22 when he solved this famous problem.

As we have seen in connection with quadratic equations, not every polynomial equation has a real root. If we permit the coefficients of such an equation to be complex numbers in general, then it is far from evident that the equation will have any root. Consequently, the Fundamental Theorem of Algebra, which states that every such equation does have a root, is of basic importance. (See Theorem 13.4a.)

▼ THEOREM 13.4a

The fundamental theorem of algebra

Every polynomial equation, $P(x) = 0$, has at least one root in the set of complex numbers.

The proof of Theorem 13.4a is far beyond the scope of this text, and we shall simply accept its validity. Note that the coefficients of $P(x)$ are not restricted to real numbers; they may be real or imaginary, and in either case, the root may be real or imaginary. The proof of the theorem is of the type known as an "existence" proof; it demonstrates that there is a root but does not show how to find it. On the basis of the Fundamental Theorem, we can derive a second important theorem, the proof of which is left for the problems.

▼ THEOREM 13.4b

Let

$$P(x) = a_n x^n + a_{n-1}x^{n-1} + \cdots + a_0 \qquad a_n \neq 0$$

Then the equation $P(x) = 0$ has exactly n roots, so that $P(x)$ can be written in the form

$$P(x) = a_n(x - r_1)(x - r_2)\cdots(x - r_n)$$

where $r_1, r_2, \ldots, r_n$ are the roots.

It should be noted that Theorem 13.4b does not assume that the roots are all distinct. In fact, two or more of the roots may be equal. If a root occurs twice, it is called a *double* root; if three times, a *triple* root; and if m times, a root of *multiplicity* m. A root that occurs more than once is often said to be a *multiple* root, and a root that occurs exactly once is said to be a *simple* root.

EXAMPLE 1

Form an equation of the fourth degree that has $1 - i$ and $1 + i$ as simple roots and has 2 as a double root.

SOLUTION:

Using Theorem 13.4b, we see that, except for a constant factor, the equation in factored form must be

$$[x - (1 - i)][x - (1 + i)][(x - 2)(x - 2)] = 0$$

Because the left side may be written as

$$[(x - 1)^2 - i^2](x - 2)^2 = (x^2 - 2x + 2)(x^2 - 4x + 4)$$

we can multiply out to find the desired equation:

$$x^4 - 6x^3 + 14x^2 - 16x + 8 = 0$$

EXAMPLE 2

The number $1 + 2i$ and its conjugate $1 - 2i$ are both roots of $x^4 + 2x^3 + x^2 + 12x + 20 = 0$. Find the remaining roots.

SOLUTION:

Since $1 + 2i$ and $1 - 2i$ are both roots of the equation, we know that the polynomial on the left side must be divisible by the product

$$\begin{aligned}[x - (1 + 2i)][x - (1 - 2i)] &= (x - 1)^2 - (2i)^2 \\ &= x^2 - 2x + 5\end{aligned}$$

If we divide the polynomial $x^4+2x^3+x^2+12x+20$ by x^2-2x+5, the result is $x^2 + 4x + 4$. (Details are left for you.) The equation

$$x^2 + 4x + 4 = 0$$

factors into

$$(x + 2)^2 = 0$$

showing that -2 is a double root. This accounts for the four roots that the given equation must have.

▼ THEOREM 13.4c

The rational root theorem

If p/q, a rational number in its lowest terms, is a root of the equation

$$a_nx^n + a_{n-1}x^{n-1} + \cdots + a_1x + a_0 = 0 \qquad a_n \neq 0$$

where the a's are all integers, then p is a factor of a_0, the constant term, and q is a factor of a_n, the coefficient of x^n.

A great many of the polynomial equations that occur in applications have integers for coefficients. It is easy to check such an equation for the presence of rational roots by means of Theorem 13.4c, whose proof is left for the exercises. As an illustration of this theorem, we can check (by synthetic division or by direct substitution) that $\frac{2}{3}$ is a root of the equation

$$3x^3 + 4x^2 + 11x - 10 = 0$$

Note that the numerator of the root is a factor of -10, the constant term, and the denominator is a factor of 3, the coefficient of x^3.

EXAMPLE 3

The equation $2x^3 + x^2 - 2x - 6 = 0$ has a rational root. Find this root, and then find the remaining roots.

SOLUTION:

If p/q is a rational root, then Theorem 13.4c tells us that p must be a factor of -6 and q must be a factor of 2. We list the possibilities:

$$p\!: \ \pm1, \pm2, \pm3, \pm6 \qquad\qquad q\!: \ 1, 2$$

$$\frac{p}{q}\!: \ \pm\tfrac{1}{2}, \pm1, \pm\tfrac{3}{2}, \pm2, \pm3, \pm6$$

It is not necessary to list both positive and negative values of q because no additional values of p/q would be obtained. The values of p/q are listed in order of increasing magnitude because it is usually simpler to test them in this order.

You may show by synthetic division that $\pm\frac{1}{2}$ and ±1 are not roots of the equation. For our trial of $\frac{3}{2}$, we have the division

$$\begin{array}{r|rrrr} \frac{3}{2} & 2 & +1 & -2 & -6 \\ & & +3 & +6 & +6 \\ \hline & 2 & +4 & +4 & 0 \end{array}$$

The zero remainder shows that $\frac{3}{2}$ is a root and $x - \frac{3}{2}$ is a factor of the left side of the equation. The second factor is $2x^2 + 4x + 4$, and the remaining roots are found from the equation $2x^2 + 4x + 4 = 0$ or

$$x^2 + 2x + 2 = 0$$

The quadratic formula applied to this equation gives $x = -1 + i$ and $x = -1 - i$, so that the three roots of the original equation are

$\frac{3}{2}$, $-1 + i$, and $-1 - i$

If a factor corresponding to a root is removed, as in the preceding example, an equation of degree 1 less than that of the given equation is obtained. This new equation is called a *depressed* equation with respect to the original equation.

EXAMPLE 4

Show that the equation $\frac{1}{6}x^3 + \frac{1}{2}x^2 + \frac{7}{6}x + 1 = 0$ has no rational roots.

SOLUTION:

We first multiply the equation by 6, the LCD of the fractional coefficients, to obtain the equivalent equation

$$x^3 + 3x^2 + 7x + 6 = 0$$

Theorem 13.4c may now be applied. Because the coefficient of x^3 is 1, the only possible rational roots are the integral factors of the constant term, 6. Furthermore, if any positive number is substituted for x, each term of the left side will be a positive number, and the sum of these numbers cannot be zero. Hence, the equation can have no positive root. This cuts down the possibilities to be checked to -1, -2, -3, and -6. We can now show by synthetic division that none of these numbers is a root. Therefore, the equation has no rational roots.

The argument whereby positive roots were excluded as possibilities for the equation in Example 4 applies in general. *A polynomial equation with all its coefficients positive numbers can have no positive root.* To apply this statement to negative roots, we replace x by $-x$ in the equation $P(x) = 0$. [Obviously, the roots of $P(-x) = 0$ are the negatives of the roots of $P(x) = 0$.] For instance, if $P(x) = x^3 - 5x^2 + 7x - 13$, then the equation $P(-x) = 0$ is $-x^3 - 5x^2 - 7x - 13 = 0$, or the equivalent $x^3 + 5x^2 + 7x + 13 = 0$, which has no positive roots. Hence, the equation $x^3 - 5x^2 + 7x - 13 = 0$ has no negative roots.

EXAMPLE 5

Show that the equation $P(x) = 0$, where $P(x) = 9x^6 + 8x^4 + x^2 + 4$ has only imaginary roots.

SOLUTION:

Since all the coefficients of $P(x)$ are positive, the equation $P(x) = 0$ can have no positive roots. Furthermore, $P(x)$ contains only even powers of x, so that $P(-x) = P(x)$. Hence, the equation $P(-x) = 0$ can have no positive roots, so that the original equation can have no negative roots. By inspection, we see that $x = 0$ is not a solution. Therefore, all of the roots of the equation must be imaginary.

In Example 4, we showed that $x^3 + 3x^2 + 7x + 6 = 0$ is an equation with no rational roots. Can we determine whether the equation has any real roots? To answer this question, we state Theorem 13.4d, whose proof is left to the problems. It follows immediately from this theorem that *if $P(x)$ is a real polynomial of odd degree, then the equation $P(x) = 0$ has at least one real root.* Hence, the cubic equation

$$x^3 + 3x^2 + 7x + 6 = 0$$

has at least one real root. Of course, we also know now (see Example 4) that this root must be an irrational number.

▼ THEOREM 13.4d

If a polynomial equation, $P(x) = 0$, with real coefficients has an imaginary root $a + bi$ $(b \neq 0)$, then it also has the root $a - bi$; that is, imaginary roots occur only in conjugate pairs.

EXAMPLE 6

It is easy to check by substituting $x = i$, that i is a root of $x^4 - x^3 + 2x^2 - x + 1 = 0$. Find the remaining three roots.

SOLUTION:

Because one root is i, Theorem 13.4d tells us that the conjugate, $-i$, is also a root. Hence, a factor of the left side of the equation is

$$(x - i)(x + i) = x^2 + 1$$

The other factor of the left side may be found by long division to be $x^2 - x + 1$, so that the depressed equation remaining to be solved is

$$x^2 - x + 1 = 0$$

For this equation, the quadratic formula gives the roots $\frac{1}{2} \pm (\sqrt{3}/2)i$. Thus, the required roots are

$$-i, \frac{1}{2} + \frac{\sqrt{3}}{2}i, \text{ and } \frac{1}{2} - \frac{\sqrt{3}}{2}i$$

EXAMPLE 7

Form a polynomial equation of lowest possible degree with real coefficients that has 3 and $2 - i$ for two of its roots.

SOLUTION:

Because the equation is to have real coefficients and $2 - i$ as a root, it must have $2 + i$ as another root. Thus, the required equation is

$$(x - 3)(x - 2 + i)(x - 2 - i) = 0$$

or

$$x^3 - 7x^2 + 17x - 15 = 0$$

Note that we could multiply this equation by any nonzero constant without changing the roots, so that the answer is not unique. The answer could be made unique by requiring the leading coefficient to have some fixed value, say 1.

EXERCISE 13.4

In Problems 1–8, form the polynomial equation of lowest degree that has the given numbers for its roots. Let the coefficient of the highest power be 1.

1 $-2, 1, 3, 4$

2 $-1, 2, 1+\sqrt{2}, 1-\sqrt{2}$

3 $-3, 0, 3, i\sqrt{2}, -i\sqrt{2}$

4 $2, 0, 4, i\sqrt{5}, -i\sqrt{5}$

5 $1+\sqrt{5}, 1-\sqrt{5}$, and -1 as a double root

6 $2+i\sqrt{3}, 2-i\sqrt{3}$, and 3 as a double root

7 -1, and $2i$ and $-2i$ both as double roots

8 $i\sqrt{7}$ and $-i\sqrt{7}$, and $\sqrt{2}$ and $-\sqrt{2}$ both as double roots

In Problems 9–22, find all the roots of the given equation.

9 $x^3 - 4x^2 + x + 6 = 0$

10 $x^3 + x^2 - 22x - 40 = 0$

11 $y^3 + 3y^2 - 5y - 39 = 0$

12 $y^3 - 6y^2 + 13y - 10 = 0$

13 $3x^3 + 19x^2 + 14x - 90 = 0$

14 $2x^3 - 15x^2 + 38x - 30 = 0$

15 $y^4 - 4y^3 - 5y^2 + 36y - 36 = 0$

16 $y^4 - 6y^3 - y^2 + 34y + 8 = 0$

17 $3z^4 - 2z^3 + 2z^2 + 10z + 3 = 0$

18 $12z^4 + 5z^3 + 10z^2 + 5z - 2 = 0$

19 $w^4 - 9w^3 + 16w^2 + 15w + 25 = 0$

20 $w^4 + 4w^3 - 18w^2 - 80w - 32 = 0$

21 $4w^5 + 4w^4 - 5w^3 + 25w^2 - 84w + 36 = 0$

22 $20z^5 - 9z^4 - 74z^3 + 30z^2 + 42z - 9 = 0$

In Problems 23–28, form the polynomial equation of lowest degree with real coefficients and 1 as the coefficient of the highest degree term if two of the roots are the given numbers.

23 $-2, 2+3i$

24 $3, 1-2i$

25 $-2i, 1+2i$

26 $i\sqrt{2}, -i\sqrt{3}$

27 $\sqrt{3}, 1+i$

28 $\sqrt{2}, 2+i$

29 One root of $3x^4 - 16x^3 + 52x^2 - 40x - 39 = 0$ is $2 + 3i$. Find the other roots.

30 One root of $5z^4 + 34z^3 + 40z^2 - 78z + 51 = 0$ is $-4 - i$. Find the other roots.

31 The volume of a rectangular tank is to be made 891 cubic feet larger by adding the same amount to each dimension. If the original dimensions are 4 feet by 6 feet by 7.5 feet, by what amount must these dimensions be increased?

32 A rectangular box is to be formed from a sheet of tin 16 inches wide by 19 inches long by cutting equal squares from each corner and turning up the sides. If the volume is to be 385 cubic inches, what must be the length of the side of the cut-out squares? (There are two answers.)

33 Prove Theorem 13.4b.

Hint: The equation has a root, say r_1, by the Fundamental Theorem, so that $P(x)$ must have a factor $x - r_1$. If the equation is depressed by dividing out this factor, the depressed equation, say $Q_1(x) = 0$, is of degree $n - 1$. The equation $Q_1(x) = 0$ has a root, say r_2, by the same theorem, so that $Q_1(x)$ must have the factor $x - r_2$, and you can depress the equation again. You should be able to continue this reasoning to show that the equation must have at least n roots. Why can't it have more than n roots?

34 Theorem 13.4c may be proved as follows: If p/q is a root of the equation, then

$$a_n\left(\frac{p}{q}\right)^n + a_{n-1}\left(\frac{p}{q}\right)^{n-1} + \cdots + a_1\left(\frac{p}{q}\right) + a_0 = 0 \tag{A}$$

Now, multiply through by q^{n-1} to obtain

$$\frac{a_np^n}{q} + a_{n-1}p^{n-1} + \cdots + a_1pq^{n-2} + a_0q^{n-1} = 0 \tag{B}$$

In Equation B, all the terms except possibly the first are integers. (Why?) But if the first term is not an integer, then Equation B cannot be true. (Why?) Therefore, the first term must also be an integer. This means that q must divide a_np^n. Recall that p and q have no common factors, so that q cannot divide p^n; it must divide a_n. Thus, we have proved that q is a factor of a_n.

Next, return to Equation B and multiply through by q/p. Then apply the same reasoning to show that p is a factor of a_0.

35 If m is a positive integer, an equation of the form

$$a_m x^{2m} + a_{m-1}x^{2m-2} + \cdots + a_1 x^2 + a_0 = 0$$

where the coefficients are all positive numbers, has exactly m pairs of conjugate imaginary roots. Explain why.

36 Theorem 13.4d may be proved as follows: Let $P(x) = 0$ be the equation. Because $a + bi$ is a root, $P(a + bi) = 0$. Form the quadratic polynomial $Q(x)$, where

$$Q(x) = (x - a - bi)(x - a + bi)$$
$$= x^2 - 2ax + a^2 + b^2$$

Now, $Q(x)$ is a real polynomial. (Why?) If we divide $P(x)$ by $Q(x)$, carrying the division to the point where the remainder is of degree less than 2, the result is of the form

$$P(x) = P_1(x)Q(x) + cx + d$$

where $P_1(x)$ is a real polynomial of degree $n - 2$, and c and d are real numbers. Since both $P(a + bi) = 0$ and $Q(a + bi) = 0$, it follows, on substituting $x = a + bi$, that

$$0 = 0 + c(a + bi) + d$$

or

$$(ac + d) + bci = 0$$

Therefore, $ac + d = 0$ and $bc = 0$. Now, recall that $b \neq 0$, so that $c = 0$, and if $c = 0$, then $d = 0$. This shows that

$$P(x) = P_1(x)Q(x)$$

You should be able to complete the proof by substituting $x = a - bi$ and recalling how $Q(x)$ was formed.

37 There is a theorem similar to Theorem 13.4d that is as follows: If a polynomial equation with *rational* coefficients has a root $a + b\sqrt{c}$, where a, b, and c are rational numbers and c is not a perfect square, then $a - b\sqrt{c}$ is also a root. The proof can be constructed along the same lines as the proof given in Problem 36. Construct this proof.

38 One root of the equation $x^4 - 4x^3 - 2x^2 + 12x - 3 = 0$ is $2 + \sqrt{3}$. Find the other roots.

Hint: See Problem 37.

39 One root of the equation $x^4 + x^3 - x^2 - 2x - 2 = 0$ is $\sqrt{2}$. Find the other roots.

Hint: See Problem 37.

40 One root of the equation $x^4 - 4x^3 - 16x^2 + 4x + 55$ is $4 - \sqrt{5}$. Find the other roots.

Hint: See Problem 37.

USING YOUR KNOWLEDGE 13.4

A polynomial $P(x)$ with real coefficients and arranged in order of descending powers of x is said to have a *variation in sign* if two consecutive terms have opposite signs. For example, the polynomial

$$x^5 - 4x^4 - 2x^3 + x - 1$$

has three variations in sign: one from x^5 to $-4x^4$; one from $-2x^3$ to $+x$; and one from $+x$ to -1. Note that some powers of x might be missing. We have already seen in the preceding section that a polynomial equation with real coefficients and with no variations in sign can have no positive roots. The theorem presented here generalizes this idea.

▼

Descartes' rule of signs[2]

The number of positive roots of a polynomial equation, $P(x) = 0$, with real coefficients, is either equal to the number of variations of sign of $P(x)$ or is less than that by an even positive integer.

According to Descartes' Rule of Signs, the equation

$$x^5 - 4x^4 - 2x^3 + x - 1 = 0$$

cannot have more than three positive roots because the polynomial has only three variations in sign. The rule does not guarantee that there are three positive roots but says that there may be three or one; further examination of the equation would be needed to decide which is the case. However, notice that there must be at least one positive root. We can see at once that if a polynomial equation with real coefficients has an odd number of variations in sign, then it must have at least one positive root.

Information about negative roots can be obtained by applying the Rule of Signs to the polynomial $P(-x)$. Thus, if $P(x)$ is the polynomial in the preceding illustration, then

$$P(-x) = -x^5 - 4x^4 + 2x^3 - x - 1$$

which has two variations in sign. Therefore, the equation $P(x) = 0$ can have no more than two negative roots, but it might have none.

The information we have obtained about the equation

$$x^5 - 4x^4 - 2x^3 + x - 1 = 0$$

can be conveniently displayed as in the accompanying table. In any row showing a possible grouping of roots, the total number of roots is five, the degree of the equation; in the column for imaginary roots, the entry is zero or an even positive integer because such roots must occur in conjugate pairs.

+	−	i
3	2	0
1	2	2
3	0	2
1	0	4

Apply the Rule of Signs to each of the following equations, and state, in tabular form, conclusions as to the type of permissible roots.

1 $2y^3 + 3y^2 + 4 = 0$ **2** $6x^3 - 7x^2 - 1 = 0$ **3** $z^4 - z^2 + 5z - 4 = 0$ **4** $8z^4 - 9z^2 + 2 = 0$
5 $x^6 - 2x^5 + 4x^3 + 9 = 0$ **6** $x^7 + x^6 - x^5 + x^2 - 3x - 5 = 0$
7 $x^{2m+1} - a^{2m+1} = 0$, m a positive integer, $a > 0$ **8** $x^{2m} - a^{2m} = 0$, m a positive integer, $a > 0$

13.5 APPROXIMATING THE ROOTS

There is a legend that the inhabitants of the Greek island of Delos were once told by their oracle that to rid themselves of a certain plague they must double the size of the altar in their Temple of Apollo. They interpreted this to mean that they must double the volume of the altar, which was a cube, and the problem attracted a great many Greek mathematicians. During the years 400–200 B.C., a number of these mathematicians were able to devise geometrical methods for constructing a cube with volume twice that of a given cube.

[2] A proof of this theorem can be found in *Algebra for College Students* by Jack R. Britton and L. Clifton Snively (New York: Holt, Rinehart and Winston, 1954), pp. 375–377.

Let's see what is involved in the problem of Delos. If the original altar is a units on an edge, then its volume is a^3; if the edge of the new altar is x units, then its volume is x^3; and we require that $x^3 = 2a^3$, which is an equation with the real solution $x = \sqrt[3]{2}a$. Now we can use tables, or logarithms, or computers to find the cube root of 2 to any reasonable number of decimal places, and, for all practical purposes, the problem is solved. What bothered the ancient Greeks? First, they were not algebraists, and, second, they were not interested in an approximate solution; they wanted an *exact* geometrical construction of the cube root of 2, a construction that is impossible with straightedge and compass alone. In this section, we shall be satisfied with approximate solutions.

Suppose that, instead of being a cube, the altar was 2 feet wide, 6 feet long, and 3 feet high and that the Greeks wanted to double the volume by increasing each edge by the same amount. If x feet is the amount of the increase, then the new dimensions are $\mathbf{2 + x}$, $\mathbf{6 + x}$, and $\mathbf{3 + x}$ feet, and, if the volume is doubled, then

$$(\mathbf{2 + x})(\mathbf{6 + x})(\mathbf{3 + x}) = 2(2 \cdot 6 \cdot 3)$$

or, in simplified form,

$$x^3 + 11x^2 + 36x - 36 = 0$$

Using the methods of Section 13.4, you can show that this equation has no rational roots. We know intuitively that the equation has at least one real positive root. (If you read Using Your Knowledge 13.4, you can prove that the equation has exactly one positive root.) Let us try to approximate this root.

Let $P(x) = x^3 + 11x^2 + 36x - 36$. Then $P(0) = -36$ and $P(1) = 12$. This change of sign from $P(0)$ to $P(1)$ is of particular interest to us for the following reason: It can be shown by the methods of calculus that the graph of a polynomial function is a continuous curve; it has no breaks and no missing points. Thus, if we are to join the point $(0, -36)$ to the point $(1, 12)$ with the graph of such a function, then we must cross the x-axis at least once between $x = 0$ and $x = 1$. This shows that the equation $P(x) = 0$ has a root between $x = 0$ and $x = 1$. A piece of the graph of $y = P(x)$ is shown in Figure 13.5a on the next page.

We shall use the method of *linear interpolation*, where we replace the graph of the polynomial function by a straight line joining the two points $(0, -36)$ and $(1, 12)$. In Figure 13.5a, we have labeled the points (x_1, y_1) and (x_2, y_2) in order to be able to obtain a general formula for use with other equations. We make use of the similar triangles ABC and AHO to find h, the amount to be added to x_1 to arrive at the intersection H of the line AB and the x-axis. Usually, $x_1 + h$ is a better approximation to the root r than either x_1 or x_2. Because corresponding sides of similar triangles are proportional, we have

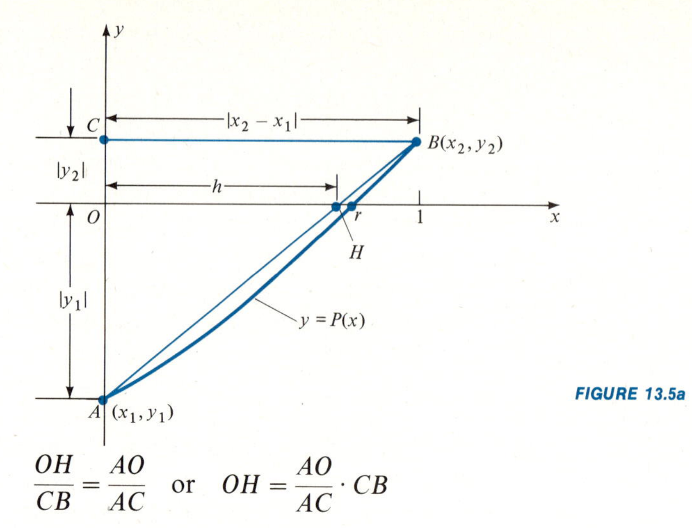

FIGURE 13.5a

$$\frac{OH}{CB} = \frac{AO}{AC} \quad \text{or} \quad OH = \frac{AO}{AC} \cdot CB$$

or, in terms of the coordinates,

$$h = \frac{|y_1|}{|y_1| + |y_2|} |x_2 - x_1| \tag{1}$$

In using this *correction formula*, we always proceed from left to right, the left-hand point being taken as (x_1, y_1), and the correction h is added to x_1.

Returning to the specific points $(0, -36)$ and $(1, 12)$, we have $x_1 = 0$, $y_1 = -36$ and $x_2 = 1$, $y_2 = 12$. Hence, $|y_1| = 36$, $|y_2| = 12$, and $|x_2 - x_1| = 1$. Substituting into Formula 1, we find

$$h = \frac{36}{36 + 12} \cdot 1 = 0.75$$

and $x_1 + h = 0 + 0.75 = 0.75$ is the next approximation to the desired root.

Using synthetic division, we obtain $P(0.75) = -2.391$, so that by comparison with $P(1) = 12$, we see that the root is now located between 0.75 and 1. We take another step in the procedure by letting $x_1 = 0.75$, $y_1 = -2.391$ and $x_2 = 1$, $y_2 = 12$. Again, using Formula 1, we obtain

$$h = \frac{2.391}{2.391 + 12}(0.25) = 0.0415$$

and $x_1 + h = 0.7915$.

Next, we find $P(0.7915) = -0.119$, and a third step in the procedure is taken by using $(x_1, y_1) = (0.7915, -0.119)$. This time we obtain

$$h = \frac{0.119}{0.119 + 12}(0.2085) = 0.00205$$

and $x_1 + h = 0.7936$ (rounded to four decimal places).

At this point, we evaluate the polynomial at $x = 0.7936$ and at 0.794, obtaining $P(0.7936) = -0.00278$ and $P(0.794) = +0.0194$. The change of sign tells us that the root is between 0.7936 and 0.794. Therefore, $r = 0.794$, correct to three decimal places. Returning to the original question, this means that the edges of the altar would each be increased by 0.794 feet or about 9.53 inches.

The preceding calculations certainly show the desirability of using a calculator or a computer to approximate the roots of an equation. With the aid of a calculator, the method is quite good and fairly rapid.

EXAMPLE 1

Find, correct to three decimal places, the largest real root of the equation $x^3 - 3x + 1 = 0$. (See Using Your Knowledge 2.1 for the hand calculator method of finding values of the polynomial.)

x	y
-2	-1 *
-1	$+3$
0	$+1$ *
1	-1 *
2	$+3$

SOLUTION:

First, we try to locate the roots between successive integers and therefore construct a table of values as shown. The asterisks indicate changes of sign, showing that the equation has three real roots, one between -2 and -1, one between 0 and 1, and one between 1 and 2. We are asked to approximate the root between 1 and 2, that is, the largest root. In order to lessen the amount of calculation with Formula 1, we evaluate the polynomial for $x = 1.5$, finding $y = -0.125$. Because $y = +3$ for $x = 2$, the root is between 1.5 and 2. We now try to locate the root between successive tenths. Hence, we evaluate the polynomial for $x = 1.6$ and find $y = +0.296$, so that the root is between 1.5 and 1.6. We now proceed to use Formula 1, arranging the work as in Table 13.5a. In the first step, we find $h_1 = 0.03$, so that the next approximation to the root is $1.5 + 0.03 = 1.53$.

TABLE 13.5a

x	y	
1.5	-0.125	$h_1 = \dfrac{0.125}{0.125 + 0.296}(0.1) = 0.03$
1.6	$+0.296$	
1.53	-0.0084	$h_2 = \dfrac{0.0084}{0.0084 + 0.0323}(0.01) = 0.002$
1.54	$+0.0323$	
1.532	-0.00036	$1.532 < r < 1.5325$
1.5325	$+0.00166$	

For $x = 1.53$, we find $y = -0.0084$, showing that the root is between 1.53 and 1.6. Hence, we try to locate the root between successive hundredths. Evaluating the polynomial for $x = 1.54$, we find $y = +0.0323$, so that the root is between 1.53 and 1.54. Next, we use Formula 1 again, obtaining $h_2 = 0.002$, so that the next approximation to the root is 1.532. For $x = 1.532$, we find $y = -0.00036$, which shows that the root is greater than 1.532. To check if 1.532 is correct to three decimal places, we evaluate the polynomial for $x = 1.5325$, obtaining $y = +0.00166$. Thus, $1.532 < r < 1.5325$, showing that the root is 1.532, correct to three decimal places.

The same method can also be used to find the other two roots of the equation in Example 1. Correct to three decimal places, these roots are 0.347 and -1.879. Another way to find these two roots is to use the approximate depressed equation $x^2 + 1.532x - 0.6530 = 0$, which gives the same values when solved by the quadratic formula. The depressed equation is obtained by dividing the given polynomial $x^3 - 3x + 1$ by the factor $x - 1.532$, corresponding to the approximate root 1.532, and discarding the remainder. Although this second scheme has the advantage of yielding a lower-degree equation to be solved for the remaining roots, it has the disadvantage of not being exact. In fact, a small inaccuracy in the first root can, in some cases, cause rather large errors in the remaining roots. In any event, the accuracy of these roots must be checked by substitution into the original equation. For instance, to check the value 0.347, we substitute 0.3465 and 0.3475 into the equation and find the values $+0.0021$ and -0.0002, respectively. The change in sign shows that the root is greater than 0.3465 and less than 0.3475, which guarantees that 0.347 is correct to three decimal places. Note that the magnitude of the two remainders is not involved in this argument; only the change of sign is important.

The linear interpolation method is not limited to polynomial equations. It can be used to solve equations that involve other types of functions just as well; this fact is illustrated next.

EXAMPLE 2

Find, correct to five decimal places, the real root of $3x + 2\log x = 6$.

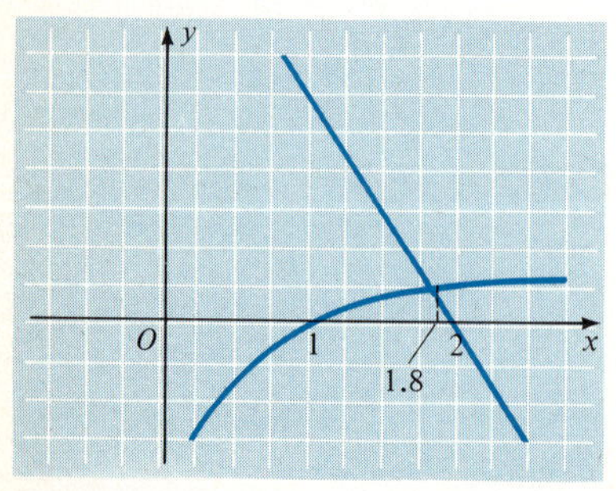

FIGURE 13.5b

SOLUTION:

If we first write the equation in the form $\log x = 3 - \frac{3}{2}x$, we see that the roots will be given by the abscissas of the points of intersection of the graphs of $y = \log x$ and $y = 3 - \frac{3}{2}x$. Figure 13.5b shows that these two graphs intersect at exactly one point whose x-coordinate is approximately 1.8.

We have used a hand calculator to do the computations indicated by the schedule in Table 13.5b, where

$$y = 3x + 2\log x - 6$$

For example, to find y for $x = 1.8$, you can proceed as follows:

Algebraic logic

[1.8] [STO] [×] [3] [+] [2] [×] [RCL] [log] [=] [−] [6] [=] ≈ -0.08945

RPN logic

[1.8] [ENTER] [STO] [3] [×] [RCL] [LOG] [2] [×] [+] [6] [−] ≈ -0.08945

Notice in the second stage of Table 13.5b that adding 0.0258 to 1.8

TABLE 13.5b

x	y	
1.8	-0.08945	$h_1 = \dfrac{(0.08945)(0.1)}{0.08945 + 0.25751} = 0.02578$
1.9	$+0.25751$	
1.8257	-0.00004117	$h_2 = \dfrac{(0.00004117)(0.0001)}{0.00004117 + 0.00030641} = 0.00001184$
1.8258	$+0.00030641$	
1.825711	-0.00000294	Therefore, $r = 1.82571$, correct to five
1.825712	$+0.00000054$	decimal places.

gives the approximation 1.8258 and that the y-value for this value of x is $+0.00030641$, so that we have gone past the root. Therefore, we back up and evaluate y for $x = 1.8257$, finding $y = -0.00004117$. Thus, the root is between 1.8257 and 1.8258. Similarly, in the next stage, where we add $h_2 = 0.000012$ to obtain the approximation 1.825712, we find $y = +0.00000054$, showing that we have gone beyond the root again. Hence, we back up and evaluate y for $x = 1.825711$, obtaining $y = -0.00000294$ and showing that the root is between 1.825711 and 1.825712. Finally, we round off to 1.82571, which is correct to five decimal places, as required.

EXERCISE 13.5

In Problems 1–8, find the indicated root correct to two decimal places.

1 $x^3 + x - 1 = 0$; the positive root

2 $x^3 + x + 1 = 0$; the negative root

3 $x^3 + x^2 + 2x - 5 = 0$; the positive root

4 $x^3 - x^2 - 11 = 0$; the positive root

5 $\sqrt{x} + \sqrt[3]{x} = 1$; the real root

6 $2x + \sqrt[3]{x} = 5$; the real root

7 $x + \log x = 2$; the real root

8 $x^2 + \log x = 4$; the real root

In Problems 9–18, find the indicated root(s) correct to three decimal places.

9 $y^3 + 2y + 28 = 0$; the negative root

10 $y^3 + 3y^2 - 10 = 0$; the positive root

11 $2z^3 - 9z^2 - 17z + 10 = 0$; the larger positive root

12 $5z^3 - 8z^2 - 11z + 15 = 0$; the negative root

13 $x^3 - 6x^2 - x + 23 = 0$; all the roots

14 $x^3 + 5x^2 - 28x - 34 = 0$; all the roots

15 $2x^4 - 5x^3 + 7 = 0$; the larger positive root

16 $2x^4 + 5x^3 - 7 = 0$; the negative root

17 The height of a rectangular box is 1 foot more than its width, and the length is 1 foot more than the height. If the volume is 20 cubic feet, what are the dimensions of the box?

18 A rectangular tank is 2 feet by 3 feet by 5 feet. A second tank is to be constructed with triple the volume by increasing each dimension by the same amount. Find the amount of the increase.

USING YOUR KNOWLEDGE 13.5

If a solid sphere is cut into two pieces by a plane, each piece is called a *segment* of the sphere. If h is the height of the segment and R is the radius of the sphere (see the accompanying diagram), then the formula for the volume of the segment is

$V = \frac{1}{3}\pi(3Rh^2 - h^3)$

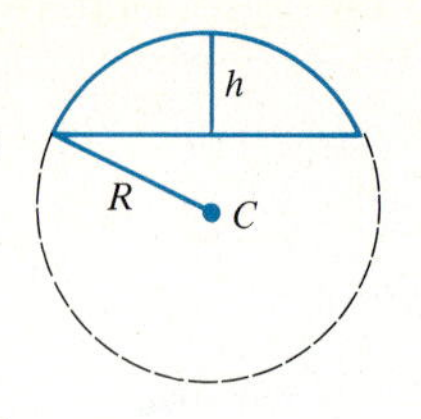

A cross-section through the center of the sphere

Use this formula to set up the following problems. In each problem find the ratio of h to R, correct to three decimal places.

C **1** Find h so that the volume of the segment is one fourth the volume of the sphere. (The volume of the sphere is $\frac{4}{3}\pi R^3$.)

C **2** Find h so that the smaller segment has half the volume of the larger segment into which a plane divides the sphere.

Hint: First find what portion of the volume of the sphere the smaller segment would be. (See Problem 1.)

SELF-TEST

1 Simplify and give the answer in standard $a + bi$ form:

a $\dfrac{5 - 6i}{-3 + 2i}$ **b** $2i^{202} - i^{-203}$

2 If $P(x) = x^3$, find $P(1 - 2i)$ in simplified standard form $a + bi$.

3 If $F(x) = (1/x) - 1/(x - 2)$, find $F(3 - i)$ in the standard $a + bi$ form.

4 If x and y are real numbers and $(x + 1) + (2x - y + 3)i = 0$, find the values of x and y.

5 a Perform the multiplication (5 cis 100°)(4 cis 140°) and then find the rectangular form of the answer.

b Perform the division (8 cis 320°)/(6 cis 170°) and then find the rectangular form of the answer.

6 a Find $(\sqrt{3} - i)^6$ in simplified polar form and then in rectangular form.

b Find $(-1 + i\sqrt{3})^5$ in simplified polar form and in rectangular form.

7 Find, in polar form, the three cube roots of 8 cis 30°.

8 Find the three cube roots of $-8i$ and express your answers in rectangular form.

9 Use synthetic division to find the quotient and the remainder when $3x^4 - x^3 + 5x - 20$ is divided by $x + 2$.

10 If $P(x) = 2x^3 + 3x - 5$, use synthetic substitution to find $P(-6)$.

11 Determine whether or not the binomial $x + 2$ is a factor of $3x^5 + 2x^4 - x^2 - 32x + 6$.

12 Find the value of the polynomial $x^4 - 2x^3 - 7x^2 - 25$ for $x = 5$.

13 If $P(x) = x^3 + x^2 + 2x - 2$, find $P(-1 + 2i)$. Use synthetic division.

14 Find the fourth-degree equation whose leading term is x^4 and that has real coefficients if $2 - i$ is a simple root and -1 is a double root.

15 The equation $x^4 - 2x^3 + 3x^2 - 2x + 2 = 0$ has $1 + i$ as a root. Find the other roots.

16 a An equation with integral coefficients has 2 as its leading coefficient and 4 as its constant term. Is there such an equation with $\frac{1}{2}$ as a root?

b In (a), if the leading coefficient is 3 rather than 2, can the equation have $\frac{1}{2}$ as a root?

17 Find the rational roots of the equation $6x^4 - x^3 + 5x^2 - x - 1 = 0$.

18 a Explain why the equation $x^4 + x^2 + 1 = 0$ can have no real roots.

b What is the least number of real roots that a fifth-degree equation with real coefficients can have?

19 Find, correct to two decimal places, the real root of the equation $x^3 + 5x - 8 = 0$.

20 Find, correct to three decimal places, the real root of the equation $xe^x = 5$.

CHAPTER 14

SEQUENCES AND SERIES

14.1 SEQUENCES

A *sequence* of numbers is simply a set of numbers written in some definite order, so that there is a first term, a second term, a third term, and so on. One of the interesting sequences in mathematics is the sequence of Fibonacci numbers:

$$1, 1, 2, 3, 5, 8, 13, 21, 34, 55, 89, \ldots,$$

which is named after the thirteenth-century mathematician, Leonardo de Pisa, better known as Fibonacci. In Example 1 we shall see how each term of this sequence is obtained.

The Fibonacci numbers occur in many places in nature. For example, the center of a certain variety of daisy displays an interesting double spiraling. Counting shows that there are 21 clockwise and 34 counterclockwise spirals. The numbers 21 and 34 are consecutive terms in the Fibonacci sequence. The Fibonacci numbers have many applications in the natural as well as in the physical sciences and in mathematics. There is even a journal, *The Fibonacci Quarterly*, that you can refer to if you wish to know more about these numbers.

We may regard a sequence of numbers as a special type of function whose domain is the set of natural numbers and whose range is another set of numbers arranged according to some given rule. If the number of elements in the domain is finite, the sequence

▼ **DEFINITION 14.1a**

A *finite sequence* is a function whose domain is the set 1, 2, 3, . . . , *n*, where *n* is a fixed natural number. An *infinite sequence* is a function whose domain is the set of *all* natural numbers.

is called a *finite sequence.* Otherwise, it is called an *infinite sequence.* (See Definition 14.1a.)

Thus, an infinite sequence is simply a function that associates to each natural number n a real number, say $a(n)$. The number $a(n)$ is usually written a_n and is called the *nth term* of the sequence. The terms of a sequence are listed by writing in order

$$a_1, a_2, a_3, \ldots, a_n, \ldots$$

In this sequence, a_1 is the first term, a_2 is the second term, and so on to a_n, the nth or *general* term. The three dots after the nth term indicate that the sequence does not terminate; it is an infinite sequence. Without these three dots, we would understand that the sequence is a finite sequence with n terms. Sometimes, the notation $\{a_n\}$ is used to denote the sequence with the general term a_n, as in the following illustrations.

1. $\{n\} = 1, 2, 3, \ldots, n, \ldots; a_n = n$
2. $\left\{\frac{1}{n}\right\} = 1, \frac{1}{2}, \frac{1}{3}, \ldots, \frac{1}{n}, \ldots; a_n = \frac{1}{n}$
3. $\{a_n\} = 1, \frac{1}{3}, 1, \frac{1}{5}, \ldots, a_n, \ldots; a_n = \begin{cases} 1 & \text{if } n \text{ is odd} \\ 1/(n+1) & \text{if } n \text{ is even} \end{cases}$
4. $\{(-1)^n\} = -1, 1, -1, 1, \ldots, (-1)^n, \ldots; a_n = (-1)^n$

As the last illustration shows, you can obtain alternating signs by using the factor $(-1)^n$ if you wish to start with a minus sign. If you wish to start with a plus sign, you should use the factor $(-1)^{n-1}$.

We can also define a sequence by giving the first few terms and a *recursion* formula that tells us how to form the succeeding terms by using one or more of the preceding ones. This idea is illustrated in Example 1.

EXAMPLE 1

(The Fibonacci Sequence)
Let $a_1 = 1$, $a_2 = 1$, and for $n > 2$, $a_n = a_{n-1} + a_{n-2}$. Find:

a. a_3

b. a_4

SOLUTION:

a. By substituting $n = 3$ in the formula for a_n, we obtain

$$a_3 = a_2 + a_1 = 1 + 1 = 2$$

b. By substituting $n = 4$ in the same formula, we obtain

$$a_4 = a_3 + a_2 = 2 + 1 = 3$$

As you can see from Example 1 and the preceding illustrations, if you are given the general term of a sequence, then you can write any desired number of terms. Is the converse of this true; that is, given n terms of a sequence, can we find the general term? The answer is *no*, as Example 2 shows.

EXAMPLE 2

Find two formulas for the nth term of the sequence whose first four terms are: 1, 3, 5, 7.

SOLUTION:

An easy guess for the next term of the sequence is 9, and $a_n = 2n - 1$ might be the general term. This would give simply the sequence of odd natural numbers. Now, consider the sequence whose general term is

$$b_n = 2n - 1 + (n-1)(n-2)(n-3)(n-4)$$

This gives $b_1 = 1$, $b_2 = 3$, $b_3 = 5$, $b_4 = 7$, but, for $n = 5$, we obtain $b_5 = 9 + (4)(3)(2)(1) = 33$.

Example 2 shows clearly that to specify a sequence unambiguously, the general term must be defined; just giving any finite number of terms is not enough.

EXAMPLE 3

Let $a_n = 6/10^n$.

a. Find the first three terms of this sequence.
Find the *sum*
$a_1 + a_2 + a_3 + \cdots$

SOLUTION:

a. For $n = 1$, we obtain $a_1 = 6/10$ or 0.6.
For $n = 2$, we obtain $a_2 = 6/10^2 = 6/100$ or 0.06.
For $n = 3$, we obtain $a_3 = 6/10^3 = 6/1000$ or 0.006.

b. $a_1 + a_2 + a_3 + \cdots = 0.6 + 0.06 + 0.006 + \cdots$
$= 0.666\ldots$ or $0.\overline{6}$,

which we recognize as the repeating decimal representation of the number $\frac{2}{3}$.

EXERCISE 14.1

In Problems 1–14, find the first four terms and the tenth term of the given sequence.

1 $\{2^n\}$

2 $\left\{\frac{1}{2^n}\right\}$

3 $a_n = 5(-1)^n$

4 $a_n = 1 + (-1)^n$

5 $a_n = \begin{cases} 0 & \text{if } n \text{ is odd} \\ 1 & \text{if } n \text{ is even} \end{cases}$

6 $a_n = \begin{cases} 1 & \text{if } n \text{ is prime} \\ 0 & \text{if } n \text{ is not prime} \end{cases}$

7 $a_n = \frac{1}{8}$

8 $a_n = 1 \cdot 2 \cdot 3 \cdots n$

9 $a_1 = \frac{3}{4}$, and for $n > 1$, $a_n = \frac{1}{2}a_{n-1}$

10 $a_1 = 1$, $a_2 = 2$, and for $n > 2$, $a_n = a_{n-1} - a_{n-2}$

11 $a_1 = 1$, and for $n > 1$, $a_n = 3a_{n-1} - 1$

12 $a_1 = 16$, and for $n > 1$, $a_n = \frac{1}{2}a_{n-1}$

13 $a_1 = 1$, and for $n > 1$, $a_n = ka_{n-1}$

14 $a_1 = \frac{1}{2}$, and for $n > 1$, $a_n = \frac{(-1)^n}{2^n}$

In Problems 15–18, find two distinct formulas for the nth term of a sequence whose first three terms are given.

15 2, 4, 8, . . .

16 $-2, 4, -8, \ldots$

17 $\frac{1}{2}, \frac{1}{4}, \frac{1}{8}, \ldots$

18 $\frac{1}{2}, \frac{2}{3}, \frac{3}{4}, \ldots$

19 In the thirteenth century, Fibonacci wrote a book called the *Liber Abaci* (the Book of Reckoning). In this book, Fibonacci proposed the following problem: How many pairs of rabbits will be produced in a year, beginning with a single pair, if in every month each pair bears a new pair which becomes productive from the second month on and none of the rabbits die?

The entire population in each month can be found by adding the number of pairs in the preceding month to the number of new pairs born. (The number of new pairs equals the number of pairs 2 months before.) The first few terms of this sequence are presented in the accompanying tables.

Month	1	2	3	4	5	...
Pairs of rabbits	1	1	2	3	5	...

a Find the number of pairs of rabbits in the sixth, seventh, and eighth months.
b It can be shown that the nth term of the Fibonacci sequence is

$$a_n = \frac{1}{\sqrt{5}}\left[\left(\frac{1+\sqrt{5}}{2}\right)^n - \left(\frac{1-\sqrt{5}}{2}\right)^n\right] \qquad n \geq 1$$

Verify this formula for $n = 1, 2, 3,$ and 4.

20 If a_n is the nth term of the Fibonacci sequence in Problem 19, show that

$$a_{n+1}a_{n-1} - a_n{}^2 = (-1)^n$$

Hint: Let $x = (1 + \sqrt{5})/2$, $y = (1 - \sqrt{5})/2$. Simplify the left side and use the fact that $xy = -1$ and $x - y = \sqrt{5}$.

USING YOUR KNOWLEDGE 14.1

Do you know that a queen bee lays two kinds of eggs, fertilized eggs that hatch into female bees (workers) and unfertilized eggs that hatch into male bees (drones)? Because the male bee comes from an unfertilized egg, he has only one parent, his mother. This curious fact leads to a peculiar family tree for the male bee. If we use M for male and F for female, then a portion of the tree appears as in the accompanying diagram. Notice that the number of ancestors in each column is a Fibonacci number.

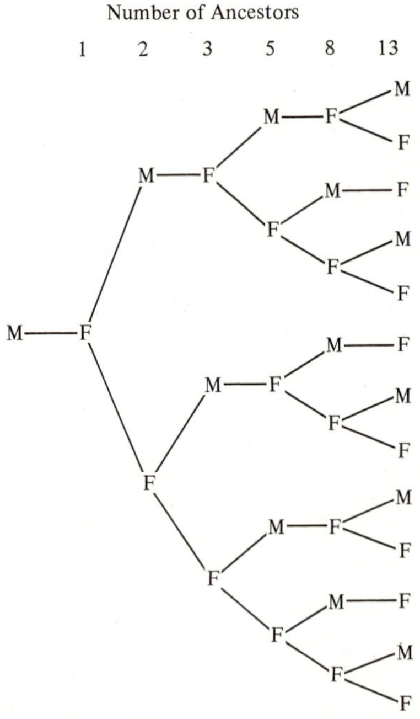

Number of Ancestors
1 2 3 5 8 13

1 Fill in the next set of ancestors in the tree diagram, and check to see if you obtain the next Fibonacci number for the number of ancestors.

2 In Example 1 of this section, we saw that the Fibonacci sequence can be defined as follows:

$$a_1 = 1, a_2 = 1, \text{ and for } n > 2, a_n = a_{n-1} + a_{n-2}$$

If we replace n by $n + 2$, we can rewrite the last formula as

$$a_{n+2} = a_{n+1} + a_n \qquad n > 0 \tag{1}$$

Solving Formula 1 for a_n, we obtain

$$a_n = a_{n+2} - a_{n+1} \tag{2}$$

Thus, $a_1 = a_3 - a_2$, $a_2 = a_4 - a_3$, $a_3 = a_5 - a_4$, $a_4 = a_6 - a_5$, and so on. Now, add the first four terms of the sequence:

$$\begin{aligned} a_4 + a_3 + a_2 + a_1 &= a_6 - a_5 \\ &\quad + a_5 - a_4 \\ &\quad + a_4 - a_3 \\ &\quad + a_3 - a_2 \\ &= a_6 - a_2 \\ &= a_6 - 1 \qquad \text{Because } a_2 = 1 \end{aligned}$$

The same procedure can be used to add the first n terms of the sequence to obtain

$$a_1 + a_2 + a_3 + \cdots + a_n = a_{n+2} - 1$$

Verify this formula by actually adding the terms of the sequence for $n = 4, 5,$ and 6.

14.2 THE SUMMATION NOTATION AND MATHEMATICAL INDUCTION

Here's an amusing anecdote about the noted mathematician Carl Friedrich Gauss, who was mentioned in the previous Chapter. When he was 10 years old, his teacher asked the class to add up all the integers from 1 to 100, expecting to keep them busy for a long time. Gauss instantly scribbled 5050 on his slate, laying it down on his desk with the proud declaration, "There it lies." When the slates were checked, the teacher found that Carl alone had the correct answer. Presumably, Gauss had discovered that each of the pairs of numbers 1 and 100, 2 and 99, 3 and 98, 4 and 97 and so on up to 50 and 51, added up to 101. Thus, the total of 50 pairs must be 50×101, as shown in the diagram.

$$1 + 2 + 3 + \cdots + 50 + 51 + \cdots + 98 + 99 + 100$$

50 pairs of 101's equal $50 \times 101 = 5050$

As you can see, the preceding problem asks for the sum of the terms of the sequence 1, 2, 3, . . . , 100. The indicated sum of the terms of a sequence is called a *series*. We define a series as in Definition 14.2a. According to this definition, there is associated with the finite sequence 1, 2, 3, . . . , 100 the finite series

▼ **DEFINITION 14.2a**

The indicated sum of the terms of a sequence is called a *series*. If the sequence is *finite*, the series is a *finite series*. If the sequence is *infinite*, then the series is an *infinite series*.

$$1 + 2 + 3 + \cdots + 100$$

Similarly, associated with the infinite sequence 1/2, 1/4, 1/8, . . . , $1/2^n$, . . . is the infinite series

$$\frac{1}{2} + \frac{1}{4} + \frac{1}{8} + \cdots + \frac{1}{2^n} + \cdots$$

SUMMATION NOTATION

It is convenient to represent a sum of terms by using the *summation notation*. In this notation, the Greek letter Σ (capital sigma), which corresponds to the English S, indicates that we are to add the given terms. Thus,

$$\sum_{i=1}^{n} a_i \qquad \text{Read "the sum of } a_i \text{ from } i = 1 \text{ to } n\text{"}$$

is defined by

$$\sum_{i=1}^{n} a_i = a_1 + a_2 + a_3 + \cdots + a_n$$

Of course, the sum need not start at $i = 1$ and end at $i = n$. Furthermore, any letter may be used in place of the *index* i. The following examples illustrate some of the possibilities.

$$\sum_{i=3}^{6} a_i = a_3 + a_4 + a_5 + a_6$$

$$\sum_{k=1}^{6} k^2 = 1^2 + 2^2 + 3^2 + 4^2 + 5^2 + 6^2$$

$$\sum_{j=1}^{5} 1 = 1 + 1 + 1 + 1 + 1$$

$$\sum_{r=1}^{4} 6x_r = 6x_1 + 6x_2 + 6x_3 + 6x_4$$

$$\sum_{n=1}^{5} (-1)^n n a_n = -a_1 + 2a_2 - 3a_3 + 4a_4 - 5a_5$$

EXAMPLE 1

Show that the given statements are true:

a. $\sum_{i=1}^{3} a_i + \sum_{i=4}^{8} a_i = \sum_{i=1}^{8} a_i$

b. $\sum_{i=1}^{3} (a_i + b_i) = \sum_{i=1}^{3} a_i + \sum_{i=1}^{3} b_i$

SOLUTION:

a.
$$\begin{aligned}\sum_{i=1}^{3} a_i + \sum_{i=4}^{8} a_i &= (a_1 + a_2 + a_3) + (a_4 + a_5 + a_6 + a_7 + a_8)\\ &= a_1 + a_2 + a_3 + a_4 + a_5 + a_6 + a_7 + a_8\\ &= \sum_{i=1}^{8} a_i\end{aligned}$$

b
$$\begin{aligned}\sum_{i=1}^{3} (a_i + b_i) &= (a_1 + b_1) + (a_2 + b_2) + (a_3 + b_3)\\ &= (a_1 + a_2 + a_3) + (b_1 + b_2 + b_3)\\ &= \sum_{i=1}^{3} a_i + \sum_{i=1}^{3} b_i\end{aligned}$$

The story about Gauss indicates that we can find the sum of the first hundred positive integers by adding the first and the last integers and then multiplying by one half the total number of integers. Let us try to use a similar procedure for the series

$$1 + 2 + 3 + \cdots + (n - 2) + (n - 1) + n$$

Because the sum of the first and last terms is $n + 1$, the sum of the second and the next to the last terms is $(n - 1) + 2 = n + 1$, and so on, the total sum would seem to be

$$\frac{n}{2}(n + 1) = \frac{n(n + 1)}{2}$$

Using the sigma notation, we can write this result as

$$\sum_{i=1}^{n} i = \frac{n(n + 1)}{2} \tag{1}$$

Have we actually proved this result? Of course not. The reasoning is not even correct if n is odd. Can we prove it by checking it for many values of n? No. No matter how many cases we check, there will always be infinitely many cases that we have not checked.

MATHEMATICAL INDUCTION

One way of proving that Equation 1 is true for all natural-number values of n is to use the *principle of mathematical induction.* This principle can be compared to the "domino effect," the toppling of a long row of dominoes standing on edge, one behind another, when the first domino is pushed over. If we assume that there are infinitely many dominoes, then one way of showing that we can topple all of them is to show two things:

1. We can topple the first domino.
2. If any domino (say number k) topples, then the next one, number $(k + 1)$, will also topple.

▼ THEOREM 14.2a

Principle of Mathematical Induction

Let $P(n)$ be a statement that depends on the natural number n. If

1. **the statement is true for $n = 1$, that is, $P(1)$ is true, and**
2. **$P(k)$ implies $P(k + 1)$, that is, if $P(k)$ is true, then $P(k + 1)$ is also true,**

then $P(n)$ is true for all natural numbers n.

These ideas are summarized in Theorem 14.2a, which we assume without formal proof. According to the theorem, to prove that a statement $P(n)$ is true by mathematical induction, you must do two things: (1) Show that the statement is true for $n = 1$, and (2) show that if the statement is true for $n = k$, then it must be true for $n = k + 1$. The next examples illustrate the procedure.

EXAMPLE 2

Prove that

$$\sum_{i=1}^{n} i = \frac{n(n+1)}{2}$$

for all natural numbers n.

SOLUTION:

We shall show the two things required by Theorem 14.2a:

1. For $n = 1$, we have

$$\sum_{i=1}^{1} i = 1 \quad \text{and} \quad \frac{1(1+1)}{2} = 1 \qquad (2)$$

Thus, the statement is true for $n = 1$.

2. We now assume that the statement is true for $n = k$; that is, we assume that

$$1 + 2 + 3 + \cdots + k = \frac{k(k+1)}{2}$$

Next we wish to show that it follows from our assumption that the statement is also true for $n = k + 1$, that is, that

$$1 + 2 + 3 + \cdots + k + (k+1) = \frac{(k+1)(k+2)}{2}$$

Examining the left side of this equation, we see that our assumed formula will sum the first k terms, so that we can write

$$(1 + 2 + 3 + \cdots + k) + \mathbf{(k+1)} = \frac{k(k+1)}{2} + \mathbf{(k+1)}$$

$$= \frac{k(k+1) + 2(k+1)}{2}$$

$$= \frac{(k+1)(k+2)}{2}$$

Thus, we have shown that if the formula is true for $n = k$, then it is also true for $n = k + 1$. This completes the proof by mathematical induction, and the original statement has been shown to be true for all natural numbers n.

EXAMPLE 3

If an amount, P dollars, is deposited in a savings account that pays interest at the rate r, compounded annually, then the amount A_n at the end of n years is given by the formula

$$A_n = P(1 + r)^n$$

Prove that this formula is correct for all natural numbers, n.

SOLUTION:

We again follow the two steps indicated by Theorem 14.2a:

1. If the amount, P dollars, is deposited for 1 year, the interest is Pr, and the amount at the end of the year is

   $$A_1 = P + Pr = P(1 + r)$$

   Thus, the desired formula is correct for $n = 1$.
2. We now assume that the formula is correct for k years; that is, we assume that

   $$A_k = P(1 + r)^k$$

is the correct accumulation at the end of k years. Then, the interest for the next year would be $P(1 + r)^k r$, and the accumulated amount at the end of $k + 1$ years would be

$$\begin{aligned} A_{k+1} &= P(1 + r)^k + P(1 + r)^k r \\ &= P(1 + r)^k(1 + r) \\ &= P(1 + r)^{k+1} \end{aligned}$$

Because this is exactly the desired formula for $n = k + 1$, we have completed the second step and have proved that

$$A_n = P(1 + r)^n$$

is correct for all natural numbers n.

At this point, we must stress the fact that to prove a formula by using the Principle of Mathematical Induction, we must show that requirements 1 and 2 of Theorem 14.2a are *both* satisfied. In the next two examples, we illustrate the consequences when we fail to verify one or the other of the two requirements.

EXAMPLE 4

A prime number is a number that has exactly two distinct divisors, itself and 1. Is the

SOLUTION:

The accompanying table shows that for the values $n = 1$, $n = 2$, and so on, up to $n = 5$, the resulting number is a prime. In fact, we would continue to obtain primes if we were to check even up to expression $n^2 + n + 41$ always prime?

n	1	2	3	4	5
$n^2 + n + 41$	43	47	53	61	71

$n = 10$, or $n = 20$, or $n = 30$. Does this prove that $n^2 + n + 41$ is always a prime? By no means. In fact, for $n = 40$, we obtain

$$\begin{aligned} n^2 + n + 41 &= 40^2 + 40 + 41 \\ &= 40(40 + 1) + 41 \\ &= 40(41) + 41 \\ &= (41)^2 \end{aligned}$$

which is obviously not a prime. In this example, it is not possible to meet the second requirement of Theorem 14.2a.

EXAMPLE 5

Is the formula $2 + 4 + 6 + \cdots + 2n = n(n + 1) + 1$ true for all natural numbers n?

SOLUTION:

Assume that the statement is true for $n = k$, so that

$$2 + 4 + 6 + \cdots + 2k = k(k + 1) + 1 \tag{2}$$

With this assumption, we now prove that it follows that the statement is also true for $n = k + 1$. To do this, we add $2(k + 1)$ to both sides of Equation (2), obtaining

$$\begin{aligned} 2 + 4 + 6 + \cdots + 2k + \mathbf{2(k + 1)} &= k(k + 1) + 1 + \mathbf{2(k + 1)} \\ &= k(k + 1) + 2(k + 1) + 1 \\ &= (k + 1)(k + 2) + 1 \end{aligned}$$

which is exactly the correct form for $n = k + 1$. Thus, if the given statement is true for $n = k$, then it is also true for $n = k + 1$. However, if we try to check the first requirement of Theorem 14.2a by substituting $n = 1$, we obtain

$$2 = (1)(2) + 1$$

or

$$2 = 3$$

which is obviously false. In fact, the proposed formula is false for all natural numbers n.

EXERCISE 14.2

In Problems 1–6, compute the indicated sums.

1 $\sum_{k=1}^{6} k^2$

2 $\sum_{i=1}^{8} i$

3 $\sum_{k=1}^{4} k^3$

4 $\sum_{n=1}^{5} 2n$

5 $\sum_{i=1}^{7} 3$

6 $\sum_{k=3}^{8} 5$

In Problems 7–14, write each expression using the sigma notation.

7 $1 + 2 + 3 + \cdots + 200$

8 $1 + 4 + 9 + 16 + \cdots + 49$

9 $1 + \frac{1}{2} + \frac{1}{3} + \frac{1}{4} + \cdots + \frac{1}{50}$

10 ${x_1}^2 + {x_2}^2 + {x_3}^2 + \cdots + {x_{100}}^2$

11 $1 - 2 + 3 - 4 + 5 - 6 + \cdots - 50$

12 $2 - 4 + 8 - 16 + 32$

13 $1 + 6 + 11 + 16 + 21$

14 $\frac{1}{2} + 1 + \frac{3}{2} + 2 + \frac{5}{2} + 3$

In Problems 15–24, use mathematical induction to prove the given statement.

15 $1 + 3 + 5 + \cdots + (2n - 1) = n^2$

16 $3 + 7 + 11 + \cdots + (4n - 1) = n(2n + 1)$

17 $1 + a + a^2 + \cdots + a^{n-1} = \dfrac{a^n - 1}{a - 1}, a \neq 1$

18 $1^3 + 2^3 + 3^3 + \cdots + n^3 = \dfrac{n^2(n + 1)^2}{4}$

19 $\sum_{i=1}^{n} i^2 = \frac{1}{6}n(n + 1)(2n + 1)$

20 $\sum_{i=1}^{n} 3^i = \frac{3}{2}(3^n - 1)$

21 $\sum_{j=1}^{n} 3j(j + 2) = \frac{1}{2}n(n + 1)(2n + 7)$

22 $\sum_{j=1}^{n} (3j - 1)(3j + 2) = n(3n^2 + 6n + 1)$

23 $\sum_{j=1}^{n} j2^{j-1} = 1 + (n - 1)2^n$

24 $\sum_{j=1}^{n} 4j3^{j-1} = 1 + (2n - 1)3^n$

25 Prove that $x^n - y^n$ is divisible by $x - y$ for all positive integer values of n.

Hint: Write $x^{k+1} - y^{k+1} = (x^{k+1} - xy^k) + (xy^k - y^{k+1})$.

26 Prove that $x^{2n} - y^{2n}$ is exactly divisible by $x + y$.

Note: See Problem 25.

USING YOUR KNOWLEDGE 14.2

In Problem 19 of Exercise 14.1, it was stated that the nth term of the Fibonacci sequence is given by

$$a_n = \frac{1}{\sqrt{5}}\left[\left(\frac{1 + \sqrt{5}}{2}\right)^n - \left(\frac{1 - \sqrt{5}}{2}\right)^n\right] \qquad n \geq 1$$

We are now prepared to prove this by mathematical induction:

1 Show that this formula is true for $n = 1$ and for $n = 2$.

2 Assume that the formula is true for $n = k$ and for $n = k + 1$. (This is slightly different from the hypothesis in step 2 of Theorem 14.2a, but the idea is essentially the same.) For convenience, let

$$x = \frac{1 + \sqrt{5}}{2} \quad \text{and} \quad y = \frac{1 - \sqrt{5}}{2}$$

Then, you will have to use the fact that $a_{k+2} = a_{k+1} + a_k$ to show that if

$$\sqrt{5}a_{k+1} = x^{k+1} - y^{k+1} \quad \text{and} \quad \sqrt{5}a_k = x^k - y^k$$

then

$$\sqrt{5}a_{k+2} = x^{k+2} - y^{k+2}$$

You should also note that $x + 1 = x^2$ and $y + 1 = y^2$.

14.3 ARITHMETIC PROGRESSIONS

When a skydiver begins plunging toward earth, it is interesting to know how far she will fall during each second before she opens her parachute. If we neglect air resistance, a free-falling body falls about 5 meters in the first second, 15 in the next, 25 in the next, and so on. In fact, the number of meters traveled each successive second up to the fifth second is given by the sequence

$$5, 15, 25, 35, 45 \tag{1}$$

The sequence specified in item 1 is an example of the type of sequence that is usually called an "arithmetic progression." An *arithmetic progression* is a sequence in which each term after the first is obtained by adding a number d, called the *common difference*, to the preceding term. Thus, an arithmetic progression can be completely defined by specifying the following numbers:

a_1, the first term;
d, the common difference; and
n, the number of terms.

▼ DEFINITION 14.3a

If a_1 and d are two real numbers, the sequence

$$a_1,\ a_1 + d,\ a_1 + 2d,\ a_1 + 3d,\ \ldots,\ a_1 + (n - 1)d \tag{2}$$

is called an *arithmetic progression of n terms* with the *common difference d*.

In the arithmetic progression defined in item 1,

$a_1 = 5$, because the first term is 5;
$d = 10$, because the difference between successive terms is 10;
$n = 5$, because the sequence has 5 terms.

In general, we define an arithmetic progression as in Definition 14.3a. As you can see from this definition, the nth term of the arithmetic progression 2 is given by

$$a_n = a_1 + (n - 1)d \tag{3}$$

EXAMPLE 1

The sixth and the eleventh terms of an arithmetic progression are 17 and 32, respectively. Find the nth term and the twenty-first term.

SOLUTION:

Using Formula 3 with $n = 6$ and then $n = 11$, we obtain

$$a_6 = a_1 + 5d = 17 \tag{4}$$
$$a_{11} = a_1 + 10d = 32 \tag{5}$$

Solving Equations 4 and 5 simultaneously yields $a_1 = 2$ and $d = 3$. By substituting these values in Formula 3, we find

$$a_n = 2 + (n - 1)(3)$$

or

$$a_n = 3n - 1$$

To find the twenty-first term, we let $n = 21$ in the preceding result to obtain

$$a_{21} = (3)(21) - 1 = 62$$

THE SUM FORMULA

Given an arithmetic progression, we can always find the sum, S_n, of the first n terms. By Definition 14.3a, we have

$$S_n = a_1 + (a_1 + d) + (a_1 + 2d) + \cdots + (a_n - 2d) + (a_n - d) + a_n \quad (6)$$

We arrange the terms on the right side of Equation 6 to obtain

$$S_n = a_n + (a_n - d) + (a_n - 2d) + \cdots + (a_1 + 2d) + (a_1 + d) + a_1 \quad (7)$$

We then add Equations 6 and 7, noticing that d, $2d$, and so on, all drop out, and we have

$$\begin{aligned} 2S_n &= (a_1 + a_n) + (a_1 + a_n) + \cdots + (a_1 + a_n) \qquad n \text{ quantities} \\ &= n(a_1 + a_n) \qquad (8) \end{aligned}$$

We obtain the formula we were seeking by dividing both sides by n:

$$\mathbf{S_n = \frac{n(a_1 + a_n)}{2}} \quad (9)$$

EXAMPLE 2

Find the twentieth term and the sum of the first 20 terms of the arithmetic progression with first term 1 and common difference 9.

SOLUTION:

To find the twentieth term, we substitute $a_1 = 1$, $d = 9$, and $n = 20$ into Formula 3, obtaining

$$a_{20} = 1 + (19)(9) = 172$$

We then substitute $n = 20$, $a_1 = 1$, and $a_{20} = 172$ into Formula 9 to find

$$S_{20} = \frac{20(1 + 172)}{2} = (10)(173) = 1730$$

EXAMPLE 3

A club raffled off a certain item by selling 1000 sealed tickets numbered in succession from 1 to 1000. These tickets were selected at random by purchasers who paid the number of cents equal to the number on the ticket. How much money did the club receive?

SOLUTION:

The number of cents the club received is

$$S_{1000} = 1 + 2 + 3 + \cdots + 1{,}000$$

Because $a_1 = 1$, $a_n = 1{,}000$, and $n = 1{,}000$, we substitute into Formula 9, obtaining

$$\begin{aligned} S_{1000} &= \frac{1{,}000(1 + 1{,}000)}{2} \\ &= (500)(1{,}001) \\ &= 500{,}500 \end{aligned}$$

Expressed in dollars, the amount the club received was \$5,005.

Notice that we could also have done the preceding problem by using the result $n(n + 1)/2$ from Example 2 of Section 14.2. On the other hand, the result of that example is easily obtainable by using the fact that $1, 2, 3, \ldots, n$ is an arithmetic progression.

EXAMPLE 4

How many terms are there in the arithmetic progression for which $a_1 = 3$, $d = 5$, and $S_n = 255$?

SOLUTION:

From Formula 9, we have

$$255 = \frac{n(3 + a_n)}{2}$$

or

$$510 = n(3 + a_n)$$

Because $a_1 = 3$ and $d = 5$, Formula 3 gives

$$a_n = 3 + (n - 1)(5) = 5n - 2$$

Substituting this expression for a_n into the preceding equation, we obtain

$$510 = n(3 + 5n - 2)$$

or

$$5n^2 + n - 510 = 0$$

By factoring, we obtain

$$(5n + 51)(n - 10) = 0$$

which gives $n = -\frac{51}{5}$ or $n = 10$. Since n must be a natural number, the correct solution is $n = 10$.

EXERCISE 14.3

In Problems 1–10, two terms of an arithmetic progression are given. Find a formula for the nth term, a_n, in terms of n.

1 $a_9 = 27, a_{15} = 45$
2 $a_{10} = 20, a_{16} = 32$
3 $a_5 = 19, a_{10} = 34$
4 $a_6 = 17, a_{10} = 33$
5 $a_{10} = -25, a_{15} = -35$
6 $a_4 = -14, a_8 = -26$
7 $a_{11} = 64, a_{29} = 100$
8 $a_{13} = 60, a_{23} = 75$
9 $a_7 = 15, a_{51} = -18$
10 $a_{27} = 238, a_{61} = 408$

In Problems 11–18, find the indicated sum for the given arithmetic progression.

11 S_8: 15, 19, 23, . . .
12 S_{19}: 31, 38, 45, . . .
13 S_{60}: 11, 1, −9, . . .
14 S_{50}: −5; −10, −15, . . .
15 S_{14}: $\frac{3}{10}, \frac{2}{5}, \frac{1}{2}, \ldots$
16 S_{19}: $\frac{3}{4}, \frac{3}{2}, \frac{9}{4}, \ldots$
17 S_{12}: $-5, -4\frac{5}{8}, -4\frac{1}{4}, \ldots$
18 S_{20}: $-2, \frac{1}{4}, \frac{5}{2}, \ldots$

In Problems 19–24, find how many terms there are in the arithmetic progression whose first term, common difference, and sum are given.

19 $a_1 = 2, d = 4, S_n = 200$
20 $a_1 = 4, d = 10, S_n = 174$
21 $a_1 = -3, d = 2, S_n = 12$
22 $a_1 = -5, d = 3, S_n = 110$
23 $a_1 = 15, d = -\frac{3}{2}, S_n = \frac{57}{2}$
24 $a_1 = 6, d = 6, S_n = 330$

25 It is estimated that a certain property, now valued at $40,000, will depreciate as follows: $1,450 the first year, $1,400 the second year, $1,350 the third year, and so on. Based on these estimates, what will be the worth of the property 15 years from now?

26 On a certain construction project, a contractor was penalized for taking more than the contractual time to finish the project. He forfeited \$750 the first day, \$900 the second day, \$1,050 the third day, and so on. How many additional days did he take to complete the project if he paid a penalty of \$12,150?

27 A rack in the form of an equilateral triangle is filled with balls so that there are 11 balls on a side. Another layer of balls is placed on this layer so that there are 10 balls on a side. A third layer, with 9 balls on a side, is then placed on the second layer, and so on. Thus, a pyramid of balls is formed with 1 ball on the top. How many balls are there in the bottom layer? How many balls are there in the pyramid?

Hint: Draw yourself a picture.

28 The sum of three numbers in arithmetic progression is 36. Find the numbers if the sum of their squares is 482.

Hint: Let $x - d$, x, and $x + d$ be the three numbers.

29 A woman is offered a position at a starting salary of \$46,000 per year with annual increases of \$2,500. How much would her total earnings amount to if she were to work 12 years under this salary schedule?

30 How far would a free-falling body fall in 5 seconds if the number of meters traveled in each second is given by Sequence 1 at the beginning of this section?

31 Show that a free-falling body traveling in accordance with Sequence 1 would fall $5n^2$ meters in n seconds.

32 A *harmonic progression* is a sequence of numbers whose reciprocals form an arithmetic progression. For example, $1, \frac{2}{3}$, and $\frac{1}{2}$ are in harmonic progression because their reciprocals $1, \frac{3}{2}$, and 2 are in arithmetic progression. If x^2, y^2, and z^2 are in arithmetic progression, show that $x + y$, $x + z$, and $y + z$ are in harmonic progression.

USING YOUR KNOWLEDGE 14.3

Money is said to earn *simple* interest if no interest is paid on previously earned interest. If a deposit of P dollars is made at the beginning of the year in a bank that pays simple interest at the rate r, expressed as a decimal, then the amount A at the end of the year is given by

$$A = P + Pr \qquad \text{or} \qquad A = P(1 + r) \tag{1}$$

For example, if \$100 is deposited at 6 percent, then at the end of the year, the accumulated amount is

$$\begin{aligned} A &= (100)(1 + 0.06) \\ &= (100)(1.06) \\ &= \$106 \end{aligned}$$

Suppose that P dollars are invested at the beginning of each year in a fund that pays dividends at the rate r, the dividends to be paid out and not reinvested in the fund. (This is equivalent to simple interest because no interest is being paid on the dividends.) How much will have been earned in dividends by the end of the nth year?

At the end of the first year, the dividends are Pr. At the end of the second year, the dividends are $2Pr$ because a second investment of P dollars is made at the beginning of the second year. Thus, the total dividends amount to

$$Pr + 2Pr$$

At the end of the third year, the dividends are $3Pr$, so that the total is

$$Pr + 2Pr + 3Pr$$

and so on. At the end of n years, the total in earned dividends is

$$Pr + 2Pr + \cdots + nPr$$

This is the sum of an arithmetic progression of n terms with $a_1 = Pr$ and $a_n = nPr$. Hence, we know that the sum is

$$S_n = \frac{n(Pr + nPr)}{2} = \frac{n(n+1)}{2} Pr$$

For example, if the amount invested at the beginning of each year is \$1000 and the dividend rate is 10 percent, then the total paid out in dividends at the end of 5 years is

$$S_5 = \frac{(5)(6)}{2}(1000)(0.10) = \$1500$$

1 If you deposit \$100 at the beginning of each month for 1 year in a fund that pays dividends at the rate of 2 percent per month, how much will you have to report to Uncle Sam as dividend income for the year?

2 A man borrows \$75,000 and signs notes agreeing to repay the loan in installments of \$5,000 at the end of 1 year, \$10,000 at the end of 2 years, \$15,000 at the end of 3 years, and so on. He also agrees to pay interest at the rate of 10 percent at the end of each year on the unpaid balance due at the beginning of the year. This means that he will pay interest on \$75,000 at the end of the first year, on \$70,000 at the end of the second year, on \$60,000 at the end of the third year, and so on. How long will it take him to repay the loan and what will be the total amount of interest he will pay?

14.4 GEOMETRIC PROGRESSIONS

There is a certain mirror trick in which a boy's image in one mirror is photographed while the boy holds a second mirror facing the first one, so that his image is reflected again and again, infinitely many times. If the mirrors are properly placed, the successive reflections are smaller and smaller, each one, after the first, being one half the height of the preceding one. Thus, if the height of the first image is h, then the heights of the successive images form the sequence

$$h, \frac{h}{2}, \frac{h}{4}, \frac{h}{8}, \ldots, \frac{h}{2^{n-1}}, \ldots \tag{1}$$

Of course, each term (except the first) of the sequence can be obtained by multiplying the preceding term by $\frac{1}{2}$.

A sequence in which each term after the first can be obtained by multiplying the preceding term by a constant, r, is called a *geometric progression*. Multiplying each term by r to produce the next term results in a fixed ratio between successive terms. This fixed ratio, r, is called the *common ratio* of the progression. Thus, we have Definition 14.4a. A geometric progression is completely defined by specifying the following numbers:

a_1, the first term;
r, the common ratio; and
n, the number of terms.

▼ DEFINITION 14.4a

If a_1 and r are two nonzero real numbers, then the sequence

$$a_1, a_1r, a_1r^2, a_1r^3, \ldots, a_1r^{n-1} \qquad r \neq 1 \tag{2}$$

where each term after the first is obtained by multiplying the preceding term by the *common ratio* r is called a *geometric progression.*

Furthermore, from Sequence 2, we can see that the nth term of a geometric progression is given by the formula

$$a_n = a_1 r^{n-1} \tag{3}$$

EXAMPLE 1

Consider Sequence 1, defined by the mirror trick. Find:

a. a_1
b. r
c. a_n

SOLUTION:

a. a_1 is the first term of the sequence, h.
b. r is the common ratio between successive terms. In this sequence, each term after the first is one half the preceding term, so that $r = \frac{1}{2}$.
c. By Formula 3,

$$a_n = a_1 r^{n-1} = h\left(\frac{1}{2}\right)^{n-1} = \frac{h}{2^{n-1}}$$

THE SUM FORMULA

As for an arithmetic progression, we can also find the sum S_n of the first n terms of a geometric progression. Using Sequence 2, we see that the sum of the first n terms is

$$S_n = a_1 + a_1 r + a_1 r^2 + \cdots + a_1 r^{n-2} + a_1 r^{n-1} \tag{4}$$

Multiplying both sides of this equation by r, we obtain

$$rS_n = a_1 r + a_1 r^2 + a_1 r^3 + \cdots + a_1 r^{n-1} + a_1 r^n \tag{5}$$

If we subtract Equation 5 from Equation 4, we have

$$S_n - rS_n = a_1 - a_1 r^n$$

or

$$S_n(1 - r) = a_1(1 - r^n) \tag{6}$$

If $r \neq 1$, we can divide both sides by $1 - r$ to obtain the desired formula

$$S_n = \frac{a_1(1 - r^n)}{1 - r} \qquad r \neq 1 \tag{7}$$

Of course, if $r = 1$, the progression has all n of its terms equal to a_1, so that $S_n = na_1$.

EXAMPLE 2

Find the tenth term and the sum of the first ten terms of the geometric progression whose first term is 1 and whose common ratio is 2.

SOLUTION:

Using Formula 3, we have

$$a_{10} = a_1 r^{10-1} = (1)(2^9) = 512$$

From Formula 7, we obtain

$$S_{10} = \frac{a_1(1 - r^n)}{1 - r} = \frac{(1)(1 - 2^{10})}{1 - 2} = \frac{-1023}{-1} = 1023$$

EXAMPLE 3

The sum of the first five terms of a geometric progression is $1\frac{3}{8}$, and the common ratio is $-\frac{1}{2}$. What is the progression?

SOLUTION:

From Formula 3, we have

$$\frac{11}{8} = \frac{a_1[1 - (-\frac{1}{2})^5]}{1 - (-\frac{1}{2})}$$

or

$$\frac{11}{8} = \frac{11}{16}a_1$$

Therefore, $a_1 = 2$, and the progression is $2, -1, \frac{1}{2}, -\frac{1}{4}, \frac{1}{8}$.

INFINITE GEOMETRIC PROGRESSIONS

We have already mentioned that in the mirror trick at the beginning of this section, each of the converging reflections is half the size of the preceding one. Can we find out how high all the reflections would reach if stacked vertically? To answer this question we need to find the sum

$$h + \frac{h}{2} + \frac{h}{4} + \cdots + \frac{h}{2^{n-1}} + \cdots \tag{8}$$

The collection of symbols in Series 8 is known as an *infinite geometric series*, where the word "infinite" means that the terms of the series proceed without end. If we examine Equation 7, we can see that if the absolute value of r is less than 1, then the term r^n in the numerator gets smaller and smaller as n gets larger and larger. In fact, we can make r^n become and remain as close to zero as we wish by taking n large enough. Thus, it appears that when $|r| < 1$, we can make S_n become and remain as close to $a_1/(1 - r)$ as we wish by taking n large enough. We express this fact by writing

$$\lim_{n \to \infty} S_n = \frac{a_1}{1 - r} \qquad \text{for } |r| < 1$$

The symbol $\lim_{n \to \infty} S_n$ is read "the limit of S_n as n increases without bound."

In case $|r| > 1$, r^n increases indefinitely in absolute value as n increases, and therefore, the value of S_n cannot be made to become and remain arbitrarily close to any fixed number. If $r = 1$, $S_n = na_1$ again increases indefinitely in absolute value if n increases. Finally, if $r = -1$, $S_n = a_1$ or 0 depending on whether n is odd or even and does not become and remain arbitrarily close to any number as n increases without limit.

A simple geometrical interpretation can be made of the behavior of S_n as n increases without limit if $r = \frac{1}{2}$. We take $a_1 = 1$ for simplicity and consider the series

$$1 + \tfrac{1}{2} + \tfrac{1}{4} + \tfrac{1}{8} + \tfrac{1}{16} + \cdots + (\tfrac{1}{2})^{n-1} + \cdots$$

For this series, we have

$S_1 = 1$

$S_2 = 1 + \frac{1}{2}$

$S_2 = 1 + \frac{1}{2} + \frac{1}{4}$

$S_4 = 1 + \frac{1}{2} + \frac{1}{4} + \frac{1}{8}$

and so on.

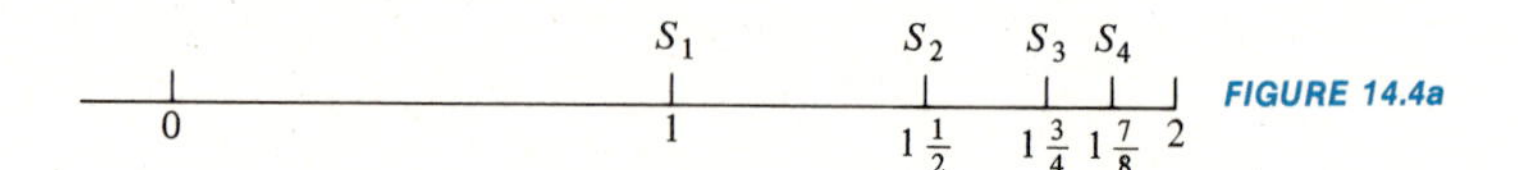

FIGURE 14.4a

In Figure 14.4a, the points corresponding to the first few S's have been marked. We see that each step cuts in half the remaining distance to the point marked 2. Thus, it appears that the value of S_n can be made to become and remain as close to 2 as we wish by taking n sufficiently large and keeping it so. For this series, Formula 9 gives

$$\lim_{n\to\infty} S_n = \frac{a_1}{1-r} = \frac{1}{1-\frac{1}{2}} = 2$$

We frequently abbreviate $\lim\limits_{n\to\infty} S_n$ by S_∞ and call it the "sum" of the geometric series. Notice, however, that S_∞ is a sum only in an extended sense of the word; it cannot be attained by adding any definite number of terms.

EXAMPLE 4

Find the sum of Series 8, that is, find the height of all the reflections in the photograph when stacked vertically.

SOLUTION:

Because the first term of Series 8 is h and the common ratio r is $\frac{1}{2}$, we have

$$\lim_{n\to\infty} S_n = \frac{a_1}{1-r} = \frac{h}{1-\frac{1}{2}} = 2h$$

Thus, if all the reflections were stacked vertically, they would be twice as high as the boy.

EXAMPLE 5

Find the sum of the geometric series:

$2 - \frac{4}{3} + \frac{8}{9} - \cdots,$

SOLUTION:

For the given series,

$a_1 = 2 \quad \text{and} \quad r = -\frac{2}{3}$

Hence,

$$S_\infty = \frac{2}{1-(-\frac{2}{3})} = \frac{2}{\frac{5}{3}} = \frac{6}{5}$$

EXERCISE 14.4

In Problems 1–6, a geometric progression is given. Find: (a) a_1, (b) r, (c) a_n, (d) S_n, and (e) S_∞, if possible.

1 $3, 6, 12, 24, \ldots$

2 $\frac{1}{3}, 1, 3, 9, \ldots$

3 $8, 24, 72, 216, \ldots$

4 $\frac{1}{5}, \frac{1}{10}, \frac{1}{20}, \frac{1}{40}, \ldots$

5 $16, -4, 1, \frac{-1}{4}, \ldots$

6 $3, -1, \frac{1}{3}, \frac{-1}{9}, \ldots$

In Problems 7–14, certain elements of a geometric progression are given. Find the elements whose values are not given. (Assume that all the elements are real numbers.)

7 $a_1 = 5, S_4 = 200; a_4; r$

8 $a_1 = 32, r = \frac{3}{2}, S_n = 422; n; a_n$

9 $r = -2, S_9 = -513; a_1; a_9$

10 $a_7 = \frac{64}{9}, r = \frac{2}{3}; a_1; S_7$

11 $a_8 = -\frac{1}{16}, r = -\frac{1}{4}; a_1; S_8$

12 $a_1 = \frac{243}{256}, a_n = 3, S_n = 9\frac{39}{256}; r; n$

13 $a_1 = \frac{7}{4}, a_n = 112, S_n = 222\frac{1}{4}; r; n$

14 $a_1 = 250, r = \frac{3}{5}, a_n = 32\frac{2}{5}; n; S_n$

In Problems 15–20, find S_∞ (if possible).

15 $\frac{7}{10} + \frac{7}{100} + \frac{7}{1000} + \cdots$

16 $\frac{17}{10^2} + \frac{17}{10^4} + \frac{17}{10^6} + \cdots$

17 $1 + \frac{1}{3} + \frac{1}{9} + \frac{1}{27} + \cdots$

18 $1 - \frac{1}{4} + \frac{1}{16} - \frac{1}{64} + \cdots$

19 $25 - 5 + 1 - \frac{1}{5} + \cdots$

20 $64 + 24 + 9 + \frac{27}{8} + \cdots$

21 The population of a certain town increases geometrically at the rate of 4 percent per year. If the present population is 200,000, what will the population be in 5 years?

22 The population of a town has increased geometrically from 59,049 to 100,000. If this growth occurred in 5 years, what is the growth rate for the town?

Hint: $9^5 = 59{,}049$

23 The number of bacteria in a culture increased geometrically from 64,000 to 729,000 in 6 days. Find the daily rate of increase if this rate is assumed to be constant.

Hint: $3^6 = 729$

24 When dropped on a hard surface, a Super Ball takes a series of bounces, each bounce being about $\frac{9}{10}$ as high as the preceding bounce. If a Super Ball is dropped from a height of 10 feet, find the approximate distance the ball travels before coming to rest.

25 Repeat Problem 24 if the ball is dropped from a height of 20 feet.

26 If the population of a certain country increases geometrically at the rate of 5 percent per year and the present population is 300,000, what will the population be 6 years from now?

27 The number of bacteria in a culture increased geometrically from 320,000 to 2,430,000 in 5 days. Find the daily rate of increase if this rate is assumed to be constant.

28 If three numbers in arithmetic progression are increased by 1, 4, and 43, respectively, the resulting numbers are in geometric progression. Find the original numbers if their sum is 36.

29 If three numbers in arithmetic progression are increased by 9, 7, and 9, respectively, the resulting numbers are in geometric progression. Find the original numbers if their sum is 3.

30 A savings bank pays interest at the rate of 5 percent, compounded annually. What will be the total accumulation at the end of 5 years if $2000 is deposited in the bank now?

USING YOUR KNOWLEDGE 14.4

In Example 3 of Section 14.2, we showed that the compound interest formula is

$$A_n = P(1 + r)^n \tag{1}$$

where P is the number of dollars deposited in a savings account that pays interest at the rate r, compounded annually, and A_n is the accumulated amount at the end of n years. Most frequently, investment problems involve interest that is calculated more than once per year and that is usually stated as "k percent, compounded m times per year." For instance, the phrase "6 percent compounded monthly" means that interest is earned at the rate of $\frac{6}{12} = 0.5$ percent each month and that at the end of the month the earned interest is made part of the principal; the new principal then earns 0.5 percent interest in the next month; and so on. In problems that involve compound interest, we may use Formula 1 by replacing r by $i = k/m$ percent and letting n be the number of interest periods. This gives the formula

$$A_n = P(1 + i)^n \tag{2}$$

where i is the interest rate *per period* and n is the *number of periods*.

For example, if \$1000 is deposited in an account that pays 6 percent, compounded monthly, then at the end of 2 years, the accumulated amount is obtained by using Formula 2 with $i = 0.005$ (0.5 percent $= 0.005$) and $n = 24$, the number of months (interest periods) in 2 years. Thus,

$$A_{24} = 1000(1.005)^{24} \qquad = \$1127.16$$

(This answer was obtained by using a calculator with a y^x key.)

We now show how the compound interest formula is used in a periodic payment type of problem that illustrates an important application of geometric progressions. Suppose that a man deposits \$100 at the beginning of each month in a fund that pays 9 percent interest, compounded monthly. How do we find the accumulated amount at the end of 5 years? We see that the man's first deposit earns interest for 60 months; his second deposit earns interest for 59 months; and so on. His last deposit, 1 month before the end of the 5 years, earns interest for 1 month. It is convenient to start from his last deposit and work backward to write the total accumulation, which we denote by S. Letting P be the periodic payment (\$100) and i be the periodic interest rate (0.09/12 or 0.0075), we obtain

$$\begin{aligned} S &= P(1 + i) + P(1 + i)^2 + \cdots + P(1 + i)^{60} \\ &= P[(1 + i) + (1 + i)^2 + \cdots + (1 + i)^{60}] \end{aligned}$$

We recognize the quantity inside the brackets as a geometric progression with $a_1 = (1 + i)$, $a_{60} = (1 + i)^{60}$, $n = 60$, and ratio $r = (1 + i)$. Thus, using the formula for the sum, we obtain

$$S = P(1 + i)\frac{1 - (1 + i)^{60}}{1 - (1 + i)} = P(1 + i)\frac{(1 + i)^{60} - 1}{i}$$

Substituting $P = 100$ and $i = 0.0075$, we obtain

$$\begin{aligned} S &= (100)(1.0075)\frac{(1.0075)^{60} - 1}{0.0075} \\ &= (100.75)(75.42414) \\ &= \$7598.98 \end{aligned}$$

(We have again used a calculator with a y^x key.)

Answers to the following problems can be conveniently obtained by using a calculator with a y^x key. If a calculator is not available, four-place logarithms can be used to give an approximate answer.

C **1** A fund is accumulated by making a deposit of \$100 at the beginning of each 6 months for 10 years. What amount is in the fund at the end of the 10 years if interest is at 6 percent, compounded semiannually?

C **2** At the end of each 3 months for 10 years, a person puts $500 into a fund that pays interest at the rate of 8 percent, compounded quarterly. How much will be in the fund just after the last deposit?

C **3** A person puts $100 into a fund at the end of each month for 4 years. If the fund pays 6 percent, compounded monthly, how much will be in the fund just after the last payment?

C **4** A fund is accumulated by making a deposit of $5000 at the beginning of each year for 20 years. What amount is in the fund at the end of the 20 years if interest is at the annual rate of 6 percent compounded continuously?

Hint: Recall that P dollars with interest at the rate r, compounded continuously, for t years accumulates to Pe^{rt} dollars.

14.5 COUNTING TECHNIQUES

TREE DIAGRAMS

If there are two questions on a true–false test, in how many different ways can these two questions be answered? One way to find out is by constructing a "tree diagram" as shown in Figure 14.5a. In this figure, the first set of branches shows the ways in which the first question can be answered (T for true, F for false), and the second set of branches shows the ways in which the second question can be answered. By tracing each path from left to right, we find that there are $2 \times 2 = 4$ ways in which the two questions can be answered. The possibilities are TT, TF, FT, and FF.

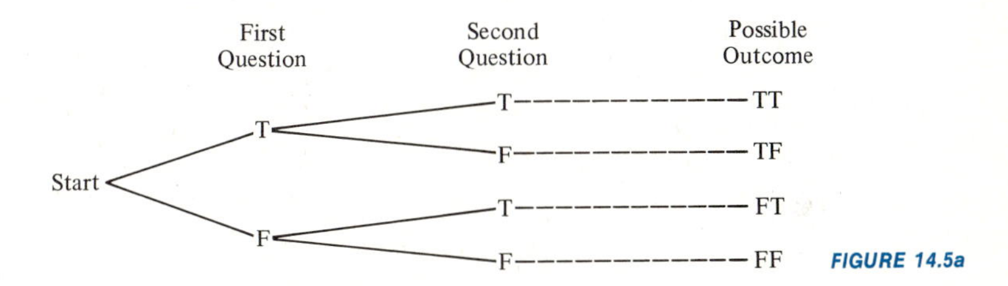

FIGURE 14.5a

There are many problems that require counting techniques related to that used in the preceding illustration. We use the tree diagram technique again in Example 1.

EXAMPLE 1

A person wishes to buy a sports car. Two body styles (convertible or hardtop), three colors (red, yellow, or green), and two styles of tires (blackwall and whitewall) are available. How many different choices of body style, color, and tires does the person have?

SOLUTION:

We can count the number of choices by using a tree diagram. (See Figure 14.5b.) The diagram shows that for each body style, convertible (c) or hardtop (h), there are available three colors, red (r), yellow (y), or green (g), and for each of the preceding choices, there are two styles of tires, blackwall (b) or whitewall (w). Thus, there are $2 \times 3 \times 2 = 12$ different choices of body style, color, and tire style possible. The figure shows all of the possible outcomes.

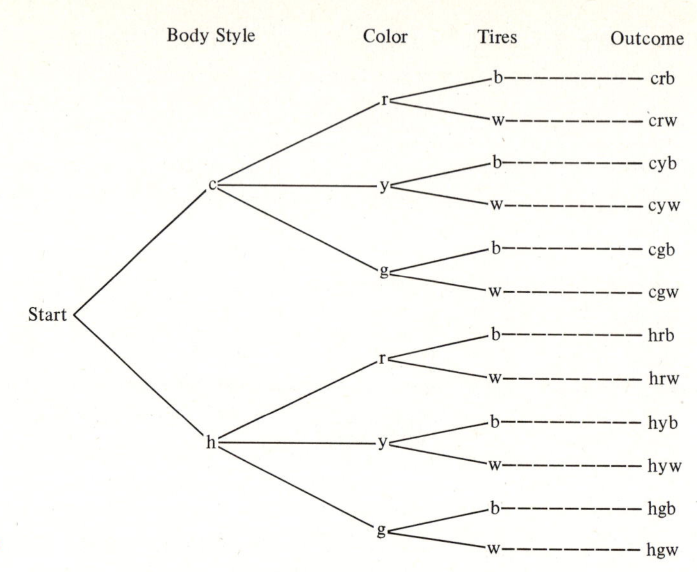

FIGURE 14.5b

THE SEQUENTIAL COUNTING PRINCIPLE

In Example 1, we found that if there are two ways to do a thing (select a body style), three ways to do a second thing (select a color), and two ways to do a third thing (select a tire style), then there are $2 \times 3 \times 2 = 12$ ways of doing the three things in the stated order. This example illustrates an important basic principle called the Sequential Counting Principle. (The validity of this principle follows from a consideration of a tree diagram as in Example 1.)

THE SEQUENTIAL COUNTING PRINCIPLE (SCP)
If one thing can occur in m ways, a second thing in n ways, a third thing in r ways, and so on, then the sequence of things can occur in the stated order in

$m \times n \times r \times \cdots$ **ways**

EXAMPLE 2

Johnny's Ice Cream Parlor lists ice cream or ice milk in 28 flavors each. Johnny serves a special consisting of one scoop of ice cream or ice milk and one of three toppings. If you want one of these specials, how many choices do you have?

SOLUTION:

Because there are two choices (ice cream or ice milk), 28 flavors for each choice, and three different toppings, the SCP gives

$2 \times 28 \times 3 = 168$

choices in all.

In some problems, it is helpful to use a diagram to represent the individual events in a sequence of events and then to apply the SCP, as in Example 3.

EXAMPLE 3

A slot machine has three dials, each with 20 symbols, as shown in Table 14.5a. If the symbols are regarded as all different,

a. how many symbol arrangements are possible on the three dials?

b. in how many ways can we obtain three bars?

c. in how many ways can we obtain the arrangement cherry–orange–lemon?

TABLE 14.5a

Symbol	Dial 1	Dial 2	Dial 3
Bar	1	3	1
Bell	1	3	3
Cherry	7	7	0
Lemon	3	0	4
Orange	3	6	7
Plum	5	1	5

SOLUTION:

a. We make three boxes representing the three dials:

☐ ☐ ☐

There are 20 choices for each box (each dial has 20 symbols). Accordingly, we enter a 20 in each box:

[20] [20] [20]

Applying the SCP, we see that the number of possibilities is

$20 \times 20 \times 20 = 8000$

b. By referring to Table 14.5a, we see that the numbers to be entered in the boxes are 1, 3, and 1, because there is one bar on the first dial, three on the second, and one on the third.

[1] [3] [1]

Thus, by the SCP, the number of ways of obtaining three bars is

$1 \times 3 \times 1 = 3$

c. Again, Table 14.5a shows that the numbers to be entered in the boxes are 7, 6, and 4, because the first dial has seven cherries, the second dial has six oranges, and the third dial has four lemons.

[7] [6] [4]

By the SCP, the number of ways of obtaining the arrangement cherry–orange–lemon is

$7 \times 6 \times 4 = 168$

PERMUTATIONS

A special case of the Sequential Counting Principle occurs when we want to count the possible arrangements of a given set of things. For example, to count the number of different orders in which we can write the letters a, b, and c in a row, with no repetitions allowed, we note that there are three choices for the first letter, two for the second, and one for the third. Thus, by the SCP, the number of different arrangements is

$3 \times 2 \times 1 = 6$

We can display these arrangements by the diagram in Figure 14.5c on the next page. There are six paths, so that the total number of arrangements is six. Notice in this illustration that abc and acb are treated as different arrangements, because the order of the letters is different. This type of arrangement, in which the order is important is called a *permutation*. (See Definition 14.5a.)

▼ DEFINITION 14.5a

An ordered arrangement of distinguishable objects, with no repetitions allowed, is called a *permutation* of the objects.

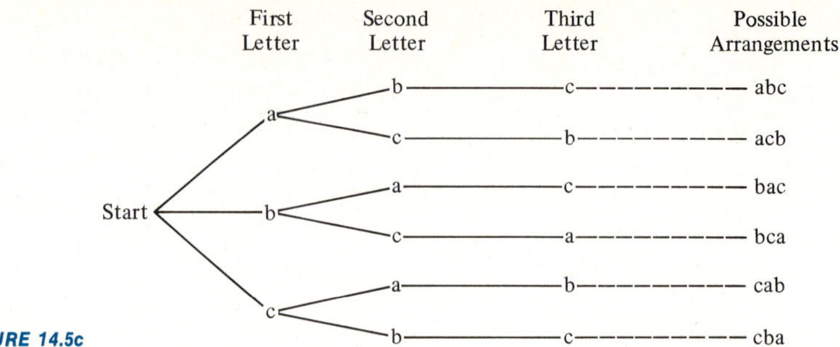

FIGURE 14.5c

In the preceding illustration, we found that the number of permutations of the three letters a, b, and c is $3 \times 2 \times 1 = 6$. A convenient way of representing the product $3 \times 2 \times 1$ is $3!$ (read "3 factorial"). The more general factorial is described in Definition 14.5b. By this definition,

▼ DEFINITION 14.5b

The symbol $n!$ (read "n factorial"), where n is a positive integer, represents the product of n and each positive integer less than n:

$$n! = n(n-1)(n-2)\cdots 3 \cdot 2 \cdot 1$$

$$5! = 5 \cdot 4 \cdot 3 \cdot 2 \cdot 1 = 120$$

and

$$6! = 6 \cdot 5 \cdot 4 \cdot 3 \cdot 2 \cdot 1 = 720$$

We shall use the notation $P(n, r)$ to represent the number of permutations of n things taken r at a time. Thus, $P(4, 4)$ stands for the number of permutations of four things taken four at a time, that is, with each permutation using all four of the objects. The symbol $P(4, 2)$ stands for the number of permutations of four things taken two at a time, that is, with each permutation using just two of the objects.

In discussing the permutations of the three letters a, b, and c, we found that $P(3, 3) = 3!$ By applying the SCP, we can see that for any positive integer n,

$$P(n, n) = n! \tag{1}$$

This result follows because if we think of arranging n things in a row, we have n choices for the first place, $n - 1$ choices for the second place, $n - 2$ choices for the third place, and so on. Thus, the SCP gives

$$P(n, n) = n(n-1)(n-2)\cdots 3 \cdot 2 \cdot 1 = n!$$

EXAMPLE 4

How many four-digit numbers can be written by using the digits 1, 2, 3, and 4 if repetition of digits is not allowed?

SOLUTION:

This question is equivalent to asking for the number of permutations of four things taken four at a time. The answer is

$$P(4, 4) = 4! = 24$$

To evaluate $P(n, r)$, where n and r are positive integers with $r < n$, we can use the same procedure as before. We think of arranging r of the n things in a row and apply the SCP. There are n choices for the first place, $n - 1$ choices for the second place, $n - 2$ for the third place, and so on, until there are $n - r + 1$ choices for the rth place. Keep in mind that $r - 1$ of the things have been used to fill the first $r - 1$ places, so that there are $n - (r - 1) = n - r + 1$ things left when we come to the rth place. By the SCP, we obtain

$$P(n, r) = n(n-1)(n-2)\cdots(n-r+1) \qquad r < n \tag{2}$$

Formula 2 is quite easy to remember. Think of the n as the number of objects available and the r as the number to be selected:

$$\underbrace{\boxed{n \text{ objects available}}}_{\downarrow} \quad \underbrace{\boxed{\text{select } r \text{ of them}}}_{\downarrow}$$
$$P(n, r)$$

You simply write down the n, then $(n - 1)$, then $(n - 2)$, and so on, until you have written r factors in all. For instance, for $P(8, 4)$, we have

8 objects available

$$P(8, 4) = \underbrace{8 \cdot 7 \cdot 6 \cdot 5}_{} = 1680$$

select 4 ⟶ 4 factors

EXAMPLE 5

How many three-letter permutations can be formed from the letters of the English alphabet? (Remember that repetition of letters is not allowed.)

SOLUTION:

Since there are 26 letters in the alphabet, we have $n = 26$ and $r = 3$. Thus, the answer is

$$P(26, 3) = 26 \cdot 25 \cdot 24 = 15{,}600$$

EXAMPLE 6

A state decides to use three letters followed by three digits for its auto license plates.

a. If no repetition of letters or of digits is to be allowed, how many different plates are possible?

b. Answer the same question if repetition of both letters and digits is allowed.

SOLUTION:

a. By the SCP, we know that the required number is the product of the number of three-letter permutations, $P(26, 3)$, and the number of three-digit permutations, $P(10, 3)$, of the ten digits 0, 1, 2, 3, 4, 5, 6, 7, 8, 9. Hence, the number of possible plates is

$$\begin{aligned} P(26, 3)P(10, 3) &= (26 \cdot 25 \cdot 24)(10 \cdot 9 \cdot 8) \\ &= (15{,}600)(720) \\ &= 11{,}232{,}000 \end{aligned}$$

b. By the SCP, the number of possible plates is

$$\begin{aligned} (26 \times 26 \times 26)(10 \times 10 \times 10) &= (26^3)(10^3) \\ &= 17{,}576{,}000 \end{aligned}$$

(Of course, not all the possibilities would be used. A car owner would probably not care to have a plate reading NUT–123.)

Notice that, for $r < n$,

$$P(n, n) = n! = n(n-1)(n-2)\cdots(n-r+1)(n-r)\cdots 3 \cdot 2 \cdot 1$$

or

$$n! = P(n, r) \cdot (n-r)!$$

By solving this equation for $P(n, r)$, we obtain the formula

$$\mathbf{P(n, r) = \frac{n!}{(n-r)!}} \qquad (3)$$

Although we have derived Formula 3 for $r < n$, it is convenient to have the formula hold for $r = n$ also. However, in this case, we have

$$P(n, n) = \frac{n!}{0!}$$

Consequently, we define

$$\mathbf{0! = 1} \qquad (4)$$

With this definition, Formula 3 holds for $r \leq n$.

EXAMPLE 7

There are 12 "face" cards (kings, queens, and jacks) in an ordinary deck of playing cards. Use Formula 3 to determine the number of ways in which these 12 cards, taken five at a time, can be arranged in a row.

SOLUTION:

We have to calculate $P(12, 5)$, and Formula 3 gives

$$\begin{aligned} P(12, 5) &= \frac{12!}{(12-5)!} \\ &= \frac{12!}{7!} \\ &= \frac{12 \cdot 11 \cdot 10 \cdot 9 \cdot 8 \cdot (7!)}{7!} \\ &= 12 \cdot 11 \cdot 10 \cdot 9 \cdot 8 \\ &= 95{,}040 \end{aligned}$$

COMBINATIONS

In the preceding work, we found the number of *ordered* arrangements that are possible with n distinguishable objects. Sometimes we wish to disregard the order in which the objects are arranged and simply count the number of subsets of a given size that can be selected from a set of n objects. Such subsets are called *combinations*. (See Definition 14.5c.)

The symbols ${}_nC_r$, $C_{n,r}$, C_n^r, and $\binom{n}{r}$ are also used to represent

▼ DEFINITION 14.5c

A subset consisting of r objects selected from a set of n objects (without regard to the order of selection) is called a *combination* of the n objects taken r at a time. The number of such combinations is denoted by the symbol $C(n, r)$ (read "the number of combinations of n objects taken r at a time").

the number of combinations of n things taken r at a time. We shall use only the symbol $C(n, r)$. As a special case, let us look at the combinations of the three letters a, b, and c, taken two at a time. This number is $C(3, 2)$. It is easy to list the combinations. They are: ab, ac, and bc, obtained by omitting one of the three letters for each combination. Thus, we have found that $C(3, 2) = 3$. It is more instructive, however, to obtain the relationship between $C(3, 2)$ and $P(3, 2)$. In Table 14.5b, we have listed the combinations and the corresponding permutations. You can see that each combination of *two* letters corresponds to $2! = 2$ permutations, so that

TABLE 14.5b

Combinations	Permutations
ab	ab, ba
ac	ac, ca
bc	bc, cb

$$P(3, 2) = (2!)C(3, 2)$$

This gives

$$C(3,2) = \frac{P(3,2)}{2!} = \frac{3 \cdot 2}{2 \cdot 1} = 3$$

as before.

The relationship between $P(3, 2)$ and $C(3, 2)$, which was just obtained, can be generalized to the corresponding relationship between $P(n, r)$ and $C(n, r)$. Any given combination of r of the n objects can be permuted in $r!$ ways, so that each of the combinations corresponds to $r!$ permutations. Thus, we have

$$P(n, r) = (r!)C(n, r)$$

and

$$C(n, r) = \frac{P(n, r)}{r!} \tag{5}$$

By using Formula 2 for $P(n, r)$, we obtain the following useful formula for $C(n, r)$.

$$C(n, r) = \frac{n(n-1)(n-2)\cdots(n-r+1)}{r!} \tag{6}$$

Notice that there are exactly r factors in the numerator and in the denominator of the right-hand member of Formula 6.

EXAMPLE 8

A set of coins consists of a penny, a nickel, a dime, a quarter, and a half-dollar. How many different sums of money can be formed if exactly three of the coins are used?

SOLUTION:

Because the order in which the coins are selected is of no importance, the required number is exactly the number of combinations of five things taken three at a time, that is, $C(5, 3)$. By Formula 6, we obtain

$$C(5, 3) = \frac{5 \cdot 4 \cdot 3}{3 \cdot 2 \cdot 1} = 10$$

Thus, ten different sums of money can be formed by using the five coins three at a time.

Another convenient formula for $C(n, r)$ can be obtained by substituting from Formula 3 into Formula 5. This gives

$$C(n, r) = \frac{P(n, r)}{r!} = \frac{n!}{r!(n-r)!} \tag{7}$$

EXAMPLE 9

In how many ways can five cards be chosen from an ordinary deck of 52 cards?

SOLUTION:

Because the order of the five cards is *not* important, this question is equivalent to, "How many combinations are there of 52 things, taken five at a time?" We use Formula 7 to obtain the answer:

$$\begin{aligned} C(52, 5) &= \frac{52!}{5!47!} \\ &= \frac{52 \cdot 51 \cdot 50 \cdot 49 \cdot 48 \cdot (47!)}{5!47!} \\ &= \frac{52 \cdot 51 \cdot 50 \cdot 49 \cdot 48}{5 \cdot 4 \cdot 3 \cdot 2 \cdot 1} \\ &= 2{,}598{,}960 \end{aligned}$$

(Note that this is the number of different five-card poker hands that can be selected from a deck of 52 cards.)

In attempting the following problems, you should keep in mind that we have considered three types of counting problems:

1. The general type of problem in which we count the total number of ways in which successive events can occur. This kind of problem is handled by the *Sequential Counting Principle* (*SCP*).
2. Problems in which we count the number of possible arrangements of the elements of the subsets of a given size formed from a specified set. Such problems involve *permutations*, because the *order* in which the elements are arranged *is important*.
3. Problems in which we count the number of subsets of a given size that can be formed from a specified set *without regard to the order* in which the elements are arranged. Such problems involve *combinations*, because the order is disregarded.

EXERCISE 14.5

1 A credit bureau classifies customers according to income (high, average, low), time spent on the present job (0–5 years, more than 5 years), and whether or not they own their own home. Draw a tree diagram showing the various possible classifications of a customer.

2 A student is taking mathematics (M), science (S), and art (A). Her three books, one for each subject, are standing together on a shelf. Draw a tree diagram to show all the possible orders in which the three books could be arranged.

3 A restaurant serves pie a la mode (pie and ice cream) for dessert. There are five flavors of ice cream (chocolate, strawberry, vanilla, peach, and coffee), and there are two kinds of pie (apple and cherry). How many different pie a la mode desserts are possible?

4 From the top of a mountain, there are three ski runs to a shelter house, and from this shelter house, there are six ski runs to the foot of the mountain. In how many ways is it possible to ski from the top to the foot of the mountain?

5 A man has three different pairs of shoes, two different suits, and four different shirts. How many different outfits does he have available?

6 Refer to the slot machine in Example 3 and find the number of ways in which one could obtain
a three bells **b** three lemons **c** three oranges **d** three plums **e** the arrangement bell–cherry–lemon

7 Compute:
a $8!$ **b** $\dfrac{10!}{7!}$ **c** $\dfrac{P(5, 2)}{2!}$

8 Use Formula 3 to compute:
a $P(10, 4)$ **b** $P(7, 2)$ **c** $P(14, 3)$ **d** $P(11, 8)$ **e** $P(n, n)$

9 A certain town has the 123 telephone exchange; that is, all the telephone numbers start with 123, which is followed by a four-digit number.
a What is the largest possible number of telephone numbers for this town?
b What is the largest possible number of numbers if repetition of digits is not allowed?

10 The area code for telephone numbers in the United States is a three-digit number.
a How many area codes are possible?
b How many area codes are possible if repetition of digits is not allowed?
c How many area codes are possible if the first digit cannot be a zero and repetition of digits is not allowed?

11 Your Social Security number is composed of nine digits.
a How many Social Security numbers are possible?
b How many if the first three digits cannot be zeros?

12 How many three-digit numbers can be formed from the digits 1, 3, 5, 6, 7, 8 if repetition of digits is not allowed?

13 Do Problem 12 if the three-digit numbers are to be even.

14 Do Problem 12 if the three-digit numbers are to be odd.

15 Calculate:
a $C(5, 2)$ **b** $C(6, 4)$ **c** $C(7, 3)$ **d** $C(5, 0)$ **e** $C(9, 6)$

16 Do Problem 12 if the three-digit numbers are to be divisible by 3.

Hint: If a number is divisible by 3, the sum of its digits must also be divisible by 3.

17 How many different sums of money can be formed from a set of coins consisting of a penny, a nickel, a dime, a quarter, a half-dollar, and a dollar if exactly three coins are to be used?

18 Do Problem 17 if four coins are to be used.

19 Do Problem 17 if *at least* two coins are to be used.

20 The Book-of-the Month Club offers a choice of three books from a list of 40 books. How many different selections of three books can be made from this list?

21 A box of light bulbs contains 95 good bulbs and five bad ones.
a In how many ways can three different bulbs (without regard to order) be selected?
b In how many ways can two good bulbs and one bad one be selected?

22 If a family has two children, the possible orders of birth are boy–boy, boy–girl, girl–boy, girl–girl. Find the possible orders for a family with three children.

Hint: Use a tree diagram.

23 A committee of three is to be selected from a group of five men and six women.
a How many choices are there for such a committee?
b How many choices are there if the committee is not to consist of all men or all women?

24 How many sets of two initials can be constructed using the letters of the English alphabet? A school has 1000 students, each with two initials. Show that at least two of these students must have the same initials.

25 Three cards are drawn in succession and without replacement from a deck of 52 cards.
a If the order of the cards is taken into account, how many different outcomes are possible?
b In how many ways can an ace, a king, and a queen be selected in that order?

26 In 1935, a chain letter fad started in Denver, Colorado. The scheme works like this: You receive a letter with a list of five names, send a dime to the person at the top of the list, cross that name out, and add your name at the bottom. Suppose that you receive one of these letters and send it to five other persons, each of whom sends it to five others, each of whom in turn sends it to five others, and so on.
a How many letters will have your name on the list?
b How much money would you receive if the chain were not broken?

27 If 20 people all shake hands with each other, how many handshakes are there?

28 A committee is to consist of three men and three women. If five men and six women are available, how many choices are there for this committee?

29 There are nine men and eight women in a tennis club. In how many ways can a doubles match be arranged if each side is composed of a man and a woman?

30 How many positive integers are exact divisors of $2^2 3^3 5^4$?

Hint: The divisors are all of the form $2^a 3^b 5^c$, where a may be 0, 1, or 2; b may be 0, 1, 2, or 3; and so on.

31 How many positive integers are exact divisors of 55,125?

Hint: First write 55,125 as a product of its prime divisors and then see the hint in Problem 30.

32 There are ten men, each capable of playing any one of the four backfield positions on a football team. In how many ways can these positions be filled from these ten men?

33 A piggy bank contains three nickels, two dimes, and two quarters. Two coins are shaken out at random.
a How many outcomes are possible if the order of the coins is taken into account?

Hint: Label the coins N_1, N_2, N_3, D_1, D_2, Q_1, Q_2, so that they may all be regarded as being different.

b In how many of the outcomes would the combined value of the two coins be 30 cents?

34 A student has two different mathematics books, three different biology books, and four different history books. In how many ways can these books be stood on a shelf so that those in the same subject are together?

35 From a squad of 20 men, three can catch and three can pitch. If these six men can play at no other positions, how many different baseball nines can be formed? (Do not distinguish among the other positions.)

36 How many different sums of money can be formed from a penny, a nickel, a dime, a quarter, and a half-dollar if at least one coin must be used?

37 A man has a $1 bill, a $5 bill, a $10 bill, and a $20 bill. How many different sums of money can the man make if he uses at least two bills each time?

38 A list consists of 25 fiction and 15 nonfiction books. How many selections of three fiction and two nonfiction books can be made from this list?

39 How many five-card poker hands consist of three aces and two kings?

40 How many diagonals does a polygon of ten sides have? Of n sides? (A diagonal is a line that joins two nonconsecutive vertices.)

USING YOUR KNOWLEDGE 14.5

One of the most interesting and important uses of the counting techniques discussed in the preceding section occurs in the subject of probability. If a penny and a nickel are tossed, they may fall in any one of four different ways: both heads up (HH); both tails up (TT); penny heads up, nickel tails up (HT); penny tails up, nickel heads up (TH). We usually assume that any one of these outcomes is as likely as any other one. Therefore, we say that the probability of two heads is $\frac{1}{4}$, the probability of two tails is $\frac{1}{4}$, and the probability of one head and one tail is $\frac{1}{2}$.

In discussing probability, we shall call the occurrence of a desired event a *success* and the nonoccurrence of that event a *failure*. A mathematical measure of probability may then be defined as follows:

MATHEMATICAL PROBABILITY

If on any trial, a desired event E can occur in s ways (s ways to obtain a $s$uccess) and can fail to occur in f ways (f ways to obtain a $f$ailure) and if these $s + f$ ways are all equally likely, then the *probability* $P(E)$ that the event E will occur is

$$P(E) = \frac{s}{s+f}$$

For example, we have seen that if two coins are tossed, they may fall in four different ways. In three of these ways, at least one head occurs, and in the remaining way, heads fails to occur. Thus, the probability that at least one head will turn up is

$$P(\text{at least one H}) = \frac{3}{3+1} = \frac{3}{4}$$

The probability that no heads turn up is

$$P(\text{no H}) = \frac{1}{1+3} = \frac{1}{4}$$

It follows from the definition that, if an event is certain to happen, its probability of happening is 1, and, if it is certain not to happen, its probability is 0. If an event must either happen or fail to happen, the sum of the probabilities of success and failure is 1. In the preceding illustration, we found that $P(\text{at least one H}) = \frac{3}{4}$ and $P(\text{no H}) = \frac{1}{4}$, and the sum of these two probabilities is 1.

As another illustration, we refer to the slot machine in Example 3 of the preceding section and assume that all outcomes are equally likely. In the solution of Example 3, we found that there are 8000 possible outcomes and three ways of obtaining three bars. Thus, the probability of obtaining three bars is

$$P(\text{Bar–Bar–Bar}) = \tfrac{3}{8000}$$

We also found that the number of ways of obtaining the arrangement cherry–orange–lemon is 168. Hence, the probability of obtaining this arrangement is

$$P(\text{cherry–orange–lemon}) = \tfrac{168}{8000} = \tfrac{21}{1000}$$

The following problems use only the probability ideas just discussed.

1 If the four-digit numbers in Example 4 of the preceding section are all written on separate slips of paper and put into a hat, what is the probability of drawing the number 1234 if a slip is drawn at random? ("At random" means all possibilities are equally likely.)

2 In the preceding problem, what is the probability of drawing a number ending in 4?

3 If the three coins in Example 8 of the preceding section are chosen at random, what is the probability that the sum will be 80 cents?

The remaining problems refer to Exercise 14.5.

4 In Problem 3, suppose the waiter brought a pie a la mode picked up at random. What is the probability that it would be apple pie with either vanilla or coffee ice cream?

5 In Problem 6, find the probability of obtaining:
a three bells **b** three lemons **c** three oranges **d** three plums **e** bell–cherry–lemon

6 In Problem 12, if one of the three-digit numbers is selected at random, what is the probability that it is an odd number?

7 In Problem 17, if the three coins are selected at random, what is the probability that their value is more than \$1.50?

8 In Problem 21, find the probability that *at least* one of the three bulbs selected is bad.

9 In Problem 23, if the committee is selected completely at random, what is the probability that it consists entirely of one sex?

10 In Problem 25, find the probability that an ace, a king, and a queen are selected in that order.

11 In Problem 33, find the probability that the value of the two coins selected is 30 cents.

12 In Problem 39, find the probability that a five-card poker hand drawn at random consists of three aces and two kings.

14.6 THE BINOMIAL THEOREM

Suppose an office has a telephone with 12 black buttons for extensions. If there is a similar telephone at each extension (13 phones in all), how many different two-person connections are possible? The answer to this question is, of course, $C(13, 2)$, the number of combinations of 13 things taken two at a time.

It is interesting to know that this problem is closely related to the problem of finding the coefficients in the expansion of $(a + b)^{13}$. In fact, the answer to the telephone question is the coefficient of the $a^{11}b^2$ term in this expansion. To see that this is so, let us write the product using subscripts so that the a's and b's all appear to be different. The product will look like this:

$$(a_1 + b_1)(a_2 + b_2)(a_3 + b_3) \cdots (a_{13} + b_{13}) \tag{1}$$

If we multiply this product out, there will be a set of terms with exactly two b's as factors. How many such terms are there? This is the same question as the telephone question but in different words. We can select two b's from the 13 b's (without regard to order) in $C(13, 2)$ ways.

Now, suppose that we ask how many three-person connections are possible in the telephone problem. This is the same as asking how many terms will have exactly three b's as factors when we multiply out the product in 1. The answer to this question is $C(13, 3)$, the number of combinations of 13 things, taken three at a time.

It follows in exactly the same way that the number of terms in the product in 1 that contain exactly k b's as factors $(1 \leq k \leq 13)$ is $C(13, k)$, the number of combinations of 13 things taken k at a time. Instead of the symbol $C(13, k)$, it is customary to use the *binomial coefficient* symbol, $\binom{13}{k}$, to stand for the preceding number; that is

$$\binom{13}{k} = C(13, k) \tag{2}$$

In terms of this symbol, one of the terms in the expansion of $(a + b)^{13}$ (obtained by multiplying out the product in 1, erasing the subscripts, and collecting the terms) is

$$\binom{13}{k} a^{13-k} b^k$$

You should recall from Section 14.5 that $C(n, k)$ or $\binom{n}{k}$ is given by

$$\binom{n}{k} = \frac{n!}{k!(n-k)!} \qquad 1 \leq k \leq n \tag{3}$$

For convenience, we also make the following definition:

$$\binom{n}{0} = 1 \tag{4}$$

Thus, we have

$$\binom{n}{k} = \frac{n!}{k!(n-k)!} \qquad k = 0, 1, 2, \ldots, n \tag{5}$$

EXAMPLE 1

Use Formula 5 to evaluate:

a. $\binom{13}{2}$

b. $\binom{n}{n}$

c. $\binom{n}{0}$

SOLUTION:

a. By Equation 5,

$$\binom{13}{2} = \frac{13!}{2!11!} = \frac{13 \cdot 12 \cdot (11!)}{2!11!} = \frac{13 \cdot 12}{2} = 78$$

b. Similarly,

$$\binom{n}{n} = \frac{n!}{n! \cdot 0!} = \frac{n!}{n! \cdot 1} = 1$$

c. Again, by Equation 5,

$$\binom{n}{0} = \frac{n!}{0!n!} = 1$$

In order to obtain the expansion of $(a + b)^n$, we can follow the same reasoning that we used when working with $(a + b)^{13}$. We first write the product

$$(a_1 + b_1)(a_2 + b_2)(a_3 + b_3) \cdots (a_n + b_n) \tag{6}$$

which we could multiply out and drop the subscripts to obtain the expansion of $(a + b)^n$. Since there are n factors in the product, each term of the expanded result must be the product of exactly n factors, all a's and b's. Thus, if exactly k of these factors are b's, then the other $n - k$ factors must be a's. As we have already seen, the number of terms of this type is $\binom{n}{k}$, so that if we multiply out the product in 6, drop the subscripts, and collect like terms, we will obtain a sum of all possible terms of the form

$$\binom{n}{k} a^{n-k} b^k \qquad 0 \le k \le n$$

This leads us to Theorem 14.6a.

▼ THEOREM 14.6a

The binomial theorem

For all natural numbers n,

$$(a + b)^n = \sum_{k=0}^{n} \binom{n}{k} a^{n-k} b^k$$
$$= \binom{n}{0} a^n + \binom{n}{1} a^{n-1} b + \cdots$$
$$+ \binom{n}{k} a^{n-k} b^k + \cdots$$
$$+ \binom{n}{n} b^n$$

EXAMPLE 2

Verify that the binomial theorem gives the known expansion

$(a + b)^4 =$
$a^4 + 4a^3b + 6a^2b^2 + 4ab^3 + b^4$

SOLUTION:

By Theorem 14.6a, we have

$$(a + b)^4 = \binom{4}{0} a^4 + \binom{4}{1} a^3 b + \binom{4}{2} a^2 b^2 + \binom{4}{3} ab^3 + \binom{4}{4} b^4$$

By Example 1, $\binom{4}{0} = 1$ and $\binom{4}{4} = 1$. By Formula 5,

$$\binom{4}{1} = \frac{4!}{1! \cdot 3!} = \frac{4 \cdot 3!}{1 \cdot 3!} = 4 \qquad \binom{4}{2} = \frac{4!}{2! \cdot 2!} = \frac{4 \cdot 3 \cdot 2!}{2 \cdot 1 \cdot 2!} = 6$$

and

$$\binom{4}{3} = \frac{4!}{3! \cdot 1!} = \frac{4 \cdot 3!}{3! \cdot 1} = 4$$

Thus, we have verified that the binomial theorem gives

$$(a + b)^4 = a^4 + 4a^3b + 6a^2b^2 + 4ab^3 + b^4$$

Notice that we found $\binom{4}{1} = \binom{4}{3}$ in Example 2. This is a special case of the general result

$$\binom{n}{k} = \binom{n}{n-k} \tag{7}$$

which follows directly from Formula 5. Details are left for you.

EXAMPLE 3

Expand $(x-2y)^5$.

SOLUTION:

We use the binomial theorem with $n = 5$, $a = x$, and $b = -2y$. This gives

$$(x-2y)^5 = \binom{5}{0}x^5 + \binom{5}{1}x^4(-2y) + \binom{5}{2}x^3(-2y)^2 + \binom{5}{3}x^2(-2y)^3 + \binom{5}{4}x(-2y)^4 + \binom{5}{5}(-2y)^5$$

We know that $\binom{5}{0} = \binom{5}{5} = 1$. By Formulas 5 and 7,

$$\binom{5}{4} = \binom{5}{1} = \frac{5!}{1!\cdot 4!} = \frac{5\cdot 4!}{1\cdot 4!} = 5$$

and

$$\binom{5}{3} = \binom{5}{2} = \frac{5!}{2!\cdot 3!} = \frac{5\cdot 4\cdot 3!}{2\cdot 1\cdot 3!} = 10$$

Thus,

$$\begin{aligned}(x-2y)^5 &= x^5 + 5x^4(-2y) + 10x^3(-2y)^2 + 10x^2(-2y)^3 \\ &\quad + 5x(-2y)^4 + (-2y)^5 \\ &= x^5 - 10x^4y + 40x^3y^2 - 80x^2y^3 + 80xy^4 - 32y^5\end{aligned}$$

EXAMPLE 4

Find the fifth term in the expansion of $(x-3)^8$.

SOLUTION:

By referring to Theorem 14.6a, we note that the terms have been numbered $k = 0, 1, 2, \ldots$, so that the rth term would be obtained by taking $k = r - 1$. Hence, to obtain the fifth term, we take $k = 4$ in the formula

$$\binom{n}{k}a^{n-k}b^k$$

With $n = 8$, $a = x$, and $b = -3$, we obtain

$$\begin{aligned}\binom{8}{4}x^4(-3)^4 &= \frac{8!}{4!4!}(81x^4) \\ &= \frac{8\cdot 7\cdot 6\cdot 5\cdot 4!}{4!\cdot 4!}(81x^4) \\ &= (70)(81x^4) \\ &= 5670x^4\end{aligned}$$

EXAMPLE 5

Find the term that contains no x in the expansion of $(x^2 - 2/x)^9$.

SOLUTION:

Because $n = 9$, we know that the general term in the expansion is

$$\binom{9}{k}(x^2)^{9-k}\left(-\frac{2}{x}\right)^k = \binom{9}{k}x^{18-3k}(-2)^k$$

Thus, we can see that for $18 - 3k = 0$, that is, for $k = 6$, the term will have no x in it. For $k = 6$, we obtain the seventh term of the expansion

$$\binom{9}{6}(x^2)^{9-6}\left(-\frac{2}{x}\right)^6 = \frac{9 \cdot 8 \cdot 7 \cdot (6!)}{3 \cdot 2 \cdot 1 \cdot (6!)}(2^6)$$
$$= (84)(64)$$
$$= 5376$$

EXERCISE 14.6

In Problems 1–8, evaluate the given expression.

1 $\binom{6}{2}$ **2** $\binom{6}{5}$ **3** $\binom{10}{3}$

4 $\binom{10}{7}$ **5** $8!$ **6** $9!$

7 $\dfrac{10!}{8!}$ **8** $\dfrac{13!}{11!}$

In Problems 9–20, expand the given expression by the binomial theorem and simplify the result.

9 $(a + 2b)^3$ **10** $(a - 3b)^3$ **11** $(x + 4)^4$

12 $(x - 4)^4$ **13** $(2x - 1)^5$ **14** $(2x + 1)^5$

15 $(2x - 3y)^4$ **16** $(2x + 3y)^4$ **17** $\left(\dfrac{1}{x} + \dfrac{x}{y}\right)^5$

18 $\left(\dfrac{1}{x} - \dfrac{x}{y}\right)^5$ **19** $(x + \frac{1}{2}y)^6$ **20** $(\frac{1}{2}x - y)^6$

21 Find the sixth term in the expansion of $(x - 3)^8$.

22 Find the fourth term in the expansion of $(x + 2)^8$.

23 Find the fifth term in the expansion of $(2x - y)^6$.

24 Find the sixth term in the expansion of $(3x - 2y)^7$.

25 Find the fourth term in the expansion of $(3x - \frac{1}{3}y)^9$.

26 Find the term that contains no x in the expansion of $(x - 1/x^3)^{12}$.

27 Find the term with the variable factor y^{-12} in the expansion of $(y + 1/y^3)^{12}$.

28 Find the term with the variable factor $x^{5/2}$ in the expansion of $(x^2 - 2/\sqrt{x})^{10}$.

29 Find the value of $(1.3)^4$ by writing it in the form $(1 + 0.3)^4$ and using the binomial theorem.

30 Find the value of $(1.2)^5$ by writing it in the form $(1 + 0.2)^5$ and using the binomial theorem.

C **31** Find the value of $(1.02)^{10}$ correct to four decimal places by expanding $(1 + 0.02)^{10}$ by the binomial theorem.

C **32** Find the value of $(1.005)^{20}$ correct to four decimal places by expanding $(1 + 0.005)^{20}$ by the binomial theorem.

33 Prove that $\binom{n}{k} = \binom{n}{n-k}$.

34 Prove that $\binom{n}{k-1} + \binom{n}{k} = \binom{n+1}{k}$.

USING YOUR KNOWLEDGE 14.6

The compound interest formula

$$A_n = P(1 + i)^n \tag{1}$$

is usually employed when n is a positive integer. However, in some financial transactions, money earns interest for a fraction of a period. In such cases, a bank may calculate the interest as simple interest for the fractional period, or it may use Formula 1 for the fractional period. In the latter case, it would be necessary to calculate the value of $(1 + i)^n$, where n is a fraction between 0 and 1. For example, if P dollars draws interest at 6 percent, compounded annually, and we want to use Formula 1 to find the interest for the first half year, then we would have to find

$$A_{1/2} = P(1.06)^{1/2}$$

to calculate the accumulated amount. This formula causes no difficulty if we have a large enough table of square roots or a calculator with a y^x key on it. Quite a large table of logarithms would be needed to give the accuracy required for most financial calculations.

To calculate an expression such as $(1.06)^{1/2}$ directly, we fall back on a generalization of the binomial theorem that is obtained in calculus. It is shown for n not an integer and for $-1 < x < 1$ that

$$(1 + x)^n = 1 + \binom{n}{1}x + \binom{n}{2}x^2 + \cdots + \binom{n}{k}x^k + \cdots \tag{2}$$

is a valid infinite series expansion in the sense that $(1 + x)^n$ can be calculated as accurately as is desired by taking enough consecutive terms of this series. In this expansion, we use the formula

$$\binom{n}{k} = \frac{n(n-1)(n-2)\cdots(n-k+1)}{k!}$$

For instance,

$$\binom{\frac{1}{2}}{3} = \frac{(\frac{1}{2})(-\frac{1}{2})(-\frac{3}{2})}{3!} = \frac{1}{16}$$

To calculate $(1.06)^{1/2}$, we take $n = \frac{1}{2}$ and $x = 0.06$, so that

$$\begin{aligned}(1.06)^{1/2} &= (1 + 0.06)^{1/2} \\ &= 1 + \binom{\frac{1}{2}}{1}(0.06) + \binom{\frac{1}{2}}{2}(0.06)^2 + \binom{\frac{1}{2}}{3}(0.06)^3 + \cdots \\ &= 1 + \frac{\frac{1}{2}}{1!}(0.06) + \frac{(\frac{1}{2})(-\frac{1}{2})}{2!}(0.06)^2 + \frac{(\frac{1}{2})(-\frac{1}{2})(-\frac{3}{2})}{3!}(0.06)^3 + \cdots \\ &= 1 + 0.03 - 0.00045 + 0.0000135 - \cdots \\ &= 1.029563, \text{ correct to six decimal places}\end{aligned}$$

The last result shows that it is slightly to the bank's advantage to use the formula rather than to pay simple interest for the fractional period. For example, if the rate is 6 percent, compounded annually, then simple interest would mean 3 percent for the half year, whereas the compound interest formula would mean 2.9563 percent. It is shown in calculus that if $0 < x < 1$ and $0 < n < 1$, then $(1 + x)^n < 1 + nx$.

Calculate each of the following, correct to six decimal places, by using the binomial series given in Formula 2.

1 $(1.06)^{1/3}$ **2** $(1.08)^{1/4}$ **3** $(1.05)^{2/5}$ **4** $(1.08)^{1/5}$ **5** $(1.05)^{1/6}$ **6** $(1.10)^{1/12}$

7 Use Formula 1 to calculate the accumulated value of \$1000 for $2\frac{1}{2}$ years at 6 percent interest compounded annually.

Hint: $(1.06)^{5/2} = (1.06)^2(1.06)^{1/2}$.

8 Use Formula 1 to find the accumulated amount of \$1000 for $2\frac{1}{4}$ years at 8 percent, compounded annually.

Note: See Problem 7.

SELF-TEST

1 A certain sequence is defined by $a_1 = 1$, $a_2 = 2$, and for $n > 2$, $a_n = a_{n-1}a_{n-2}$. Find:
a a_4 **b** a_5 **c** a_6

2 The general term of a sequence is $a_n = 3/(10^n)$.
a Write a_1, a_2, and a_3 in decimal form.
b Write the series $a_1 + a_2 + a_3 + \cdots$ as a common fraction.

3 a Use summation notation to write the series

$$1^2 + 2^2 + 3^2 + \cdots + n^2$$

b Write as a single summation:

$$\sum_{i=1}^{4} a_i + \sum_{i=5}^{9} a_i$$

c Write as the difference of the sum of the a's and the sum of the b's:

$$\sum_{i=1}^{n} (a_i - b_i)$$

4 Does the formula

$$1^2 + 2^2 + 3^2 + \cdots + n^2 = \frac{8n^3 - 33n^2 + 67n - 36}{6}$$

hold for all natural numbers n?

5 Use mathematical induction to prove that

$$1 + 3 + 5 + \cdots + (2n - 1) = n^2$$

for all natural numbers n.

6 The fifth and the eleventh terms of an arithmetic progression are 2 and 14, respectively. Find:
a the nth term of the progression;
b the fifteenth term;
c the sum of the first 15 terms.

7 For a certain arithmetic progression with n terms, $a_1 = 7$, $d = 6$, and $S_n = 480$. Find the value of n.

8 An interesting type of sale is conducted by a certain store. The price of an article is decreased by 5 cents the first day, by 10 additional cents the second day, by 15 additional cents the third day, and so on, until either the article is sold or the sale ends. Until what day of the sale would you have to wait to buy an article originally priced at \$22.25 if you were willing to pay \$20 for it?

9 In the geometric progression $2, \frac{1}{2}, \frac{1}{8}, \ldots$, find the seventh term and the sum of the first seven terms.

10 The general term of a geometric progression is $(\frac{3}{2})^{n-1}$. Find the sum of the first eight terms.

11 The sum of the first three terms of a geometric progression is 248, and the common ratio is 5. Find the first three terms of this progression.

12 Find the sum of the infinite geometric series $-1 + \frac{1}{3} - \frac{1}{9} + \cdots$.

13 There are no direct roads from Ashton to Clayville, but there are three roads from Ashton to Baytown and five roads from Baytown to Clayville. If you wish to drive from Ashton to Clayville and return without driving over the same road twice, in how many ways could you do this?

14 Find the number of four-digit positive integers that can be formed from the digits 1, 3, 5, 6, 7, 8,
a if repetition of digits is not allowed;
b if repetition of digits is allowed.

15 A committee of three is to be formed from a group of ten people. How many different choices are possible for this committee?

16 How many different 13-card hands can be selected from a deck of 52 cards? Leave your answer in the $C(n, k)$ form without evaluating it.

17 A committee of four is to be formed from a group of four men and six women. If the committee is to consist of two men and two women, how many choices are there for this committee?

18 Evaluate:

a $\binom{12}{4}$ **b** $\binom{14}{0}$ **c** $\binom{10}{5}$

19 a Expand and write in simplified form:

$$\left(2x - \frac{1}{2x}\right)^4$$

b Find the sixth term in the expansion of $(y - 2)^9$.

20 Find the term whose variable factor is x^{-4} in the expansion of

$$\left(x - \frac{1}{2x}\right)^{10}$$

APPENDIX

TABLE A1 *Powers, Roots, Reciprocals*

N	N^2	$\sqrt{N}$	$\sqrt{10N}$	N^3	$\sqrt[3]{N}$	$\sqrt[3]{10N}$	$\sqrt[3]{100N}$	$1000/N$
1	1	1.00 000	3.16 228	1	1.00 000	2.15 443	4.64 159	1000.00
2	4	1.41 421	4.47 214	8	1.25 992	2.71 442	5.84 804	500.00 0
3	9	1.73 205	5.47 723	27	1.44 225	3.10 723	6.69 433	333.33 3
4	16	2.00 000	6.32 456	64	1.58 740	3.41 995	7.36 806	250.00 0
5	25	2.23 607	7.07 107	125	1.70 998	3.68 403	7.93 701	200.00 0
6	36	2.44 949	7.74 597	216	1.81 712	3.91 487	8.43 433	166.66 7
7	49	2.64 575	8.36 660	343	1.91 293	4.12 129	8.87 904	142.85 7
8	64	2.82 843	8.94 427	512	2.00 000	4.30 887	9.28 318	125.00 0
9	81	3.00 000	9.48 683	729	2.08 008	4.48 140	9.65 489	111.11 1
10	100	3.16 228	10.00 00	1 000	2.15 443	4.64 159	10.00 00	100.00 0
11	121	3.31 662	10.48 81	1 331	2.22 398	4.79 142	10.32 28	90.90 91
12	144	3.46 410	10.95 45	1 728	2.28 943	4.93 242	10.62 66	83.33 33
13	169	3.60 555	11.40 18	2 197	2.35 133	5.06 580	10.91 39	76.92 31
14	196	3.74 166	11.83 22	2 744	2.41 014	5.19 249	11.18 69	71.42 86
15	225	3.87 298	12.24 74	3 375	2.46 621	5.31 329	11.44 71	66.66 67
16	256	4.00 000	12.64 91	4 096	2.51 984	5.42 884	11.69 61	62.50 00
17	289	4.12 311	13.03 84	4 913	2.57 128	5.53 966	11.93 48	58.82 35
18	324	4.24 264	13.41 64	5 832	2.62 074	5.64 622	12.16 44	55.55 56
19	361	4.35 890	13.78 40	6 859	2.66 840	5.74 890	12.38 56	52.63 16
20	400	4.47 214	14.14 21	8 000	2.71 442	5.84 804	12.59 92	50.00 00

TABLE A1 *(Continued)*

N	N^2	$\sqrt{N}$	$\sqrt{10N}$	N^3	$\sqrt[3]{N}$	$\sqrt[3]{10N}$	$\sqrt[3]{100N}$	$1000/N$
21	441	4.58 258	14.49 14	9 261	2.75 892	5.94 392	12.80 58	47.61 90
22	484	4.69 042	14.83 24	10 648	2.80 204	6.03 681	13.00 59	45.45 45
23	529	4.79 583	15.16 58	12 167	2.84 387	6.12 693	13.20 01	43.47 83
24	576	4.89 898	15.49 19	13 824	2.88 450	6.21 446	13.38 87	41.66 67
25	625	5.00 000	15.81 14	15 625	2.92 402	6.29 961	13.57 21	40.00 00
26	676	5.09 902	16.12 45	17 576	2.96 250	6.38 250	13.75 07	38.46 15
27	729	5.19 615	16.43 17	19 683	3.00 000	6.46 330	13.92 48	37.03 70
28	784	5.29 150	16.73 32	21 952	3.03 659	6.54 213	14.09 46	35.71 43
29	841	5.38 516	17.02 94	24 389	3.07 232	6.61 911	14.26 04	34.48 28
30	900	5.47 723	17.32 05	27 000	3.10 723	6.69 433	14.42 25	33.33 33
31	961	5.56 776	17.60 68	29 791	3.14 138	6.76 790	14.58 10	32.25 81
32	1 024	5.65 685	17.88 85	32 768	3.17 480	6.83 990	14.73 61	31.25 00
33	1 089	5.74 456	18.16 59	35 937	3.20 753	6.91 042	14.88 81	30.30 30
34	1 156	5.83 095	18.43 91	39 304	3.23 961	6.97 953	15.03 69	29.41 18
35	1 225	5.91 608	18.70 83	42 875	3.27 107	7.04 730	15.18 29	28.57 14
36	1 296	6.00 000	18.97 37	46 656	3.30 193	7.11 379	15.32 62	27.77 78
37	1 369	6.08 276	19.23 54	50 653	3.33 222	7.17 905	15.46 68	27.02 70
38	1 444	6.16 441	19.49 36	54 872	3.36 198	7.24 316	15.60 49	26.31 58
39	1 521	6.24 500	19.74 84	59 319	3.39 121	7.30 614	15.74 06	25.64 10
40	1 600	6.32 456	20.00 00	64 000	3.41 995	7.36 806	15.87 40	25.00 00
41	1 681	6.40 312	20.24 85	68 921	3.44 822	7.42 896	16.00 52	24.39 02
42	1 764	6.48 074	20.49 39	74 088	3.47 603	7.48 887	16.13 43	23.80 95
43	1 849	6.55 744	20.73 64	79 507	3.50 340	7.54 784	16.26 13	23.25 58
44	1 936	6.63 325	20.97 62	85 184	3.53 035	7.60 590	16.38 64	22.72 73
45	2 025	6.70 820	21.21 32	91 125	3.55 689	7.66 309	16.50 96	22.22 22
46	2 116	6.78 233	21.44 76	97 336	3.58 305	7.71 944	16.63 10	21.73 91
47	2 209	6.85 565	21.67 95	103 823	3.60 883	7.77 498	16.75 07	21.27 66
48	2 304	6.92 820	21.90 89	110 592	3.63 424	7.82 974	16.86 87	20.83 33
49	2 401	7.00 000	22.13 59	117 649	3.65 931	7.88 374	16.98 50	20.40 82
50	2 500	7.07 107	22.36 07	125 000	3.68 403	7.93 701	17.09 98	20.00 00
51	2 601	7.14 143	22.58 32	132 651	3.70 843	7.98 957	17.21 30	19.60 78
52	2 704	7.21 110	22.80 35	140 608	3.73 251	8.04 145	17.32 48	19.23 08
53	2 809	7.28 011	23.02 17	148 877	3.75 629	8.09 267	17.43 51	18.86 79
54	2 916	7.34 847	23.23 79	157 464	3.77 976	8.14 325	17.54 41	18.51 85
55	3 025	7.41 620	23.45 21	166 375	3.80 295	8.19 321	17.65 17	18.18 18
56	3 136	7.48 331	23.66 43	175 616	3.82 586	8.24 257	17.75 81	17.85 71
57	3 249	7.54 983	23.87 47	185 193	3.84 850	8.29 134	17.86 32	17.54 39
58	3 364	7.61 577	24.08 32	195 112	3.87 088	8.33 955	17.96 70	17.24 14
59	3 481	7.68 115	24.28 99	205 379	3.89 300	8.38 721	18.06 97	16.94 92
60	3 600	7.74 597	24.49 49	216 000	3.91 487	8.43 433	18.17 12	16.66 67
N	N^2	$\sqrt{N}$	$\sqrt{10N}$	N^3	$\sqrt[3]{N}$	$\sqrt[3]{10N}$	$\sqrt[3]{100N}$	$1000/N$

TABLE A1 (*Continued*)

N	N^2	$\sqrt{N}$	$\sqrt{10N}$	N^3	$\sqrt[3]{N}$	$\sqrt[3]{10N}$	$\sqrt[3]{100N}$	$1000/N$
61	3 721	7.81 025	24.69 82	226 981	3.93 650	8.48 093	18.27 16	16.39 34
62	3 844	7.87 401	24.89 98	238 328	3.95 789	8.52 702	18.37 09	16.12 90
63	3 969	7.93 725	25.09 98	250 047	3.97 906	8.57 262	18.46 91	15.87 30
64	4 096	8.00 000	25.29 82	262 144	4.00 000	8.61 774	18.56 64	15.62 50
65	4 225	8.06 226	25.49 51	274 625	4.02 073	8.66 239	18.66 26	15.38 46
66	4 356	8.12 404	25.69 05	287 496	4.04 124	8.70 659	18.75 78	15.15 15
67	4 489	8.18 535	25.88 44	300 763	4.06 155	8.75 034	18.85 20	14.92 54
68	4 624	8.24 621	26.07 68	314 432	4.08 166	8.79 366	18.94 54	14.70 59
69	4 761	8.30 662	26.26 79	328 509	4.10 157	8.83 656	19.03 78	14.49 28
70	4 900	8.36 660	26.45 75	343 000	4.12 129	8.87 904	19.12 93	14.28 57
71	5 041	8.42 615	26.64 58	357 911	4.14 082	8.92 112	19.22 00	14.08 45
72	5 184	8.48 528	26.83 28	373 248	4.16 017	8.96 281	19.30 98	13.88 89
73	5 329	8.54 400	27.01 85	389 017	4.17 934	9.00 411	19.39 88	13.69 86
74	5 476	8.60 233	27.20 29	405 224	4.19 834	9.04 504	19.48 70	13.51 35
75	5 625	8.66 025	27.38 61	421 875	4.21 716	9.08 560	19.57 43	13.33 33
76	5 776	8.71 780	27.56 81	438 976	4.23 582	9.12 581	19.66 10	13.15 79
77	5 929	8.77 496	27.74 89	456 533	4.25 432	9.16 566	19.74 68	12.98 70
78	6 084	8.83 176	27.92 85	474 552	4.27 266	9.20 516	19.83 19	12.82 05
79	6 241	8.88 819	28.10 69	493 039	4.29 084	9.24 434	19.91 63	12.65 82
80	6 400	8.94 427	28.28 43	512 000	4.30 887	9.28 318	20.00 00	12.50 00
81	6 561	9.00 000	28.46 05	531 441	4.32 675	9.32 170	20.08 30	12.34 57
82	6 724	9.05 539	28.63 56	551 368	4.34 448	9.35 990	20.16 53	12.19 51
83	6 889	9.11 043	28.80 97	571 787	4.36 207	9.39 780	20.24 69	12.04 82
84	7 056	9.16 515	28.98 28	592 704	4.37 952	9.43 539	20.32 79	11.90 48
85	7 225	9.21 954	29.15 48	614 125	4.39 683	9.47 268	20.40 83	11.76 47
86	7 396	9.27 362	29.32 58	636 056	4.41 400	9.50 969	20.48 80	11.62 79
87	7 569	9.32 738	29.49 58	658 503	4.43 105	9.54 640	20.56 71	11.49 43
88	7 744	9.38 083	29.66 48	681 472	4.44 796	9.58 284	20.64 56	11.36 36
89	7 921	9.43 398	29.83 29	704 969	4.46 475	9.61 900	20.72 35	11.23 60
90	8 100	9.48 683	30.00 00	729 000	4.48 140	9.65 489	20.80 08	11.11 11
91	8 281	9.53 939	30.16 62	753 571	4.49 794	9.69 052	20.87 76	10.98 90
92	8 464	9.59 166	30.33 15	778 688	4.51 436	9.72 589	20.95 38	10.86 96
93	8 649	9.64 365	30.49 59	804 357	4.53 065	9.76 100	21.02 94	10.75 27
94	8 836	9.69 536	30.65 94	830 584	4.54 684	9.79 586	21.10 45	10.63 83
95	9 025	9.74 679	30.82 21	857 375	4.56 290	9.83 048	21.17 91	10.52 63
96	9 216	9.79 796	30.98 39	884 736	4.57 886	9.86 485	21.25 32	10.41 67
97	9 409	9.84 886	31.14 48	912 673	4.59 470	9.89 898	21.32 67	10.30 93
98	9 604	9.89 949	31.30 50	941 192	4.61 044	9.93 288	21.39 97	10.20 41
99	9 801	9.94 987	31.46 43	970 299	4.62 607	9.96 655	21.47 23	10.10 10
100	10 000	10.00 000	31.62 28	1 000 000	4.64 159	10.00 000	21.54 43	10.00 00
N	N^2	$\sqrt{N}$	$\sqrt{10N}$	N^3	$\sqrt[3]{N}$	$\sqrt[3]{10N}$	$\sqrt[3]{100N}$	$1000/N$

TABLE A2 *Four-Place Common Logarithms*

x	0	1	2	3	4	5	6	7	8	9
1.0	.0000	.0043	.0086	.0128	.0170	.0212	.0253	.0294	.0334	.0374
1.1	.0414	.0453	.0492	.0531	.0569	.0607	.0645	.0682	.0719	.0755
1.2	.0792	.0828	.0864	.0899	.0934	.0969	.1004	.1038	.1072	.1106
1.3	.1139	.1173	.1206	.1239	.1271	.1303	.1335	.1367	.1399	.1430
1.4	.1461	.1492	.1523	.1553	.1584	.1614	.1644	.1673	.1703	.1732
1.5	.1761	.1790	.1818	.1847	.1875	.1903	.1931	.1959	.1987	.2014
1.6	.2041	.2068	.2095	.2122	.2148	.2175	.2201	.2227	.2253	.2279
1.7	.2304	.2330	.2355	.2380	.2405	.2430	.2455	.2480	.2504	.2529
1.8	.2553	.2577	.2601	.2625	.2648	.2672	.2695	.2718	.2742	.2765
1.9	.2788	.2810	.2833	.2856	.2878	.2900	.2923	.2945	.2967	.2989
2.0	.3010	.3032	.3054	.3075	.3096	.3118	.3139	.3160	.3181	.3201
2.1	.3222	.3243	.3263	.3284	.3304	.3324	.3345	.3365	.3385	.3404
2.2	.3424	.3444	.3464	.3483	.3502	.3522	.3541	.3560	.3579	.3598
2.3	.3617	.3636	.3655	.3674	.3692	.3711	.3729	.3747	.3766	.3784
2.4	.3802	.3820	.3838	.3856	.3874	.3892	.3909	.3927	.3945	.3962
2.5	.3979	.3997	.4014	.4031	.4048	.4065	.4082	.4099	.4116	.4133
2.6	.4150	.4166	.4183	.4200	.4216	.4232	.4249	.4265	.4281	.4298
2.7	.4314	.4330	.4346	.4362	.4378	.4393	.4409	.4425	.4440	.4456
2.8	.4472	.4487	.4502	.4518	.4533	.4548	.4564	.4579	.4594	.4609
2.9	.4624	.4639	.4654	.4669	.4683	.4698	.4713	.4728	.4742	.4757
3.0	.4771	.4786	.4800	.4814	.4829	.4843	.4857	.4871	.4886	.4900
3.1	.4914	.4928	.4942	.4955	.4969	.4983	.4997	.5011	.5024	.5038
3.2	.5051	.5065	.5079	.5092	.5105	.5119	.5132	.5145	.5159	.5172
3.3	.5185	.5198	.5211	.5224	.5237	.5250	.5263	.5276	.5289	.5302
3.4	.5315	.5328	.5340	.5353	.5366	.5378	.5391	.5403	.5416	.5428
3.5	.5441	.5453	.5465	.5478	.5490	.5502	.5514	.5527	.5539	.5551
3.6	.5563	.5575	.5587	.5599	.5611	.5623	.5635	.5647	.5658	.5670
3.7	.5682	.5694	.5705	.5717	.5729	.5740	.5752	.5763	.5775	.5786
3.8	.5798	.5809	.5821	.5832	.5843	.5855	.5866	.5877	.5888	.5899
3.9	.5911	.5922	.5933	.5944	.5955	.5966	.5977	.5988	.5999	.6010
4.0	.6021	.6031	.6042	.6053	.6064	.6075	.6085	.6096	.6107	.6117
4.1	.6128	.6138	.6149	.6160	.6170	.6180	.6191	.6201	.6212	.6222
4.2	.6232	.6243	.6253	.6263	.6274	.6284	.6294	.6304	.6314	.6325
4.3	.6335	.6345	.6355	.6365	.6375	.6385	.6395	.6405	.6415	.6425
4.4	.6435	.6444	.6454	.6464	.6474	.6484	.6493	.6503	.6513	.6522
4.5	.6532	.6542	.6551	.6561	.6571	.6580	.6590	.6599	.6609	.6618
4.6	.6628	.6637	.6646	.6656	.6665	.6675	.6684	.6693	.6702	.6712
4.7	.6721	.6730	.6739	.6749	.6758	.6767	.6776	.6785	.6794	.6803
4.8	.6812	.6821	.6830	.6839	.6848	.6857	.6866	.6875	.6884	.6893
4.9	.6902	.6911	.6920	.6928	.6937	.6946	.6955	.6964	.6972	.6981
5.0	.6990	.6998	.7007	.7016	.7024	.7033	.7042	.7050	.7059	.7067
5.1	.7076	.7084	.7093	.7101	.7110	.7118	.7126	.7135	.7143	.7152
5.2	.7160	.7168	.7177	.7185	.7193	.7202	.7210	.7218	.7226	.7235
5.3	.7243	.7251	.7259	.7267	.7275	.7284	.7292	.7300	.7308	.7316
5.4	.7324	.7332	.7340	.7348	.7356	.7364	.7372	.7380	.7388	.7396
x	0	1	2	3	4	5	6	7	8	9

TABLE A2 *(Continued)*

x	0	1	2	3	4	5	6	7	8	9
5.5	.7404	.7412	.7419	.7427	.7435	.7443	.7451	.7459	.7466	.7474
5.6	.7482	.7490	.7497	.7505	.7513	.7520	.7528	.7536	.7543	.7551
5.7	.7559	.7566	.7574	.7582	.7589	.7597	.7604	.7612	.7619	.7627
5.8	.7634	.7642	.7649	.7657	.7664	.7672	.7679	.7686	.7694	.7701
5.9	.7709	.7716	.7723	.7731	.7738	.7745	.7752	.7760	.7767	.7774
6.0	.7782	.7789	.7796	.7803	.7810	.7818	.7825	.7832	.7839	.7846
6.1	.7853	.7860	.7868	.7875	.7882	.7889	.7896	.7903	.7910	.7917
6.2	.7924	.7931	.7938	.7945	.7952	.7959	.7966	.7973	.7980	.7987
6.3	.7993	.8000	.8007	.8014	.8021	.8028	.8035	.8041	.8048	.8055
6.4	.8062	.8069	.8075	.8082	.8089	.8096	.8102	.8109	.8116	.8122
6.5	.8129	.8136	.8142	.8149	.8156	.8162	.8169	.8176	.8182	.8189
6.6	.8195	.8202	.8209	.8215	.8222	.8228	.8235	.8241	.8248	.8254
6.7	.8261	.8267	.8274	.8280	.8287	.8293	.8299	.8306	.8312	.8319
6.8	.8325	.8331	.8338	.8344	.8351	.8357	.8363	.8370	.8376	.8382
6.9	.8388	.8395	.8401	.8407	.8414	.8420	.8426	.8432	.8439	.8445
7.0	.8451	.8457	.8463	.8470	.8476	.8482	.8488	.8494	.8500	.8506
7.1	.8513	.8519	.8525	.8531	.8537	.8543	.8549	.8555	.8561	.8567
7.2	.8573	.8579	.8585	.8591	.8597	.8603	.8609	.8615	.8621	.8627
7.3	.8633	.8639	.8645	.8651	.8657	.8663	.8669	.8675	.8681	.8686
7.4	.8692	.8698	.8704	.8710	.8716	.8722	.8727	.8733	.8739	.8745
7.5	.8751	.8756	.8762	.8768	.8774	.8779	.8785	.8791	.8797	8802
7.6	.8808	.8814	.8820	.8825	.8831	.8837	.8842	.8848	.8854	.8859
7.7	.8865	.8871	.8876	.8882	.8887	.8893	.8899	.8904	.8910	.8915
7.8	.8921	.8927	.8932	.8938	.8943	.8949	.8954	.8960	.8965	.8971
7.9	.8976	.8982	.8987	.8993	.8998	.9004	.9009	.9015	.9020	.9026
8.0	.9031	.9036	.9042	.9047	.9053	.9058	.9063	.9069	.9074	.9079
8.1	.9085	.9090	.9096	.9101	.9106	.9112	.9117	.9122	.9128	.9133
8.2	.9138	.9143	.9149	.9154	.9159	.9165	.9170	.9175	.9180	.9186
8.3	.9191	.9196	.9201	.9206	.9212	.9217	.9222	.9227	.9232	.9238
8.4	.9243	.9248	.9253	.9258	.9263	.9269	.9274	.9279	.9284	.9289
8.5	.9294	.9299	.9304	.9309	.9315	.9320	.9325	.9330	.9335	.9340
8.6	.9345	.9350	.9355	.9360	.9365	.9370	.9375	.9380	.9385	.9390
8.7	.9395	.9400	.9405	.9410	.9415	.9420	.9425	.9430	.9435	.9440
8.8	.9445	.9450	.9455	.9460	.9465	.9469	.9474	.9479	.9484	.9489
8.9	.9494	.9499	.9504	.9509	.9513	.9518	.9523	.9528	.9533	.9538
9.0	.9542	.9547	.9552	.9557	.9562	.9566	.9571	.9576	.9581	.9586
9.1	.9590	.9595	.9600	.9605	.9609	.9614	.9619	.9624	.9628	.9633
9.2	.9638	.9643	.9647	.9652	.9657	.9661	.9666	.9671	.9675	.9680
9.3	.9685	.9689	.9694	.9699	.9703	.9708	.9713	.9717	.9722	.9727
9.4	.9731	.9736	.9741	.9745	.9750	.9754	.9759	.9763	.9768	.9773
9.5	.9777	.9782	.9786	.9791	.9795	.9800	.9805	.9809	.9814	.9818
9.6	.9823	.9827	.9832	.9836	.9841	.9845	.9850	.9854	.9859	.9863
9.7	.9868	.9872	.9877	.9881	.9886	.9890	.9894	.9899	.9903	.9908
9.8	.9912	.9917	.9921	.9926	.9930	.9934	.9939	.9943	.9948	.9952
9.9	.9956	.9961	.9965	.9969	.9974	.9978	.9983	.9987	.9991	.9996
x	0	1	2	3	4	5	6	7	8	9

TABLE A3 *Four-Place Natural Logarithms*

To find ln N when N is beyond the range of this table, write N in the form $P \times 10^m$, where P lies within the range of the table, and m is a positive or negative integer. Then use

$$\ln N = \ln(P \times 10^m) = \ln P + m \ln 10$$

ln 10 = 2.3026	3 ln 10 = 6.9076	5 ln 10 = 11.5129	7 ln 10 = 16.1181	9 ln 10 = 20.7233
2 ln 10 = 4.6052	4 ln 10 = 9.2103	6 ln 10 = 13.8155	8 ln 10 = 18.4207	10 ln 10 = 23.0259

N	.00	.01	.02	.03	.04	.05	.06	.07	.08	.09
1.0	0.0000	0.0100	0.0198	0.0296	0.0392	0.0488	0.0583	0.0677	0.0770	0.0862
1.1	0.0953	0.1044	0.1133	0.1222	0.1310	0.1398	0.1484	0.1570	0.1655	0.1740
1.2	0.1823	0.1906	0.1989	0.2070	0.2151	0.2231	0.2311	0.2390	0.2469	0.2546
1.3	0.2624	0.2700	0.2776	0.2852	0.2927	0.3001	0.3075	0.3148	0.3221	0.3293
1.4	0.3365	0.3436	0.3507	0.3577	0.3646	0.3716	0.3784	0.3853	0.3920	0.3988
1.5	0.4055	0.4121	0.4187	0.4253	0.4318	0.4383	0.4447	0.4511	0.4574	0.4637
1.6	0.4700	0.4762	0.4824	0.4886	0.4947	0.5008	0.5068	0.5128	0.5188	0.5247
1.7	0.5306	0.5365	0.5423	0.5481	0.5539	0.5596	0.5653	0.5710	0.5766	0.5822
1.8	0.5878	0.5933	0.5988	0.6043	0.6098	0.6152	0.6206	0.6259	0.6313	0.6366
1.9	0.6419	0.6471	0.6523	0.6575	0.6627	0.6678	0.6729	0.6780	0.6831	0.6881
2.0	0.6931	0.6981	0.7031	0.7080	0.7129	0.7178	0.7227	0.7275	0.7324	0.7372
2.1	0.7419	0.7467	0.7514	0.7561	0.7608	0.7655	0.7701	0.7747	0.7793	0.7839
2.2	0.7885	0.7930	0.7975	0.8020	0.8065	0.8109	0.8154	0.8198	0.8242	0.8286
2.3	0.8329	0.8372	0.8416	0.8459	0.8502	0.8544	0.8587	0.8629	0.8671	0.8713
2.4	0.8755	0.8796	0.8838	0.8879	0.8920	0.8961	0.9002	0.9042	0.9083	0.9123
2.5	0.9163	0.9203	0.9243	0.9282	0.9322	0.9361	0.9400	0.9439	0.9478	0.9517
2.6	0.9555	0.9594	0.9632	0.9670	0.9708	0.9746	0.9783	0.9821	0.9858	0.9895
2.7	0.9933	0.9969	1.0006	1.0043	1.0080	1.0116	1.0152	1.0188	1.0225	1.0260
2.8	1.0296	1.0332	1.0367	1.0403	1.0438	1.0473	1.0508	1.0543	1.0578	1.0613
2.9	1.0647	1.0682	1.0716	1.0750	1.0784	1.0818	1.0852	1.0886	1.0919	1.0953
3.0	1.0986	1.1019	1.1053	1.1086	1.1119	1.1151	1.1184	1.1217	1.1249	1.1282
3.1	1.1314	1.1346	1.1378	1.1410	1.1442	1.1474	1.1506	1.1537	1.1569	1.1600
3.2	1.1632	1.1663	1.1694	1.1725	1.1756	1.1787	1.1817	1.1848	1.1878	1.1909
3.3	1.1939	1.1969	1.2000	1.2030	1.2060	1.2090	1.2119	1.2149	1.2179	1.2208
3.4	1.2238	1.2267	1.2296	1.2326	1.2355	1.2384	1.2413	1.2442	1.2470	1.2499
3.5	1.2528	1.2556	1.2585	1.2613	1.2641	1.2669	1.2698	1.2726	1.2754	1.2782
3.6	1.2809	1.2837	1.2865	1.2892	1.2920	1.2947	1.2975	1.3002	1.3029	1.3056
3.7	1.3083	1.3110	1.3137	1.3164	1.3191	1.3218	1.3244	1.3271	1.3297	1.3324
3.8	1.3350	1.3376	1.3403	1.3429	1.3455	1.3481	1.3507	1.3533	1.3558	1.3584
3.9	1.3610	1.3635	1.3661	1.3686	1.3712	1.3737	1.3762	1.3788	1.3813	1.3838
4.0	1.3863	1.3888	1.3913	1.3938	1.3962	1.3987	1.4012	1.4036	1.4061	1.4085
4.1	1.4110	1.4134	1.4159	1.4183	1.4207	1.4231	1.4255	1.4279	1.4303	1.4327
4.2	1.4351	1.4375	1.4398	1.4422	1.4446	1.4469	1.4493	1.4516	1.4540	1.4563
4.3	1.4586	1.4609	1.4633	1.4656	1.4679	1.4702	1.4725	1.4748	1.4770	1.4793
4.4	1.4816	1.4839	1.4861	1.4884	1.4907	1.4929	1.4951	1.4974	1.4996	1.5019
N	.00	.01	.02	.03	.04	.05	.06	.07	.08	.09

TABLE A3 (*Continued*)

N	.00	.01	.02	.03	.04	.05	.06	.07	.08	.09
4.5	1.5041	1.5063	1.5085	1.5107	1.5129	1.5151	1.5173	1.5195	1.5217	1.5239
4.6	1.5261	1.5282	1.5304	1.5326	1.5347	1.5369	1.5390	1.5412	1.5433	1.5454
4.7	1.5476	1.5497	1.5518	1.5539	1.5560	1.5581	1.5602	1.5623	1.5644	1.5665
4.8	1.5686	1.5707	1.5728	1.5748	1.5769	1.5790	1.5810	1.5831	1.5851	1.5872
4.9	1.5892	1.5913	1.5933	1.5953	1.5974	1.5994	1.6014	1.6034	1.6054	1.6074
5.0	1.6094	1.6114	1.6134	1.6154	1.6174	1.6194	1.6214	1.6233	1.6253	1.6273
5.1	1.6292	1.6312	1.6332	1.6351	1.6371	1.6390	1.6409	1.6429	1.6448	1.6467
5.2	1.6487	1.6506	1.6525	1.6544	1.6563	1.6582	1.6601	1.6620	1.6639	1.6658
5.3	1.6677	1.6696	1.6715	1.6734	1.6752	1.6771	1.6790	1.6808	1.6827	1.6845
5.4	1.6864	1.6882	1.6901	1.6919	1.6938	1.6956	1.6974	1.6993	1.7011	1.7029
5.5	1.7047	1.7066	1.7084	1.7102	1.7120	1.7138	1.7156	1.7174	1.7192	1.7210
5.6	1.7228	1.7246	1.7263	1.7281	1.7299	1.7317	1.7334	1.7352	1.7370	1.7387
5.7	1.7405	1.7422	1.7440	1.7457	1.7475	1.7492	1.7509	1.7527	1.7544	1.7561
5.8	1.7579	1.7596	1.7613	1.7630	1.7647	1.7664	1.7681	1.7699	1.7716	1.7733
5.9	1.7750	1.7766	1.7783	1.7800	1.7817	1.7834	1.7851	1.7867	1.7884	1.7901
6.0	1.7918	1.7934	1.7951	1.7967	1.7984	1.8001	1.8017	1.8034	1.8050	1.8066
6.1	1.8083	1.8099	1.8116	1.8132	1.8148	1.8165	1.8181	1.8197	1.8213	1.8229
6.2	1.8245	1.8262	1.8278	1.8294	1.8310	1.8326	1.8342	1.8358	1.8374	1.8390
6.3	1.8405	1.8421	1.8437	1.8453	1.8469	1.8485	1.8500	1.8516	1.8532	1.8547
6.4	1.8563	1.8579	1.8594	1.8610	1.8625	1.8641	1.8656	1.8672	1.8687	1.8703
6.5	1.8718	1.8733	1.8749	1.8764	1.8779	1.8795	1.8810	1.8825	1.8840	1.8856
6.6	1.8871	1.8886	1.8901	1.8916	1.8931	1.8946	1.8961	1.8976	1.8991	1.9006
6.7	1.9021	1.9036	1.9051	1.9066	1.9081	1.9095	1.9110	1.9125	1.9140	1.9155
6.8	1.9169	1.9184	1.9199	1.9213	1.9228	1.9242	1.9257	1.9272	1.9286	1.9301
6.9	1.9315	1.9330	1.9344	1.9359	1.9373	1.9387	1.9402	1.9416	1.9430	1.9445
7.0	1.9459	1.9473	1.9488	1.9502	1.9516	1.9530	1.9544	1.9559	1.9573	1.9587
7.1	1.9601	1.9615	1.9629	1.9643	1.9657	1.9671	1.9685	1.9699	1.9713	1.9727
7.2	1.9741	1.9755	1.9769	1.9782	1.9796	1.9810	1.9824	1.9838	1.9851	1.9865
7.3	1.9879	1.9892	1.9906	1.9920	1.9933	1.9947	1.9961	1.9974	1.9988	2.0001
7.4	2.0015	2.0028	2.0042	2.0055	2.0069	2.0082	2.0096	2.0109	2.0122	2.0136
7.5	2.0149	2.0162	2.0176	2.0189	2.0202	2.0215	2.0229	2.0242	2.0255	2.0268
7.6	2.0281	2.0295	2.0308	2.0321	2.0334	2.0347	2.0360	2.0373	2.0386	2.0399
7.7	2.0412	2.0425	2.0438	2.0451	2.0464	2.0477	2.0490	2.0503	2.0516	2.0528
7.8	2.0541	2.0554	2.0567	2.0580	2.0592	2.0605	2.0618	2.0631	2.0643	2.0656
7.9	2.0669	2.0681	2.0694	2.0707	2.0719	2.0732	2.0744	2.0757	2.0769	2.0782
8.0	2.0794	2.0807	2.0819	2.0832	2.0844	2.0857	2.0869	2.0882	2.0894	2.0906
8.1	2.0919	2.0931	2.0943	2.0956	2.0968	2.0980	2.0992	2.1005	2.1017	2.1029
8.2	2.1041	2.1054	2.1066	2.1078	2.1090	2.1102	2.1114	2.1126	2.1138	2.1150
8.3	2.1163	2.1175	2.1187	2.1199	2.1211	2.1223	2.1235	2.1247	2.1258	2.1270
8.4	2.1282	2.1294	2.1306	2.1318	2.1330	2.1342	2.1353	2.1365	2.1377	2.1389
N	.00	.01	.02	.03	.04	.05	.06	.07	.08	.09

TABLE A3 (*Continued*)

N	.00	.01	.02	.03	.04	.05	.06	.07	.08	.09
8.5	2.1401	2.1412	2.1424	2.1436	2.1448	2.1459	2.1471	2.1483	2.1494	2.1506
8.6	2.1518	2.1529	2.1541	2.1552	2.1564	2.1576	2.1587	2.1599	2.1610	2.1622
8.7	2.1633	2.1645	2.1656	2.1668	2.1679	2.1691	2.1702	2.1713	2.1725	2.1736
8.8	2.1748	2.1759	2.1770	2.1782	2.1793	2.1804	2.1815	2.1827	2.1838	2.1849
8.9	2.1861	2.1872	2.1883	2.1894	2.1905	2.1917	2.1928	2.1939	2.1950	2.1961
9.0	2.1972	2.1983	2.1994	2.2006	2.2017	2.2028	2.2039	2.2050	2.2061	2.2072
9.1	2.2083	2.2094	2.2105	2.2116	2.2127	2.2138	2.2148	2.2159	2.2170	2.2181
9.2	2.2192	2.2203	2.2214	2.2225	2.2235	2.2246	2.2257	2.2268	2.2279	2.2289
9.3	2.2300	2.2311	2.2322	2.2332	2.2343	2.2354	2.2364	2.2375	2.2386	2.2396
9.4	2.2407	2.2418	2.2428	2.2439	2.2450	2.2460	2.2471	2.2481	2.2492	2.2502
9.5	2.2513	2.2523	2.2534	2.2544	2.2555	2.2565	2.2576	2.2586	2.2597	2.2607
9.6	2.2618	2.2628	2.2638	2.2649	2.2659	2.2670	2.2680	2.2690	2.2701	2.2711
9.7	2.2721	2.2732	2.2742	2.2752	2.2762	2.2773	2.2783	2.2793	2.2803	2.2814
9.8	2.2824	2.2834	2.2844	2.2854	2.2865	2.2875	2.2885	2.2895	2.2905	2.2915
9.9	2.2925	2.2935	2.2946	2.2956	2.2966	2.2976	2.2986	2.2996	2.3006	2.3016
10.0	2.3026	2.3036	2.3046	2.3056	2.3066	2.3076	2.3086	2.3096	2.3106	2.3115

TABLE A4 *Exponential Functions*

x	e^x	e^{-x}	x	e^x	e^{-x}	x	e^x	e^{-x}
0.00	1.0000	1.00000	**0.20**	1.2214	.81873	**0.40**	1.4918	.67032
0.01	1.0101	0.99005	**0.21**	1.2337	.81058	**0.41**	1.5068	.66365
0.02	1.0202	.98020	**0.22**	1.2461	.80252	**0.42**	1.5220	.65705
0.03	1.0305	.97045	**0.23**	1.2586	.79453	**0.43**	1.5373	.65051
0.04	1.0408	.96079	**0.24**	1.2712	.78663	**0.44**	1.5527	.64404
0.05	1.0513	.95123	**0.25**	1.2840	.77880	**0.45**	1.5683	.63763
0.06	1.0618	.94176	**0.26**	1.2969	.77105	**0.46**	1.5841	.63128
0.07	1.0725	.93239	**0.27**	1.3100	.76338	**0.47**	1.6000	.62500
0.08	1.0833	.92312	**0.28**	1.3231	.75578	**0.48**	1.6161	.61878
0.09	1.0942	.91393	**0.29**	1.3364	.74826	**0.49**	1.6323	.61263
0.10	1.1052	.90484	**0.30**	1.3499	.74082	**0.50**	1.6487	.60653
0.11	1.1163	.89583	**0.31**	1.3634	.73345	**0.51**	1.6653	.60050
0.12	1.1275	.88692	**0.32**	1.3771	.72615	**0.52**	1.6820	.59452
0.13	1.1388	.87810	**0.33**	1.3910	.71892	**0.53**	1.6989	.58860
0.14	1.1503	.86936	**0.34**	1.4049	.71177	**0.54**	1.7160	.58275
0.15	1.1618	.86071	**0.35**	1.4191	.70469	**0.55**	1.7333	.57695
0.16	1.1735	.85214	**0.36**	1.4333	.69768	**0.56**	1.7507	.57121
0.17	1.1853	.84366	**0.37**	1.4477	.69073	**0.57**	1.7683	.56553
0.18	1.1972	.83527	**0.38**	1.4623	.68386	**0.58**	1.7860	.55990
0.19	1.2092	.82696	**0.39**	1.4770	.67706	**0.59**	1.8040	.55433

TABLE A4 *(Continued)*

x	e^x	e^{-x}
0.60	1.8221	.54881
0.61	1.8404	.54335
0.62	1.8589	.53794
0.63	1.8776	.53259
0.64	1.8965	.52729
0.65	1.9155	.52205
0.66	1.9348	.51685
0.67	1.9542	.51171
0.68	1.9739	.50662
0.69	1.9937	.50158
0.70	2.0138	.49659
0.71	2.0340	.49164
0.72	2.0544	.48675
0.73	2.0751	.48191
0.74	2.0959	.47711
0.75	2.1170	.47237
0.76	2.1383	.46767
0.77	2.1598	.46301
0.78	2.1815	.45841
0.79	2.2034	.45384
0.80	2.2255	.44933
0.81	2.2479	.44486
0.82	2.2705	.44043
0.83	2.2933	.43605
0.84	2.3164	.43171
0.85	2.3396	.42741
0.86	2.3632	.42316
0.87	2.3869	.41895
0.88	2.4109	.41478
0.89	2.4351	.41066
0.90	2.4596	.40657
0.91	2.4843	.40252
0.92	2.5093	.39852
0.93	2.5345	.39455
0.94	2.5600	.39063
0.95	2.5857	.38674
0.96	2.6117	.38289
0.97	2.6379	.37908
0.98	2.6645	.37531
0.99	2.6912	.37158
1.00	2.7183	.36788
1.01	2.7456	.36422
1.02	2.7732	.36059
1.03	2.8011	.35701
1.04	2.8292	.35345
1.05	2.8577	.34994
1.06	2.8864	.34646
1.07	2.9154	.34301
1.08	2.9447	.33960
1.09	2.9743	.33622
1.10	3.0042	.33287
1.11	3.0344	.32956
1.12	3.0649	.32628
1.13	3.0957	.32303
1.14	3.1268	.31982
1.15	3.1582	.31664
1.16	3.1899	.31349
1.17	3.2220	.31037
1.18	3.2544	.30728
1.19	3.2871	.30422
1.20	3.3201	.30119
1.21	3.3535	.29820
1.22	3.3872	.29523
1.23	3.4212	.29229
1.24	3.4556	.28938
1.25	3.4903	.28650
1.26	3.5254	.28365
1.27	3.5609	.28083
1.28	3.5966	.27804
1.29	3.6328	.27527
1.30	3.6693	.27253
1.31	3.7062	.26982
1.32	3.7434	.26714
1.33	3.7810	.26448
1.34	3.8190	.26185
1.35	3.8574	.25924
1.36	3.8962	.25666
1.37	3.9354	.25411
1.38	3.9749	.25158
1.39	4.0149	.24908
1.40	4.0552	.24660
1.41	4.0960	.24414
1.42	4.1371	.24171
1.43	4.1787	.23931
1.44	4.2207	.23693
1.45	4.2631	.23457
1.46	4.3060	.23224
1.47	4.3492	.22993
1.48	4.3929	.22764
1.49	4.4371	.22537
1.50	4.4817	.22313
1.51	4.5267	.22091
1.52	4.5722	.21871
1.53	4.6182	.21654
1.54	4.6646	.21438
1.55	4.7115	.21225
1.56	4.7588	.21014
1.57	4.8066	.20805
1.58	4.8550	.20598
1.59	4.9037	.20393
1.60	4.9530	.20190
1.61	5.0028	.19989
1.62	5.0531	.19790
1.63	5.1039	.19593
1.64	5.1552	.19398
1.65	5.2070	.19205
1.66	5.2593	.19014
1.67	5.3122	.18825
1.68	5.3656	.18637
1.69	5.4195	.18452
1.70	5.4739	.18268
1.71	5.5290	.18087
1.72	5.5845	.17907
1.73	5.6407	.17728
1.74	5.6973	.17552
1.75	5.7546	.17377
1.76	5.8124	.17204
1.77	5.8709	.17033
1.78	5.9299	.16864
1.79	5.9895	.16696
x	e^x	e^{-x}

TABLE A4 *(Continued)*

x	e^x	e^{-x}	x	e^x	e^{-x}	x	e^x	e^{-x}
1.80	6.0496	.16530	**2.20**	9.0250	.11080	**2.60**	13.464	.07427
1.81	6.1104	.16365	**2.21**	9.1157	.10970	**2.61**	13.599	.07354
1.82	6.1719	.16203	**2.22**	9.2073	.10861	**2.62**	13.736	.07280
1.83	6.2339	.16041	**2.23**	9.2999	.10753	**2.63**	13.874	.07208
1.84	6.2965	.15882	**2.24**	9.3933	.10646	**2.64**	14.013	.07136
1.85	6.3598	.15724	**2.25**	9.4877	.10540	**2.65**	14.154	.07066
1.86	6.4237	.15567	**2.26**	9.5831	.10435	**2.66**	14.296	.06995
1.87	6.4883	.15412	**2.27**	9.6794	.10331	**2.67**	14.440	.06925
1.88	6.5535	.15259	**2.28**	9.7767	.10228	**2.68**	14.585	.06856
1.89	6.6194	.15107	**2.29**	9.8749	.10127	**2.69**	14.732	.06788
1.90	6.6859	.14957	**2.30**	9.9742	.10026	**2.70**	14.880	.06721
1.91	6.7531	.14808	**2.31**	10.074	.09926	**2.71**	15.029	.06654
1.92	6.8210	.14661	**2.32**	10.176	.09827	**2.72**	15.180	.06587
1.93	6.8895	.14515	**2.33**	10.278	.09730	**2.73**	15.333	.06522
1.94	6.9588	.14370	**2.34**	10.381	.09633	**2.74**	15.487	.06457
1.95	7.0287	.14227	**2.35**	10.486	.09537	**2.75**	15.643	.06393
1.96	7.0993	.14086	**2.36**	10.591	.09442	**2.76**	15.800	.06329
1.97	7.1707	.13946	**2.37**	10.697	.09348	**2.77**	15.959	.06266
1.98	7.2427	.13807	**2.38**	10.805	.09255	**2.78**	16.119	.06204
1.99	7.3155	.13670	**2.39**	10.913	.09163	**2.79**	16.281	.06142
2.00	7.3891	.13534	**2.40**	11.023	.09072	**2.80**	16.445	.06081
2.01	7.4633	.13399	**2.41**	11.134	.08982	**2.81**	16.610	.06020
2.02	7.5383	.13266	**2.42**	11.246	.08892	**2.82**	16.777	.05961
2.03	7.6141	.13134	**2.43**	11.359	.08803	**2.83**	16.945	.05901
2.04	7.6906	.13003	**2.44**	11.473	.08716	**2.84**	17.116	.05843
2.05	7.7679	.12873	**2.45**	11.588	.08629	**2.85**	17.288	.05784
2.06	7.8460	.12745	**2.46**	11.705	.08543	**2.86**	17.462	.05727
2.07	7.9248	.12619	**2.47**	11.822	.08458	**2.87**	17.637	.05670
2.08	8.0045	.12493	**2.48**	11.941	.08374	**2.88**	17.814	.05613
2.09	8.0849	.12369	**2.49**	12.061	.08291	**2.89**	17.993	.05558
2.10	8.1662	.12246	**2.50**	12.182	.08208	**2.90**	18.174	.05502
2.11	8.2482	.12124	**2.51**	12.305	.08127	**2.91**	18.357	.05448
2.12	8.3311	.12003	**2.52**	12.429	.08046	**2.92**	18.541	.05393
2.13	8.4149	.11884	**2.53**	12.554	.07966	**2.93**	18.728	.05340
2.14	8.4994	.11765	**2.54**	12.680	.07887	**2.94**	18.916	.05287
2.15	8.5849	.11648	**2.55**	12.807	.07808	**2.95**	19.106	.05234
2.16	8.6711	.11533	**2.56**	12.936	.07730	**2.96**	19.298	.05182
2.17	8.7583	.11418	**2.57**	13.066	.07654	**2.97**	19.492	.05130
2.18	8.8463	.11304	**2.58**	13.197	.07577	**2.98**	19.688	.05079
2.19	8.9352	.11192	**2.59**	13.330	.07502	**2.99**	19.886	.05029
x	e^x	e^{-x}	x	e^x	e^{-x}	x	e^x	e^{-x}

TABLE A5 *Circular Functions* (*Trigonometric Functions for Angles in Radians*)

Rad.	Sin	Tan	Cot	Cos
0.00	.00000	.00000	—	1.00000
.01	.01000	.01000	99.997	0.99995
.02	.02000	.02000	49.993	.99980
.03	.03000	.03001	33.323	.99955
.04	.03999	.04002	24.987	.99920
.05	.04998	.05004	19.983	.99875
.06	.05996	.06007	16.647	.99820
.07	.06994	.07011	14.262	.99755
.08	.07991	.08017	12.473	.99680
.09	.08988	.09024	11.081	.99595
0.10	.09983	.10033	9.9666	.99500
.11	.10978	.11045	9.0542	.99396
.12	.11971	.12058	8.2933	.99281
.13	.12963	.13074	7.6489	.99156
.14	.13954	.14092	7.0961	.99022
.15	.14944	.15114	6.6166	.98877
.16	.15932	.16138	6.1966	.98723
.17	.16918	.17166	5.8256	.98558
.18	.17903	.18197	5.4954	.98384
.19	.18886	.19232	5.1997	.98200
0.20	.19867	.20271	4.9332	.98007
.21	.20846	.21314	4.6917	.97803
.22	.21823	.22362	4.4719	.97590
.23	.22798	.23414	4.2709	.97367
.24	.23770	.24472	4.0864	.97134
.25	.24740	.25534	3.9163	.96891
.26	.25708	.26602	3.7591	.96639
.27	.26673	.27676	3.6133	.96377
.28	.27636	.28755	3.4776	.96106
.29	.28595	.29841	3.3511	.95824
0.30	.29552	.30934	3.2327	.95534
.31	.30506	.32033	3.1218	.95233
.32	.31457	.33139	3.0176	.94924
.33	.32404	.34252	2.9195	.94604
.34	.33349	.35374	2.8270	.94275
.35	.34290	.36503	2.7395	.93937
.36	.35227	.37640	2.6567	.93590
.37	.36162	.38786	2.5782	.93233
.38	.37092	.39941	2.5037	.92866
.39	.38019	.41105	2.4328	.92491
Rad.	Sin	Tan	Cot	Cos

Rad.	Sin	Tan	Cot	Cos
0.40	.38942	.42279	2.3652	.92106
.41	.39861	.43463	2.3008	.91712
.42	.40776	.44657	2.2393	.91309
.43	.41687	.45862	2.1804	.90897
.44	.42594	.47078	2.1241	.90475
.45	.43497	.48306	2.0702	.90045
.46	.44395	.49545	2.0184	.89605
.47	.45289	.50797	1.9686	.89157
.48	.46178	.52061	1.9208	.88699
.49	.47063	.53339	1.8748	.88233
0.50	.47943	.54630	1.8305	.87758
.51	.48818	.55936	1.7878	.87274
.52	.49688	.57256	1.7465	.86782
.53	.50553	.58592	1.7067	.86281
.54	.51414	.59943	1.6683	.85771
.55	.52269	.61311	1.6310	.85252
.56	.53119	.62695	1.5950	.84726
.57	.53963	.64097	1.5601	.84190
.58	.54802	.65517	1.5263	.83646
.59	.55636	.66956	1.4935	.83094
0.60	.56464	.68414	1.4617	.82534
.61	.57287	.69892	1.4308	.81965
.62	.58104	.71391	1.4007	.81388
.63	.58914	.72911	1.3715	.80803
.64	.59720	.74454	1.3431	.80210
.65	.60519	.76020	1.3154	.79608
.66	.61312	.77610	1.2885	.78999
.67	.62099	.79225	1.2622	.78382
.68	.62879	.80866	1.2366	.77757
.69	.63654	.82534	1.2116	.77125
0.70	.64422	.84229	1.1872	.76484
.71	.65183	.85953	1.1634	.75836
.72	.65938	.87707	1.1402	.75181
.73	.66687	.89492	1.1174	.74517
.74	.67429	.91309	1.0952	.73847
.75	.68164	.93160	1.0734	.73169
.76	.68892	.95045	1.0521	.72484
.77	.69614	.96967	1.0313	.71791
.78	.70328	.98926	1.0109	.71091
.79	.71035	1.0092	.99084	.70385
Rad.	Sin	Tan	Cot	Cos

TABLE A5 (*Continued*)

Rad.	Sin	Tan	Cot	Cos
0.80	.71736	1.0296	.97121	.69671
.81	.72429	1.0505	.95197	.68950
.82	.73115	1.0717	.93309	.68222
.83	.73793	1.0934	.91455	.67488
.84	.74464	1.1156	.89635	.66746
.85	.75128	1.1383	.87848	.65998
.86	.75784	1.1616	.86091	.65244
.87	.76433	1.1853	.84365	.64483
.88	.77074	1.2097	.82668	.63715
.89	.77707	1.2346	.80998	.62941
0.90	.78333	1.2602	.79355	.62161
.91	.78950	1.2864	.77738	.61375
.92	.79560	1.3133	.76146	.60582
.93	.80162	1.3409	.74578	.59783
.94	.80756	1.3692	.73034	.58979
.95	.81342	1.3984	.71511	.58168
.96	.81919	1.4284	.70010	.57352
.97	.82489	1.4592	.68531	.56530
.98	.83050	1.4910	.67071	.55702
.99	.83603	1.5237	.65631	.54869
1.00	.84147	1.5574	.64209	.54030
1.01	.84683	1.5922	.62806	.53186
1.02	.85211	1.6281	.61420	.52337
1.03	.85730	1.6652	.60051	.51482
1.04	.86240	1.7036	.58699	.50622
1.05	.86742	1.7433	.57362	.49757
1.06	.87236	1.7844	.56040	.48887
1.07	.87720	1.8270	.54734	.48012
1.08	.88196	1.8712	.53441	.47133
1.09	.88663	1.9171	.52162	.46249
1.10	.89121	1.9648	.50897	.45360
1.11	.89570	2.0143	.49644	.44466
1.12	.90010	2.0660	.48404	.43568
1.13	.90441	2.1198	.47175	.42666
1.14	.90863	2.1759	.45959	.41759
1.15	.91276	2.2345	.44753	.40849
1.16	.91680	2.2958	.43558	.39934
1.17	.92075	2.3600	.42373	.39015
1.18	.92461	2.4273	.41199	.38092
1.19	.92837	2.4979	.40034	.37166
1.20	.93204	2.5722	.38878	.36236
1.21	.93562	2.6503	.37731	.35302
1.22	.93910	2.7328	.36593	.34365
1.23	.94249	2.8198	.35463	.33424
1.24	.94578	2.9119	.34341	.32480
1.25	.94898	3.0096	.33227	.31532
1.26	.95209	3.1133	.32121	.30582
1.27	.95510	3.2236	.31021	.29628
1.28	.95802	3.3413	.29928	.28672
1.29	.96084	3.4672	.28842	.27712
1.30	.96356	3.6021	.27762	.26750
1.31	.96618	3.7471	.26687	.25785
1.32	.96872	3.9033	.25619	.24818
1.33	.97115	4.0723	.24556	.23848
1.34	.97348	4.2556	.23498	.22875
1.35	.97572	4.4552	.22446	.21901
1.36	.97786	4.6734	.21398	.20924
1.37	.97991	4.9131	.20354	.19945
1.38	.98185	5.1774	.19315	.18964
1.39	.98370	5.4707	.18279	.17981
1.40	.98545	5.7979	.17248	.16997
1.41	.98710	6.1654	.16220	.16010
1.42	.98865	6.5811	.15195	.15023
1.43	.99010	7.0555	.14173	.14033
1.44	.99146	7.6018	.13155	.13042
1.45	.99271	8.2381	.12139	.12050
1.46	.99387	8.9886	.11125	.11057
1.47	.99492	9.8874	.10114	.10063
1.48	.99588	10.983	.09105	.09067
1.49	.99674	12.350	.08097	.08071
1.50	.99749	14.101	.07091	.07074
1.51	.99815	16.428	.06087	.06076
1.52	.99871	19.670	.05084	.05077
1.53	.99917	24.498	.04082	.04079
1.54	.99953	32.461	.03081	.03079
1.55	.99978	48.078	.02080	.02079
1.56	.99994	92.621	.01080	.01080
1.57	1.00000	1255.8	.00080	.00080
1.58	.99996	−108.65	−.00920	−.00920
1.59	.99982	−52.067	−.01921	−.01920
1.60	.99957	−34.233	−.02921	−.02920
Rad.	Sin	Tan	Cot	Cos

TABLE A6 *Trigonometric Functions for Angles in Degrees*

Deg.	Sin	Tan	Cot	Cos	Deg.
0.0	.00000	.00000	—	1.00000	**90.0**
.1	.00175	.00175	572.96	1.00000	**.9**
.2	.00349	.00349	286.48	0.99999	**.8**
.3	.00524	.00524	190.98	.99999	**.7**
.4	.00698	.00698	143.24	.99998	**.6**
.5	.00873	.00873	114.59	.99996	**.5**
.6	.01047	.01047	95.489	.99995	**.4**
.7	.01222	.01222	81.847	.99993	**.3**
.8	.01396	.01396	71.615	.99990	**.2**
.9	.01571	.01571	63.657	.99988	**.1**
1.0	.01745	.01746	57.290	.99985	**89.0**
.1	.01920	.01920	52.081	.99982	**.9**
.2	.02094	.02095	47.740	.99978	**.8**
.3	.02269	.02269	44.066	.99974	**.7**
.4	.02443	.02444	40.917	.99970	**.6**
.5	.02618	.02619	38.188	.99966	**.5**
.6	.02792	.02793	35.801	.99961	**.4**
.7	.02967	.02968	33.694	.99956	**.3**
.8	.03141	.03143	31.821	.99951	**.2**
.9	.03316	.03317	30.145	.99945	**.1**
2.0	.03490	.03492	28.636	.99939	**88.0**
.1	.03664	.03667	27.271	.99933	**.9**
.2	.03839	.03842	26.031	.99926	**.8**
.3	.04013	.04016	24.898	.99919	**.7**
.4	.04188	.04191	23.859	.99912	**.6**
.5	.04362	.04366	22.904	.99905	**.5**
.6	.04536	.04541	22.022	.99897	**.4**
.7	.04711	.04716	21.205	.99889	**.3**
.8	.04885	.04891	20.446	.99881	**.2**
.9	.05059	.05066	19.740	.99872	**.1**
3.0	.05234	.05241	19.081	.99863	**87.0**
.1	.05408	.05416	18.464	.99854	**.9**
.2	.05582	.05591	17.886	.99844	**.8**
.3	.05756	.05766	17.343	.99834	**.7**
.4	.05931	.05941	16.832	.99824	**.6**
.5	.06105	.06116	16.350	.99813	**.5**
.6	.06279	.06291	15.895	.99803	**.4**
.7	.06453	.06467	15.464	.99792	**.3**
.8	.06627	.06642	15.056	.99780	**.2**
.9	.06802	.06817	14.669	.99768	**.1**
Deg.	**Cos**	**Cot**	**Tan**	**Sin**	**Deg.**

Deg.	Sin	Tan	Cot	Cos	Deg.
4.0	.06976	.06993	14.301	.99756	**86.0**
.1	.07150	.07168	13.951	.99744	**.9**
.2	.07324	.07344	13.617	.99731	**.8**
.3	.07498	.07519	13.300	.99719	**.7**
.4	.07672	.07695	12.996	.99705	**.6**
.5	.07846	.07870	12.706	.99692	**.5**
.6	.08020	.08046	12.429	.99678	**.4**
.7	.08194	.08221	12.163	.99664	**.3**
.8	.08368	.08397	11.909	.99649	**.2**
.9	.08542	.08573	11.664	.99635	**.1**
5.0	.08716	.08749	11.430	.99619	**85.0**
.1	.08889	.08925	11.205	.99604	**.9**
.2	.09063	.09101	10.988	.99588	**.8**
.3	.09237	.09277	10.780	.99572	**.7**
.4	.09411	.09453	10.579	.99556	**.6**
.5	.09585	.09629	10.385	.99540	**.5**
.6	.09758	.09805	10.199	.99523	**.4**
.7	.09932	.09981	10.019	.99506	**.3**
.8	.10106	.10158	9.8448	.99488	**.2**
.9	.10279	.10334	9.6768	.99470	**.1**
6.0	.10453	.10510	9.5144	.99452	**84.0**
.1	.10626	.10687	9.3572	.99434	**.9**
.2	.10800	.10863	9.2052	.99415	**.8**
.3	.10973	.11040	9.0579	.99396	**.7**
.4	.11147	.11217	8.9152	.99377	**.6**
.5	.11320	.11394	8.7769	.99357	**.5**
.6	.11494	.11570	8.6427	.99337	**.4**
.7	.11667	.11747	8.5126	.99317	**.3**
.8	.11840	.11924	8.3863	.99297	**.2**
.9	.12014	.12101	8.2636	.99276	**.1**
7.0	.12187	.12278	8.1443	.99255	**83.0**
.1	.12360	.12456	8.0285	.99233	**.9**
.2	.12533	.12633	7.9158	.99211	**.8**
.3	.12706	.12810	7.8062	.99189	**.7**
.4	.12880	.12988	7.6996	.99167	**.6**
.5	.13053	.13165	7.5958	.99144	**.5**
.6	.13226	.13343	7.4947	.99122	**.4**
.7	.13399	.13521	7.3962	.99098	**.3**
.8	.13572	.13698	7.3002	.99075	**.2**
.9	.13744	.13876	7.2066	.99051	**.1**
Deg.	**Cos**	**Cot**	**Tan**	**Sin**	**Deg.**

TABLE A6 *(Continued)*

Deg.	Sin	Tan	Cot	Cos	Deg.
8.0	.13917	.14054	7.1154	.99027	**82.0**
.1	.14090	.14232	7.0264	.99002	**.9**
.2	.14263	.14410	6.9395	.98978	**.8**
.3	.14436	.14588	6.8548	.98953	**.7**
.4	.14608	.14767	6.7720	.98927	**.6**
.5	.14781	.14945	6.6912	.98902	**.5**
.6	.14954	.15124	6.6122	.98876	**.4**
.7	.15126	.15302	6.5350	.98849	**.3**
.8	.15299	.15481	6.4596	.98823	**.2**
.9	.15471	.15660	6.3859	.98796	**.1**
9.0	.15643	.15838	6.3138	.98769	**81.0**
.1	.15816	.16017	6.2432	.98741	**.9**
.2	.15988	.16196	6.1742	.98714	**.8**
.3	.16160	.16376	6.1066	.98686	**.7**
.4	.16333	.16555	6.0405	.98657	**.6**
.5	.16505	.16734	5.9758	.98629	**.5**
.6	.16677	.16914	5.9124	.98600	**.4**
.7	.16849	.17093	5.8502	.98570	**.3**
.8	.17021	.17273	5.7894	.98541	**.2**
.9	.17193	.17453	5.7297	.98511	**.1**
10.0	.17365	.17633	5.6713	.98481	**80.0**
.1	.17537	.17813	5.6140	.98450	**.9**
.2	.17708	.17993	5.5578	.98420	**.8**
.3	.17880	.18173	5.5026	.98389	**.7**
.4	.18052	.18353	5.4486	.98357	**.6**
.5	.18224	.18534	5.3955	.98325	**.5**
.6	.18395	.18714	5.3435	.98294	**.4**
.7	.18567	.18895	5.2924	.98261	**.3**
.8	.18738	.19076	5.2422	.98229	**.2**
.9	.18910	.19257	5.1929	.98196	**.1**
11.0	.19081	.19438	5.1446	.98163	**79.0**
.1	.19252	.19619	5.0970	.98129	**.9**
.2	.19423	.19801	5.0504	.98096	**.8**
.3	.19595	.19982	5.0045	.98061	**.7**
.4	.19766	.20164	4.9594	.98027	**.6**
.5	.19937	.20345	4.9152	.97992	**.5**
.6	.20108	.20527	4.8716	.97958	**.4**
.7	.20279	.20709	4.8288	.97922	**.3**
.8	.20450	.20891	4.7867	.97887	**.2**
.9	.20620	.21073	4.7453	.97851	**.1**
Deg.	**Cos**	**Cot**	**Tan**	**Sin**	**Deg.**

Deg.	Sin	Tan	Cot	Cos	Deg.
12.0	.20791	.21256	4.7046	.97815	**78.0**
.1	.20962	.21438	4.6646	.97778	**.9**
.2	.21132	.21621	4.6252	.97742	**.8**
.3	.21303	.21804	4.5864	.97705	**.7**
.4	.21474	.21986	4.5483	.97667	**.6**
.5	.21644	.22169	4.5107	.97630	**.5**
.6	.21814	.22353	4.4737	.97592	**.4**
.7	.21985	.22536	4.4373	.97553	**.3**
.8	.22155	.22719	4.4015	.97515	**.2**
.9	.22325	.22903	4.3662	.97476	**.1**
13.0	.22495	.23087	4.3315	.97437	**77.0**
.1	.22665	.23271	4.2972	.97398	**.9**
.2	.22835	.23455	4.2635	.97358	**.8**
.3	.23005	.23639	4.2303	.97318	**.7**
.4	.23175	.23823	4.1976	.97278	**.6**
.5	.23345	.24008	4.1653	.97237	**.5**
.6	.23514	.24193	4.1335	.97196	**.4**
.7	.23684	.24377	4.1022	.97155	**.3**
.8	.23853	.24562	4.0713	.97113	**.2**
.9	.24023	.24747	4.0408	.97072	**.1**
14.0	.24192	.24933	4.0108	.97030	**76.0**
.1	.24362	.25118	3.9812	.96987	**.9**
.2	.24531	.25304	3.9520	.96945	**.8**
.3	.24700	.25490	3.9232	.96902	**.7**
.4	.24869	.25676	3.8947	.96858	**.6**
.5	.25038	.25862	3.8667	.96815	**.5**
.6	.25207	.26048	3.8391	.96771	**.4**
.7	.25376	.26235	3.8118	.96727	**.3**
.8	.25545	.26421	3.7848	.96682	**.2**
.9	.25713	.26608	3.7583	.96638	**.1**
15.0	.25882	.26795	3.7321	.96593	**75.0**
.1	.26050	.26982	3.7062	.96547	**.9**
.2	.26219	.27169	3.6806	.96502	**.8**
.3	.26387	.27357	3.6554	.96456	**.7**
.4	.26556	.27545	3.6305	.96410	**.6**
.5	.26724	.27732	3.6059	.96363	**.5**
.6	.26892	.27921	3.5816	.96316	**.4**
.7	.27060	.28109	3.5576	.96269	**.3**
.8	.27228	.28297	3.5339	.96222	**.2**
.9	.27396	.28486	3.5105	.96174	**.1**
Deg.	**Cos**	**Cot**	**Tan**	**Sin**	**Deg.**

TABLE A6 (*Continued*)

Deg.	**Sin**	**Tan**	**Cot**	**Cos**	**Deg.**	**Deg.**	**Sin**	**Tan**	**Cot**	**Cos**	**Deg.**
16.0	.27564	.28675	3.4874	.96126	**74.0**	**20.0**	.34202	.36397	2.7475	.93969	**70.0**
.1	.27731	.28864	3.4646	.96078	**.9**	**.1**	.34366	.36595	2.7326	.93909	**.9**
.2	.27899	.29053	3.4420	.96029	**.8**	**.2**	.34530	.36793	2.7179	.93849	**.8**
.3	.28067	.29242	3.4197	.95981	**.7**	**.3**	.34694	.36991	2.7034	.93789	**.7**
.4	.28234	.29432	3.3977	.95931	**.6**	**.4**	.34857	.37190	2.6889	.93728	**.6**
.5	.28402	.29621	3.3759	.95882	**.5**	**.5**	.35021	.37388	2.6746	.93667	**.5**
.6	.28569	.29811	3.3544	.95832	**.4**	**.6**	.35184	.37588	2.6605	.93606	**.4**
.7	.28736	.30001	3.3332	.95782	**.3**	**.7**	.35347	3.7787	2.6464	.93544	**.3**
.8	.28903	.30192	3.3122	.95732	**.2**	**.8**	.35511	.37986	2.6325	.93483	**.2**
.9	.29070	.30382	3.2914	.95681	**.1**	**.9**	.35674	.38186	2.6187	.93420	**.1**
17.0	.29237	.30573	3.2709	.95630	**73.0**	**21.0**	.35837	.38386	2.6051	.93358	**69.0**
.1	.29404	.30764	3.2506	.95579	**.9**	**.1**	.36000	.38587	2.5916	.93295	**.9**
.2	.29571	.30955	3.2305	.95528	**.8**	**.2**	.36162	.38787	2.5782	.93232	**.8**
.3	.29737	.31147	3.2106	.95476	**.7**	**.3**	.36325	.38988	2.5649	.93169	**.7**
.4	.29904	.31338	3.1910	.95424	**.6**	**.4**	.36488	.39190	2.5517	.93106	**.6**
.5	.30071	.31530	3.1716	.95372	**.5**	**.5**	.36650	.39391	2.5386	.93042	**.5**
.6	.30237	.31722	3.1524	.95319	**.4**	**.6**	.36812	.39593	2.5257	.92978	**.4**
.7	.30403	.31914	3.1334	.95266	**.3**	**.7**	.36975	.39795	2.5129	.92913	**.3**
.8	.30570	.32106	3.1146	.95213	**.2**	**.8**	.37137	.39997	2.5002	.92849	**.2**
.9	.30736	.32299	3.0961	.95159	**.1**	**.9**	.37299	.40200	2.4876	.92784	**.1**
18.0	.30902	.32492	3.0777	.95106	**72.0**	**22.0**	.37461	.40403	2.4751	.92718	**68.0**
.1	.31068	.32685	3.0595	.95052	**.9**	**.1**	.37622	.40606	2.4627	.92653	**.9**
.2	.31233	.32878	3.0415	.94997	**.8**	**.2**	.37784	.40809	2.4504	.92587	**.8**
.3	.31399	.33072	3.0237	.94943	**.7**	**.3**	.37946	.41013	2.4383	.92521	**.7**
.4	.31565	.33266	3.0061	.94888	**.6**	**.4**	.38107	.41217	2.4262	.92455	**.6**
.5	.31730	.33460	2.9887	.94832	**.5**	**.5**	.38268	.41421	2.4142	.92388	**.5**
.6	.31896	.33654	2.9714	.94777	**.4**	**.6**	.38430	.41626	2.4023	.92321	**.4**
.7	.32061	.33848	2.9544	.94721	**.3**	**.7**	.38591	.41831	2.3906	.92254	**.3**
.8	.32227	.34043	2.9375	.94665	**.2**	**.8**	.38752	.42036	2.3789	.92186	**.2**
.9	.32392	.34238	2.9208	.94609	**.1**	**.9**	.38912	.42242	2.3673	.92119	**.1**
19.0	.32557	.34433	2.9042	.94552	**71.0**	**23.0**	.39073	.42447	2.3559	.92050	**67.0**
.1	.32722	.34628	2.8878	.94495	**.9**	**.1**	.39234	.42654	2.3445	.91982	**.9**
.2	.32887	.34824	2.8716	.94438	**.8**	**.2**	.39394	.42860	2.3332	.91914	**.8**
.3	.33051	.35020	2.8556	.94380	**.7**	**.3**	.39555	.43067	2.3220	.91845	**.7**
.4	.33216	.35216	2.8397	.94322	**.6**	**.4**	.39715	.43274	2.3109	.91775	**.6**
.5	.33381	.35412	2.8239	.94264	**.5**	**.5**	.39875	.43481	2.2998	.91706	**.5**
.6	.33545	.35608	2.8083	.94206	**.4**	**.6**	.40035	.43689	2.2889	.91636	**.4**
.7	.33710	.35805	2.7929	.94147	**.3**	**.7**	.40195	.43897	2.2781	.91566	**.3**
.8	.33874	.36002	2.7776	.94088	**.2**	**.8**	.40355	.44105	2.2673	.91496	**.2**
.9	.34038	.36199	2.7625	.94029	**.1**	**.9**	.40514	.44314	2.2566	.91425	**.1**
Deg.	**Cos**	**Cot**	**Tan**	**Sin**	**Deg.**	**Deg.**	**Cos**	**Cot**	**Tan**	**Sin**	**Deg.**

TABLE A6 *(Continued)*

Deg.	Sin	Tan	Cot	Cos	Deg.	Deg.	Sin	Tan	Cot	Cos	Deg.
24.0	.40674	.44523	2.2460	.91355	**66.0**	**28.0**	.46947	.53171	1.8807	.88295	**62.0**
.1	.40833	.44732	2.2355	.91283	**.9**	**.1**	.47101	.53395	1.8728	.88213	**.9**
.2	.40992	.44942	2.2251	.91212	**.8**	**.2**	.47255	.53620	1.8650	.88130	**.8**
.3	.41151	.45152	2.2148	.91140	**.7**	**.3**	.47409	.53844	1.8572	.88048	**.7**
.4	.41310	.45362	2.2045	.91068	**.6**	**.4**	.47562	.54070	1.8495	.87965	**.6**
.5	.41469	.45573	2.1943	.90996	**.5**	**.5**	.47716	.54296	1.8418	.87882	**.5**
.6	.41628	.45784	2.1842	.90924	**.4**	**.6**	.47869	.54522	1.8341	.87798	**.4**
.7	.41787	.45995	2.1742	.90851	**.3**	**.7**	.48022	.54748	1.8265	.87715	**.3**
.8	.41945	.46206	2.1642	.90778	**.2**	**.8**	.48175	.54975	1.8190	.87631	**.2**
.9	.42104	.46418	2.1543	.90704	**.1**	**.9**	.48328	.55203	1.8115	.87546	**.1**
25.0	.42262	.46631	2.1445	.90631	**65.0**	**29.0**	.48481	.55431	1.8040	.87462	**61.0**
.1	.42420	.46843	2.1348	.90557	**.9**	**.1**	.48634	.55659	1.7966	.87377	**.9**
.2	.42578	.47056	2.1251	.90483	**.8**	**.2**	.48786	.55888	1.7893	.87292	**.8**
.3	.42736	.47270	2.1155	.90408	**.7**	**.3**	.48938	.56117	1.7820	.87207	**.7**
.4	.42894	.47483	2.1060	.90334	**.6**	**.4**	.49090	.56347	1.7747	.87121	**.6**
.5	.43051	.47698	2.0965	.90259	**.5**	**.5**	.49242	.56577	1.7675	.87036	**.5**
.6	.43209	.47912	2.0872	.90183	**.4**	**.6**	.49394	.56808	1.7603	.86949	**.4**
.7	.43366	.48127	2.0778	.90108	**.3**	**.7**	.49546	.57039	1.7532	.86863	**.3**
.8	.43523	.48342	2.0686	.90032	**.2**	**.8**	.49697	.57271	1.7461	.86777	**.2**
.9	.43680	.48557	2.0594	.89956	**.1**	**.9**	.49849	.57503	1.7391	.86690	**.1**
26.0	.43837	.48773	2.0503	.89879	**64.0**	**30.0**	.50000	.57735	1.7321	.86603	**60.0**
.1	.43994	.48989	2.0413	.89803	**.9**	**.1**	.50151	.57968	1.7251	.86515	**.9**
.2	.44151	.49206	2.0323	.89726	**.8**	**.2**	.50302	.58201	1.7182	.86427	**.8**
.3	.44307	.49423	2.0233	.89649	**.7**	**.3**	.50453	.58435	1.7113	.86340	**.7**
.4	.44464	.49640	2.0145	.89571	**.6**	**.4**	.50603	.58670	1.7045	.86251	**.6**
.5	.44620	.49858	2.0057	.89493	**.5**	**.5**	.50754	.58905	1.6977	.86163	**.5**
.6	.44776	.50076	1.9970	.89415	**.4**	**.6**	.50904	.59140	1.6909	.86074	**.4**
.7	.44932	.50295	1.9883	.89337	**.3**	**.7**	.51054	.59376	1.6842	.85985	**.3**
.8	.45088	.50514	1.9797	.89259	**.2**	**.8**	.51204	.59612	1.6775	.85896	**.2**
.9	.45243	.50733	1.9711	.89180	**.1**	**.9**	.51354	.59849	1.6709	.85806	**.1**
27.0	.45399	.50953	1.9626	.89101	**63.0**	**31.0**	.51504	.60086	1.6643	.85717	**59.0**
.1	.45554	.51173	1.9542	.89021	**.9**	**.1**	.51653	.60324	1.6577	.85627	**.9**
.2	.45710	.51393	1.9458	.88942	**.8**	**.2**	.51803	.60562	1.6512	.85536	**.8**
.3	.45865	.51614	1.9375	.88862	**.7**	**.3**	.51952	.60801	1.6447	.85446	**.7**
.4	.46020	.51835	1.9292	.88782	**.6**	**.4**	.52101	.61040	1.6383	.85355	**.6**
.5	.46175	.52057	1.9210	.88701	**.5**	**.5**	.52250	.61280	1.6319	.85264	**.5**
.6	.46330	.52279	1.9128	.88620	**.4**	**.6**	.52399	.61520	1.6255	.85173	**.4**
.7	.46484	.52501	1.9047	.88539	**.3**	**.7**	.52547	.61761	1.6191	.85081	**.3**
.8	.46639	.52724	1.8967	.88458	**.2**	**.8**	.52696	.62003	1.6128	.84989	**.2**
.9	.46793	.52947	1.8887	.88377	**.1**	**.9**	.52844	.62245	1.6066	.84897	**.1**
Deg.	**Cos**	**Cot**	**Tan**	**Sin**	**Deg.**	**Deg.**	**Cos**	**Cot**	**Tan**	**Sin**	**Deg.**

TABLE A6 *(Continued)*

Deg.	Sin	Tan	Cot	Cos	Deg.	Deg.	Sin	Tan	Cot	Cos	Deg.
32.0	.52992	.62487	1.6003	.84805	58.0	36.0	.58779	.72654	1.3764	.80902	54.0
.1	.53140	.62730	1.5941	.84712	.9	.1	.58920	.72921	1.3713	.80799	.9
.2	.53288	.62973	1.5880	.84619	.8	.2	.59061	.73189	1.3663	.80696	.8
.3	.53435	.63217	1.5818	.84526	.7	.3	.59201	.73457	1.3613	.80593	.7
.4	.53583	.63462	1.5757	.84433	.6	.4	.59342	.73726	1.3564	.80489	.6
.5	.53730	.63707	1.5697	.84339	.5	.5	.59482	.73996	1.3514	.80386	.5
.6	.53877	.63953	1.5637	.84245	.4	.6	.59622	.74267	1.3465	.80282	.4
.7	.54024	.64199	1.5577	.84151	.3	.7	.59763	.74538	1.3416	.80178	.3
.8	.54171	.64446	1.5517	.84057	.2	.8	.59902	.74810	1.3367	.80073	.2
.9	.54317	.64693	1.5458	.83962	.1	.9	.60042	.75082	1.3319	.79968	.1
33.0	.54464	.64941	1.5399	.83867	57.0	37.0	.60182	.75355	1.3270	.79864	53.0
.1	.54610	.65189	1.5340	.83772	.9	.1	.60321	.75629	1.3222	.79758	.9
.2	.54756	.65438	1.5282	.83676	.8	.2	.60460	.75904	1.3175	.79653	.8
.3	.54902	.65688	1.5224	.83581	.7	.3	.60599	.76180	1.3127	.79547	.7
.4	.55048	.65938	1.5166	.83485	.6	.4	.60738	.76456	1.3079	.79441	.6
.5	.55194	.66189	1.5108	.83389	.5	.5	.60876	.76733	1.3032	.79335	.5
.6	.55339	.66440	1.5051	.83292	.4	.6	.61015	.77010	1.2985	.79229	.4
.7	.55484	.66692	1.4994	.83195	.3	.7	.61153	.77289	1.2938	.79122	.3
.8	.55630	.66944	1.4938	.83098	.2	.8	.61291	.77568	1.2892	.79016	.2
.9	.56775	.67197	1.4882	.83001	.1	.9	.61429	.77848	1.2846	.78908	.1
34.0	.55919	.67451	1.4826	.82904	56.0	38.0	.61566	.78129	1.2799	.78801	52.0
.1	.56064	.67705	1.4770	.82806	.9	.1	.61704	.78410	1.2753	.78694	.9
.2	.56208	.67960	1.4715	.82708	.8	.2	.61841	.78692	1.2708	.78586	.8
.3	.56353	.68215	1.4659	.82610	.7	.3	.61978	.78975	1.2662	.78478	.7
.4	.56497	.68471	1.4605	.82511	.6	.4	.62115	.79259	1.2617	.78369	.6
.5	.56641	.68728	1.4550	.82413	.5	.5	.62251	.79544	1.2572	.78261	.5
.6	.56784	.68985	1.4496	.82314	.4	.6	.62388	.79829	1.2527	.78152	.4
.7	.56928	.69243	1.4442	.82214	.3	.7	.62524	.80115	1.2482	.78043	.3
.8	.57071	.69502	1.4388	.82115	.2	.8	.62660	.80402	1.2437	.77934	.2
.9	.57215	.69761	1.4335	.82015	.1	.9	.62796	.80690	1.2393	.77824	.1
35.0	.57358	.70021	1.4281	.81915	55.0	39.0	.62932	.80978	1.2349	.77715	51.0
.1	.57501	.70281	1.4229	.81815	.9	.1	.63068	.81268	1.2305	.77605	.9
.2	.57643	.70542	1.4176	.81714	.8	.2	.63203	.81558	1.2261	.77494	.8
.3	.57786	.70804	1.4124	.81614	.7	.3	.63338	.81849	1.2218	.77384	.7
.4	.57928	.71066	1.4071	.81513	.6	.4	.63473	.82141	1.2174	.77273	.6
.5	.58070	.71329	1.4019	.81412	.5	.5	.63608	.82434	1.2131	.77162	.5
.6	.58212	.71593	1.3968	.81310	.4	.6	.63742	.82727	1.2088	.77051	.4
.7	.58354	.71857	1.3916	.81208	.3	.7	.63877	.83022	1.2045	.76940	.3
.8	.58496	.72122	1.3865	.81106	.2	.8	.64011	.83317	1.2002	.76828	.2
.9	.58637	.72388	1.3814	.81004	.1	.9	.64145	.83613	1.1960	.76717	.1
Deg.	Sin	Tan	Cot	Cos	Deg.	Deg.	Sin	Tan	Cot	Cos	Deg.

TABLE A6 *(Continued)*

Deg.	Sin	Tan	Cot	Cos	Deg.	Deg.	Sin	Tan	Cot	Cos	Deg.
40.0	.64279	.83910	1.1918	.76604	**50.0**	**42.5**	.67559	.91633	1.0913	.73728	**47.5**
.1	.64412	.84208	1.1875	.76492	**.9**	**.6**	.67688	.91955	1.0875	.73610	**.4**
.2	.64546	.84507	1.1833	.76380	**.8**	**.7**	.67816	.92277	1.0837	.73491	**.3**
.3	.64679	.84806	1.1792	.76267	**.7**	**.8**	.67944	.92601	1.0799	.73373	**.2**
.4	.64812	.85107	1.1750	.76154	**.6**	**.9**	.68072	.92926	1.0761	.73254	**.1**
.5	.64945	.85408	1.1708	.76041	**.5**	**43.0**	.68200	.93252	1.0724	.73135	**47.0**
.6	.65077	.85710	1.1667	.75927	**.4**	**.1**	.68327	.93578	1.0686	.73016	**.9**
.7	.65210	.86014	1.1626	.75813	**.3**	**.2**	.68455	.93906	1.0649	.72897	**.8**
.8	.65342	.86318	1.1585	.75700	**.2**	**.3**	.68582	.94235	1.0612	.72777	**.7**
.9	.65474	.86623	1.1544	.75585	**.1**	**.4**	.68709	.94565	1.0575	.72657	**.6**
41.0	.65606	.86929	1.1504	.75471	**49.0**	**.5**	.68835	.94896	1.0538	.72537	**.5**
.1	.65738	.87236	1.1463	.75356	**.9**	**.6**	.68962	.95229	1.0501	.72417	**.4**
.2	.65869	.87543	1.1423	.75241	**.8**	**.7**	.69088	.95562	1.0464	.72297	**.3**
.3	.66000	.87852	1.1383	.75126	**.7**	**.8**	.69214	.95897	1.0428	.72176	**.2**
.4	.66131	.88162	1.1343	.75011	**.6**	**.9**	.69340	.96232	1.0392	.72055	**.1**
.5	.66262	.88473	1.1303	.74896	**.5**	**44.0**	.69466	.96569	1.0355	.71934	**46.0**
.6	.66393	.88784	1.1263	.74780	**.4**	**.1**	.69591	.96907	1.0319	.71813	**.9**
.7	.66523	.89097	1.1224	.74664	**.3**	**.2**	.69717	.97246	1.0283	.71691	**.8**
.8	.66653	.89410	1.1184	.74548	**.2**	**.3**	.69842	.97586	1.0247	.71569	**.7**
.9	.66783	.89725	1.1145	.74431	**.1**	**.4**	.69966	.97927	1.0212	.71447	**.6**
42.0	.66913	.90040	1.1106	.74314	**48.0**	**.5**	.70091	.98270	1.0176	.71325	**.5**
.1	.67043	.90357	1.1067	.74198	**.9**	**.6**	.70215	.98613	1.0141	.71203	**.4**
.2	.67172	.90674	1.1028	.74080	**.8**	**.7**	.70339	.98958	1.0105	.71080	**.3**
.3	.67301	.90993	1.0990	.73963	**.7**	**.8**	.70463	.99304	1.0070	.70957	**.2**
.4	.67430	.91313	1.0951	.73846	**.6**	**.9**	.70587	.99652	1.0035	.70834	**.1**
						45.0	.70711	1.00000	1.0000	.70711	**45.0**
Deg.	**Cos**	**Cot**	**Tan**	**Sin**	**Deg.**	**Deg.**	**Cos**	**Cot**	**Tan**	**Sin**	**Deg.**

ANSWERS TO ODD-NUMBERED PROBLEMS AND SELF-TESTS

EXERCISE 1.1

1a $\{1\}$ **b** $\{x|x$ is a natural number less than $2\}$ **3a** $\{9, 10\}$ **b** $\{x|x$ is a natural number between 8 and $11\}$ **5a** $\{5, 6, 7, 8\}$ **b** $\{x|x$ is a natural number less than 9 and greater than $4\}$ **7a** $\{-1, -2, -3, \ldots\}$ **b** $\{x|x$ is an integer less than $0\}$ **9a** $\{0, 1, 2, \ldots\}$ **b** $\{x|x$ is an integer and x is not negative$\}$ **11a** $\{$r, o, t$\}$ **b** $\{x|x$ is a letter in the word "root"$\}$ **13** $\varnothing$ **15** $\{1, 2, 3, 4, 5, 6, 7, 8, 9, 10\}$ **17** $\{1, 4, 5, 6, 9\}$ **19** $\{1, 2, 3, 4, 5, 6, 7, 8, 9, 10\}$ **21** R **23** R **25** R **27** Rational, $0.\overline{45}$ **29** Rational, $0.\overline{296}$ **31** Rational, $1.1\overline{6}$ **33** Irrational **35** Rational, 0.8

USING YOUR KNOWLEDGE 1.1

1 0.46875 **3** 0.252 inch

EXERCISE 1.2

1 Reflexive Law **3** Substitution Law **5** Substitution Law **7** Closure, A-1 **9** Commutative Law, A-2 **11** Associative Law, A-3 **13** Commutative Law, M-2 **15** Commutative Law, M-2 **17** Distributive Law, D-1 **19** Multiplicative Inverse Law, M-5

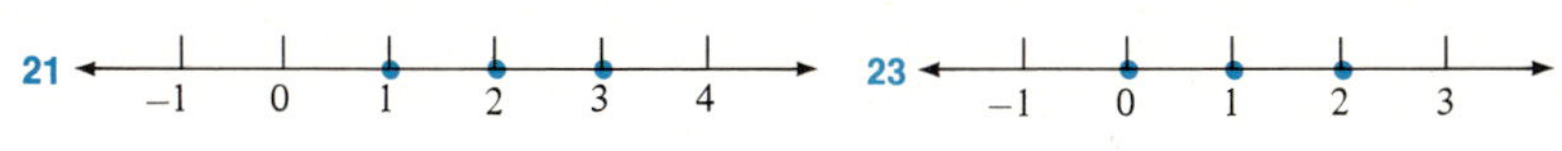

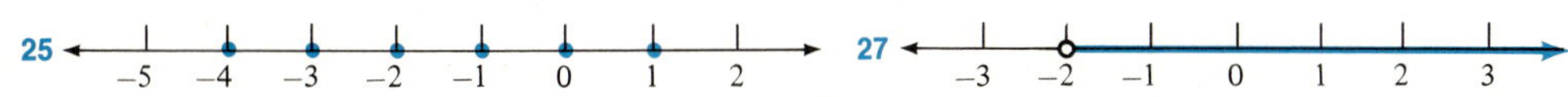

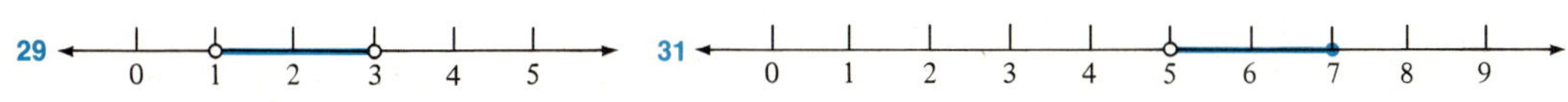

37 $0 < x < 3$ **39** $x > 0$ **41** $x \geq 0$ **43** $-3 \leq x \leq 4$ **45** $3 < 2x < 5$

USING YOUR KNOWLEDGE 1.2

1a $(3, 8)$ **b** $[-2, 0)$ **c** $[-1, 1]$ **d** $(2, 5]$ **3a** $(2, 3)$ **b** $[-1, 5]$

EXERCISE 1.3

1 -15 **3** -9 **5** $-\frac{3}{5}$ **7** 2 **9** -16 **11** $\frac{2}{5}$ **13** -0.7 **15** -1.1 **17** -36 **19** 18 **21** 24 **23** -24 **25** -9 **27** 2 **29** -1 **31** -2
33 $-\frac{1}{10}$ **35** $-\frac{23}{24}$ **37** $\frac{1}{24}$ **39** $\frac{2}{3}$ **41** $-\frac{2}{9}$ **43** $\frac{2}{3}$ **45** $-\frac{3}{4}$ **47** $\frac{4}{5}$ **49** $\frac{9}{20}$

51
$(-a) + a = 0$ — $-a$ and a are opposites. (A-5)
$a = -(-a)$ — The additive inverse of a number is unique. (A-5)

53
$a + (-1)a = 1 \cdot a + (-1)a$ — 1 is the additive identity.
$= [1 + (-1)]a$ — Distributive Law (D-2)
$= 0 \cdot a$ — 1 and -1 are additive inverses. (A-5)
$= 0$ — The product of 0 and any number is 0.
$(-1)a = -a$ — The additive inverse of a number is unique. (A-5)

USING YOUR KNOWLEDGE 1.3

1 0 **3** 0 **5** 7

SELF-TEST, CHAPTER 1

1a $\{0, 1, 2\}$ **b** $\{x \mid x$ is a whole number less than $3\}$ **2a** $\{1, 3, 5, 6, 7, 8\}$ **b** $\{1\}$ **c** $\{1, 7\}$ **3a** 0.4 **b** $0.\overline{90}$ **4a** Rational **b** Rational **c** Irrational **d** Rational **5a** Substitution Axiom **b** Symmetry Axiom **6a** Commutative Axiom for Addition (A-2) **b** Identity for Addition (A-4) **7a** Multiplicative Inverse Axiom (M-5) **b** Commutative Axiom for Multiplication (M-2)

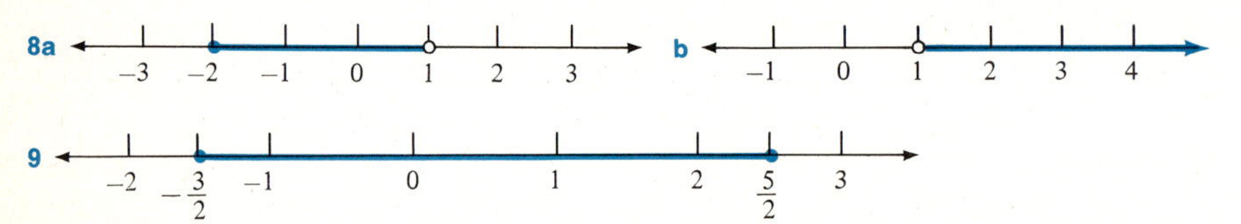

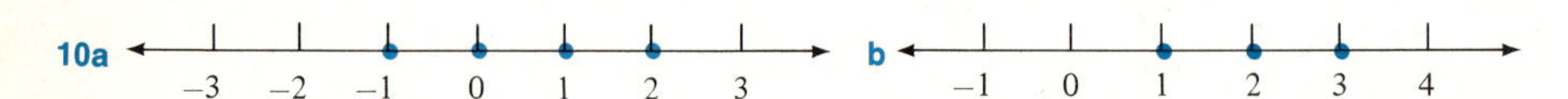

11a $2 < x < 5$ **b** $x < 0$ **12a** -12 **b** 8 **c** -2 **d** 2 **e** 5 **f** -2 **13a** 7 **b** 12 **c** -6 **d** 2 **e** -1 **14a** -18 **b** 24 **c** -35 **d** -36 **e** 9 **15a** -2 **b** -3 **c** 0 **d** 3 **e** -2 **16a** $\frac{8}{9}$ **b** $-\frac{7}{6}$ **17a** 30 **b** -28 **18a** $\frac{13}{10}$ **b** $\frac{17}{15}$ **c** $\frac{27}{40}$ **d** $-\frac{8}{99}$ **e** $\frac{27}{40}$ **f** $\frac{2}{15}$ **19a** $\frac{4}{3}$ **b** $-\frac{3}{2}$ **20a** -1 **b** $-\frac{3}{8}$

EXERCISE 2.1

1 Trinomial, degree 2 **3** Binomial, degree 4 **5** Monomial, degree 0 **7** The zero polynomial, no degree assigned **9** Binomial, degree 8 **11** -5 **13** -28 **15a** $6x^2 + 2x - 1$ **b** $-2x^2 + 8x - 5$ **17a** $16x^3 + 7x^2 + 12x - 11$ **b** $7x^2 - 2x + 1$ **19a** $5x^2 - 8x - 7$ **b** $3x^2 - 6x - 3$ **21** $5x - 7$ **23** $2x^3 - x^2 + 4x - 7$ **25** $-2a^2 + 5a$ **27** $x^3 + 3x^2 + 4y^2$ **29a** -3 **b** -6 **c** 15 **31a** -1 **b** 1 **c** -1 **33** 27

USING YOUR KNOWLEDGE 2.1

1 -9.317184 **3** 3.423744 **5** 0.90709936

EXERCISE 2.2

1 $3x^2 + 21x$ **3** $x^7 + x^6$ **5** $x^3y - x^2y^2 + 3x^2y - 3xy^2 + 2x - 2y$ **7** $x^2 + x - 8$ **9** $2a^4 - a^3 - 6a^2 + 7a - 2$ **11** $2t^4 - 3t^3 - 21t^2 - 2t + 24$ **13** $x^2 + 7x - 18$ **15** $x^2 + 7x + 12$ **17** $4x^2 + 16x + 16$ **19** $x^2 - 10x + 25$

21 $4x^2 - 20xy + 25y^2$ **23** $\frac{x^2}{4} - \frac{xy}{3} + \frac{y^2}{9}$ **25** $9x^2 - y^2$ **27** $3x^2 - xy - 10y^2$ **29** $6x^2 - 19xy + 15y^2$ **31** $8x^2 - 6xy - 9y^2$

33 $-3x^3 + 18x^2 - 24x$ **35** $2x^3 - 4x^2 + 2x$ **37** $x^3 + 5x^2 + 8x + 4$ **39** $-3x^3 - 9x^2 - 6x$ **41** $3x^3 + 16x^2 + 17x + 4$

43 $\frac{x^2}{2} + \frac{x}{4} - \frac{2}{3} + \frac{1}{x}$ **45** $\frac{2u}{v} + \frac{3v}{u}$ **47** $\frac{a^4}{b^4} - 6a^2 + 6b^2 - \frac{b^4}{a^4}$ **49** $x - 2$, remainder 0 **51** $y - 5$, remainder 1

53 $x^3 - x^2y + xy^2 - y^3$, remainder 0 **55** $x + 1$, remainder 0 **57** $x^2 + 4y^2 + 4z^2 + 2xy + 2xz - 4yz$, remainder 0

59
$(x^m)^n = x^m \cdot x^m \cdots x^m$ (n factors, each x^m) — Definition 2.1
$= x \cdot x \cdots x$ (mn factors, because each x^m has m x's as factors)
$= x^{mn}$ — Definition 2.1

61 $\frac{x^m}{x^n} = \frac{x \cdot x \cdots x\ (m \text{ factors})}{x \cdot x \cdots x\ (n \text{ factors})}$ Definition 2.1

$= \frac{x \cdot x \cdots x\ (m-n \text{ factors})\ x \cdot x \cdots x\ (n \text{ factors})}{x \cdot x \cdots x\ (n \text{ factors})}$ $(m > n$, given)

$= x \cdot x \cdots x\ (m-n \text{ factors})$ Simplifying

$= x^{m-n}$ Definition 2.1

USING YOUR KNOWLEDGE 2.2

For answers, see Exercise 2.2

EXERCISE 2.3

1 $2(3x+5)$ **3** $7x^2(3+x^2)$ **5** $5a^2b^2(a-2b+3a^2b^2)$ **7** $(a+b)(x+y)$ **9** $(a+3)(x+2y)$ **11** $(2x-5)(b-3n)$ **13** $(x+2)(x+4)$ **15** $(c-3d)(c-5d)$ **17** $(x-7y)^2$ **19** $(y+1)^2$ **21** $(3x+5y)^2$ **23** $(x-7)^2$ **25** $(a-b)^2$ **27** $(7x-2y)^2$ **29** $(m^2-6)^2$

31 $\left(\frac{x}{2}-y\right)^2$ **33** $(2x-1)(x+2)$ **35** $(5y-2)(y+3)$ **37** $(3x^2-2)(2x^2+3)$ **39** $(5x-2a)(3x+a)$ **41** $(x-7)(x+7)$

43 $(3xy-7z)(3xy+7z)$ **45** $(a-b-c)(a+b+c)$ **47** $(b^4+y^8)(b^2+y^4)(b+y^2)(b-y^2)$ **49** $(4x^2+9y^2)(2x+3y)(2x-3y)$ **51** $(x+5)(x^2-5x+25)$ **53** $(x^2+1)(x^4-x^2+1)$ **55** $2a(a^2+3b^2)$ **57** $-y(3x^2+3xy+y^2)$ **59** $(3a^2-b-3c)(9a^4+3a^2b+9a^2c+b^2+6bc+9c^2)$

USING YOUR KNOWLEDGE 2.3

1 $(x^2-2x+2)(x^2+2x+2)$ **3** $(x^2-x+1)(x^2+x+1)$ **5** $(x^2-x+2)(x^2+x+2)$ **7** $(b^4-b^2+1)(b^2-b+1)(b^2+b+1)$ **9** $(c^2-5c-10)(c^2+5c-10)$

EXERCISE 2.4

1 $\frac{2y}{3x^2}$ **3** $\frac{4a^3b^3}{5a^2+3b}$ **5** $\frac{5}{c+d}$ **7** $r(r-1)$ **9** $-a^2b^2$ **11** -1 **13** -1 **15** $(3x+y)^2$ **17** $\frac{2x+3}{5(x+2)}$ **19** $\frac{5y+10}{(y-5)(y+5)}$

21 $\frac{6x^2-28xy}{(x-2y)(x+2y)(x-3y)}$ **23** $\frac{2(a+b)}{a-b}$ **25** $\frac{-4a}{a-b}$ **27** $\frac{2}{x(x-1)(x+1)}$ **29** $\frac{-(x+3)}{(x-1)(x-2)(x-3)}$ **31** $\frac{-4(x^3-x^2+x+1)}{(2x-1)(2x+1)}$

33 0 **35** $\frac{x+2}{2x-3}$ **37** $\frac{a}{a-b} = \frac{-a}{b-a} = -\frac{a}{b-a} = -\frac{-a}{a-b}$ **39** Not true for $x = 2$

USING YOUR KNOWLEDGE 2.4

1 $\frac{-h}{x(x+h)}$ **3** $\frac{h(2x+h)}{(1-x^2)[1-(x+h)^2]}$

EXERCISE 2.5

1 $\frac{2ab^2}{3}$ **3** $\frac{a+b}{a-b}$ **5** $\frac{49}{6x^3y^3}$ **7** $\frac{2xy^3(x-1)}{x+4}$ **9** $\frac{(y+2)(y+3)}{y+6}$ **11** 1 **13** $\frac{x^2+ax+a^2}{x-2a}$ **15** $\frac{a^2+ax+x^2}{a^2-ax+x^2}$ **17** 1 **19** $\left(\frac{x-y-z}{x-y+z}\right)^2$

21 $\frac{x^2-81}{a}$ **23** $\frac{x}{y-1}$ **25** $\frac{b+a}{b-a}$ **27** $-\frac{1}{ax}$ **29** $\frac{-1}{(x-1)(a-1)}$ **31** $x(x+1)$ **33** $\frac{2x+4}{3x+2}$ **35** $\frac{c+v}{c-v}$ **37** $\frac{5x+7y}{x+y}\%$

USING YOUR KNOWLEDGE 2.5

1 $\frac{2m_0c^2eV+e^2V^2}{(m_0c^2+eV)^2}$ or $\frac{eV(2m_0c^2+eV)}{(m_0c^2+eV)^2}$

SELF-TEST, CHAPTER 2

1a Trinomial, degree 2 **b** Trinomial, degree 3 **c** Monomial, no degree assigned **d** Binomial, degree 7 **e** Monomial, degree 1 **2a** $7x^2-2x-10$ **b** $-x^3+4x^2-6x+6$ **3a** $4x^2-4x+2$ **b** $2x^2-6x+8$ **c** -3 **d** 13 **e** 10 **4** $x^5+3x^4-x^3-11x+6$ **5a** $x^2-3x-10$ **b** $6x^2+x-12$ **c** $25x^2-20xy+4y^2$ **d** $25x^2-4y^2$ **e** x^3+y^3 **6a** $3x^2+6x+2$ **b** $2a-2$, remainder $6a-4$ **7a** $4a^2b(3ab-2)$ **b** $(a-b)(8c-3)$ **c** $(b-5)(b+5)$ **d** $(4-z)(16+4z+z^2)$ **e** $(3b+2c)(9b^2-6bc+4c^2)$ **8a** $(x-3)^2$ **b** $(x-4)^2$ **c** $(6u-5v)^2$ **9a** $(2x-5)(x+3)$ **b** $(3x+4)(2x-3)$ **10a** $(x^2+y^4)(x^4-x^2y^4+y^8)$ **b** $(a^2+4b^2)(a+2b)(a-2b)$

11a $\frac{4x}{y^2}$ **b** $-\frac{2x+3}{x+5}$ **12a** $\frac{2}{x+2}$ **b** $\frac{2x+1}{(x-2)(x+3)}$ **13a** $\frac{x+6}{2x+2}$ **b** $\frac{1}{2-x}$ **14** $\frac{-4x^3+4x^2+8x+2}{4x^2-1}$ **15** $\frac{-3x+1}{(2x-1)(x+2)}$

16a a^2b^2 **b** $a^2-2ab+4b^2$ **17** -1 **18** 1 **19** 3 **20** $\frac{a-2}{a+2}$

EXERCISE 3.1

1 $\frac{1}{125}$ **3** $-\frac{27}{8}$ **5** 16 **7** $\frac{1}{9}$ **9** $\frac{x^2}{y^4}$ **11** a^8b^4 **13** $\frac{2v^7}{m^6}$ **15** $\frac{a^2}{b^3}$ **17** $\frac{1}{m^7}$ **19** $\frac{b}{a^6}$ **21** $\frac{27x}{4y^3}$ **23** $\frac{1}{x^{2n}}$ **25** x^{2n+2} **27** $\frac{1}{x^3}$ **29** x^ny^3 **31** $\frac{y+x}{y-x}$

33 $\frac{b^2-a^2}{ab}$ **35** 8.7×10 **37** 1.2×10^{-3} **39** 2.9×10^{-7} **41** 1.0×10^4 **43** 1.2×10^{-1} **45** 8×10 **47** 2×10^3 **49** 1.58×10^{79}

51 1.674×10^{-24} gm **53** 2.992×10^{-23} gm

USING YOUR KNOWLEDGE 3.1

1 2.2×10^9 **3** 31 years **5a** 4.86×10^{11} **b** 794 years

EXERCISE 3.2

1 -5 **3** $-\frac{1}{3}$ **5** x^4y **7** $\frac{4}{3}x^3y^4$ **9** $-4a^3b$ **11** $5y^3\sqrt{5x}$ **13** $8x^2y^3\sqrt[3]{2x}$ **15** $m^3\sqrt{3am}$ **17** $(x+y)\sqrt{x-y}$ **19** $-\frac{\sqrt{6ab}}{4ab^2}$

21 $\frac{x^2}{5y^2}\sqrt{35xy}$ **23** $-\frac{a}{7b}\sqrt[3]{42a^2b}$ **25** x^2y **27** $7xy\sqrt[3]{y^2}$ **29** $\frac{a\sqrt{b}}{b^2}$ **31** $x\sqrt{x}$ **33** x^2y^7 **35** $\frac{\sqrt{14}}{7}$ **37** $\frac{\sqrt[3]{12xy}}{4x}$ **39** $\frac{x\sqrt[3]{4t^2x}}{t^2}$

41 $\frac{\sqrt[5]{x^4y}}{x}$ **43** $(10^7)\sqrt[3]{75.3}$ **45** $\frac{m_0c\sqrt{c^2-v^2}}{c^2-v^2}$ **47** $|x-2|\sqrt{2}$ **49** $|z-4|\sqrt{3}$ **51** $\sqrt[n]{x}$ is an nth root of x. This is the meaning of $\sqrt[n]{\ }$.
$(\sqrt[n]{x})^n = x$ Definition 3.2a

53a $x^n = x^n$ Reflexive Property of Equality
If $x \ge 0$, then $\sqrt[n]{x^n} = x$. Definition 3.2b for n odd and $x \ge 0$
If $x < 0$, then $\sqrt[n]{x^n} = x$. Definition 3.2b for n odd and $x < 0$
Thus, for n odd, $\sqrt[n]{x^n} = x$.

b If n is even, then
$(\pm x)^n = x^n$ $(\pm 1)^n = 1$ for n even
Thus, x and $-x$ are both nth roots of x^n. Because the principal root is the nonnegative root,
$\sqrt[n]{x^n} = |x|$ Definition 3.2b for n even

USING YOUR KNOWLEDGE 3.2

1 $|x||y| = (\sqrt{x^2})(\sqrt{y^2})$ Theorem 3.2a, Part IV(b)
$= \sqrt{x^2y^2}$ Theorem 3.2a, Part II
$= \sqrt{(xy)^2}$ Laws of Exponents
$= |xy|$ Theorem 3.2a, Part IV(b)

EXERCISE 3.3

1 2 **3** $\frac{1}{2}$ **5** 9 **7** $\frac{1}{4}$ **9** $\frac{27}{64}$ **11** 11 **13** $6x^2$ **15** $-\frac{4}{a^{5/2}}$ **17** $\frac{x^{5/4}}{y^{3/2}}$ **19** $\frac{4x}{27y}$ **21** $\frac{z^4}{x^4y}$ **23** x^n **25** x^{2n} **27** $\frac{1}{x}$ **29** $x+x^{4/3}$ **31** $x^{1/5}+\frac{1}{x}$

33 $|x|(x-5)^{1/2}$ **35** $\frac{1}{|x+3|(x-1)^{1/2}}$

37 $[(-2)^6]^{1/2}$ cannot be negative because the exponent $\frac{1}{2}$ calls for the *principal* square root. Thus, $[(-2)^6]^{1/2} \ne (-2)^{6\cdot(1/2)}$. In fact, $(-2)^{6/2}$ is *not* defined because the exponent is not in lowest terms.

39a 0.15368 **b** 0.14582

USING YOUR KNOWLEDGE 3.3

1a 7805 m/sec **b** 87.6 min **3a** 5.20 A.U. **b** 4.84×10^8 miles

EXERCISE 3.4

1 $-4\sqrt{3}$ **3** $\sqrt[3]{2}$ **5** $7ck\sqrt{k}$ **7** $(7-a)\sqrt[3]{3a}$ **9** $\frac{7\sqrt[3]{6}}{12}$ **11** $\frac{(e-2f)\sqrt{3ef}}{ef}$ **13** $x-\sqrt{xy}$ **15** $x-5y$ **17** $4(1+\sqrt{3})$ **19** $\sqrt[3]{4}$ **21** $\frac{\sqrt[6]{81x^5}}{x}$

23 $\frac{8\sqrt[6]{2x^2y^2}}{x}$ **25** $x\sqrt[6]{16x}$ **27** $-3x\sqrt[12]{27x^5y^{11}}$ **29** $4-\sqrt{15}$ **31** $\frac{x^2-2x\sqrt{3y}+3y}{x^2-3y}$

USING YOUR KNOWLEDGE 3.4

1 $\dfrac{2}{\sqrt{2x+1}+\sqrt{2a+1}}$ **3** $\dfrac{x+a}{\sqrt{x^2+b^2}+\sqrt{a^2+b^2}}$

SELF-TEST, CHAPTER 3

1a $\dfrac{1}{x^9}$ **b** $\dfrac{x^{12}}{y^8}$ **2** $\dfrac{y^9}{x^{12}}$ **3** $\dfrac{x^2y-x}{y+xy^2}$ or $\dfrac{x(xy-1)}{y(xy+1)}$ **4** 3×10^7 **5** 6.02×10^{26} **6a** $\frac{4}{5}$ **b** $-\frac{1}{2}$ **7a** $3x^2y^3\sqrt{y}$ **b** $2xy^2z\sqrt[3]{2x^2}$

8a $\dfrac{x^2}{4y^2}\sqrt{6xy}$ **b** $\dfrac{3x}{yz}\sqrt[3]{2xyz^2}$ **9** $\dfrac{\sqrt[3]{x+y}}{x+y}$ **10** $|x-2y|\sqrt{x}$ **11a** 9 **b** $\frac{1}{8}$ **12** $4xy^{1/3}$ **13** $x^{5/4}-3$ **14** $x^{7/12}$ **15a** $8\sqrt{3}$ **b** $16xy\sqrt[3]{x^2y}$

16a $2\sqrt{2}+3$ **b** $x+4\sqrt{xy}-21y$ **17** $2^{5/6}$ or $\sqrt[6]{32}$ **18** $|x|y\sqrt[6]{x^2y}$ **19** $\dfrac{y(\sqrt{x}+\sqrt{y})}{x-y}$ **20** $\dfrac{1}{\sqrt{x}-\sqrt{y}}$

EXERCISE 4.1

1 2 **3** 6 **5** $\frac{1}{8}$ **7** 1 **9** 4 **11** $-\frac{45}{4}$ **13** $-\frac{10}{3}$ **15** $\frac{11}{3}$ **17** $\frac{5}{4}$ **19** 6 **21** No solution **23** -2 **25** $\dfrac{2A}{a}$ **27** $\dfrac{v}{t}$ **29** $\dfrac{P-2W}{2}$ **31** $\dfrac{bf}{b-f}$

33 $\dfrac{a+1}{a-1}$ **35** $-\dfrac{m^2+n^2}{2m}$ **37** $c=1$ **39** $x=2$ and $x^2=4$, or $x^2=4$ and $x=\sqrt{4}$

USING YOUR KNOWLEDGE 4.1

1 No **3** Yes **5** 5 feet, 8.4 inches

EXERCISE 4.2

5 $x \le 2$ (−3 −2 −1 0 1 2 3) **7** $x > -1$ (−3 −2 −1 0 1 2 3)

9 $x > \frac{12}{5}$ (−1 0 1 2 3 4) **11** $-\frac{1}{2} \le x < 2$ (−1 0 1 2 3)

13 $-2 \le x < 4$ (−3 −2 −1 0 1 2 3 4 5)

15 $1 < x \le 6$ (−1 0 1 2 3 4 5 6 7)

17 $5 \le x \le 7$ (0 1 2 3 4 5 6 7 8 9)

19 $-2 \le x < 7$ (−3 −2 −1 0 1 2 3 4 5 6 7 8 9) **21** $x > 8$ **23** $x \ge 167$

25 a and b are positive numbers and

$b \ne 3a$	Given
$b - 3a \ne 0$	Subtract 3a from both sides.
$(b-3a)^2 > 0$	The square of any nonzero real number is positive.
$b^2 - 6ab + 9a^2 > 0$	Multiplying out
$b^2 + 9a^2 > 6ab$	Add $6ab$ to both sides.
$\dfrac{b}{3a} + \dfrac{3a}{b} > 2$	Divide both sides by $3ab$.

27 n and v are positive numbers and

$n \ne v$	Given
$\sqrt{n} \ne \sqrt{v}$	Take square roots.
$\sqrt{n} - \sqrt{v} \ne 0$	Subtract $\sqrt{v}$ from both sides.
$(\sqrt{n}-\sqrt{v})^2 > 0$	The square of any nonzero real number is positive.
$n + v - 2\sqrt{nv} > 0$	Multiply out.
$n + v > 2\sqrt{nv}$	Add $2\sqrt{nv}$ to both sides.
$\dfrac{\sqrt{n}}{\sqrt{v}} + \dfrac{\sqrt{v}}{\sqrt{n}} > 2$	Divide both sides by $\sqrt{nv}$.

29 $x \ne 0$ and $x \ge -3$

USING YOUR KNOWLEDGE 4.2

1 $-3 < x < -1$ **3** $-1 < x \le 2$ **5** $x = 3$

EXERCISE 4.3

1 $x = 5$ or $x = -5$

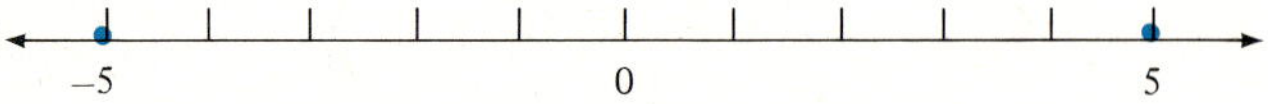

3 $-1 \le x \le 1$

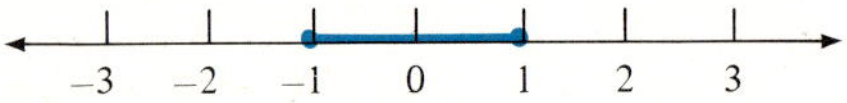

5 $x < -3$ or $x > 3$ 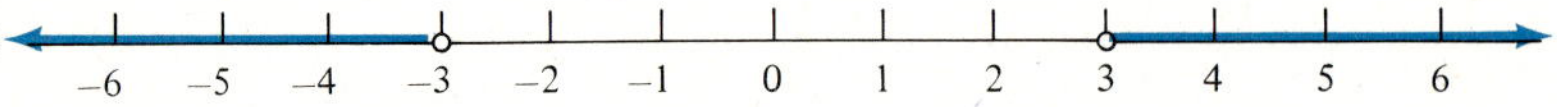 **7** $|x| \le 4$ **9** $|x| > 3$

11 $|x - 1| \ge 3$ **13** $x = -1$ or $x = 5$

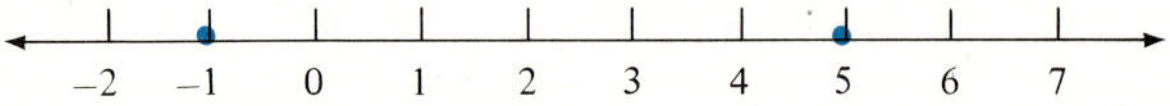

15 $-5 < x < 1$

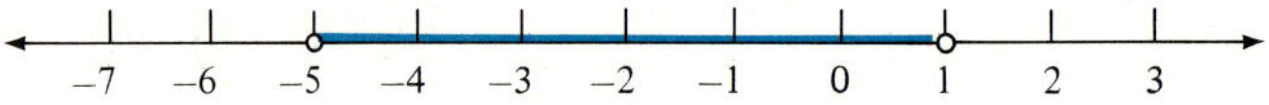

17 $-5 \le x \le 2$

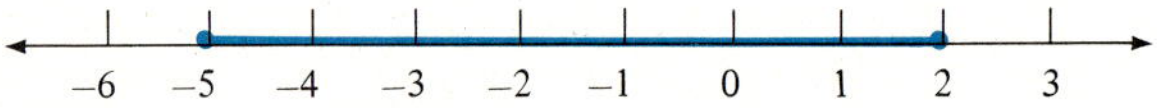

19 $-1 < x < \frac{5}{2}$

21 $x \le -1$ or $x \ge 3$

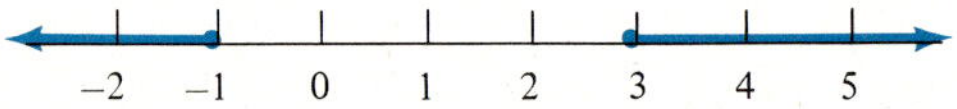

23 $x < -1$ or $x > 2$

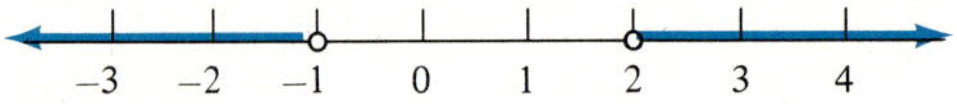

25 $x < -1$ or $x > \frac{5}{2}$

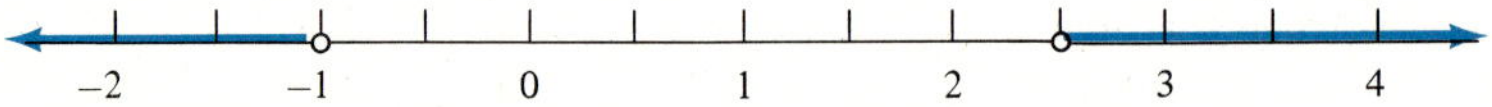

27 $x = -\frac{5}{3}$ or $x = 3$

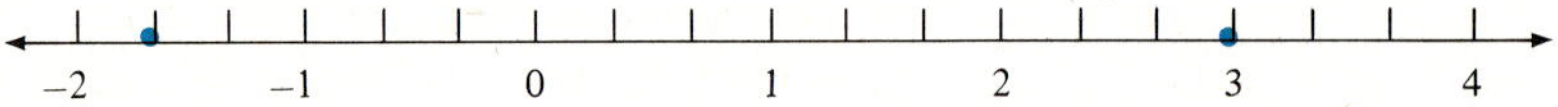

29 $-2 \le x \le 6$

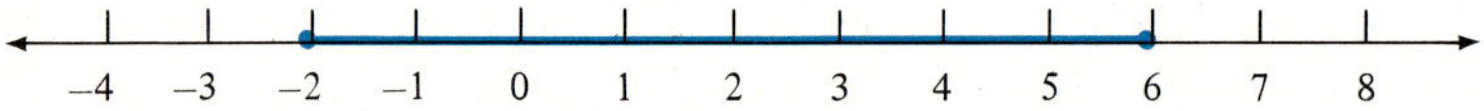

31 $x < \frac{7}{2}$ or $x > \frac{9}{2}$

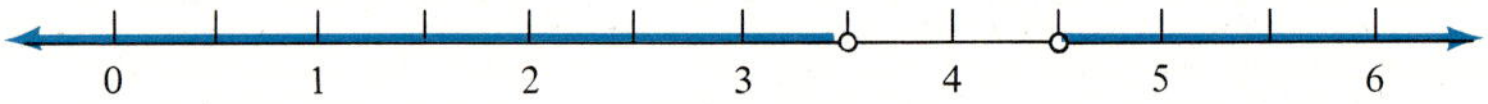

33 $x \le -\frac{7}{2}$ or $x \ge -\frac{5}{2}$

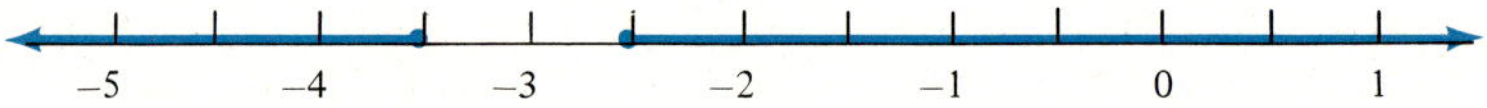

35 $\frac{7}{2} < x < \frac{9}{2}$

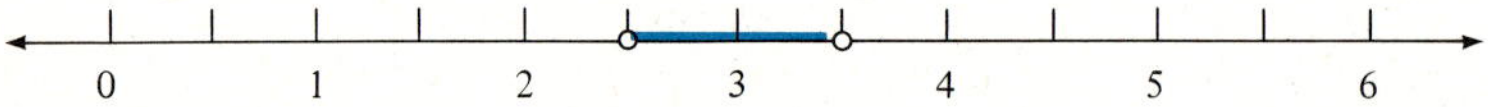

37 $x \le \frac{3}{2}$ **39** $x \ge \frac{2}{3}$ **41** $x = \frac{1}{3}$ or $x = 1$ **43** There is no solution. **45** $x = \frac{5}{3}$

47

$xy \le \lvert xy \rvert$	A real number is less than or equal to its own absolute value.
$x^2 + 2xy + y^2 \le x^2 + 2\lvert xy \rvert + y^2$	Multiply both sides by 2 and add $x^2 + y^2$ to both sides.
$\lvert x + y \rvert^2 \le (\lvert x \rvert + \lvert y \rvert)^2$	Factor both sides.
$\lvert x + y \rvert \le \lvert x \rvert + \lvert y \rvert$	Take the positive square root of both sides.

USING YOUR KNOWLEDGE 4.3

1 $|(x^2 + 4) - 8| = |x^2 - 4| = |x + 2|\,|x - 2|$

If x is restricted to an interval of unit radius centered at $x = 2$, then $1 < x < 3$, so that $3 < x + 2 < 5$. Thus, $|x + 2| < 5$ and, if $x \ne 2$,

$|(x^2 + 4) - 8| < 5|x - 2|$

3 $\left|\dfrac{10}{x+5} - 1\right| = \left|\dfrac{5-x}{x+5}\right| = \dfrac{|x-5|}{|x+5|}$

If x is restricted to an interval of unit radius centered at $x = 5$, then $4 < x < 6$, so that $9 < x + 5 < 11$. This gives

$$\frac{1}{11} < \frac{1}{x+5} < \frac{1}{9}$$

Therefore,

$$\left|\frac{10}{x+5} - 1\right| < \frac{1}{9}|x-5|$$

(If $x = 5$, both sides are zero.)

EXERCISE 4.4

1 $x = -3, x = -1$ **3** $x = -2$ **5** $x = 3, x = 7$ **7** $x = 8$ **9** $x = -3, x = -\frac{1}{2}$ **11** $x = \frac{2}{5}, x = \frac{3}{5}$ **13** $x = -4, x = 6$ **15** $x = 1 \pm \sqrt{2}$ **17** $x = -1, x = -\frac{1}{2}$ **19** $x = a \pm 5$ **21** $y = -5 \pm \sqrt{7}$ **23** $x = -3, x = \frac{2}{3}$ **25** $y = 3 \pm \sqrt{2}$ **27** $x = -2 \pm \sqrt{3}$ **29** $t = -\dfrac{\sqrt{3}}{2}, t = \sqrt{3}$ **31** $x = -\sqrt{2}$ **33** $y = -5, y = 0$ **35** $x = -\frac{3}{2}, x = \frac{1}{3}$ **37** $x = 1$ **39** $x = -\frac{5}{3}, x = -1$ **41** $y = -\dfrac{1}{2} \pm \dfrac{\sqrt{2}}{2}$ **43** $x = -\sqrt{3} \pm 4$ **45** $x = -\frac{1}{2}, x = 3$ **47** $t = -\sqrt{5}, t = \dfrac{\sqrt{5}}{5}$ **49** Rational **51** Irrational **53** $k = 1$ **55a** $k < \frac{5}{4}$ **57** $x = -1.368, x = 0.782$ **59** $x = -3.159, x = 1.104$

USING YOUR KNOWLEDGE 4.4

1 $r_1 + r_2 = \dfrac{-b - \sqrt{b^2 - 4ac}}{2a} + \dfrac{-b + \sqrt{b^2 - 4ac}}{2a} = \dfrac{-2b}{2a} = -\dfrac{b}{a}$ **3a** $r_1 + r_2 = 4, r_1 r_2 = 2$ **b** $r_1 + r_2 = \frac{2}{3}, r_1 r_2 = -\frac{8}{3}$ **c** $r_1 + r_2 = 6, r_1 r_2 = \frac{5}{2}$ **5** -30

EXERCISE 4.5

1 $x = 8$ **3** $x = 6$ **5** $y = 4$ **7** $y = -1, y = 0$ **9** $x = 16$ **11** $y = 0$ **13** $y = 1$ **15** $y = 1, y = 9$ **17** $x = 9$ **19** $x = -4, x = 1, x = 2 \pm 2\sqrt{2}$ **21** $x = -5, x = -1, x = \dfrac{9 \pm \sqrt{61}}{2}$ **23** $x = -4, x = -1, x = 2$ **25** $x = -3, x = -\frac{4}{3}$ **27** $x = -5, x = -1, x = 3$ **29** $w = -\frac{3}{2}, w = 3$ **31** $y = \pm 13, y = \pm\sqrt{29}$ **33** $x = \pm 2, x = \pm 5$ **35** $y = \pm\frac{1}{2}, y = \pm 1$ **37** $y = -1, y = 8$

USING YOUR KNOWLEDGE 4.5

1 16 feet **3** 64 pounds per square inch

EXERCISE 4.6

1 $x < -1$ or $x > 3$ **3** $-4 \le x \le 0$ **5** $-1 \le x \le 2$ **7** $x \le 0$ or $x \ge 3$ **9** No solution **11** $-1 - \sqrt{5} < x < -1 + \sqrt{5}$ **13** $x < 0$ or $x \ge \frac{1}{2}$ **15** $x > 2$ **17** $1 < x < 7$ **19** $\frac{1}{2} < x < 3$ **21** $-2 < x < 1$ **23** $x < -\frac{1}{2}$ or $x > 0$ **25** $x \le -3$ or $x \ge 3$ **27** $x \le 1$ or $x \ge 5$

29 a and b are nonnegative and

$a^2 > b^2$	Given
$a + b > 0$	$b \ge 0$ and $a \ge 0$, but $a^2 > b^2$, so that $a > 0$.
$a^2 - b^2 > 0$	Subtract b^2 from both sides of the first inequality.
$(a - b)(a + b) > 0$	Factor the left side.
$a - b > 0$	Divide both sides by $a + b$, which is positive.
$a > b$	Add b to both sides.

31

$(x - y)^2 \ge 0$	The square of a real number is nonnegative.
$x^2 - 2xy + y^2 \ge 0$	Multiply out.
$2x^2 + 2y^2 - x^2 - 2xy - y^2 \ge 0$	Replace x^2 by $2x^2 - x^2$ and y^2 by $2y^2 - y^2$.
$2(x^2 + y^2) - (x + y)^2 \ge 0$	Factor.

$$\frac{x^2 + y^2}{2} - \frac{(x + y)^2}{4} \geq 0 \qquad \text{Divide both sides by 4.}$$

$$\frac{x^2 + y^2}{2} \geq \left(\frac{x + y}{2}\right)^2 \qquad \text{Add } \left[\frac{(x + y)}{2}\right]^2 \text{ to both sides.}$$

33 $0 < x < 4$

USING YOUR KNOWLEDGE 4.6

1 $23.5 \leq v < 26.5$ **3** $49.1 \leq v < 50.9$

EXERCISE 4.7

1 20, 21, 22 **3** $\frac{5}{12}$ **5** Tower, 1454 feet; antenna, 346 feet **7** 50 mph and 65 mph **9** \$8000 in Stock A, \$7000 in Stock B **11** 6.5% on \$8000; 6% on \$12,000 **13** 10 **15** 5 **17** $12\frac{1}{2}$ **19** 12 inches by 24 inches **21** 17 inches by 19 inches **23** 12 **25** $3\frac{3}{7}$ days **27** $x > 3$ **29** $2 \leq x < 4$ **31** ≥ 86

USING YOUR KNOWLEDGE 4.7

Harry could not have been at the scene of the crime, because point B is 6 miles south of A.

SELF-TEST, CHAPTER 4

1a $x = \frac{3}{2}$ **b** $x = 3$ **2a** $x = 1$ **b** No solution **3a** $r = \dfrac{A - P}{Pt}$ **b** $a = \dfrac{2S - nb}{n}$ **4** $y = -9$ **5a** $x < -\frac{1}{2}$ **b** $x \geq -\frac{1}{6}$

6 $-3 \leq x < 2$

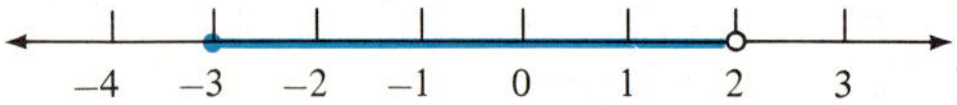

7 $x < 400$ **8a** $x = -\frac{1}{2}$ or $x = 1$ **b** $x = -\frac{1}{4}$

9a $-2 < x < \frac{4}{3}$

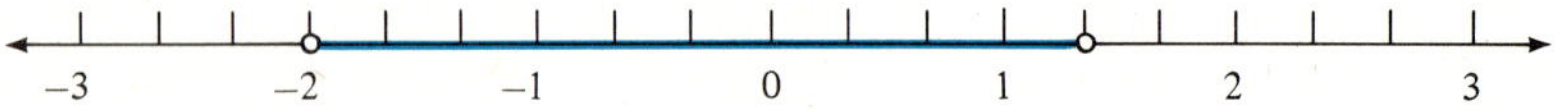

b $x \leq -1$ or $x \geq \frac{1}{3}$

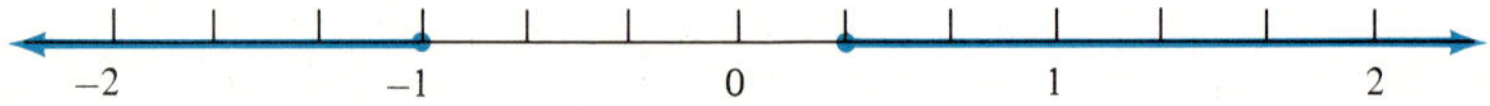

10a $x = -1$ or $x = \frac{1}{3}$ **b** $x = -\frac{2}{3}$ or $x = \frac{5}{2}$ **11a** $x = \pm\frac{6}{5}\sqrt{5}$ **b** $x = -2 \pm \sqrt{2}$ **12a** $x = \dfrac{-7 \pm \sqrt{109}}{6}$ **b** No real-number solution

13a $k > \frac{1}{2}$ **b** $\sqrt{b^2 - 4ac} = \frac{9}{4} = (\frac{3}{2})^2$, so that the roots are rational numbers. **14** $x = 1$ **15** $x = \pm\sqrt{3}$ or $x = \pm 2$ **16** $x = \pm 5$

17a $x \leq -3$ or $x \geq 2$ **b** $x < 0$ or $x > 3$ **18** 150 mph **19** $2\frac{3}{4}$ liters of 10% and $2\frac{1}{4}$ liters of 50% solution **20** $x > 5$

EXERCISE 5.1

1 **a** $(-3, \frac{5}{2})$ is in Q II. **b** $(0, -3)$ is on the y-axis. **c** $(4, -\frac{2}{3})$ is in Q IV. **d** $(-5, -\frac{1}{3})$ is in Q III.

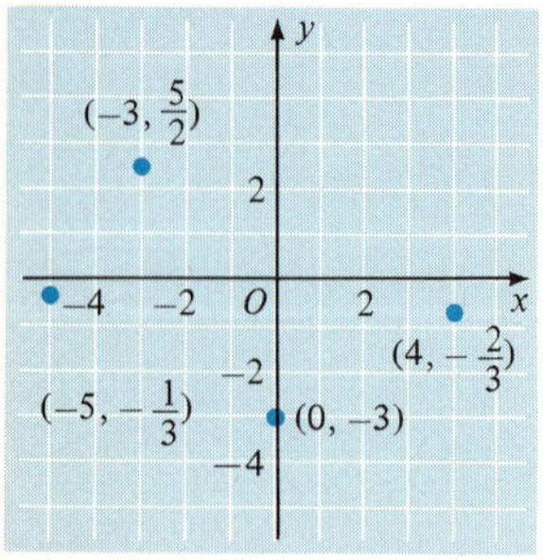

3a 5 **b** $(4, \frac{7}{2})$ **5a** 5 **b** $(\frac{3}{2}, 3)$ **7a** 6 **b** $(1, 3)$ **9a** $2\sqrt{17}$ **b** $(2, -3)$ **11a** 8 **b** $(-4, 1)$ **13** $(x - 2)^2 + (y - 5)^2 = 9$ **15** $(x + 3)^2 + (y + 4)^2 = 1$ **17** $C(2, 4)$, $r = 6$ **19** $C(-1, -3)$, $r = 4$ **21** Not a right triangle; scalene **23** An isosceles right triangle **25** Collinear **27** Collinear

29

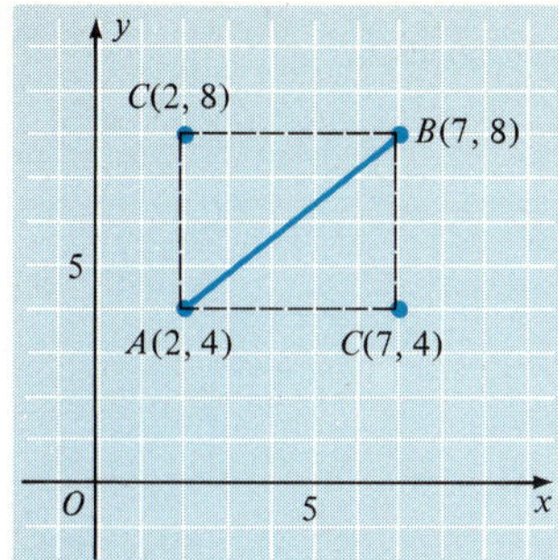

31

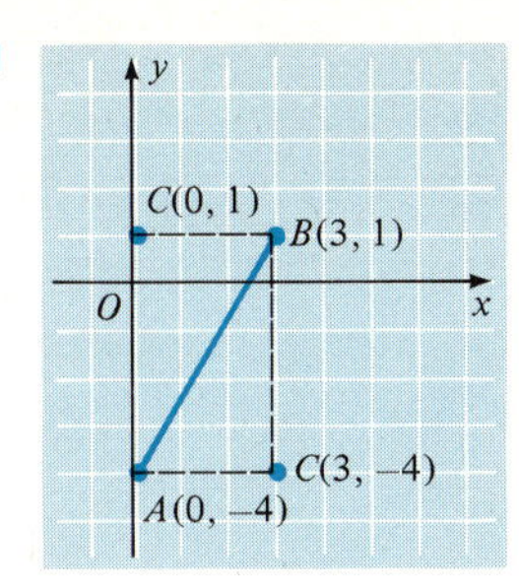

33 $x = -6$ or $x = 2$ **35** $x + y = 4$ **37** 9.6 **39** Yes, they do lie on a straight line. **41** Approx. 340 miles.

USING YOUR KNOWLEDGE 5.1

1 13 inches

EXERCISE 5.2

1a $\{1\}$ **b** $\{2, 3, 4\}$

c

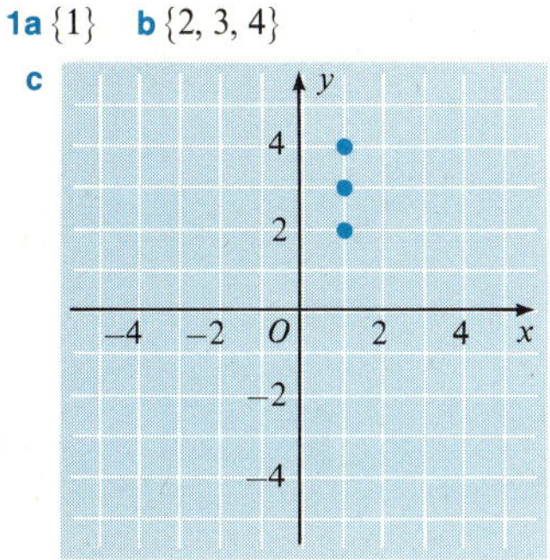

3a $\{x \mid -1 < x < 2\}$ **b** $\{y \mid y > 0\}$

c

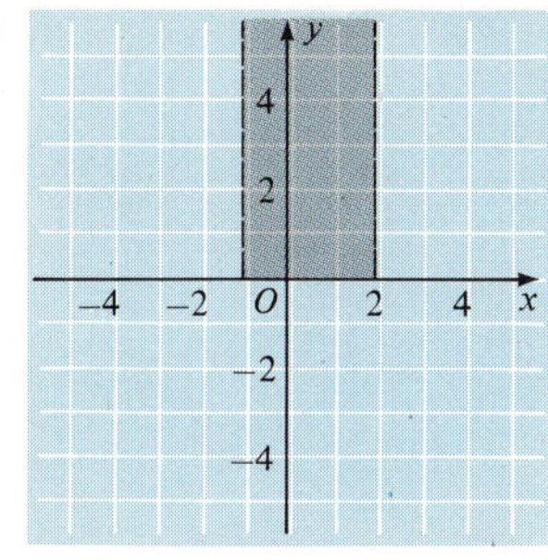

5a $\{-2\}$ **b** $\{y \mid y$ is a real number$\}$

c

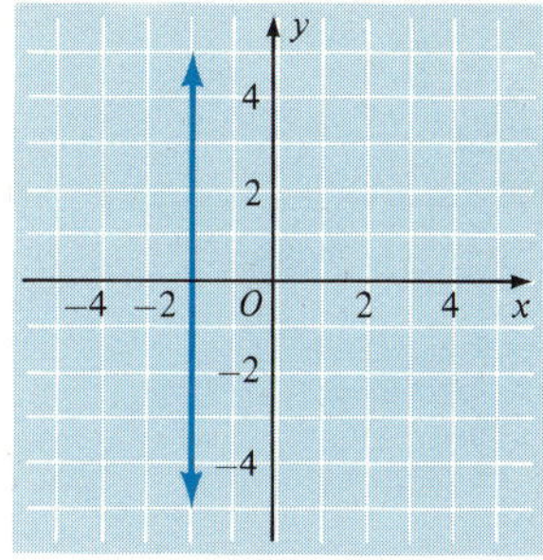

7a $\{x \mid -4 < x < 4\}$
b $\{y \mid -1 < y < 1\}$

c

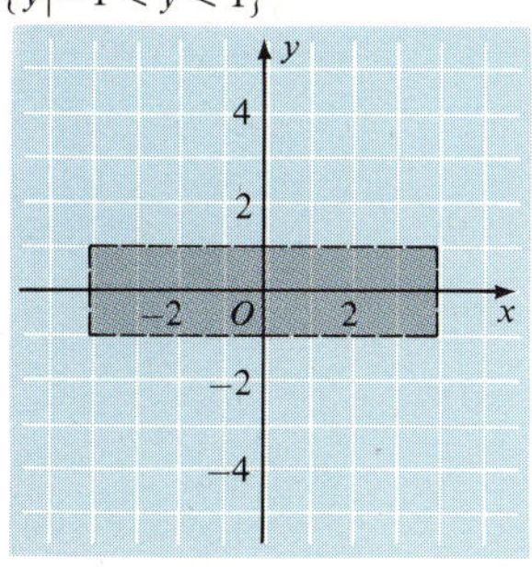

9a $\{x \mid x$ is a real number$\}$
b $\{y \mid y$ is a real number$\}$

c

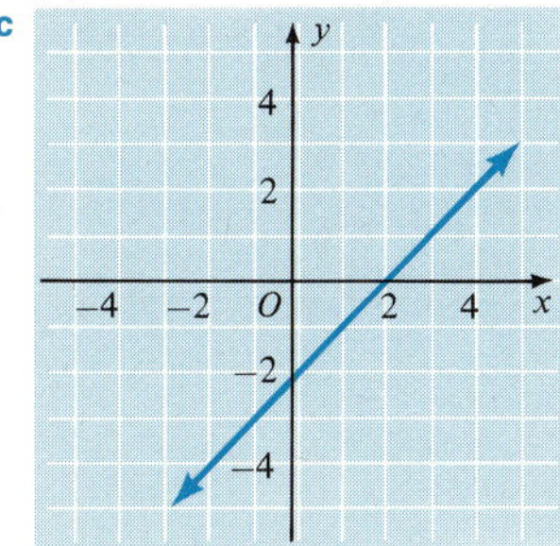

11a $\{x \mid x$ is a real number$\}$
b $\{y \mid y$ is a real number$\}$

c

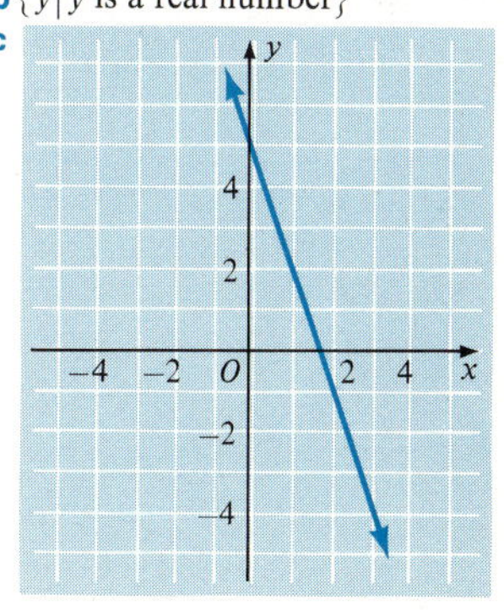

13a $\{x \mid x$ is a real number$\}$
b $\{y \mid y$ is a real number$\}$

c

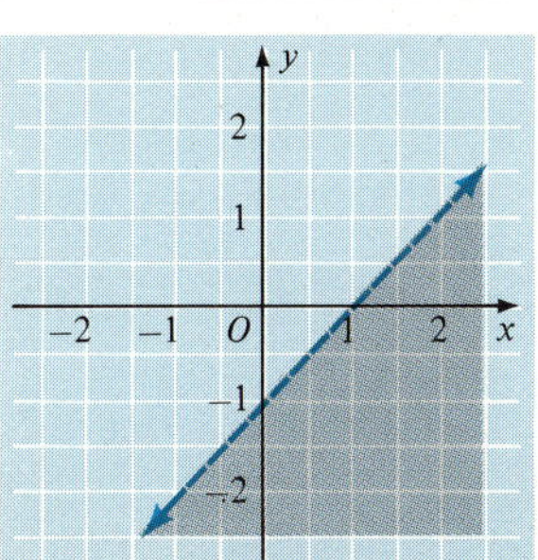

15a $\{x \mid x \text{ is a real number}\}$
b $\{y \mid y \text{ is a real number}\}$
c

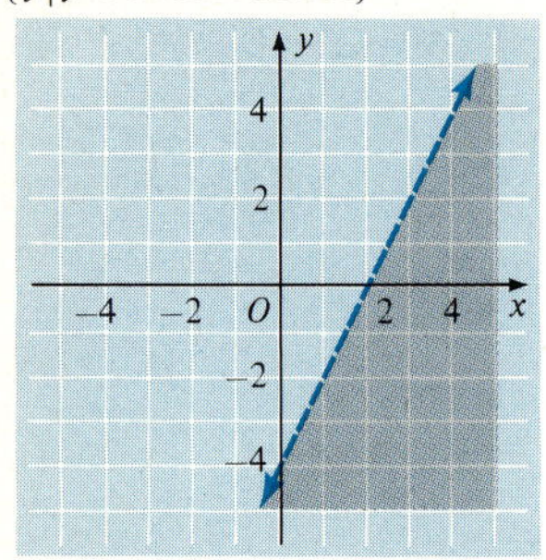

17a $\{x \mid x \geq -1\}$
b $\{y \mid y \text{ is a real number}\}$
c

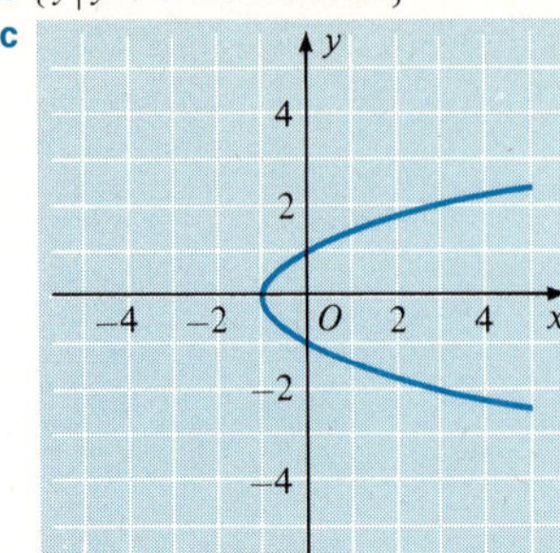

19a $\{x \mid x \leq 4\}$
b $\{y \mid y \text{ is a real number}\}$
c

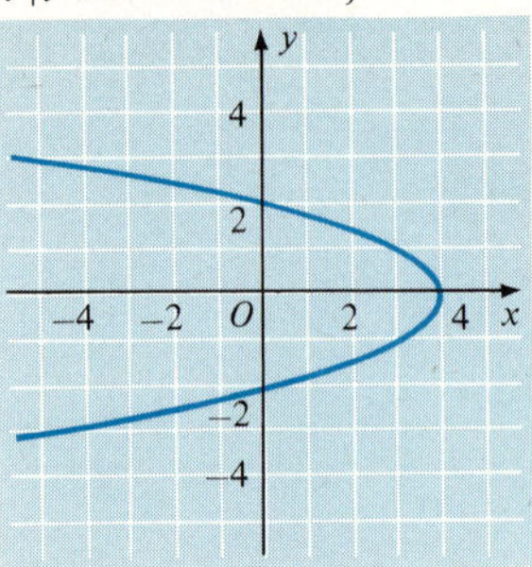

21 $(x-2)^2 + (y-1)^2 = 16$ **23** $(x+4)^2 + (y-3)^2 = 1$ **25** $6x + 4y - 19 = 0$ **27** $x^2 = 4y - 12$ **29** $3x^2 + 3y^2 - 4x - 8y = 0$

31

x	y
0.0	± 2.00
0.5	± 2.06
1.0	± 2.24
1.5	± 2.50
2.0	± 2.83
2.5	± 3.20
3.0	± 3.61
3.5	± 4.03
4.0	± 4.47
4.5	± 4.92
5.0	± 5.39

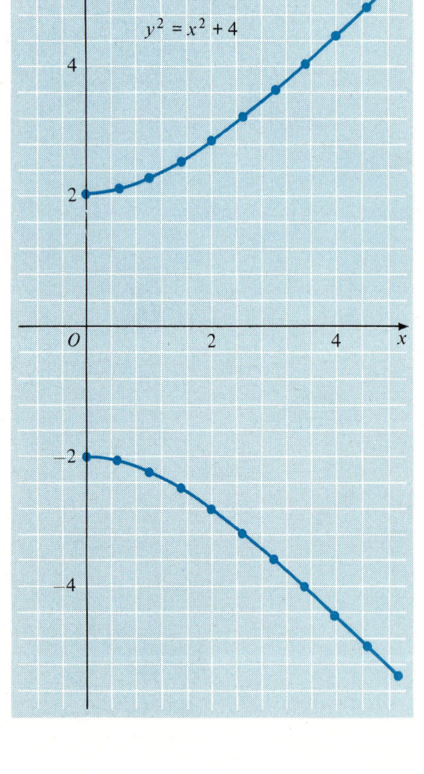

33
$$\begin{aligned} d(B, C) &= \sqrt{(x_3 - x_2)^2 + (y_3 - y_2)^2} \\ &= \sqrt{(x_3 - x_2)^2 + (mx_3 + b - mx_2 - b)^2} \\ &= (x_3 - x_2)\sqrt{1 + m^2} \end{aligned}$$

$$\begin{aligned} d(A, C) &= \sqrt{(x_3 - x_1)^2 + (y_3 - y_1)^2} \\ &= \sqrt{(x_3 - x_1)^2 + (mx_3 + b - mx_1 - b)^2} \\ &= (x_3 - x_1)\sqrt{1 + m^2} \end{aligned}$$

Because

$$(x_3 - x_1)\sqrt{1 + m^2} = (x_2 - x_1)\sqrt{1 + m^2} + (x_3 - x_2)\sqrt{1 + m^2}$$

we have

$$d(A, C) = d(A, B) + d(B, C)$$

so that the points A, B, and C are on a straight line. Because A, B, and C are any three points on the graph, the graph must be a straight line.

USING YOUR KNOWLEDGE 5.2

1

3 The temperature at the top of Mt. Rainier is about $-20°C$. Other altitudes less than 100 kilometers with the same temperature are 40 kilometers and 62 kilometers.

EXERCISE 5.3

1a $\{x \mid x \text{ is a real number}\}$
b $\{y \mid y \text{ is a real number}\}$
c

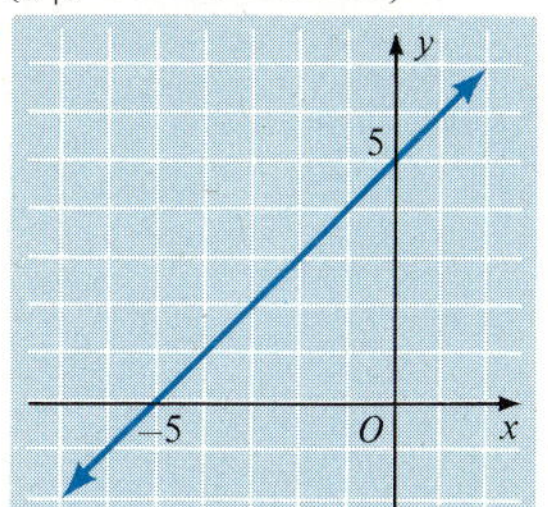

3a $\{x \mid x \text{ is a real number}\}$
b $\{y \mid y \text{ is a real number}\}$
c

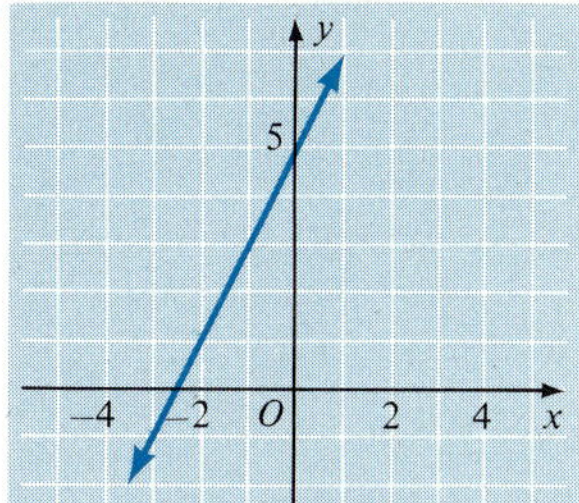

5a $\{x \mid x \text{ is a real number}\}$
b $\{y \mid y \geq 3\}$
c

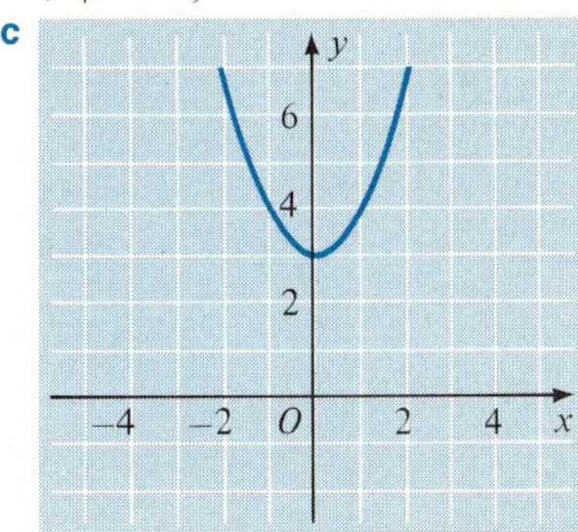

7a $\{x \mid x \text{ is a real number}\}$ **b** $\{y \mid y \leq 4\}$
c

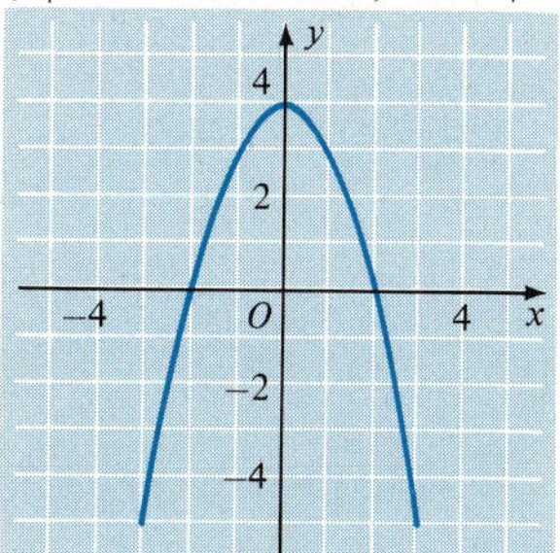

9a $\{x \mid x \text{ is a real number}\}$ **b** $\{y \mid y \text{ is a real number}\}$
c

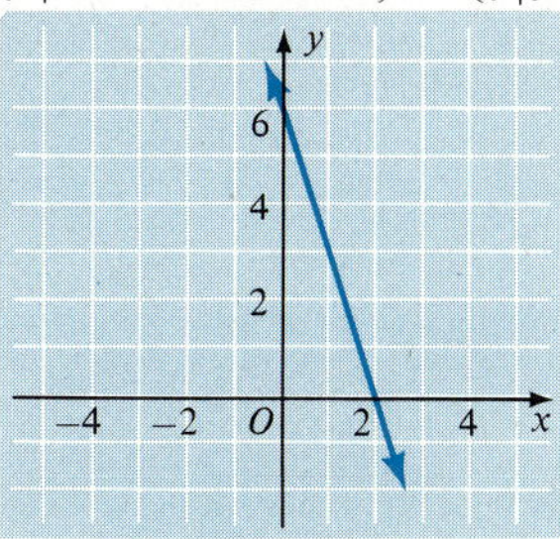

11a $\{x \mid x \text{ is a real number}\}$ **b** $\{5\}$
c

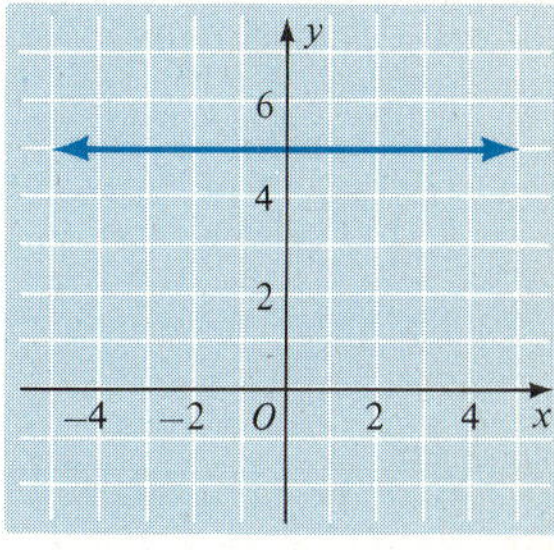

13a $\{x \mid x \text{ is a real number}\}$ **b** $\{y \mid y \geq -1\}$
c

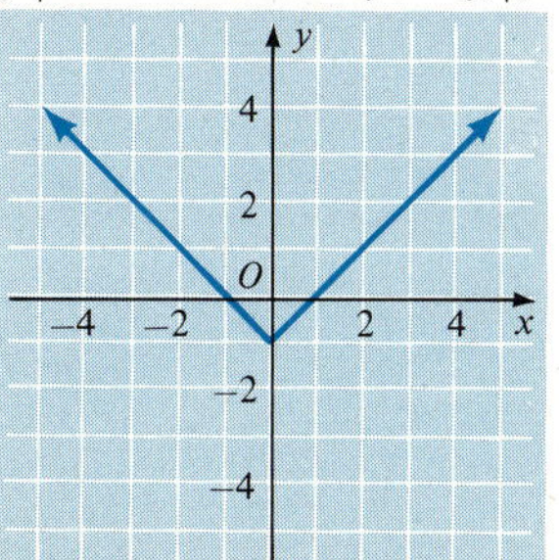

15a $\{x \mid x \text{ is a real number}\}$ **b** $\{y \mid y \geq 0\}$
c

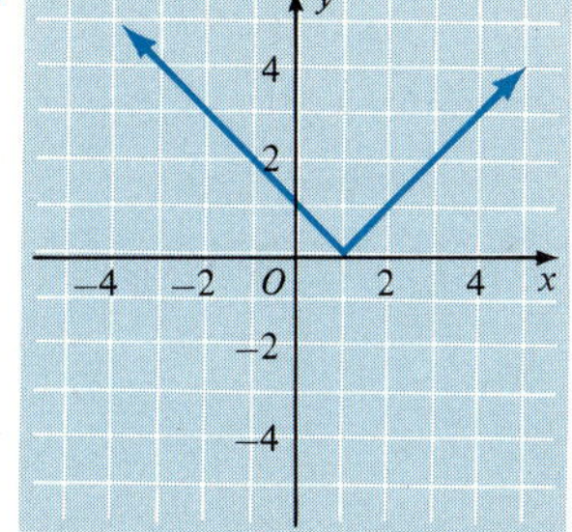

17a $\{x \mid x \text{ is a real number}\}$ **b** $\{-1, 1\}$
c

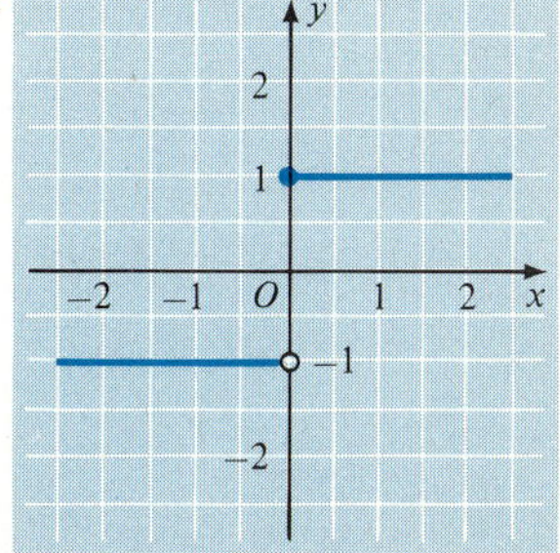

19a $\{x \mid x \text{ is a real number}\}$
b $\{y \mid -2 \leq y \leq 2\}$
c

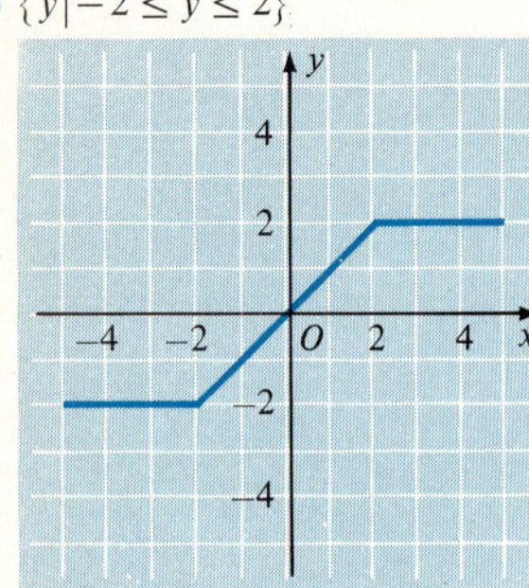

21a $\{x \mid x \text{ is a real number}\}$
b $\{y \mid y \text{ is an integer}\}$
c

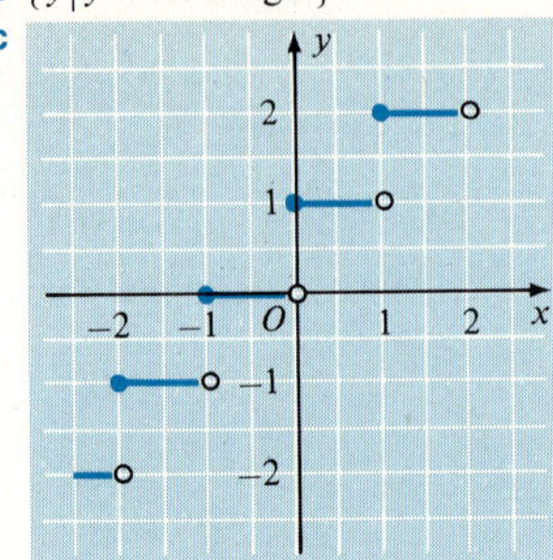

23a $\{1, 2, 5\}$
b A function
c

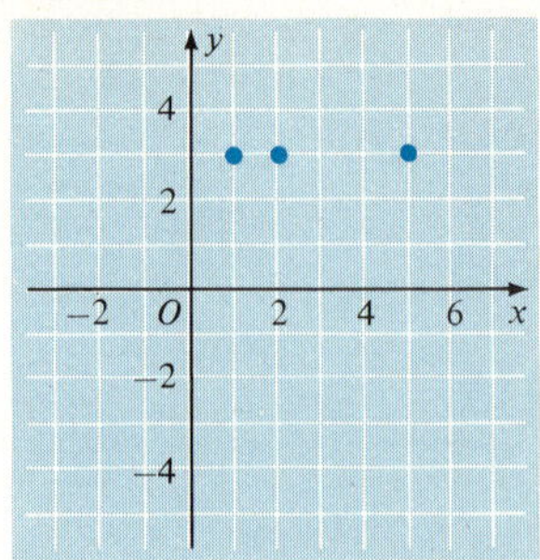

25a $\{-4, 2, 3\}$ **b** Not a function
c

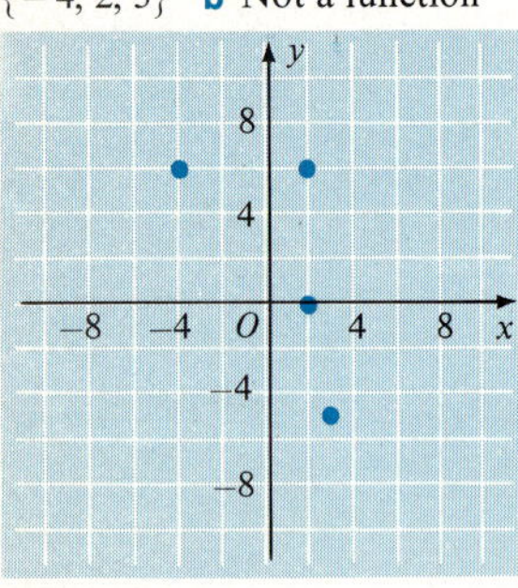

27a $\{x \mid x \text{ is a real number}\}$ **b** A function
c

29a $\{x \mid x \geq 0\}$ **b** Not a function
c

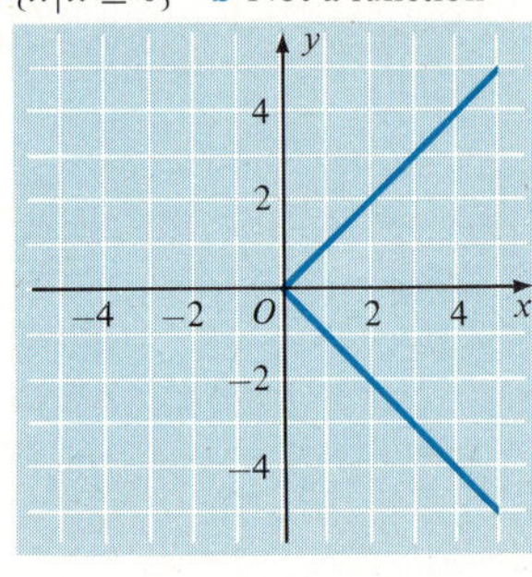

31a $2 - a$ **b** $-a$ **c** $1 - x + y$ **d** $1 - x - y$ **33a** $y + \frac{1}{y}$ **b** $\frac{1}{y} + y$ **c** $\frac{y}{y^2 + 1}$ **35a** $x \geq 2$ **b** $x = 18$ **37a** $x + h + 2$ **b** 1

39a $\sqrt{x + h + 1}$ **b** $\frac{1}{\sqrt{x + h + 1} + \sqrt{x + h}}$ **41a** \$22.45 **b** \$29.95 **c** $C(m) = 14.95 + 0.15m$

43

x	y	
100	18,000	
105	18,125	
110	18,200	
115	18,225	← A price of \$115 gives the greatest profit.
120	18,200	
125	18,125	
130	18,000	

45 $v = 8\sqrt{h}$ (ft/sec) **47** $d = \frac{1}{4}\sqrt{1 + t^2}$ (miles)

49 Let n be the whole number such that $n < x \leq n + 1$. Then,

$-n > -x \geq -n - 1$ Multiply each member by -1.

and

$1 - n > 1 - x \geq -n$ Add 1 to to each member.

Therefore,

$-n \leq 1 - x < -n + 1$ Read the preceding inequality from right to left.

Hence,

$[\![1 - x]\!] = -n$ and $p(x) = 15 + 13n$

which is the correct postage.

USING YOUR KNOWLEDGE 5.3

1 An even function **3** An even function **5** An odd function **7** An even function

EXERCISE 5.4

1 $f^{-1} = \{(3, 4), (2, 3), (1, 2)\}$ **3** $f^{-1} = \left\{(x, y) \mid y = \dfrac{x+2}{3}\right\}$ **5** $f^{-1} = \{(x, y) \mid y = \sqrt{x+2}\}$ **7** $f^{-1}(x) = 2 - \dfrac{1}{x}, x \neq 0$

9 $g^{-1}(x) = \sqrt[3]{x+1}$ **11** $h^{-1}(x) = \dfrac{x^2 - 4}{3}, x \geq 0$

13 $f(g(x)) = \dfrac{(4x-2)+2}{4} = x$ for all real values of x.

$g(f(x)) = 4\left(\dfrac{x+2}{4}\right) - 2 = x$ for all real values of x.

Therefore, f and g are inverses of each other.

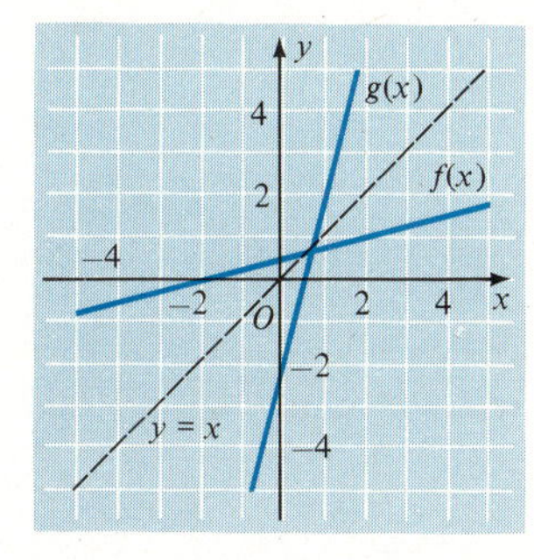

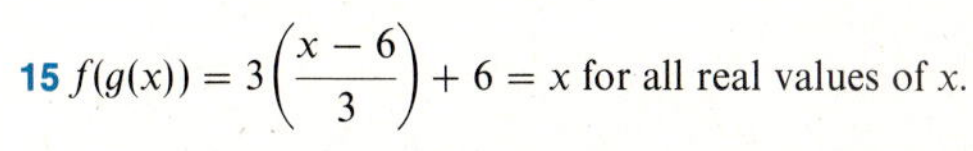

15 $f(g(x)) = 3\left(\dfrac{x-6}{3}\right) + 6 = x$ for all real values of x.

$g(f(x)) = \dfrac{3x+6-6}{3} = x$ for all real values of x.

Therefore, f and g are inverses of each other.

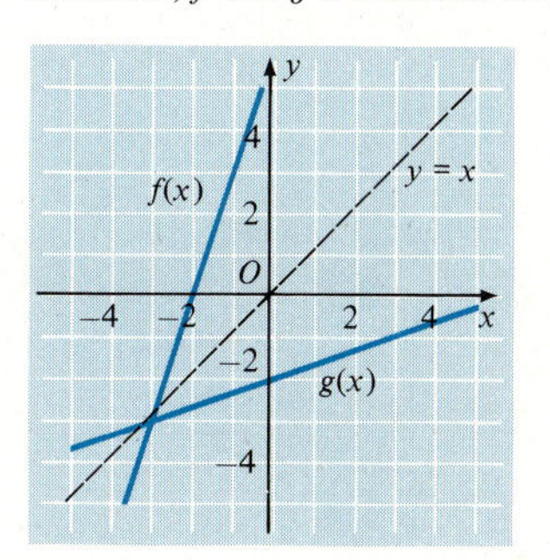

17 $f(g(x)) = f(x) = x$ for all real values of x.
$g(f(x)) = g(x) = x$ for all real values of x.
Therefore, f and g are inverses of each other.

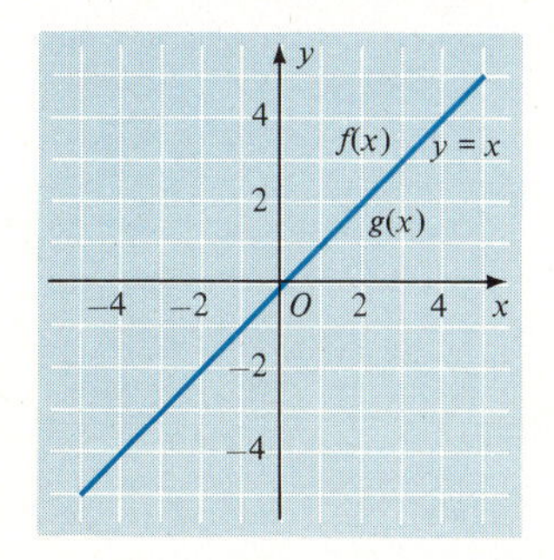

19a $x^2 - x - 3$ **b** $x^2 + x - 11$ **c** $-x^3 + 4x^2 + 7x - 28$ **d** $-\dfrac{x^2 - 7}{x - 4}$

21a $-x^2 + 3x + 4$ **b** $-x^2 + x - 2$ **c** $-x^3 - x^2 + 7x + 3$ **d** $-\dfrac{x^2 - 2x - 1}{x + 3}$ **23a** $f(g(x)) = 3x + 1$ **b** $g(f(x)) = 3x - 1$

25a $f(g(x)) = |x|$ **b** $g(f(x)) = x$ **27a** $f(g(x)) = 3$ **b** $g(f(x)) = -1$ **29** $f(g(x)) = (x^2 - x - 1)^3$, where $f(x) = x^3$, $g(x) = x^2 - x - 1$

31 $f(g(x)) = \sqrt{x+1} + 5$, where $f(x) = \sqrt{x} + 5$, $g(x) = x + 1$ **33a** $m = 0.0254i$ **b** $m = g(f(i))$

35 $i = \dfrac{100m}{2.54} \approx 39.37m$ can be used to convert meters to inches. **37a** $G^{-1}(S) = \dfrac{S^{3/2}}{6\pi^{1/2}}$

b This formula can be used to find the volume if the surface area is known.

39 $f_1(10) = 1.7782794$, $f_2(10) = 1.1547820$, $f_3(10) = 1.0366329$, $f_4(10) = 1.0090350$. $f_n(10)$ gets closer and closer to 1 as n gets larger and larger.

USING YOUR KNOWLEDGE 5.4

1a $g^{-1}(n) = \frac{n}{4} + 40$ **b** $f^{-1}(d) = 6d + 4$ **c** $h^{-1}(F) = \frac{5}{9}(F - 32)$

SELF-TEST, CHAPTER 5

1a

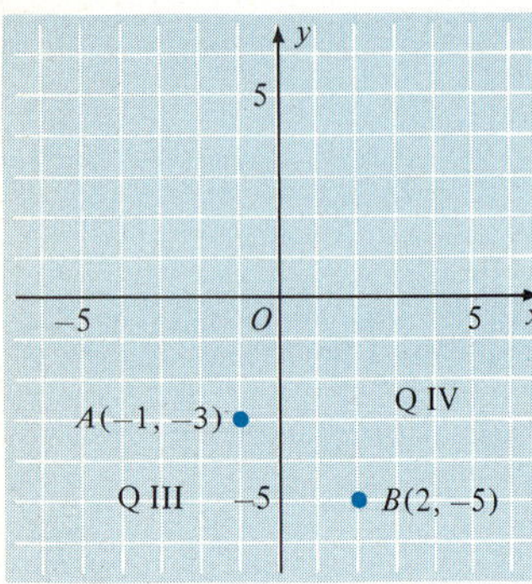

b $\sqrt{13}$ **2a** $(x - 2)^2 + (y + 3)^2 = 16$ **b** The distance from the center of the circle to the point is $\sqrt{18}$ (or $3\sqrt{2}$), which is greater than the radius. Therefore, the point is outside the circle.

3 $d(A, B) = \sqrt{50}$; $d(B, C) = \sqrt{85}$; $d(A, C) = \sqrt{125}$
$[d(A, B)]^2 + [d(B, C)]^2 = 50 + 85 = 135 \neq [d(A, C)]^2$
Thus, ABC is not a right triangle.

4 $\sqrt{29}$

5a Domain: $\{x \mid -1 < x < 2\}$
Range: $\{y \mid -2 < y < 3\}$

b

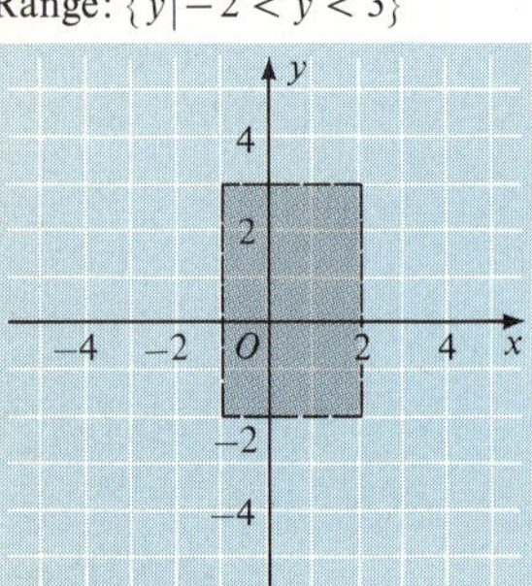

6

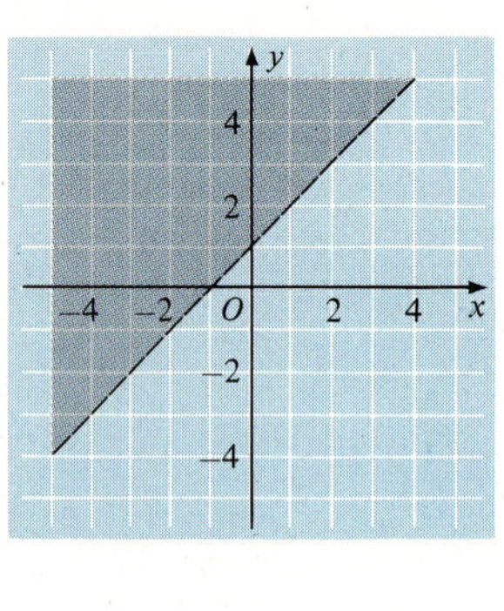

7 $(x + 2)^2 + (y - 1)^2 = 4$ **8a**

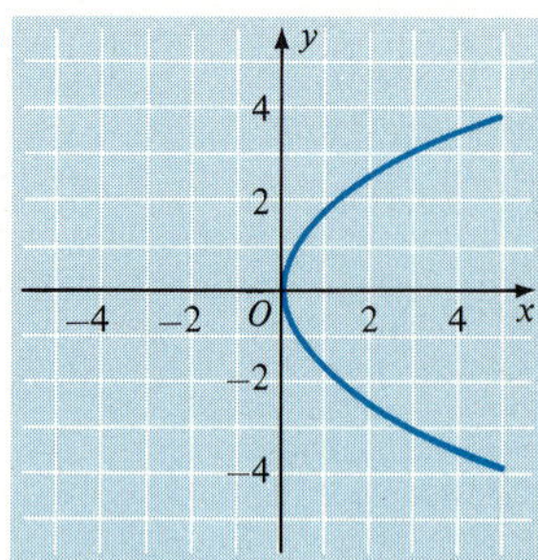

b Domain: $\{x \mid x \geq 0\}$
Range: the set of all real numbers

9 $x^2 + y^2 - 9x + 18 = 0$

10a Domain: the set of all real numbers
Range: the set of all real numbers

b

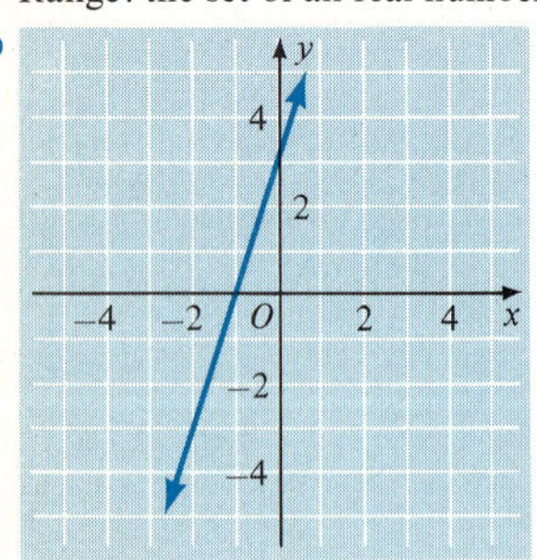

11a Domain: the set of all real numbers
Range: the set of all nonpositive real numbers

b

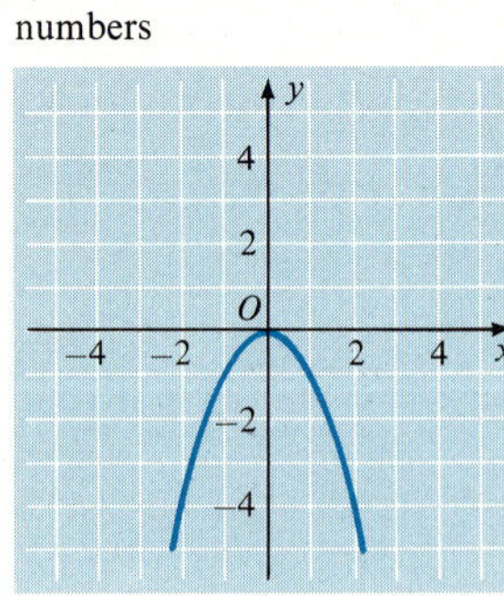

12a Domain: $\{x \mid x \leq -3 \text{ or } x \geq 3\}$ **b** ± 5

13a Domain: the set of all real numbers
Range: the set of all real numbers ≥ 1

b

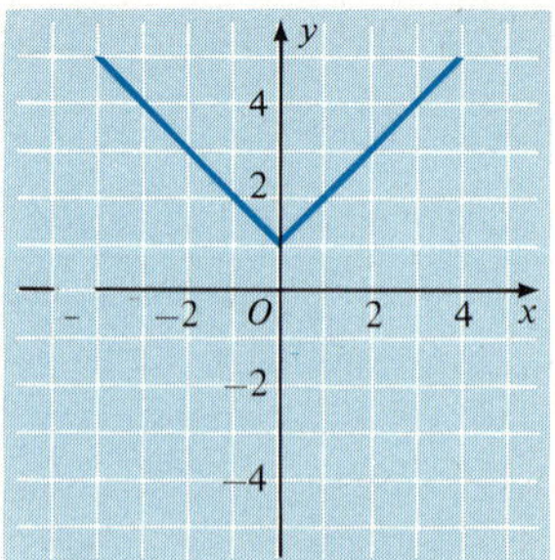

14a \$23.50 **b** $C(x) = 16 + 0.15x$ **15a** $f(3) = 5$ **b** $f(x+h) = \sqrt{(x+h)^2 + 16}$ **16** $f^{-1} = \{(x, y) | y = x + 4\}$
17a $f_r(x) = x^2 - 1, x \geq 0$ **b** $f_r^{-1}(x) = \sqrt{x+1}, x \geq -1$ **18a** $x^2 + 2x$ **b** $x^2 - 2x + 2$ **c** $2x^3 - x^2 + 2x - 1$ **d** $\dfrac{x^2 - 1}{2x - 1}$
19a $\sqrt{x^2 + 1}$ **b** Domain: the set of all real numbers **20** $f(x) = \sqrt[3]{x}, g(x) = x^2 - 1$ (Other answers are possible.)

EXERCISE 6.1

1 -1 **3** Undefined **5** $\frac{5}{2}$ **7** $-\sqrt{6}$ **9** $x - 2y - 8 = 0$ **11** $5x - y + 3 = 0$ **13** $3x - 8y - 6 = 0$ **15** $3x - 2y + 1 = 0$
17 $3x - y + 9 = 0$ **19** $2x - y - 2 = 0$ **21** $y + 1 = 0$ **23a** $x + 2 = 0$ **b** $y - 5 = 0$ **25** $x + 2 = 0$
27a $-\frac{3}{2}$ **b** 3 **c** 2

d

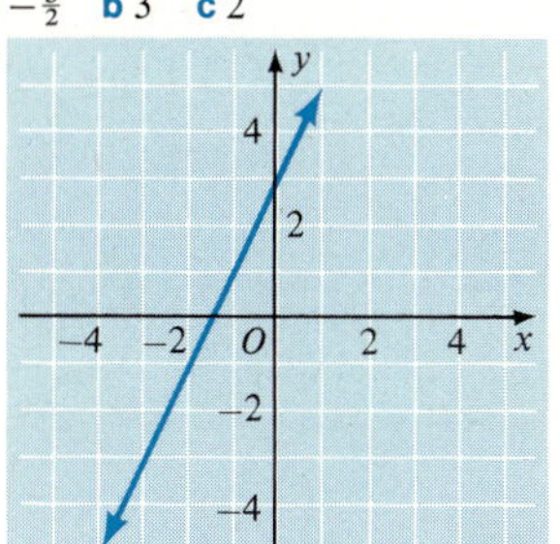

29a -4 **b** 2 **c** $\frac{1}{2}$

d

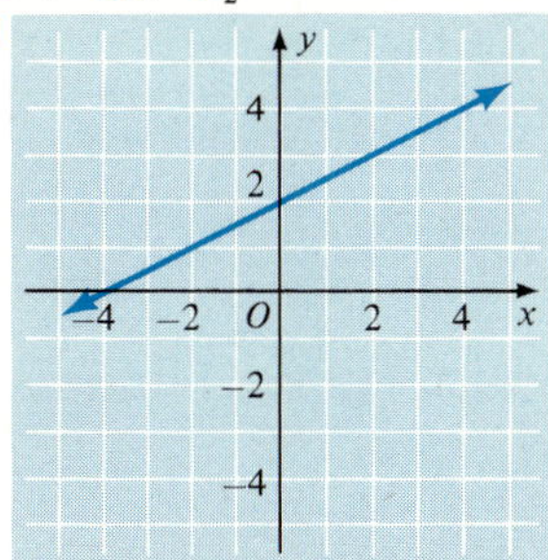

31a 3 **b** None **c** Undefined

d

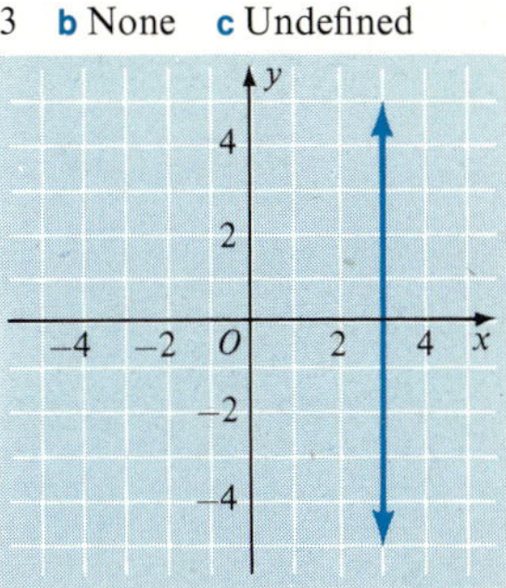

33a $y = 2x - 2$ **b** $y = -\frac{1}{2}x + \frac{1}{2}$ **35a** $y = 0$ **b** The required line has no slope and thus no slope-intercept equation. An equation is $x - 1 = 0$. **37a** $y = x - 1$ **b** $y = -x + 1$ **39a** $y = \frac{1}{4}x - \frac{1}{4}$ **b** $y = -4x + 4$ **41** $a = 7$ **43** $b = 4$
45 $m \approx 10$. This means there was an average increase of 10% per year in the on-time performance. **47** -3.65
49 x-intercept $= -2.307$; y-intercept $= 0.515$
51 The slope of the line is

$$m = \frac{y_2 - y_1}{x_2 - x_1}$$

If (x, y) is any point on the line other than one of the given points, then the slope is also

$$m = \frac{y - y_1}{x - x_1}$$

Thus, an equation of the line is

$$\frac{y - y_1}{x - x_1} = \frac{y_2 - y_1}{x_2 - x_1}$$

53 If $B \neq 0$, then both lines have slope $m = -A/B$. Thus, since $C \neq D$, the lines are distinct and parallel. If $B = 0$, then both lines are vertical and therefore parallel.
55 If L_1 and L_2 are parallel, then the two indicated angles in Figure 6.11 are equal, so that the two right triangles are similar and the corresponding sides are proportional. Thus,

$$\frac{y_3 - y_1}{y_4 - y_1} = \frac{x_3 - x_1}{x_4 - x_1}$$

or

$$\frac{y_3 - y_1}{x_3 - x_1} = \frac{y_4 - y_1}{x_4 - x_1}$$

that is, the slopes are equal. This argument can be repeated in the reverse direction to show that if the slopes are equal, then the lines are parallel.

USING YOUR KNOWLEDGE 6.1

1 $C(x) = 3x + 8000$ **3a** \$4 **b** \$9000 **c** \$13,000

EXERCISE 6.2

1
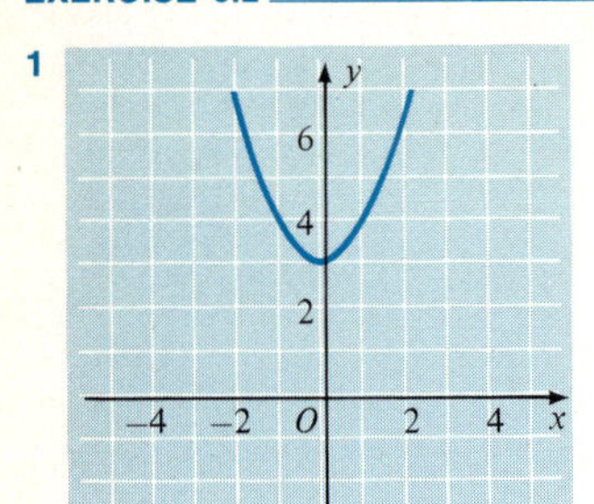

3
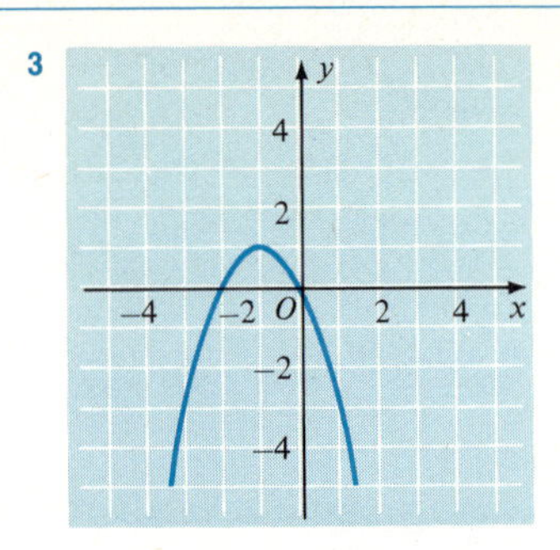

5
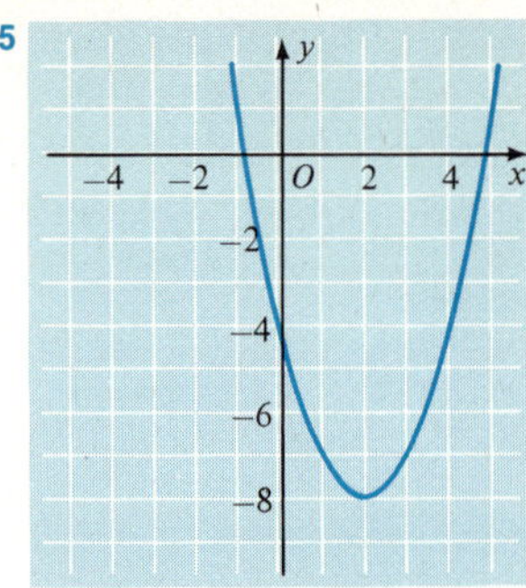

7
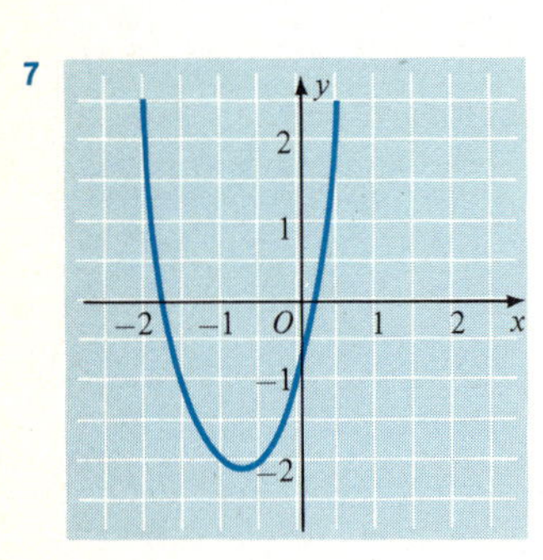

9
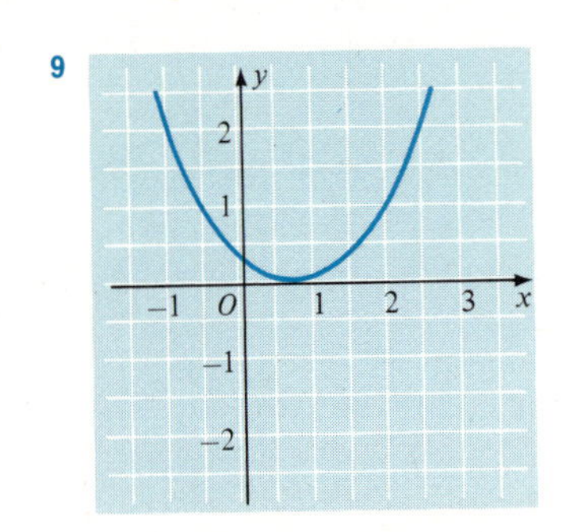

11
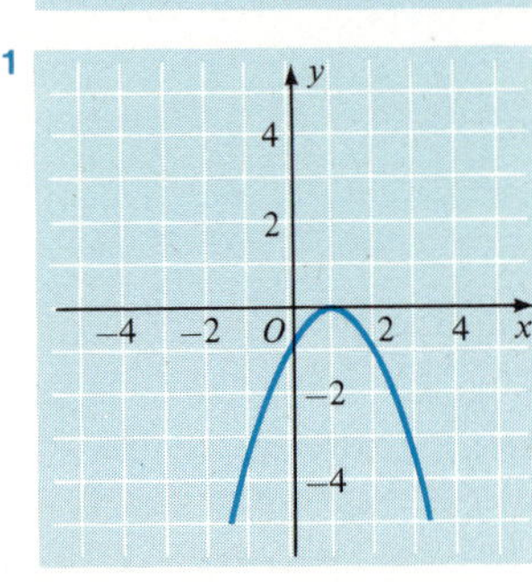

13 True for all real values of x. **15** $x \leq -2$ or $x \geq 0$ **17** $x < 2 - 2\sqrt{2}$ or $x > 2 + 2\sqrt{2}$ **19** $\dfrac{-3 - \sqrt{17}}{4} \leq x \leq \dfrac{-3 + \sqrt{17}}{4}$

21 True for no values of x. **23** True for $x = 1$ only. **25** There are no real roots. **27** There are two real roots.
29 There are two real roots.

31
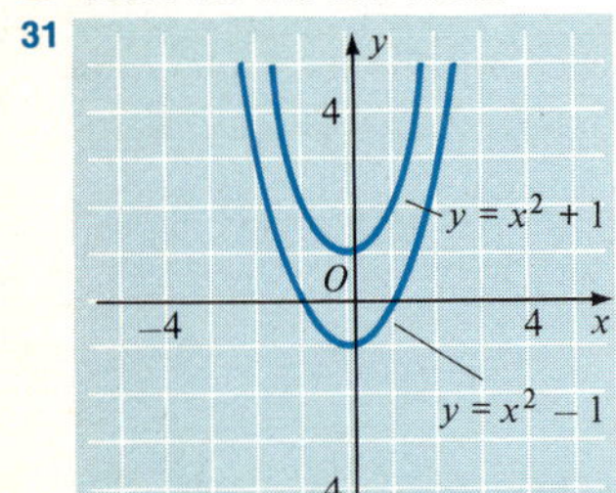

33
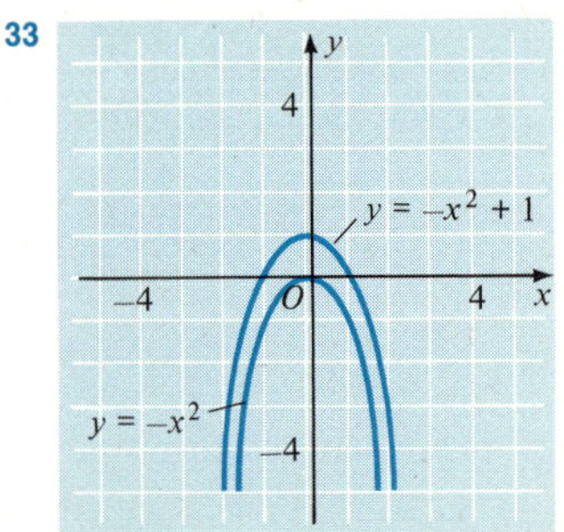

35
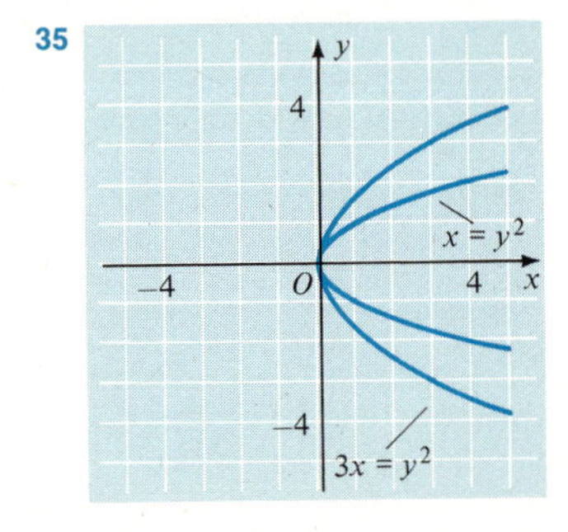

41 5 inches **43** 100 m by 100 m **45** 25 feet

37
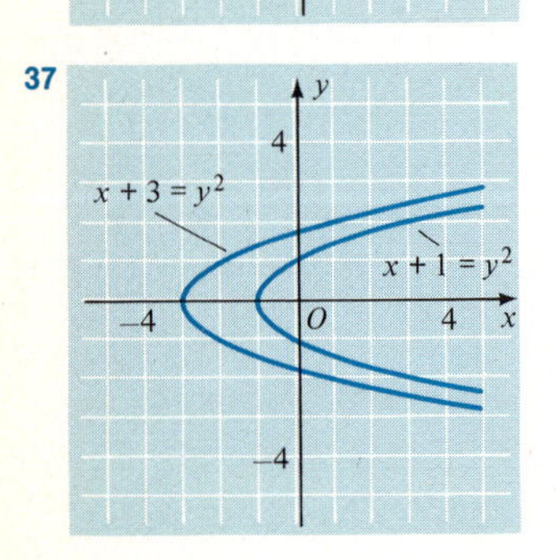

39
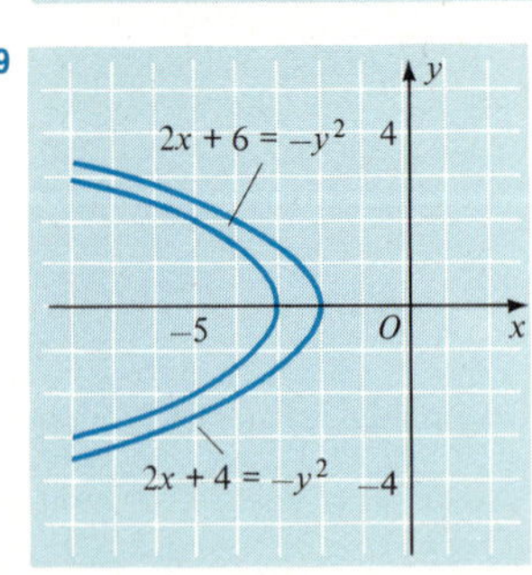

47 Vertex $(1.02, -173.02)$; x-intercepts: 2.96 and -0.92 **49** \$40.19, \$40.75, \$40.91, \$40.75

USING YOUR KNOWLEDGE 6.2

1 $y = -\frac{1}{10}x^2$

EXERCISE 6.3

1

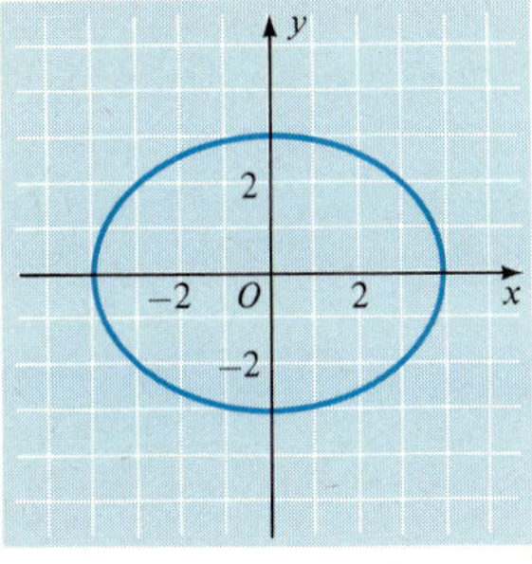

3

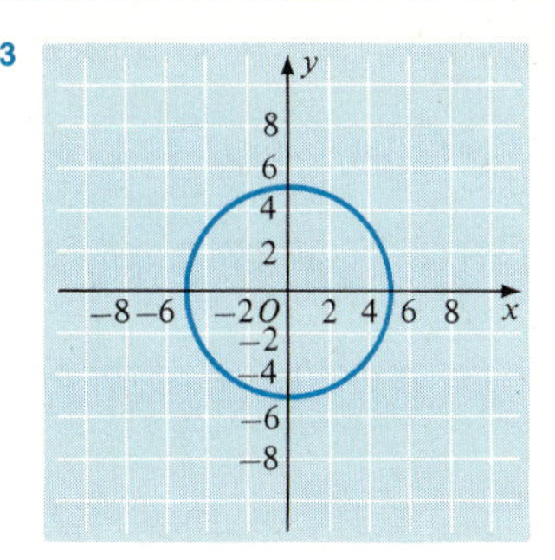

5

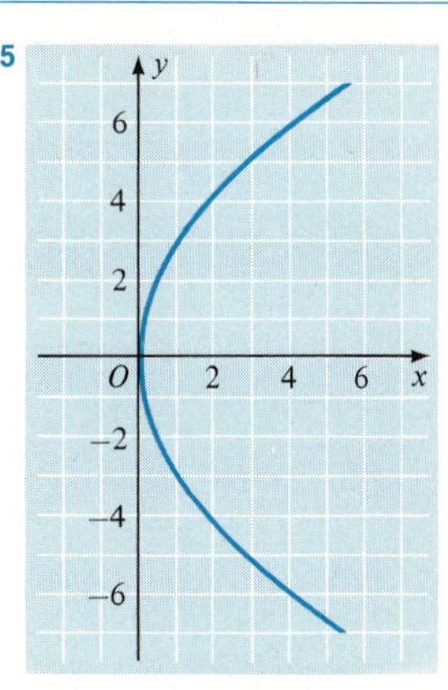

7

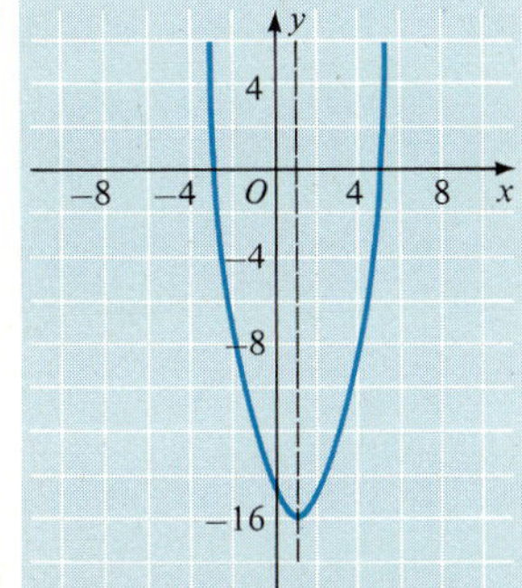

9

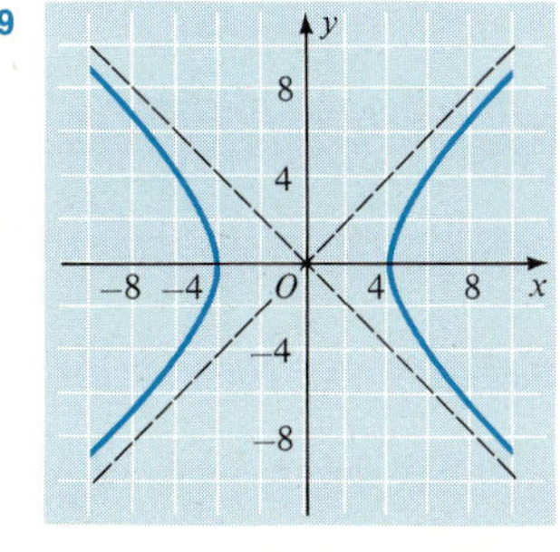

11

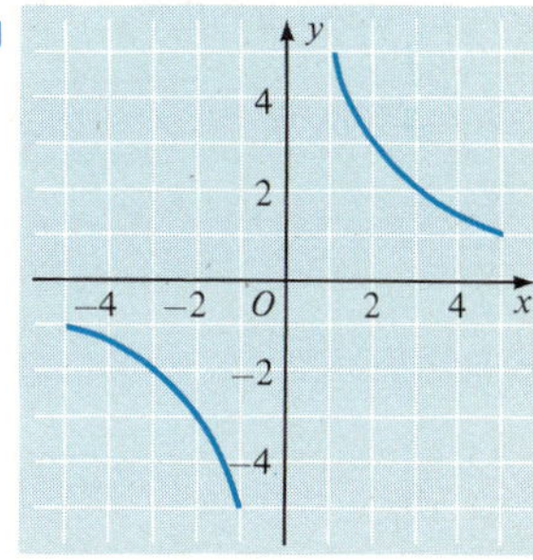

13 x-intercepts: $\pm\sqrt{\frac{145}{42}} \approx \pm 1.9$; y-intercepts: $\pm\sqrt{\frac{145}{29}} \approx \pm 2.4$ **15** $(-99.2, -4.8)$

USING YOUR KNOWLEDGE 6.3

1 $x^2 = \frac{25}{2}y$. The focus is 3.125 feet from the vertex.

EXERCISE 6.4

1

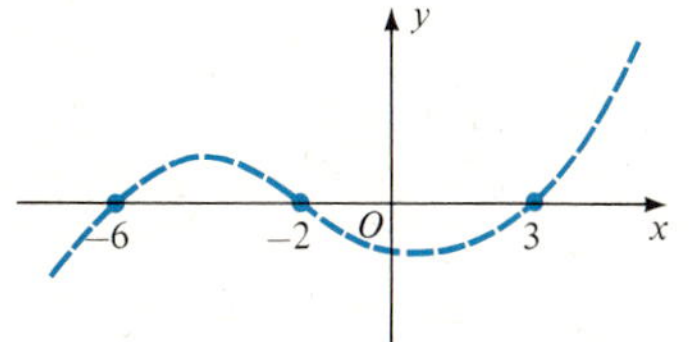

3

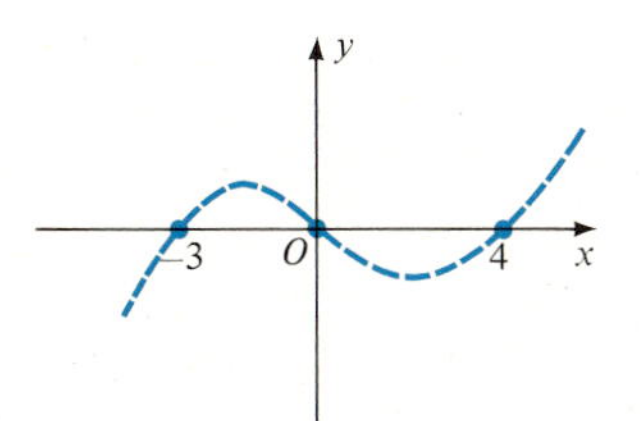

5

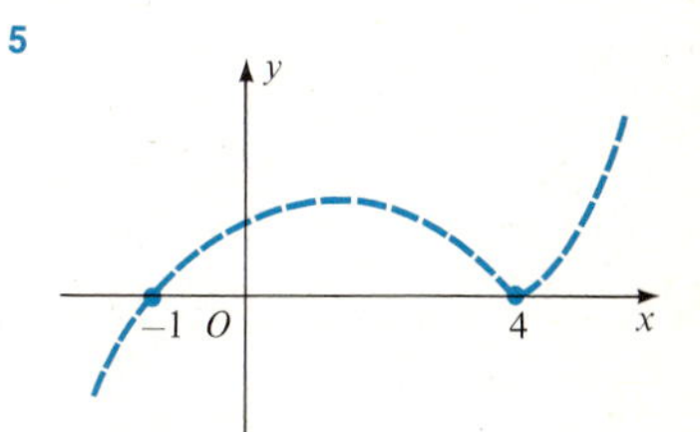

7

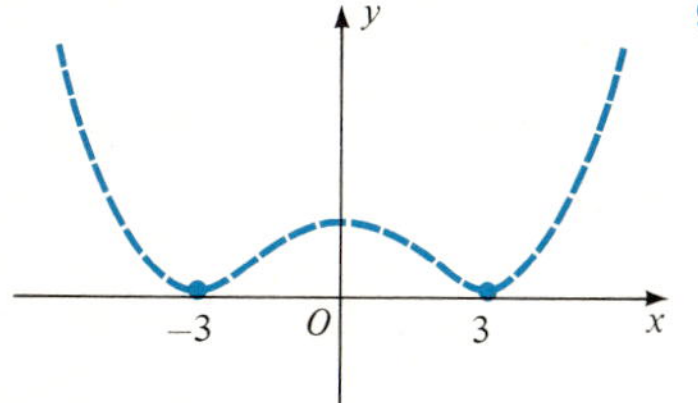

9

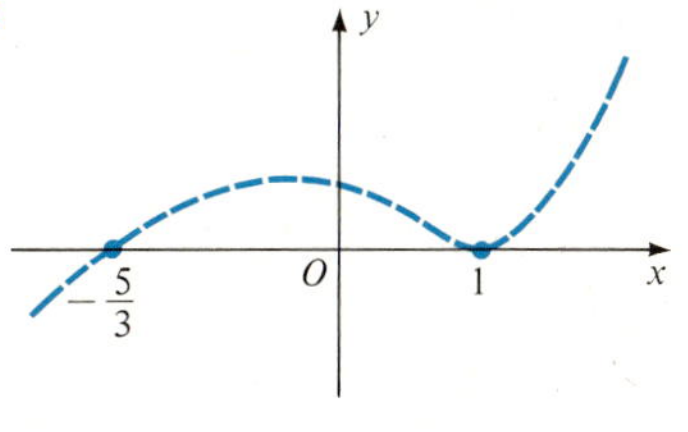

11

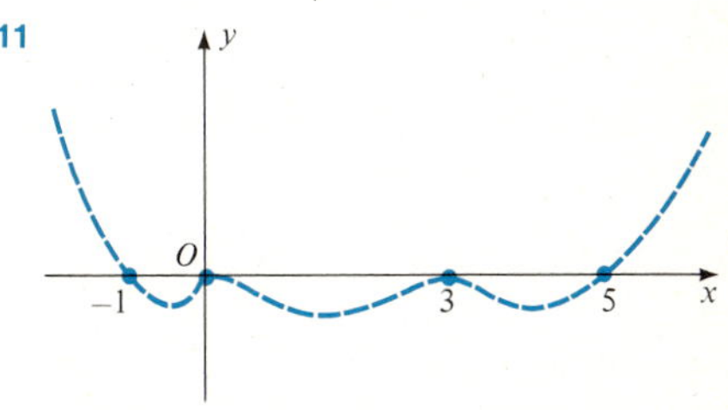

13

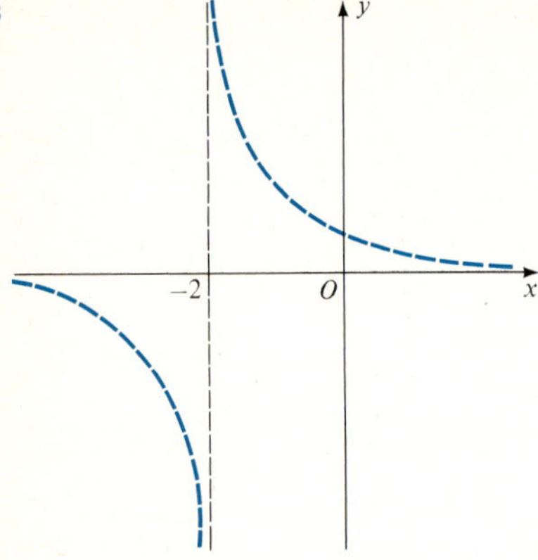

15

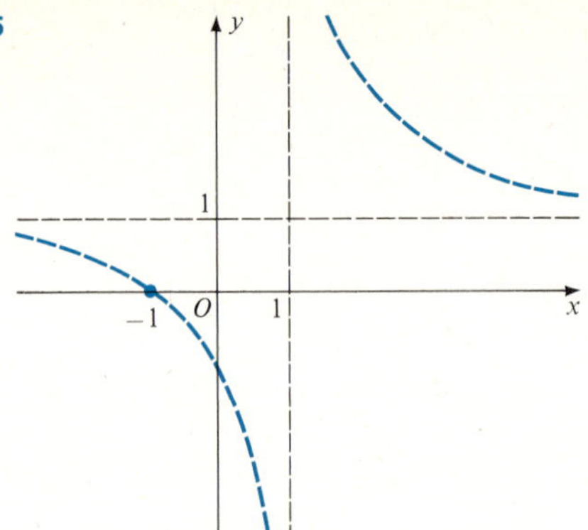

17

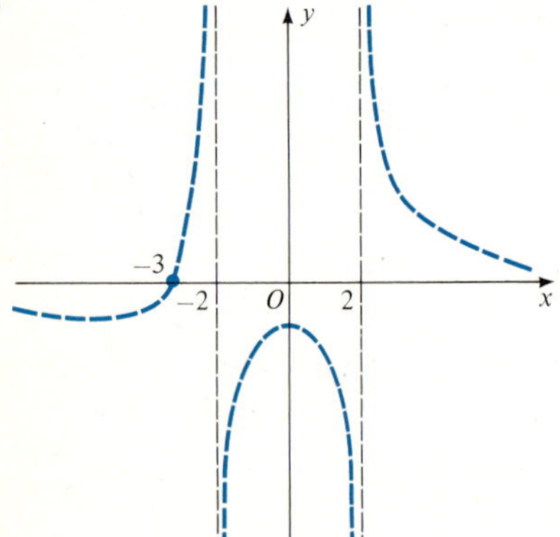

19

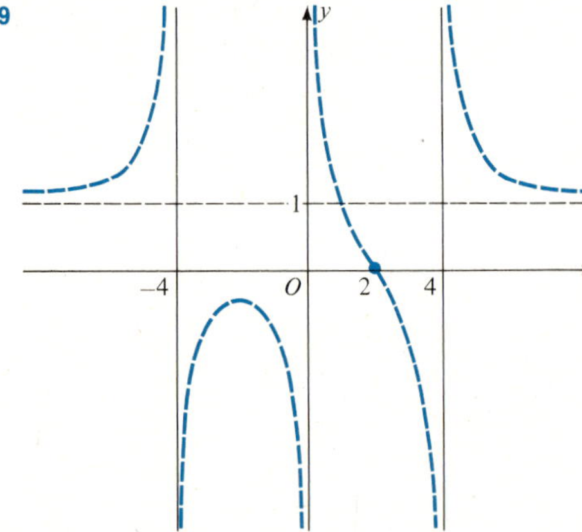

21 $-3 < z < 0$ or $z > 4$ **23** $x \leq -1$ or $x = 0$ or $x = 3$ or $x \geq 5$ **25** $-3 \leq x < -2$ or $x > 2$ **27** $-4 < x < 0$ or $2 < x < 4$
29a $x > 20$ cm **b** 40 cm **c** $x > f$ **d** $2f$ **31** Vertical asymptotes: $x = -1$ and $x = 1$
Oblique asymptote: $y = 2x + 1$

33

x	50	100	500	1000	5000	10,000
y	2.4510	2.2002	2.0367	2.0182	2.0036	2.0018

USING YOUR KNOWLEDGE 6.4

1

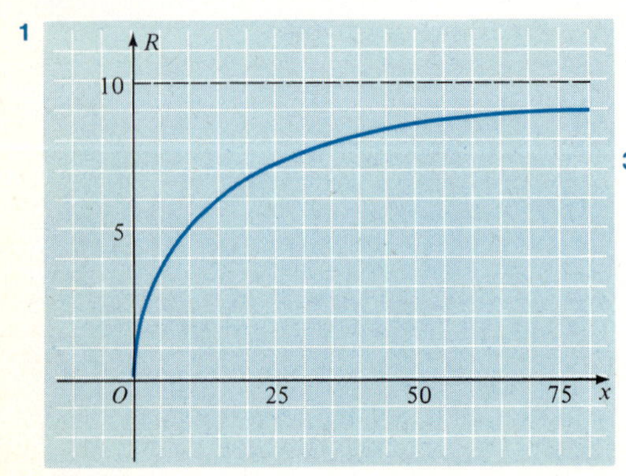

As x gets larger and larger, the equivalent resistance R gets closer and closer to 10 ohms. The current would tend to flow through the 10-ohm resistance alone.

3 $N(x)$ increases by about 2.67 tasks per additional practice task.

EXERCISE 6.5

1 $P = kd$ **3** $A = kr^2$ **5** $R = k/d^2$ **7** $P = kI^2R$ **9** $f = \frac{k}{L}\sqrt{\frac{T}{m}}$ **11a** $k = \frac{1}{40}$ **b** 5 lb **13** 36,864 lb **15a** $P = \frac{k}{V}$ **b** 320 psi

17a $F = k\frac{m_1 m_2}{d^2}$ **b** 40 dynes **19** $\frac{4}{3}$ ft-c **21** 65.2 lb

SELF-TEST, CHAPTER 6

1 $4x + y - 19 = 0$ **2** $y = -6x + 32$ **3a** -3 **b** $\frac{5}{3}$ **4** $2x - y + 4 = 0$ **5** $x + 2y + 5 = 0$ **6a**

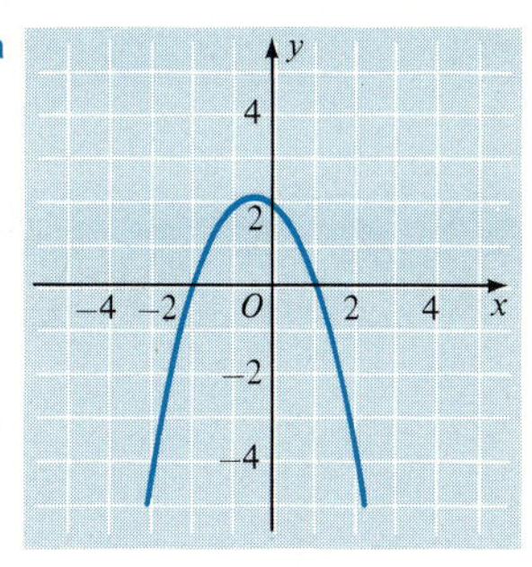

b $-2 < x < 1$ **c** Two

7 72 sq. in. **8** Ellipse

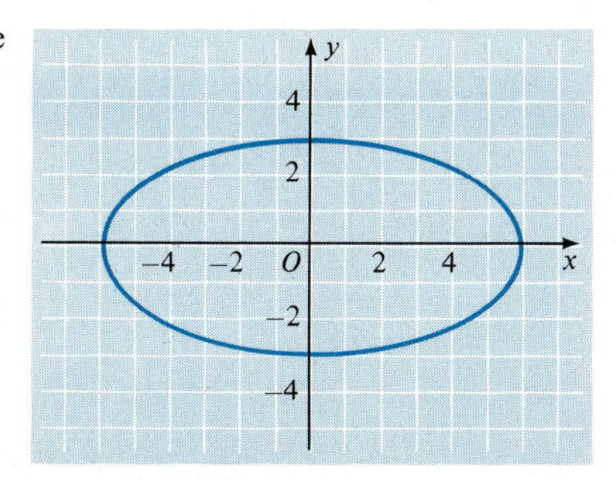

9 Hyperbola

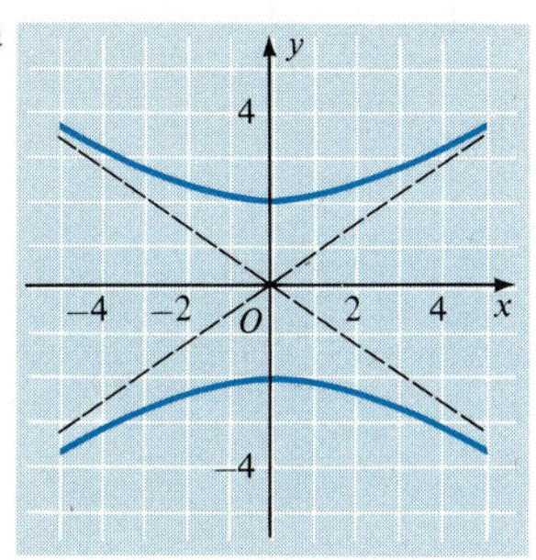

10 Circle

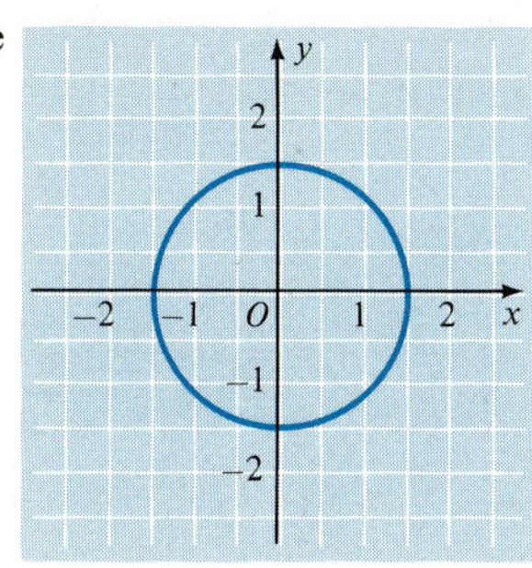

11 Parabola

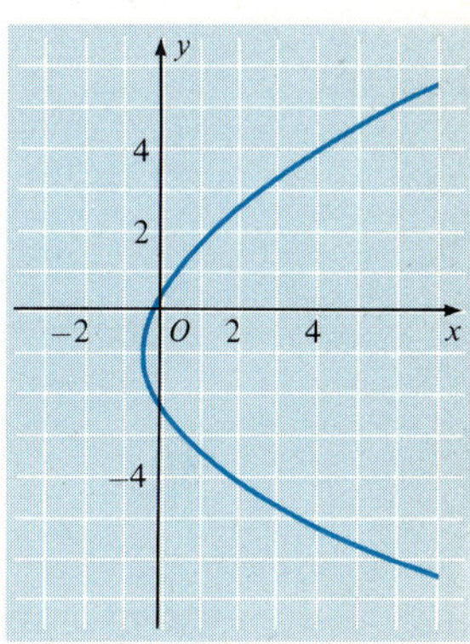

12

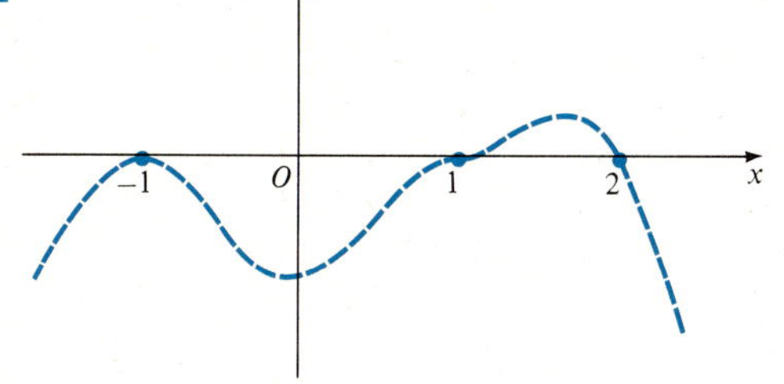

13 $1 \le x \le 2$ or $x = -1$ **14** $x = 1, x = -2$ **15** $y = 2$

16

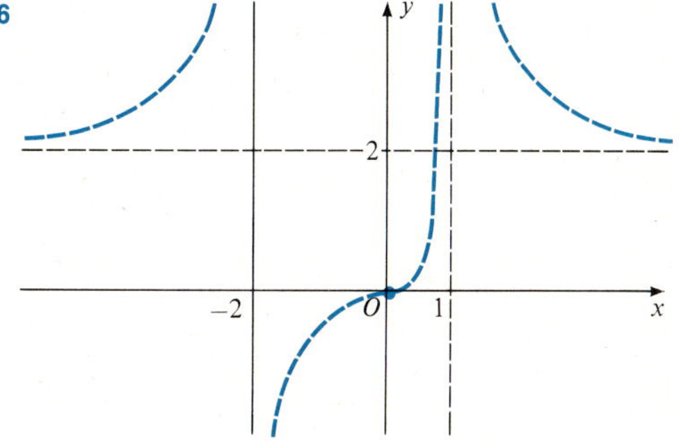

17 $P = \frac{T}{92}$ **18** $d = 16t^2$ **19** $P = \frac{10}{3}$ psi **20** $a = \frac{kF}{m}$

EXERCISE 7.1

1

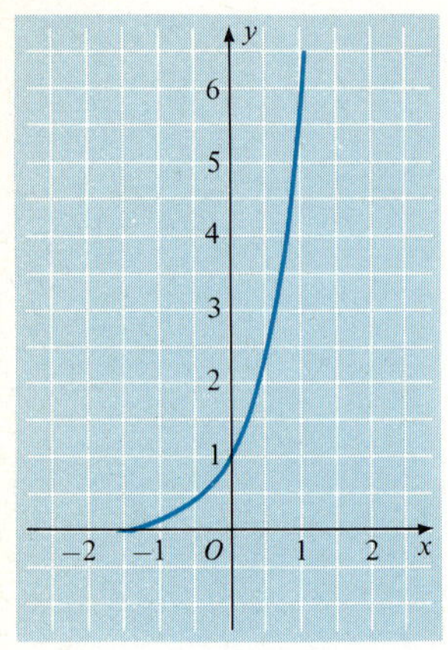

3

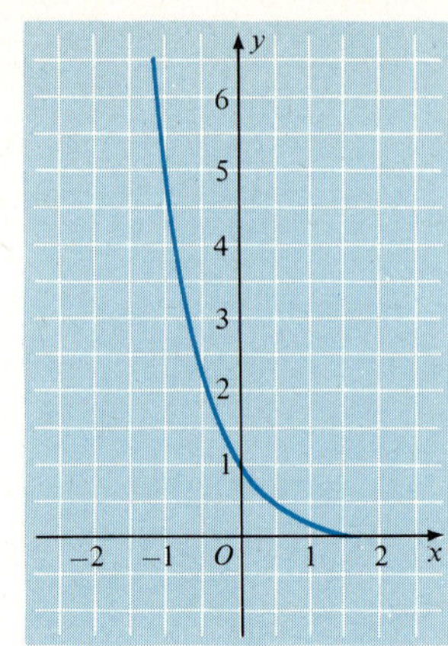

5

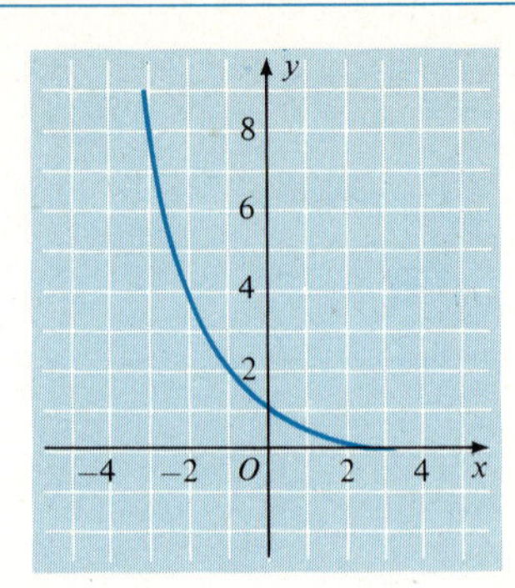

7

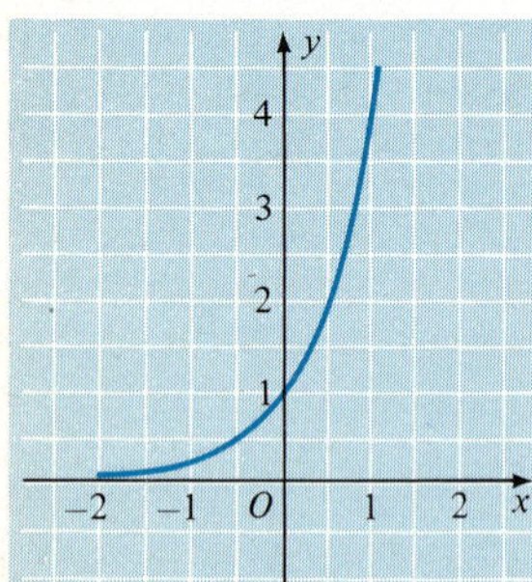

9

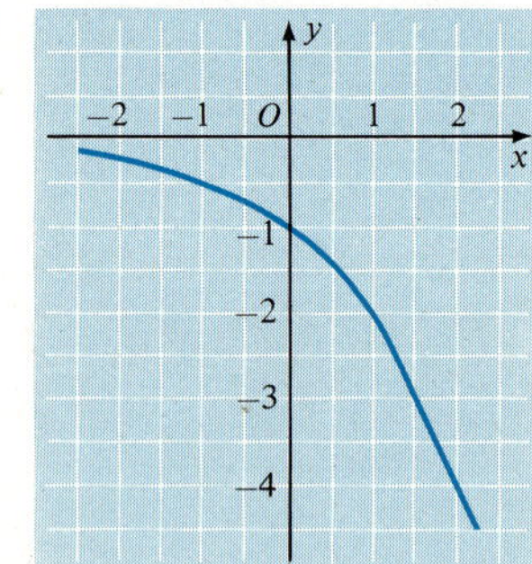

11

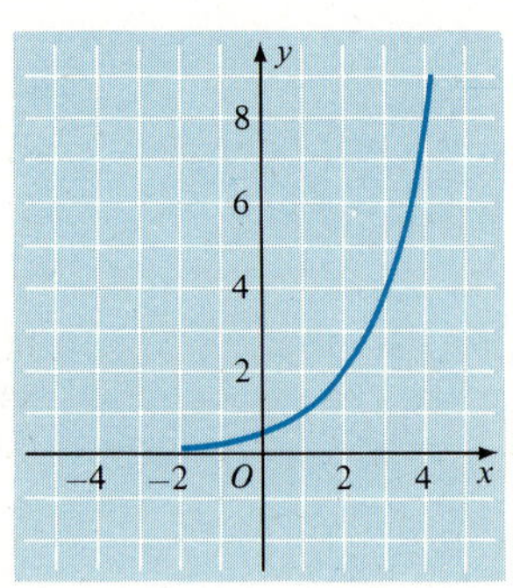

13

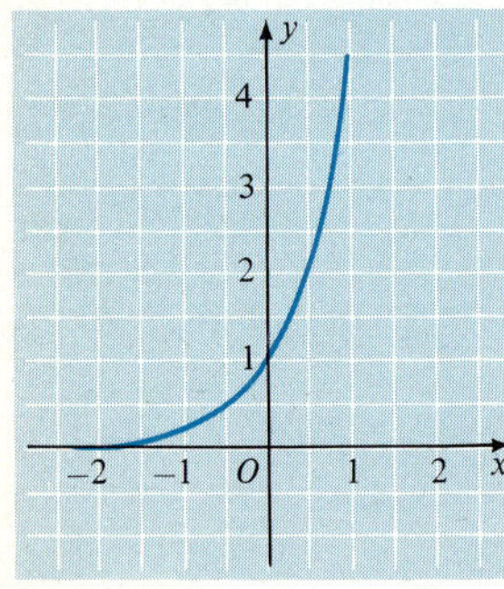

15 $x = -1$ **17** $x = -2$ **19** $x = -2$ **21** $x = \frac{1}{3}$ **23** $k = \frac{1}{22}$ **25** $k = \frac{1}{5600}$
27a 93.3% **b** 87.1% **c** 70.7% **d** 35.4% **e** 4.4%

USING YOUR KNOWLEDGE 7.1

1 $x = 2.906$

EXERCISE 7.2

1 $9^3 = 729$ **3** $2^{-8} = \frac{1}{256}$ **5** $10^{2.4771} = 300$ **7** $\log_2 128 = 7$ **9** $\log_{64} 2 = \frac{1}{6}$ **11** $\log_{10} 0.001 = -3$ **13** 8 **15** $\frac{4}{3}$ **17** 3 **19** $\frac{1}{5}$
21 $x = 243$ **23** $x = 8$ **25** $y = 100$ **27** $x = 11$ **29** $x = 4$ **31** $x = \frac{1}{7}$ **33** $x = 3$ **35** $x = 2$ **37** $x = 4$ **39** $x = -3$ or $x = -1$
41 1.556 **43** 0.681 **45** 0.175 **47** log 3 **49** $\log(\sqrt{2}b^4)$
57 Let $x = \log_b M$ and $y = \log_b N$. Then $b^x = M$ and $b^y = N$. Thus,

$$\frac{M}{N} = \frac{b^x}{b^y} = b^{x-y}$$

so that $\log_b(M/N) = x - y = \log_b M - \log_b N$.
59 The logarithm function is a one-to-one function. Therefore, $\log_b M = \log_b N$ if and only if $M = N$.

USING YOUR KNOWLEDGE 7.2

1 3.5 **3** 6.4 **5** 8.3

EXERCISE 7.3

1a 0 **b** 1 **c** 2 **3a** 0.4771 **b** 1.4771 **c** 2.4771 **5a** 9.4914 − 10 **b** 8.4914 − 10 **c** 6.4914 − 10 **7a** 1.6232 **b** 2.6232 **c** 3.6232 **9a** 7.0899 − 10 **b** 6.0899 − 10 **c** 5.0899 − 10 **11** 18.5 **13** 4.93 **15** 0.0608 **17** 0.0000000998 **19** 0.6735 **21** 3.7602 **23** 9.8881 − 10 **25** 2.249 **27** 2.518 **29** 0.03864 **31** 0.0001650 **33** 140 **35** 210 **37** 40

USING YOUR KNOWLEDGE 7.3

1 5.01×10^7 **3** 3.16×10^4

EXERCISE 7.4

1 0.002912 **3** −0.03882 **5** 29.52 **7** −28.01 **9** 12.98 **11** 7.197 **13** −4.919 **15** 0.7726 **17** 0.1450 **19** 2.402 **21** 1.760 **23** 2.771 **25** 6.44 psi **27** $196.72 **29** $10,765.16

USING YOUR KNOWLEDGE 7.4

1a 11 years, 7 months **b** 8 years, 9 months **c** 5 years, 10 months **3** About 493 years

EXERCISE 7.5

1 0.186% **3** 0.194% **5** 0.201% **7** 11.6 yr **9** 6.93 yr **11** 19,800 yr **13** 5.30 yr **15** 1590 yr **17** 0.6045 **19** 1.054 **21** −0.2619 **23** 5.301 **25** $\ln x - x$ **27** $x - \ln(e^x + 1)$ **29a** 5.82 billion **b** 34.66 yr **31** 64,000,000 **33** About 6610 years ago **35** 1.5 ml **37** $1.54 \times 10^{-8}\%$

USING YOUR KNOWLEDGE 7.5

1 About 75

SELF-TEST, CHAPTER 7

1a

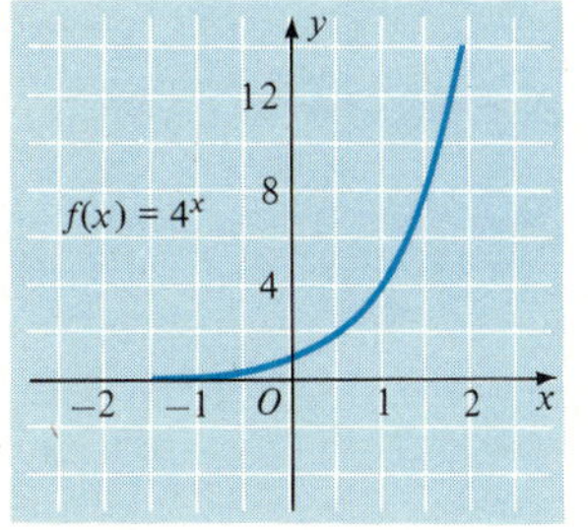

b

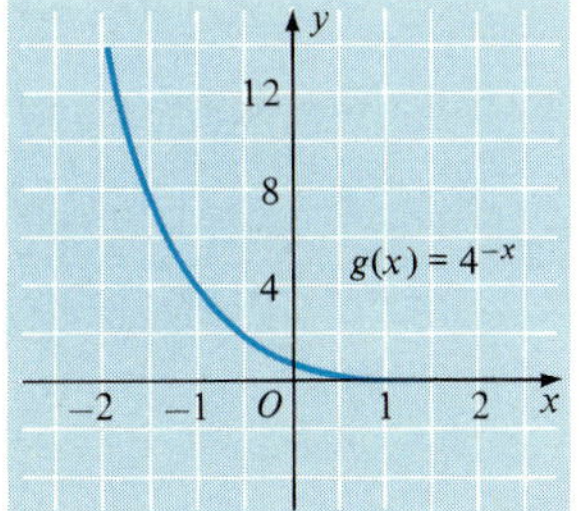

c

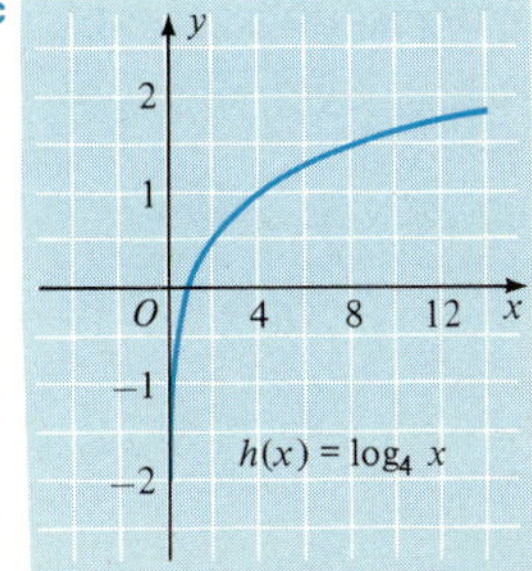

2a $k = \frac{1}{23{,}100} \approx 0.00004329$ **b** 25 mg **3a** $6^2 = 36$ **b** $\log_{32} 4 = \frac{2}{5}$ **4a** 4 **b** $\frac{3}{2}$ **5** $x = -1$ **6** $x = -\frac{14}{29}$ **7a** 1.6 **b** $\frac{4}{3} \log_b |x|$

8 $\log \sqrt{\frac{a^3}{2}}$ **9a** 1 **b** 4 **c** −3 **d** −1 **10a** 0.6243 **b** 0.9523 **c** 0.3636 **11a** 1.6649 **b** 8.9839 − 10 **12a** 430.5 **b** 0.04623

13 1.568 **14** 0.4947 **15** 0.0892 **16** 2.0437 **17** $\frac{\log 15}{\log 9} \approx 1.232$ **18** About 30.1 yr **19** $\frac{\ln 2}{5760} \approx 0.000120$ **20** $\frac{\ln 2}{0.08} \approx 8.66$ yr

EXERCISE 8.1

1 $\cos t = \dfrac{\sqrt{3}}{2}$ **3** $\sin t = -\frac{1}{2}$ **5** $\sin t = \dfrac{2\sqrt{2}}{3}$ **7** $\cos t = -\dfrac{\sqrt{2}}{2}$ **9** $\sin t = -\dfrac{\sqrt{3}}{2}$ **11** $\sin(3000\pi + t) = \sin(1500 \cdot 2\pi + t) = \sin t$

13 $\sin(1316\pi - t) = \sin(658 \cdot 2\pi - t) = \sin(-t) = -\sin t$ **15** $\cos\left(-\dfrac{\pi}{4}\right) = \cos\dfrac{\pi}{4}$ **17** $\sin\left(\pi + \dfrac{\pi}{3}\right) = -\sin\dfrac{\pi}{3}$

19 $\cos\left(\pi - \dfrac{\pi}{6}\right) = -\cos\dfrac{\pi}{6}$ **21** $\sin\dfrac{7\pi}{4} = -\sin\dfrac{\pi}{4}$ **23** $\cos\dfrac{13\pi}{3} = \cos\dfrac{\pi}{3}$ **25** $\cos\left(-\dfrac{19\pi}{5}\right) = \cos\dfrac{\pi}{5}$

27 $\sin 12 = -\sin(4\pi - 12) \approx -\sin 0.566$ **29** $\cos(-7) = \cos(2\pi - 7) \approx \cos 0.717$

31 Let $P(\pi/4) = (x, y)$. Then $x = y$, $\sin \pi/4 = \cos \pi/4$, and, because $x^2 + y^2 = 1$, $2x^2 = 1$. Thus, $x = \sqrt{2}/2$ and $\sin \pi/4 = \cos \pi/4 = \sqrt{2}/2$.

33

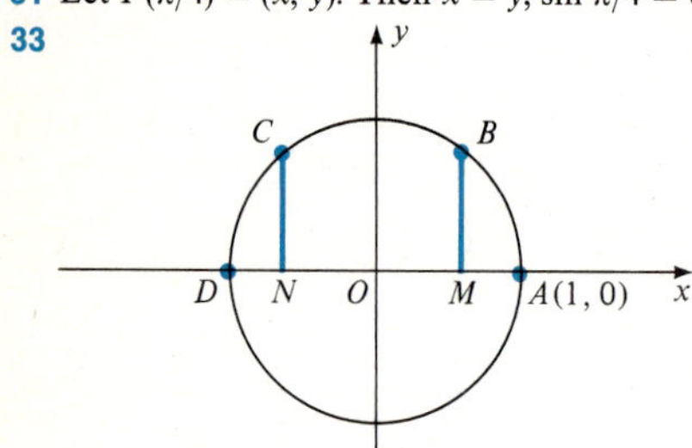

Suppose $\sin 2 = \sin 1$. Then $MB = NC$ and arc $AB =$ arc $DC = 1$. Because arc $BC = 1$, arc $ABCD = 3$, so that $3 = \pi$, which is *false*. Therefore, $\sin 2 \neq \sin 1$.

35 If $\sin s = \sin t$, then $t = 2m\pi + s$, m an integer, or $t = (2m + 1)\pi - s$, m an integer.

37 We need to know the quadrant of $P(s)$ or whether $\cos s$ is positive or negative.

39 We may take $p = ac$ and $q = bc$, where a and b are relatively prime integers. Then $f(x + abc) = f(x + bp) = f(x)$ because f has the period p, and $g(x + abc) = g(x + aq) = g(x)$ because g has the period q. Thus, f and g both have the period abc, so that $f + g$ also has the period abc.

USING YOUR KNOWLEDGE 8.1

1a $\dfrac{1}{2}$ **b** $\dfrac{\sqrt{3}}{2}$ **3a** $\sin\dfrac{2\pi}{3} = \dfrac{\sqrt{3}}{2}$; $\cos\dfrac{2\pi}{3} = -\dfrac{1}{2}$ **b** $\sin\dfrac{5\pi}{3} = -\dfrac{\sqrt{3}}{2}$; $\cos\dfrac{5\pi}{3} = \dfrac{1}{2}$ **c** $\sin\left(-\dfrac{11\pi}{3}\right) = \dfrac{\sqrt{3}}{2}$; $\cos\left(-\dfrac{11\pi}{3}\right) = \dfrac{1}{2}$

EXERCISE 8.2

1

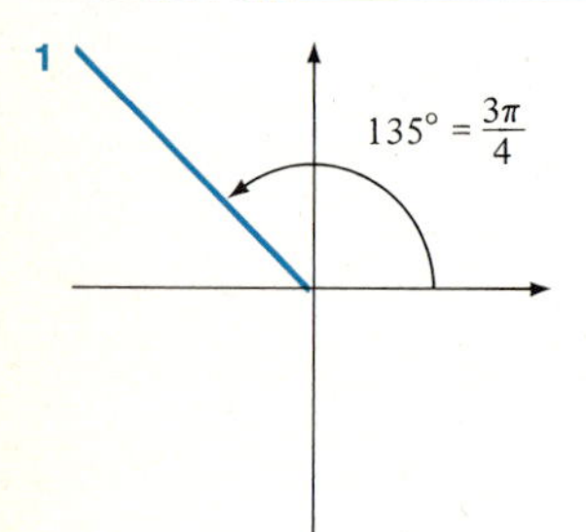

3

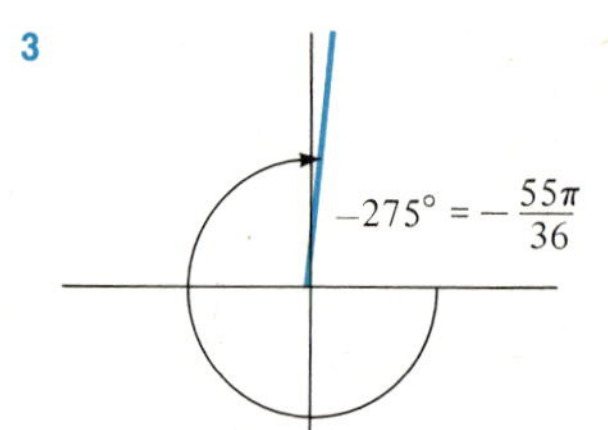

5

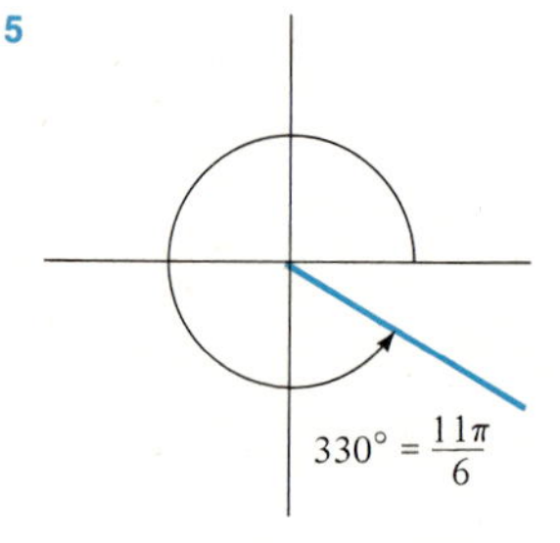

7

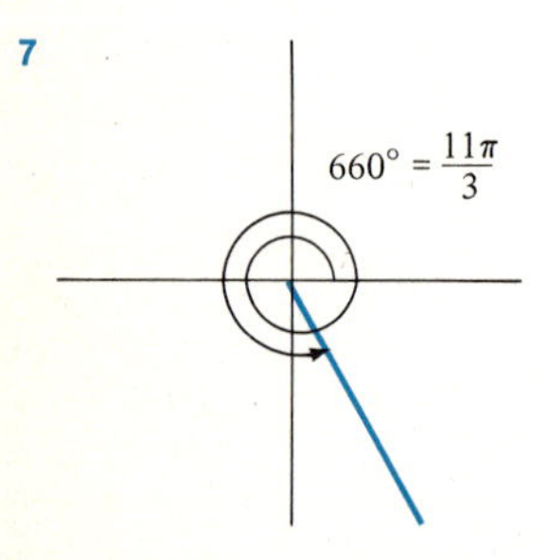

9

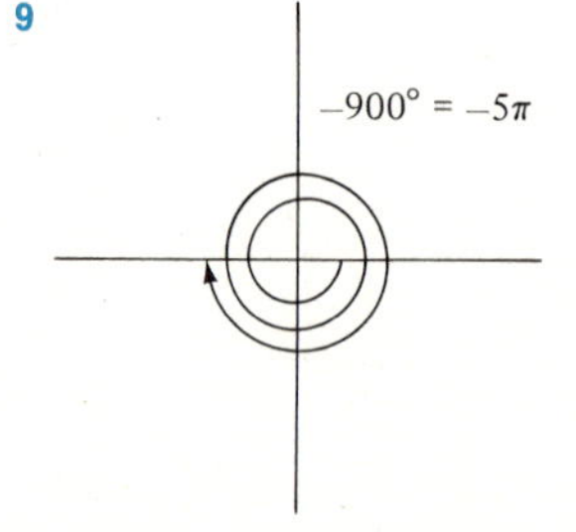

11

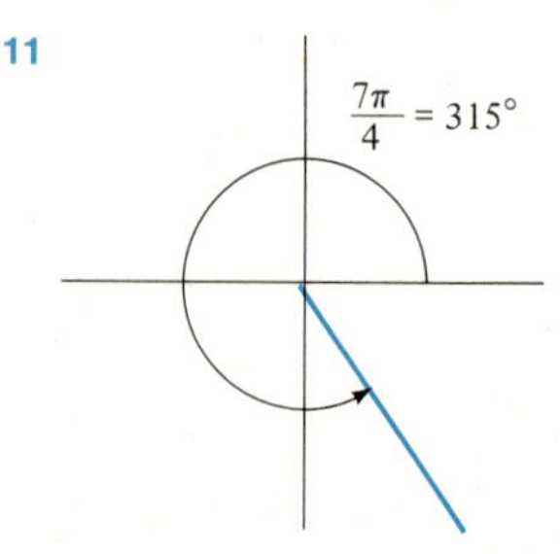

13

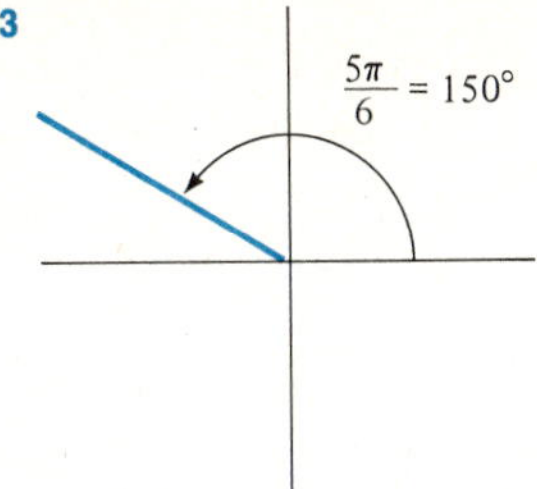

15

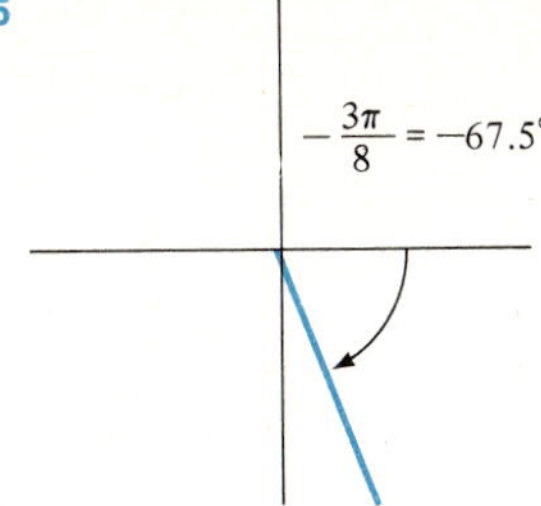

17

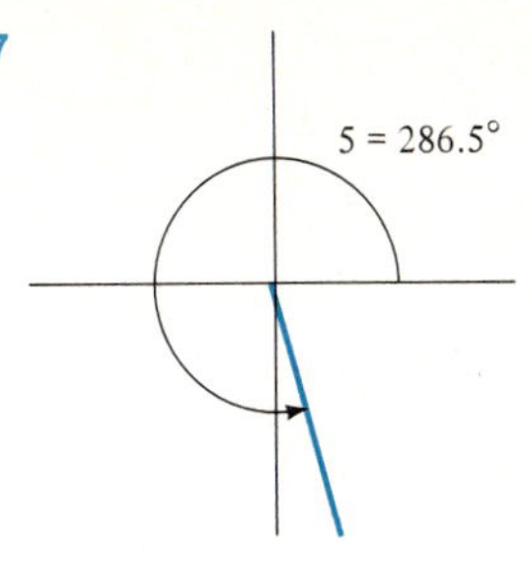

19

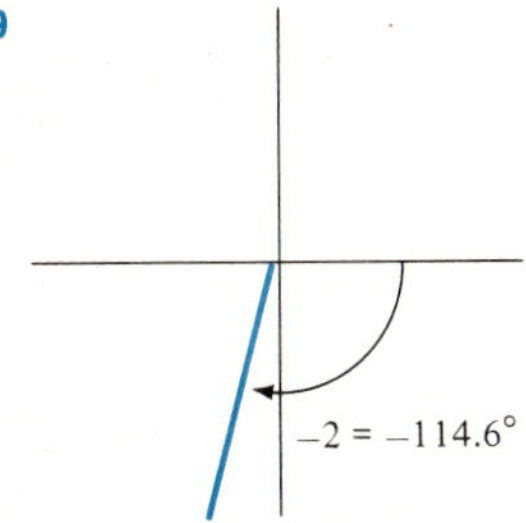

21

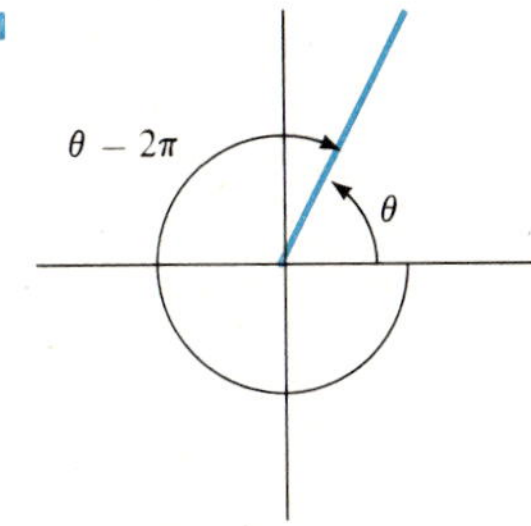

23

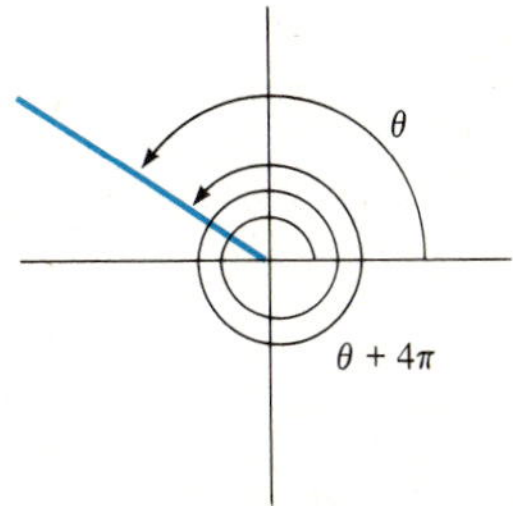

25 96.426° **27** -142.248° **29** 1.495

USING YOUR KNOWLEDGE 8.2

1 20 radians per second **3** $73\frac{1}{3}$ radians per second

EXERCISE 8.3

1 $\cos\theta = \frac{3}{5}, \sin\theta = \frac{4}{5}$ **3** $\cos\theta = \dfrac{1}{\sqrt{5}}, \sin\theta = -\dfrac{2}{\sqrt{5}}$ **5** $\cos\theta = -\frac{5}{13}, \sin\theta = \pm\frac{12}{13}$ **7** $\cos 23\pi = -1, \sin 23\pi = 0$

9 $\cos 100\pi = 1, \sin 100\pi = 0$ **11** $\cos\dfrac{9\pi}{2} = 0, \sin\dfrac{9\pi}{2} = 1$ **13** $\cos\left(-\dfrac{21\pi}{2}\right) = 0, \sin\left(-\dfrac{21\pi}{2}\right) = -1$ **15** $\cos\dfrac{23\pi}{6} = \dfrac{\sqrt{3}}{2}, \sin\dfrac{23\pi}{6} = -\dfrac{1}{2}$

17 $\cos\left(-\dfrac{31\pi}{3}\right) = \dfrac{1}{2}, \sin\left(-\dfrac{31\pi}{3}\right) = -\dfrac{\sqrt{3}}{2}$ **19** $\cos\dfrac{59\pi}{6} = \dfrac{\sqrt{3}}{2}, \sin\dfrac{59\pi}{6} = -\dfrac{1}{2}$ **21** $\cos\dfrac{19\pi}{4} = -\dfrac{\sqrt{2}}{2}, \sin\dfrac{19\pi}{4} = \dfrac{\sqrt{2}}{2}$

23 $\cos 120^\circ = -\dfrac{1}{2}, \sin 120^\circ = \dfrac{\sqrt{3}}{2}$ **25** $\cos 315^\circ = \dfrac{\sqrt{2}}{2}, \sin 315^\circ = -\dfrac{\sqrt{2}}{2}$ **27** $\cos 300^\circ = \dfrac{1}{2}, \sin 300^\circ = -\dfrac{\sqrt{3}}{2}$

29 $\cos(-450^\circ) = 0, \sin(-450^\circ) = -1$ **31** $\cos(-600^\circ) = -\dfrac{1}{2}, \sin(-600^\circ) = \dfrac{\sqrt{3}}{2}$ **33** $\cos 990^\circ = 0, \sin 990^\circ = -1$

35 $\cos 765^\circ = \dfrac{\sqrt{2}}{2}, \sin 765^\circ = \dfrac{\sqrt{2}}{2}$ **37** $\cos(-240^\circ) = -\dfrac{1}{2}, \sin(-240^\circ) = \dfrac{\sqrt{3}}{2}$ **39** $-\cos 55^\circ, \sin 55^\circ$ **41** $-\cos 70^\circ, -\sin 70^\circ$

43 $\cos 55^\circ, -\sin 55^\circ$ **45** $\cos 15^\circ, \sin 15^\circ$ **47** $-\cos 80^\circ, -\sin 80^\circ$ **49** $-\cos 22^\circ, -\sin 22^\circ$ **51** $-\cos 1.14, \sin 1.14$

53 $\cos 1.28, -\sin 1.28$ **55** $-\cos 0.58, -\sin 0.58$ **57** $\cos 1.28, \sin 1.28$ **59** $\dfrac{3\pi}{4}$ and $\dfrac{7\pi}{4}$

61 $\cos t = \dfrac{1}{\sqrt{5}}, \sin t = \dfrac{2}{\sqrt{5}}$ or $\cos t = -\dfrac{1}{\sqrt{5}}, \sin t = -\dfrac{2}{\sqrt{5}}$ **63** $-\dfrac{\pi}{2} < t < \dfrac{\pi}{4}$

USING YOUR KNOWLEDGE 8.3

1 $(6400)(50)\left(\dfrac{\pi}{180}\right) \approx 5600$ km

EXERCISE 8.4

	sin	cos	tan	cot	sec	csc
1	$-\frac{\sqrt{3}}{2}$	$\frac{1}{2}$	$-\sqrt{3}$	$-\frac{\sqrt{3}}{3}$	2	$-\frac{2\sqrt{3}}{3}$
3	$-\frac{\sqrt{3}}{2}$	$-\frac{1}{2}$	$\sqrt{3}$	$\frac{\sqrt{3}}{3}$	-2	$-\frac{2\sqrt{3}}{3}$
5	$-\frac{\sqrt{2}}{2}$	$\frac{\sqrt{2}}{2}$	-1	-1	$\sqrt{2}$	$-\sqrt{2}$
7	0	-1	0	—	-1	—
9	$\frac{1}{2}$	$-\frac{\sqrt{3}}{2}$	$-\frac{\sqrt{3}}{3}$	$-\sqrt{3}$	$-\frac{2\sqrt{3}}{3}$	2
11	$\frac{\sqrt{2}}{2}$	$-\frac{\sqrt{2}}{2}$	-1	-1	$-\sqrt{2}$	$\sqrt{2}$
13	$-\frac{1}{2}$	$-\frac{\sqrt{3}}{2}$	$\frac{\sqrt{3}}{3}$	$\sqrt{3}$	$-\frac{2\sqrt{3}}{3}$	-2
15	$-\frac{\sqrt{2}}{2}$	$\frac{\sqrt{2}}{2}$	-1	-1	$\sqrt{2}$	$-\sqrt{2}$
17	$\frac{\sqrt{2}}{2}$	$\frac{\sqrt{2}}{2}$	1	1	$\sqrt{2}$	$\sqrt{2}$
19	$\frac{1}{2}$	$-\frac{\sqrt{3}}{2}$	$-\frac{\sqrt{3}}{3}$	$-\sqrt{3}$	$-\frac{2\sqrt{3}}{3}$	2
21	$\frac{\sqrt{2}}{2}$	$-\frac{\sqrt{2}}{2}$	-1	-1	$-\sqrt{2}$	$\sqrt{2}$
23	1	0	—	0	—	1

25 210° **27** 300° **29** 240° **31** 150° **33** 225° **35** 240° **37** 240°, 300° **39** 45°, 225° **41** 0°, 180°, 360° **43** 90°, 270°

USING YOUR KNOWLEDGE 8.4

1

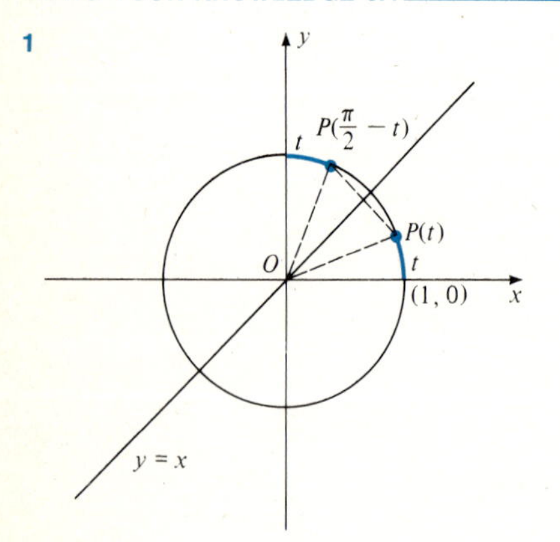

As the figure shows, the two points $P(t)$ and $P(\pi/2 - t)$ are symmetric with respect to the line $y = x$. Thus, if the coordinates of $P(t)$ are $(\cos t, \sin t)$, then the coordinates of $P(\pi/2 - t)$ are $(\sin t, \cos t)$. Therefore,

$$\cos\left(\frac{\pi}{2} - t\right) = \sin t$$

$$\sin\left(\frac{\pi}{2} - t\right) = \cos t$$

3 $\cos\frac{5\pi}{8} = \sin\left(\frac{\pi}{2} - \frac{5\pi}{8}\right) = \sin\left(-\frac{\pi}{8}\right) = -\sin\frac{\pi}{8}$ **5** $\tan\frac{7\pi}{12} = \cot\left(\frac{\pi}{2} - \frac{7\pi}{12}\right) = \cot\left(-\frac{\pi}{12}\right) = -\cot\frac{\pi}{12}$

7 $\sin 72° = \cos(90° - 72°) = \cos 18°$ **9** $\tan 265° = \tan(180° + 85°) = \tan 85° = \cot(90° - 85°) = \cot 5°$

EXERCISE 8.5

1 $\sin\theta\cot\theta = \sin\theta\dfrac{\cos\theta}{\sin\theta} = \cos\theta$ **3** $\sec\theta\cot\theta = \dfrac{1}{\cos\theta}\cdot\dfrac{\cos\theta}{\sin\theta} = \dfrac{1}{\sin\theta} = \csc\theta$

5 $\cos^2\theta - \sin^2\theta = \cos^2\theta - (1-\cos^2\theta) = \cos^2\theta - 1 + \cos^2\theta = 2\cos^2\theta - 1$ **7** $\cot^2\theta + \csc^2\theta = (\csc^2\theta - 1) + \csc^2\theta = 2\csc^2\theta - 1$

9 $\tan t + \sec t = \dfrac{\sin t}{\cos t} + \dfrac{1}{\cos t} = \dfrac{1+\sin t}{\cos t}\cdot\dfrac{1-\sin t}{1-\sin t} = \dfrac{1-\sin^2 t}{\cos t(1-\sin t)} = \dfrac{\cos^2 t}{\cos t(1-\sin t)} = \dfrac{\cos t}{1-\sin t}$

11 $\tan\alpha + \cot\alpha = \dfrac{\sin\alpha}{\cos\alpha} + \dfrac{\cos\alpha}{\sin\alpha} = \dfrac{\sin^2\alpha + \cos^2\alpha}{\cos\alpha\sin\alpha} = \dfrac{1}{\cos\alpha\sin\alpha} = \sec\alpha\csc\alpha$

13 $\dfrac{\cos\theta}{1-\sin\theta} - \dfrac{1-\sin\theta}{\cos\theta} = \dfrac{\cos\theta\,(1+\sin\theta)}{1-\sin^2\theta} - \dfrac{1-\sin\theta}{\cos\theta} = \dfrac{\cos\theta\,(1+\sin\theta)}{\cos^2\theta} - \dfrac{1-\sin\theta}{\cos\theta} = \dfrac{1+\sin\theta-1+\sin\theta}{\cos\theta} = \dfrac{2\sin\theta}{\cos\theta} = 2\tan\theta$

15 $\dfrac{\sin A + \sin B}{\sin A - \sin B} = \dfrac{(1/\csc A) + (1/\csc B)}{(1/\csc A) - (1/\csc B)} = \dfrac{\csc B + \csc A}{\csc B - \csc A}$

17 $\csc^4 A - \cot^4 A = (\csc^2 A - \cot^2 A)(\csc^2 A + \cot^2 A) = \csc^2 A + \cot^2 A = (1 + \sin^2 A\cot^2 A)\csc^2 A = (1+\cos^2 A)\csc^2 A = [(\sin^2 A + \cos^2 A) + \cos^2 A]\csc^2 A = (\sin^2 A + 2\cos^2 A)\csc^2 A$

19 $\dfrac{1+\cos\theta}{1-\cos\theta} = \dfrac{(1+\cos\theta)^2}{1-\cos^2\theta} = \dfrac{(1+\cos\theta)^2}{\sin^2\theta} = \left(\dfrac{1+\cos\theta}{\sin\theta}\right)^2 = \left(\dfrac{1}{\sin\theta} + \dfrac{\cos\theta}{\sin\theta}\right)^2 = (\csc\theta + \cot\theta)^2$

21 For $t = \pi/6$, $\sin 3t = 1$ and $3\sin t = \frac{3}{2}$. Thus, the given equation is not an identity.
23 For $t = \pi/3$, $\tan t/2 = \sqrt{3}/3$ and $\frac{1}{2}\tan t = \sqrt{3}/2$. Thus, the given equation is not an identity.
25 For $t = \pi/4$, $\cos 2t = 0$ and $1 + \sin^2 t = 1 + (\sqrt{2}/2)^2 = \frac{3}{2}$. Thus, the given equation is not an identity. **27** $\dfrac{\pi}{3}, \dfrac{5\pi}{3}$ **29** $\dfrac{\pi}{6}, \dfrac{5\pi}{6}$

31 $\dfrac{\pi}{4}, \dfrac{3\pi}{4}, \dfrac{5\pi}{4}, \dfrac{7\pi}{4}$ **33** $\dfrac{3\pi}{4}, \dfrac{7\pi}{4}$ **35** $\dfrac{\pi}{6}, \dfrac{7\pi}{6}$ **37** $\dfrac{\pi}{3}, \dfrac{2\pi}{3}, \dfrac{4\pi}{3}, \dfrac{5\pi}{3}$ **39** $0, \dfrac{2\pi}{3}, \dfrac{4\pi}{3}, 2\pi$ **41** $m\pi + (-1)^m\dfrac{\pi}{6}, m = 0, \pm 1, \pm 2, \ldots$

43 $m\pi + \dfrac{\pi}{6}, m\pi + \dfrac{\pi}{3}, m = 0, \pm 1, \pm 2, \ldots$ **45** $m\pi + (-1)^m\dfrac{\pi}{6}, m = 0, \pm 1, \pm 2, \ldots$ **47** $\dfrac{\pi}{4} < t < \dfrac{\pi}{2}$ **49** $\dfrac{\pi}{4} < t < \dfrac{\pi}{2}$ **51** $0 < t < \dfrac{\pi}{4}$

USING YOUR KNOWLEDGE 8.5

1 About 25.5° toward the normal. **3** No. This would require $\sin\alpha = 2\sqrt{3}/3$, which is greater than 1, and this is impossible.

EXERCISE 8.6

1 0.50553 **3** 7.6018 **5** 3.6133 **7** 1.2572 **9** 2.4023 **11** 1.2218 **13** 3.7321 **15** 0.60876 **17** 3.0777 **19** 0.86179 **21** 0.13664 **23** 0.27732 **25** 0.50362 **27** 1.2028 **29** 0.47516 **31** 0.17 **33** 2.42 **35** 0.97 **37** 16.44 **39** 10.37 **41** 32.63° **43** 277.47° **45** 134.48° **47** 38.25° **49** 133.40° **51** 0.79, 2.36 **53** 1.57 **55** 0.23, 2.91 **57** About 64.7 feet

USING YOUR KNOWLEDGE 8.6

1 0.50726 **3** 0.58861 **5** 0.27732 **7** 0.13664 **9** 2.86152 **11** −0.73060 **13** 0.50362 **15** 1.20280 **17** 2.10454 **19** −0.36748 **21** −1.55706 **23** 1.04824 **25** 0.55557 **27** 0.25676 **29** 1.05604 **31** 0.17 **33** 2.42 **35** 0.81 **37** 16.44 **39** 10.37 **41** 32.63° **43** 277.47° **45** 134.48° **47** 38.25° **49** 126.35° **51** 0.79, 2.36 **53** 0.57 **55** 0.23 **57** 0.97, 5.31 **59** 0.64, 2.50, 4.71 **61** 41.81°

SELF-TEST, CHAPTER 8

1a $\frac{2\pi}{5}$ **b** 2 **2a** Negative **b** Negative **c** Negative **d** Negative **e** Positive **3a** $-\frac{\sqrt{2}}{2}$ **b** $\frac{\pi}{6}$ **4** $\frac{4\pi}{5}$ **5a** $m \cdot 360°$, m an integer

b $2m\pi$, m an integer **6a**

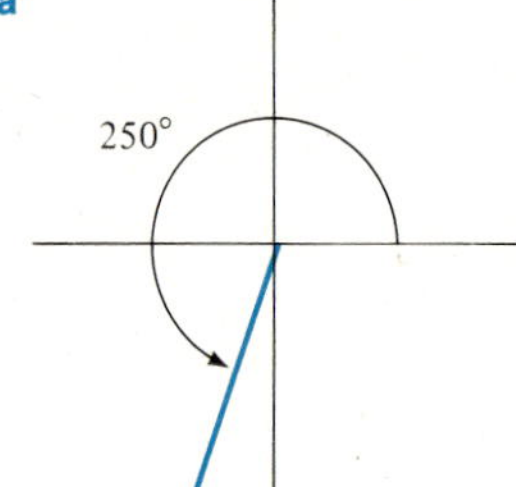

b

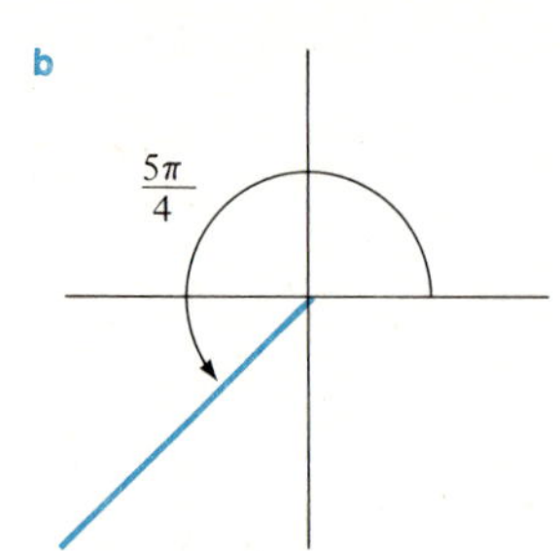

7a $\frac{7\pi}{6}$ **b** $\left(\frac{540}{\pi}\right)^\circ \approx 171.9°$

8a $\sin\theta = \frac{1}{\sqrt{5}}$, $\cos\theta = -\frac{2}{\sqrt{5}}$, $\tan\theta = -\frac{1}{2}$ **b** $\sin\theta = \pm\frac{3}{5}$, $\cos\theta = \frac{4}{5}$, $\tan\theta = \pm\frac{3}{4}$ **9a** 0 **b** $-\frac{1}{2}$ **c** 0 **10a** $-\frac{\sqrt{2}}{2}$ **b** $-\frac{1}{2}$ **c** $\sqrt{3}$

11a 34° **b** $\sin 1226° = \sin 34°$, $\cos 1226° = -\cos 34°$, $\tan 1226° = -\tan 34°$ **12** (b) and (d) **13a** $\frac{1}{\sqrt{3}}$ **b** $-\frac{2}{\sqrt{3}}$ **c** $-\frac{2}{\sqrt{3}}$ **d** 0

e -1 **f** Undefined **14a** $\sqrt{3}$ **b** 2 **c** 2 **d** $\sqrt{3}$ **e** $\frac{1}{2}$ **f** $\frac{1}{2}$ **15** (a), (c), (e), and (f)

16 $\tan t + \cot t = \frac{\sin t}{\cos t} + \frac{\cos t}{\sin t} = \frac{\sin^2 t + \cos^2 t}{\cos t \sin t} = \frac{1}{\cos t \sin t} = \sec t \csc t$ **17** $m\pi \pm \frac{\pi}{3}$, $m = 0, \pm 1, \pm 2, \ldots$ **18** $\frac{\pi}{6}, \frac{5\pi}{6}, \frac{7\pi}{6}, \frac{11\pi}{6}$

19a 3.84 **b** 5.79 **20a** 143.13° **b** 212.17°

EXERCISE 9.1

1

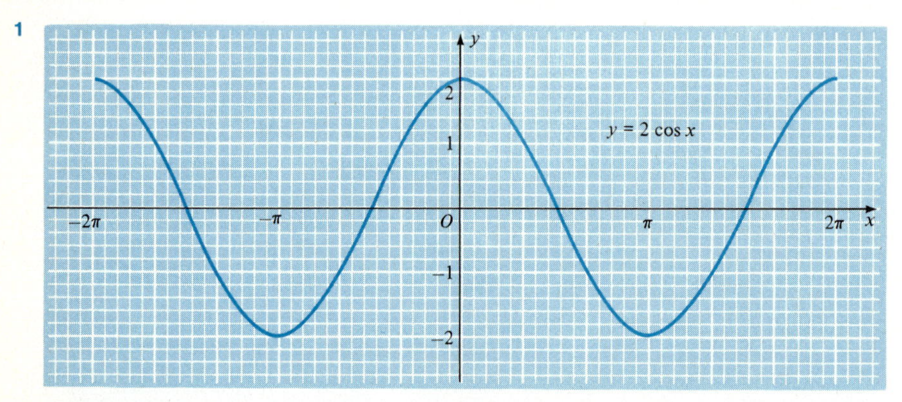

3

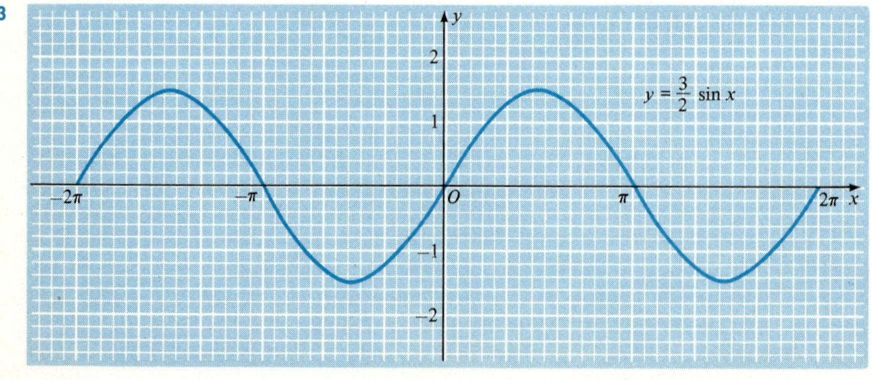

5

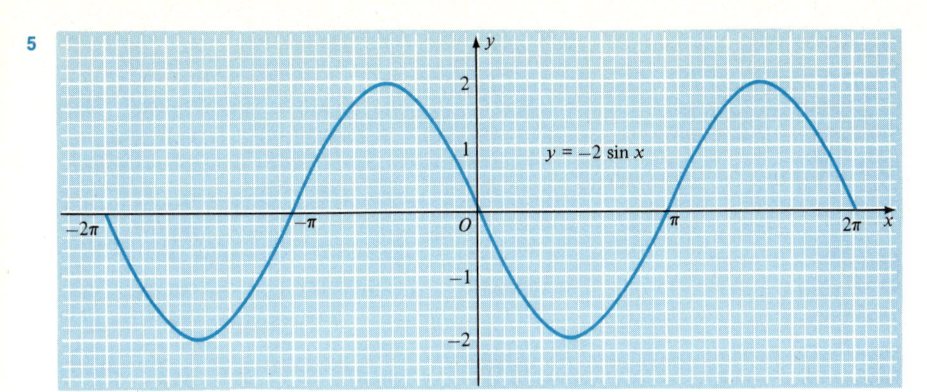
y
2
1
$y = -2 \sin x$
-2π
$-\pi$
O
π
2π
x
−1
−2

7

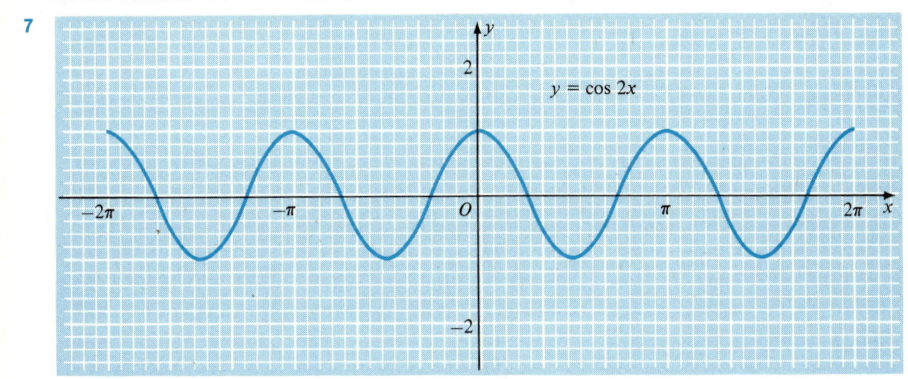
y
2
$y = \cos 2x$
-2π
$-\pi$
O
π
2π
x
−2

9

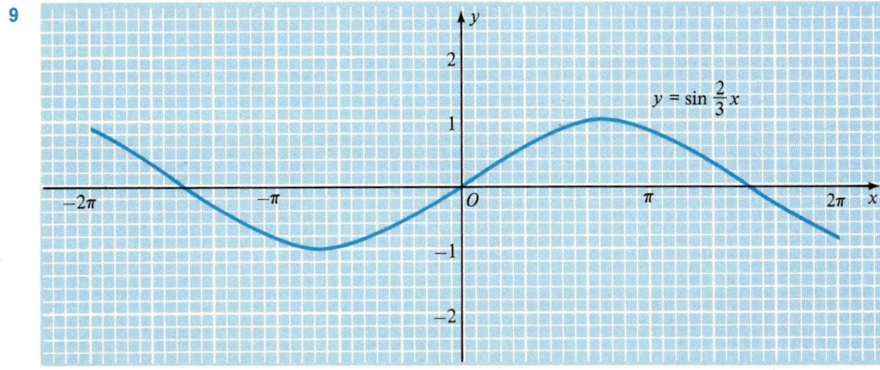
y
2
$y = \sin \frac{2}{3} x$
1
-2π
$-\pi$
O
π
2π
x
−1
−2

11

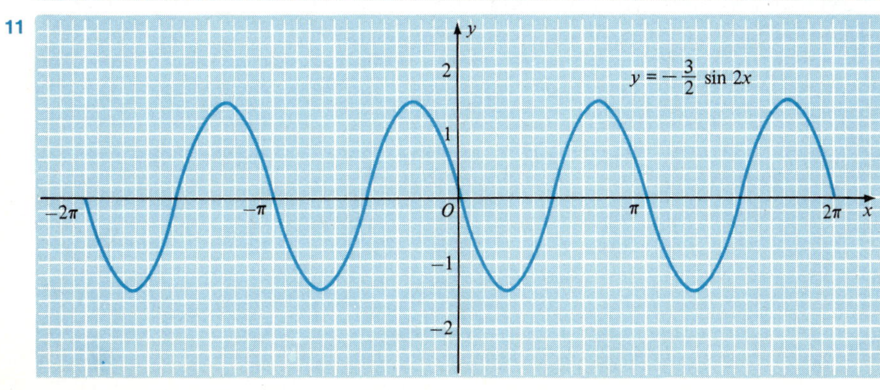
y
2
$y = -\frac{3}{2} \sin 2x$
1
-2π
$-\pi$
O
π
2π
x
−1
−2

13

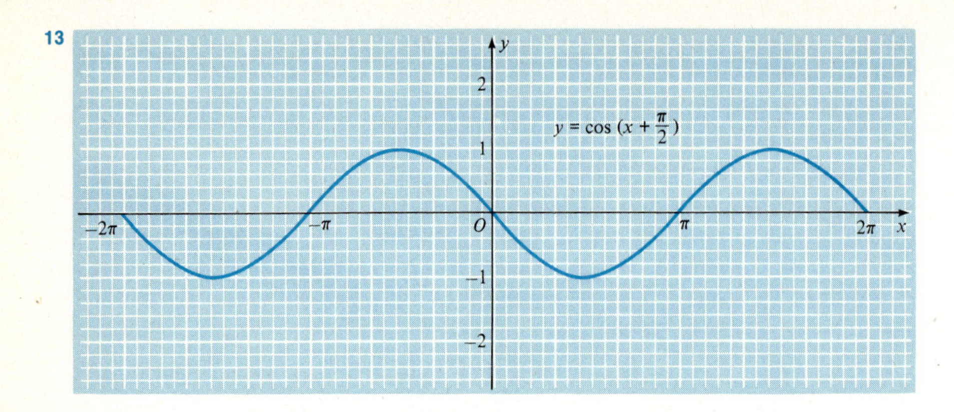
y
2
1
O
−1
−2
−2π
−π
π
2π
x
$y = \cos(x + \frac{\pi}{2})$

15

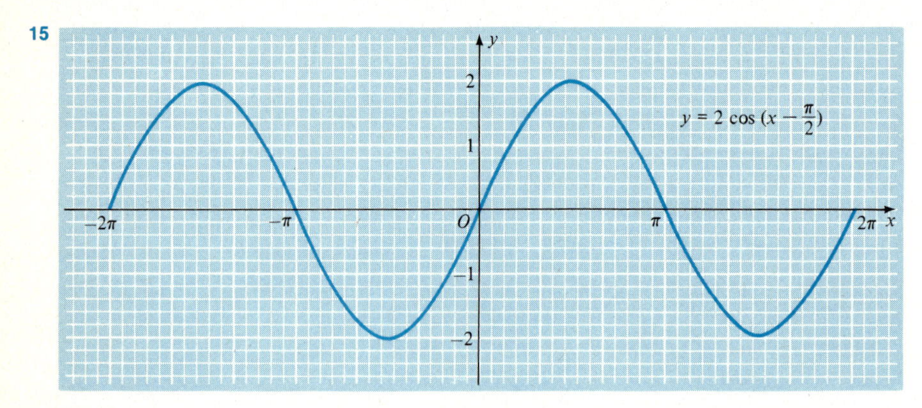
y
2
1
O
−1
−2
−2π
−π
π
2π
x
$y = 2\cos(x - \frac{\pi}{2})$

17

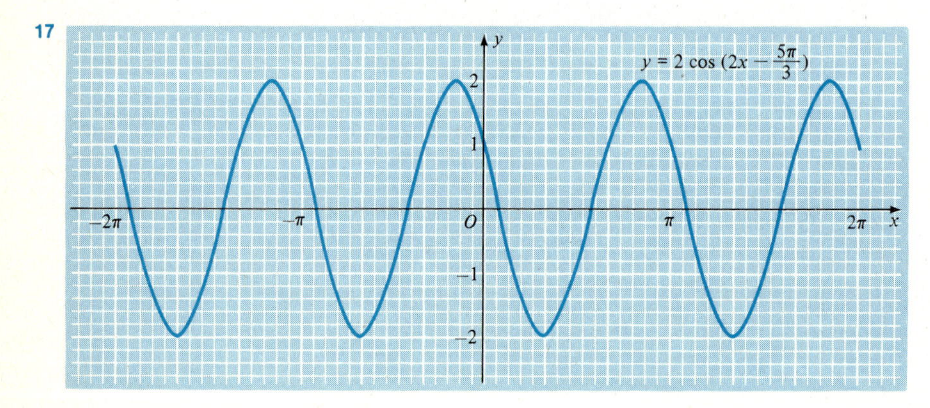
y
2
1
O
−1
−2
−2π
−π
π
2π
x
$y = 2\cos(2x - \frac{5\pi}{3})$

19

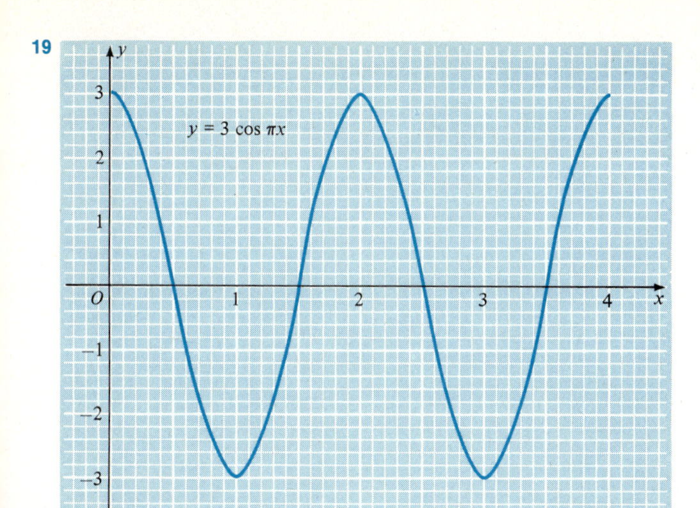

21

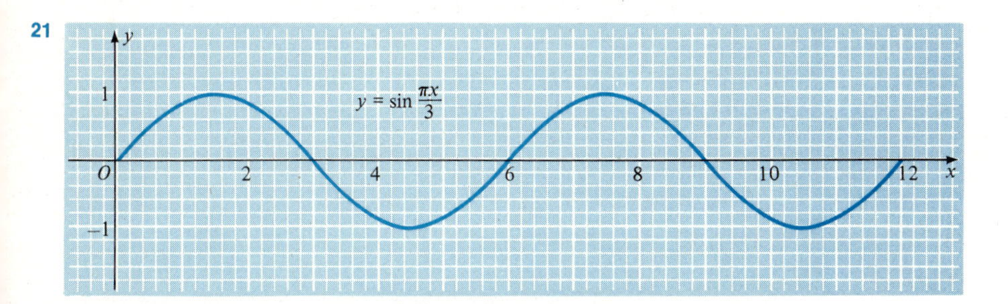

23

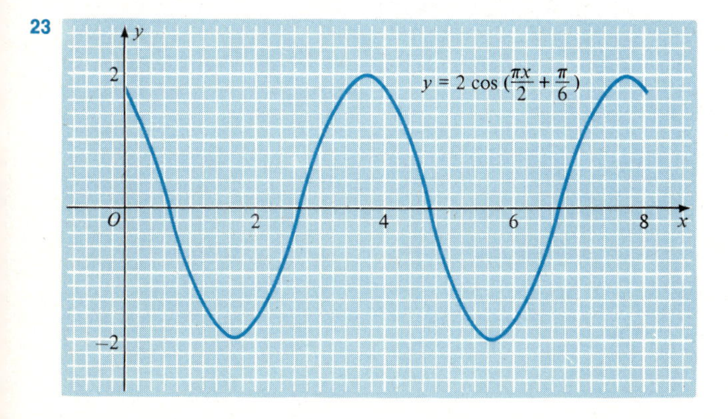

25 $0, \pm\frac{\pi}{4}, \pm\frac{\pi}{2}, \pm\frac{3\pi}{4}\ \pm\pi, \pm\frac{5\pi}{4}, \pm\frac{3\pi}{2}, \pm\frac{7\pi}{4}, \pm 2\pi$ **27** $-\frac{7\pi}{4}, -\frac{3\pi}{4}, \frac{\pi}{4}, \frac{5\pi}{4}$ **29** 0, 3, 6, 9, 12

31

t	x	y
0	0	0
$\pi/8$	0.02	0.15
$2\pi/8$	0.16	0.59
$3\pi/8$	0.51	1.23
$4\pi/8$	1.14	2
$5\pi/8$	2.08	2.77
$6\pi/8$	3.30	3.41
$7\pi/8$	4.73	3.85
$8\pi/8$	6.28	4

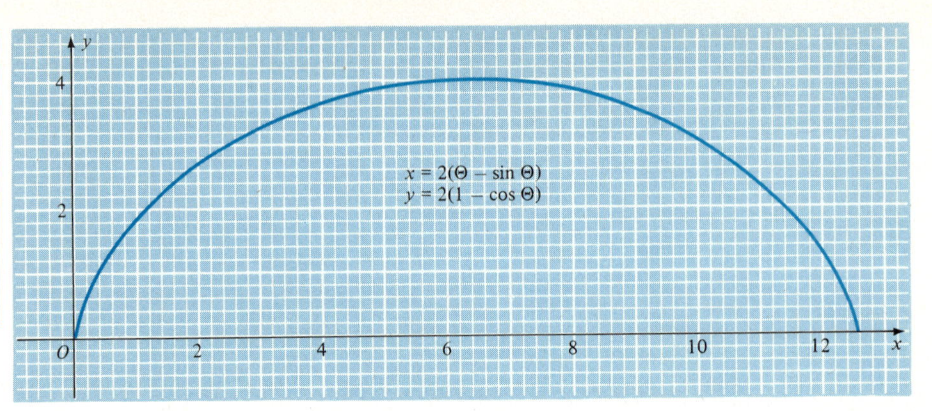

USING YOUR KNOWLEDGE 9.1

1

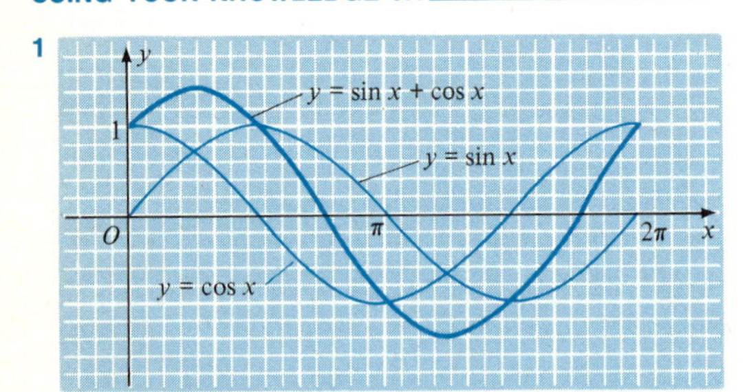

3

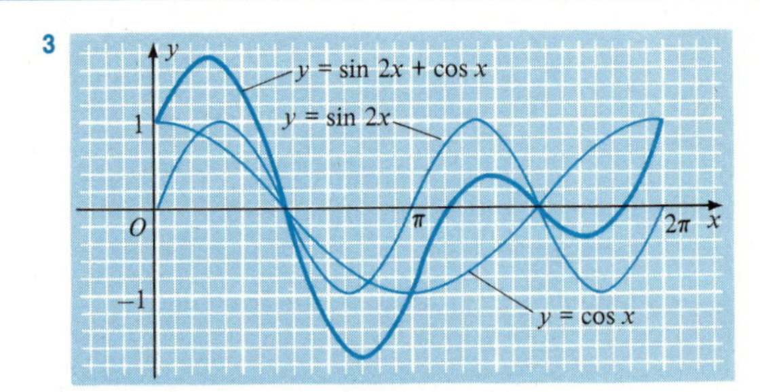

5

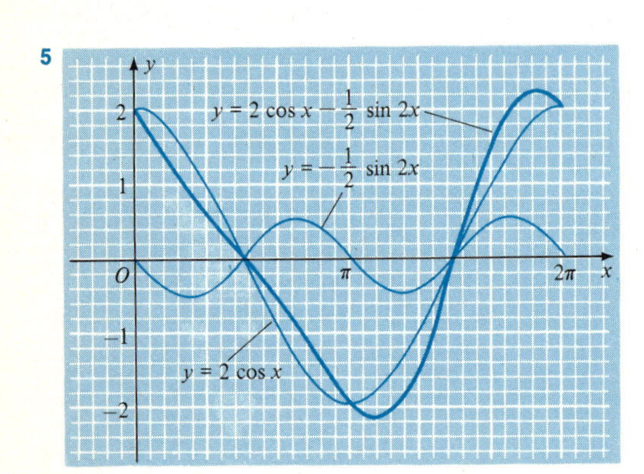

7

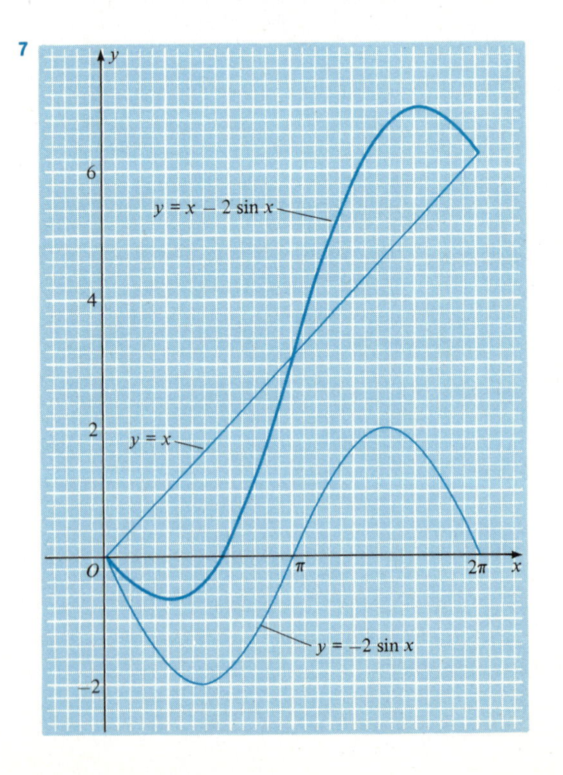

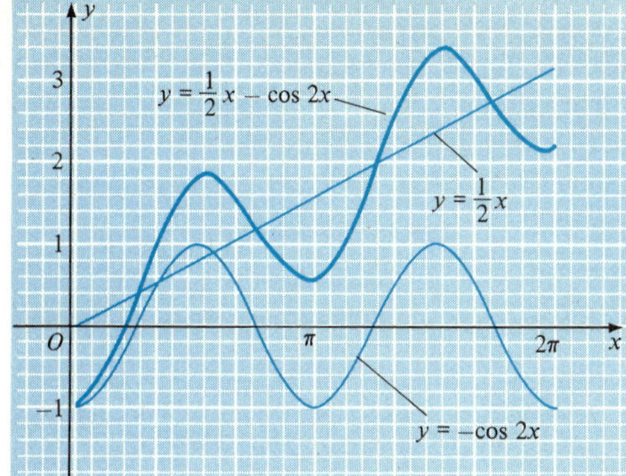

EXERCISE 9.2

1

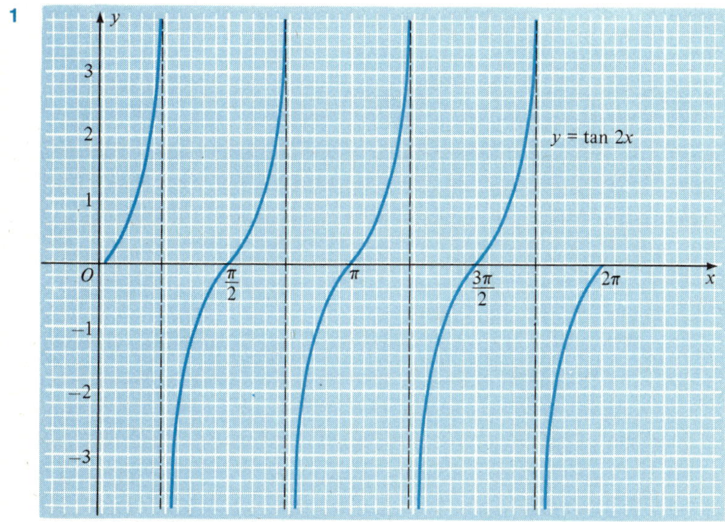

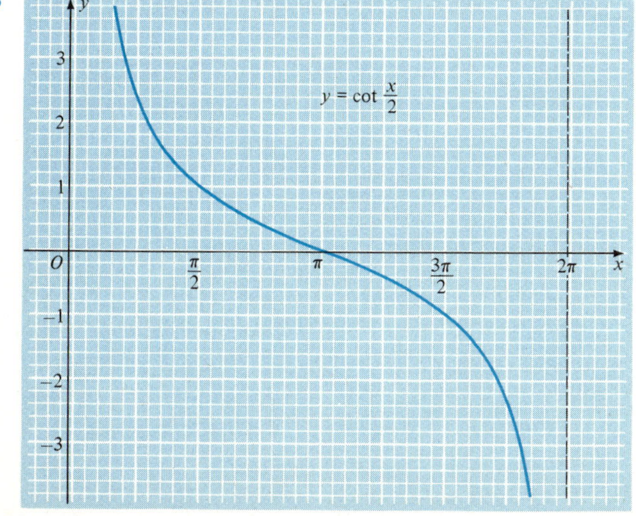

5

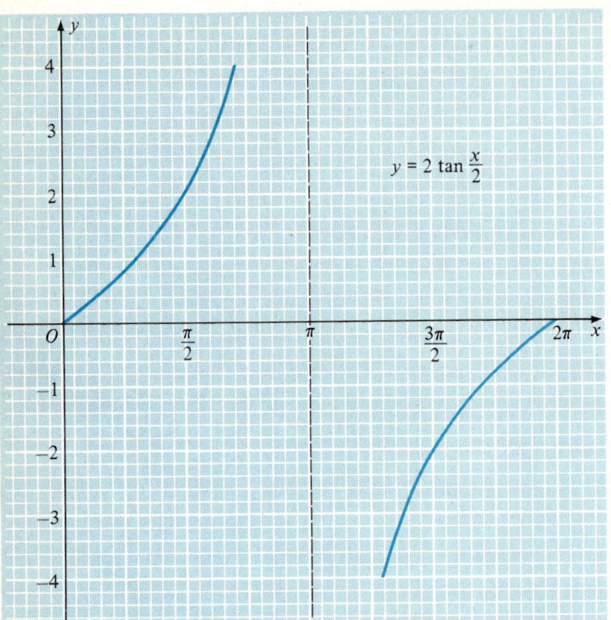
y
4
3
2
1
O
−1
−2
−3
−4
π/2
π
3π/2
2π
x
y = 2 tan x/2

7

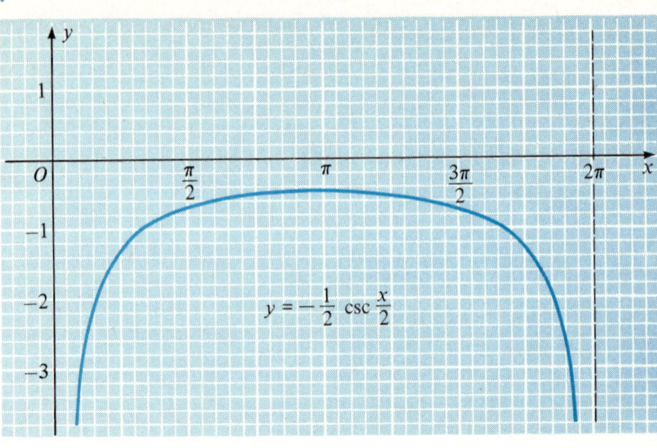
y
1
O
−1
−2
−3
π/2
π
3π/2
2π
x
y = −1/2 csc x/2

9

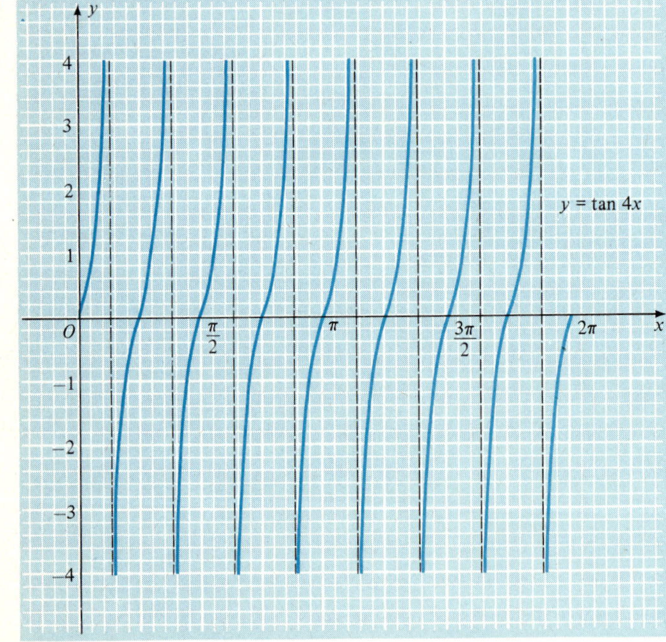
y
4
3
2
1
O
−1
−2
−3
−4
π/2
π
3π/2
2π
x
y = tan 4x

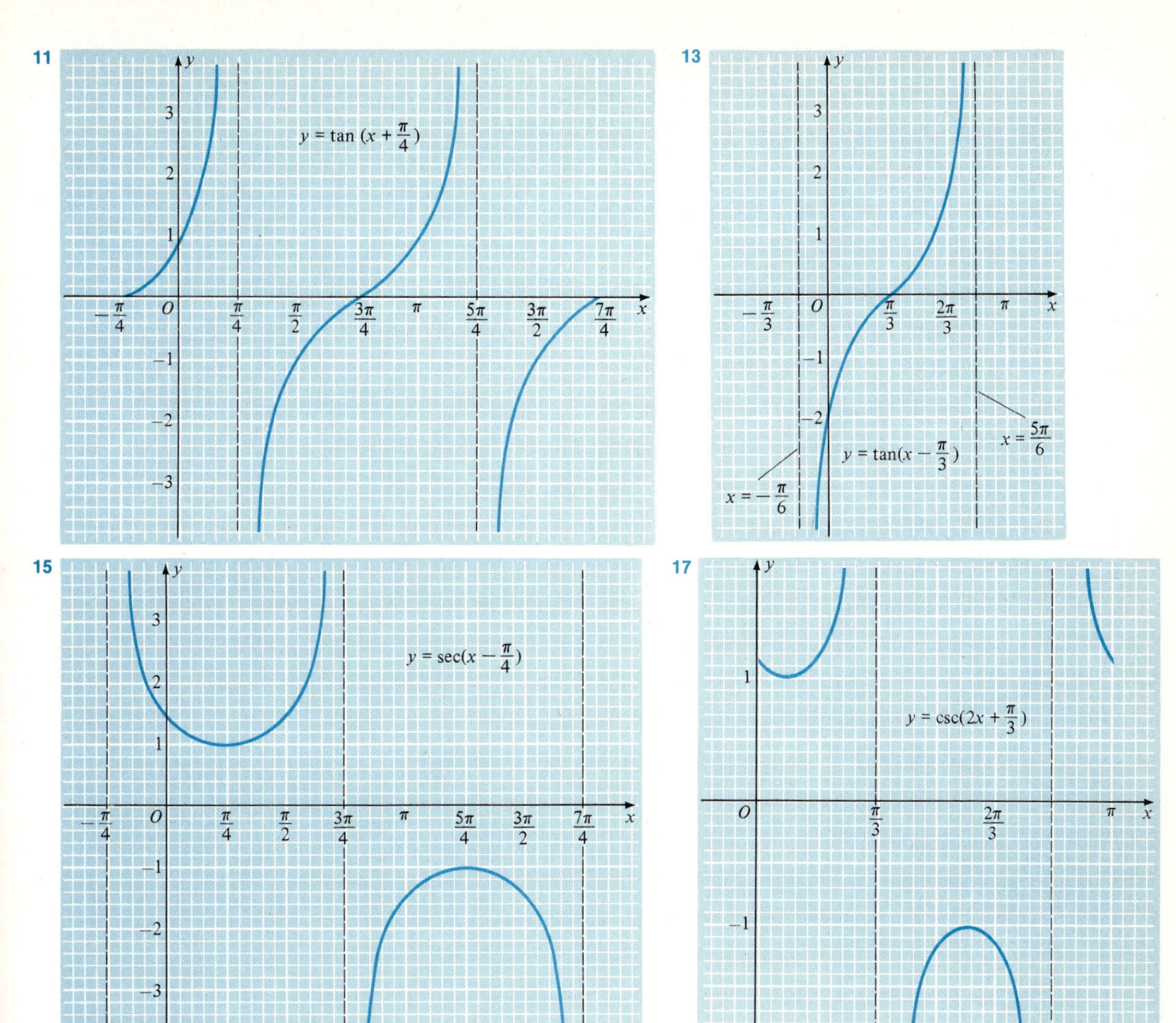
11
y = tan (x + π/4)
13
y = tan(x − π/3)
x = −π/6
x = 5π/6
15
y = sec(x − π/4)
17
y = csc(2x + π/3)

19

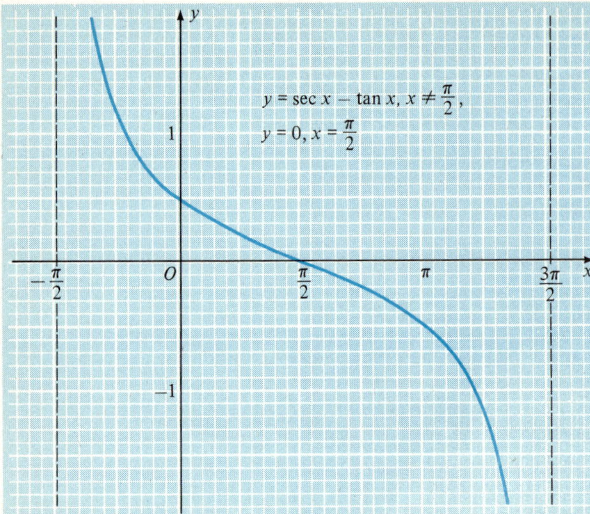

21 $x = 0.9$

USING YOUR KNOWLEDGE 9.2

1 $x = 0.860$ **3** $x = 0.741$

EXERCISE 9.3

1 $\cos(180^\circ - \theta) = \cos 180^\circ \cos\theta + \sin 180^\circ \sin\theta$
$= (-1)\cos\theta + (0)\sin\theta$
$= -\cos\theta$

3 $\sin(90^\circ + A) = \sin 90^\circ \cos A + \cos 90^\circ \sin A$
$= (1)\cos A + (0)\sin A$
$= \cos A$

5 $\tan(3\pi - x) = \dfrac{\tan 3\pi - \tan x}{1 + \tan 3\pi \tan x} = \dfrac{0 - \tan x}{1 + (0)\tan x} = -\tan x$

7 $\sin(270^\circ + C) = \sin 270^\circ \cos C + \cos 270^\circ \sin C$
$= (-1)\cos C + (0)\sin C$
$= -\cos C$

9
$$\begin{aligned}\sin\left(2m\pi + \frac{\pi}{2} + \theta\right) &= \sin 2m\pi \cos\left(\frac{\pi}{2} + \theta\right) + \cos 2m\pi \sin\left(\frac{\pi}{2} + \theta\right)\\ &= (0)\cos\left(\frac{\pi}{2} + \theta\right) + (1)\sin\left(\frac{\pi}{2} + \theta\right)\\ &= \sin\left(\frac{\pi}{2} + \theta\right)\\ &= \sin\frac{\pi}{2}\cos\theta + \cos\frac{\pi}{2}\sin\theta\\ &= (1)\cos\theta + (0)\sin\theta\\ &= \cos\theta\end{aligned}$$

11 $\dfrac{\sqrt{6} + \sqrt{2}}{4}$ **13** $2 + \sqrt{3}$ **15** $\sqrt{3} - 2$ **17** $\dfrac{\sqrt{2} - \sqrt{6}}{4}$ **19** $2 + \sqrt{3}$ **21** $\dfrac{\sqrt{2} - \sqrt{6}}{4}$ **23** $\dfrac{1}{2}\cos\theta + \dfrac{\sqrt{3}}{2}\sin\theta$ **25** $\dfrac{\tan\theta - 1}{\tan\theta + 1}$

27 $-\dfrac{1}{2}\cos\theta - \dfrac{\sqrt{3}}{2}\sin\theta$ **29** $-\frac{56}{65}$ **31** $-\frac{120}{169}$

33 $\cos A = \cos[(A + B) - B]$
$= \cos(A + B)\cos B + \sin(A + B)\sin B$

35
$$\begin{aligned}\sin\left(\theta + \frac{\pi}{3}\right) - \cos\left(\theta - \frac{\pi}{6}\right) &= \sin\theta\cos\frac{\pi}{3} + \cos\theta\sin\frac{\pi}{3} - \cos\theta\cos\frac{\pi}{6} - \sin\theta\sin\frac{\pi}{6}\\ &= \frac{1}{2}\sin\theta + \frac{\sqrt{3}}{2}\cos\theta - \frac{\sqrt{3}}{2}\cos\theta - \frac{1}{2}\sin\theta\\ &= 0\end{aligned}$$

37 $\sin A + \sin B = \sin(s+t) + \sin(s-t)$
$= \sin s \cos t + \cos s \sin t + \sin s \cos t - \cos s \sin t$
$= 2 \sin s \cos t$
$= 2 \sin \dfrac{A+B}{2} \cos \dfrac{A-B}{2}$

39 $\cos A + \cos B = \cos(s+t) + \cos(s-t)$
$= \cos s \cos t - \sin s \sin t + \cos s \cos t + \sin s \sin t$
$= 2 \cos s \cos t$
$= 2 \cos \dfrac{A+B}{2} \cos \dfrac{A-B}{2}$

41 $\cot(s+t) = \dfrac{1}{\tan(s+t)}$
$= \dfrac{1 - \tan s \tan t}{\tan s + \tan t} \cdot \dfrac{\cot s \cot t}{\cot s \cot t}$
$= \dfrac{\cot s \cot t - 1}{\cot s + \cot t}$

43 $\dfrac{\pi}{4}, \dfrac{5\pi}{4}$ **45** $\dfrac{3\pi}{4}, \dfrac{7\pi}{4}$

47

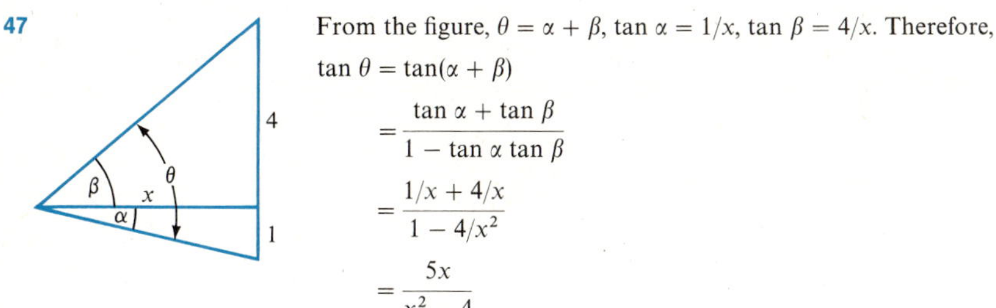

From the figure, $\theta = \alpha + \beta$, $\tan \alpha = 1/x$, $\tan \beta = 4/x$. Therefore,

$\tan \theta = \tan(\alpha + \beta)$
$= \dfrac{\tan \alpha + \tan \beta}{1 - \tan \alpha \tan \beta}$
$= \dfrac{1/x + 4/x}{1 - 4/x^2}$
$= \dfrac{5x}{x^2 - 4}$

USING YOUR KNOWLEDGE 9.3

1 $\sqrt{2} \sin\left(t + \dfrac{\pi}{4}\right)$ **3** $2 \sin\left(t - \dfrac{\pi}{6}\right)$

EXERCISE 9.4

1 $\cos 2\theta = \frac{7}{25}$, $\sin 2\theta = -\frac{24}{25}$ **3** $\cos 2\theta = -\frac{161}{289}$, $\sin 2\theta = \frac{240}{289}$ **5** $\frac{1}{2}\sqrt{2+\sqrt{3}}$ **7** $1 + \sqrt{2}$ **9** $\frac{1}{2}\sqrt{2-\sqrt{3}}$ **11** $\frac{1}{2}\sqrt{2+\sqrt{3}}$
13 $\frac{1}{2}\sqrt{2-\sqrt{2}}$

15 Let $s = t = \theta$ in the formula $\cos(s+t) = \cos s \cos t - \sin s \sin t$ to obtain

$\cos 2\theta = \cos \theta \cos \theta - \sin \theta \sin \theta$
$= \cos^2 \theta - \sin^2 \theta$

Also

$\cos 2\theta = \cos^2 \theta - (1 - \cos^2 \theta) = 2 \cos^2 \theta - 1$

and

$\cos 2\theta = (1 - \sin^2 \theta) - \sin^2 \theta = 1 - 2 \sin^2 \theta$

17 From $\cos 2\alpha = 1 - 2 \sin^2 \alpha$, we obtain

$2 \sin^2 \alpha = 1 - \cos 2\alpha$
$\sin^2 \alpha = \dfrac{1 - \cos 2\alpha}{2}$
$\sin \alpha = \pm\sqrt{\dfrac{1 - \cos 2\alpha}{2}}$

Let $\alpha = \theta/2$ to obtain the formula

$\sin \dfrac{\theta}{2} = \pm\sqrt{\dfrac{1 - \cos \theta}{2}}$

19 $\tan B + \cot B = \dfrac{\sin B}{\cos B} + \dfrac{\cos B}{\sin B} = \dfrac{\sin^2 B + \cos^2 B}{\sin B \cos B} = \dfrac{1}{\sin B \cos B}$
$= \dfrac{2}{2 \sin B \cos B} = \dfrac{2}{\sin 2B} = 2 \csc 2B$

21 $\tan \dfrac{x}{2} + \cot \dfrac{x}{2} = \dfrac{1 - \cos x}{\sin x} + \dfrac{1 + \cos x}{\sin x} = \dfrac{2}{\sin x} = 2 \csc x$

23 $\dfrac{1 - \tan^2 C}{1 + \tan^2 C} = \dfrac{1 - (\sin^2 C/\cos^2 C)}{1 + (\sin^2 C/\cos^2 C)} = \dfrac{\cos^2 C - \sin^2 C}{\cos^2 C + \sin^2 C} = \cos^2 C - \sin^2 C = \cos 2C$

25 $\cot x - \tan x = \dfrac{1 + \cos 2x}{\sin 2x} - \dfrac{1 - \cos 2x}{\sin 2x} = \dfrac{2 \cos 2x}{\sin 2x} = 2 \cot 2x$ **27** $\cos t = 1 - 2 \sin^2\left(\dfrac{t}{2}\right)$, which gives $2 \sin^2\left(\dfrac{t}{2}\right) = 1 - \cos t$

29 $\sin 3t = 3 \sin t - 4 \sin^3 t$ **31** $\cos^4 x = \frac{3}{8} + \frac{1}{2} \cos 2x + \frac{1}{8} \cos 4x$ **33** $\dfrac{\pi}{3}, \dfrac{2\pi}{3}, \dfrac{4\pi}{3}, \dfrac{5\pi}{3}$ **35** $\dfrac{\pi}{12}, \dfrac{5\pi}{12}, \dfrac{7\pi}{12}, \dfrac{11\pi}{12}, \dfrac{13\pi}{12}, \dfrac{17\pi}{12}, \dfrac{19\pi}{12}, \dfrac{23\pi}{12}$

37 $\dfrac{\pi}{2}, \dfrac{3\pi}{4}, \dfrac{3\pi}{2}, \dfrac{7\pi}{4}$ **39** $0, \dfrac{\pi}{2}, \dfrac{3\pi}{2}, 2\pi$ **41** $\dfrac{7\pi}{6}, \dfrac{11\pi}{6}$ **43** $\dfrac{n\pi}{2} \pm \dfrac{\pi}{6}, n = 0, \pm 1, \pm 2, \ldots$ **45** $n\pi, 2n\pi \pm \dfrac{\pi}{3}, n = 0, \pm 1, \pm 2, \ldots$

47 $n\pi, n\pi \pm \dfrac{\pi}{3}, n = 0, \pm 1, \pm 2, \ldots$ **49** $\dfrac{n\pi}{2}, n = 0, \pm 1, \pm 2, \ldots$ **51** $2n\pi \pm \dfrac{\pi}{6}, n = 0, \pm 1, \pm 2, \ldots$

USING YOUR KNOWLEDGE 9.4

1
$$\begin{aligned}
\sin 6^\circ &= \sin(36^\circ - 30^\circ) \\
&= \sin 36^\circ \cos 30^\circ - \cos 36^\circ \sin 30^\circ \\
&= (\tfrac{1}{4}\sqrt{10 - 2\sqrt{5}})(\sqrt{3}/2) - [\sqrt{1 - \tfrac{1}{16}(10 - 2\sqrt{5})}](\tfrac{1}{2}) \\
&= \tfrac{1}{8}\sqrt{30 - 6\sqrt{5}} - \tfrac{1}{8}\sqrt{16 - 10 + 2\sqrt{5}} \\
&= \tfrac{1}{8}\sqrt{30 - 6\sqrt{5}} - \tfrac{1}{8}\sqrt{6 + 2\sqrt{5}} \\
&= \tfrac{1}{8}\sqrt{30 - 6\sqrt{5}} - \tfrac{1}{8}(\sqrt{5} + 1) \\
&= \tfrac{1}{8}(\sqrt{30 - 6\sqrt{5}} - \sqrt{5} - 1)
\end{aligned}$$

3 Let $\theta = 36^\circ$. Then $16 \sin^5 \theta - 20 \sin^3 \theta + 5 \sin \theta = 0$. Since $\sin 36^\circ \neq 0$, we may divide by $\sin \theta$. This gives the quadratic type equation

$$16 \sin^4 \theta - 20 \sin^2 \theta + 5 = 0$$

Using the quadratic formula, we find

$$\sin^2 \theta = \frac{10 \pm 2\sqrt{5}}{16}$$

We know that $0 < \sin 36^\circ < \sin 45^\circ \approx 0.71$, but

$$\frac{10 + 2\sqrt{5}}{16} \approx 0.9$$

Hence, for $\theta = 36^\circ$,

$$\sin^2 \theta = \frac{10 - 2\sqrt{5}}{16}$$

and

$$\sin \theta = \sqrt{\frac{10 - 2\sqrt{5}}{16}} = \frac{1}{4}\sqrt{10 - 2\sqrt{5}}$$

SELF-TEST, CHAPTER 9

1a Period $= \pi$, amplitude $= 3$ **b** $\dfrac{\pi}{4}$ **c** $\dfrac{\pi}{8}$ **2**

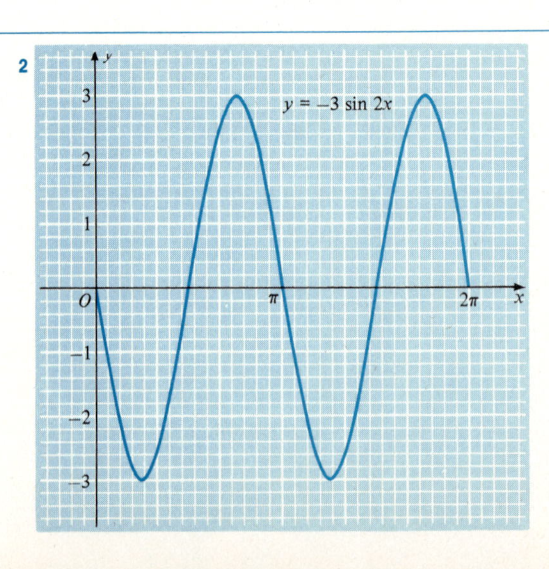

3

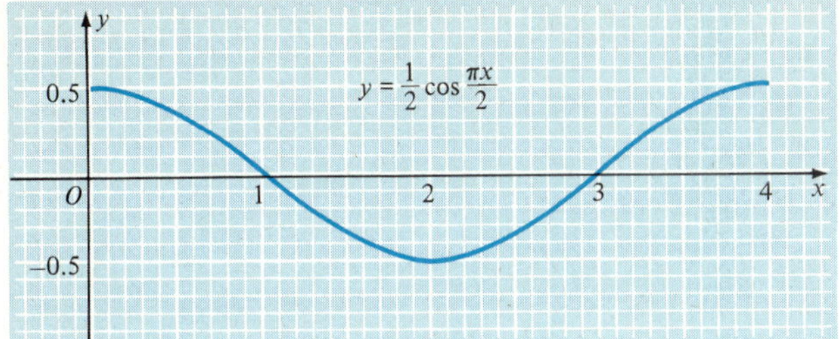
y
0.5
$y = \frac{1}{2}\cos\frac{\pi x}{2}$
O
1
2
3
4
x
−0.5

4

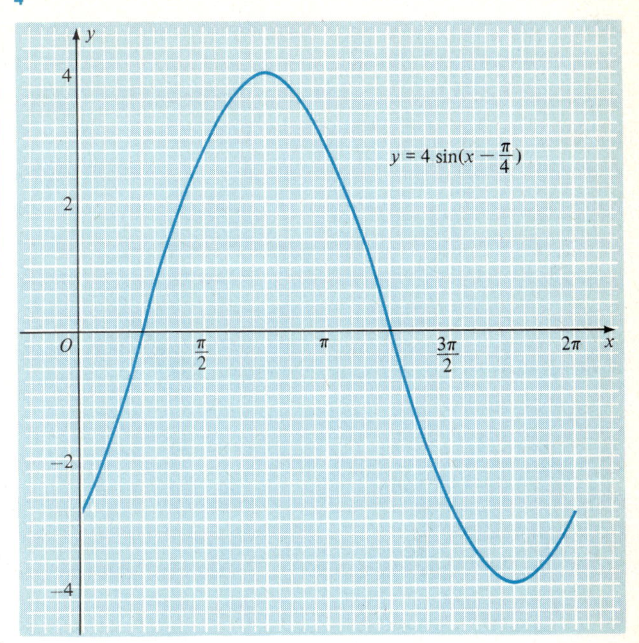
y
4
$y = 4\sin(x - \frac{\pi}{4})$
2
O
$\frac{\pi}{2}$
π
$\frac{3\pi}{2}$
2π
x
−2
−4

5

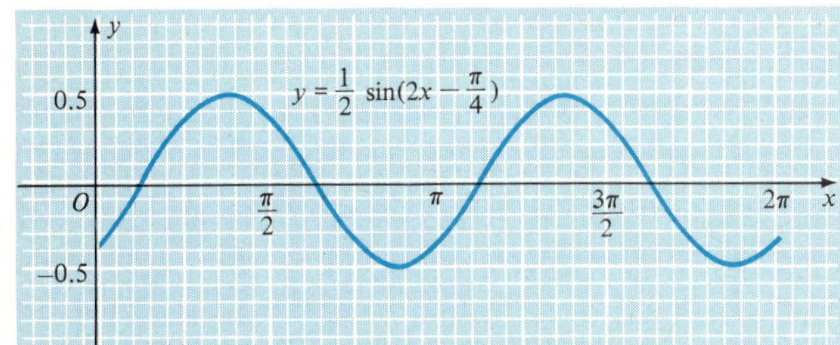
y
0.5
$y = \frac{1}{2}\sin(2x - \frac{\pi}{4})$
O
$\frac{\pi}{2}$
π
$\frac{3\pi}{2}$
2π
x
−0.5

6

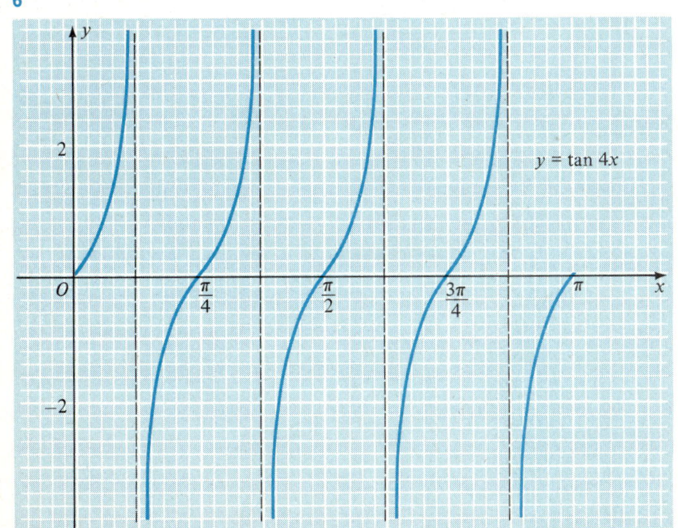
y
2
$y = \tan 4x$
O
$\frac{\pi}{4}$
$\frac{\pi}{2}$
$\frac{3\pi}{4}$
π
x
−2

7

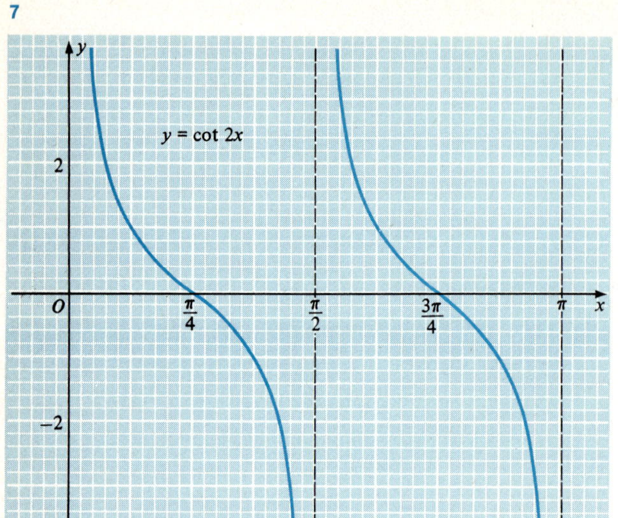

8

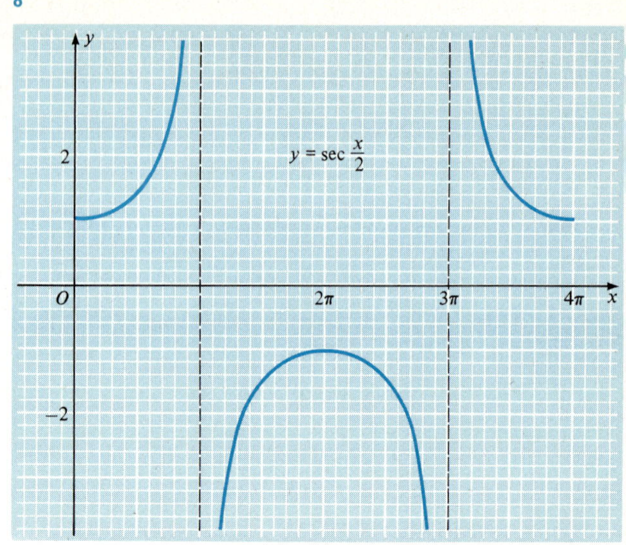

9

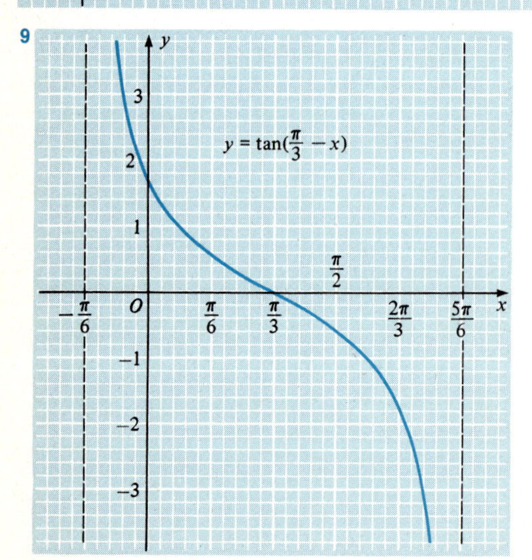

10a $\dfrac{\sqrt{2}-\sqrt{6}}{4}$ **b** $2-\sqrt{3}$ **11a** $\dfrac{\sqrt{3}}{2}\cos\theta - \dfrac{1}{2}\sin\theta$ **b** $\dfrac{1+\tan x}{1-\tan x}$ **12a** $\sin 27^\circ$ **b** $-\cot 10^\circ$ **13** $\dfrac{4\sqrt{2}-\sqrt{5}}{9}$ **14** $\dfrac{\pi}{4}, \dfrac{5\pi}{4}$ **15** $-\dfrac{4\sqrt{2}}{9}$

16 $\cos 4\theta = 8\cos^4\theta - 8\cos^2\theta + 1$ **17** $-\dfrac{\sqrt{14}}{6}$ **18** $\dfrac{2\pi}{3}, \pi, \dfrac{4\pi}{3}$ **19** $n\pi, (2n+1)\pi \pm \dfrac{\pi}{3}, n = 0, \pm 1, \pm 2, \dots$

20
$$\csc 2x - \cot 2x = \frac{1}{\sin 2x} - \frac{\cos 2x}{\sin 2x} = \frac{1-\cos 2x}{\sin 2x} = \tan \tfrac{1}{2}(2x) = \tan x$$

or

$$\tan x = \tan \tfrac{1}{2}(2x) = \frac{1-\cos 2x}{\sin 2x} = \frac{1}{\sin 2x} - \frac{\cos 2x}{\sin 2x} = \csc 2x - \cot 2x$$

EXERCISE 10.1

1 $\alpha = 53.1°, \beta = 36.9°, c = 15$ **3** $\beta = 38°, b = 15.6, c = 25.4$ **5** $\beta = 47.4°, b = 6.3, c = 8.6$ **7** $\beta = 65°, a = 7.6, b = 16.3$
9 $b = 13.7, \alpha = 23.6°, \beta = 66.4°$ **11** 30.0 m **13** 2.0 km **15** 245.1 in^2 **17** 34.9° **19** $17\frac{6}{7}$ in. **21** 39.8° **23** 239,000 miles
25 240.0 m **27**

$$x = (d + y)\tan\beta$$
$$x = y\tan\alpha$$
$$x = d\tan\beta + y\tan\beta$$
$$= d\tan\beta + \frac{x\tan\beta}{\tan\alpha}$$

$$x\left(1 - \frac{\tan\beta}{\tan\alpha}\right) = d\tan\beta$$
$$x(\tan\alpha - \tan\beta) = d\tan\alpha\tan\beta$$
$$x\left(\frac{\sin\alpha}{\cos\alpha} - \frac{\sin\beta}{\cos\beta}\right) = d\,\frac{\sin\alpha\sin\beta}{\cos\alpha\cos\beta}$$
$$x(\sin\alpha\cos\beta - \cos\alpha\sin\beta) = d\sin\alpha\sin\beta$$
$$x\sin(\alpha - \beta) = d\sin\alpha\sin\beta$$
$$x = \frac{d\sin\alpha\sin\beta}{\sin(\alpha - \beta)}$$

USING YOUR KNOWLEDGE 10.1

1 2694.8 ft **3** 500.3 ft

EXERCISE 10.2

1 867.5 **3** 7.99 **5** 2.299 **7** 1.60×10^4 **9** 2.28×10^0 **11** 4.00×10^2 **13** 4.0×10^3 **15** 65.0 **17** 3.14 **19** 2.96 **21** 0.225
23 0.584 **25** 6.091 **27** $\beta = 51.9°, b = 47.1, c = 59.8$ **29** $\beta = 70.17°, a = 202.6, b = 561.8$ **31** $c = 256, \alpha = 31.9°, \beta = 58.1°$
33 $b = 73, \alpha = 46°, \beta = 44°$ **35** $\beta = 59.9°, a = 22.9, b = 39.5$

USING YOUR KNOWLEDGE 10.2

1
$$(37.15)(2.555) \le (37.2)(2.56) < (37.25)(2.565)$$
$$94.91825 \le 95.235 < 95.54625$$
The properly rounded answer is 95.2.

3
$$\frac{37.15}{2.565} < \frac{37.2}{2.56} < \frac{37.25}{2.555}$$
$$14.483431 < 14.53125 < 14.579256$$
The properly rounded answer is 14.5.

5
$$(52.25)\sin 63.65° \le (52.3)\sin 63.7° < (52.35)\sin 63.75°$$
$$46.821196 \le 46.88624 < 46.951288$$
The properly rounded answer is 46.9.

7
$$37.15 + 2.555 \le 37.2 + 2.56 < 37.25 + 2.565$$
$$39.705 \le 39.76 < 39.815$$
The properly rounded answer is 39.8.

9
$$37.15 - 2.565 < 37.2 - 2.56 < 37.25 - 2.555$$
$$34.585 < 34.64 < 34.695$$
The properly rounded answer is 34.6.

EXERCISE 10.3

1 $150\sqrt{3}$ sq units **3** 14 sq units **5** $15\sqrt{2}$ sq units **7** 628 sq units **9** 3.55×10^3 sq units **11** 1.5×10^2 sq units
13 $\alpha = 70°, b = 4.08, c = 4.61$ **15** $\beta = 45°, a = 55.1, c = 57.1$ **17** $\gamma = 47°, a = 150, b = 112$ **19** $\alpha = 62.0°, b = 75.9, c = 79.7$
21 $\beta = 45.5°, a = 137, b = 98.8$ **23** $\alpha = 25°, a = 36.7, c = 76.3$ **25** $\alpha = 24.57°, a = 14.44, b = 25.44$ **27** 11.8 cm

29 $A = \frac{1}{2}bc\sin\alpha,\ b = \frac{c\sin\beta}{\sin\gamma} = \frac{c\sin\beta}{\sin(180° - \alpha - \beta)} = \frac{c\sin\beta}{\sin(\alpha + \beta)}$

Thus,

$$A = \frac{1}{2}\frac{c\sin\beta}{\sin(\alpha + \beta)}c\sin\alpha = \frac{c^2\sin\alpha\sin\beta}{2\sin(\alpha + \beta)}$$

31 1.08×10^3 ft **33** 200 ft **35** 3996 sq m

USING YOUR KNOWLEDGE 10.3

1 $R = 584.9$ ft, $OC = 593.9$ ft

EXERCISE 10.4

1 $c = \sqrt{7}$ **3** $a = \sqrt{31}$ **5** $c = \sqrt{146}$ **7** $c = \sqrt{7} \approx 2.65, \alpha = 49.1°, \beta = 100.9°$ **9** $a = \sqrt{31} \approx 5.57, \beta = 68.9°, \gamma = 51.1°$
11 $c = \sqrt{146} \approx 12.1, \alpha = 24.4°, \beta = 20.6°$ **13** $c = 12, \alpha = 46°, \beta = 102°$ **15** $b = 71, \alpha = 28°, \gamma = 46°$

17 $c = 17.2$, $\alpha = 51.0°$, $\beta = 108.3°$ **19** $a = 110.6$, $\beta = 59.90°$, $\gamma = 70.85°$ **21** $a\sqrt{2-\sqrt{3}}$ **23** $d_1 = 27.4$ ft, $d_2 = 58.3$ ft
25 She would be about 234 yd in the direction S 10° E from her starting point.
27 The target is about 8.9 km in the direction N 13.8° E from A.

USING YOUR KNOWLEDGE 10.4

1 About 2000 miles

EXERCISE 10.5

1 $\alpha = 26.38°$, $\beta = 36.34°$, $\gamma = 117.28°$ **3** $\alpha = 82.82°$, $\beta = 41.41°$, $\gamma = 55.77°$ **5** $\alpha = 46.4°$, $\beta = 57.6°$, $\gamma = 76.0°$
7 $\alpha = 37.3°$, $\beta = 84.1°$, $\gamma = 58.6°$ **9** $\alpha = 40.23°$, $\beta = 124.12°$, $\gamma = 15.65°$ **11** S 64° E **13** N 75.1° E **15** 10.67 in² **17** 4482 in²

19 From the Law of Cosines,

$$\cos\beta = \frac{a^2 + c^2 - b^2}{2ac}$$

Thus,

$$1 - \cos\beta = 1 - \frac{a^2 + c^2 - b^2}{2ac}$$
$$= \frac{b^2 - (a^2 - 2ac + c^2)}{2ac}$$
$$= \frac{b^2 - (a-c)^2}{2ac}$$
$$= \frac{(b + a - c)(b - a + c)}{2ac}$$
$$= \frac{(a + b - c)(b + c - a)}{2ac}$$

21 Join the vertices to the center of the circle as shown in the figure. This divides the triangle into three triangles, each with r as altitude and one side of the original triangle as base. Thus,

$$A = \tfrac{1}{2}ra + \tfrac{1}{2}rb + \tfrac{1}{2}rc = \tfrac{1}{2}r(a + b + c) = rs$$

Since $A = \sqrt{s(s-a)(s-b)(s-c)}$, we have

$$rs = \sqrt{s(s-a)(s-b)(s-c)}$$

and

$$r = \sqrt{\frac{(s-a)(s-b)(s-c)}{s}}$$

23 See Problem 9.

25 In the accompanying figure,

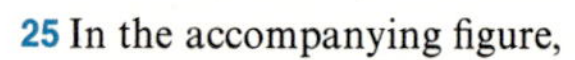

$$\alpha = \angle A = \tfrac{1}{2}\widehat{BC} \quad \text{and} \quad \angle BOC = \widehat{BC}$$

Hence, $\angle BOC = 2\alpha$, so that

$$\angle BOP = \alpha$$

From triangle BOP, we find

$$\tfrac{1}{2}a = R\sin\alpha$$

Thus,

$$R = \frac{a}{2\sin\alpha}$$

Similarly, we obtain

$$R = \frac{b}{2\sin\beta} \quad \text{and} \quad R = \frac{c}{2\sin\gamma}$$

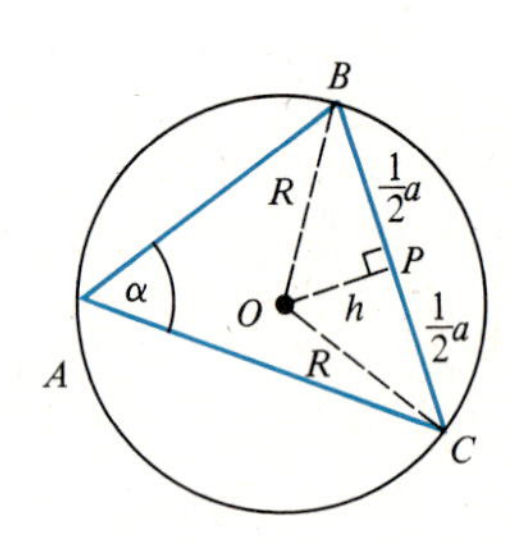

USING YOUR KNOWLEDGE 10.5

1 Because $\tan\frac{1}{2}\alpha = r/(s-a)$, we find

$$\sin\tfrac{1}{2}\alpha = \frac{r}{\sqrt{r^2 + (s-a)^2}} \quad \text{and} \quad \cos\tfrac{1}{2}\alpha = \frac{s-a}{\sqrt{r^2 + (s-a)^2}}$$

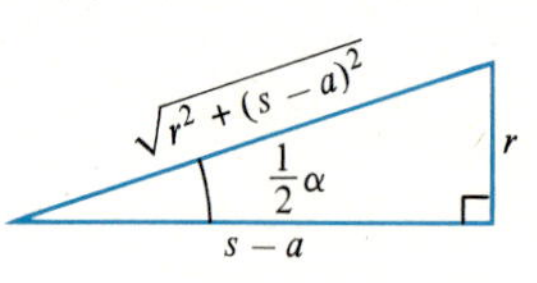

(See the accompanying figure.) From the double-angle formulas,

$$\sin\alpha = 2\sin\tfrac{1}{2}\alpha\cos\tfrac{1}{2}\alpha = \frac{2r(s-a)}{r^2 + (s-a)^2}$$

Also,

$$r^2 + (s-a)^2 = \frac{(s-a)(s-b)(s-c)}{s} + (s-a)^2$$
$$= \frac{s-a}{s}[(s-b)(s-c) + s(s-a)]$$

$$(s-b)(s-c)+s(s-a)=\left(\frac{a+c-b}{2}\right)\left(\frac{a+b-c}{2}\right)+\left(\frac{a+b+c}{2}\right)\left(\frac{b+c-a}{2}\right)$$

$$=\frac{a^2-(b-c)^2}{4}+\frac{(b+c)^2-a^2}{4}$$

$$=\frac{(b+c)^2-(b-c)^2}{4}$$

$$=bc$$

Thus,

$$\sin\alpha=\frac{2r(s-a)}{[(s-a)bc]/s}=\frac{2rs}{bc}$$

3 $r=\frac{4\sqrt{6}}{3}$, $R=\frac{35\sqrt{6}}{12}$

EXERCISE 10.6

1 There is one triangle. $\beta=90°$, $\gamma=60°$, $c=4\sqrt{3}$ **3** There are two triangles. $\beta=101.41°$, $\gamma=48.59°$, $b=7.84$
$\beta'=18.59°$, $\gamma'=131.41°$, $b'=2.55$

5 There is no triangle. **7** There is no triangle. **9** There is one triangle. $\gamma=14.86°$, $\beta=5.14°$, $b=2.09$

11 There are two triangles. $\alpha=54.4°$, $\beta=87.4°$, $b=26.7$
$\alpha'=125.6°$, $\beta'=16.2°$, $b'=7.45$

13 There is one triangle. $\alpha=80.27°$, $\beta=47.03°$, $a=171.6$ **15** $\frac{\sqrt{14}}{7}\approx 0.53$

17 Two parallelograms are possible. $A=77\text{ in}^2$; $A'=13\text{ in}^2$ **19** About 2.2 miles

USING YOUR KNOWLEDGE 10.6

1 $4\sqrt{3}$ **3** There is no triangle possible. **5** $b=2.09$

SELF-TEST, CHAPTER 10

1a $\frac{AE}{AF}$ **b** $\frac{BF}{AF}$ **c** $\frac{DE}{AD}$ **d** $\frac{AF}{AE}$ **2a** 26.6° **b** 9.7 **c** 6.6 **d** 72.5° **3** $x=\frac{d\cos\beta}{\cos\alpha}$, $y=\frac{d\sin\beta}{\cos\alpha}$ **4a** 73.5 **b** 5.4×10^5

5a 1.15×10^4 m **b** 23.0 in. **6a** 50.4 ft **b** 40.0° **7a** 30 sq units **b** $2\sqrt{13}$ **8a** $\frac{15\sqrt{2}}{2}$ **b** $20\sqrt{2}$ **9** $50\sqrt{2}$ m

10 $x=b(\tan\theta-\tan\alpha)$ or $x=\frac{b\sin(\theta-\alpha)}{\cos\theta\cos\alpha}$ **11a** $\sqrt{21}$ **b** $\frac{2\sqrt{7}}{7}$ **c** $\frac{5\sqrt{7}}{14}$ **12a** 1 **b** $\alpha=105°$, $\beta=15°$ **13** $5\sqrt{7}$ miles **14** 71.8°

15 0.51 **16** S 46.6° W **17a** None **b** Two **c** One **18** $\beta=24.6°$, $\gamma=5.4°$, $c=1.13$ **19** $\alpha=59.56°$, $b=34.83$ m, $c=27.05$ m
20 235 sq ft

EXERCISE 11.1

1a $\frac{\pi}{3}$ **b** $-\frac{\pi}{3}$ **3a** $-\frac{\pi}{4}$ **b** $-\frac{\pi}{2}$ **5a** $\frac{\pi}{4}$ **b** $\frac{\pi}{3}$ **7**

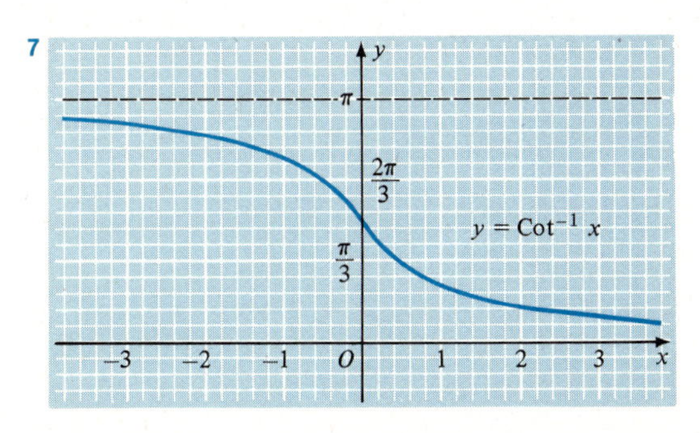

9 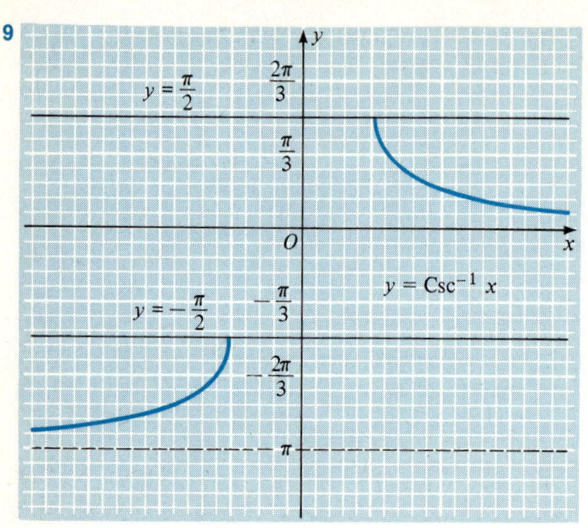

11 Let $y = \text{Tan}^{-1} x$. Then, for all real values of x, $\tan y = x$, by the definition of $\text{Tan}^{-1} x$. Thus, $\tan(\text{Tan}^{-1} x) = \tan y = x$ for all real values of x.

13 No. For $-1 \le x \le 1$, $\text{Cos}^{-1}(-x) = \pi - \text{Cos}^{-1} x$ (Theorem 11.1b). For example,

$$\text{Cos}^{-1}\left(-\frac{1}{2}\right) = \frac{2\pi}{3} \quad \text{and} \quad \text{Cos}^{-1}\frac{1}{2} = \frac{\pi}{3}$$

15 Let $y = \text{Cos}^{-1} x$. Then, $\cos y = x$, $0 \le y \le \pi$ and $\cos(\pi - y) = -x$. Therefore, $\text{Cos}^{-1}(-x) = \pi - y = \pi - \text{Cos}^{-1} x$

17 $-\frac{\pi}{2} < x < \frac{\pi}{2}$

19 One solution is $\theta_1 = \text{Cos}^{-1} c$. All the solutions are then given by $2m\pi \pm \theta_1$ or $2m\pi \pm \text{Cos}^{-1} c$, $m = 0, \pm 1, \pm 2, \ldots$.

21 Let $\alpha = \text{Tan}^{-1} \frac{1}{2}$, $\beta = \text{Tan}^{-1} \frac{1}{3}$. Then

$$\tan(\alpha + \beta) = \frac{\tan \alpha + \tan \beta}{1 - \tan \alpha \tan \beta} = \frac{\frac{1}{2} + \frac{1}{3}}{1 - (\frac{1}{2})(\frac{1}{3})} = \frac{\frac{5}{6}}{\frac{5}{6}} = 1$$

From the values of $\tan \alpha$ and $\tan \beta$, it follows that

$$0 < \alpha < \frac{\pi}{4} \quad \text{and} \quad 0 < \beta < \frac{\pi}{4}$$

Therefore, $\tan(\alpha + \beta) = 1$ implies that $\alpha + \beta = \pi/4$.

23 Let $\alpha = \text{Sin}^{-1} \frac{3}{5}$, $\beta = \text{Sin}^{-1} \frac{4}{5}$. Then,

$$0 < \alpha < \frac{\pi}{2} \quad \text{and} \quad 0 < \beta < \frac{\pi}{2}$$

so that $0 < \alpha + \beta < \pi$. Because

$$\begin{aligned} \sin(\alpha + \beta) &= \sin \alpha \cos \beta + \cos \alpha \sin \beta \\ &= (\tfrac{3}{5})(\tfrac{3}{5}) + (\tfrac{4}{5})(\tfrac{4}{5}) \\ &= \tfrac{9}{25} + \tfrac{16}{25} \\ &= 1 \end{aligned}$$

it follows that $\alpha + \beta = \pi/2$.

25 We know that $0 < \alpha < \pi/2$ and $0 < \beta < \pi/2$, so that $0 < \alpha + \beta < \pi$. Since $\tan(\alpha + \beta) = -1$, we know that $\pi/2 < \alpha + \beta < \pi$, which means that we cannot use $\text{Tan}^{-1}(-1)$. The correct solution is $\alpha + \beta = 3\pi/4$, which also satisfies $\tan(\alpha + \beta) = -1$.

27 $\frac{119}{169}$ 29 $\frac{\sqrt{26}}{26}$ 31 $-\frac{2}{11}$ 33 0 35 $\frac{5}{13}$

USING YOUR KNOWLEDGE 11.1

1 Let $\alpha = \text{Tan}^{-1}\left(\frac{a + h}{x}\right)$ and $\beta = \text{Tan}^{-1}\left(\frac{a}{x}\right)$. Then,

$$\begin{aligned} \tan \theta &= \tan(\alpha - \beta) \\ &= \frac{\tan \alpha - \tan \beta}{1 + \tan \alpha \tan \beta} \\ &= \frac{(a + h)/x - a/x}{1 + [(a^2 + ah)/x^2]} \\ &= \frac{hx}{x^2 + a^2 + ah} \end{aligned}$$

Therefore,

$$\theta = \text{Tan}^{-1}\left(\frac{hx}{x^2 + a^2 + ah}\right)$$

3 $F(\sqrt{a^2 + ah}) = \dfrac{h\sqrt{a^2 + ah}}{(\sqrt{a^2 + ah})^2 + a^2 + ah} = \dfrac{h}{2\sqrt{a^2 + ah}}$

5 From Problem 4, it follows that $F(\sqrt{a^2 + ah}) - F(x) \geq 0$ for all real values of x. Therefore, $F(x) \leq F(\sqrt{a^2 + ah})$. Because $F(x) = F(\sqrt{a^2 + ah})$ for $x = \sqrt{a^2 + ah}$, this value of x makes $F(x)$ a maximum and, thus, θ a maximum.

EXERCISE 11.2

1

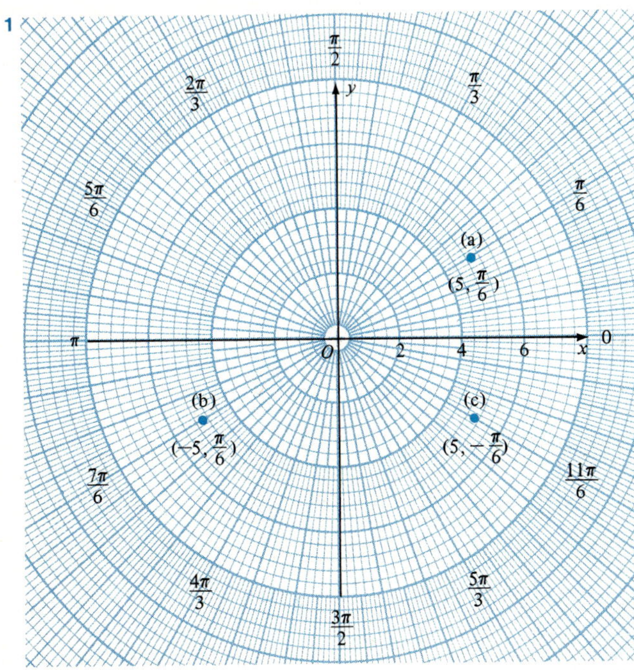

a $\left(\frac{5\sqrt{3}}{2}, \frac{5}{2}\right)$ **b** $\left(-\frac{5\sqrt{3}}{2}, -\frac{5}{2}\right)$ **c** $\left(\frac{5\sqrt{3}}{2}, -\frac{5}{2}\right)$

3

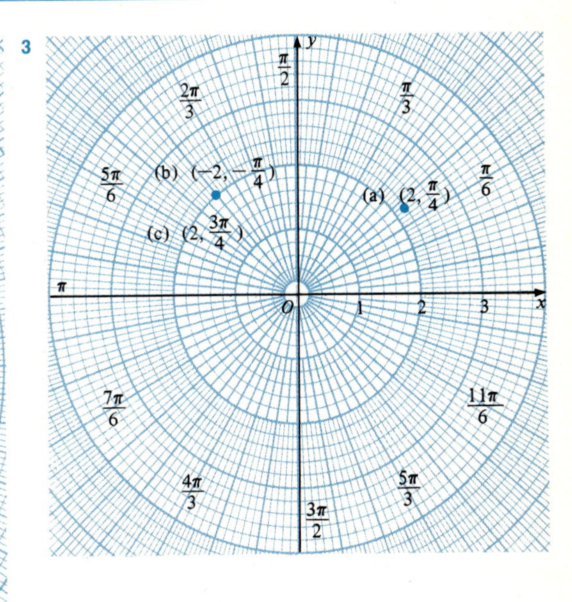

a $(\sqrt{2}, \sqrt{2})$ **b** $(-\sqrt{2}, \sqrt{2})$ **c** $(-\sqrt{2}, \sqrt{2})$

5

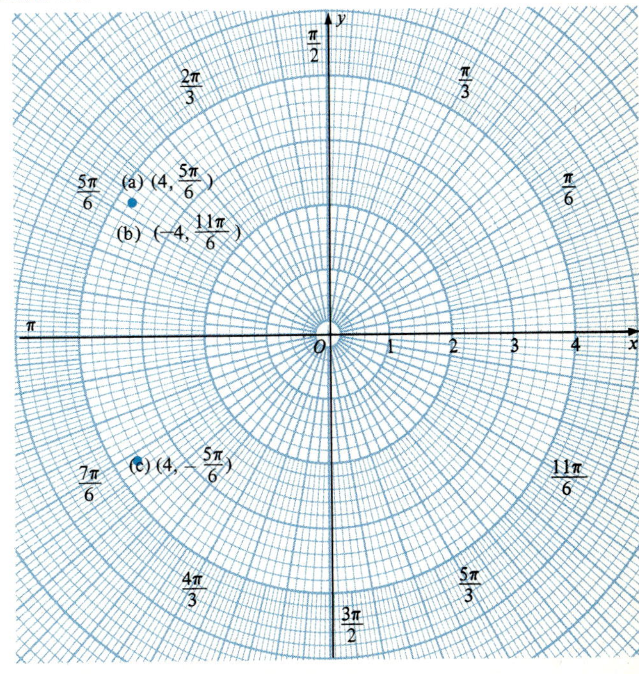

a $(-2\sqrt{3}, 2)$ **b** $(-2\sqrt{3}, 2)$ **c** $(-2\sqrt{3}, -2)$

7 **a** $\left(\sqrt{2}, \frac{7\pi}{4}\right)$ and $\left(-\sqrt{2}, \frac{11\pi}{4}\right)$ **b** $\left(\sqrt{2}, \frac{3\pi}{4}\right)$ and $\left(-\sqrt{2}, \frac{7\pi}{4}\right)$ **c** $\left(\sqrt{2}, \frac{5\pi}{4}\right)$ and $\left(-\sqrt{2}, \frac{9\pi}{4}\right)$

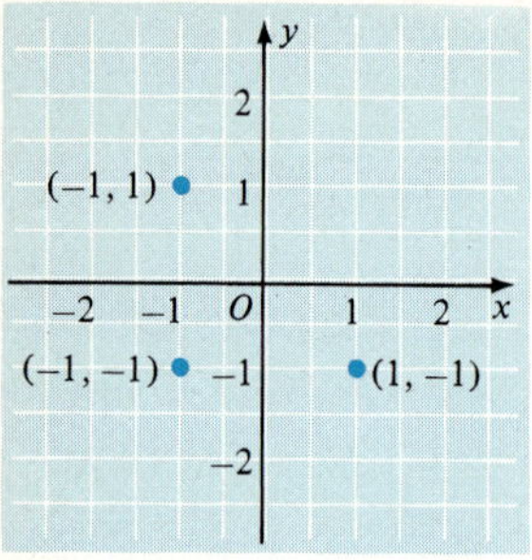

9 **a** $\left(2, \frac{\pi}{2}\right)$ and $\left(-2, \frac{3\pi}{2}\right)$ **b** $(2, \pi)$ and $(-2, 2\pi)$ **c** $\left(2, \frac{3\pi}{2}\right)$ and $\left(-2, \frac{5\pi}{2}\right)$

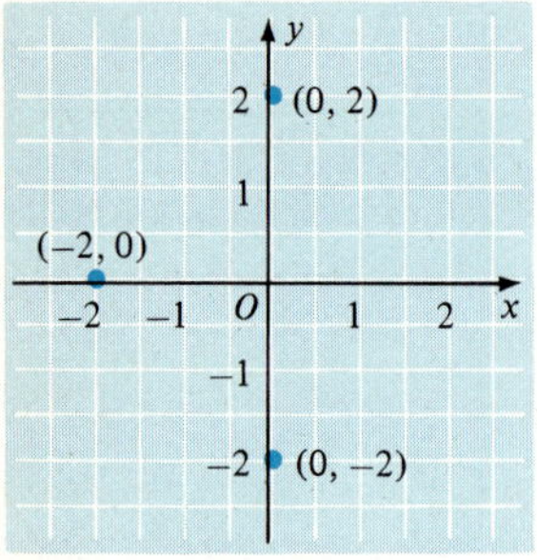

11a No **b** Yes **13a** No **b** No **15a** Yes **b** Yes **17a** No **b** Yes **19** $2\sqrt{3}$ **21** $\sqrt{37}$ **23** $r = 2 \sin \theta$ **25** $r = 2 \csc \theta \cot \theta$
27 $(x^2 + y^2)^2 = 4(x^2 - y^2)$ **29** $(x^2 + y^2)^3 = y^2$

31 The radius of the circle is 2. If $P(r, \theta)$ is a point on the circle, then the square of the distance of P from $(2, \pi/3)$ is 4. Thus, $r^2 + 4 - 4r \cos(\theta - \pi/3) = 4$, so that $r = 4 \cos(\theta - \pi/3)$.

33 $r^2 + 25 - 10r \cos(\theta - \text{Tan}^{-1} \frac{3}{4}) = 25$, so that

$$r = 10 \cos(\theta - \text{Tan}^{-1} \tfrac{3}{4})$$

Let $\phi = \text{Tan}^{-1} \frac{3}{4}$. Then,

$$\begin{aligned} r &= 10 \cos(\theta - \phi) \\ &= 10(\cos \theta \cos \phi + \sin \theta \sin \phi) \\ &= 10(\tfrac{4}{5} \cos \theta + \tfrac{3}{5} \sin \theta) \\ &= 8 \cos \theta + 6 \sin \theta \end{aligned}$$

USING YOUR KNOWLEDGE 11.2

1 From Figure 11.2j, $DP = AO + OB$. $AO = p$ and $OB = r \cos \theta$. Thus, $DP = p + r \cos \theta$.

3 In the definition of Problem 17, Exercise 6.3, the distance of P from the focus equals its distance from the directrix. Thus, $e = 1$.

5 $e > 1$ and $r = \dfrac{ep}{1 - e \cos \theta}$

Because $e > 1$, we can make $1 - e \cos \theta$ either positive or negative and arbitrarily close to zero. This means that r can be arbitrarily large in absolute value, so that the range of r is infinite.

EXERCISE 11.3

1 $r \sin\theta = -2$ or $y = -2$

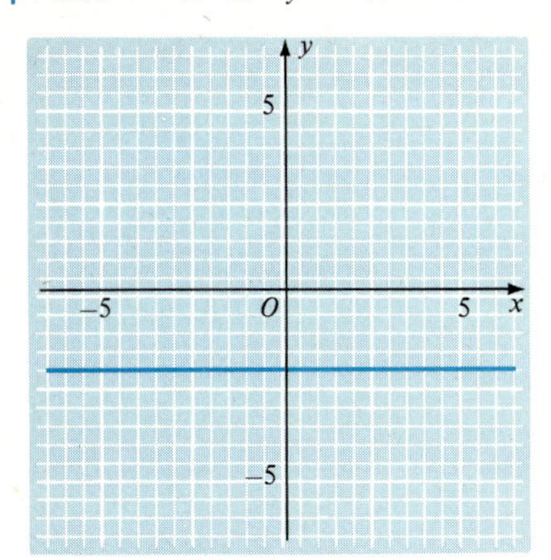

3 $r = 5$ or $x^2 + y^2 = 25$

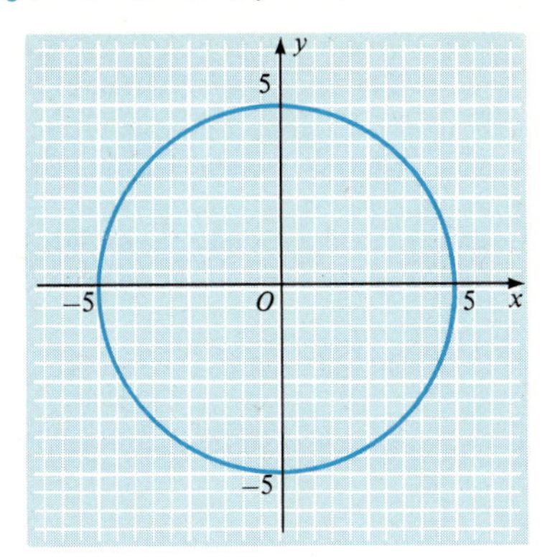

5 $r = 4\cos\theta$ or $x^2 + y^2 = 4x$

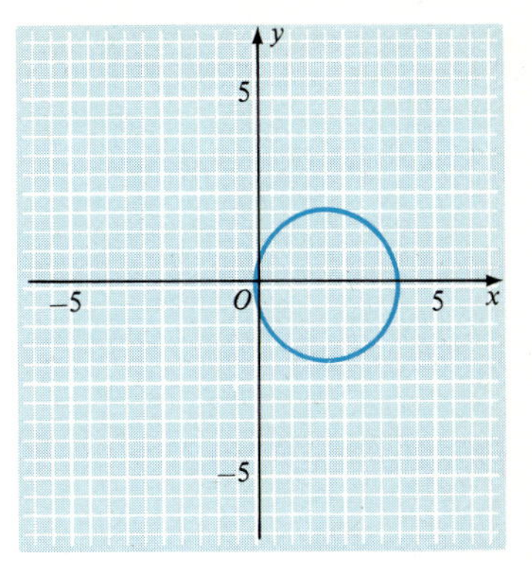

7 $r = \cos\theta - \sin\theta$ or $x^2 + y^2 = x - y$

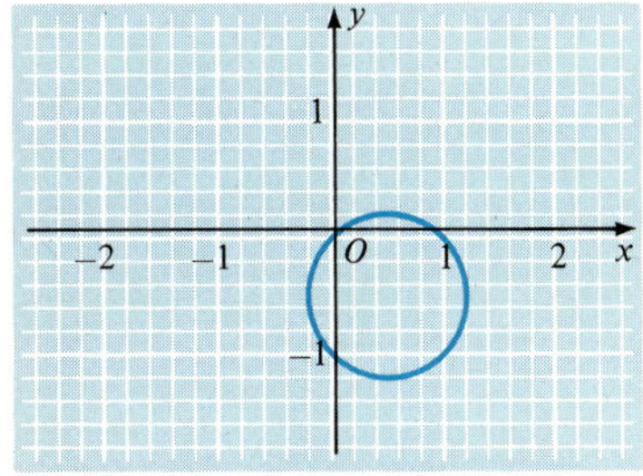

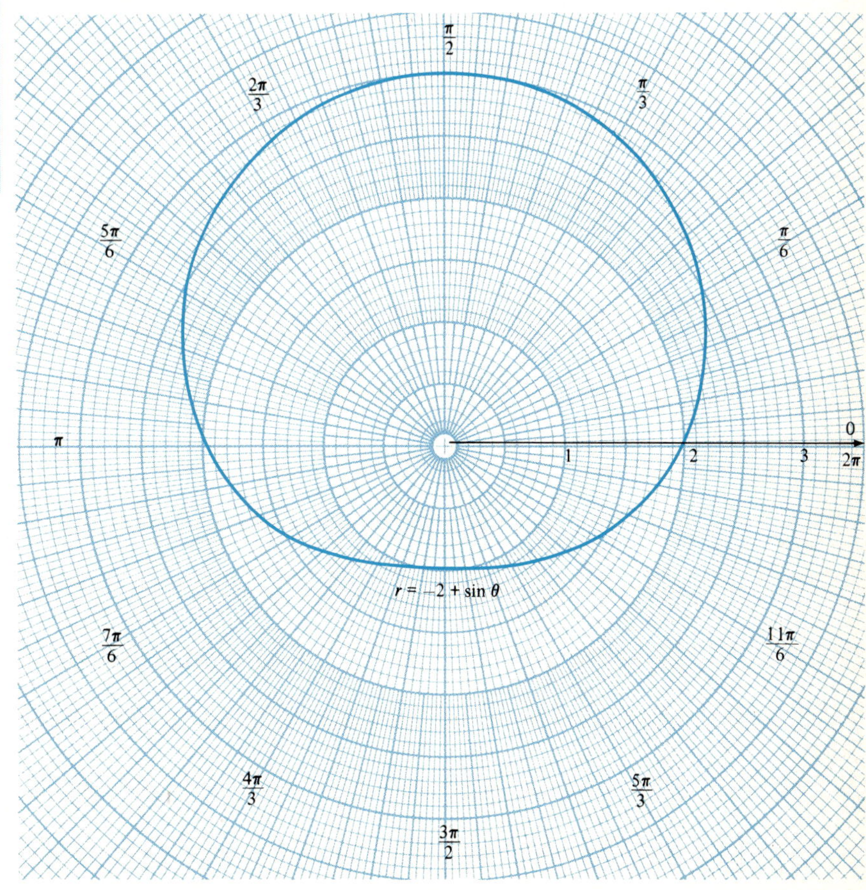

9 1. $\theta = 0$, $r = -2$; $\theta = \pi/2$, $r = -1$; $\theta = \pi$, $r = -2$; $\theta = 3\pi/2$, $r = -3$. r cannot be zero.
2. Symmetric to y-axis
3. There is no restriction on θ.
4. As θ increases from 0 to $\pi/2$, r increases from -2 to -1. As θ increases from $\pi/2$ to π, r decreases from -1 to -2. As θ increases from π to $3\pi/2$, r decreases from -2 to -3. As θ increases from $3\pi/2$ to 2π, r increases from -3 to -2.

11 1. For $\theta = 0$, $r = -3$; for $\theta = \pi/2$, $r = 3$; for $\theta = \pi$, $r = 9$; for $\theta = 3\pi/2$, $r = 3$. $r = 0$ for $\theta = \pi/3$ and for $\theta = 5\pi/3$.

2. The graph is symmetric to the x-axis.
3. There is no restriction on θ.
4. As θ increases from 0 to $\pi/2$, r increases from -3 to 3. As θ increases from $\pi/2$ to π, r increases from 3 to 9. As θ increases from π to $3\pi/2$, r decreases from 9 to 3. As θ increases from $3\pi/2$ to 2π, r decreases from 3 to -3.

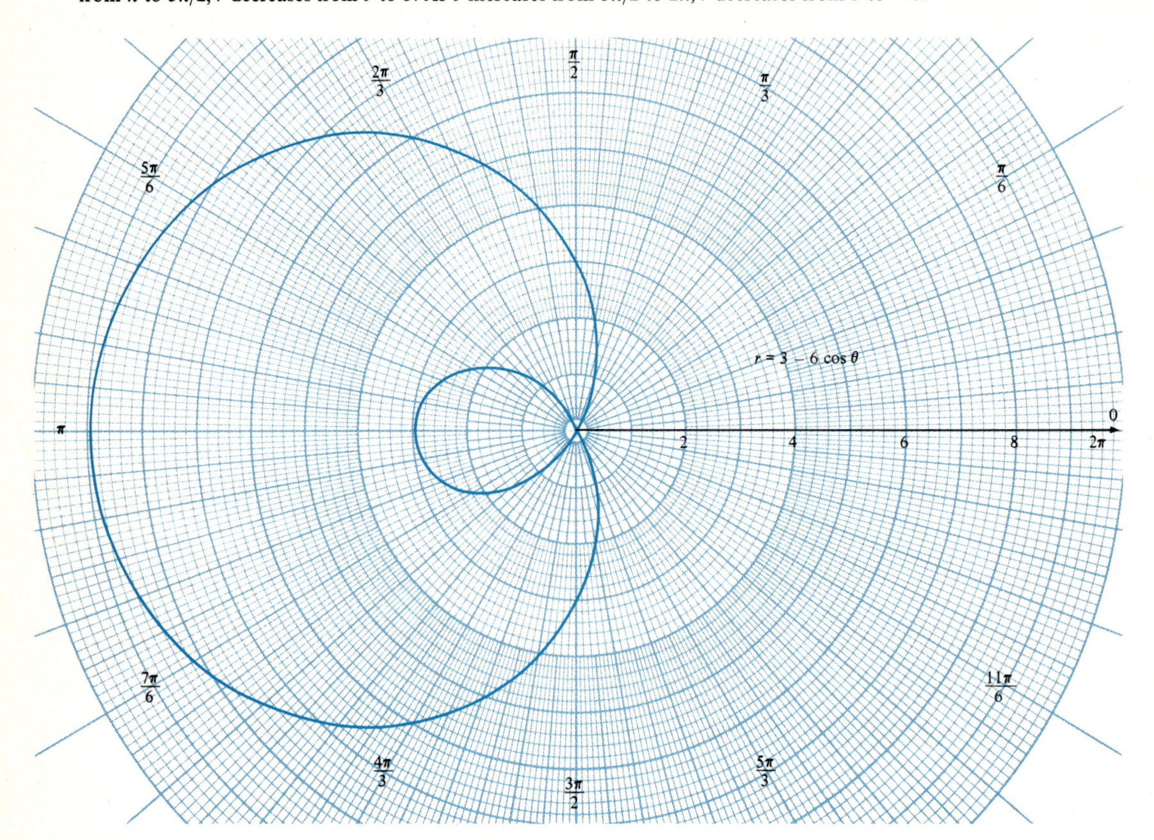

13 1. For $\theta = 0$, $r = 5$; for $\theta = \pi/2$, $r = 3$; for $\theta = \pi$, $r = 1$; for $\theta = 3\pi/2$, $r = 3$. r cannot be zero.

2. The graph is symmetric to the x-axis.

3. There is no restriction on θ.

4. As θ increases from 0 to $\pi/2$, r decreases from 5 to 3. As θ increases from $\pi/2$ to π, r decreases from 3 to 1. As θ increases from π to $3\pi/2$, r increases from 1 to 3. As θ increases from $3\pi/2$ to 2π, r increases from 3 to 5.

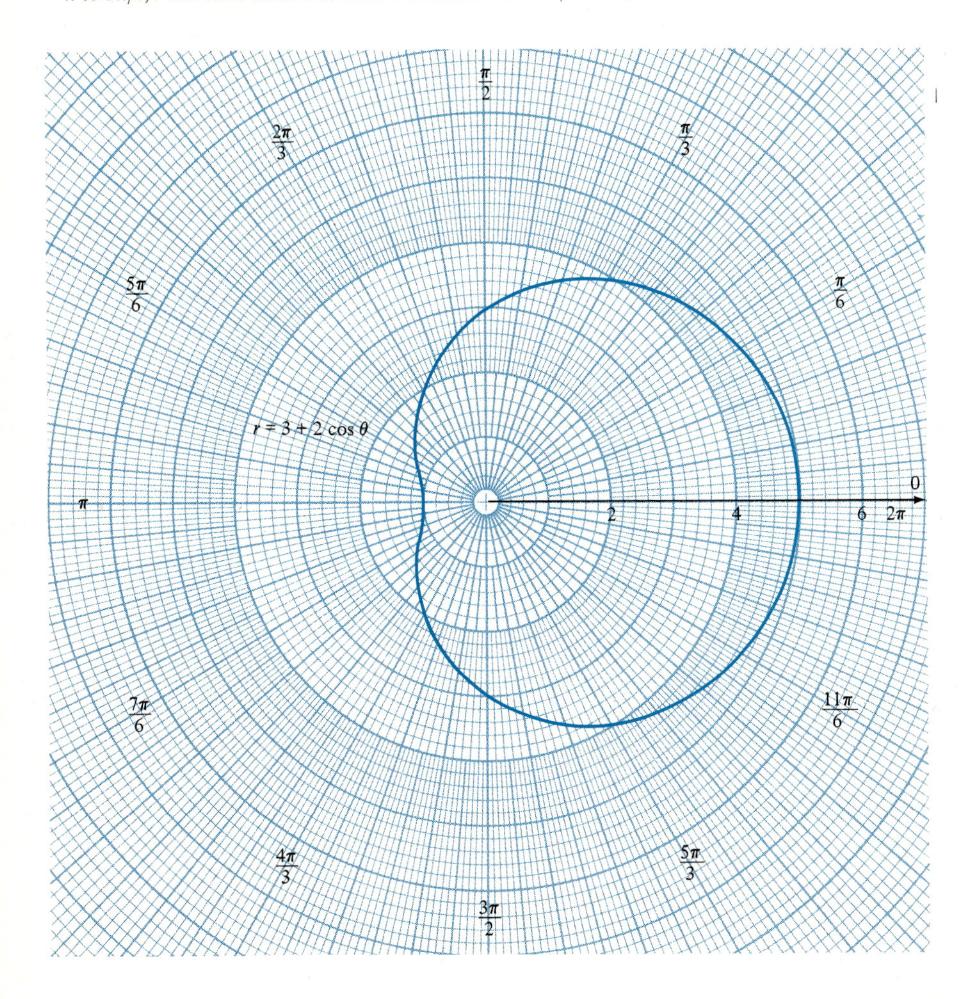

15 1. There is no intersection with the axes except at $r = 0$. For $r = 0$, $\theta = 0$, $\pi/4$, $\pi/2$, $3\pi/4$, π, $5\pi/4$, $3\pi/2$, $7\pi/4$.
2. The graph is symmetric to both axes and the pole.
3. There is no restriction on θ.
4. As θ increases from 0 to $\pi/8$, r increases from 0 to 1. As θ increases from $\pi/8$ to $\pi/4$, r decreases from 1 to 0. This is one loop of the graph, and the remaining loops can be obtained by symmetry. (The curve is an eight-leaved rose curve.)

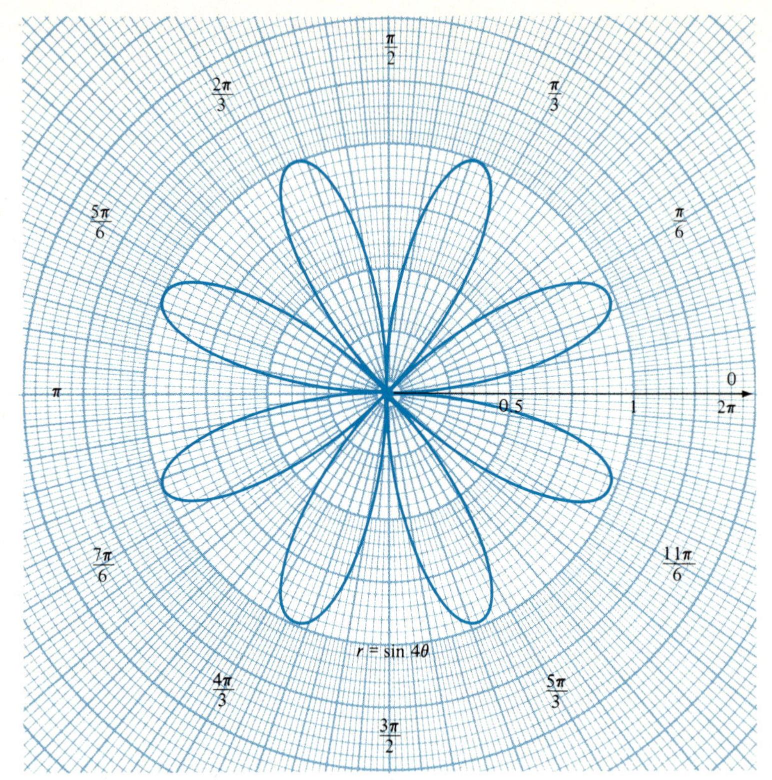

17 1. For $\theta = 0$, $r = 2$; for $\theta = \pi/2$, $r = 4$; for $\theta = \pi$, $r = 2$; for $\theta = 3\pi/2$, $r = 0$. For $r = 0$, the graph is tangent to the line $\theta = 3\pi/2$.
2. The graph is symmetric to the y-axis.
3. There is no restriction on θ.
4. As θ increases from 0 to $\pi/2$, r increases from 2 to 4. As θ increases from $\pi/2$ to π, r decreases from 4 to 2. As θ increases from π to $3\pi/2$, r decreases from 2 to 0. As θ increases from $3\pi/2$ to 2π, r increases from 0 to 2.

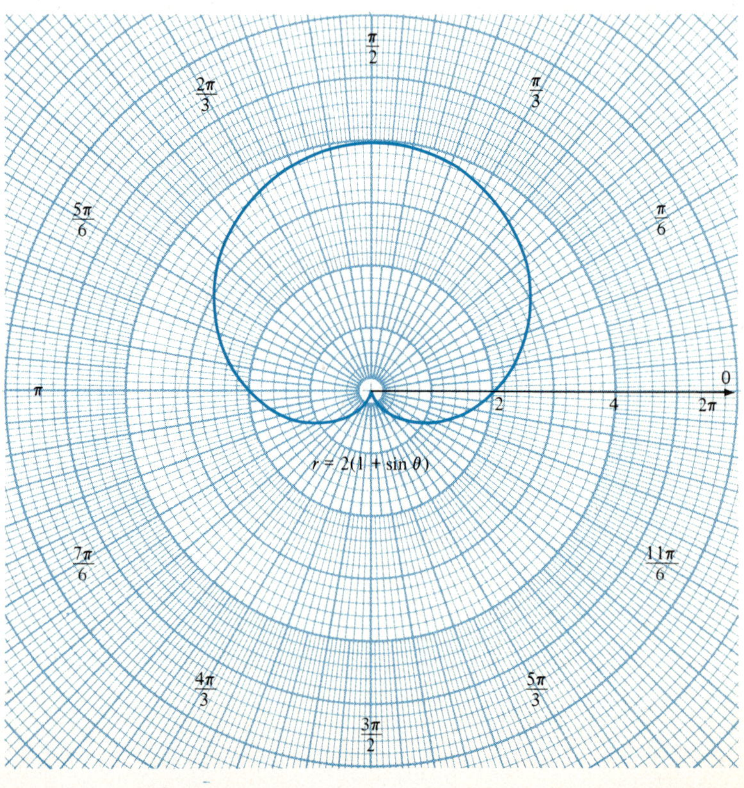

19 1. For $\theta = 0$, $r = 1$ or -1. For $r = 0$, $\theta = \pi/4, 3\pi/4, 5\pi/4, 7\pi/4$.

2. The graph is symmetric to both axes and the pole.
3. First-quadrant values of θ are restricted to lie between 0 and $\pi/4$. By symmetry, it follows that the graph lies between the lines $\theta = -\pi/4$ and $\theta = \pi/4$, and between the lines $\theta = 3\pi/4$ and $\theta = 5\pi/4$.
4. As θ increases from 0 to $\pi/4$, $|r|$ decreases from 1 to 0. As θ increases from $3\pi/4$ to π, $|r|$ increases from 0 to 1. (The curve is a lemniscate.)

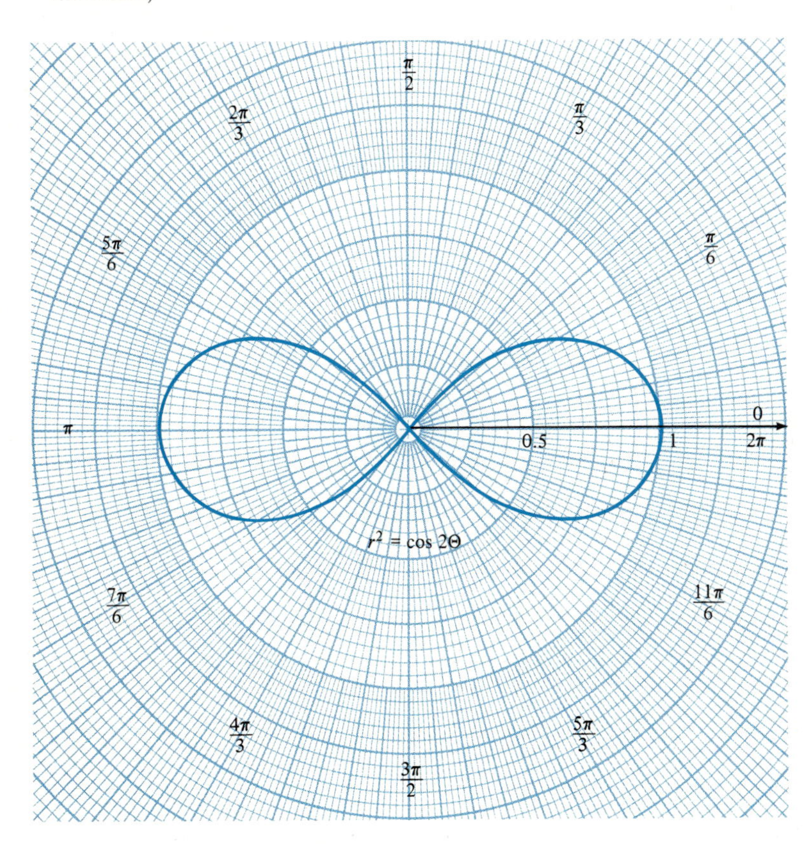

21 1. For $\theta = m\pi/2$, $r = 2/m\pi$, $m = \pm 1, \pm 2, \ldots$, $r \neq 0$.

2. The graph is symmetric to the y-axis. (The portion for $\theta < 0$ is symmetric to that for $\theta > 0$ with respect to the y-axis. The graph for $\theta < 0$ is not shown in the figure.)
3. $\theta \neq 0$ and $r \neq 0$.
4. As $|\theta|$ increases, $|r|$ decreases and gets closer and closer to 0. As $|\theta|$ decreases and gets closer and closer to 0, $|r|$ increases without bound.

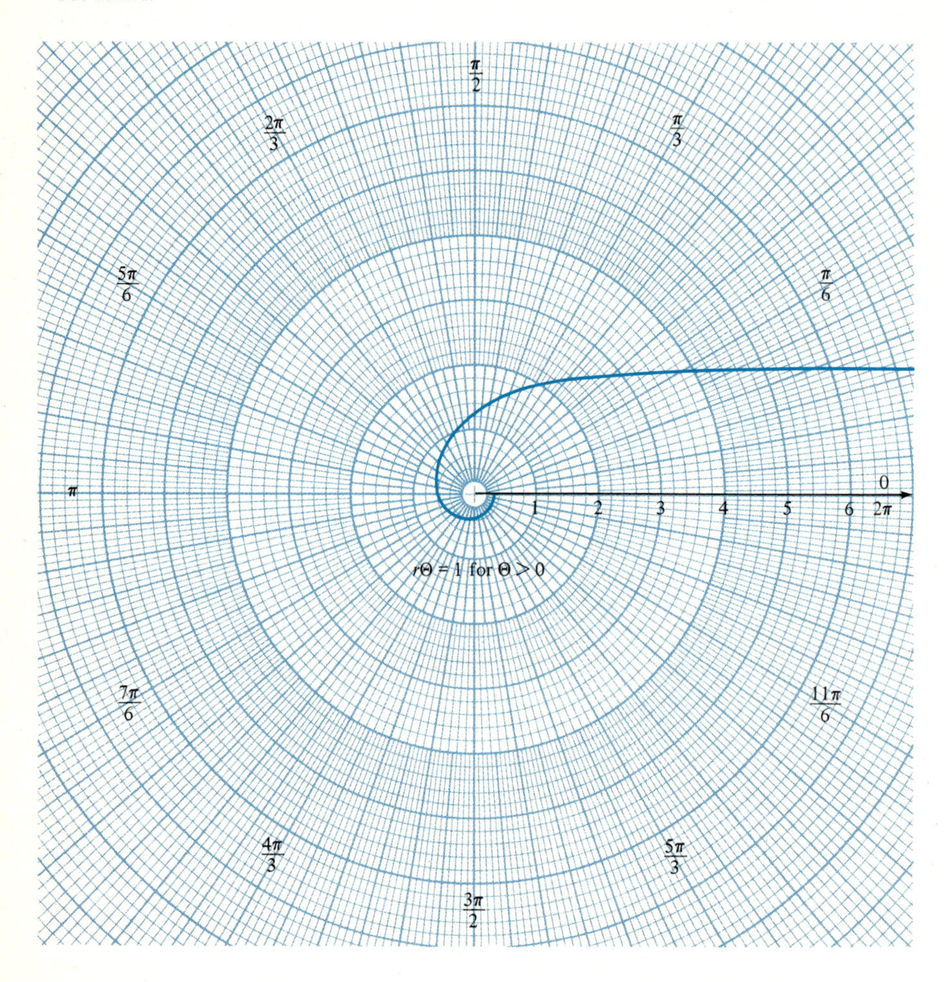

(Graph for $\theta > 0$ only. See Item 2.)

23 1. For $\theta = m\pi/2$, $r = e^{m\pi/2}$, $m = 0, \pm 1, \pm 2, \ldots, r \neq 0$.
2. There is no symmetry.
3. There is no restriction on θ, but r must be positive.
4. As θ increases, r increases. If θ is negative, then as $|\theta|$ increases, r decreases and gets closer and closer to 0.

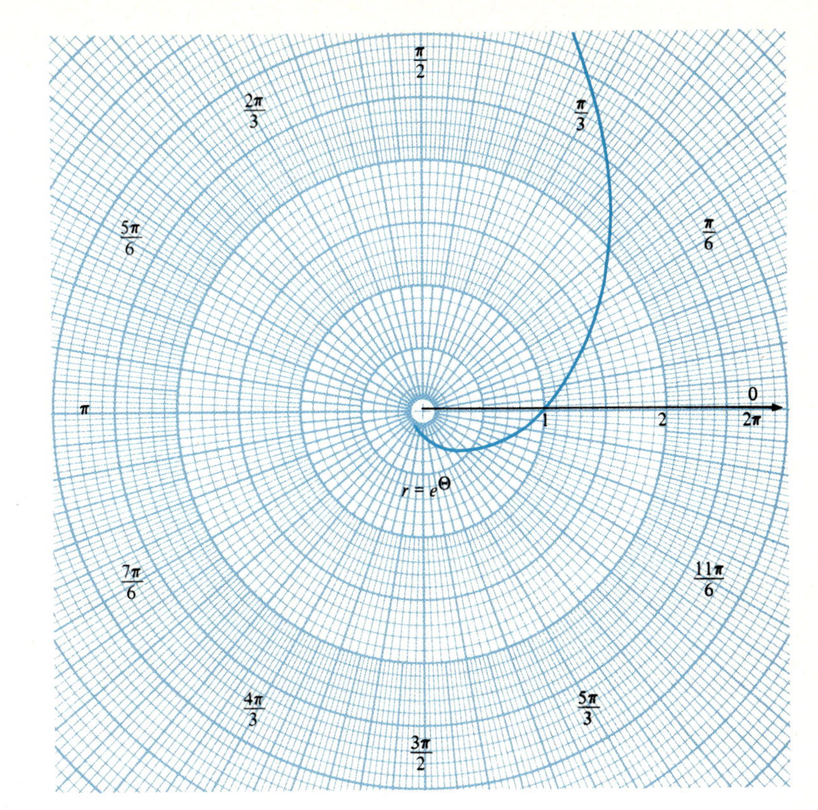

25 $r = \cos 3\theta$
1. For $\theta = 0$, $r = 1$; for $\theta = \pi/2$, $r = 0$; for $\theta = \pi$, $r = -1$. This brings us back to the starting point, and the entire curve is retraced as θ takes on values greater than π. For $r = 0$, $\theta = \pi/6, \pi/2, 5\pi/6, 7\pi/6, 3\pi/2, 11\pi/6$.
2. The graph is symmetric to the x-axis.
3. There is no restriction on θ.
4. As θ increases from 0 to $\pi/6$, r decreases from 1 to 0. As θ increases from $\pi/6$ to $\pi/3$, r decreases from 0 to -1. As θ increases from $\pi/3$ to $\pi/2$, r increases from -1 to 0. The remainder of the graph can be obtained by symmetry. The curve is a three-leaved rose curve.

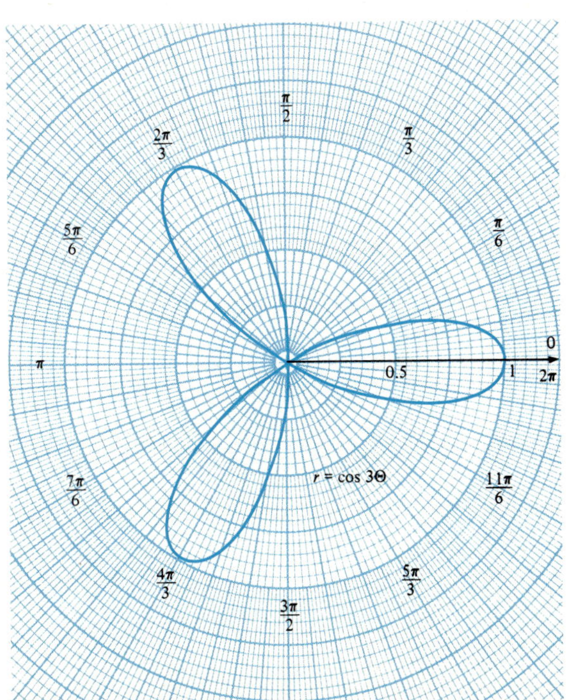

27 Let (r, θ) be any point on the curve $r^2 = a^2 \cos 2\theta$. Let d_1 be the distance from $(a/\sqrt{2}, 0)$ to (r, θ) and d_2 be the distance from $(a/\sqrt{2}, \pi)$ to (r, θ). Then, using the distance formula, we find

$$\begin{aligned}
d_1{}^2 d_2{}^2 &= \left[r^2 + \frac{a^2}{2} - \sqrt{2}ar\cos(\pi - \theta)\right]\left[r^2 + \frac{a^2}{2} - \sqrt{2}ar\cos\theta\right] \\
&= \left(r^2 + \frac{a^2}{2} + \sqrt{2}ar\cos\theta\right)\left(r^2 + \frac{a^2}{2} - \sqrt{2}ar\cos\theta\right) \\
&= \left(r^2 + \frac{a^2}{2}\right)^2 - 2a^2r^2\cos^2\theta \\
&= \left(a^2\cos 2\theta + \frac{a^2}{2}\right)^2 - 2a^4\cos 2\theta\cos^2\theta \qquad r^2 = a^2\cos 2\theta \\
&= a^4[(\cos 2\theta + \tfrac{1}{2})^2 - 2\cos 2\theta\cos^2\theta] \\
&= a^4[(2\cos^2\theta - 1 + \tfrac{1}{2})^2 - 2(2\cos^2\theta - 1)\cos^2\theta] \\
&= a^4[(2\cos^2\theta - \tfrac{1}{2})^2 - 2(2\cos^2\theta - 1)\cos^2\theta] \\
&= a^4(4\cos^4\theta - 2\cos^2\theta + \tfrac{1}{4} - 4\cos^4\theta + 2\cos^2\theta) \\
&= \frac{a^4}{4}
\end{aligned}$$

Thus,

$$d_1 d_2 = \frac{a^2}{2}$$

USING YOUR KNOWLEDGE 11.3

1 Intersections at the pole, at $(\frac{1}{2}, \pi/6)$, and at $(\frac{1}{2}, 5\pi/6)$ **3** Intersections at the pole, at $(1, 0)$, at $(-\frac{1}{2}, 2\pi/3)$, and at $(-\frac{1}{2}, 4\pi/3)$

EXERCISE 11.4

1a $(6, 1)$ **b** $\sqrt{37}$ **3a** $(2, -7)$ **b** $\sqrt{53}$ **5** $(16, -1)$ **7** $(9, -1)$ **9** $\sqrt{229}$
11 $\mathbf{v} = \mathbf{b} - \mathbf{a}$. If the position of $\mathbf{v}$ does not matter, then $\mathbf{v} = \mathbf{u}$. In any event, $\mathbf{v}$ has the same magnitude and direction as $\mathbf{u}$.
13 The zero vector. **15** $6\mathbf{i} + 2\mathbf{j}$ **17** $-8\mathbf{i} + 9\mathbf{j}$ **19** $2\sqrt{10}$ **21** $\sqrt{145}$ **23** $\mathbf{r} = 5\mathbf{i} + 45\sqrt{3}\mathbf{j}$, $|\mathbf{r}| = 10\sqrt{61}$, $\theta = \text{Tan}^{-1}\, 9\sqrt{3} \approx 86.33^\circ$
25 $\mathbf{r} = (-5\sqrt{2}, 35\sqrt{2})$, $|\mathbf{r}| = 50$, $\theta = 180^\circ - \text{Tan}^{-1}\, 7 \approx 98.1^\circ$ **27** $\mathbf{r} = (2.5, 39.6)$, $|\mathbf{r}| = 39.6$, $\theta = 86.3^\circ$
29 The vectors $\mathbf{a}$, $\mathbf{b}$, and $\mathbf{c} = -(\mathbf{a} + \mathbf{b})$ form a closed triangle because their sum is the zero vector. Since one side of a triangle cannot be greater than the sum of the other two sides, it follows that $|\mathbf{c}| \le |\mathbf{a} + \mathbf{b}|$. Equality holds if $\mathbf{a}$ and $\mathbf{b}$ are collinear and have the same sense.

31 $\dfrac{5}{\sqrt{41}}\mathbf{i} + \dfrac{4}{\sqrt{41}}\mathbf{j}$ **33**

$$\begin{aligned}
k(\mathbf{a} + \mathbf{b}) &= k[(a_1, a_2) + (b_1, b_2)] \\
&= k(a_1 + b_1, a_2 + b_2) \\
&= [k(a_1 + b_1), k(a_2 + b_2)] \\
&= (ka_1 + kb_1, ka_2 + kb_2) \\
&= (ka_1, ka_2) + (kb_1, kb_2) \\
&= k(a_1, a_2) + k(b_1, b_2) \\
&= k\mathbf{a} + k\mathbf{b}
\end{aligned}$$

35 $\mathbf{u} = \dfrac{2}{\sqrt{13}}\mathbf{i} + \dfrac{3}{\sqrt{13}}\mathbf{j}$ or $\mathbf{u} = -\dfrac{2}{\sqrt{13}}\mathbf{i} - \dfrac{3}{\sqrt{13}}\mathbf{j}$

USING YOUR KNOWLEDGE 11.4

1 $-4\mathbf{i} + 2\mathbf{j}$

EXERCISE 11.5

1 $\sqrt{113}$ lb, making an angle of 41.2° with the positive x direction **3** 11.74 lb, S 83.1° E
5 $5\sqrt{394}$ lb, downward on a line of slope $-\frac{13}{15}$ **7** 701.8 mph, S 85.9° W **9** 43.3 lb **11** $10\sqrt{61}$ lb, N 50.2° E **13** N 76.4° E
15 5196 lb, right-hand cable; 3000 lb, left-hand cable **17a** 1.06 tons **b** 0.34 ton

USING YOUR KNOWLEDGE 11.5

About 999 lb

SELF-TEST, CHAPTER 11

1a $\frac{\pi}{4}$ **b** $-\frac{\pi}{4}$ **c** 0.95 **d** -0.95 **2a** $\frac{\pi}{3}$ **b** $\frac{2\pi}{3}$ **c** 0.55 **d** 2.59 **3a** $\frac{\pi}{6}$ **b** $-\frac{\pi}{6}$ **c** 1.30 **d** -1.30 **4a** $-2\sqrt{5}/5$

b No. $-\pi/2 \le \text{Sin}^{-1}(\sin x) \le \pi/2$, but x does not have to be in this interval. For example, if $x = 5\pi/6$, then $\sin x = \frac{1}{2}$ and $\text{Sin}^{-1}\frac{1}{2} = \pi/6 \ne 5\pi/6$.

5 $\text{Tan}^{-1}\left(\frac{x-y}{x+y}\right)$ **6a** $\left(2, \frac{2\pi}{3}\right)$ and $\left(-2, \frac{5\pi}{3}\right)$ **b** $\left(2, 2m\pi + \frac{2\pi}{3}\right)$ and $\left(-2, 2m\pi + \frac{5\pi}{3}\right)$, $m = 0, \pm 1, \pm 2, \ldots$ **7a** $\left(-3, \frac{7\pi}{6}\right)$ **b** 7 units

8a $\left(2, \frac{2\pi}{3}\right)$ or $\left(-2, \frac{5\pi}{3}\right)$ **b** $(1, -\sqrt{3})$ **9a** $r = 2\cos\theta$ **b** $(x^2 + y^2)^2 = x^2 - y^2$ **10** See Problem 19, Exercise 11.3.

11 1. The graph crosses the axes at $(1, 0)$, $(3, \pi/2)$, $(5, \pi)$, and $(3, 3\pi/2)$. The entire curve is traced out as θ goes from 0 to 2π. Because $3 - 2\cos\theta$ cannot be zero, the curve does not pass through the origin.
2. If we replace θ by $-\theta$, the equation is unchanged. Thus, the curve is symmetric to the x-axis. There is no other symmetry.
3. All values of θ are permissible.
4. As θ increases from 0 to $\pi/2$, r increases from 1 to 3. As θ increases from $\pi/2$ to π, r increases from 3 to 5. As θ increases from π to $3\pi/2$, r decreases from 5 to 3. As θ increases from $3\pi/2$ to 2π, r decreases from 3 to 1.

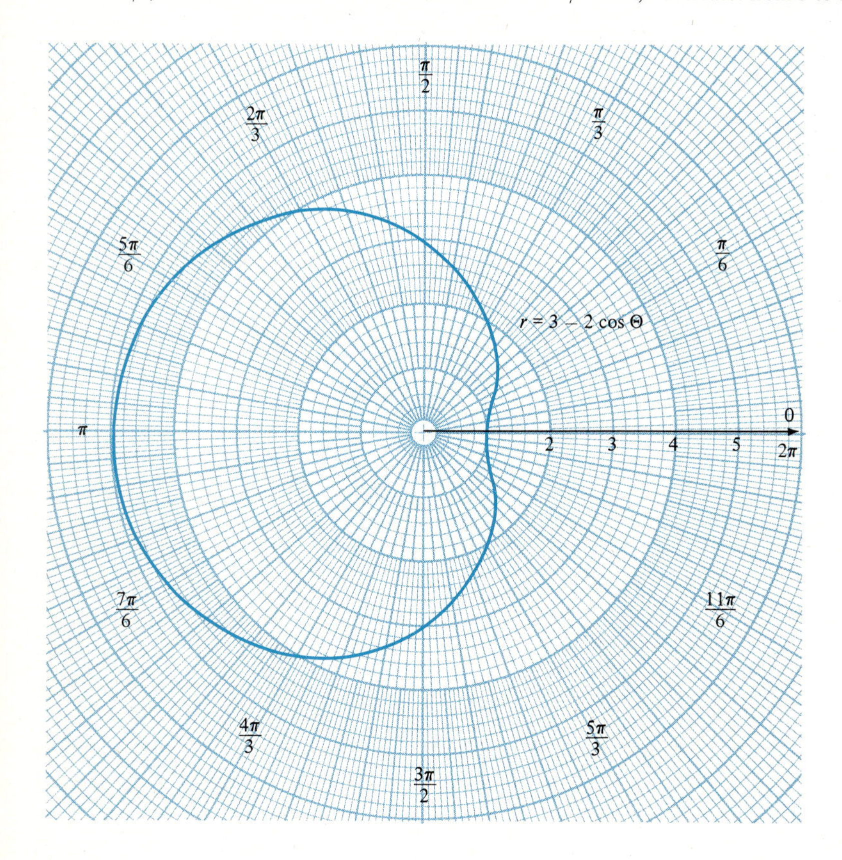

12 1. The graph crosses the axes at $(1, 0)$, $(-1, \pi/2)$, $(1, \pi)$, and $(3, 3\pi/2)$. The entire curve is traced out as θ goes from 0 to 2π. The curve passes through the origin at $\theta = \pi/6$ and $\theta = 5\pi/6$.

2. If we replace θ by $\pi - \theta$, the equation is unchanged. Thus, the curve is symmetric to the y-axis. There is no other symmetry.
3. All values of θ are permissible.
4. As θ increases from 0 to $\pi/6$, r decreases from 1 to 0. As θ increases from $\pi/6$ to $\pi/2$, r decreases from 0 to -1. As θ increases from $\pi/2$ to $5\pi/6$, r increases from -1 to 0. As θ increases from $5\pi/6$ to π, r increases from 0 to 1. As θ increases from π to $3\pi/2$, r increases from 1 to 3. As θ increases from $3\pi/2$ to 2π, r decreases from 3 to 1.

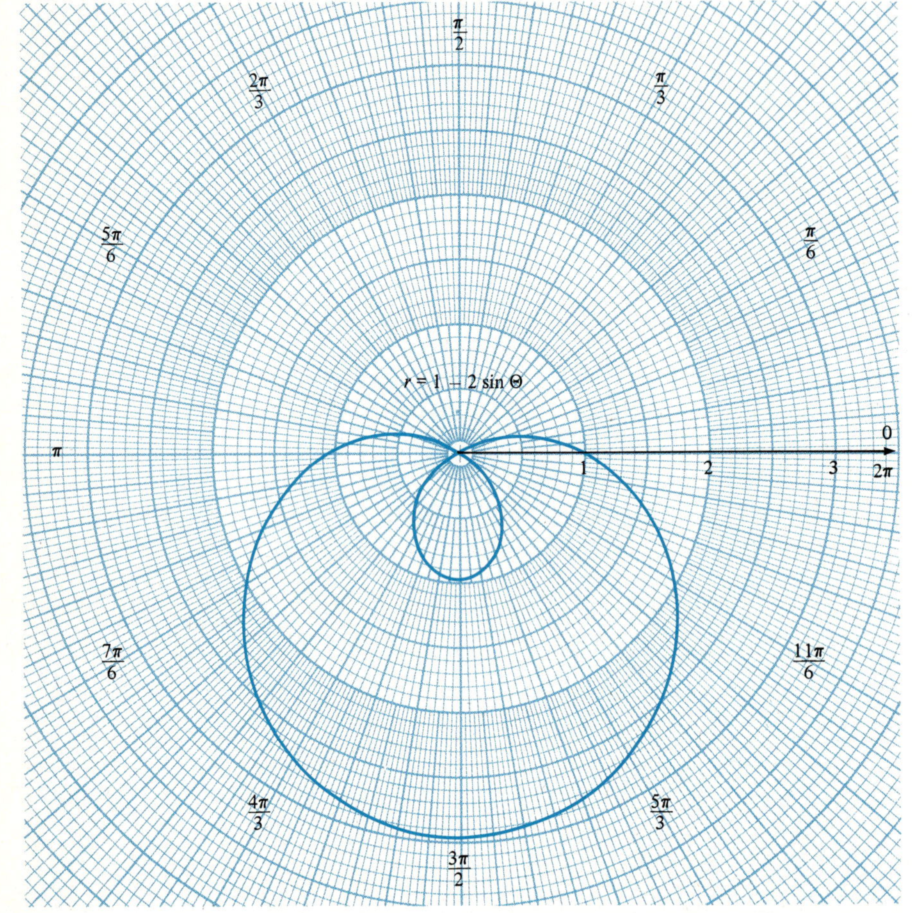

13a $(-4, 3)$ **b** 5 **14a** 5, 126.87° **b** $\sqrt{5}$, 63.43° **15a** $-2\mathbf{i} + 7\mathbf{j}$ **b** $2\sqrt{2}$ **c** $\sqrt{13} + \sqrt{17}$ **16** 75 yd **17** N 67.4° E **18** $100\sqrt{5}$ m, S 18.4° W **19** 11.2 lb, S 63.4° W **20** 500 lb in the right-hand cable and $500\sqrt{3}$ lb in the left-hand cable

EXERCISE 12.1

1 $(2, 1)$ **3** $(5, 3)$ **5** Inconsistent, no solution **7** $(3, -4)$ **9** $\left(k, \dfrac{4k-8}{3}\right)$, dependent **11** $(5, -2)$ **13** $(-\frac{1}{2}, -\frac{1}{3})$

15 Inconsistent, no solution **17** $(1, 2)$ **19** Inconsistent, no solution **21** $(15, 12)$ **23** $(-3, -4)$ **25** $(8, 20)$ **27** $(2, 4)$ **29** $(u, v) = (\frac{5}{3}, \frac{1}{3})$ **31** $(7, -4)$ **33** $(b + a, b - a)$ **35** Antenna 222 ft; building 1250 ft **37** 8600 million nonreturnables, 460 million returnables **39** $\frac{3}{16}$ **41** 30 miles; 40 minutes

USING YOUR KNOWLEDGE 12.1

1 $6d + 4 = \frac{5}{9}(\frac{1}{4}n + 40 - 32)$ **3** $2\frac{2}{3}$ cm/sec

EXERCISE 12.2

1 (2, 3, −4) **3** (1, 0, −2) **5** Inconsistent, no solution **7** Inconsistent, no solution **9** (0, 10, −19) **11** (1, 2, 3)
13 $(k, 2k-3, 5k-7)$ **15** (2, 4, 5) **17** Inconsistent, no solution **19** $(-3, 4, \frac{3}{2})$ **21** (6, 12, 18) **23** (2, 4, −10) **25** $(\frac{9}{2}, \frac{1}{2}, \frac{5}{2})$
27 (−2, −3, −4) **29** $(-\frac{1}{20}, \frac{1}{3}, \frac{1}{35})$ **31** $2a + b - c = 0$ **33** $k = 7$ **35** 210 dimes, 60 quarters, 24 half-dollars
37 A: 40 days, B: 120 days, C: 60 days **39** $a = -\frac{3}{8}, b = \frac{11}{4}, c = -5$

USING YOUR KNOWLEDGE 12.2

1 (2.57, −6.29, 5.69)

EXERCISE 12.3

For answers to Problems 1–20, see Exercise 12.2.
21 (−1, −1, 3) **23** Inconsistent, no solution **25** Inconsistent, no solution

USING YOUR KNOWLEDGE 12.3

1 (1, −2, 3, 2)

EXERCISE 12.4

1 1 **3** $\frac{1}{24}$ **5** $-\frac{17}{12}$ **7** $-\frac{3}{10}$ **9** −1 **11** −4 **13** 77 **15** $a - a^3$ **17** 166 **29** $x = -2, x = 1$ **31** $x = \pm 2$ **33** $x = -3$

USING YOUR KNOWLEDGE 12.4

1 $2x - y + 3 = 0$ **3** $2x + 9y - 34 = 0$ **5** $bx + ay - ab = 0$

EXERCISE 12.5

1 (5, −2) **3** $(-\frac{1}{2}, -\frac{1}{3})$ **5** Inconsistent, no solution **7** (4, 9) **9** (2, 3, −4) **11** (1, 0, −2) **13** Inconsistent, no solution
15 Inconsistent, no solution **17** (−1, −1, 3) **19** Inconsistent, no solution **21** (5, 2, −4) **23** $w = 2, x = -1, y = -2, z = 3$
25 The system reduces to

$$\begin{aligned} x + 2y - z &= 4 \\ -5y + 3z &= -2 \\ 0z &= 0 \end{aligned}$$

and has the solutions

$$\left(\frac{16-k}{5}, \frac{2+3k}{5}, k\right) \quad \text{where } k \text{ is any number}$$

USING YOUR KNOWLEDGE 12.5

1 For $k = 6$, the system has the solution (3, −5).

SELF-TEST, CHAPTER 12

1 (2, −4) **2** (−7, 5) **3** $(k, 2k + 2)$, k any number **4** Inconsistent, no solution **5** (−5, 3) **6** (1, 2, −1)
7 $(k, -2k, k)$, k any number **8** 15 pennies, 5 nickels, 6 dimes **9** (2, −3, −2) **10** $(3k + 4, -2k - 7, k)$, k any number
11 (5, −2, −5) **12** Inconsistent, no solution **13** 10 **14a** $0(0 + 5) - (-1)(-3 + 4) + 4(3 - 0) = 13$
b $-0(8 - 0) + 3(4 - 0) - (-1)(-3 + 4) = 13$ **15** 24 **16** $k = -\frac{1}{2}, k = 2$ **17** (−1, 2)
18 $D = 0$, $D_x = 22$. Therefore, the system is inconsistent. **19** $(-2, \frac{17}{3}, -\frac{5}{3})$ **20** $x = -2, x = 4$

EXERCISE 13.1

1a $8 - i$ **b** $-2 + 5i$ **c** $21 + i$ **d** $\frac{9}{34} + \frac{19}{34}i$ **3a** $10 + 6\sqrt{2}i$ **b** $-4 + 4\sqrt{2}i$ **c** $11 + 38\sqrt{2}i$ **d** $\dfrac{31}{51} + \dfrac{32\sqrt{2}}{51}i$ **5a** $5 + 4i$
b $1 - 2i$ **c** $3 + 11i$ **d** $\frac{9}{13} - \frac{7}{13}i$ **7a** $-5 + 19i$ **b** $-1 - i$ **c** $-84 - 48i$ **d** $\frac{12}{13} + \frac{3}{26}i$ **9a** $5 - 2\sqrt{3}i$ **b** $-1 + 4\sqrt{3}i$ **c** $15 - 3\sqrt{3}i$
d $-\frac{1}{12} + \frac{1}{4}\sqrt{3}i$ **11a** $3 - \frac{7}{2}i$ **b** $-3 + \frac{1}{2}i$ **c** $-3 - \frac{9}{2}i$ **d** $\frac{3}{13} - \frac{9}{26}i$ **13** $x = -3 \pm i$ **15** $x = 1 \pm \frac{1}{2}\sqrt{2}i$ **17** $t = \frac{1}{3} \pm \frac{1}{3}\sqrt{2}i$ **19** −2
21 0 **23** 1 **25** $2 + 2i; 2\sqrt{2}$ **27** $(\sqrt{6} + 1) + (\sqrt{3} - \sqrt{2})i; 2\sqrt{3}$ **29** $1 - \sqrt{3}i; 2$ **31** $4 + 2i; 2\sqrt{5}$ **33** $-\frac{6}{5}; \frac{6}{5}$
35 $-16 + 16i; 16\sqrt{2}$ **37** $P(1 + 2i) = -11 - 2i; P(1 - 2i) = -11 + 2i$ **39** $-5 - 23i$ **43** $x = -1, y = 2$ **45** $x = \frac{3}{2}, y = -\frac{5}{2}$

47 $x = -1, y = 2$ or $x = 3, y = -5$ **49** Because $|c + di| = |c - di| = \sqrt{c^2 + d^2}$, we have **51** $-0.311 \pm 1.144i$ **53** $1.043 \pm 1.841i$

$$\left|\frac{a + bi}{c + di}\right| = \left|\frac{(a + bi)(c - di)}{c^2 + d^2}\right|$$
$$= \frac{|a + bi|\,|c - di|}{c^2 + d^2}$$
$$= \frac{|a + bi|\,|c + di|}{|c + di|^2}$$
$$= \frac{|a + bi|}{|c + di|}$$

USING YOUR KNOWLEDGE 13.1

3 The vector from the marker to point Q is
$-i[(-1 - a) - bi] = -b + (1 + a)i$

EXERCISE 13.2

1 8 cis 180°, 8 cis(180° + $k \cdot 360°$) **3** 6 cis 315°, 6 cis(315° + $k \cdot 360°$) **5** 20 cis 300°, 20 cis(300° + $k \cdot 360°$) **7** 5 cis 323.13°, 5 cis(323.13° + $k \cdot 360°$) **9** 13 cis 157.38°, 13 cis(157.38° + $k \cdot 360°$) **11** $3 + 3\sqrt{3}i$ **13** $-17i$ **15** $-3 + \sqrt{3}i$ **17** $1.37 + 3.76i$ **19** $7.73 - 2.07i$ **21** $6\sqrt{3} + 6i$ **23** $2\sqrt{3} + 2i$ **25** $-8 + 8\sqrt{3}i$ **27** $-32i$ **29** $\frac{1}{2} + \frac{1}{2}\sqrt{3}i$ **31** i **33** $-16\sqrt{3} + 16i$

35 2 cis 60°, 2 cis 240° **37** 2 cis 45°, 2 cis 165°, 2 cis 285° **39** 2 cis 60°, 2 cis 150°, 2 cis 240°, 2 cis 330°

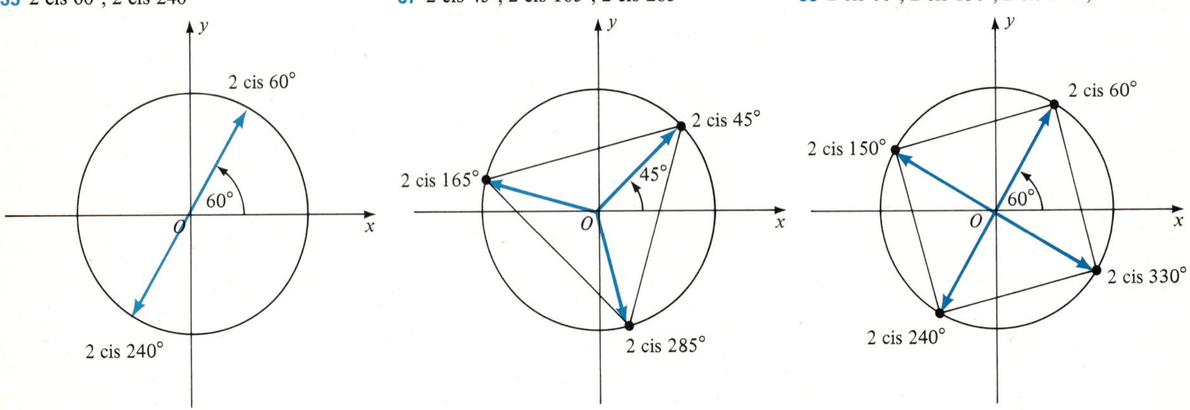

41 2 cis 45°, 2 cis 117°, 2 cis 189°, 2 cis 261°, 2 cis 333° **43** cis 45°, cis 165°, cis 285°

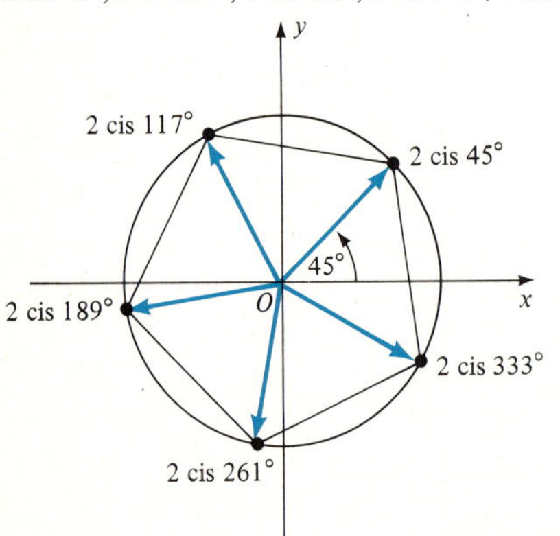

y
cis 45°
cis 165°
O
x
cis 285°

45 2 cis 60°, 2 cis 150°, 2 cis 240°, 2 cis 330°

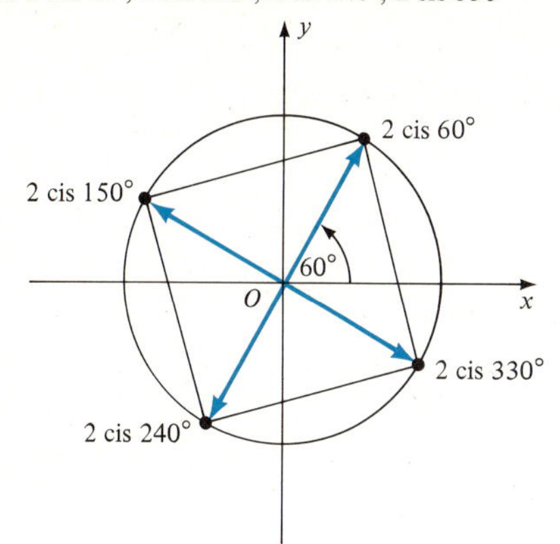

47 2 cis 60°, 2 cis 132°, 2 cis 204°, 2 cis 276°, 2 cis 348°

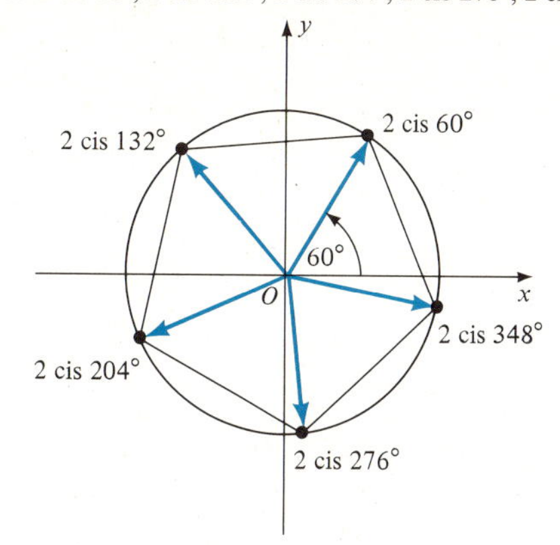

49 $3 \text{ cis } 0° = 3$
$3 \text{ cis } 120° = -\frac{3}{2} + \frac{3\sqrt{3}}{2}i$
$3 \text{ cis } 240° = -\frac{3}{2} - \frac{3\sqrt{3}}{2}i$

51 $3 \text{ cis } 90° = 3i$
$3 \text{ cis } 210° = -\frac{3\sqrt{3}}{2} - \frac{3}{2}i$
$3 \text{ cis } 330° = \frac{3\sqrt{3}}{2} - \frac{3}{2}i$

53 $\text{cis } 0° = 1$
$\text{cis } 72° = 0.309 + 0.951i$
$\text{cis } 144° = -0.809 + 0.588i$
$\text{cis } 216° = -0.809 - 0.588i$
$\text{cis } 288° = 0.309 - 0.951i$

55 $2 \text{ cis } 0° = 2$
$2 \text{ cis } 60° = 1 + \sqrt{3}i$
$2 \text{ cis } 120° = -1 + \sqrt{3}i$
$2 \text{ cis } 180° = -2$
$2 \text{ cis } 240° = -1 - \sqrt{3}i$
$2 \text{ cis } 300° = 1 - \sqrt{3}i$

USING YOUR KNOWLEDGE 13.2

1 $I_s = 2.999 \cos(400t - 2.889)$

EXERCISE 13.3

1 Degree 4, $a_4 = 5$, $a_3 = -7$, $a_2 = 0$, $a_1 = 20$, $a_0 = -3$
3 Degree 7, $a_7 = \sqrt{2}$, $a_6 = 0$, $a_5 = -3$, $a_4 = 0$, $a_3 = 0$, $a_2 = 0$, $a_1 = 3\sqrt{2}$, $a_0 = -4$ **5** $x^2 + 3x + 1$, remainder 0
7 $x^2 + 6x + 5$, remainder 15 **9** $2y^3 + y^2 + 3y + 3$, remainder 0 **11** $x^3 + (4 + \sqrt{2})x^2 + (4\sqrt{2} - 7)x - 7\sqrt{2}$, remainder 0
13 $x^3 + bx^2 + 5b^2x + 3b^3$, remainder 0 **15** $2x^2 + (3 + 2i)x + 3i$, remainder 0
17 $P(-2) = -32 + 32 = 0$. Therefore, $x + 2$ is a factor of $x^5 + 32$.
19 $P(-2) = -8 + 22 - 14 = 0$. Therefore, $x + 2$ is a factor of $x^3 - 11x - 14$.
21 $P(-1) = -1 + 1 - 2 + 5 + 2 - 5 = 0$. Therefore, $y + 1$ is a factor of $y^5 + y^4 + 2y^3 + 5y^2 - 2y - 5$.
23 $P(-i) = -1 + 1 = 0$. Therefore, $z + i$ is a factor of $z^6 + 1$. **25** $P(2) = -2$, $P(3) = 64$
27 $P(6) = -591$, $P(-\frac{1}{2}) = -\frac{191}{8} = -23.875$ **29** $P(2i) = -11 - 12i$, $P(-i) = -2 + 3i$ **31** $\frac{1}{5} \pm \frac{1}{5}\sqrt{34}i$ **33** $-5 \pm 3i$ **35** $\frac{2}{5} \pm \frac{1}{5}\sqrt{11}i$

USING YOUR KNOWLEDGE 13.3

1 -6.541, -0.459

EXERCISE 13.4

1 $x^4 - 6x^3 + 3x^2 + 26x - 24 = 0$ **3** $x^5 - 7x^3 - 18x = 0$ **5** $x^4 - 7x^2 - 10x - 4 = 0$ **7** $x^5 + x^4 + 8x^3 + 8x^2 + 16x + 16 = 0$
9 $-1, 2, 3$ **11** $3, -3 \pm 2i$ **13** $\frac{5}{3}, -4 \pm \sqrt{2}i$ **15** $-3, 2, 2, 3$ **17** $-1, -\frac{1}{3}, 1 \pm \sqrt{2}i$ **19** $5, 5, -\frac{1}{2} \pm \frac{1}{2}\sqrt{3}i$ **21** $-3, \frac{1}{2}, \frac{3}{2}, \pm 2i$

23 $x^3 - 2x^2 + 5x + 26 = 0$ **25** $x^4 - 2x^3 + 9x^2 - 8x + 20 = 0$ **27** $x^3 - (2 + \sqrt{3})x^2 + (2 + 2\sqrt{3})x - 2\sqrt{3} = 0$ **29** $2 - 3i, \frac{2 \pm \sqrt{13}}{3}$

31 4.5 ft
33 Let r_1 denote a root whose existence is assured by the Fundamental Theorem of Algebra. Then, by the Factor Theorem, $x - r_1$ is a factor of $P(x)$, so that

$$P(x) = (x - r_1)P_1(x)$$

where $P_1(x)$ is a polynomial of degree $n - 1$ in x.

By the Fundamental Theorem, $P_1(x) = 0$ also has a root, say r_2, so that $x - r_2$ is a factor of $P_1(x)$ and

$P_1(x) = (x - r_2)P_2(x)$

where $P_2(x)$ is a polynomial of degree $n - 2$ in x.

Thus, the original polynomial may be written

$P(x) = (x - r_1)(x - r_2)P_2(x)$

Using this procedure n times, we obtain $P(x)$ in the form

$P(x) = (x - r_1)(x - r_2) \cdots (x - r_n)P_n(x)$

where $P_n(x)$ must be of degree $n - n = 0$; that is, $P_n(x)$ is a constant. By comparison with $P(x) = a_n x^n + \cdots + a_0$, we see that $P_n(x) = a_n$. Hence,

$P(x) = a_n(x - r_1)(x - r_2) \cdots (x - r_n)$

as was to be shown.

It follows that the equation $P(x) = 0$ has n roots, for $P(x)$ has the value zero if x equals any of the numbers $r_1, r_2, \ldots, r_n$. Furthermore, the equation cannot have more than n roots, because, if it did, then $P(x)$ would have more than n first-degree factors, one for each root. In that case, $P(x)$ would be of degree higher than n, which contradicts the hypothesis that $P(x)$ is of degree n. This completes the proof of the theorem.

35 Let $P(x) = a_m x^{2m} + a_{m-1}x^{2m-2} + \cdots + a_1 x^2 + a_0$. Then, if r is a real number, $P(r) \neq 0$, because all of the terms are positive numbers and the sum of positive numbers cannot be zero. Thus, all the roots are imaginary, and because these roots must occur in conjugate pairs, the equation has exactly m pairs of conjugate imaginary roots.

37 Let $Q(x) = (x - a - b\sqrt{c})(x - a + b\sqrt{c}) = x^2 - 2ax + a^2 - b^2c$, where a, b, c are rational numbers and c is not a perfect square. Divide $P(x)$ by $Q(x)$ to obtain

$$P(x) = P_1(x)Q(x) + Ax + B \qquad (1)$$

where A and B are rational numbers. Since $P(a + b\sqrt{c}) = 0$ and $Q(a + b\sqrt{c}) = 0$, we can substitute $x = a + b\sqrt{c}$ into Equation 1 to obtain

$0 = 0 + A(a + b\sqrt{c}) + B$

Since A and B are rational numbers, the preceding equation gives

$Ab\sqrt{c} = 0 \quad \text{and} \quad Aa + B = 0$

Because $b\sqrt{c} \neq 0$, we have $A = 0$, and from $Aa + B = 0$, we have $B = 0$. Thus, $P(x) = P_1(x)Q(x)$, so that

$P(a - b\sqrt{c}) = P_1(a - b\sqrt{c})Q(a - b\sqrt{c})$

Because $P_1(a - b\sqrt{c})$ is simply some real number and $Q(a - b\sqrt{c}) = 0$, we see that $P(a - b\sqrt{c}) = 0$, which shows that $a - b\sqrt{c}$ is a root of the equation $P(x) = 0$.

39 $-\sqrt{2}, -\frac{1}{2} \pm \frac{1}{2}\sqrt{3}i$

USING YOUR KNOWLEDGE 13.4

1

+	−	i
0	1	2

3

+	−	i
3	1	0
1	1	2

5

+	−	i
2	2	2
2	0	4
0	2	4
0	0	6

7

+	−	i
1	0	2m

EXERCISE 13.5

1 0.68 **3** 1.13 **5** 0.19 **7** 1.76 **9** −2.817 **11** 5.814 **13** −1.784, 2.389, 5.395 **15** 2.146 **17** 1.837 ft by 2.837 ft by 3.837 ft

USING YOUR KNOWLEDGE 13.5

1 $\dfrac{h}{R} = 0.653$

SELF-TEST, CHAPTER 13

1a $-\frac{27}{13} + \frac{8}{13}i$ **b** $-2 - i$ **2** $-11 + 2i$ **3** $-\frac{1}{5} - \frac{2}{5}i$ **4** $x = -1, y = 1$ **5a** $20 \text{ cis } 240° = -10 - 10\sqrt{3}i$ **b** $-\frac{2\sqrt{3}}{3} + \frac{2}{3}i$

6a $64 \text{ cis } 180° = -64 + 0i = -64$ **b** $32 \text{ cis } 240° = -16 - 16\sqrt{3}i$ **7** $2 \text{ cis } 10°, 2 \text{ cis } 130°, 2 \text{ cis } 250°$ **8** $2i, -\sqrt{3} - i, \sqrt{3} - i$

9
$$\begin{array}{r|rrrrr} -2 & 3 & -1 & 0 & +5 & -20 \\ & & -6 & +14 & -28 & +46 \\ \hline & 3 & -7 & +14 & -23 & +26 \end{array}$$
Quotient is $3x^3 - 7x^2 + 14x - 23$.
Remainder is 26.

10
$$\begin{array}{r|rrrr} -6 & 2 & 0 & +3 & -5 \\ & & -12 & +72 & -450 \\ \hline & 2 & -12 & +75 & -455 \end{array}$$
$P(-6) = -455$

11
$$\begin{array}{r|rrrrrr} -2 & 3 & +2 & 0 & -1 & -32 & +6 \\ & & -6 & +8 & -16 & +34 & -4 \\ \hline & 3 & -4 & +8 & -17 & +2 & +2 \end{array}$$
Since the remainder is not zero, $x + 2$ is not a factor.

12 175 **13**
$$\begin{array}{r|rrrr} -1 + 2i & 1 & +1 & +2 & -2 \\ & & -1 + 2i & -4 - 2i & +6 - 2i \\ \hline & 1 & +2i & -2 - 2i & +4 - 2i \end{array}$$
Therefore, $P(-1 + 2i) = 4 - 2i$.

14 $x^4 - 2x^3 - 2x^2 + 6x + 5 = 0$ **15** $1 - i, i, -i$ **16a** Yes. For example, $2x^2 - 9x + 4 = 0$ has $\frac{1}{2}$ as one of its roots. **b** No
17 $-\frac{1}{3}, \frac{1}{2}$ **18a** If x is a real number, then x^4 and x^2 are both positive numbers, and $x^4 + x^2 + 1 \neq 0$. **b** One **19** 1.23 **20** 1.327

EXERCISE 14.1

1 2, 4, 8, 16; $a_{10} = 1024$ **3** $-5, 5, -5, 5$; $a_{10} = 5$ **5** 0, 1, 0, 1; $a_{10} = 1$ **7** $\frac{1}{8}, \frac{1}{8}, \frac{1}{8}, \frac{1}{8}$; $a_{10} = \frac{1}{8}$ **9** $\frac{3}{4}, \frac{3}{8}, \frac{3}{16}, \frac{3}{32}$; $a_{10} = \frac{3}{2048}$
11 1, 2, 5, 14; $a_{10} = 9842$ **13** $1, k, k^2, k^3$; $a_{10} = k^9$ **15** $2^n, 2^n + (n - 1)(n - 2)(n - 3)$ (Other answers are possible.)
17 $\frac{1}{2^n}, \frac{1}{2^n} + (n - 1)(n - 2)(n - 3)$ (Other answers are possible.) **19a** 8, 13, and 21

USING YOUR KNOWLEDGE 14.1

1 Yes. You get 21.

EXERCISE 14.2

1 91 **3** 100 **5** 21 **7** $\sum_{k=1}^{200} k$ **9** $\sum_{k=1}^{50} \frac{1}{k}$ **11** $\sum_{k=1}^{50} (-1)^{k-1}k$ **13** $\sum_{k=1}^{5} (5k - 4)$

15 1. For $n = 1$, we have $1 = 1^2$, which is true.
2. Assume that the formula is true for $n = k$. Then,

$$1 + 3 + 5 + \cdots + (2k - 1) = k^2$$

We then wish to show that

$$1 + 3 + 5 + \cdots + (2k - 1) + (2k + 1) = (k + 1)^2$$

The assumed formula sums the first k terms on the left, so that

$$\begin{aligned} 1 + 3 + 5 + \cdots + (2k - 1) + (2k + 1) &= k^2 + (2k + 1) \\ &= k^2 + 2k + 1 \\ &= (k + 1)^2 \end{aligned}$$

Thus, the formula is true for $n = k + 1$ if it is true for $n = k$. This completes the proof.

17 1. For $n = 1$, we have

$$1 = \frac{a - 1}{a - 1} \qquad a \neq 1$$

which is true.
2. Assume that the formula is true for $n = k$. Then

$$1 + a + a^2 + \cdots + a^{k-1} = \frac{a^k - 1}{a - 1}$$

We wish to show that

$$1 + a + a^2 + \cdots + a^{k-1} + a^k = \frac{a^{k+1} - 1}{a - 1}$$

The assumed formula sums the first k terms on the left, so that

$$\begin{aligned} 1 + a + a^2 + \cdots + a^{k-1} + a^k &= \frac{a^k - 1}{a - 1} + a^k \\ &= \frac{a^k - 1 + a^{k+1} - a^k}{a - 1} \\ &= \frac{a^{k+1} - 1}{a - 1} \end{aligned}$$

Thus, the formula is true for $n = k + 1$ if it is true for $n = k$. This completes the proof.

19 1. For $n = 1$, we have

$$\sum_{i=1}^{1} i^2 = 1^2 = \tfrac{1}{6}(1)(2)(3)$$

which is true.

2. Assume that the formula is true for $n = k$. Then,

$$\sum_{i=1}^{k} i^2 = \tfrac{1}{6}k(k+1)(2k+1)$$

We wish to show that

$$\sum_{i=1}^{k+1} i^2 = \tfrac{1}{6}(k+1)(k+2)(2k+3)$$

The assumed formula sums the first k terms on the left to give

$$\begin{aligned}\sum_{i=1}^{k+1} i^2 &= \sum_{i=1}^{k} i^2 + (k+1)^2\\ &= \tfrac{1}{6}k(k+1)(2k+1) + (k+1)^2\\ &\tfrac{1}{6}(k+1)[k(2k+1) + 6(k+1)]\\ &= \tfrac{1}{6}(k+1)(2k^2+7k+6)\\ &= \tfrac{1}{6}(k+1)(k+2)(2k+3)\end{aligned}$$

Thus, the formula is true for $n = k + 1$ if it is true for $n = k$. This completes the proof.

21 1. For $n = 1$, we have

$$\sum_{j=1}^{1} 3j(j+2) = (3)(1)(3) = \tfrac{1}{2}(1)(2)(9)$$

which is true.

2. Assume the formula is true for $n = k$. Then,

$$\sum_{j=1}^{k} 3j(j+2) = \tfrac{1}{2}k(k+1)(2k+7)$$

We wish to show that

$$\sum_{j=1}^{k+1} 3j(j+2) = \tfrac{1}{2}(k+1)(k+2)(2k+9)$$

The assumed formula sums the first k terms on the left, giving

$$\begin{aligned}\sum_{j=1}^{k+1} 3j(j+2) &= \sum_{j=1}^{k} 3j(j+2) + 3(k+1)(k+3)\\ &= \tfrac{1}{2}k(k+1)(2k+7) + 3(k+1)(k+3)\\ &= \tfrac{1}{2}(k+1)[k(2k+7) + 6(k+3)]\\ &= \tfrac{1}{2}(k+1)(2k^2+13k+18)\\ &= \tfrac{1}{2}(k+1)(k+2)(2k+9)\end{aligned}$$

Thus, the formula is true for $n = k + 1$ if it is true for $n = k$. This completes the proof.

23 1. For $n = 1$, we have

$$\sum_{j=1}^{1} j2^{j-1} = (1)(2^0) = 1 + (0)(2^0)$$

which is true.

2. Assume the formula is true for $n = k$. Then,

$$\sum_{j=1}^{k} j2^{j-1} = 1 + (k-1)2^k$$

We wish to show that

$$\sum_{j=1}^{k+1} j2^{j-1} = 1 + k \cdot 2^{k+1}$$

The assumed formula sums the first k terms on the left, giving

$$\begin{aligned}\sum_{j=1}^{k+1} j2^{j-1} &= \sum_{j=1}^{k} j2^{j-1} + (k+1)2^k\\ &= 1 + (k-1)2^k + (k+1)2^k\\ &= 1 + 2k \cdot 2^k = 1 + k \cdot 2^{k+1}\end{aligned}$$

Thus, the formula is true for $n = k + 1$ if it is true for $n = k$. This completes the proof.

25 1. For $n = 1$, we have $x - y = (x - y) \cdot 1$, which is true.

2. Assume the result is true for $n = k$. Then,

$$x^k - y^k = (x - y)Q_1$$

where Q_1 is a polynomial in x and y. We wish to show that

$$x^{k+1} - y^{k+1} = (x - y)Q_2$$

where Q_2 is a polynomial in x and y. We write

$$\begin{aligned}x^{k+1} - y^{k+1} &= (x^{k+1} - xy^k) + (xy^k - y^{k+1})\\ &= x(x^k - y^k) + (x - y)y^k\\ &= x(x - y)Q_1 + (x - y)y^k\\ &= (x - y)(xQ_1 + y^k)\end{aligned}$$

Because $xQ_1 + y^k$ is a polynomial in x and y, we may write

$$x^{k+1} - y^{k+1} = (x - y)Q_2$$

where $Q_2 = xQ_1 + y^k$. This shows that the result is true for $n = k + 1$ if it is true for $n = k$. The proof is thus complete.

USING YOUR KNOWLEDGE 14.2

1 For $n = 1$, we have

$$a_1 = \frac{1}{\sqrt{5}}\left[\left(\frac{1+\sqrt{5}}{2}\right) - \left(\frac{1-\sqrt{5}}{2}\right)\right] = \left(\frac{1}{\sqrt{5}}\right)(\sqrt{5}) = 1$$

which is the first term of the Fibonacci sequence. For $n = 2$, we have

$$\begin{aligned}a_2 &= \frac{1}{\sqrt{5}}\left[\left(\frac{1+\sqrt{5}}{2}\right)^2 - \left(\frac{1-\sqrt{5}}{2}\right)^2\right]\\ &= \frac{1}{\sqrt{5}}\left[\left(\frac{1+\sqrt{5}}{2} - \frac{1-\sqrt{5}}{2}\right)\left(\frac{1+\sqrt{5}}{2} + \frac{1-\sqrt{5}}{2}\right)\right]\end{aligned}$$

$$= \frac{1}{\sqrt{5}}(\sqrt{5})(1) = 1$$

which is the second term of the Fibonacci sequence.

EXERCISE 14.3

1 $a_n = 3n$ **3** $a_n = 3n + 4$ **5** $a_n = -2n - 5$ **7** $a_n = 2n + 42$ **9** $a_n = -\frac{1}{4}(3n - 81)$ **11** 232 **13** $-17{,}040$ **15** $\frac{133}{10}$ **17** $-\frac{141}{4}$ **19** 10 **21** 6 **23** 2 or 19 **25** \$23,500 **27** 66 balls in the bottom layer; 286 balls in the pyramid **29** \$717,000 **31** $a_1 = 5$, $d = 10$, so that $a_n = 5 + 10(n - 1) = 10n - 5$ and $S_n = (n/2)(5 + 10n - 5) = 5n^2$.

USING YOUR KNOWLEDGE 14.3

1 \$156

EXERCISE 14.4

1a 3 **b** 2 **c** $3 \cdot 2^{n-1}$ **d** $3(2^n - 1)$ **e** Not possible **3a** 8 **b** 3 **c** $8 \cdot 3^{n-1}$ **d** $4(3^n - 1)$ **e** Not possible **5a** 16 **b** $-\frac{1}{4}$ **c** $16(-\frac{1}{4})^{n-1}$ **d** $\frac{64}{5}[1 - (-\frac{1}{4})^n]$ **e** $\frac{64}{5}$ **7** $r = 3$, $a_4 = 135$ **9** $a_1 = -3$, $a_9 = -768$ **11** $a_1 = 1024$, $S_8 = \frac{13{,}107}{16}$ **13** $r = 2$, $n = 7$ **15** $\frac{7}{9}$ **17** $\frac{3}{2}$ **19** $\frac{125}{6}$ **21** About 243,000 **23** 50% **25** 380 ft **27** 50% **29** -5, 1, and 7 or 7, 1, and -5

USING YOUR KNOWLEDGE 14.4

1 \$2767.65 **3** \$5409.78

EXERCISE 14.5

1

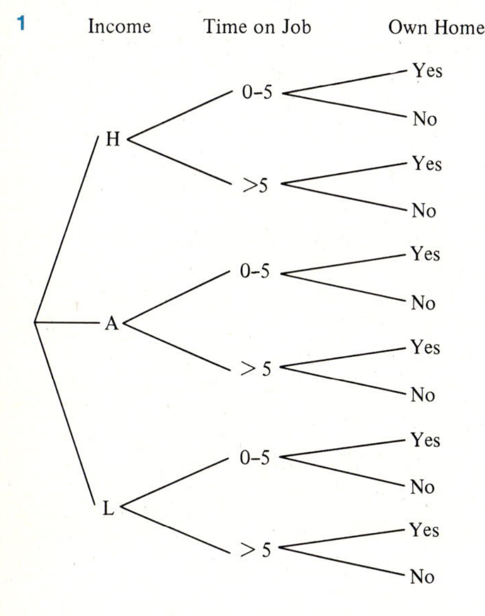

3 10 **5** 24 **7a** 40,320 **b** 720 **c** 10 **9a** 10,000 **b** 5040 **11a** $10^9 = 1{,}000{,}000{,}000$ **b** 729,000,000 **13** $2 \cdot 5 \cdot 4 = 40$ **15a** 10 **b** 15 **c** 35 **d** 1 **e** 84 **17** 20 **19** 57 **21a** 161,700 **b** 22,325 **23a** 165 **b** 135 **25a** $52 \cdot 51 \cdot 50 = 132{,}600$ **b** $4 \cdot 4 \cdot 4 = 64$ **27** 190 **29** 2016 **31** 36 **33a** 42 **b** 12 **35** $3 \cdot 3 \cdot C(14, 7) = 30{,}888$ **37** 11 **39** 24

USING YOUR KNOWLEDGE 14.5

1 $\frac{1}{24}$ **3** $\frac{1}{10}$ **5a** $\frac{9}{8000}$ **b** 0 **c** $\frac{63}{4000}$ **d** $\frac{1}{320}$ **e** $\frac{7}{2000}$ **7** $\frac{1}{5}$ **9** $\frac{2}{11}$ **11** $\frac{2}{7}$

EXERCISE 14.6

1 15 **3** 120 **5** 40,320 **7** 90 **9** $a^3 + 6a^2b + 12ab^2 + 8b^3$ **11** $x^4 + 16x^3 + 96x^2 + 256x + 256$ **13** $32x^5 - 80x^4 + 80x^3 - 40x^2 + 10x - 1$ **15** $16x^4 - 96x^3y + 216x^2y^2 - 216xy^3 + 81y^4$ **17** $\frac{1}{x^5} + \frac{5}{x^3y} + \frac{10}{xy^2} + \frac{10x}{y^3} + \frac{5x^3}{y^4} + \frac{x^5}{y^5}$

19 $x^6 + 3x^5y + \frac{15}{4}x^4y^2 + \frac{5}{2}x^3y^3 + \frac{15}{16}x^2y^4 + \frac{3}{16}xy^5 + \frac{1}{64}y^6$ **21** $-13{,}608x^3$ **23** $60x^2y^4$ **25** $-2268x^6y^3$ **27** $924y^{-12}$ **29** 2.8561

31 1.2190 **33** $\binom{n}{k} = \dfrac{n!}{k!(n-k)!}$ and $\binom{n}{n-k} = \dfrac{n!}{(n-k)!k!}$

Therefore,

$$\binom{n}{k} = \binom{n}{n-k}$$

USING YOUR KNOWLEDGE 14.6

1 1.019613 **3** 1.019708 **5** 1.008165 **7** \$1156.82

SELF-TEST, CHAPTER 14

1a 4 **b** 8 **c** 32 **2a** 0.3, 0.03, 0.003 **b** $\frac{1}{3}$ **3a** $\sum_{k=1}^{n} k^2$ **b** $\sum_{i=1}^{9} a_i$ **c** $\sum_{i=1}^{n} a_i - \sum_{i=1}^{n} b_i$ **4** No. It does not hold for $n \geq 4$.

5 1. For $n = 1$, we have $1 = 1^2$, which is true.
2. Assume the formula holds for $n = k$. Then,

$$1 + 3 + 5 + \cdots + (2k - 1) = k^2$$

We wish to show that

$$1 + 3 + 5 + \cdots + (2k - 1) + (2k + 1) = (k + 1)^2$$

The assumed formula sums the first k terms on the left to give

$$\begin{aligned} 1 + 3 + 5 + \cdots + (2k - 1) + (2k + 1) &= k^2 + (2k + 1) \\ &= k^2 + 2k + 1 \\ &= (k + 1)^2 \end{aligned}$$

This shows that the formula holds for $n = k + 1$ if it holds for $n = k$. Thus, the proof is complete.

6a $a_n = 2n - 8$ **b** 22 **c** 120 **7** $n = 12$ **8** The ninth day **9** $a_7 = \frac{1}{2048}$, $S_7 = \frac{5461}{2048}$ **10** $\frac{6305}{128}$ **11** 8, 40, 200 **12** $-\frac{3}{4}$ **13** 120

14a 360 **b** $6^4 = 1296$ **15** 120 **16** $C(52, 13)$ **17** 90 **18a** 495 **b** 1 **c** 252 **19a** $16x^4 - 16x^2 + 6 - \dfrac{1}{x^2} + \dfrac{1}{16x^4}$ **b** $-4032y^4$

20 $-\frac{15}{16}x^{-4}$

INDEX

82 83 84 85 9 8 7 6 5 4 3 2 1

(The radian measure of θ in the figure is s/r.)

$\cos\theta = \cos\frac{s}{r} = \frac{x}{r}$ $\quad$ $\sec\theta = \sec\frac{s}{r} = \frac{r}{x}$

$\sin\theta = \sin\frac{s}{r} = \frac{y}{r}$ $\quad$ $\csc\theta = \csc\frac{s}{r} = \frac{r}{y}$

$\tan\theta = \tan\frac{s}{r} = \frac{y}{x}$ $\quad$ $\cot\theta = \cot\frac{s}{r} = \frac{x}{y}$

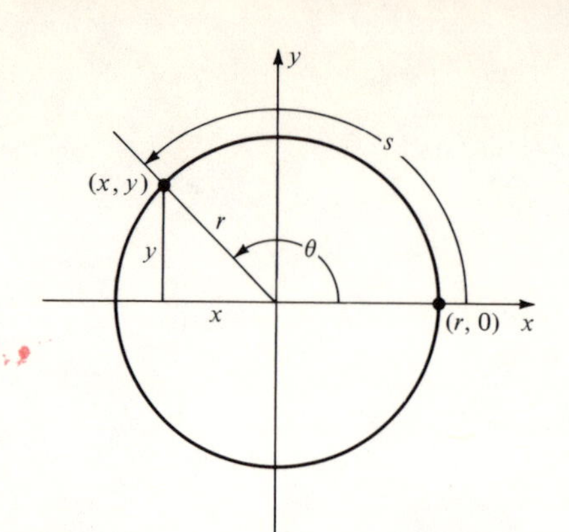

Some Exact Values

Degrees	Radians	sin θ	cos θ	tan θ	cot θ	sec θ	csc θ
0	0	0	1	0	***	1	***
30	$\pi/6$	$1/2$	$\sqrt{3}/2$	$1/\sqrt{3}$	$\sqrt{3}$	$2/\sqrt{3}$	2
45	$\pi/4$	$\sqrt{2}/2$	$\sqrt{2}/2$	1	1	$\sqrt{2}$	$\sqrt{2}$
60	$\pi/3$	$\sqrt{3}/2$	$1/2$	$\sqrt{3}$	$1/\sqrt{3}$	2	$2/\sqrt{3}$
90	$\pi/2$	1	0	***	0	***	1
180	π	0	-1	0	***	-1	***
270	$3\pi/2$	-1	0	***	0	***	-1
360	2π	0	1	0	***	1	***

IDENTITIES

Basic Identities

1 $\tan\theta = \dfrac{\sin\theta}{\cos\theta}$ $\quad$ **2** $\cot\theta = \dfrac{\cos\theta}{\sin\theta}$

3 $\sec\theta = \dfrac{1}{\cos\theta}$ $\quad$ **4** $\csc\theta = \dfrac{1}{\sin\theta}$

5 $\cot\theta = \dfrac{1}{\tan\theta}$ $\quad$ **6** $\cos^2\theta + \sin^2\theta = 1$

7 $1 + \tan^2\theta = \sec^2\theta$ $\quad$ **8** $\cot^2\theta + 1 = \csc^2\theta$

Functions of Sums and Differences

9 $\cos(\alpha+\beta) = \cos\alpha\cos\beta - \sin\alpha\sin\beta$

10 $\cos(\alpha-\beta) = \cos\alpha\cos\beta + \sin\alpha\sin\beta$

11 $\sin(\alpha+\beta) = \sin\alpha\cos\beta + \cos\alpha\sin\beta$

12 $\sin(\alpha-\beta) = \sin\alpha\cos\beta - \cos\alpha\sin\beta$

13 $\tan(\alpha+\beta) = \dfrac{\tan\alpha + \tan\beta}{1 - \tan\alpha\tan\beta}$ $\quad$ **14** $\tan(\alpha-\beta) = \dfrac{\tan\alpha - \tan\beta}{1 + \tan\alpha\tan\beta}$

Double-Angle Formulas

15 $\cos 2\theta = \cos^2\theta - \sin^2\theta$
$= 2\cos^2\theta - 1$
$= 1 - 2\sin^2\theta$

16 $\sin 2\theta = 2\sin\theta\cos\theta$

17 $\tan 2\theta = \dfrac{2\tan\theta}{1 - \tan^2\theta}$